P. Wriggers · W. Wagner (Eds.)

Nonlinear Computational Mechanics

State of the Art

Springer-Verlag
Berlin Heidelberg New York
London Paris Tokyo
Hong Kong Barcelona Budapest

Prof. Dr.-Ing. habil. P. Wriggers
Technische Hochschule Darmstadt
Institut für Mechanik IV
Hochschulstraße 1
W-6100 Darmstadt 1

PD Dr.-Ing. habil. W. Wagner
Universität Hannover
Institut für Baumechanik
und Numerische Mechanik
Appelstraße 9A
W-3000 Hannover 1

ISBN 3-540-54254-X Springer-Verlag Berlin Heidelberg NewYork
ISBN 0-387-54254-X Springer-Verlag NewYork Berlin Heidelberg

Typesetting: Camera ready by authors;
Offsetprinting: Color-Druck Dorfi GmbH, Berlin; Bookbinding: Lüderitz & Bauer, Berlin
61/3020-543210 – Printed on acid-free paper.

Dedication

Professor Erwin Stein's contributions to the field of finite elements have been inspiring and substantial, and have earned him growing recognition from colleagues which come from mechanics and from mathematics. In total he has authored over one hundred and fifty professional papers containing engineering and mathematical aspects of the finite element method. His never ending and always constructive engagement and service for the finite element community is celebrated with this special anniversary volume.

Erwin Stein was born in Altendiez, Germany, on 5 July 1931. He studied Civil Engineering and Mathematics at the Technische Hochschule Darmstadt. After this he worked for a one year period as a structural engineer in a design office,where he constructed pre-stressed concrete bridges, before he went back to the university. He received the Doctor of Engineering 1964 and his habilitation 1969, respectively, from the University of Stuttgart. In 1971 he accepted a call to the University of Hannover. Since then he helds the chair of the Institute for Mechanics and Computational Mechanics at the University of Hannover. From 1978 to 1981 he was the Dean of the faculty of Civil Engineering. He is a member of the Braunschweigische Wissenschaftliche Gesellschaft, the board of the GAMM, the Leibniz–Society in Hannover, a founding member of the IACM, a member of the International Society for the Interaction of Mathematics and Mechanics (ISIMM), and is secretary of DEKOMECH (German Committee for Mechanics. He serves as a managing director of the international journal "Computational Mechanics", as a co–editor of the journal "Bauingenieur" and on the editorial boards of several international publications in the field of finite element methods.

We are glad to celebrate Professor Stein's 60th birthday with this special anniversary volume which contains over 40 very stimulating and interesting papers contributed by his colleagues and former student from all over the world.

P. Wriggers W. Wagner

Preface

The field of Computational Mechanics has grown during the last decade very rapidly. This is due to the fact that modern engineering design needs complex models which can only be analyzed and simulated on powerful computers and workstations using numerical methods like finite element, boundary element or finite difference techniques. This volume presents an overview of current research areas and thus a state of the art in the field of nonlinear computational mechanics.

In many engineering problems different types of nonlinearities have to be considered at the same time which is true for e.g. forming processes exhibiting large deformations, inelasticity and contact or stability problems of composite shell structures including damage behaviour. Furthermore coupling of various types of nonlinear partial differential equations is essential to model problems like fluid–structure–interaction. Due to the complexity of such simulated processes accurate, robust and reliable formulations, algorithms and computer programs can only be designed when mechanical models are well founded by the mathematical theory. Furthermore deep knowledge of computer science is needed to obtain efficient algorithms for different types of computer architecture, like scalar, vector and parallel processors. All these aspects of Computational Mechanics are combined in this volume.

We are overwhelmed and thankful that so many outstanding scientists in this field have taken their time to contribute fresh and impressive papers to this volume which covers areas considering

- the mathematical theory and numerical algorithms,
- nonlinear finite element procedures,
- boundary element techniques,
- beam, plate and shell formulations,
- inelastic constitutive models and
- contact formulations.

This list shows also the wide range of applications within Computationals Mechanics.

W. Rust has produced the figures on the cover depicting the adaptive mesh refinement of a cylindrical shell, which is gratefully acknowleged.

Finally we would like to thank the Springer Verlag, especially A. v. Hagen, for the good cooperation during the editorial work.

P. Wriggers — W. Wagner

Darmstadt — Hannover

Contents

I. Mathematical Theory of Finite Elements

1. **D. Braess**
Kirchhoff– and Mindlin–Reissner Plates: Combinations from the Viewpoint of Finite Elements 1

2. **R. Rannacher**
On the Convergence of the Newton–Raphson Method for Strongly Nonlinear FE–Equations 11

3. **W. C. Rheinboldt, J.-L. Liu**
A Posteriori Error Estimates for Parametrized Nonlinear Equations. 31

II. Numerical Methods in Nonlinear Finite Element Procedures

1. **M. A. Crisfield, J. Shi**
A Review of Solution Procedures and Path–Following Techniques in Relation to the Non-Linear Finite Element Analysis of Structures 47

2. **W. Wagner, P. Wriggers**
Calculation of Bifurcation Points via Fold Curves 69

3. **W. B. Krätzig, Y.-Y. Quian**
On Stability Conditions for Nonlinear Static and Dynamic Buckling Responses of Arbitrary Structures 85

4. **B. Kröplin, D. Dinkler**
An Energy–Based Concept of Structural Stability 98

5. **O. C. Zienkiewicz, B. A. Schrefler, L. Simoni, Y. M. Xie, X. Y. Zhan**
Two and Three Phase Behaviour in Soil Dynamics 103

6. **R. W. Lewis, A. F. Usmani, J. T. Cross**
Adaptive Finite Element Analysis of Heat Transfer and Flow Problems 137

7. **E. Ramm, K.–U. Bletzinger, S. Kimmich**
Strategies in Shape Optimization of Free Form Shells 163

III. Boundary Element Techniques

1. **H. Antes, B. Steinfeld**
Unilateral Contact with Friction by a Time Domain BEM. 193

2. **H. Schmitz, G. Schneider, W. L. Wendland**
Boundary Element Methods for Problems Involving Unilateral Boundary Conditions 212

3. **L. C. Wrobel, C. A. Brebbia**
An Overview of Boundary Element Applications to Nonlinear Heat Transfer Problems 226

IV. Beam, Plate and Shell Formulations

1. **O. C. Zienkiewicz, E. Oñate**
Finite Volumes versus Finite Elements. Is there really a Choice? 240

2. **T. H. H. Pian, C.–C. Wu**
Use of Additional Incompatible Displacements for Finite Element Formulation 255

3. **A. K. Noor, C. M. Andersen**
Hybrid Analytical Technique for the Nonlinear Analysis of Curved Beams .. 267

4. **C. Felippa, L. Crivelli**
A Congruential Formulation of Nonlinear Finite Elements 283

5. **P. G. Bergan, K. M. Mathisen**
Large Displacement Analysis of Highly Flexible Off–Shore Structures 303

6. **F. G. Rammerstorfer, K. Dorninger, A. Starlinger**
Nonlinear Finite Element Analysis of Composite and Sandwich Shell Structures 332

7. **F. G. Kollmann, D. Cordts, H.–P. Hackenberg**
A Family of Mixed and Hybrid Finite Elements for the Numerical Analysis of Viscoplastic Shells of Revolution 360

V. Inelastic Constitutive Models and Numerical Algorithms

Metal Plasticity

1. **I. Babuška, Y. Li, K. L. Jerina**
Reliability of Computational Analysis of Plasticity Problems 375

2. **O. T. Bruhns**
Shear Band Formation in Impact Problems 399

3. **H. Okada, T. Tamura, N. Ramakrishnan, S. N. Atluri, J. S. Epstein**
On the Improvement of Fracture Toughness due to Transformation Induced Plasticity 418

4. **C. Comi, A. Corigliano, G. Maier**
Extremum Properties of Finite–Step Solutions in Elastoplasticity with Nonlinear Mixed Hardening 429

5. **D. Perić, D. R. J. Owen**
Computational Model for Large Elastic–Plastic Deformations at Moderate Elastic Strains 449

6. **E. Massoni, J. L. Chenot, M. Bellet**
A 3D Finite Element Approach for Predicting the Deformation of an Anisotropic Elasto–Plastic Membrane 473

7. **L. G. Popov, J. R. Whiteman**
Models of Plasticity in Strain Space 497

8. **L. Bernspång, K. Mattiasson, A. Samuelsson**
Sheet Metal Forming Modelled as a Quasi–Dynamic Process 520

9. **J. Appeltauer, O. Mahrenholtz**
Some Thoughts about Metal Forming Simulation Models and Numerical Performance 530

10. **T. Niezgoda, M. Kleiber, J. A. König**
On Thermomechanical Shakedown Analysis using Incremental Procedures .. 537

11. **H. Rothert, N. Gebbeken, V. Brandt**
Computational Aspects of a Rate Dependent Yield Hinge Theory 549

General Inelastic Models

1. **H. G. Matthies**
Computation of Constitutive Response 573

2. **P. Steinmann, K. Willam**
A Critical Assessment of FE–Performance in Localized Failure Analysis 586

3. **G. Hofstetter, H. A. Mang, G. Meschke**
3D Inelastic Constitutive Modeling for FE Analysis of Concrete Structures . 606

4. **W. Wunderlich, E. Heins**
Nonlinear Analysis of Cracked Concrete Beams under Bending and Torsion 627

5. **A. Matzenmiller, K. Schweizerhof**
Crashworthiness Simulations of Composite Structures - a first Step with Explicit Time Integration 642

VI. Contact Formulations

1. **C.-Y. Lee, J. T. Oden, M. Ainsworth**
Local A Posteriori Error Estimates and Numerical Results for Contact Problems and Problems of Flow Through Porous Media. 671

2. **R. L. Taylor, P. Papadopoulos**
On a Patch Test for Contact Problems in Two Dimensions 690

3. **A. L. Eterovic, K.–J. Bathe**
An Interface Interpolation Scheme for Quadratic Convergence in the Finite Element Analysis of Contact Problems 703

4. **T. A. Laursen, J. C. Simo**
On the Formulation and Numerical Treatment of Finite Deformation Frictional Contact Problems 716

5. **F. D. Fischer, E. Hinteregger, F. G. Rammerstorfer**
A Computational Study of the Residual Stress Distribution in Thermally Loaded Beams of Arbitrary Cross–Section on Frictional Support 737

Kirchhoff- and Mindlin-Reissner Plates: Combinations from the Viewpoint of Finite Elements

Dietrich Braess

Mathematisches Institut
Ruhr-Universität Bochum
D 4630 Bochum 1

In the theory of plates usually two different limit cases are considered: the Kirchhoff plate and the Mindlin-Reissner plate.

The *Kirchhoff plate* model leads to an elliptic differential equation of fourth order. In most cases it is treated by using nonconforming elements or mixed methods.

The *Mindlin-Reissner plate* model is obtained under less restrictive hypotheses. Since it is modelled by partial differential equations of second order, formally this plate may be treated by conforming H^1-elements. Therefore, at first glance its numerical treatment seems to be much easier than that of the Kirchhoff plate.

On the contrary, it turns out that the computations for the Mindlin-plate are more difficult. The problems known for the Kirchhoff-plate are not eliminated but only concealed. This is not the end of the story, the hide-and-seek is observed in various forms: Some finite element calculations make use of combinations or refer in some way to both models. Since partial differential equations often give rise to surprises, it is necessary to ascertain (i.e. to prove) that in this way the advantages and not the disadvantages of the models are added.

It is the aim of this contribution to elucidate this point. In doing so we encounter another source of confusion, (perhaps only from the viewpoint of a mathematician): When some finite element methods are analysed, it is not easy to decide whether they are *mixed methods* or whether they are better counted to *other nonconforming methods.* This question is not only of academic interest, but it is substantial for the development of fast multigrid algorithms, see e.g. Peisker, Rust and Stein [1990], Braess and Blömer [1990].

The Variational Problem

In plate theory the energy

$$\Pi(w,\theta) = \frac{1}{2}a(\theta,\theta) + \frac{t^{-2}}{2}\int_\Omega |\nabla w - \theta|^2 dx - \int_\Omega fw\, dx \tag{1}$$

is to be minimized where

$$a(\theta,\psi) := \int_\Omega [\mu \sum_{i,j} D_{ij}\theta\, D_{ij}\psi + \lambda \operatorname{div}\theta \operatorname{div}\psi] dx, \tag{2}$$

$$D_{ij}\theta := \frac{1}{2}(\frac{\partial}{\partial x_i}\theta_j + \frac{\partial}{\partial x_j}\theta_i).$$

We are given the Lamé constants λ, μ, the load f, and the parameter t which equals the thickness of the plate multiplied by a correction factor. The bending $w : \Omega \to \mathbf{R}$ and the rotation $\theta : \Omega \to \mathbf{R}^2$ are to be determined. Since we will restrict our attention to the clamped plate, w and θ satisfy zero boundary conditions. When the Mindlin plate model is used, the variational problem is directly solved with the variables w and θ as stated in (1) and (2). The problem is H^1-elliptic. Here $H^m = H^m(\Omega)$ refers to the usual Sobolev space of the functions having square-integrable derivatives up to order m. The norms $\|\cdot\|_m$ and semi-norms $|\cdot|_m$ refer to these spaces. — The famous Korn inequalities imply that

$$a(\psi,\psi) \geq c\|\psi\|_1^2 \quad \text{for } \psi \in H_0^1(\Omega). \tag{3}$$

Set $c_1 = \min(c, 1)$. The quadratic term in (1) may now be estimated from below by making use of Friedrichs' inequality

$$\begin{aligned}\frac{1}{2}a(\psi,\psi) + \frac{t^{-2}}{2}\|\nabla v - \psi\|_0^2 &\geq \frac{1}{2}c\|\psi\|_1^2 + \frac{1}{4}\|\nabla v - \psi\|_0^2 \\ &\geq \frac{1}{4}c\|\psi\|_1^2 + \frac{1}{4}c_1(\|\psi\|_0^2 + \|\nabla v - \psi\|_0^2) \\ &\geq \frac{1}{8}c_1(\|\psi\|_1^2 + \|\nabla v\|_0^2) \\ &\geq c'(\|\psi\|_1^2 + \|v\|_1^2).\end{aligned}$$

Since the right hand side of (1) is also continuous with respect to the H^1-norm, it follows from the general theory that the variational problem is solvable. Therefore one might think of computations with standard conforming C^0-elements. Unfortunately the Céa lemma tells us that the error of the finite element solution may be augmented by a factor of

$$t^{-2}$$

when compared to best approximation. Indeed, this factor is not only found in theoretical results, it implies that poor results are obtained in actual computations unless specific devices are applied, see e.g. Brezzi and Fortin [1986], Arnold and Falk [1989], Peisker and Braess [*].

The theory of the Kirchhoff plate is based on the hypothesis that in the limit case $(t \to 0)$

$$\nabla w - \theta = 0 \tag{4}$$

holds, where (4) is understood as an equality in H^1. In this case one minimizes

$$\Pi = \frac{1}{2}a(\theta,\theta) - \int_\Omega f w \, dx \tag{5}$$

subject to the constraint (4). The restriction admits to eliminate the rotation θ and we have to minimize

$$\frac{1}{2}a(\nabla w, \nabla w) - (f, w)_0 \; . \tag{6}$$

For convenience, the integral from (5) has been rewritten as an L_2-inner product $(.,.)_0$. From Korn's and Friedrichs' inequalities it follows that

$$\begin{aligned} a(\nabla w, \nabla w) &\geq c\|\nabla w\|_1^2 \\ &\geq \frac{1}{2}c\|\nabla w\|_1^2 + \frac{1}{2}c\|\nabla w\|_0^2 \geq c'\|w\|_2^2. \end{aligned} \tag{7}$$

Hence, we obtain an H^2-elliptic problem as already mentioned. Therefore, one often hears that this variational problem is the more difficult one. This is true, however, only from a formal viewpoint.

Mixed Methods for the Kirchhoff Plate

While both plate theories seem to be very different when the displacement formulation is used, the common features become more apparent with the mixed methods. In this context it does not matter that the interest in the mixed formulation has very different reasons in both cases.

We start with the discussion of the Kirchhoff plate. The minimization of the expression (5) with the constraint (4)

$$\nabla v = \psi \quad \text{in } \Omega$$

leads to the following saddle point problem. Find $(w, \theta) \in X$ and $\gamma \in M$ such that

$$\begin{aligned} a(\theta, \psi) + (\nabla v - \psi, \gamma)_0 &= (f, v) \quad &&\text{for } (v, \psi) \in X, \\ (\nabla w - \theta, \eta)_0 \qquad\qquad &= 0 &&\text{for } \eta \in M \; . \end{aligned} \tag{8}$$

The last equation is the weak formulation of the constraint (4), and therefore (8) is considered as the standard saddle point problem for a variational problem with constraints.

The spaces X and M are written as abstract spaces in (8). The canonical choice of the spaces for the clamped plate is

$$X := H_0^2(\Omega) \times H_0^1(\Omega)^2, \quad M := H^{-1}(\Omega)^2. \tag{9}$$

Obviously the bilinear forms in (8) are continuous. From Korn's inequality we conclude with the same arguments as in (7)

$$\begin{aligned} a(\psi,\psi) &\geq c\|\psi\|_1^2 \\ &= \frac{c}{2}\|\psi\|_1^2 + \frac{c}{2}\|\nabla v\|_1^2 \\ &\geq c'(\|\psi\|_1^2 + \|v\|_2^2), \end{aligned}$$

whenever $\nabla v = \psi$ holds. Consequently, a satisfies the ellipticity condition for saddle point problems. Furthermore,

$$\sup_{v,\psi} \frac{(\nabla v - \psi, \eta)_0}{\|v\|_2 + \|\psi\|_1} \geq \sup_{\psi} \frac{-(\psi,\eta)_0}{\|\psi\|_1} = \|\eta\|_{-1}.$$

Therefore the famous inf-sup condition of Ladyšenskaja, Babuška and Brezzi holds (see Brezzi [1973]). The solution of the mixed problem can now be estimated by arguments from the general theory

$$\|w\|_2 + \|\theta\|_1 + \|\gamma\|_{-1} \leq c\|f\|_{-2}.$$

The following regularity result for convex domains Ω of Blum and Rannacher [1980] will be crucial in the analysis of finite elements

$$\|w\|_3 + \|\theta\|_2 + \|\gamma\|_0 \leq c\|f\|_{-1}\,. \tag{10}$$

It follows from (10) that both sides of (8) are H^1-continuous expressions of v. Since $H_0^2(\Omega)$ is dense in $H_0^1(\Omega)$, the first equation of (8) holds for all $v \in H_0^1(\Omega)$, whenever Ω is convex and $f \in H^{-1}(\Omega)$.

We move already into the direction of the other plate model, when we consider the mixed formulation of the Kirchhoff plate with another pair of spaces:

$$X := H_0^1(\Omega) \times H_0^1(\Omega)^2, \quad M := H^{-1}(\mathrm{div}, \Omega). \tag{11}$$

Here $H^{-1}(\mathrm{div},\Omega)$ is the completion with respect to the norm

$$\|\eta\|_{H^{-1}(\mathrm{div},\Omega)} := (\|\eta\|_{-1}^2 + \|\mathrm{div}\,\eta\|_{-1}^2)^{\frac{1}{2}}.$$

If the domain Ω satisfies a cone condition, $H^{-1}(\mathrm{div},\Omega)$ may be understood as the set of vector-valued H^{-1}-functions whose divergence also belongs to H^{-1}.

The pairing (11) plays an important role, since w is no longer restricted to the set of H^2-functions. We abandon the proof that ellipticity and the Ladyšenskaja-Babuška-Brezzi condition hold. — We note that the variational problem (8) need not have the same solutions for the two choices of spaces. It is guaranteed when the regularity result (10) holds which in particular implies that $\gamma \in L_2(\Omega)$.

A Mixed Formulation for the Mindlin Plate

The variational problem contains a small parameter, and the problem (1) does not make sense in the limit case $t = 0$ unless special rearrangements are performed. Typically, a penalty term enters into the theory when a mixed formulation for this kind of problems is chosen.

The starting point is the weak formulation of (1): Find $w \in H_0^1(\Omega)$, $\theta \in H_0^1(\Omega)^2$ such that

$$a(\theta,\psi) + t^{-2}(\nabla w - \theta, \nabla v - \psi)_0 = (f,v)_0 \quad \text{for } v \in H_0^1(\Omega), \psi \in H_0^1 \,. \tag{12}$$

Next a shear term with appropriate scaling is introduced:

$$\gamma := t^{-2}(\nabla w - \theta). \tag{13}$$

When equation (12) is combined with the weak formulation of (13), we obtain the system

$$\begin{aligned} a(\theta,\psi) \quad &+ (\nabla v - \psi, \gamma)_0 = (f,v)_0 \qquad && \text{for } (v,\psi) \in X, \\ (\nabla w - \theta, \eta)_0 \quad &- t^2(\gamma,\eta)_0 \qquad = 0 && \text{for } \eta \in M. \end{aligned} \tag{14}$$

There is a substantial difference to (8), namely a penalty term $-t^2(\gamma,\eta)_0$ is encountered. This bilinear form is not continuous, although the spaces (11) are closer to those spaces in which we would have continuity. The appropriate norm is here t-dependent:

$$\|w\|_1 + \|\theta\|_1 + \|\gamma\|_{H^{-1}(\mathrm{div},\Omega)} + t\|\gamma\|_0. \tag{15}$$

The penalty term is a singular perturbation, see Arnold and Falk [1990]. The singular behaviour is more than a coincidence, it is the mathematical counterpart to an observation for actual plates: A substantial deviation between the rotation θ and the normal ∇w is possible only in the neighbourhood of the boundary $\partial\Omega$.

The general theory of saddle point problems with penalty term is only applicable since Brezzi and Fortin [1986] decomposed the shear term γ into a gradient and a rotation. In this framework one can also deal with those finite elements for which the mathematical theory is most developed, namely the MITC-elements, see Bathe, Brezzi, and Cho [1989], Brezzi, Fortin, and Stenberg [*], Peisker and Braess [*].

Finally we note that the variable γ has a different meaning in both theories. It enters in Kirchhoff plate theory as a Lagrangian parameter, while it is a normalized shear term in the other model. Nevertheless, we have convergence to the right value as $t \to 0$.

DKT-Elements

Finite element calculations are obviously simpler than with C^1-elements, if the Kirchoff hypothesis $\nabla w = \theta$ is not postulated in the whole domain Ω, but only at the vertices of a triangulation. This has led to the triangular elements with discrete Kirchhoff constraints short DKT-elements.

To be precise, the DKT-elements are nonconforming elements for the displacement formulation. Moreover the quadratic form a in (6) is replaced by a mesh-dependent quadratic form a_h. Nevertheless, the connection with the mixed formulation (8) facilitates the mathematical analysis. On the other hand, it is often very difficult to recognize in the proofs, in which model the actual discussion is performed. We hope that the framework and the viewpoint will be clear in the following discussion.

We will consider one DKT-element with details of the proof. At the end we will briefly mention another element which has to be treated by quite different arguments.

Both elements refer to reduced cubic polynomials. Let a_1, a_2, a_3 be the vertices of a triangle and a_0 its center of mass:

$$\mathcal{P}_{3,red} := \Big\{p \in \mathcal{P}_3;\ 6p(a_0) - \sum_{i=1}^{3}[2p(a_i) - \nabla p(a_i)(a_i - a_0)] = 0\Big\}.$$

Here $\mathcal{P}_n$ is the set of polynomials of degree n. The bubble function has been eliminated by the linear constraint (which on the other hand is often added to polynomials of lower degree). When compared with standard Lagrangian (cubic) elements, the nodal function to the interior point is missing. Only the 9 points on the boundary are incorporated. — We may also use the function values and the first derivatives at the vertices instead of the 9 Lagrangian values mentioned above.

The DKT-element due to Batoz, Bathe und Ho [1980] refers to the two spaces

$$\begin{aligned} W_h &:= \{w \in H_0^1(\Omega);\ \nabla w \text{ is continuous at all vertices of } \mathcal{T}_h, \\ &\qquad w_{|T} \in \mathcal{P}_{3,red} \text{ for } T \in \mathcal{T}_h\}, \\ \Theta_h &:= \{\theta \in H_0^1(\Omega)^2;\ \theta_{|T} \in (\mathcal{P}_2)^2 \text{ and } \theta \cdot n \in \mathcal{P}_1(e) \\ &\qquad \text{for each edge } e \in \partial T \text{ and } T \in \mathcal{T}_h\}. \end{aligned} \tag{17}$$

As usual, T denotes a triangle of the triangulation $\mathcal{T}_h$; the normal vector n and the tangential vector t are defined on the boundary of T. Furthermore, $a_{ij} := \frac{1}{2}(a_i + a_j)$.

Now, a rotation is associated to each bending by a *discrete Kirchhoff condition*:

$$\begin{aligned} w_h \longmapsto \theta_h(w_h):\quad \theta_h(a_i) &= \nabla w_h(a_i) && \text{for the vertices } a_i, \\ \theta_h(a_{ij}) \cdot t &= \nabla w_h(a_{ij}) \cdot t && \text{for the midpoints } a_{ij}. \end{aligned} \tag{18}$$

The mapping (18) is well defined since the normal component of the rotation for each point a_{ij} is already given by the rotations at the vertices. Indeed, it follows from the definition of the space Θ_h that

$$\theta_h(a_{ij}) \cdot n = \frac{1}{2}[\nabla w_h(a_i) + \nabla w_h(a_j)] \cdot n.$$

Originally, a mapping from W_h to Θ_h is defined by (18). The domain can be canonically extended to $W_h \oplus H^3(\Omega)$. Since $H^3(\Omega) \subset C^1(\Omega)$, we just apply (18) to the larger set. The extension will be important later when we will apply the lemmas of Strang.

Now we are in a position to define the associate finite element approximation: Find $w_h \in W_h$ such that

$$a_h(w_h, v_h) = (f, v_h)_0 \quad \text{for } v_h \in W_h \tag{19}$$

where

$$a_h(w_h, v_h) := a(\theta_h(w_h), \theta_h(v_h)). \tag{20}$$

Following Pitkäranta [1988] we introduce the mesh dependent norms

$$\begin{aligned} |v|_{s,h} &:= \Big(\sum_{T\in\mathcal{T}_h} |v|_{s,T}^2\Big)^{\frac{1}{2}}, \\ \|v\|_{2,h} &:= (\|v\|_1^2 + |v|_{2,h}^2)^{\frac{1}{2}}. \end{aligned} \tag{21}$$

Remark. The norms and seminorms

$$\begin{aligned} &\|\nabla v_h\|_0 \text{ and } \|\theta_h(v_h)\|_0, \\ &|\nabla v_h|_{1,h} \text{ and } |\theta_h(v_h)|_1, \text{ resp.} \\ &\|\nabla v\|_{1,h} \text{ and } \|\theta_h(v_h)\|_1, \text{ resp.} \end{aligned}$$

are equivalent on W_h, provided that the triangulations $\mathcal{T}_h$ are uniform.

We abandon the proof which is based on the Bramble-Hilbert-Lemma and the standard arguments for affine families.

Corollary. There is an h-independent constant $c > 0$ for the DKT-elements (17) such that

$$a(\theta_h(v_h), \theta_h(v_h)) \geq c\|v_h\|_{2,h}^2 \quad \text{for } v_h \in W_h.$$

Proof. From Korn's inequality we conclude that

$$a(\theta_h(v_h), \theta_h(v_h)) \geq c\|\theta_h(v_h)\|_1^2 = c\,(\|\theta_h(v_h)\|_0^2 + |\theta_h(v_h)|_1^2).$$

Now it follows from the remark above and Friedrichs' inequality that

$$\begin{aligned} a(\theta_h(v_h), \theta_h(v_h)) &\geq c'(\|\nabla v_h\|_{0,h}^2 + \sum_{T\in\mathcal{T}_h} |\nabla v_h|_{1,T}^2) \\ &\geq c''(\|v_h\|_1^2 + \sum_{T\in\mathcal{T}_h} |v_h|_{2,T}^2) \\ &= c''\|v_h\|_{2,h}^2. \end{aligned}$$

□

By similar arguments we obtain

$$\|\theta_h(v_h) - \nabla v_h\|_0 \leq c \cdot h|v_h|_{2,h} \text{ for } v_h \in W_h. \tag{22}$$

As usual, first we get an estimate with $\|v_h\|_{2,h}$ on the right hand side from the theory of affine families. Since the interpolation for gradients of linear functions is exact, the expression is reduced to $|v_h|_{2,h}$ by the aid of the Bramble-Hilbert lemma.

We want to estimate the consistency error which enters into the second Strang lemma. The lemma provides the bound

$$\|w - w_h\|_h \le c(\inf_{v_h \in W_h} \|w - v_h\|_h + \sup_{v_h \in W_h} \frac{|a_h(w, v_h) - (f, v_h)_0|}{\|v_h\|_h}). \tag{23}$$

We emphasize that here a_h must be defined on $W_h \oplus H^3(\Omega)$. This was already done in conncetion with the extension of the mapping (18).

Next, we note that we have in general $\theta_h(w) \neq \nabla w$. Since $\theta_h(w)$ is an interpolant, we obtain an error estimate from the standard approximation theorems

$$\|\theta_h(w) - \nabla w\|_0 \le ch^2 \|w\|_3. \tag{24}$$

Now we make use of the results for the mixed variational problem. We set $v = v_h$ and $\psi = \theta_h(v_h)$ in the first equation of (8) and obtain

$$(f, v_h)_0 = a(\theta, \theta_h(v_h)) + (\nabla v_h - \theta_h(v_h), \gamma)_0 \,.$$

This and the preceding estimates imply that

$$\begin{aligned} a_h(w, v_h) - (f, v_h)_0 &= a(\theta_h(w) - \theta, \theta_h(v_h)) + (\nabla v_h - \theta_h(v_h), \gamma) \\ &\le c\|\theta_h(w) - \theta\|_1 \cdot \|\theta_h(v_h)\|_1 + \|\nabla v_h - \theta_h(v_h)\|_0 \cdot \|\gamma\|_0 \\ &\le ch\|w\|_3 \|\nabla v_h\|_1 + c'\|f\|_{-1} \cdot |v_h|_{2,h} \\ &\le c'' h\|f\|_{-1} \cdot \|v_h\|_{2,h}. \end{aligned}$$

Since the approximation error $\inf\{\|w - \psi_h\|_{2,h};\ \psi_h \in W_h\}$ is also bounded by $ch \cdot \|w\|_3$, the lemma of Strang yields the

Theorem. Let $\mathcal{T}_h$ be a family of quasi-uniform triangulations, and let w_h be the finite element approximation by the DKT-element due to Batoz, Bathe and Ho. Then we have

$$\|w - w_h\|_{2,h} \le c \cdot h\|f\|_{-1}$$

where c is a constant not depending of h.

The discussion of the DKT-element above might give the impression that each discrete Kirchhoff condition guarantees convergence. This is not true, however, and the following example shows that one cannot do without convergence analysis.

The DKT-element which is based on the Zienkiewicz triangle uses a smaller space for the rotations in order to reduce the computational effort:

$$\begin{aligned} W_h &:= \{w \in H_0^1(\Omega);\ \nabla w \text{ is continuous at all vertices of } \mathcal{T}_h, \\ &\qquad\qquad w_{|T} \in \mathcal{P}_{3,red} \text{ for } T \in \mathcal{T}_h\}, \\ \Theta_h &= \{\theta \in H_0^1(\Omega)^2;\ \theta_{|T} \in (P_1)^2 \text{ for } T \in \mathcal{T}_h\}. \end{aligned}$$

Also in this case a rotation is associated to a bending by a discrete Kirchhoff condition

$$w_h \longmapsto \theta_h(w_h): \quad \theta_h(a_i) = \nabla w_h(a_i) \quad \text{for the vertices } a_i.$$

It follows already from a counting of the dimension, that $\theta_h = 0$ may be true in some triangle T although $\nabla w_h \neq 0$ holds in that triangle. Therefore, a quadratic form a_h is chosen, which differs from that in (19):

$$a_h(w_h, v_h) := \sum_{T \in \mathcal{T}_h} a(\nabla w_h, \nabla v_h)_T .$$

It is interesting that theoretical considerations and actual computations show that convergence of this nonconforming method is restricted to 3-direction-meshes, see Lasceaux and Lesaint [1975].

In order to eliminate the restriction to 3-direction-meshes, a penalty term is added to the quadratic form

$$a_h(v_h, v_h) := a(\theta_h(v_h), \theta_h(v_h)) + \sum_{T \in \mathcal{T}_h} \frac{1}{h_T^2} \int_T |\nabla v_h - \theta_h(v_h)|^2 dx.$$

Here h_T is the diameter of the element T. The penalty term has obviously be designed by looking at the Mindlin-Reissner theory, although we have not left the Kirchhoff theory.

This completes the circle in our discussion.

References

D.N. Arnold and R.S. Falk [1989]: A uniformly accurate finite element method for the Mindlin-Reissner plate. SIAM J. Numer. Anal. 26, 1276–1290

D.N. Arnold and R.S. Falk [1990]: The boundary layer for the Reissner-Mindlin plate model. SIAM J. Math. Anal. 21, 281–312

K.-J. Bathe, F. Brezzi and S.W. Cho [1989]: The MITC7 and MITC9 plate bending elements. J. Computers and Structures 32, 797–814

J.-L. Batoz, K.-J. Bathe and L.W. Ho [1980]: A study of three-node triangular plate bending elements. Int. J. Num. Math. Eng. 15, 1771–1812

D. Braess [*]: Vorlesungen über Finite Elemente. (Kap. VI. Finite Elemente in der Mechanik elastischer Körper.) Springer-Verlag (in preparation)

D. Braess und C. Blömer [1990]: A multigrid method for a parameter dependent problem in solid mechanics. Numer. Math. 57, 747–761

F. Brezzi [1974]: On the existence, uniqueness and approximation of saddle-point problems arising from Lagrangian multipliers. RAIRO Anal. Numér. 8, R-2, 129–151

F. Brezzi and M. Fortin [1986]: Numerical approximation of Mindlin-Reissner plates. Math. Comp. 47, 151–158

F. Brezzi, M. Fortin and R. Stenberg [*]: Error analysis of mixed-interpolated elements for Reissner-Mindlin plates. (to appear)

P. Peisker and D. Braess [*]: Uniform convergence of mixed interpolated elements for Reissner-Mindlin plates. RAIRO Anal. Numér. (to appear)

P. Peisker, W. Rust and E. Stein [1990]: Iterative solution methods for plate bending problems: multi-grid and preconditioned cg algorithm. SIAM J. Numer. Anal. 27, 1450–1465

J. Pitkäranta [1988]: Analysis of some low-order finite element schemes for Mindlin-Reissner and Kirchhoff plates. Numer. Math. 53, 237–254

On the Convergence of the Newton-Raphson Method for Strongly Nonlinear

ROLF RANNACHER

Institut für Angewandte Mathematik
Universität Heidelberg, Im Neuenheimer Feld 293/294,
D-6900 Heidelberg, Germany

Summary

Primal as well as mixed finite element discretizations are considered for second order quasilinear elliptic boundary value problems. It is proved that the convergence properties of the Newton-Raphson method for solving the discrete problems are nearly independent of the mesh size. For starting values within an $O(\log(1/h)^{-1})$-neighborhood of the true solution the accuracy of the discretization error is reached in at most $O(\log\log(1/h)^{-1})$ iteration steps. Typical examples to which this analysis applies are the subsonic full potential equation and the minimal surface problem.

1 Introduction

The finite element Galerkin discretization of quasilinear elliptic boundary value problems leads to strongly nonlinear algebraic systems which may be solved by the Newton-Raphson method. In general, the quadratic convergence of this method cannot be guaranteed on the continuous level because this would require to have pointwise control on the gradients of the iterates. However, it will be shown below that, even for this situation, the accuracy of the discretization error is reached in at most $O(\log\log(1/h))$ iteration steps, provided that the starting value is taken in an $O(\log(1/h)^{-1})$-neighborhood of the true solution. The proof is based on homotopy arguments and on sharp a priori estimates with respect to L^∞-norms, which are available from [10], [23], [25], [16], and [31]. This analysis is applied to standard primal as well as to mixed discretizations of scalar elliptic problems. Extensions to nonlinear systems, e.g., like those of nonlinear elasticity are also discussed. A simple numerical example confirms the theoretical results.

We consider the scalar quasi-linear elliptic boundary value problem

$$- \operatorname{div} F[\cdot,\nabla u] = f, \text{ in } \Omega, \quad u|_{\Gamma_0} = b, \quad n\cdot F(\cdot,\nabla u)|_{\Gamma_1} = g, \tag{1.1}$$

where Ω is a bounded domain in $\mathbf{R}^n$, $n\in\{2,3\}$, with the boundary $\partial\Omega$ consisting of two disjoint parts Γ_1 and Γ_2. It is assumed that the vector function $F = (F_i)_{i=1}^n$ and the boundary $\partial\Omega$ are sufficiently regular and that Γ_1 is non-trivial. Problems of this type arise in glaciological models and in seepage problems for porous media. Another example is the (subsonic) full potential equation, where

$$F[\eta] = (1-\tfrac{\gamma-1}{2}\eta^2)^{1/(\gamma-1)}\eta,$$

with $\gamma<-1$; the limit case with $\gamma=-1$ is the classical minimal surface problem.

For the ease of presentation it is assumed that Ω is an n-polyhedron, and that $b = 0$ and $g = 0$. With the appropriate technical modifications, the results presented below extend to more general problems, including also additional lower order terms, linear or nonlinear, in (1.1). The Lebesgue and Sobolev spaces of n-vector functions are denoted by $L^p(\Omega)$ and $W^{m,p}(\Omega)$, provided with the norms $\|v\|_p \equiv \|v\|_{L^p(\Omega)}$ and $\|v\|_{m,p} \equiv \|v\|_{W^{m,p}(\Omega)}$, respectively. The L^2-inner product and norm are shortly written as $(\cdot,\cdot)$ and $\|\cdot\|$, respectively. Further, we set

$$H = \left\{v\in W^{1,2}(\Omega),\ v_{|\Gamma_1}=0\right\}, \qquad a(v;w) = (F[\cdot,\nabla v],\nabla w).$$

Then, the usual variational formulation of problem (1.1) reads as follows.

(P) Find $u\in H\cap W^{1,\infty}(\Omega)$, such that

$$a(u;v) = (f,v), \quad \forall\ v\in H. \tag{1.2}$$

We are not interested here in the question of the existence of solutions of problem (P) but in their numerical approximation by the finite element Galerkin method. Therefore, we simply presuppose the existence of a solution $u\in H$ which is in $W^{2,p}(\Omega)$, for some $p>n$, and which is unique in a certain $W^{1,\infty}$-ball

$$B_R(u) = \{v\in H,\ |||u-v|||_\infty \le R\}, \qquad |||w|||_\infty = \operatorname{ess\,sup}_{x\in\Omega}|\nabla w(x)|$$

The local uniqueness of the solution u is related to the assumption that the eigenvalues of the Jacobian matrix $F'[x,\eta] \equiv \nabla_\eta F[x,\eta]$, evaluated at any $v\in B_R(u)$, are bounded away from zero, uniformly for $x\in\Omega$. This is equivalent to the property

$$\sup\left\{|F'[x,\eta]^{-1}|,\ \eta=\nabla v(x),\ v\in B_R(u),\ x\in\Omega\right\} > 0, \tag{1.3}$$

which particularly implies that the Frechét derivative of the semilinear form $a(\cdot;\cdot)$, $a'(u;\cdot,\cdot) = (F'[\cdot,\nabla u]\nabla\cdot,\nabla\cdot)$, is coercive on H,

$$\sup\left\{a'(u;v,w)\ ,\ w\in H,\ \|\nabla w\|=1\right\} \geq \alpha\,\|\nabla v\|\ , \quad v\in H\ . \tag{1.4}$$

The data of problem (1.1), particularely the domain Ω should be sufficiently regular such that the linear operator $A'(u)$: $H\cap W^{2,q}(\Omega)\to L^q(\Omega)$ generated by the bilinear form $a'(u;\cdot,\cdot)$ is onto and satisfies the a priori estimate

$$\|v\|_{2,q} \leq c\,\|A'(u)v\|_q\ , \tag{1.5}$$

for some $q>n$. We emphasize that we have avoided making any assumptions which would guarantee the global monotonicity of the problem (1.1). Only local regularity and continuity properties are required in a $W^{1,\infty}$-neighborhood of a sufficiently regular solution. For related results based on monotonicity arguments, we refer to [8], [9], and [20], and to the literature cited therein. Our assumptions exclude problems with solutions having unbounded gradients, e.g., in the neighborhood of reentrant corners or at points where the ellipticity degenerates.

2 Primal Finite Element Discretization

For the discretization of problem (P) we consider a very simple finite element method. Let $\{T_h\}_{h>0}$ be a "regular" family of decompositions of Ω into simplices T of width h, $h\to 0$, satisfying the usual "uniform size and shape" condition (see, e.g., [8]). Corresponding to T_h, we introduce the finite element spaces

$$H_h = \left\{v_h\in H\ ,\ v_{h|T}\in P_1\ ,\ \forall\ T\in T_h\right\} \subset W^{1,\infty}(\Omega)\ ,$$

where, for $k\in \mathbf{N}_0$, P_k is the space of all polynomials of degree less or equal k. Then, the discrete analogue of problem (P) reads as follows.
(P_h) Find $u_h\in H_h$, such that

$$a(u_h;v_h) = (f,v_h)\ ,\ \forall\ v_h\in H_h\ . \tag{2.1}$$

Introducing the usual nodal basis of H_h, problem (P_h) may be written as a nonlinear algebraic system in $\mathbf{R}^{N_h}$ where $N_h=\dim(H_h)$.

Theorem 1. *For sufficiently small* h, $h<h_0$, *the discrete problems* (P_h) *possess locally unique solutions* $u_h\in H_h\cap B_R(u)$, *and there holds*

$$|||u-u_h|||_\infty \leq ch^{1-n/p}\ . \tag{2.2}$$

Proof. The proof uses a homotopy argument and bases on the stability properties of the linear Ritz operator $R_h: H_h^* \to H_h$, corresponding to the regular bilinear form $a'(u;\cdot,\cdot)$, which, for a functional $g \in H_h^*$ is defined through the relation

$$a'(u;R_h g,v_h) = \langle g,v_h \rangle , \quad \forall\, v_h \in H_h .$$

This notion is a natural extension of the classical Ritz projection if one identifies a function $v \in W^{1,2}(\Omega)$ with the functional $<v,\cdot> = a'(u;v,\cdot)$. Under our assumptions, there holds (see [14] and [23] for a proof in $\mathbf{R}^2$)

$$|||R_h g|||_\infty \le c \log(1/h)\, ||g||_{\infty;h} , \tag{2.3}$$

where the discrete norm $||\cdot||_{\infty;h}$ is defined through

$$||g||_{\infty;h} = \sup\left\{<g,v_h> ,\ v_h \in H_h,\ ||v_h||_{1,1}=1\right\} .$$

In view of the well-known approximation estimate

$$\inf\{|||u-v_h|||_\infty,\ v_h \in H_h\} \le ch^{1-n/p}||u||_{2,p} ,$$

this bound for R_h implies the error estimate

$$|||u-R_h u|||_\infty \le \hat{c}h^{1-n/p}||u||_{2,p} . \tag{2.4}$$

In order to carry this result over to the nonlinear problem (P), we use an argument introduced in [15] and [10]. We set $a_\theta(v;w) \equiv \theta a(v;w) + (1-\theta)a'(u;v-u,w)$, for a parameter $\theta \in [0,1]$, and consider the auxiliary problems

(P^θ) Find $u^\theta \in V$, such that: $\quad a_\theta(u_\theta;v) = \theta(f,v) , \quad \forall\, v \in H ,$

(P_h^θ) Find $u_h^\theta \in V_h$, such that: $\quad a_\theta(u_h^\theta;v_h) = \theta(f,v_h) , \quad \forall\, v_h \in H_h .$

Clearly, $u_h^1 = u_h$, and $u_h^0 = R_h u$. Further, for all $\theta \in [0,1]$, $u_\theta = u$ is a solution of (P^θ), and there holds $a_\theta'(u;\cdot,\cdot) \equiv a'(u;\cdot,\cdot)$. We introduce the set

$$\Xi_h = \left\{\theta \in [0,1],\ (P_h^\theta) \text{ has a solution } u_h^\theta \in H_h \cap B_R(u), \text{ s.t. } |||u-u_h^\theta|||_\infty \le 2\hat{c}h^{1-n/p}\right\} ,$$

where $\hat{c}$ is the constant in the error estimate (2.4). For $\theta \in \Xi_h$, we find after some rearrangement of terms that

$$\begin{aligned} a'(u;R_h u-u_h^\theta,v_h) &= a'(u;u-u_h^\theta,v_h) + \theta\left\{a(u;v_h)-a(u_h^\theta,v_h)\right\} - \theta\left\{a(u;v_h)-a(u_h^\theta,v_h)\right\} \\ &= \theta a'(u;u-u_h^\theta,v_h) - \theta \int_0^1 a'(u_h^\theta+s(u-u_h^\theta);u-u_h^\theta,v_h)\, ds \quad . \end{aligned}$$

The right hand side of this equation can be viewed as a functional $v_h^* \in H_h^*$ being applied to v_h. In this sense, there holds $R_h u - u_h^\theta = R_h v_h^*$, and, consequently,

$$|||R_h u - u_h^\theta|||_\infty \le c \log(1/h)\, \|v_h^*\|_{\infty;h} \le c\theta \log(1/h)\, |||u - u_h^\theta|||_\infty^2 .$$

Hence, by the error estimate (2.4), we obtain that

$$|||u - u_h^\theta|||_\infty \le c' h^{1-n/p} + c''\theta \log(1/h)\, |||u - u_h^\theta|||_\infty^2 ,$$

with constants c', c" independent of h and θ. Then, for sufficiently small h, $h<h_0$, a standard continuation argument implies that the set Ξ_h is open as well as closed in $[0,1]$. Further, by definition, $0 \in \Xi_h$. Then, it follows that $\Xi_h = [0,1]$, and therefore that $1 \in \Xi_h$, which implies the assertion. ♦

Remark 2.1. The following example shows that the logarithm in the stability estimate (2.3) cannot be removed in. On the circle $\Omega = \{|x|<1\}$, the function $v(x) = (x_1+x_2)\log(|x|)$ is the unique solution in $H = W_0^{1,2}(\Omega)$ of the variational equation $(\nabla v, \nabla\varphi) = (g, \nabla\varphi)$, $\forall\ \varphi \in H$, where the right hand side

$$g(x) = |x|^{-2}(x_1^2+2x_1x_2-x_2^2, -x_1^2+2x_1x_2+x_2^2)^T$$

is bounded while ∇v has a logarithmic singularity. Hence, for the Ritz projection $R_h g$ the estimate (2.2) is best possible. However, if the functional $g \in H_h^*$ is given in the form $\langle g, \varphi_h \rangle = a'(u; v, \varphi)$, with some function $v \in H \cap W^{1,\infty}(\Omega)$, then the logarithm in the estimate (5) can be dropped, $|||R_h v|||_\infty \le c|||v|||_\infty$ (see [25]). This sharper result will be used below whenever possible.

Remark 2.2. We remark that, under the above assumptions, one can derive an improved estimate for the pointwise error of the form

$$\|u - u_h\|_\infty = O(h^{2-n/p}\log(1/h)) . \tag{2.5}$$

The proof is by a "duality argument" using the adjoint of the regular operator A'(u) (see [10] and [15]). We omit the technical details since this result does not contribute much to the objective of this paper.

For the computation of the solutions u_h we consider the Newton-Raphson method. Starting with a suitable initial guess $u_h^0 \in V_h \cap B_R(u)$, one obtains a sequence of iterates $u_h^t \in H_h$, $t \in \mathbb{N}$, by successively solving the linear problems

$$a'(u_h^{t-1}; u_h^t, v_h) = a'(u_h^{t-1}; u_h^{t-1}, v_h) - a(u_h^{t-1}; v_h) + (f, v_h), \quad \forall\ v_h \in H_h . \tag{2.6}$$

In view of the results of Theorem 1, the quadratic convergence of this iteration is guaranteed for any fixed h which is sufficiently small, say $h<h_1<h_0$; see the general theory in [21]. But, since our assumptions do not ensure the convergence of the Newton iteration for the "continuous" problem (P), it cannot be expected that the speed of this convergence is uniform, as $h\to 0$, i.e., the "mesh-independence principle" of [1] does not hold in the strict sense in this situation. However, we have the following somewhat weaker result.

Theorem 2. *For sufficiently small* h , $h\le h_0$, *there exist positive constants* c_1, c_2, c_3 *independent of* h , *such that for any starting value satisfying*

$$u_h^0\in H_h\cap B_R(u)\,, \qquad |||u-u_h^0|||_\infty \le c_1\log(1/h)^{-1}\,, \tag{2.7}$$

the Newton iterates $u_h^t\in H_h\cap B_R(u)$ *are well defined and the accuracy of the discretization error is reached after at most* $t_h\le c_2\log\log(1/h)$ *steps,*

$$|||u-u_h^t|||_\infty \le c_3h^{1-n/p}\,, \quad t\ge t_h\,. \tag{2.8}$$

Proof. The proof is similar to that of Theorem 1. First, assuming the existence of the iterate $u_h^t\in H_h\cap B_R(u)$, for some $t>0$, we derive the identity

$$\begin{aligned} a'(u;R_hu-u_h^t,v_h) &= a'(u;u-u_h^t,v_h) - a'(u_h^{t-1};u-u_h^t,v_h) \\ &\quad + a'(u_h^{t-1};u-u_h^{t-1},v_h) - \int_0^1 a'(u_h^{t-1}+s(u-u_h^{t-1});u-u_h^{t-1},v_h)\,ds\,. \end{aligned}$$

Then, analogously as in the proof of Theorem 1 we conclude the estimate

$$|||u-u_h^t|||_\infty \le c'h^{1-n/p} + c''\log(1/h)\Big(|||u-u_h^t|||_\infty+|||u-u_h^{t-1}|||_\infty\Big)\,|||u-u_h^{t-1}|||_\infty\,. \tag{2.9}$$

Setting $a_t=\|u-u_h^t\|_{1,\infty}$, $\delta=c'h^{1-n/p}$, and $\beta=c''\log(1/h)$, this reads

$$a_t\le\delta+\beta(a_t+a_{t-1})a_{t-1}\,.$$

We want to show that, for sufficiently small h ,

$$a_t\le 3\delta+\frac{1}{3\beta}(3\beta a_0)^{2^t}\,, \quad t\ge 1\,. \tag{2.10}$$

From this one easily obtains that, for starting values u_h^0 satisfying (2.7), all iterates $u_h^t\in H_h\cap B_R(u)$ exist and satisfy (2.8) after $t\le t_h=c_2\log\log(1/h)$ steps. To proof (2.10), we set $b_t=3\beta a_t$, and require h to be sufficiently small, such that $c'c''\log(1/h)h^{1-n/p}\le 1/32$. By induction we find that $b_t\le 1$, and consequently,

$b_t \le 9\beta\delta/2 + b_{t-1}^2$. Hence, (2.10) holds true for $t = 1$. Suppose that it holds for some t-1 . Since $b_0 \le 1$, and $1+18\beta\delta+4b_0 \le 2$, for $c_1 = (24c'')^{-1}$, we obtain that

$$b_t \le \tfrac{9}{2}\beta\delta + (9\beta\delta+b_0^{2^{t-1}})^2 = \tfrac{9}{2}\beta\delta(1+18\beta\delta+4b_0^{2^t}) + b_0^{2^t} \le 9\beta\delta + b_0^{2^t} ,$$

and (2.10) follows by induction. ♦

Next, we consider a multi-level variant of the Newton method (2.6) which works on successively refined meshes; see, e.g., [29] and [30]. Let H_t be finite element subspaces of H , with decreasing mesh sizes $h_0 > \cdots > h_t \to 0$ $(t\to 0)$. Starting with an initial guess $u_0 \in H_0$, on the coarsest mesh, one determines approximations $u_t \in H_t$, $t \in \mathbf{N}$, through successively solving the linear problems

$$a'(u_{t-1};u_t,v) = a'(u_{t-1};u_{t-1},v) - a(u_{t-1};v) + (f,v) , \quad \forall\ v \in H_t . \tag{2.11}$$

The question is now how rapidly the meshsize h_t may be decreased in this process without loosing the convergence properties of the Newton method. In [24] the local quadratic convergence was shown under the condition that $h_t \ge \kappa h_{t-1}$. With the present approach this result can be sharpened as follows.

Theorem 3. *Suppose that the coarsest mesh size* h_0 *is sufficiently small and that the initial value* $u_0 \in H_0 \cap B_R(u)$ *satisfies*

$$|||u-u_0|||_\infty \le c_0 h_0^{1-n/p} . \tag{2.13}$$

Further, let the mesh sizes h_t *be chosen such that*

$$h_t \ge \kappa h_{t-1}^2 \log(1/h_t)^{p/(p-n)}, \tag{2.14}$$

with a sufficiently large constant $\kappa > 0$. *Then, the iterates* $u_t \in H_t \cap B_R(u)$ *are well defined and there holds*

$$|||u-u_t|||_\infty \le c_1 h_t^{1-n/p} . \tag{2.15}$$

Proof. The proof follows the same line of argument as the proof of Theorem 2. Assuming that the iterate $u_{t-1} \in H_{t-1} \cap B_R(u)$ exists, we obtain the estimate

$$|||u-u_t|||_\infty \le c' h_t^{1-n/p} + c'' \log(1/h_t) \left(|||u-u_t|||_\infty + |||u-u_{t-1}|||_\infty\right) |||u-u_{t-1}|||_\infty .$$

Setting $a_t = |||u-u_t|||_\infty$, $\delta_t = c' h_t^{1-n/p}$, and $\beta_t = c'' \log(1/h_t)$, this inequality reads

$$a_t \le \delta_t + \beta_t (a_t + a_{t-1}) a_{t-1} .$$

Next, we set $\sigma = \sup_{t\geq 1} \beta_t/\beta_{t-1} \geq 1$, and fix a number $m\geq 2\sigma+1$. Then, for h_0 sufficiently small, such that $a_0\leq(m\beta_1)^{-1}$ and $\delta_t\leq(m^2\beta_t)^{-1}$, an elementary induction argument shows that $a_t\leq(m\sigma\beta_t)^{-1}$. This implies that

$$a_t \leq \gamma\delta_t+\gamma\beta_t a_{t-1}^2 ,$$

where $\gamma=m/(m-1)$. To show (2.16), we note that it clearly holds for $t=1$, with $c_1=2\gamma\sigma$. Suppose that it holds true for some $t-1 \geq 2$. Then,

$$a_t \leq \gamma\delta_t + \gamma\beta_t a_{t-1}^2 \leq \gamma\delta_t + \kappa^{-2}\gamma c'' c_1^2\left(\log(1/h_t)^{p/(p-n)} h_{t-1}^2\right)^{1-n/p} ,$$

and, consequently, $a_t \leq c_1\delta_t$, for sufficiently large κ . From this the asserted estimate follows by induction. ♦

3 Mixed Finite Element Discretization

If mainly the approximations to the stresses $\sigma \equiv F[\cdot,\nabla u]$ in problem (1.1) are of interest, one may use a "mixed" scheme basing on the variational principle of Prange-Hellinger-Reissner. To this end, one introduces the function spaces

$$V = L^2(\Omega) , \quad W = \left\{\mu\in [L^2(\Omega)]^n,\ \nabla\cdot\mu\in L^2(\Omega),\ n\cdot\mu_{|\Gamma_2}=0\right\} ,$$

and the corresponding norms

$$\|v_h\|_V = \|v_h\|_2 , \quad \|\mu_h\|_W = \|\mu_h\|_2 + \|\nabla\cdot\mu_h\|_2 .$$

In view of our assumptions, the function $F[x,\cdot]$ is invertible in a R-neighborhood of $\eta=\nabla u(x)$, $x\in\Omega$, with the inverse denoted by $G = F^{-1}$. Then, setting

$$a(\mu;\psi) = (G[\cdot,\mu],\psi) , \quad b(v,\mu) = (v,\nabla\cdot \mu) , \quad v\in V ,\ \mu,\psi\in W ,$$

a mixed formulation of problem (1.1) reads

$(\widetilde{P})$ Find $\{u,\sigma\}\in V\times W$, such that

$$a(\sigma;\mu) + b(u,\mu) = 0 , \qquad \forall\ \mu\in W , \tag{3.1}$$

$$b(v,\sigma) = -(v,f) , \quad \forall\ v\in V . \tag{3.2}$$

Clearly, the locally unique solution $u\in H\cap W^{1,\infty}(\Omega)$ of the primal problem (1.1) corresponds to a solution $\{u,\sigma\}\in V\times W$ of the saddle point problem $(\widetilde{P})$, where $\sigma \equiv F[\cdot,\nabla u]$. By our assumptions , the Jacobian matrix of $G[x,\cdot]$,

$$G'[x,\cdot] = \nabla_\mu G[x,\cdot] = F'[x,F^{-1}[\cdot]]^{-1} ,$$

evaluated at any $\mu=F[x,\nabla v(x)]$, $v\in B_R(u)$, is uniformly regular for $x\in\Omega$,

$$\sup\left\{\|G'[x,\mu]^{-1}\|\,,\ \mu=\nabla v,\ v\in B_R(u),\ x\in\Omega\right\} > 0\,. \tag{3.3}$$

This particularly implies that the solution $\{u,\sigma\}\in V\times W$ of problem $(\widetilde{P})$ is unique in a certain ball

$$B_{R'}(\sigma) = \{\mu\in L^\infty(\Omega)^n,\ \|\mu-\sigma\|_\infty\le R'\}\,,$$

and that the bilinear form $a'(\sigma;\cdot,\cdot) = (G'[\cdot,\sigma]\cdot,\cdot)$ is coercive on the null-space,

$$Z=\left\{\mu\in W\,,\ b(v,\mu)=0\,,\ \forall\ v\in V\right\},$$

of the bilinear form b ,

$$\sup\left\{a'(\sigma;\mu,\psi)\,,\ \psi\in W,\ \|\psi\|_2=1\right\} \ge \gamma\,\|\mu\|_W\,,\quad \mu\in Z\,. \tag{3.4}$$

For this it is crucial that $\mu\in Z$ necessarily implies that $\nabla\cdot\mu=0$, which is usually called the "equilibrium condition". Furthermore, there holds the stability estimate

$$\sup\left\{b(v,\psi)\,,\ \psi\in W\right\} \ge \beta\,\|v\|_2\,,\quad v\in V\,. \tag{3.5}$$

In view of the abstract theory in [5], the estimates (3.4) and (3.5) guarantee that, for any given functionals $g\in V^*$ and $\rho\in W^*$, the linear saddle point problem

$$a'(\sigma;\mu,\psi) + b(v,\psi) + b(w,\mu) = \langle g,w\rangle + \langle\rho,\psi\rangle\,, \tag{3.6}$$

for all $\{w,\psi\}\in V\times W$ possesses a unique solution $\{v,\mu\}\in V\times W$.

For discretizing problem $(\widetilde{P})$, let $V_h\subset V$ and $W_h\subset W$ be proper finite element subspaces. Then, the discrete problem reads

$(\widetilde{P}_h)$ Find $\{u_h,\sigma_h\}\in V_h\times W_h$, such that

$$a(\sigma_h;\mu_h) + b(u_h,\mu_h) = 0\,,\qquad \forall\ \mu_h\in W_h\,, \tag{3.6}$$

$$b(v_h,\sigma_h) = -(v_h,f)\,,\quad \forall\ v_h\in V_h\,. \tag{3.7}$$

Appropriate examples of lowest order elements are due to Raviart-Thomas [26],

$$V_h = \left\{v_h\in V\,,\ v_{h|T}\in P_0,\ \forall\ T\in \mathbf{T}_h\right\},$$

$$\widetilde{W}_h = \left\{\mu_h\in [L^2(\Omega)]^n,\ \mu_{h|T}\in [P_0]^n\oplus\mathrm{span}\{x\},\ \forall\ T\in\mathbf{T}_h\right\},$$

$$W_h = \left\{\mu_h\in \widetilde{W}_h\,,\ n\cdot\mu_h \text{ continuous across interelement boundaries}\right\},$$

and to Brezzi, Douglas and Marini [6],

$$V_h = \left\{ v_h \in V \,,\ v_{h|T} \in P_0, \forall\, T \in \mathbf{T}_h \right\},$$

$$\widetilde{W}_h = \left\{ \mu_h \in [L^2(\Omega)]^n,\ \mu_{h|T} \in [P_1]^n, \forall\, T \in \mathbf{T}_h \right\},$$

$$W_h = \left\{ \mu_h \in \widetilde{W}_h \,,\ n \cdot \mu_h \text{ continuous across interelement boundaries} \right\}.$$

The continuity requirement on the normal components guarantees that $W_h \subset W$. Below, we will use the specific properties of these finite element spaces, particularly that they satisfy the discrete "equilibrium condition",

$$Z_h = \left\{ \psi_h \in W_h \,,\ b(v_h, \psi_h) = 0 \right\} \subset Z \,. \tag{3.8}$$

Both classes of elements are well known to converge with optimal orders in the linear case, i.e., the approximations for both variables σ and u are of the order $O(h)$, and a certain post-processing yields improved approximations to u of the order $O(h^2)$; see [2]. In particular, these elements satisfy the uniform stability condition

$$\sup\left\{ b(v_h, \mu_h) \,,\ \mu_h \in W_h \,,\ \|\mu_h\|_W = 1 \right\} \geq \gamma \|v_h\|_V \,,\quad v_h \in V_h \,, \tag{3.9}$$

which is the discrete analogue of the relation (3.4). On the discrete level, the existence of solutions of the primal problem (P_h) does not directly imply also the existence of solutions of the nonlinear saddlepoint problems $(\widetilde{P}_h)$. For this we shall employ an argument analogous to that used in the proof of Theorem 1.

Theorem 4. *For sufficiently small* h, *the discrete mixed problems* $(\widetilde{P}_h)$ *possess locally unique solutions* $\{u_h, \sigma_h\} \in V_h \times W_h$, *and there holds*

$$\|\sigma - \sigma_h\|_\infty \leq c h^{1-n/p} \,. \tag{3.10}$$

Proof. Again we employ the L^∞-stability of the "mixed" finite element approximation in the linear case. By a standard perturbation argument, for sufficiently small h (say for $h < h_0$), the bilinear form $c'(\sigma;\cdot,\cdot)$ is also coercive on the discrete null-spaces Z_h. This together with the stability estimate (3.9) guarantees that then, for any fixed functional $\rho \in W_h^*$, there exist unique solutions $\{v_h, \mu_h\} \in V_h \times W_h$ of the linear saddle point problems

$$a'(\sigma; \mu_h, \psi_h) + b(v_h, \psi_h) + b(w_h, \mu_h) = \langle \rho, \psi_h \rangle \,, \tag{3.11}$$

for all $\{w_h, \psi_h\} \in V_h \times W_h$. For this we postulate the a priori estimate

$$\|\mu_h\|_\infty \le c \log(1/h)\, \|\rho\|_{\infty;h}\,, \tag{3.12}$$

uniformly for $0<h\le h_0$, where the discrete norm $\|\cdot\|_{\infty;h}$ is defined through

$$\|\rho\|_{\infty;h} = \sup\{<\rho,\psi_h>\,,\ \psi_h\in W_h,\ \|\psi_h\|_1=1\}\ .$$

For the particular elements, mentioned above, this stability result follows from the error analysis in [11] and [31]. Again, if the functional ρ is given through a pair $\{v,\mu\}\in V\times W$, satisfying $\mu=F[\cdot,\nabla v]$, in the form $<\rho,\cdot> = a'(\sigma;\mu,\cdot) + b(v,\cdot)$, then the logarithm in (3.12) can be dropped, i.e., there holds $\|\mu_h\|_\infty \le c\|\mu\|_\infty$. This particularly implies the error estimate

$$\|\mu_h-\mu\|_\infty \le \hat{c}h^{1-n/p}\|\mu\|_{1,p}\,. \tag{3.13}$$

Next, we introduce the semi-linear form $a_\theta(\mu;\psi) \equiv \theta a(\mu;\psi) + (1-\theta)a'(\sigma;\mu-\sigma,\psi)$, for a parameter $\theta\in[0,1]$, and consider the auxiliary problems

$(\tilde{P}^\theta)$ Find $\{u^\theta,\sigma^\theta\}\in V\times W$, such that

$$a_\theta(\sigma^\theta;\psi) + b(u^\theta,\psi) + b(v,\sigma^\theta) = -(v,f)\,,\quad \forall\ \{v,\psi\}\in V\times W\,,$$

$(\tilde{P}_h^\theta)$ Find $\{u_h^\theta,\sigma_h^\theta\}\in V_h\times W_h$, such that

$$a_\theta(\sigma_h^\theta;\psi_h) + b(u_h^\theta,\psi_h) + b(v_h,\sigma_h^\theta) = -(v_h,f)\,,\quad \forall\ \{v_h,\psi_h\}\in V_h\times W_h\,.$$

Again, by construction, $\{u_h^1,\sigma_h^1\} = \{u_h,\sigma_h\}$, and $\{u_h^0,\sigma_h^0\}$ is the solution of the equation (3.11), for $<\rho,\cdot> \equiv a'(\sigma;\sigma,\cdot)+b(u,\cdot)$. Further, $\{u^\theta,\sigma^\theta\}=\{u,\sigma\}$ is a solution of $(\tilde{P}^\theta)$, for all $\theta\in[0,1]$. We consider again the set

$$\Xi_h = \left\{\theta\in[0,1]\,,\ (\tilde{P}_h^\theta) \text{ has a solution } \{u_h^\theta,\sigma_h^\theta\}\in V_h\cap W_h, \text{ s.t. } \|\sigma-\sigma_h^\theta\|_\infty \le 2\hat{c}h^{1-n/p}\right\},$$

where $\hat{c}$ is the constant in the error estimate (3.13). Let us, for some $\theta\in[0,1]$, assume that $\{u_h^\theta,\sigma_h^\theta\}\in V_h\times\{W_h\cap B_{R'}(\sigma)\}$. Then, a simple calculation leads us to the equations

$$a'(\sigma;\sigma_h^0-\sigma_h^\theta,\psi_h) + b(u_h^0-u_h^\theta,\psi_h) = \theta a'(\sigma;\sigma-\sigma_h^\theta,\psi_h) - \theta\int_0^1 a'(\sigma_h^\theta+s(\sigma-\sigma_h^\theta);\sigma-\sigma_h^\theta,\psi_h)\,ds$$

and

$$b(v_h,\sigma_h^0-\sigma_h^\theta) = b(v_h,\sigma-\sigma_h^\theta) = 0\,,$$

for all $\{v_h,\psi_h\}\in V_h\times W_h$. The right hand side of the first equation can be viewed as a functional $\rho_h\in W_h^*$, applied to ψ_h, satisfying

$$\|\rho_h\|_{1;h} \le c\theta\|\sigma-\sigma_h^\theta\|_\infty^2\,. \tag{3.14}$$

Hence, by the stability estimate (3.12) and the error estimate (3.13), we obtain that

$$\|\sigma-\sigma_h^\theta\|_\infty \le c'h^{1-n/p} + c''\theta \log(1/h) \|\sigma-\sigma_h^\theta\|_\infty^2 , \tag{3.15}$$

with constants c' , c'' independent of h and θ . Clearly, for $\theta=0$, the discrete solution $\{u_h^0,\sigma_h^0\}$ exists and has all the desired properties. Then, for sufficiently small h , $h<h_0$, one again concludes that Ξ_h is open as well as closed in $[0,1]$, and hence $\Xi_h = [0,1]$. Therefore, for $h<h_0$, there exist locally unique solutions $\{u_h^1,\sigma_h^1\}\in V_h\times\{W_h\cap B_R(\sigma)\}$ of the problems $(\tilde{P}_h^1)$ satisfying

$$\|\sigma^1-\sigma_h^1\|_\infty \le 2c'h^{1-n/p} , \tag{3.16}$$

which completes the proof. ♦

Remark 3.1. We remark that, under the above assumptions, one can derive also a pointwise error estimate for the displacement component u_h ,

$$\|u-u_h\|_\infty = O(h^{1-n/p}\log(1/h)) . \tag{3.17}$$

By a certain local postprocessing (see, e.g., [2] and [16]) one may obtain approximations $\bar{u}_h$ which even satisfy

$$\|u-\bar{u}_h\|_\infty = O(h^{2-n/p}\log(1/h)) , \tag{3.18}$$

i.e., the mixed finite element scheme has the same asymptotic accuracy for both variables u and $\sigma = F[\cdot,\nabla u]$ as its primal counterpart. We again omit the technical details since they are not relevant for the purpose of this paper.

For the computation of the solutions $\{u_h,\sigma_h\}\in V_h\times\{W_h\cap B_{R'}(\sigma)\}$ we consider again the Newton-Raphson method. The t-th iteration step reads

(P_h^t) Find $\{u_h^t,\sigma_h^t\}\in V_h\times W_h$, such that

$$a'(\sigma_h^{t-1};\sigma_h^t,\mu_h) + b(u_h^t,\mu_h) = a'(\sigma_h^{t-1};\sigma_h^{t-1},\mu_h) - a(\sigma_h^{t-1};\mu_h) , \quad \forall\, \mu_h\in W_h , \tag{3.19}$$

$$b(v_h,\sigma_h^t) = -(v_h,f) , \forall\, v_h\in V_h . \tag{3.20}$$

Starting with a suitable initial pair $\{u_h^0,\sigma_h^0\}\in V_h\times\{W_h\cap B_R(\sigma)\}$, one obtains a sequence of iterates $\{u_h^t,\sigma_h^t\}$ by successively solving linear saddlepoint problems. In analogy to the result of Theorem 2, we have the following convergence result.

Theorem 5. *Let* h *be sufficiently small, such that the assertions of Theorem 4 hold. Then, there exist positive constants* c_1, c_2, c_3 *independent of* h , *such that for any starting value* $\sigma_h^0 \in W_h \cap B_R(\sigma)$ *satisfying*

$$\|\sigma-\sigma_h^0\|_\infty \le c_1 \log(1/h)^{-1} , \tag{3.21}$$

the Newton iterates $\{u_h^t, \sigma_h^t\} \in V_h \cap \{W_h \cap B_R(\sigma)\}$ *exist and the accuracy of the discretization error is reached after at most* $t_h \le c_2 \log|\log(1/h)|$ *steps,*

$$\|\sigma-\sigma_h^t\|_\infty \le c_3 h^{1-n/p}, \quad t \ge t_h . \tag{3.22}$$

Proof. The proof is similar to that of Theorem 4. Let $\{\tilde{u}_h, \tilde{\sigma}_h\} \in V_h \times W_h$ denote the solution of the auxiliary linear problem (3.11), with the special right hand side $\langle\rho,\cdot\rangle = a'(\sigma;\sigma,\cdot) + b(u,\cdot)$. Then, assuming again the existence of a solution $\{u_h^t, \sigma_h^t\} \in V_h \times \{H_h \cap B_R(\sigma)\}$, a simple calculation leads us to

$$a'(\sigma;\tilde{\sigma}_h-\sigma_h^t,\psi_h) + b(\tilde{u}_h-u_h^t,\psi_h) = a'(\sigma;\sigma-\sigma_h^t,\psi_h) + b(u-u_h^t,\psi_h)$$

$$= a'(\sigma;\sigma-\sigma_h^t,\psi_h) - a'(\sigma_h^{t-1};\sigma-\sigma_h^t,\psi_h)$$

$$+ a'(\sigma_h^{t-1};\sigma-\sigma_h^{t-1},\psi_h) - \int_0^1 a'(\sigma_h^{t-1}+s(\sigma-\sigma_h^{t-1});\sigma-\sigma_h^{t-1},\psi_h)\, ds$$

From this we conclude analogously as in the proof of Theorem 4 that

$$\|\sigma-\sigma_h^t\|_\infty \le c'h^{1-n/p} + c''\log(1/h)\left(\|\sigma-\sigma_h^t\|_\infty + \|\sigma-\sigma_h^{t-1}\|_\infty\right)\|\sigma-\sigma_h^{t-1}\|_\infty . \tag{3.23}$$

Then, as in the proof of Theorem 2, this implies that all iterates are well defined and satisfy (3.22) after at most $t < t_h = c_2 \log\log(1/h)$ steps. ♦

Remark 3.2. Following the line of argument of the above proof, one also obtains a convergence result analogous to Theorem 3 for the multilevel Newton method applied to the mixed approximation scheme. We omit the technical details.

Remark 3.3. We address the question of how to evaluate the form $c'(\cdot;\cdot,\cdot)$ in the Newton iteration in the case that only the function $F[\cdot,\eta]$ is explicitly given. Notice that $G'[\cdot,\eta] = F'[\cdot,F^{-1}[\cdot,\eta]]^{-1}$. In order to compute, for example, the contribution to $a'(\sigma_h^{t-1};\sigma_h^{t-1},\psi_h)$, on an element $T \in \mathbf{T}_h$, one has first to evaluate $\eta_T = F^{-1}[\cdot,\sigma_h^{t-1}]_{|T}$, and then to invert the matrix function $F'[\cdot,\eta_T]$. This needs to be done only at the center z_T of the element T , which generally results in a quadrature error. This error, however, does not affect the asymptotic accuracy of the scheme.

4 Extensions to Systems in Nonlinear Elasticity

The application of the analysis of the preceding sections to the elliptic systems of nonlinear elasticity is possible if all the needed technical prerequisites can be made available for the finite element discretization. In this case the nonlinear problem (1.1) formally reads

$$- \nabla\cdot\sigma = f\,, \quad \sigma = F[\cdot,\varepsilon(u)]\,, \quad \text{in } \Omega\,, \tag{4.1}$$

with the boundary conditions

$$u_{|\Gamma_1} = b\,, \quad n\cdot\sigma_{|\Gamma_2} = g\,. \tag{4.2}$$

Here, $\varepsilon(u) = (\nabla u + \nabla u^T)/2$ and $\sigma(u)$ are the symmetric strain and stress tensors, respectively, of the displacement field u, and $F = (F_{ik})_{i,k=1}^n$ is a nonlinear symmetric $n\times n$-tensor function. If we restrict ourselves to the special case $b = 0$, and $g = 0$, the variational formulation of problem (4.1), (4.2) reads like problem (P), where in this case

$$H = \left\{ v\in [W^{1,2}(\Omega)]^n,\ v_{|\Gamma_1}=0 \right\}, \qquad a(v;w) = (F[\cdot,\varepsilon(v)],\varepsilon(w))\,.$$

The condition for local uniqueness for a solution $u\in H\cap[W^{1,\infty}(\Omega)]^n$ is now that the eigenvalues of the Jacobian $(n\times n)\times(n\times n)$-tensor $F'[x,\eta] = \nabla_\eta F[x,\eta]$, taken at any $\eta = \varepsilon(v)$, with $v\in B_R(u)$, are uniformly bounded away from zero. This is equivalent to the property

$$\sup\left\{ |F'[x,\eta]^{-1}|\,,\ \eta=\varepsilon(v(x)),\ v\in B_R(u),\ x\in\Omega \right\} > 0\,, \tag{4.3}$$

which particularly implies for the derivative

$$a'(u;v,w) = (F'[\cdot,\varepsilon(u)]\varepsilon(v),\varepsilon(w))$$

of the semi-linear form $a(\cdot;\cdot)$, at u, the estimate

$$\sup\left\{ a'(u;v,w)\,,\ w\in H,\ \|\nabla w\|=1 \right\} \geq \alpha\,\|\varepsilon(v)\|\,, \qquad v\in H\,. \tag{4.4}$$

Hence, by the second Korn inequality, there holds

$$\sup\left\{ a'(u;v,w)\,,\ w\in H,\ \|\nabla w\|=1 \right\} \geq \alpha\,\|\nabla v\|\,, \qquad v\in H\,. \tag{4.5}$$

If this is made an assumption, then the statements of Theorems 1, 2, and 3 for the primal finite element discretization (P_h) carry over to this situation without problems. The required L^∞-estimates for the linearized problem are available through the analysis in [24] and [10]. We omit the technical details since the argument follows closely the line of that already used above for the scalar case.

For the mixed finite element discretization the situation is much more complicated. This is due to the fact that even in the linear case the problem of constructing stable elements for general two- or three-dimensional elasticity problems is not satisfactorily solved yet. The mixed formulation of problem (4.1), (4.2) and its discrete analogues look formally like problems $(\widetilde{P})$ and $(\widetilde{P}_h)$, respectively. In this case the notation is

$$V = [L^2(\Omega)]^n, \quad W = \left\{\mu \in [L^2(\Omega)]^{n\times n},\ \mu=\mu^T,\ \nabla\cdot\mu \in [L^2(\Omega)]^n,\ n\cdot\mu_{|\Gamma_2}=0\right\},$$

and, setting $G = F^{-1}$,

$$a(\sigma;\psi) = (G[\cdot,\sigma],\psi), \qquad b(v,\sigma) = (v,\nabla\cdot\sigma).$$

As in the scalar case, the finite element subspaces $V_h \subset V$ and $W_h \subset W$ have to satisfy the uniform stability condition (3.9). But now also the regularity of the linearized form $a'(\sigma;\cdot,\cdot)$ on W_h is a problem, since it is difficult to construct pairs of spaces $\{V_h,W_h\}$ which satisfy simultaneously the stability condition (3.9) as well as the equilibrium condition (3.8), i.e.,

$$\mu_h \in W_h:\ b(v_h,\mu_h) = 0,\ \forall\ v_h \in V_h \quad \Rightarrow \quad \nabla\cdot\mu_h = 0. \tag{4.6}$$

One possible solution of this problem goes back to an idea of deVeubeke [13] . If the symmetry of the stress tensor σ_h is only approximately imposed through the use of Lagrangian multipliers, one gains additional flexibility in the construction of the trial spaces. This concept has succesfully been used in [3] and [4] in the construction of convergent elements for elasticity problems; see also the earlier literature [17], [18], [19], [22], and the discussion in [12]. This approach is based on the identity $\varepsilon(u) = \nabla u - \omega(u)$, where $\omega(u) \equiv (\nabla u - \nabla u^T)/2$ is the skew symmetric rotation tensor. Then, with the function spaces

$$\overline{W} = \left\{\mu \in [L^2(\Omega)]^{n\times n},\ \nabla\cdot\mu \in [L^2(\Omega)]^n,\ n\cdot\mu_{|\Gamma_2}=0\right\},$$

$$X = \left\{\lambda \in [L^2(\Omega)]^{n\times n},\ \lambda+\lambda^T=0\right\},$$

the modified mixed formulation of the elasticity problem (4.1), (4.2) reads

$(\overline{P})$ Find $\{u,\sigma,\omega\} \in V\times\overline{W}\times X$, such that

$$a(\sigma;\psi) + b(u,\psi) + (\omega,\psi) = 0, \qquad \forall\ \psi \in \overline{W}, \tag{4.7}$$

$$b(v,\sigma) + (\lambda,\sigma) = -(v,f), \qquad \forall\ \{v,\lambda\} \in V\times X. \tag{4.8}$$

Clearly, a solution $\{u,\sigma\}\in V\times W$ of problem $(\tilde{P})$ corresponds to a solution $\{u,\sigma,\omega\}\in V\times\overline{W}\times X$ of problem $(\overline{P})$, if one sets $\omega = \omega(u)$. The finite element approximation of problem $(\overline{P})$ works with subspaces $V_h\times\overline{W}_h\times X_h \subset V\times\overline{W}\times X$ and reads as follows.

$(\overline{P}_h)$ Find $\{u_h,\sigma_h,\omega_h\}\in V_h\times\overline{W}_h\times X_h$, such that

$$a(\sigma_h;\psi_h) + b(u_h,\psi_h) + (\omega_h,\psi_h) = 0\,, \qquad \forall\ \psi_h\in\overline{W}_h\,, \tag{4.9}$$

$$b(v_h,\sigma_h) + (\lambda_h,\sigma_h) = -(v_h,f_h)\,, \quad \forall\ \{v_h,\lambda_h\}\in V_h\times X_h\,. \tag{4.10}$$

Such schemes can be constructed on the basis of the the Raviart-Thomas as well of the Brezzi-Douglas-Marini elements. For example, the PEERS-element introduced in [3] uses the following spaces in $n=3$ dimensions:

$$V_h = \left\{v_h\in V\,,\ v_{h|T}\in[P_0]^3,\ \forall\ T\in\mathbf{T}_h\right\},$$

$$\overline{W}_h = \left\{\mu_h\in\overline{W},\ \mu_{h|T}\in RT_0\oplus B_0(T),\ \forall\ T\in\mathbf{T}_h\right\},$$

$$X_h = \left\{\lambda_h\in[C(\overline{\Omega})]^{3\times3},\ \lambda_{h|T}\in[P_1]^{3\times3},\ \forall\ T\in\mathbf{T}_h,\ \lambda_h+\lambda_h^T=0\right\},$$

where

$$RT_0 = \left\{\sigma\in[P_1]^{3\times3}\,,\ \sigma_i\in[P_0]^3\oplus\mathrm{span}\{x\},\ i=1,2,3\right\},$$

$$B_0(T) = \left\{\sigma\in[P_2]^{3\times3},\ \sigma_i=\mathrm{curl}(b_Ta)\,,\ i=1,2,3,\ a\in[P_0]^3\right\},$$

and b_T is the cubic "bubble function" on T which vanishes along $\partial\Omega$ and has mean value one. For this particular element the stability estimate (3.9) and the equilibrium property (3.8) have been established in [27] and [28]. In a standard way, for linear problems, these imply the optimal order L^2-error estimate

$$\|P_hu-u_h\|_2 + h\|\sigma-\sigma_h\|_2 \le ch^2\left\{\|\sigma\|_{1,2}+\|\omega(u)\|_{1,2}\right\}, \tag{4.11}$$

where P_h denotes the L^2-projection onto V_h , and the force is required to be piecewise constant, $f\in V_h$. It is very likely that also sharp L^∞-error bounds can be obtained for this scheme which would then lead to convergence results analogous to those stated in Theorems 4 and 5 for the scalar nonlinear case. This, however, must be left as a subject of further research.

5 A Numerical Example

A simple numerical test confirms our theoretical result on the convergence properties of the finite element Newton method. We consider the scalar boundary value problem

$$- \nabla \cdot \left\{ (1+|\nabla u|^2)^{p/2-1} \nabla u \right\} = f \,, \quad \text{in } \Omega \,, \qquad u_{|\partial\Omega} = 0 \,, \tag{5.1}$$

on the unit square $\Omega = (0,1)\times(0,1)$. We take the value $p = 1$, corresponding to the classical minimal surface problem, i.e., $F[\eta] = (1+|\eta|^2)^{-1/2}$. The right hand side is adjusted, such that the solution is $u(x) = \sin(\pi x_1)\sin(\pi x_2)/2$. The application of the primal finite element scheme leads to a nonlinear algebraic equation of the form

$$K(\xi)\xi = b \,, \tag{5.2}$$

for the nodal vector $\xi \in \mathbf{R}^N$ corresponding to u_h . On a rectangular uniform mesh of width h , the matrix $K(\xi)$ is penta-diagonal. The dimension of the system (5.2) is $n = (h^{-1}-1)^2$. The Newton iteration for its solution reads

$$\left(\xi_{t-1}^T K'(\xi_{t-1}) + K(\xi_{t-1}) \right) \Delta\xi_t = K(\xi_{t-1})\xi_{t-1} - b \,, \quad \xi_t = \xi_{t-1} + \Delta\xi_t \,, \tag{5.3}$$

where $K'(\cdot)$ denotes the Jacobian of $K(\cdot)$. The starting value is taken as $\xi_0 = 0$.

In the special case $p = 1$, the function $\mu = F[\cdot,\eta]$ can be explicitly inverted,

$$\eta = G[\cdot,\mu] \equiv F^{-1}[\cdot,\mu] = (1-|\mu|^2)^{-1/2}\mu \,,$$

for $|\mu| = (1+|\eta|^2)^{-1}|\eta| < 1$. Hence, the mixed scheme $(\widetilde{P}_h)$ for solving problem (5.1) is easily formulated. The resulting nonlinear system has the form

$$\begin{pmatrix} A(\eta) & B \\ B^T & 0 \end{pmatrix} \begin{pmatrix} \eta \\ \xi \end{pmatrix} = \begin{pmatrix} 0 \\ -b \end{pmatrix} , \tag{5.4}$$

for the nodal vectors $\{\eta,\xi\} \in R^M \times R^N$ corresponding to $\{\sigma_h, u_h\}$. For the lowest order element from the Raviart-Thomas class, on a rectangular uniform mesh, the matrices $A(\eta)$ and B^T are tridiagonal. The dimension of the mixed system (5.4) is $n = h^{-1}(5h^{-1}+2)$. The Newton iteration for its solution reads

$$\begin{pmatrix} \eta_{t-1}^T A'(\eta_{t-1}) + A(\eta_{t-1}) & B \\ B^T & O \end{pmatrix} \begin{pmatrix} \Delta\eta_t \\ \Delta\xi_t \end{pmatrix} = - \begin{pmatrix} A(\eta_{t-1})\eta_{t-1} + B\xi_{t-1} \\ B^T\eta_{t-1} + b \end{pmatrix} , \tag{5.5}$$

$\eta_t = \eta_{t-1}+\Delta\eta_t$, $\xi_t = \xi_{t-1}+\Delta\xi_t$, where again $A'(\cdot)$ is the Jacobian of $A(\cdot)$. The matrix $\eta_{t-1}^T A'(\eta_{t-1})+A(\eta_{t-1})$ is symmetric and positive definite, with a condition number of the order $O(1)$, as $h\to 0$, and is easily inverted. Consequently, the linear system (5.5) may be solved through a conjugate gradient iteration applied to a reduced equation, for $\Delta\xi_t$ alone, with the matrix $B^T[\eta_{t-1}^T A'(\eta_{t-1})+A(\eta_{t-1})]^{-1}B$. The starting value is again taken as $(\eta_0,\xi_0) = (0,0)$.

The following tables show the accuracy of these schemes as well as the convergence behavior of the Newton iterations. The quantity T_δ is the number of iteration steps needed for reaching the level of the discretization error $\delta_h = \|u-u_h\|_\infty$, and Δ_h indicates the size of $|||u-u_h|||_\infty$ and $\|\sigma-\sigma_h\|_\infty$, respectively, at which the transition from linear to quadratic convergence occurs in the Newton iteration.

Table 5.1. *Convergence behavior of the primal scheme*

h	n	$h^{-1}\|\|\|u-u_h\|\|\|_\infty$	$h^{-2}\|\|u-u_h\|\|_\infty$	T_δ	Δ_h
2^{-3}	49	2.726	1.202	4	.3393
2^{-4}	225	2.548	1.101	4	.2049
2^{-5}	961	2.493	1.083	4	.1937
2^{-6}	3969	2.474	1.079	4	.1907
2^{-7}	16129	2.469	1.078	5	.1900

Table 5.2. *Convergence behavior of the mixed scheme*

h	n	$h^{-2}\|\|\sigma-\sigma_h\|\|_\infty$	$h^{-2}\|\|\bar{u}-u_h\|\|_\infty$	T_δ	Δ_h
2^{-2}	88	1.280	.9371	3	.1366
2^{-3}	336	1.194	.8046	4	.1057
2^{-4}	1312	1.303	.8243	4	.1101
2^{-5}	5184	1.326	.8634	4	.1109
2^{-6}	20608	1.332	.8752	4	.1108

($\bar{u}$ denotes the L^2-projection of u into V_h .)

These results confirm the mesh independence of the Newton iteration as predicted by our theory. They also show the expected super approximation property of the mixed scheme.

References

1. Allgower,E.L., et al.: A mesh-independence principle for operator equations and their discretizations, SIAM J.Numer.Anal. 23,160-169 (1986).
2. Arnold,D.N., Brezzi,F.: Mixed and nonconforming finite element methods: implementation, postprocessing and error estimates, M^2AN 19, 7-32 (1985).
3. Arnold,D.N., Brezzi,F., Douglas,J.,Jr.: PEERS: A new finite element for plane elasticity, Jap.J.Appl.Math. 1, 347-367 (1984).
4. Arnold,D.N., Falk,R.S.: A new mixed formulation for elasticity, preprint 1989.
5. Brezzi,F.: On the existence, uniqueness and approximation of saddlepoint problems arising from Lagrangian multipliers, R.A.I.R.O.Anal.Numér. 8, 129-151 (1974).
6. Brezzi,F., Douglas,J.,Jr., Marini,L.D.: Two families of mixed finite element methods for second order elliptic problems, Numer.Math. 47, 217-235 (1985).
7. Brezzi,F., Douglas,J.,Jr., Marini,L.D.: Recent results on mixed finite element methods for second order elliptic problems, in "Vistas in Applied Mathematics", pp. 25-43, Springer: Heidelberg 1986.
8. Ciarlet,P.G.: The Finite Element Method for Elliptic Problems, North-Holland, Amsterdam 1978.
9. Chow,S.-S.: Finite element error estimates for non-linear elliptic equations of monotone type, Numer.Math. 54, 373-393 (1989).
10. Dobrowolski,M., Rannacher,R.: Finite element methods for nonlinear elliptic systems of second order, Math.Nachr. 94, 155-172 (1980).
11. Durán,R.: Error analysis in L^p for mixed finite element methods for linear and quasi-linear elliptic problems, R.A.I.R.O.Anal.Numér. 22, 371-387 (1988).
12. Falk,R.S., Morley, M.E.: Equivalence of finite element methods for problems in elasticity, preprint 1985.
13. Fraeijs deVeubeke,B.X.: Stress function approach, Bournemouth, World Conference in Finite Elements 1975, pp. J.1-J.51.
14. Frehse,J., Rannacher,R.: Eine L^1-Fehlerabschätzung für diskrete Grundlösungen in der Methode der finiten Elemente, in "Finite Elemente", J.Frehse et al., eds., Bonn.Math.Schr. 89, 92-114 (1976).
15. Frehse,J., Rannacher,R.: Asymptotic L^∞-error estimates for linear finite element approximations of quasi-linear boundary value problems, SIAM J.Numer.Anal. 15, 418-431 (1978).

16. Gastaldi,L., Nochetto,R.: Optimal L^∞-error estimates for nonconforming and mixed finite element methods of lowest order, Numer.Math. 50, 567-611 (1987).

17. Mirza,F.A., Olson,M.D.: The mixed finite element method in plane elasticity, Int.J.Numer.Methods Eng. 15, 273-290 (1980).

18. Nedelec,J.C.: Mixed finite elements in $\mathbf{R}^3$, Numer.Math. 35, 315-341 (1980).

19. Johnson,C., Mercier,B.: Some equilibrium finite element methods for two-dimensional elasticity problems, Numer.Math. 30, 103-116 (1978).

20. Noor,M.A.: Finite element approximation theory for strongly nonlinear problems, Comment.Math.Univ.St.Paul 31, 1-7 (1982).

21. Ortega,J.M., Rheinboldt,W.C.: Iterative Solution of Nonlinear Equations in Several Variables, Academic Press, New York-London 1970.

22. Pitkäranta,J., Stenberg,R.: Analysis of some mixed finite element methods for plane elasticity equations, Math.Comp. 41, 399-423 (1983).

23. Rannacher,R.: Zur L^∞-Konvergenz linearer finiter Elemente beim Dirichlet-Problem, Math.Z. 149, 69-77 (1976).

24. Rannacher,R.: On finite element approximation of general boundary value problems in nonlinear elasticity, Calcolo 17, 174-193 (1980).

25. Rannacher,R., Scott,R.: Some optimal error estimates for piecewise linear finite element approximation, Math.Comp. 38, 437-445 (1982).

26. Raviart,P.A., Thomas,J.M.: A mixed finite element method for second order elliptic problems, in Mathematical Aspects of the Finite Element Method, Lecture Notes in Mathematics 606, Springer, Berlin, 1977.

27. Stenberg,R.: On the construction of optimal mixed finite element methods for the elasticity problem, Numer.Math. 48, 447-462 (1986).

28. Stenberg,R.: A family of mixed finite elements for the elasticity problem, Numer.Math. 53, 513-538 (1988).

29. Vainikko,G.M.: Galerkin's perturbation method and the general theory of approximation methods for nonlinear equations, USSR Comp.Math.Phys. 7, 1-41 (1967).

30. Witsch,K.: Projektive Newton-Verfahren und Anwendungen auf nichtlineare Randwertaufgaben, Numer.Math. 31, 209-230 (1978/79).

31. Wang,J.P.: Asymptotic expansions and L^∞-error estimates for mixed finite element methods for second order elliptic problems, Numer.Math. 55, 401-430 (1989).

A Posteriori Error Estimates for Parametrized Nonlinear Equations

Werner C. Rheinboldt and Jinn-Liang Liu
Institute for Computational Mathematics and Applications
University of Pittsburgh, Pittsburgh, PA 15260

Abstract: *A new approach to the construction of a posteriori error estimates for finite element solutions of multi-parameter nonlinear problems is presented. The estimates are derived from local, element-by-element solutions of linearizations of the problems; they turn out to be very effective, computationally rather inexpensive, and insensitive to the choice of the local coordinate system on the solution manifold.*

Introduction

Frequently, in practical computations in engineering and science, the aim is to obtain results which are sufficiently accurate and reliable to allow for a decision about the physical system under study. A posteriori error estimates play a very important role in achieving this aim. Such estimates are needed not only for judging the reliability of the computed results but also for controlling adaptive processes designed to achieve desired error tolerances at minimal cost or best possible solutions within allowable cost ranges.

Many structural mechanics problems are modelled by parameter dependent nonlinear equations. The parameters may characterize, for instance, load points and load directions, material properties, geometrical data, etc. In general, the set of all solutions forms a differentiable manifold in the space of the state variables and parameters. This is often called the equilibrium surface of the structure and its form and characteristic features can provide considerable insight into the behavior and stability properties of the structure. Not surprisingly, the quality assessment and control of approximate solutions of such parameterized nonlinear problems is much more difficult and expensive than that of linear problems. In particular, the parameter-dependence causes the discretization error to become a local concept which

---

This work was supported in part by ONR-grant N-00014-90-J-1025 and NSF-grant CCR-8907654.

depends on the choice of the local coordinate system on the equilibrium manifold, [17].

In this paper, we present a new approach for computing a posteriori error estimates of finite element solutions of nonlinear equations involving several real parameters. The approach is insensitive to the choice of the local coordinate system on the solution manifold. Moreover, it has already shown itself to be highly effective and computationally inexpensive.

Many error estimations for nonlinear problems work with linearizations of the given equations (see e.g. [16] [19]) and apply known a posteriori error estimators for linear problems to approximate the solution-norms of these linearizations. So far, these techniques have been very limited in the permissible choices of the local coordinate systems on the solution manifold. Another approach consists in the defect-based estimations proposed, for instance, in [10], [11], [12], [13]. However, their widespread applications appears to have been prevented by a rather high computational cost.

Our approach here combines both approaches. It is based on the use of linearizations of the original problem, but determines the required norms of their solution by solving the linearized equations only locally, element-by-element. In essence, the defect-correction is performed locally on the linearized problem. As a result, the computational cost is reduced to very acceptable levels that are comparable to the cost of typical a posteriori estimates for linear problems. Moreover, the resulting estimates vary smoothly on the solution manifold of the given problem and apply to large classes of discretizations and error norms.

In Section 2, we discuss briefly some properties of solution manifolds of parameterized nonlinear problems and their discretizations. Then Section 3 presents the general construction of the error estimates and Section 4 concerns algorithmic aspects. Finally, some results of numerical experiments for a model problem are given in Section 5.

Solution Manifolds and Their Discretizations

As noted before, equilibrium problems for many physical systems are modelled by parameter dependent nonlinear equations

$$F(x) = 0, \quad x = (z, \lambda) \tag{2.1}$$

where z represents a state variable and λ a vector of parameters. More specifically, suppose that the nonlinear mapping F satisfies the condition

(Fcon) : *$F : S \subset X \mapsto Y$ is a Fredholm mapping of class C^r, $r \geq 1$ and index $p \geq 1$ from an open subset $S \subset X$ of a real Banach space X into another such space Y.*

Then it is well-known that the set of all regular solutions,

$$M = \{x;\ x \in S,\ F(x) = 0,\ DF(x)X = Y\} \tag{2.2}$$

is either empty or a p-dimensional C^r-manifold in X without boundary. We assume always that $M \neq \emptyset$

For the numerical analysis of the solution manifold (2.2), we need a computationally feasible scheme for fixing local coordinate systems at a given point $x_0 \in M$. For this suppose that a splitting

$$X = T \oplus W,\ \dim T = p \tag{2.3}$$

of X has been chosen such that

$$W \cap \ker DF(x_0) = \{0\}. \tag{2.4}$$

Evidently (Fcon) and (2.4) together imply that the partial derivative $D_W F(x_0)$ is an isomorphism between W and Y. Now the following result (see [17]) shows some neighborhood of the origin of T to be diffeomorphic to a (relative) neighborhood of x_0 in M:

Theorem 2.1: *Under the condition (Fcon), suppose that at $x_0 \in M$, a splitting (2.3), (2.4) has been given. Then there exist an open ball $B = B(0, r) \subset T$ centered at $0 \in T$, an open neighborhood $U \subset X$ of x_0, and a unique C^r-function $\eta : B \to W$ such that $\eta(0) = 0$ and the local coordinate mapping*

$$\Phi : B \subset T \mapsto X,\ \Phi(x) = x(t) = x_0 + t + \eta(t), \quad \forall t \in B \tag{2.5}$$

is a C^r-diffeomorphism from B onto $M \cap U$.

A point $x_0 \in M$ is called a foldpoint with respect to the splitting (2.3) if the condition (2.4) is violated; that is, if (2.3) does not induce a local coordinate system. Note that the finite-dimensional subspace $T = \ker DF(x_0)$ always admits a splitting (2.3) for which (2.4) holds.

In applications the equation (2.1) usually represents some boundary value problem and, as indicated in (2.1), we have a natural parameter splitting $X = Z \oplus \Lambda$, $\dim \Lambda = p$ of X into a state space Z and p-dimensional parameter space Λ. Evidently, this natural parameter splitting may be considered for the definition of a local coordinate system. We will assume that there is at least one point $x \in M$ where this is possible; that is, where $Z \cap \ker DF(x) = \{0\}$. Then, as noted earlier, Z and Y are isomorphic, and accordingly there exists an operator $Q \in L(X, Y)$ such that $\ker Q = \Lambda$ and the restriction $Q_0 = Q|_Z \in L(Z, Y)$ is an isomorphism.

For the computation, it is necessary to replace (2.1) by some finite dimensional approximating equation. Since the parameter space Λ is already finite dimensional, only the state space Z has to be discretized. We follow the approach in [17] in defining a discretization of (2.1).

Let $\{\Pi_h\}$ be a family of finite-rank projections $\Pi_h \in L(Z)$ with range spaces $Z_h = \Pi_h Z$, indexed by a positive real number $h > 0$. Correspondingly we introduce the extensions $\hat{\Pi}_h \in L(X)$ defined by $\hat{\Pi}_h x = \Pi_h z + \lambda$ for $x = z + \lambda \in X$ which have the ranges $X_h = Z_h \oplus \Lambda$. With the earlier isomorphism Q_0 between Z and Y define now the projections $P_h \in L(Y)$, $P_h = Q_0 \Pi_h (Q_0)^{-1}$ with the ranges $Y_h = P_h Y = Q_0 Z_h$. Then the desired approximate equations are given by

$$F_h(x_h) = 0, \ x_h = z_h + \lambda \in X_h \tag{2.6}$$

where

$$F_h : S_h \subset X_h \mapsto Y_h, \ F_h(x) = P_h F(x), \ \forall x \in S_h = S \cap X_h. \tag{2.7}$$

Hence, the corresponding discrete solution manifold is

$$M_h = \{x; x \in S_h, \ F_h(x) = 0, \ DF_h(x) X_h = Y_h\}. \tag{2.8}$$

For the convergence theory additional assumptions about these discretizations are needed the first of which will be the following consistency condition:

$$\lim_{h \to 0} P_h y = y, \quad \forall y \in Y. \tag{2.9}$$

Clearly, any comparison of the solution manifolds M and M_h must be done locally. Conceptually, we expect that for sufficiently small h the local coordinate system at a point $x_0 \in M$ established by Theorem 2.1 also constitutes a local coordinate system at the corresponding approximate solution $x_h \in M_h$. For this to be correct we require some stability condition.

For the analysis of the approximate problems, the discrete operator (2.7) can be extended to all of $S \subset X$ as follows:

$$\hat{F}_h : S \mapsto Y, \ \hat{F}_h(x) = (I_Y - P_h) Q x + P_h F(x), \quad x \in S. \tag{2.10}$$

Then the mentioned stability condition at $x_0 \in M$ assumes the form:

$$\|D\hat{F}_h(x_0) w\|_Y \geq \gamma \|w\|_W, \ \forall w \in W, \text{ and sufficiently small } h > 0, \tag{2.11}$$

where $\gamma > 0$ is a positive constant independent of w and h. As shown in [17] this condition is relatively easy to verify in many practical situations. The existence and convergence of solutions of the approximate problems (2.6) and the a priori error estimates between M and M_h can now be summarized in the form of the following theorem which was proved in [17].

Theorem 2.2: *(i) Under the condition (Fcon) consider the discretized problem (2.6) and assume that the consistency condition (2.9) holds. At a given point $x_0 \in M$, suppose that the splitting (2.3), (2.4) induces the local C^r-coordinate-mapping (2.5) and that the stability condition (2.11) is satisfied. Then for all sufficiently small $h > 0$ there exists a unique point $x_h \in M_h$ and we have $\lim_{h\to 0} x_h = x_0$. Moreover, the splitting $X_h = T \oplus W_h$ with $W_h = W \cap X_h$ defines at $x_h \in M_h$ a local C^r-coordinate-mapping $\Phi_h : B \subset T \mapsto X_h$. (ii) If, in addition, DF is Lipschitz continuous on bounded sets, then there exists a closed ball $\bar{B}(0, r_0) \subset B$, such that for all sufficiently small $h > 0$ the estimate*

$$\|x_0(t) - x_h(t)\|_X \leq C\|(I_Y - P_h)Qx_0(t)\|_Y, \quad \forall t \in \bar{B}(0, r_0),$$

holds with a constant C that is independent of h.

A Posteriori Error Estimates

For finite element discretizations of various linear problems the theory and application of reliable a posteriori error estimates has advanced considerably in recent years, we refer only to [3] and the two proceedings [1], [5] where many other references are given. These results are now also being extended to nonlinear problems, see e.g. [2], [4], [16], [8], [19].

In [2] and [4] it has been shown that effective a posteriori error estimates and adaptive procedures can be constructed and incorporated into a general continuation process for tracing paths on the equilibrium surface. But the applicability of these results is somewhat restricted due to the relatively high computational cost of the estimates. In [16] the estimates use a linearized form of the problem and hence can be computed about as rapidly as in the case of linear problems. However, these estimates were based on a fixed local coordinate system, such as that induced by the natural coordinate splitting $X = Z \oplus \Lambda$. Hence, by definition they are not valid at any foldpoint of that local coordinate system, and in fact, as already simple examples indicate, the estimates may become unduly large near any such point.

As noted earlier, our new construction of a posteriori error estimates is also utilizes local linearizations of the given nonlinear mapping. For the relation between the solutions of the nonlinear and these linearized problems, we present first a simple result based on tools from the theory of Newton's method (see e.g.[14], [15])

Theorem 3.1: *Let X, $\hat{X}$ be real Banach spaces and $G : S \mapsto \hat{X}$ a C^r-map, $r \geq 1$, on the open subset $S \subset X$ such that DG is Lipschitz continuous on bounded subsets. Consider any $x_0 \in S$ where $G(x_0) = 0$ and $DG(x_0) \in \mathrm{Isom}(X, \hat{X})$. Then there exists a closed ball $\bar{B} = \bar{B}(x_0, r) \subset S$, $r > 0$, such that for any $x \in \bar{B}$ the linearized problem*

$$G(x) + DG(x)w = 0 \tag{3.1}$$

has a unique solution $w = w(x) \in X$ and

$$\|x_0 - x\| = (1 + \mathrm{O}(\|x - x_0\|))\|w(x)\|, \quad \text{as } x \to x_0, \ x \in \bar{B}. \tag{3.2}$$

Proof: There exists $r_0 > 0$ such that $\bar{B}(x_0, r_0) \subset S$ and $DG(x)$ is an isomorphism for each x in this ball. Let γ be the Lipschitz constant of DG on the ball and with $\beta = \|DG(x_0)^{-1}\|$ and $\alpha = \beta\gamma$ set $r = \min(r_0, 1/(2\alpha))$. Then for all $x \in \bar{B} = \bar{B}(x_0, r)$ the standard perturbation lemma ensures that $DG(x) \in \mathrm{Isom}(X, \hat{X})$ and

$$\|DG(x)^{-1}\| \leq \frac{\beta}{1 - \alpha\|x - x_0\|} \leq 2\beta, \ \forall x \in \bar{B}.$$

Hence, $w(x) = -DG(x)^{-1}G(x)$ is uniquely defined for $x \in \bar{B}$ and with $x_1 = x + w(x)$ we have

$$\begin{aligned} \|x_1 - x_0\| &\leq \|DG(x)^{-1}\|\|G(x_0) - G(x) - DG(x)(x_0 - x)\| \\ &\leq \alpha\|x - x_0\|^2. \end{aligned}$$

Thus

$$\|w(x)\| \leq \|x - x_0\| + \|x_0 - x_1\| \leq (1 + \alpha\|x - x_0\|)\|x - x_0\|$$

together with

$$\|x - x_0\| \leq \|x - x_1\| + \|x_1 - x_0\| \leq \|w(x)\| + \alpha\|x - x_0\|^2$$

shows that

$$\frac{1}{1 + \alpha\|x - x_0\|}\|w(x)\| \leq \|x - x_0\| \leq \frac{1}{1 - \alpha\|x - x_0\|}\|w(x)\|$$

which proves (3.2).

Now suppose that we are in the setting of both parts of Theorem 2.2. In other words, consider the problem (2.1) where the mapping F satisfies the condition (Fcon) and DF is Lipschitz continous on bounded sets. Suppose that a discretization has been introduced for which the consistency condition (2.9) holds. At the currently given point $x_0 \in M$ we choose a splitting $X = T \oplus W$, $\dim T = p$ for which x_0 is not a foldpoint; that is, for which

(2.4) is satisfied. Let $\tau \in L(X)$ denote the natural projection of X onto T along W. For the mapping

$$G : S \subset X \mapsto Y \times T,\ G(x) = (F(x), \tau(x - x_0)),\ \forall x \in S, \tag{3.3}$$

$DG(x_0)v = 0$ is equivalent with $v \in kerDF(x_0) \cap W$ and thus $v = 0$ and $DG(x_0)$ is injective. By (Fcon) the mapping is also surjective, and hence we find that $DG(x_0) \in \mathrm{Isom}(X, Y \times T)$ which means that Theorem 3.1 is applicable to G.

Since $x_0 \in M$ represents here the desired solution of the original equations (2.1), we assume accordingly that x is the corresponding solution x_h of the discretized equations (2.6) for which, by definition $x_0 - x_h \in W$. Thus by Theorem 3.1, the problem of estimating the error $x_0 - x_h$ is asymptotically equivalent to solving the constrained linearized problem

$$F(x_h) + DF(x_h)w = 0, \quad \tau(w) = 0. \tag{3.4}$$

Of course, (3.4) is still an infinite dimensional problem which can only be solved approximately. For this we use the same discretization as in the computation of x_h. In other words, we approximate (3.4) by the constrained discretized problem

$$F_h(x_h) + DF_h(x_h)w_h = 0, \quad \tau(\hat{\Pi}_h w_h) = 0, \tag{3.5}$$

where the second equation evidently requires that $w_h \in W_h$ with $W_h = X_h \cap W$. Because of $F_h(x_h) = 0$, the solution of (3.5) is $w_h = 0$. Hence the norm $\|w\|$ of the solution w of (3.4) is exactly the error arising in the discretization (3.5) of (3.4). In other words, by applying one of the known, linear a posteriori error estimates to (3.5) we obtain asymptotically valid estimates of the discretization error $\|w\|$ and therefore, by Theorem 3.1, also of the error $\|x_0 - x_h\|$.

This was the approach taken in [16] and [17]. However, note that (3.4) and (3.5) incorporate constraints involving the complement W of the local coordinate space T. The available a posteriori error estimates cannot handle arbitrary choices of such spaces W; in fact, many of them work only when W is the given state space Z. Clearly, it is important to allow for a flexible choice of the local coordinate system during the computation of points on the solution manifold of (2.1). In other words, this approach is severely limitated by that requirement.

Since the discretized problems (3.5) only produce the trivial approximation $w_h = 0$ of the solution w of the infinite-dimensional linearized problem (3.4), the question arises how to obtain better approximations of w. Evidently, the difficulty arises when we construct (3.5) by means of the same discretization used in the computation of x_h. In other words, we should work with more accurate discretizations of (3.4). This is the basic concept of the defect correction principle. For matrix equations that principle is also called the method of iterative improvements and was first described by Wilkinson [21]. Extensions to ordinary differential equations were proposed in [20] and since then the principle has been used extensively in many settings, see e.g., [6], [7], [9], [10], [11], [12] , [13].

We shall not use the full iterative form of the defect correction principle, but work instead only with one specific improved discretization of (3.4):

$$F_{\bar{h}}(x_h) + DF_{\bar{h}}(x_h)w_{\bar{h}} = 0, \quad w_{\bar{h}} \in W_{\bar{h}}, \tag{3.6}$$

where $W_h \subset W_{\bar{h}} \subset W$. In other words, the system (3.6) has more degrees of freedom than used in the computation of x_h.

For ease of notation we assume that the spaces W_h increase monotonically when h decreases. Then the convergence of the solutions $w_{\bar{h}}$ of (3.6) to that of (3.4) for $\bar{h} \to 0$ may be formulated in the form of the following result:

Theorem 3.2: *Suppose that all conditions of Theorem 2.2 hold. Then there is some sufficiently small ε such that for all $0 < \bar{h} < h < \varepsilon$, the solutions w and $w_{\bar{h}}$ of (3.4) and (3.6), respectively, exist and that*

$$\lim_{\bar{h} \to 0} w_{\bar{h}} = w. \tag{3.7}$$

Proof: Since W and Y are isomorphic, we may choose some $A \in \mathrm{Isom}(Y, W)$. Then, by construction of the local coordinate system, $DF(x)A$ is an isomorphism in some ball $B_\rho = \bar{B}(x_0, \rho) \subset S$. Moreover, by the Lipschitz continuity of DF on bounded sets, we may choose $\rho > 0$ small enough such that the stability condition (2.11) holds for all $x \in B_\rho$ and $0 < h < \varepsilon$ with a suitably small $\varepsilon > 0$; in other words, that

$$\|D\hat{F}_h(x)Ay\|_Y \geq \gamma_0 \|y\|_Y, \quad \forall y \in Y \tag{3.8}$$

with some $\gamma_0 > 0$ that is independent of x and h. Finally, by shrinking, if necessary, ρ and ε further, we may assume that for all $0 < h < \varepsilon$ the approximate solution $x_h \in M_h$ of Theorem 2.2 exists and that $x_h \in B_\rho$.

Clearly (3.8) implies that $D\hat{F}_h(x)A \in \mathrm{Isom}(Y)$ whence for any $y_1 \in Y$ there is a unique $y_2 \in Y$ for which $D\hat{F}_h(x)Ay_2 = y_1$. For $y_1 \in Y_h$ it follows that

$$QAy_2 = P_hQAy_2 - P_hDF(x)Ay_2 + y_1 \in Y_h$$

and therefore $Ay_2 \in X_h$ and $(I_Y - P_h)QAy_2 = 0$; that is, $P_hDF(x)Ay_2 = y_1$. Hence we obtain from (3.8) that

$$\|DF_h(x)Ay\|_Y \geq \gamma_0 \|y\|_Y, \quad \forall y \in Y_h.$$

In other words, $(DF_h(x)A)^{-1} \in L(Y_h)$ exists and is uniformly bounded by $\delta = 1/\gamma_0$ for all $x \in B_\rho$ and $0 < h < \varepsilon$.

Now let h with $0 < h < \varepsilon$ be fixed and choose any $\bar{h}$ such that $0 < \bar{h} < h$. By construction of B_ρ we know that $DF(x_h)A$ is an isomorphism and hence that (3.4) has a unique solution $w = Ay \in W$. Moreover, the inverse $(DF_{\bar{h}}(x_h)A)^{-1} \in L(Y_{\bar{h}})$ exists and thus also (3.6) has a unique solution $w_{\bar{h}} = Ay_{\bar{h}}, y_{\bar{h}} \in Y_h$. By subtraction we find from (3.4) and (3.6) that

$$\|w_{\bar{h}} - w\| \leq \delta(\|(I_Y - P_{\bar{h}})DF(x_h)w\| + \|(I_Y - P_{\bar{h}})F(x_h)\|)$$

which by the consistency condition (2.9) implies that (3.7) holds.

Algorithm

As shown in the previous section, for the a posteriori estimation of the discretization error of the computed point $x_h \in M_h$ we should solve the more accurate equation (3.6). Obviously, if $X_{\bar{h}} = X_h$ then the solution is $w_h = 0$; hence for any reasonable approximation of the discretization error w, we require $X_{\bar{h}}$ to be a sufficiently larger subspace than X_h. In practice, the dimension of X_h is already expected to be large and thus a full solution of the equation (3.6) will be more expensive than the computation of x_h itself. This constitutes a severe restriction for the applicability of the defect-correction approach.

Without any further assumptions about the discretization process for (2.1) there appears to be little chance of resolving this problem. However, if we suppose that (2.1) is a differential equation to which a finite element approximation can be applied, then it turns out that an estimate of $\|w_{\bar{h}}\|$ can be obtained by solving (3.6) locally on each element. A detailed theory of this approximation process would exceed the space limitations for this paper and will be given elsewhere. Instead, we shall sketch here only the general computational procedure.

We consider a nonlinear problem in a generic weak form requiring the determination of $z \in Z$ such that for given $\lambda \in \Lambda$

$$b(z, \lambda, v) = g(\lambda, v), \quad \forall v \in V \tag{4.1}$$

where Z and V are real Hilbert spaces and Λ is a p-dimensional inner-product parameter-space. In (4.1), b stands for a form on $Z \times \Lambda \times V$ which is nonlinear in the first two variables but linear in the third one, and g is a functional on $\Lambda \times V$ which is linear in v but may be nonlinear in the parameter-variable. We introduce also the Hilbert space $X = Z \times \Lambda$ with the usual inner product for product spaces denoted by $\langle .,. \rangle$.

The discretizations are now specified by the choice of finite dimensional linear subspaces $Z_h \subset Z$ and $V_h \subset V$ and the corresponding orthogonal projections $\Pi_h \in L(Z)$ onto Z_h. In other words, the resulting (Galerkin-type) discretization (2.6) becomes the problem

determine $z_h \in Z_h$ such that

$$b(z_h, \lambda, v_h) = g(\lambda, v_h), \quad v_h \in V_h. \tag{4.2}$$

If $z_h^j,\ j = 1, 2, \dots, m_h$ and $v_h^j,\ j = 1, 2, \dots, m_h$ denote bases of Z_h and V_h, respectively, then (4.2) assumes the usual form

$$b(\sum_{j=1}^{m_h} \zeta_j z_h^j, \lambda, v_h^k) = g(\lambda, v_h^k), \quad k = 1, 2, \dots, m_h \tag{4.3}$$

of a system of m_h nonlinear equations in $m_h + p$ variables.

At any solution $x = (z, \lambda)$ of (4.1) consider now a local coordinate system defined by a p-dimensional subspace $T \subset X$ together with its orthogonal complement $W = T^\perp$ in X. Under appropriate smoothness assumptions on b and g the linearized problem (3.4) at a computed approximation $x_h \in X_h = Z_h \oplus \Lambda$ of x then becomes

determine $u \in Z,\ \mu \in \Lambda$ such that

$$\begin{aligned} &a_h(u, \mu, v) = f(v), \quad \forall v \in V, \\ &\langle t_k, (u, \mu)\rangle = 0, \quad k = 1, \dots, p \end{aligned} \tag{4.4}$$

where $t_k \in X,\ k = 1, 2, \dots, p$ represents a basis of T and

$$\begin{aligned} a_h(u, \mu, v) &= D_z b(z_h, \lambda_h, v)u + D_\lambda b(z_h, \lambda_h, v)\mu - D_\lambda g(\lambda_h, v)\mu \\ f(v) &= -b(z_h, \lambda_h, v) + g(\lambda_h, v) \end{aligned} \tag{4.5}$$

As we saw, we have to solve (4.4) with a discretization induced by some larger subspaces $Z_{\bar{h}} = Z_h \oplus Z_h^c$ and $V_{\bar{h}} = V_h \oplus V_h^c$ where Z_h^c and V_h^c denote here certain complements of Z_h and V_h in $Z_{\bar{h}}$ and $V_{\bar{h}}$,respectively . Hence, the resulting discretization is

determine $u_h \in Z_h,\ u_h^c \in Z_h^c,\ \mu \in \Lambda$ such that

$$a_h(u_h, \mu, v_h) = f(v_h), \quad \forall v_h \in V_h, \tag{4.6a}$$

$$a_h(u_h^c, \mu, v_h) = 0, \quad \forall v_h \in V_h, \tag{4.6b}$$

$$a_h(u_h + u_h^c, \mu, v_h^c) = f(v_h^c), \quad \forall v_h^c \in V_h^c, \tag{4.6c}$$

$$\langle t_k, (u_h, \mu)\rangle = 0, \quad k = 1, \dots, p \tag{4.6d}$$

$$\langle t_k, (u_h^c, \mu)\rangle = 0, \quad k = 1, \dots, p \tag{4.6e}$$

But (4.6a) and (4.6d) constitute exactly the discretization of (4.4) defined by the restriction $u_h \in Z_h$ and $v_h \in V_h$, for which we found the solution to be zero. Hence, it remains only to solve the reduced problem

determine $u_h^c \in Z_h^c,\ \mu \in \Lambda$ such that

$$a_h(u_h^c, \mu, v_h^c) = f(v_h^c), \quad \forall v_h^c \in V_h^c, \tag{4.7a}$$

$$a_h(u_h^c, \mu, v_h) = 0, \quad \forall v_h \in V_h, \tag{4.7b}$$

$$\langle t_k, (u_h^c, \mu)\rangle = 0, \quad k = 1, \dots, p \tag{4.7c}$$

where (4.7b) represents a boundary condition for u_h^c.

For the practical application we have to choose a suitable space T. Since the computed point x_h belongs to the manifold M_h of the discretized problem (4.3) it is natural to select T as the null-space of the Jacobian of(4.2); that is, as the subspace of X_h corresponding to the tangent vector of M_h at x_h. If Λ is one-dimensional and M_h is computed by a standard continuation process then a normalized basis vector of T is usually available at each computed point. Analogously, in the multi-parameter case, if a triangulation of M_h is being computed (see e.g. [18]) then again orthonormal "tangent-bases" t_h^j, $j = 1, \ldots, p$ are readily available at the computed points on the manifold.

Let Ω_j^h be a typical element of the mesh on the underlying domain Ω of the original problem (4.1) corresponding to the finite element space Z_h. In order to obtain $Z_{\bar{h}} \supset Z_h$, we may subdivide Ω_j^h into sub-elements or increase the order of the element or both, in other words we may use either a local h-version, p-version, or h-p-version of the finite element method. This has the advantage of being applicable to large classes of error norms for u. Of course, for the solution of (4.7a,b,c) it now will become necessary to "extend" the computed point x_h and its corresponding "tangent"-basis t_h^j, $j = 1, \ldots, p$ to $X_{\bar{h}}$ by suitable interpolation. The norm of the computed local solution on Ω_j^h is called the local error indicator on that element; that is,

$$\eta_j = \|u_h^c\|_{\Omega_j^h}, \tag{4.8}$$

and the error estimate for the computed point x_h is the sum

$$\varepsilon = (\sum_j \eta_j^2)^{1/2}, \tag{4.9}$$

over the error-indicators of all elements of Ω.

Altogether, for the computation of the error estimate at any computed point of the solution manifold M_h of the discreterized problem we have now the following algorithm:

1. Let $x_h \in X_h$ on M_h be the current solution of (4.2). Select the tangent space $T_h = span\{t_h^1, \ldots, t_h^p\}$of M_h at x_h to define the local coordinate system.
2. Loop over all elements Ω_j^h of the domain Ω
   - 2.1 Subdivide the element Ω_j^h into sub-elements corresponding to the desired choice of a finite dimensional subspace $Z_{\bar{h}}^j$ on Ω_j^h.
   - 2.2 On the refined element solve the local linearized finite element problem (4.7a,b,c) for $(u_h^c, \mu) \in X_{\bar{h}}^j = Z_{\bar{h}}^j \oplus \Lambda$. This involves the interpolation of x_h and $t_1^h, \cdots, t_p^h$ from X_h^j into $X_{\bar{h}}^j$.
   - 2.3 Compute the local error indicator (4.8) for the element.

3. Compute the desired a posteriori estimate (4.9) for the current point $x_h \in M_h$.

Numerical Example

As a model problem for numerical experiments we consider the two-dimensional nonlinear boundary value problem

$$-\Delta z = \lambda e^z, \quad z = z(x,y), \quad \forall (x,y) \in \Omega = (0,1) \times (0,1), \qquad (5.1)$$
$$z = 0 \quad \text{on } \partial\Omega.$$

A weak formulation is

$$\int_\Omega (z_x v_x + z_y v_y - \lambda e^z v)\, dx\, dy = 0, \quad \forall v \in H_0^1(\Omega). \qquad (5.2)$$

and we assume that $\lambda \in R^1$ which means that the solutions of (5.2) are expected to form a one-dimensional manifold M.

For the discretization we use a uniform mesh of sixteen bi-quadratic elements on Ω. For the computation of the one-dimensional solution manifold M_h of the discretized problem a continuation process (PITCON) is applied starting from $(z_h^0, \lambda_h^0) = (0,0)$ and our aim is to determine a posteriori estimates of the error between M and M_h at all computed solutions $(z_h, \lambda_h) \in M_h$.

In line with (4.4), the linearized problem at (z, λ) has the form

$$\int_\Omega [u_x v_x + u_y v_y - (\lambda u + \mu) e^z v]\, dx\, dy$$
$$= -\int_\Omega (z_x v_x + z_y v_y - \lambda e^z v)\, dx\, dy, \quad \forall v \in H_0^1(\Omega). \qquad (5.3)$$

As before the point (z, λ) stands here always for one of the computed points $(z_h, \lambda_h) \in M_h$. We apply now the algorithm of Section 4. and approximate the error $\|u\|$ by solving (5.3) locally on a larger finite element subspace $Z_{\bar{h}}$ of $H_0^1(\Omega)$ together with the auxiliary condition

$$\langle t_h, u \rangle = 0, \qquad (5.4)$$

Here, as noted in Section 4., we choose t_h as a normalized tangent vector on M_h at (z_h, λ_h). Such a tangent vector is available at each step of the continuation process and hence the equation (5.4) will involve little additional computational cost.

For the local solution each one of the 16 elements of Ω were divided into $(k+1)^2$ biquadratic sub-elements with $k = 1, 2, 3, 4$. The resulting error estimates are shown in Table 5.1 where $\|u_{\bar{h}_k}\|$ denotes the computed error norms for the four cases of k. The computations are very cost-effective, since each local problem involves only a fixed number of degrees of freedom depending on the value of k. If $\|u_{\bar{h}_4}\|$ is taken as the exact error then the effectivity index of the estimates is surprisingly good even for $k = 1$. The table also shows that, as expected, the estimated errors vary smoothly along the solution path M_h and show no sudden increases near the limit point $\lambda = 6.804524$.

As mentioned earlier, if the natural coordinate system induced by the parameter space Λ is chosen then we expect the resulting error estimates to become unduly large near the limit point. This is indeed the case as the last column of Table 5.2 shows. At the same time, it should be noted that the computational cost of the two approaches are practically identical.

Table 5.1

step	λ	$\|u_{\bar{h}_1}\|$	$\|u_{\bar{h}_2}\|$	$\|u_{\bar{h}_3}\|$	$\|u_{\bar{h}_4}\|$
1	0.118710	0.000516	0.000538	0.000542	0.000544
2	0.473781	0.002040	0.002128	0.002146	0.002152
3	0.946226	0.004010	0.004184	0.004220	0.004231
4	1.415882	0.005898	0.006155	0.006208	0.006225
5	1.882269	0.007695	0.008031	0.008101	0.008123
6	2.344791	0.009389	0.009801	0.009887	0.009914
7	2.785973	0.010913	0.011395	0.011495	0.011527
8	3.220584	0.012313	0.012861	0.012976	0.013011
9	3.647333	0.013578	0.014188	0.014315	0.014355
10	4.064580	0.014694	0.015361	0.015500	0.015544
11	4.470212	0.015648	0.016368	0.016519	0.016555
12	4.861489	0.016433	0.017200	0.017362	0.017413
13	5.234830	0.017050	0.017861	0.018032	0.018087
14	5.585562	0.017522	0.018373	0.018554	0.018611
15	5.907655	0.017914	0.018805	0.018994	0.019054
16	6.193552	0.018368	0.019300	0.019499	0.019562
17	6.434373	0.019144	0.020128	0.020337	0.020404
18	6.620862	0.020640	0.021692	0.021916	0.021986
19	6.745397	0.023317	0.024468	0.024710	0.024787
20	6.804524	0.027539	0.028832	0.029100	0.029185
21	6.800451	0.033461	0.034945	0.035250	0.035345
22	6.740221	0.041066	0.042795	0.043146	0.043255
23	6.633252	0.050283	0.052305	0.052712	0.052838
24	6.489009	0.061065	0.063424	0.063896	0.064041
25	6.328924	0.072480	0.075190	0.075729	0.075894
26	6.144920	0.085598	0.088703	0.089318	0.089506
27	5.941748	0.100500	0.104046	0.104745	0.104959
28	5.723487	0.117289	0.121324	0.122117	0.122359
29	5.493577	0.136117	0.140694	0.141591	0.141865
30	5.254879	0.157154	0.162338	0.163350	0.163659

Table 5.2

step	λ	$\|u_{\bar{h}_4}\|$	$\|u\|$
15	5.907655	0.019054	0.022905
16	6.193552	0.019562	0.024439
17	6.434373	0.020404	0.027792
18	6.620862	0.021986	0.041992
19	6.745397	0.024787	0.159595
20	6.804524	0.029185	0.176171
21	6.800451	0.035345	0.086353
22	6.740221	0.043255	0.069906
23	6.633252	0.052838	0.065171
24	6.489009	0.064041	0.065099
25	6.328924	0.075894	0.067672

References

1 Babuška, I.; Chandra, J. ; Flaherty, J. E. eds.: Adaptive Computational Methods for Partial Differential Equations. SIAM Publications, Philadelphia (1983)

2 Babuška, I.; Rheinboldt, W. C.: Computational error estimates and adaptive processes for some nonlinear structural problems. Comp. Meth. in Appl. Meth. and Eng. 34 (1982) 895-937

3 Babuška, I.; Rheinboldt, W. C.: A survey of a posteriori error estimates and adaptive approaches in the finite element method. Proc. China-France Symposium on Finite Element Methods, F. Kang and J.L. Lions, eds., Gordon & Breach Inc., New York (1983) 895-937

4 Babuška, I.; Rheinboldt, W. C.: Adaptive finite element processes in structural mechanics. Elliptic Problem Solvers II, G. Birkhoff and A. Schoenstadt, eds., Academic Press, New York (1984) 345-378

5 Babuška, I.; Zienkiewicz, O. C.; Gago, J.; Oliviera, E. R. eds.: Accuracy Estimates and Adaptive Refinements in Finite Elements Computations. J. Wiley & Sons, New York (1986)

6 Böhmer, K.: Discrete Newton method and iterated defect corrections. Numer. Math. 37 (1981) 167-192

7 Böhmer, K.; Hemker, P.; Stetter, H. J.: The defect correction approach. Defect correction methods: theory and applications, K. Böhmer and H.J. Steller, eds., Computing Suppl. 5, Wien (1984) 1-32

8 Demkowicz, L.; Oden, J. T.: An adaptive characteristic Petrov-Galerkin finite element method for convection-dominated linear and nonlinear parabolic problems in one space variable. J. Comp. Phys. 67 (1986) 188-213

9 Kaucher, E. W.; Miranker, W. L.: Self-Validating Numerics for Function Space Problems. Academic Press, Orlando (1984)

10 Lippold, G.: Diskretisierungsfehler und Defektkorrekturen bei der Lösung von Gleichungen in Banach-Räumen. ZAMM 60 (1980) 459-468

11 Lippold, G.: Schätzung von Diskretisierungsfehler bei der Lösung von Gleichungen in Banach-Räumen. ZAMM 61 (1981) 487-493

12 Lippold, G.: Fehlerschätzung und Nachiteration bei Galerkin-Verfahren. ZAMM 62 (1982) 435-440

13 Lippold, G.: Error estimates for the approximate solution of linear fixed point equations. ZAMM 69 (1989) 8564-8572

14 Ortega, J. M.; Rheinboldt, W. C.: Iterative Solution of Nonlinear Equations in Several Variables. Academic Press, New York (1970)

15 Rheinboldt, W. C.: An adaptive continuation process for solving systems of nonlinear equations. Proc. of the Semester on Mathematical Models and Numerical Methods, Stefan Banach Center Publ. Vol. 3, Polish Academy of Science, Warsaw, Poland (1977) 129-142

16 Rheinboldt, W. C.: Error estimates for nonlinear finite element computations. Computers and Structures 20 (1985) 91-98

17 Rheinboldt, W. C.: Numerical Analysis of Parametrized Nonlinear Equations. J. Wiley & Sons, New York (1986)

18 Rheinboldt, W. C.: On the computation of multi-dimensional solution manifolds of parametrized equations. Num. Math. 53 (1988) 165-181

19 Rank, E.; Werner, H.: An adaptive finite-element approach to a nonlinear seepage problem. Int. J. for Num. Meth. in Engng.

20 Stetter, H. J.: The defect correction principle and discretization methods. Numer. Math. 29 (1978) 425-443

21 Wilkinson, J. H.: Rounding Errors in Algebraic Processes. Prentice Hall, Englewood Cliffs, NJ (1963)

A Review of Solution Procedures and Path-Following Techniques in Relation to the Non-Linear Finite Element Analysis of Structures

by M.A. Crisfield & J. Shi
Department of Aeronautics, Imperial College, London

Abstract

The paper reviews the use of path-following techniques in relation to the non-linear static analysis of structures. The review starts with the early techniques that were largely motivated by structural analogies and moves on to some of the latest techniques that have often stemmed from the mathematical literatures.

Introduction

The user of a non-linear finite element computer program will usually be interested in tracing the complete load/deflection response of a structure up to its collapse. He or she may also be interested in the post-collapse regime. The latter response may be required because the analysis may involve a component rather than the complete structure so that the nature of the post-collapse response may be required for a later overall analysis. Also, the computer simulation should lead to a physical understanding of the collapse process. This is best achieved if the analysis can be taken to the stage beyond collapse with plots of the deformed structure, principal stresses and other features then being available to the engineer. In many instances, there can be local limit points so that the analysis must be able to proceed to the final "maximum".

For imperfection sensitive, thin-plated structures, the analyst may wish to start by analysing the "perfect structure" before introducing imperfections. The former response will often involve "bifurcations" and "post-buckling paths" and as a consequence the analyst will require "branch-switching techniques". Even when analysing "imperfect structures", the analyst can encounter severe numerical difficulties because of the alternative equilibrium states that stem from the "unfolding" of the perfect response. Similar difficulties stem from "softening materials" [C7]. The challenge for research workers and code developers lies in the development of robust and reliable analysis techniques to handle these problems.

In part, these analysis techniques can be taken from the mathematical literature[U1]; in particular from the field of "mathematical programming". However, there are many special problems that relate specifically to structural problems. In particular, the number of variables is usually very large and the banded-nature of the equations is of great importance. In addition, many problems involve "plasticity" so that the paths are "irreversible".

Early work

The earliest paper on non-linear finite elements appears to be that by Turner et al [T2] which dates from 1960 and, significantly stems from the aircraft industry. In this and subsequent papers [A4,A5,D7,M7,M9], the analysts adopted an "incremental procedure" in which they effectively solved a series of conventional linear analyses although the stiffness matrix, $\mathbf{K}$, was now replaced by the tangential stiffness matrix $\mathbf{K}_t$. Similar procedures were used for plasticity [M10,Z1,Z2]. The analysts were using a "forward Euler" technique which inevitably drifted from equilibrium (usually in a unsafe manner).

As an alternative, some analysts applied energy search methods [F9,M8,S10] which stemmed from the mathematical programming literature. Others iterated to equilibrium using either the full or modified Newton-Raphson method in which the stiffness matrix was not continuously up-dated [H1,M7,O2,O3,S7,S8,Z1,Z2]. Some of the earliest uses of "predictor-corrector methods" were due to Brebbia and Connor [B16] and Murray and Wilson [M11,M12]. The "predictor" involved a forward Euler or incremental solution while the "corrector" involved full or modified Newton iterations. Such techniques still provide the basis for most (implicit) non-linear finite element codes.

Work involving alternative pridictors and correctors

As we have indicated, most nonlinear analysis uses the tangential vector as the predictor. However, higher order terms can also be used [E3,R13,W10]. This approach has not often been adopted perhaps because the main difficulties relate to sudden non-linearities for which standard methods of producing a better predictor would be inaccurate. Den Herjer and Rheinboldt [D5] argued that "high-order predictions are very rarely effective ". Nevertheless, Wagner[W10] found that such a refined predictor leads to a reduction of 20-30% in the required number of corrector iterations. He also found that the resulting algorithm was more robust. Higher order terms are probably of most relevance in relation to " branch-switching ", following bifurcation. This topic will be discussed later.

Much work has been concentrated on the use of alternative "correctors" which have often stemmed from "mathematical programming" [B12,F5,W6]. In particular, there have been two main developments; firstly the use of "line-searches" and, secondly, the introduction of "quasi-Newton" techniques. With the former, the nodal displacements, **p**, would be up-dated according to:

$$\mathbf{p}_n = \mathbf{p}_o + \eta\delta\mathbf{p} \tag{1}$$

where the subscript n means "new" and the subscript o means "old". The iterative "direction" is $\delta\mathbf{p}$ and would usually come from a Newton or Newton-like technique. The scalar η in (1) is the "line-search" parameter. In the mathematical programming literature, η would be chosen so as to ensure a "sufficient reduction in energy" [W6,G3] and also to give a small enough "energy slope" via:

$$\|s_n\| = \|\mathbf{g}_n^t(\eta)\delta\mathbf{p}\| \leq TOL\| \mathbf{g}^t(\eta=0)\delta\mathbf{p}\| = TOL\|s_o\| \tag{2}$$

where **g** are the out-of-balance forces (or gradient of the total potential energy). In finite element work, the energy is often not available and hence (2) is often used on its own. In addition, for finite element work, with sudden non-linearities (as with plasticity or concrete-cracking [C11, C12]), it may not be worth using a refined interpolation involving the energies to obtain η. Instead, the first author [C8,C9,C12] uses a simple (safe-guarded) linear interpolation of the slopes (see (2)).

With a "slack line-search" [C9,C13,G3], a typical value for TOL would be 0.8. Lee [L3] uses 0.5. He notes that the method is efficient for preventing divergence in complex non-linear problems. However he finds that "indiscriminate use causes adverse effects on efficiency". The authors agree that such inefficiency can result but would argue that, in general, with a slack tolerance (say $\eta \approx 0.8$), this inefficiency will not often occur.

As originally introduced, the line-search method was applied to problems in which $\mathbf{K}_t$ was positive definite. In these circumstances, s_o in (2) will be negative and the iterative direction will be "down-hill". Various techniques have been devised for directly modifying $\mathbf{K}_t$ or otherwise changing the iterative procedure to ensure that it is positive definite [A1,B1,F6,L2]. However, as previously indicated, for non-linear structural analysis, we may aim at convergence on an unstable equilibrium state. Even in these circumstances, the author has found that the line-search technique can be beneficial [C9,C11]. However, much more work is required.

The second development from mathematical programming involves the quasi-Newton techniques [B11,B13,B14,B15,D1,D2,D6,F5,G1,G4,L2,S4] for which a good review has been given by Dennis and Moore[D6]. Such techniques are usually combined with line-searches. However, in contrast to the full Newton-Raphson method, $\mathbf{K}_t$ (or its inverse) is not directly formed but, instead, is up-dated using a secant-related formula involving previous iterative changes and previous residuals (or out-of-balance force vectors). Direct implementations of the method were not very successful because, away from equilibrium, they destroyed the banded nature of the (approximate) tangent stiffness matrix. Matthies and Strang [M1] showed that one could use an indirect vector-based method whereby only the factors of the original approximation are directly used. This approach has much common with work in the mathematical programming literature due to Buckley [B14], Nazareth [N1] and Nocedal [N2]. The method has been applied to non-linear finite element analysis by a number of workers [B2,B3,G2,H4,P9].

Matthies and Strang's method was based on the rank-two BFGS up-date [F5,B12,S4] for which positive-definiteness is a basic assumption. For systems that are not necessarily positive definite, it might be more appropriate to use a rank-one up-date such as that due to Davidon [D2] and Broyden [B13] or to Broyden as given by Dennis and Moore [D6] - the latter is unsymmetric and might therefore be appropriate for non-symmetric systems such as those arising from non-associative plasticity.

The quasi-Newton method is but one of the many iterative methods that can be used to solve non-linear equations. When taken to the limit, it will not only converge to the solution of $\mathbf{g}(\mathbf{p}) = \mathbf{0}$ but will also obtain $\mathbf{K}_t = \partial\mathbf{g}/\partial\mathbf{p}$. In contrast iterative techniques such as dynamic relaxation [D3,F4,K8,O1], conjugate gradients [C3,F7,H3,P10,S3] and the Lanczos method [L1,P8] involve no direct approximation to $\mathbf{K}_t$. However, without some pre-conditioner [C1,C3,C10, H5,H6,J1,K4,N3,P2] they can produce very slow convergence characteristics. There is considerable scope for using the finite element idealisation itself to provide such a pre-conditioner [C1,C10,H5,H6]. For non-linear analysis one can then embed the iterative technique in an inner-outer loop whereby the outer-loop is a form of Newton-procedure for the non-linearities and the inner-loop involves the iterative solution of the linear equations [N4,P3-P5]. Because of the outer loop, the inner loop need not be solved accurately for the early outer iterations. There would seem to be considerable scope for the development of these methods, particularly in relation to parallel computing.

Arc-length and similar techniques

Early finite element techniques for passing limit-points involved artificial springs [S9,W9], a switch from load to displacement control [A5,B4] or the abandonment of equilibrium iterations in the close vicinity of the limit point [B7,B8]. In to-day's finite element computer programs, some form of "arc-length method" would usually be used instead.

In relation to structural analysis, the arc-length method was originally introduced by Riks [R8,R9] and Wempner [W5] with later modifications being made by the first author [C5,C6, C11] and others [B5,F8,G2,M2-M4,R1,R2,S2,S6]. A recent comparative study of the mostly used arc-length methods has been done by Clarke and Hanncock [C2]. Closely related work can be found in the mathematical literature [K1-K3,R6] with the first such paper appearing to be due to Haslegrove [H2].

As a starting point to various "continuation methods", we can write the equilibrium equations as:

$$\mathbf{g}(\mathbf{p}, \lambda) = \mathbf{q}_i(\mathbf{p}) - \lambda \mathbf{q}_{ef} = \mathbf{0} \tag{3}$$

where $\mathbf{q}_i$ are the internal forces which are functions of the displacements, $\mathbf{p}$, the vector $\mathbf{q}_{ef}$ is a "fixed external loading vector" and the scalar λ is a "load level parameter" that multiplies $\mathbf{q}_{ef}$. Equation (3) defines a state of "proportional loading" in which the loading pattern is kept fixed. Non-proportional loading can be covered by making a simple modification [C13].

Conventional finite element solution procedure would apply (3) by incrementing the load parameter, λ. However, near a limit-point, there may be no intersection between the equilibrium path of (3) and the plane λ = constant which represents the next "load level". Instead of applying (3), with λ = constant, one may aim to find the intersection of (3) with "s = constant" where s is the "arc-length" defined by:

$$s = \int ds \tag{4}$$

and

$$ds = \sqrt{d\mathbf{p}^t d\mathbf{p} + d\lambda^2 \psi^2 \mathbf{q}_{ef}^t \mathbf{q}_{ef}} \tag{5}$$

The scaling parameter ψ is required in (5) because the load contribution depends on the adopted scaling between the load and displacement terms. Having introduced the arc-length ,s, one may attempt to directly solve:

$$\mathbf{g}(s) = \mathbf{q}_i(\mathbf{p}(s)) - \lambda(s)\mathbf{q}_{ef} = \mathbf{0} \tag{6}$$

using a higher order ODE method [W3]. However, with this approach it is often very difficult to successfully limit the "drift from equilibrium" and hence "predictor-corrector" methods are usually used. To this end, one would effectively replace the differential form of (5) with an incremental form:

$$a = (\Delta \mathbf{p}^t \Delta \mathbf{p} + \Delta\lambda^2 \psi^2 \mathbf{q}_{ef}{}^t \mathbf{q}_{ef}) - \Delta l^2 = 0 \tag{7}$$

where Δl is the fixed "radius" for the desired intersection. The vector $\Delta \mathbf{p}$ and scalar $\Delta\lambda$ are incremental (not iterative for which we will use δ's) and relate back to the last converged equilibrium state.

The main essence of the arc-length methods is that the load parameter, λ,becomes a variable. Hence, together with the n displacement variables, we have a total of n+1 variables. To solve for these, we have the n equilibrium equations of (3) and the constraint equation of (7). Following Riks [R8,R9] and Wempner[W5], we can solve for these n+1 variables by directly applying the Newton-Raphson method to (3) and (7). This leads to:

$$\begin{pmatrix} \delta \mathbf{p} \\ \delta\lambda \end{pmatrix} = - \begin{pmatrix} \mathbf{K}_t & -\mathbf{q}_{ef} \\ 2\Delta \mathbf{p}^t & 2\Delta\lambda\psi^2 \mathbf{q}_{ef}^t \mathbf{q}_{ef} \end{pmatrix}^{-1} \begin{pmatrix} \mathbf{g}_o \\ a_o \end{pmatrix} \tag{8}$$

The augmented "Jacobian" or "stiffness matrix" within the square brackets in (8) remains non-singular even when $\mathbf{K}_t = \partial \mathbf{g}/\partial \mathbf{p}$ is singular. Equation of the form of (8) are also known as "bordered equations" [R7]. Other interesting structural uses of bordered equations have been given by Kroplin et al [K5,K7,K9]). Equations (8) can be used directly to find the changes $\delta \mathbf{p}$ and $\delta\lambda$. However, in contrast to $\mathbf{K}_t$, the augmented stiffness matrix in (8) is neither symmetric nor banded.

Instead of solving (8), one may directly introduce the constraint of (7) by, following Batoz and Dhatt [B4] for displacement-control at a single point. To this end, the iterative displacement, $\delta \mathbf{p}$, is split into two parts. Hence the Newton change at the new unknown load level, $\lambda_n = \lambda_o + \delta\lambda$, becomes:

$$\delta \mathbf{p} = -\mathbf{K}_t^{-1}(\mathbf{g}(\mathbf{p}_o,\lambda_o) - \delta\lambda \mathbf{q}_{ef}) = \delta \mathbf{p}_1 + \delta\lambda\delta \mathbf{p}_t \tag{9}$$

Here, $\delta \mathbf{p}_1$ is the iterative change that would stem from the standard load-controlled Newton-Raphson method (at a fixed load level, λ_o) while $\delta \mathbf{p}_t$ is the displacement vector corresponding to the fixed load vector, $\mathbf{q}_{ef}$. Having obtained $\delta \mathbf{p}$ from (9) (with $\delta\lambda$ still unknown), the new incremental displacements $\Delta \mathbf{p}_n = \Delta \mathbf{p}_o + \delta \mathbf{p}$ can be substituted into the constraint of (7) to obtain a scalar quadratic equation for $\delta\lambda$. To prevent the solution algorithm doubling back on itself, the root of the equation can be chosen so as to obtain the minimum angle between $\Delta \mathbf{p}_n$ and $\Delta \mathbf{p}_o$ [C6,C13].

In theory, this method suffers from the limitation, that precisely at the limit-point, $\mathbf{K}_t$ will be singular and hence cannot be factorised. In reality, the authors have not found this to be a significant problem, because one appears never to arrive precisely at the limit point (see also [B4]). Nonetheless, a number of authors have addressed the issue of "stabilising" the stiffness matrix near to limit points. Riks and Rankin [R7] have proposed two techniques, one of which requires knowledge of the lowest eigenmode of $\mathbf{K}_t$. Felippa [F1,F2] has combined a partitioning device of Rheinboldt [R6] with the original "fictitious spring approach" of Sharifi and Popov [S9]. As already discussed, in the mathematical programming literature there are approaches to modify $\mathbf{K}_t$ in the presence of "uphill directions" [F6,L2]. These methods must also be relevant.

With regard to the "scaling parameter", ψ, in the constraint of (7), both the author [C6] and Ramm [R1,R2] independently concluded that, for practical problems involving a realistic number of variables, the "loading terms" (those involving ψ) had little effect and hence advocated setting ψ to zero. Padovan et al [P1] and Park [P7] proposed adopting a variable ψ which should be large in the initial stages (so that the technique then tends towards load-control) and small when the limit points were approached. The first author [C13] has discussed a procedure that produces a similar effect by switching to "arc-length control" as the limit point is reached. Felippa [F1,F2] and Simo et al [S6] have taken the scaling or weighting further by replacing the $\Delta \mathbf{p}^t \Delta \mathbf{p}$ term in (7) $\Delta \mathbf{p}^t \mathbf{S} \Delta \mathbf{p}$ where one suggestion for the scaling matrix $\mathbf{S}$ is Diag($\mathbf{K}_t$). Weighting schemes have also been advocated by de Borst [D4] and Gierlinski et al [G2].

A number of authors [F8,F10,R1,R2,R8,R9,S2,W5] have advocated linearised forms of arc-length method. Differentiation of (7) leads to:

$$f(\Delta \mathbf{p}_o, \Delta\lambda_o) = \Delta \mathbf{p}_o^t \delta \mathbf{p} + \delta\lambda\Delta\lambda_o \psi^2 \mathbf{q}_{ef}^t \mathbf{q}_{ef} = -a_o/2 \tag{10}$$

where a_o is the "old" value of the arc-length mismatch (see (7)). If a_o is taken as zero, we have Ramm's approach [R1,R2] which ensures that the iterative change $(\delta\mathbf{p}, \delta\lambda\psi\mathbf{q}_{ef})$ is orthogonal to the "secant change" $(\Delta\mathbf{p}_o, \Delta\lambda_o\psi\mathbf{q}_{ef})$. Using (9), this leads to:

$$\delta\lambda(\Delta\mathbf{p}_o, \Delta\lambda_o) = (-a_o/2 - \Delta\mathbf{p}_o{}^t\,\delta\mathbf{p}_1)/(\Delta\mathbf{p}_o{}^t\,\delta\mathbf{p}_t + \Delta\lambda_o\psi^2\mathbf{q}_{ef}{}^t\,\mathbf{q}_{ef}) \tag{11}$$

This technique is closely related to the original procedures due to Riks [R8,R9] and Wempner [W5] which involves making the iterative change $(\delta\mathbf{p}, \delta\lambda\psi\mathbf{q}_{ef})$ orthogonal to the predictor solution $(\Delta\mathbf{p}_p, \Delta\lambda_p)$ (again with $a_o = 0$). The relevant formulae are simply obtained by replacing the "old incremental" $(\Delta\mathbf{p}_o, \Delta\lambda_o)$ with the "initial predictor", $(\Delta\mathbf{p}_p, \Delta\lambda_p)$ in (10) and (11). A further variation has been given by Fried [F10] which uses $(\delta\mathbf{p}_t, (\psi^2\mathbf{q}_{ef}{}^t\,\mathbf{q}_{ef})^{-1})$ in place of $(\Delta\mathbf{p}_o, \Delta\lambda_o)$ with also $a_o = 0$. In this approach, which is related to a procedure by Haselgrove [H2], the solution process does not depend on the predictor solution. The idea of including the a_o terms in (10) and (11) is due to Schweizerhoff et al [S2] and Riks [R11] with a further modification by Forde et al [F8]

The previous methods can be considered as forms of generalised displacement control which can be applied to problems that, physically, may not involve displacement-control. This has effectively been achieved with the spherical arc-length method (with $\psi = 0$) by constraining the Euclidian norm of the incremental displacement to a fixed quantity (equation (7)). Instead, following Batoz and Dhatt [B4], one may constrain the displacement increment at a particular variable to a specified quantity so that in place of (7) one would have:

$$\Delta\mathbf{p}_n(k) = \Delta\mathbf{p}_o(k) + \delta\mathbf{p}(k) = \Delta p_k \tag{12}$$

where Δp_k is the prescribed magnitude of the the kth incremental displacement variable. Using, as before, (9) for the iterative change $\delta\mathbf{p}$, the constraint (12) then leads to:

$$\delta\lambda = (\Delta p_k - \Delta\mathbf{p}_o(k) - \delta\mathbf{p}_1(k))/\,\delta\mathbf{p}_t(k) \tag{13}$$

The tangential, predictor solution would simply be achieved by finding

$$\delta \mathbf{p}_t = \mathbf{K}_t^{-1} \mathbf{q}_{ef} \tag{14}$$

with $\mathbf{K}_t$ as the tangent stiffness matrix at the beginning of the increment and then from:

$$\Delta\lambda \delta \mathbf{p}_t(k) = \Delta p_k \tag{15}$$

Rheinboldt [R3,R6] has adopted this approach with the variable k being changed for each increment so as to relate to the largest tangential (predictor) component.

Simons and Bergan [S5], expanding on the work of Powell and Simons [P11] and Bergan and Mollestad [B10], have advocated a "hyper-plane control method" which effectively involves an extension of Batoz and Dhatt's method with a weighted linear combination of "individual specified displacements" with say half-a-dozen key displacement variables and weights being specified by the user. A procedure lying between the arc-length method and this "hyper-plane control" has been proposed by Gierlinski et al [G3]

The arc-length method can be combined with the quasi-Newton methods[C6,S2] but, as already discussed, further work is required in relation to the line-search. Special techniques can also be devised for contact algorithms [S11].

When using a "predictor/corrector" based "continuation method" for path-following an important issue is the step-size (be it related to load or arc-length). A number of procedures have been devised [B7,B9,C4,C6,D5,R1,R2,S1] many of which relate to some form of "path-parameter" [B6-B8,E2] such as the "current stiffness parameter" [B6-B8,C13]. Probably the simplest method relates the step-size to the inverse of the number of iterations (or square root of it) at the previous step [C6,C13,R1,R2]

Bifurcations, branch-switching and direct computation of singular points

For the previous arc-length methods, the predictor can be computed from:

$$\Delta \mathbf{p}_p = \Delta\lambda_p \delta \mathbf{p}_t = \Delta\lambda_p \mathbf{K}_t^{-1} \mathbf{q}_{ef} \tag{16}$$

where $\mathbf{K}_t$ is the tangent stiffness matrix and in order to satisfy the constraint of (9), the scalar $\Delta\lambda_p$ is given by:

$$\Delta\lambda_p = \pm \frac{\Delta l}{\sqrt{\delta \mathbf{p}_t^t \delta \mathbf{p}_t}} = s \frac{\Delta l}{\sqrt{\delta \mathbf{p}_t^t \delta \mathbf{p}_t}} \tag{17}$$

Because of the plus or minus sign in (17), we have two possible predictors. Following [C6], we let s be + 1 when $\mathbf{K}_t$ (at the beginning of the increment) is positive definite and let s be -1 when we encounter a negative pivot in the factorisation of $\mathbf{K}_t$. This approach presupposes that the latter occurs when we have just passed a limit point. However a negative pivot can also result when a bifurcation point is passed and, in these circumstances, the previous solution algorithm will lead to an oscillation about the bifurcation point [C13]. Such a bifurcation point can be distinguished from a limit point because, in the former case, $\mathbf{z}^t \mathbf{q}_{ef} = 0$ while, in the latter case, $\mathbf{z}^t \mathbf{q}_{ef} \neq 0$ [A3,R8,W2]. In the previous expressions, $\mathbf{z}$ is the eigenvector corresponding to the lowest eigenvalue of $\mathbf{K}_t$ (zero at the singular point). Having detected a bifurcation point, some form of bracketing procedure (based, say, on the determinant of $\mathbf{K}_t$ or the lowest pivot in the factorisation of $\mathbf{K}_t$) can be used to "home-in" on the bifurcation point [A3,W2,R8] (or close to it so as to avoid the exact singularity). With plasticity, care must be taken not to reverse the loading path during this process.

Having reached the bifurcation point, there are a number of strategies that can be implemented for subsequent "branch switching" [C7,E1,K6,M6,R5,R11,T1,W1,W7,W8]. With the most common symmetric bifurcation, the simplest strategy [R10,W1,W2,W7,W8] is to apply one of the arc-length methods with the predictor of (16) and (17) being replaced by:

$$\Delta \mathbf{p}_p = \Delta l \, \mathbf{z}; \qquad \Delta\lambda_p = 0 \tag{18}$$

If the linearised arc-length method is used, with the scaling parameter ψ as zero and $a_o = 0$, then (10) gives $\delta \mathbf{p}^t \mathbf{z} = 0$ from which (see (11)):

$$\delta\lambda = - (\delta \mathbf{p}_1{}^t \mathbf{z})/(\delta \mathbf{p}_t{}^t \mathbf{z}) \tag{19}$$

which enforces the iterative change $\delta \mathbf{p}$ to be orthogonal to $\mathbf{z}$. As a consequence, there should be no problems with the solution returning to the fundamental path. The quadratic form could also be used with the root for now being chosen so that $\delta \mathbf{p}$ is such as to impose the minimum angle between $\Delta \mathbf{p}_n$ and $\Delta \mathbf{p}_o = \Delta l \mathbf{z}$. One could also use the "specific displacement approach" of (12)-(13) with the variable k again being chosen to involve the largest component of the predictor (now proportional to $\mathbf{z}$). There are strong links between the method of (19) and a method due to Rheinboldt[R5] although Rheinboldt also uses some higher-order data concerning the fundamental path and instead of enforcing $\delta \mathbf{p}^t \mathbf{z} = 0$ appears to enforce $\mathbf{g}^t \mathbf{z} = 0$.

The previous methods can also be used for asymmetric bifurcations although in this case the predictor of (18) may not be very accurate. To obtain a better predictor, one can introduce the higher-order terms [A2,R10,K6]:

$$B_1 = \mathbf{z}^t \left\{ \frac{\partial^2 g}{\partial p_i \partial p_j} z_i z_j \right\} \tag{20a}$$

$$B_2 = \mathbf{z}^t \left\{ \frac{\partial^2 g}{\partial p_i \partial p_j} z_i y_j \right\} \tag{20b}$$

$$B_3 = \mathbf{z}^t \left\{ \frac{\partial^2 g}{\partial p_i \partial p_j} y_i y_j \right\} \tag{20c}$$

$$B_4 = -\mathbf{q}^t\mathbf{z} \tag{20d}$$

where:

$$\mathbf{y} = (\mathbf{K}_t + \beta \mathbf{z}\mathbf{z}^t)^{-1}\mathbf{q} \tag{21}$$

is orthogonal to **z**. β ensures that the modified tangent stiffness matrix is non-singular.

At a bifurcation point, $B_4 = 0$ and at a symmetric bifurcation point $B_1 = 0$. In addition, for an asymmetric bifurcation, one can solve [A2]:

$$B_1 \Delta\bar{s}^2 + 2B_2 \Delta\bar{s}\Delta\lambda + B_3 \Delta\lambda^2 = 0 \tag{22}$$

to obtain the tangents to the paths:

$$\Delta\mathbf{p} = \Delta\bar{s}\mathbf{z} + \Delta\lambda\mathbf{y} \tag{23}$$

thereby obtaining a better predictor than (18). The higher order terms in (20) $\partial^2 g/\partial p_i \partial p_j$ are not available in conventional finite element codes. However, they can be computed [E3,W7] for simple elements. As an alternative, one can use finite difference approximations [E3,R10,W8].

Such terms have been used by Wriggers and co-workers[W7,W8] in an alternative procedure which follows on from the mathematical work in [M5,M6] so as to directly compute singular points (be they limit or bifurcation points) and to compute the eigenmode as a by-product. To this end, they supplement the equilibrium equations of (3) with the relationship:

$$\mathbf{K}_t\mathbf{z} = \mathbf{0} \tag{24}$$

which only applies at the singular points (because the eigenvalue is then zero).

A truncated Taylor series then leads to :

$$(\mathbf{K}_t\mathbf{z})_n = (\mathbf{K}_t\mathbf{z})_o + \mathbf{K}_t\delta\mathbf{z} + \nabla_{\mathbf{p}}(\mathbf{K}_t\mathbf{z}) = \mathbf{0} \tag{25}$$

where:

$$\nabla_{\mathbf{p}}(\mathbf{K}_t\mathbf{z}) = \nabla_{\mathbf{p}}(\delta\mathbf{p}, \mathbf{z}) = \partial^2 g/(\partial p_i \partial p_j)\, \delta p_i z_j \tag{26}$$

is the "directional derivative" of $\mathbf{K}_t\mathbf{z}$ with respect to $\mathbf{p}$ (with $\mathbf{z}$ fixed). Equation (26) involves similar vectors to those in (20) and again, for simple elements, the vector $\nabla_{\mathbf{p}}(\delta\mathbf{p}, \mathbf{z})$ is easily computed at the element level [E3,W7] provided that $\mathbf{z}$ and are known. (For other elements, one can use a finite difference approximation[E3,W8,R12]). In order to obtain the vector $\nabla_{\mathbf{p}}(\delta\mathbf{p}, \mathbf{z})$, it is best to use the "two vector form" of (9) for $\delta\mathbf{p}$ so that:

$$\nabla_{\mathbf{p}}(\delta\mathbf{p}, \mathbf{z}) = \nabla_{\mathbf{p}}(\delta\mathbf{p}_1, \mathbf{z}) + \delta\lambda\nabla_{\mathbf{p}}(\delta\mathbf{p}_t, \mathbf{z}) \tag{27}$$

Hence, from (25), one can obtain:

$$\begin{aligned}\delta\mathbf{z} &= -\,\mathbf{K}_t^{-1}\{\,(\mathbf{K}_t\mathbf{z})_o + \nabla_{\mathbf{p}}(\delta\mathbf{p}_1, \mathbf{z}) + \delta\lambda\nabla_{\mathbf{p}}(\delta\mathbf{p}_t, \mathbf{z})\} \\ &= \delta\mathbf{z}_1 + \delta\lambda\mathbf{z}_2\end{aligned} \tag{28}$$

Finally to eliminate the scaler $\delta\lambda$, one can use a constraint such as $\mathbf{z}^t\mathbf{z} = 1$or $z_k=1$(the latter being similar to (12)). As with the conventional arc-length method, these procedures only involve one factorization of $\mathbf{K}_t$ at each iteration.

If the iterations are taken "too far", the iterative performance degenerates as det($\mathbf{K}_t$) tends to zero. Wriggers and Simo [W8] have adapted the technique to overcome this limitation by ammending a procedure due to Felippa[F3] which involves a "penalty spring".

Special indirect solution methods can be devised to avoid directly solving the resulting enlarged system of equations [M5,M6]. In its pure form, this method is only directly applicable for elastic (reversible) materials and as indicated in [W8], special care is required for plasticity (although few details are given).

Conclusions

The paper has reviewed a range of solution procedures that can be used for nonlinear finite element analysis. With the aid of these techniques, today's finite element systems are often efficient and robust. However, many difficulties remain; particularly for problems involving instability that can be considered as being " unfolded " from a perfect system (and with softening materials). In these circumstances, the solution procedures will often branch-switch and oscillate [C13]. As a result methods are required to prevent this behaviour and to converge to the "lowest energy state". However, in some instances, one can take advantage of this phenomenon and provide an energy perturbation that will induce branch-switching thereby providing some form of stability measure [C14,K10].

A new challenge for nonlinear solution algorithms involves their integration within an enviroment that is suitable for parallel computing. To this end, it is likely that we will see a resurgence of interest in iterative techniques that do not directly involve a stiffness matrix. However, as previously discussed, an effective preconditioner will be required.

References

A1. ALLMAN, D.J., An iterative method to locate minima of a function of several variables, Royal Aircraft Estab. Tech. Report 84038, (1984).

A2. ALLMAN, D.J., Calculation of the stable equilibrium paths of discrete conservative systems with singular points, Royal Aircraft Estab. Tech. Report 86053, (1986) also Computers & Structures, 32, pp.1045-1054, (1989)

A3. ALLMAN, D.J., On the general theory of the stability of equilibrium of discrete conservative systems, Aeronautical Journal, pp. 29-35, (Jan. 1989).

A4. ARGYRIS, J.H., Recent advances in matrix methods of structural analysis, Pergamon Press, (1964).

A5. ARGYRIS, J.H., Continua and discontinua, Proc. Conf. Matrix Methods in Struct. Mech. Air Force Inst. of Tech., Wright Patterson Air Force Base, Ohio, (Oct. 1965).

B1. BARTHOLOMEW, P., A simple iterative method for the minimisation of an unconstrained function of many variables, Royal Aircraft Estab., Tech. Memo MAT/STR 1021, Farnborough, Hants., (1983).

B2. BATHE, K-J & DVORKIN, E.N., On the automatic solution of nonlinear finite element equations, Computers & Structures, 17, pp.871-879, (1983).

B3. BATHE, K.J. & CIMENTO, A.P., Some practical procedures for the solution of nonlinear finite element equations, Comp. Meth. in Appl. Mech. & Engng., 22, pp.59-85, (1980).

B4. BATOZ, J.L. & DHATT, G., Incremental displacement algorithms for non-linear problems, Int. J. for Num. Meth. in Engng., 14, pp. 1262-1266, (1979).

B5 BELLINI, P.X. & CHULYA, A., An improved automatic incremental algorithm for the efficient solution of nonlinear finite element equations, Computers & Structures, 26, pp.99-110, (1987).

B6. BERGAN, P.G., Solution algorithms for non-linear structural problems, Engineering Applications of the Finite Element Method, Computer & Structures, 12, pp.497-509, (1980).

B7. BERGAN, P.G. & SOREIDE, T., Solution of large displacement and instability problems using the current stiffness parameter, Finite Elements in Non-linear Mechanics, Tapir Press, pp. 647-669, (1978)

B8. BERGAN, P.G., HORRIGMOE, G., KRAKELAND, B. & SOREIDE, T.H., Solution techniques for non-linear finite element problems, Int. J. for Num. Meth. in Engng., 12, pp.1677-1696. (1978)

B9. BERGAN, P.G. & SOREIDE, T., A comparative study of different numerical techniques as applied to a nonlinear structural problem, Comp. Meth. in Appl. Mech. & Engng., 2, pp.185-201, (1973).

B10 BERGAN, P.G. & MOLLESTAD, E., Static and dynamic solution strategies in nonlinear analysis, Numerical Methods for Non-linear Problems, Vol. 2., ed. C. Taylor et al., Pineridge, Swansea, pp.3-17, (1984).

B11. BRODLIE, K.W., GOURLAY, A.R. & GREENSTADT, J., Rank-one and rank-two corrections to positive definite matrices expressed in product form, J. Inst. Math., Appl., 11, pp.73-82, (1973).

B12 BROYDEN, C.G., The convergence of a double-rank minimisation 2: the new algorithm, J. Inst. Math. Appl., 6, 222-231, (1970) (BFGS)

B13. BROYDEN,C.G., Quasi-Newton methods and their application to function minimisation, Math. Comp., 21, pp.368-381, (1967)

B14. BUCKLEY, A.G., A combined conjugate-gradient quasi-Newton minimisation algorithm, Math. Programm., 15, pp.200-210, (1978)

B15 BUCKLEY, A. & LENIR, A., QN-like variable storage conjugate gradients, Math. Programming, 27, pp.155-175, (1983).

B16. BREBBIA, C. & CONNOR, J., Geometrically non-linear finite element analysis, Proc. ASCE, J. of Eng. Mech. Div., Proc. paper 6516, (April 1969).

C1 CAREY, G.F. & BO-NAN, J., Element-by-element preconditioned conjugate gradient algorithm for compressible flow, Innovative Methods for Nonlinear Problems, Pineridge, Swansea, pp.41-49, (1984).

C2. CLARKE, M.J. & HANCOCK, G.J., A study of incremental-iterative strategies for non-linear analysis, Int. J. for Num. Meth. in Engng., 29, pp.1365-1391, (1990).

C3. CRISFIELD, M.A., Finite elements and solution procedures for structural analysis, vol 1: linear analysis, Pineridge Press, Swansea, 1986.

C4. CRISFIELD, M.A. & WILLS, J., Criteria for a reliable non-linear finite element system, in Reliability of Methods for Engineering Analysis, ed. K.J. Bathe et al., Pineridge Press, Swansea, pp. 159-179, (1986).

C5. CRISFIELD, M.A., Incremental/iterative solution procedures for non-linear structural analysis, Num. Meth. for Non-linear Problems, Vol. 1, ed. C. Taylor et al, Pineridge Press, Swansea, pp.261-290, (1980)

C6 CRISFIELD, M.A., A fast incremental/iterative solution procedure that handles "snap-through", Computers & Structures, 13, pp.55-62, (1981).

C7 CRISFIELD, M.A., Willis, J., Solution strategies and softening materials, Comp. Meth. in Appl. Mech. & Engng., 66, pp.267-289, (1988).

C8 CRISFIELD, M.A. Accelerating and damping the modified Newton-Raphson method, Computers & Struct., 18, pp. 395-407, (1984)

C9 CRISFIELD, M.A., Variable step-lengths for non-linear structural analysis, Transport and Road Res. Lab. Report LR1049, Crowthorne, Berks., England, (1982).

C10 CRISFIELD, M.A., New solution procedures for linear and nonlinear finite element analysis, The Mathematics of Finite Elements and Applications V, ed. J. Whiteman, Academic Press, pp.49-81, (1985).

C11 CRISFIELD, M.A., An arc-length method including line-searches and accelerations, Int. J. for Num. Meth. in Engng., 19, pp.1269-1289, (1983).

C12 CRISFIELD, M.A., Accelerated solution techniques and concrete cracking, Comp. Meth. in Appl. Mech. & Engng., 33, pp.585-607, (1982).

C13.CRISFIELD, M.A, Nonlinear Finite Element Analysis of Solids and Structures, Vol 1, Essentials, Vol2, Advanced Topics, to be published by John-Wiley Ltd

C14. CRISFIELD, M.A., SHI, J. & PENG, X., Some recent developments with shell and stability analysis, Proc. of Europ. Conf. on New Advances in Comp. Struct. Mech., pp521-529, April, 1991.

D1. DAVIDON, W.C., Variable metric method for minimisation, Argonne Nat. Lab. Report ANL-5990, (1959)

D2. DAVIDON, W.C., Variance algorithms for minimisation, Computer J., 10, pp.406-410, (1968).

D3. DAY, A.S., Introduction to dynamic relaxation, The Engineer, 219, pp.218-221, (1965)

D4 DE BORST, R., Computation of post-bifurcation and post-failure behaviour of strain-softening solids, Computers & Structures, 25, pp.211-224, (1987).

D5. DEN HEIJER, C. & RHEINBOLDT, W.C., On steplength algorithms for a class of continuation methods, SIAM J. Num. Analysis, 18, pp.925-948, (1981).

D6. DENNIS, J.E. & MORE, J., Quasi-Newton methods, motivation and theory, SIAM Rev., 19, pp.46-89, (1977).

D7. DUPIUS, G.A., HIBBIT, H.D., MCNAMARA, S.F. & MARCAL, P.V., Non-linear material and geometric behaviour of shell structures, Comp. & Struct., 1, pp. 223-239, (1971).

E1. ERIKSSON, A., On linear constraints for Newton-Raphson corrections and critical point searches in structural f.e. problems, Int. J. for Num. Meth. in Engng., 28, pp.1317, 1334, (1989)

E2. ERIKSSON, A., On some path-related measures for non-linear structural f.e. problems, Int. J. for Num. Meth. in Engng., 26, pp.1791-1803, (1988).

E3. Eriksson, A., On the numerical usage of differentiated tangent stiffness matrices for elastic instability problems, Extended Abstracts from 2nd World Congress on Comp. Mech., Stuttgart, August 1990, pp.67-70

F1. FELIPPA, C.A., Solution of nonlinear static equations, Chapt. prepared for North-Holland Series on Comp. Meth. in Mechanics, Vol. on Large deflection and stability of structures, ed. K-J Bathe, Dept. of Mech. Engng., MIT, (1986).

F2. FELIPPA, C.A., Dynamic relaxation and quasi-Newton methods, 2nd. Int. Conf. on Numerical Meth. for Nonlinear Problems, Barcelona, (1984).

F3 FELIPPA, C.A., Transversing critical points with penalty springs, in Pande and Middleton (ed.), Transient/Dynamic Analysis and Constitutive Laws for Engineering Materials, Nijhoff, Dordrecht, 1987, pp.C2/1-C2/8.

F4 FELIPPA, C.A., Dynamic relaxation under general increment control, Innovative methods for Nonlinear Problems, ed. W.K Liu et al., Pineridge Press Int. Ltd., Swansea pp.103-134, (1984)

F5. FLETCHER, R., A new approach to variable metric algorithms, Computer J., 13, pp. 317-322,, (1970)

F6. FLETCHER,R. Practical methods of optimisation, 2nd Ed., John Wiley & Sons., (1987).

F7. FLETCHER, R. & REEVES, C.M., Function minimisation by conjugate gradients, Comp. J., 6, pp.149-154, (1964).

F8. FORDE, B.W.R. & STIEMER, S.F., Improved arc length orthogonality methods for nonlinear finite element analysis, Computers & Structures, 27, pp.625-630, (1987).

F9 FOX, L. & STANTON, E., Developments in structural analysis by direct energy minimisation, AIAA J., 6, pp.1036-1042, (1968).

F10 FRIED, I., Orthogonal trajectory accession to the nonlinear equilibrium curve, Comp. Meth. in Appl. Mech. & Engng. Comp. Meth. in Appl.Mech. & Engng., 47, pp.283-298, (1984).

G1. GERADIN, M., IDELSOHN, S. & HOGGE, M., Computational strategies for the solution of large non-linear problems via quasi-Newton methods, Computational Methods in Nonlinear Structural and Solid Mechanics, ed. A.K.Noor et al, Pergammon Press, pp.73-82, (1980).

G2. GIERLINSKI, J.T. & GRAVES-SMITH, T.R., A variable load iteration procedure for thin-walled structures, Computers & Structures, 21, pp.1085-1094, (1985).

G3. GILL, P.E. & MURRAY, W., Safeguarded step-length algorithms for optimisation using descent methods, National Physical Laboratory Report NAL 37, (1974).

G4. GILL, P.E. & MURRAY, W., Quasi-Newton methods for unconstrained optimisation, J. Inst. Math. Appl., 9, pp.91-108, (1972).

H1. HAISLER, W.E., STRICKLIN, J.E. & STEBBINS, F.J., Development and evaluation of solution procedures for geometrically non-linear structural analysis by the discrete stiffness method, AIAA/ASME 12th Structure, Structural Dynamics & Materials Conf., Anaheim, California, (April 1971).

H2 HASELGROVE, C.B., The solution of non-linear equations and of differential equations with two-point boundary conditions, Computer J., pp.255-259, (1961).

H3. HESTENES, M. & STEIFEL, E., Method of conjugate gradients for solving linear systems, J. Res. Nat. Bur. Stand., 49, pp.409-436, (1952).

H4. HINTON, E., ABDAL-RAHMAN, H.H. & ZIENKIEWICZ, O.C., Computational models for reinforced concrete slab systems, IABSE Colloq. on Advanced Mechanics of Reinforced Concrete, Final Report, IABSE, Delft, pp.303-314, (1981).

H5. HUGHES, T.J.R., LEVIT, I. & WINGET, J., An element-by-element algorithm for problems of structural and solid mechanics, Comp. Meth. in Appl. Mech. & Engng., 36, pp.241-254, (1983).

H6. HUGHES, T.J.R., FERENCZ, R.M. & HALLQUIST, J.O., Large-scale vectorised implicit calculations in solid mechanics on a CRAY-XMP/48 utilising EBE preconditioned conjugate gradients, Comp. Meth. in Appl. Mech. & Engng., 61, pp.215-248, (1987).

J1. JENNINGS, A., Development of an ICCG algorithm for large sparse systems, Preconditioning Techniques in Numerical Solution of Partial Differential Equations, ed. D.J. Evans, Gordon & Breach, New York, pp.426-438, (1983).

K1. KEARFOTT, R.B., A derivative free arc continuation method and a bifurcation technique, Numerical Solution of Nonlinear Equations, ed. E.L Allgower et al., Springer-Verlag, Berlin, pp.183-198, (1981).

K2. KELLER, H.B., Global homotopies and Newton methods, Recent Advances in Numerical Analysis, ed. C. de Boor G.H. Golub, Academic Press, New York, pp.73-94, (1978).

K3. KELLER, H.B., Numerical solution of bifurcation and nonlinear eigenvalue problems, in Applications of Bifurcation Theory, ed. P. Rabinowitz, Academic Press, pp.359-384, (1977).

K4. KERSHAW, E., The incomplete Cholesky conjugate gradient method for iterative solution of systems of linear equations, J. Comp. Phys., 26, pp.43-65, (1978).

K5 KROPLIN, B. & DINKLER, D., Global constraints applied to static instability problems, in Numerical Meth. for Nonlinear problems, ed. C. Taylor et al., Vol. 3., Pineridge, Swansea, pp.41-60, (1986).

K6. KOUHIA, R. & MIKKOLA, M., Tracing the equilibrium path beyond simple critical points, Int. J. for Num. Meth. in Engng., 28, pp.2923-2941, (1989).

K7. KROPLIN, B., DINKLER, D. & HILLMANN, J., An energy perturbation applied to nonlinear structural analysis, Comp. Meth. in Appl. Mech. & Engng., 52, pp.885-897, (1985).

K8. KROPLIN, B.H., A viscous approach to post buckling analysis, Engng. Struct., 3, pp.187-189, (1981).

K9. KROPLIN, B., DINKLER, D. & HILLMAN, J., Global constraints in nonlinear solution strategies, Finite Element Methods for Nonlinear Problems, ed. P.G. Bergan et al., Springer, Berlin, (1986).

K10. KROPLIN, B. & DINKLER, D., An energy-based concept of structural stability, Proc. of Europ. Conf. on New Advances in Comp. Struct. Mech., pp649-654, April, 1991.

L1. LANCZOS, C, Applied analysis, Prentice Hall, New York, (1956).

L2. LUENBERGER, D.G., Linear and nonlinear programming, 2nd. Ed., Addison-Wesley Pub. Co., Reading, Massach., (1984).

L3. LEE, S.H., Rudimentary considerations for effective line search method in nonlinear finite element analysis, Computers and Structures, 32,6, 1989, pp.1287,1301 Page 29

M1. MATTHIES, H. & STRANG, G., The solution of non-linear finite element equations, Int. J. Num. Meth. in Engng., 14, pp.1613-1626, (1979).

M2. MEEK, J.L. & LOGANATHAN, S., Large displacement analysis of space-frame structures, Comp. Meth. in Appl. Mech. & Engng., 72, pp.57-75 (1989),

M3. MEEK, J.L. & LOGANATHAN, S., Geometrically non-linear behaviour of space frame structures, Computers & Structures, 31, pp.35-45, (1989).

M4. MEEK, J.L. & TAN, H.S., Geometrically nonlinear analysis of space frames by an incremental iterative technique, Comp. Meth. in Appl. & Engng., 47, pp.261-282, (1984).

M5. MOORE, G. The numerical treatment of non-trivial bifurcation points, Numer. Funct. Anal. & Optimiz., 2, 6, pp.441-472, (1980).

M6. MOORE, G. & Spence, A., The calculation of turning points of nonlinear equations, SIAM J. Num. Anal., 17, 1980, pp.567-576.

M7. MALLET, R.H. & MARCAL, P.V., Finite element analysis of non-linear structures, Proc. ASCE, J. of Struct. Div., 94, ST9, pp.2081-2105, (1968).

M8 MALLET, R.H. & SCHMIDT, L.A., Non-linear structural analysis by energy search, Proc. ASCE, J. of Struct. Div., 93, ST3, pp.221-234, (1967).

M9. MARCAL, P.V. Finite element analysis of combined problems of non-linear material and geometric behaviour, Proc. Am. Soc. Mech. Conf. on Comp. Approaches in Appl. Mech., (June 1969).

M10. MARCAL, P.V., Finite element analysis with material non-linearities - theory and practice, Recent Advances in Matrix Meth. of Structural Analysis & Design, ed. R.H. Gallagher et al., The Univ. of Alabama Press, pp.257-282, (1971).

M11. MURRAY, D.W. & WILSON, E.L., Finite element postbuckling analysis of thin elastic plates, Proc. ASCE, J. of Engng. Mech. Div., 95, EM1, pp.143-165, (1969).

M12. MURRAY, D.W. & WILSON, E.L., Finite element postbuckling analysis of thin elastic plates, AIAA. J., 7, pp.1915-1930, (1969).

N1. NAZARETH, L., A relationship between the BFGS and conjugate gradient algorithms and its implications for new algorithms, SIAM J. on Num. Analysis, 16, pp.794-800, (1979).

N2 NOCEDAL, J., Up-dating quasi-newton matrices with limited storage, Math. of Computation, 35, pp.773-782, (1980).

N3. NOUR-OMID, B., A preconditioned conjugate gradient method for solution of finite element equations, Innovative Methods for Nonlinear Problems, Pineridge, Swansea, pp. 17-40, (1984).

N4. NOUR-OMID, B., PARLETT. B.N. & TAYLOR, R.L., A Newton-Lanczos method for solution of nonlinear finite element equations, Computers & Structs., 16, pp.241-252, (1983).

O1. OTTER, J.R.H. & DAY, A.S., Tidal computations, The Engineer, 209, pp.177-182, (1960).

O2. ODEN, J.T., Numerical formulation of non-linear elasticity problems, Proc. ASCE, J. of Struct. Div., 93, ST3, paper 5290, (1967).

O3. ODEN, J.T., Finite element applications in non-linear structural analysis, Proc. Conf. on Finite Element Meth., Vanderbilt Univ., Tenesee, (Nov. 1969).

P1. PADOVAN, J.P. & ARECHAGA, T., Formal convergence characteristics of elliptically constrained incremental Newton-Raphson algorithms, Int. J. Engng. Sci., 20, pp. 1077-1097, (1982).

P2. PAPADRAKAKIS, M., Accelerating vector iteration methods, J. Appl. Mech., ASME, 53, pp.291-297, (1986).

P3. PAPADRAKAKIS, M. & GANTES, C.J., Preconditioned conjugate- and secant-Newton methods for non-linear problems, Int. J. for Num. Meth. in Engng., 28, pp.1299-1316, (1989).

P4. PAPADRAKAKIS, M., A truncated Newton-Lanczos method for overcoming limit and bifurcation points, Int. J. for Num. Meth. In Engng., 29, pp.1065-1077, (1990).

P5 PAPADRAKAKIS, M. & NOMIKOS, N., Automatic non-linear solution with arc-length and Newton-Lanczos methods, Engineering Comp., 7,1, (1980)

P6. PARK, K.C. & RANKIN, C.C., A semi-implicit dynamic relaxation algorithm for static nonlinear structural analysis, Research in Structural and Sold Mechanics - 1982, NASA Conf. Pub. 2245, (1982).

P7 PARK. K.C., A family of solution algorithms for nonlinear structural analysis based on the relaxation equations, Int. J. for Num. Meth. in Engng., 18, pp.1337-1347, (1982).

P8. PARLETT, B.N., A new look at the Lanczos algorithm for solving symmetric systems of linear equations, Linear Algebraic Applications, 29, pp.323-346, (1980).

P9. PICA, A & HINTON, E., The quasi-Newton BFGS method in large deflection analysis of plates, Num. Meth. for Nonlinear Problems, ed. C. Taylor, et al., Vol. 1, Pineridge, Swansea, pp.355-366, (1980).

P10 POLAK,E. & RIBIERE, Nore sur la convergence de methodes de methodes de direction conjugees, Rev. Fr. Inform Rech. Oper., 16-R1, pp.35-43, (1969).

P11 POWELL, G. & SIMONS, J., Improved iteration strategy for nonlinear structures, Int. J. for Num. Meth. in Engng., 17, pp.1455-1467, (1981).

R1 RAMM, E., Strategies for tracing the no-linear response near limit-points, Non-linear Finite Element Analysis in Structural Mechanics, ed. W. Wunderlich, Springer-Verlag, Berlin, pp.63-89, (1981).

R2 RAMM, E., The Riks/Wempner approach - an extension of the displacement control method in non-linear analysis, Non-linear Computational Mechanics, ed. E. Hinton et al., Pineridge, Swansea, pp. 63-86, (1982).

R3. RHEINBOLDT, W.C., Numerical analysis of parametrised nonlinear equations, Vol. 7, John Wiley & Sons, (1986)

R4. RHEINBOLDT, W.C. & RIKS, E., A survey of solution techniques for nonlinear finite element equations, Nat. Aerospace Lab. Report NLR 82027U, (1982). also State of the Art Surveys on Finite Element Techniques, ed. A.K. Noor et al., ASME, (1983).

R5. RHEINBOLDT, W.C., Numerical methods for a class of finite dimensional bifurcation problems, SIAM J. Num. Analysis, 15, 1, pp. 1-11, (1978).

R6. RHEINBOLDT, W.C., Numerical analysis of continuation methods for nonlinear structural problems, Computers and Structures, 13, pp.103-113, (1981).

R7. RIKS, E. & RANKIN, C.C., Bordered equations in continuation methods: an improved solution technique, Nat. Aero Lab. Report NLR MP 82057 U, (1987)

R8. RIKS, E., An incremental approach to the solution of snapping and buckling problems, Int. J. Solids & Structs. Vol. 15, pp. 529-551, (1979)

R9. RIKS, E., The application of Newton's method to the problem of elastic stability, J. of Appl. Mech., 39, pp. 1060-1066, 1972.

R10 RIKS, E., Bifurcation and Stability, a numerical approach, Innovative methods for Nonlinear Problems, ed. W.K Liu et al., Pineridge Press Int. Ltd., Swansea pp.313-344, (1984)

R11 RIKS, E., Progress in collapse analysis, in: Collapse Analysis of Structures, ASME -Pressure Vessel and Piping Conf., -084, (San Antonio, Texas) ed. L.H. Sobel et al, New York, (1984).

R12 RIKS, E., Some computational aspects of the stability analysis of nonlinear structures, Nat. Aero Lab. Report NLR MP 82034 U, (1981) also Comp. Meth. in Appl. Mech. & Engng., 47, pp.219-259, (1984).

R13 RIKS, E., A unified method for the computation of critical equilibrium states of nonlinear elastic systems, Acta Technica Scientiarum Hungaricae, 87, pp.121-141, (1978).

S1. SCHMIDT, W.F., Adaptive step size selection for use with the continuation method, Int. J. for Num. Meth. in Engng., 12, pp.677-694, (1978).

S2. SCHWEIZERHOF, K. & WRIGGERS, P., Consistent linearisation for path following methods in nonlinear f.e. analysis, Comp. Meth. in Appl. Mech. & Engng., 59, pp.261-279, (1986).

S3. SHANNO, D.F., Conjugate gradient methods with inexact searches, Math. of O.R., 13, 3, pp.244-255, (1978)

S4. SHANNO, D.F., Conditioning of quasi-Newton methods for function minimisation, Maths. of Computation, 24, pp.322-334, (1970).

S5. SIMONS, J. & BERGAN, P.G., Hyperplane displacement control methods in nonlinear analysis, Innovative methods for Nonlinear Problems, ed. W.K Liu et al., Pineridge Press Int. Ltd., Swansea pp.345-364, (1984)

S6. SIMO, J.C., WRIGGERS, P., SCHWEIZERHOFF, K.H. & TAYLOR, R.L., Finite deformation postbuckling analysis involving inelasticity and contact constraints, Innovative methods for Nonlinear Problems, ed. W.K Liu et al., Pineridge Press Int. Ltd., Swansea pp.365-388, (1984)

S7. STRICKLIN, J.A., HAISLER, W.E. & VON RIESEMANN, W.A., Evaluation of solution procedures for nonlinear structural analysis, AIAA J., 11, pp.292-299, (1973).

S8. STRICKLIN, J.A. & HAISLER, W.E., Formulation and solution procedures for nonlinear structural analysis, Computers and Structures, 7, pp.125-136, (1977).

S9. SHARIFI, P. & POPOV, E.P., Nonlinear buckling analysis of sandwich arches, Proc. ASCE, J. Engng. Mech. Div., 97, pp.1397-1411, (1971)

S10. SCHMIDT, F.K., BOGNOR, F.K. & FOX, R.L., Finite deflection structural analysis using plate and shell discrete elements, AIAA J., 6, 5, pp.781-791, (1968).

S11. STEIN, E., WAGNER., W. & WRIGGERS, P., Finite element postbuckling analysis of shells with nonlinear contact constraints, in: P.G. Bergan, et al, Ed., Finite Element Method for Nonlinear Problems (Springer, Berlin, 1986), pp719-744.

T1. THURSTON, G.A., BROGAN, F.A. & STEHLIN, P., Postbuckling analysis using a general purpose code, AIAA Paper 85-0719-CP, presented at the AIAA/ASME/ASCE/AHS 26th Structures, Structural Dynamics and Materials Conf., Orlando, Florida, (1985). also J. AIAA, V24, N6, 1986.

T2. TURNER, M.J., DILL, E.H., MARTIN, H.C. & MELOSH, R.J., Large deflection of structures subject to heating and external load, J. of Aero. Sci., v27, n2,p97, (1960).

U1. ULRICH, K., State of the art in numerical methods for continuation and bifurcation problems with applications in continuum mechanics - a survey and comparative study, Report No. 031/88, Laboratorio Nacional de Computacao Cientifica, Rio de Janeiro, Brazil, (1988).

W1. WAGNER, W. & WRIGGERS, P., A simple method for the calculation of postcritical branches, Engineering Computations, 5, pp.103-110, (1988).

W2. WASZCZYSZYN, Z., Numerical problems of nonlinear stability analysis of elastic structures, Computers and Structures, 17,1, pp.13-24, (1983).

W3. WATSON, L.T., An algorithm that is globally convergent with probability one for a class of non-linear two-point boundary value problems, SIAM J. Num. Anal., 16, pp.394-401, (1979). W4 WATSON, L.T. & HOLZER, M., Quadratic convergence of Crisfield's method, 17, pp.69-72, (1983).

W5. WEMPNER, G.A., Discrete approximations related to nonlinear theories of solids, Int. J. Solids & Structs., 7, pp.1581-1599, (1971)

W6. WOLFE, M.A., Numerical methods for unconstrained optimisation - an introduction, van Nostrand Reinhold Co., (1978).

W7. WRIGGERS, P., WAGNER, W. & MIEHE, C., A quadratically convergent procedure for the calculation of stability points in finite element analysis, Comp. Meth. in Appl. Mech. & Engng., 70, pp. 329-347, (1988).

W8. WRIGGERS, P. & SIMO, J.C., A general procedure for the direct computation of turning and bifurcation points, Int. J. for Num. Meth. in Engng., 30, pp.155-176, (1990)

W9. WRIGHT, E.W. & GAYLORD, E.H., Analysis of unbraced multistory steel rigid frames, Proc. ASCE, J. of Struct. Div., 94, pp.1143-1163, (1968).

W10. WAGNER, W., A path-following algorithm with quadratic predictor, Comp. & Struc., V39, N3/4, pp. 339-348, 1991.

Z1. ZIENKIEWICZ, O.C., VALLIAPAN, S. & KING, I.P., Elasto-plastic solutions of engineering problems. Initial stress, finite element approach, Int. J. for Num. Meth. in Engng., 1, pp.75-100, (1969).

Z2. ZIENKIEWICZ, O.C., The finite element in engineering science, McGraw-Hill, London, (1971).

Calculation of Bifurcation Points via Fold Curves

W. WAGNER

Institut für Baumechanik und Numerische Mechanik,
Universität Hannover, D-3000 Hannover 1, FRG

and

P. WRIGGERS

Institut für Mechanik, Technische Hochschule Darmstadt, D-6100 Darmstadt, FRG

Summary

In this paper we study the application of two parameter systems to engineering structures. Especially a finite element formulation is given which allows a quite general usage of the derived methods. The considerations lead to an extended system which can be solved by an arc–length scheme to obtain the fold curve describing the stability behaviour of a perturbed system. A numerical example shows the relevance of the method for engineering structures.

1 Introduction

Bifurcation analysis in engineering design starts in general from the perfect unperturbed structure. This topic is due to its importance well established and several algorithms have been developed for the computation of limit and bifurcation points of the nonlinear equations governing solid mechanics, see e.g. Koiter [9], Keener, Keller [7], Moore, Spence [10], Wagner, Wriggers [17], Wriggers, Wagner, Miehe [20], Stein, Wagner, Wriggers [15].

Due to the fact that many engineering structures exhibit imperfections it is often of great interest how these imperfection influence the stability behaviour. Very expensive investigations have been performed to measure experimentally the shape of imperfection e.g. for shell structures, see e.g. Thompson, Hunt [16] and the references in this book. However it would be preferable to find the most dangerous form or amplitude of an imperfection by pure simulations using the underlying equations.

One step in this direction is the analysis of the behaviour of a structure based on two parameter systems which introduce additionally to the load parameter also a geometrical parameter. This parameter can describe e.g. the amplitude of an imperfection, a thickness or an imperfection in the loading conditions. The variation of this parameter yields then a so called fold curve which under certain conditions connects all singular points due to the parameter variation. This curve is then very useful to estimate the influence of the magnitude of an imperfection on the stability behaviour of a structure.

In this paper we like to introduce the concept of two parameter systems which is well known in the mathematical theory, see e.g. Jepson, Spence [5], in combination with finite elements. For this purpose we first recall some known results concerning singular points of one parameter and two parameter systems in detail. Then we formulate the associated extended system and state an arc-length scheme for the computation of the fold curve. Finally a numerical example shows the applicability of the method to engineering structures.

2 Basic Equations of Nonlinear Finite Elements

This section describes briefly the basic equations for the finite element formulation used in this paper. All equations are given for the three-dimensional elasticity. For beam or shell structures the corresponding equations have to be used. The restriction to elastic deformations is sufficient for most purposes concerning stability problems.

For elastic bodies the principle of virtual work can be stated with respect to the reference configuration in the following form

$$D\Pi(\mathbf{u}) \cdot \boldsymbol{\eta}_1 = \int_B \mathbf{S} \cdot \mathbf{E}'(\mathbf{u}, \boldsymbol{\eta}_1)\, dv - \int_B \rho\hat{\mathbf{b}} \cdot \boldsymbol{\eta}_1\, dv - \int_{\partial B_\sigma} \hat{\mathbf{t}} \cdot \boldsymbol{\eta}_1\, da = 0 \tag{1}$$

with the Green–Lagrangian strain tensor $\mathbf{E} = 1/2(\mathbf{F}^T\mathbf{F} - \mathbf{1})$, the variation of $\mathbf{E}$ given by $\mathbf{E}'(\mathbf{u}, \boldsymbol{\eta}_1) = \frac{1}{2}(\mathbf{F}^T \mathrm{Grad}\, \boldsymbol{\eta}_1 + \mathrm{Grad}^T \boldsymbol{\eta}_1\, \mathbf{F})$, and the 2nd Piola–Kirchhoff stress tensor $\mathbf{S}$. $\rho\hat{\mathbf{b}}$ are the body forces and $\hat{\mathbf{t}}$ describes the surface tractions acting on ∂B_σ whereas $\boldsymbol{\eta}_1$ are the virtual displacements or test functions with $\boldsymbol{\eta}_1 = \mathbf{0}$ on ∂B_u.

For the further developments within this paper we need depart from the weak form also the tangent operator and its derivative with respect to the displacement field. The tangent operator follows from the linearization of the principle of virtual work, eq. (1) for St. Venant material

$$D[\, D\Pi(\mathbf{u}) \cdot \boldsymbol{\eta}_1\,] \cdot \boldsymbol{\eta}_2 = \int_B \{ \mathbf{E}'(\mathbf{u}, \boldsymbol{\eta}_1) \cdot \mathbb{D}\, [\mathbf{E}'(\mathbf{u}, \boldsymbol{\eta}_2\,] + \mathbf{S} \cdot \mathbf{E}''(\boldsymbol{\eta}_1, \boldsymbol{\eta}_2) \} dv\,, \tag{2}$$

with

$$\begin{aligned} \mathbb{D} &= \frac{\partial \mathbf{S}}{\partial \mathbf{E}}\,, \\ \mathbf{E}'(\mathbf{u}, \boldsymbol{\eta}_2) &= \frac{1}{2}(\, \mathbf{F}^T\, \mathrm{Grad}\, \boldsymbol{\eta}_2 + \mathrm{Grad}^T \boldsymbol{\eta}_2\, \mathbf{F}\,)\,, \\ \mathbf{E}''(\boldsymbol{\eta}_1, \boldsymbol{\eta}_2) &= \frac{1}{2}(\, \mathrm{Grad}^T \boldsymbol{\eta}_1\, \mathrm{Grad}\, \boldsymbol{\eta}_2 + \mathrm{Grad}^T \boldsymbol{\eta}_2\, \mathrm{Grad}\, \boldsymbol{\eta}_1\,)\,. \end{aligned} \tag{3}$$

Eqs. (1)–(3) are standard within nonlinear finite element methods. For the later introduced extended system we define also the next derivative by repeating the linearization procedure in a formal way. We obtain

$$\begin{aligned} D\{\, D[\, D\Pi(\mathbf{u}) \cdot \boldsymbol{\eta}_1\,] \cdot \boldsymbol{\eta}_2\, \} \cdot \boldsymbol{\eta}_3 = \int_B \{\, & \mathbf{E}''(\boldsymbol{\eta}_1, \boldsymbol{\eta}_3) \cdot \mathbb{D}\, [\, \mathbf{E}'(\mathbf{u}, \boldsymbol{\eta}_2)\,] \\ & + \mathbf{E}'(\mathbf{u}, \boldsymbol{\eta}_1) \cdot \mathbb{D}\, [\, \mathbf{E}''(\boldsymbol{\eta}_2, \boldsymbol{\eta}_3)\,] + \mathbf{E}''(\boldsymbol{\eta}_1, \boldsymbol{\eta}_2) \cdot \mathbb{D}\, [\mathbf{E}'(\mathbf{u}, \boldsymbol{\eta}_3)\,]\, \} dv \end{aligned} \tag{4}$$

with

$$
\begin{aligned}
\mathbf{E}'(\mathbf{u},\boldsymbol{\eta}_3) &= \frac{1}{2}\,(\,\mathbf{F}^T\,\mathrm{Grad}\,\boldsymbol{\eta}_3 + \mathrm{Grad}^T\,\boldsymbol{\eta}_3\,\mathbf{F}\,)\,,\\
\mathbf{E}''(\boldsymbol{\eta}_1,\boldsymbol{\eta}_3) &= \frac{1}{2}\,(\,\mathrm{Grad}^T\,\boldsymbol{\eta}_1\,\mathrm{Grad}\,\boldsymbol{\eta}_3 + \mathrm{Grad}^T\,\boldsymbol{\eta}_3\,\mathrm{Grad}\,\boldsymbol{\eta}_1\,)\,,\\
\mathbf{E}''(\boldsymbol{\eta}_1,\boldsymbol{\eta}_1) &= \frac{1}{2}\,(\,\mathrm{Grad}^T\,\boldsymbol{\eta}_1\,\mathrm{Grad}\,\boldsymbol{\eta}_2 + \mathrm{Grad}^T\,\boldsymbol{\eta}_2\,\mathrm{Grad}\,\boldsymbol{\eta}_1\,)\,,\\
\mathbf{E}''(\boldsymbol{\eta}_2,\boldsymbol{\eta}_3) &= \frac{1}{2}\,(\,\mathrm{Grad}^T\,\boldsymbol{\eta}_2\,\mathrm{Grad}\,\boldsymbol{\eta}_3 + \mathrm{Grad}^T\,\boldsymbol{\eta}_3\,\mathrm{Grad}\,\boldsymbol{\eta}_2\,)\,.
\end{aligned}
\tag{5}
$$

The finite element discretization of B is given by

$$
B^h = \bigcup_{e=1}^{n_e} \Omega_e
$$

with n_e elements. Within a single element Ω_e the displacement field $\mathbf{u}$, the test functions $\boldsymbol{\eta}_1$ and the higher order displacement fields $\boldsymbol{\eta}_2$ and $\boldsymbol{\eta}_3$ are approximated by

$$
\mathbf{u}^h = \mathbf{N}\,\mathbf{v}_e, \quad \boldsymbol{\eta}_1^h = \mathbf{N}\,\mathbf{q}_{1e}, \quad \boldsymbol{\eta}_2^h = \mathbf{N}\,\mathbf{q}_{2e}, \quad \boldsymbol{\eta}_3^h = \mathbf{N}\,\mathbf{q}_{3e}. \tag{6}
$$

Here the matrix $\mathbf{N}$ contains the shape functions whereas $\mathbf{v}_e$, $\mathbf{q}_{ie} (i = 1,3)$ are the nodal values. Within standard finite element formulations we introduce the matrices $\mathbf{B}^l$ und $\mathbf{B}^{nl}$ containing derivations of the shape functions, which have to be specified for the chosen theory. The components of the Green–Lagrangian strains $\mathbf{E} = 1/2(\mathbf{F}^T\mathbf{F} - \mathbf{1})$ are approximated by the vectors

$$
\begin{aligned}
\mathbf{E}_e^h &= [\mathbf{B}^l + \frac{1}{2}\mathbf{B}^{nl}(\mathbf{v}_e)]\,\mathbf{v}_e = \bar{\mathbf{B}}(\mathbf{v}_e)\,\mathbf{v}_e\,,\\
\mathbf{E}_e'^h &= [\mathbf{B}^l + \quad \mathbf{B}^{nl}(\mathbf{v}_e)]\,\mathbf{q}_{1e} = \mathbf{B}(\mathbf{v}_e)\,\mathbf{q}_{1e}\,.
\end{aligned}
\tag{7}
$$

With (7) the discrete form of the virtual work is given for an element by

$$
G_e^h\,(\mathbf{v}_e, \mathbf{q}_{1e}) = \mathbf{q}_{1e}^T\{\int_{\Omega_e} \mathbf{B}^T(\mathbf{v}_e)\mathbf{S}(\mathbf{v}_e)\,d\Omega - \int_{\Omega_e} \mathbf{N}^T\rho_R\hat{\mathbf{b}}\,d\Omega - \int_{\Gamma_e} \mathbf{N}^T\hat{\mathbf{t}}\,d\Gamma\} \tag{8}
$$

where the vector $\mathbf{S}$ contains the components of the 2nd Piola-Kirchhoff stress tensor.

The assembling procedure to the global system leads to an algebraic system of equations. The operator $\bigcup$ is used to describe the assembling process which yields the global residual $\mathbf{G}$

$$
\mathbf{q}_1^T\,\mathbf{G}\,(\,\mathbf{v}\,) = \bigcup_{e=1}^{n_e} G_e^h\,. \tag{9}
$$

Based on (9) we obtain for an arbitrary test functions $\mathbf{q}_1$ the algebraic equilibrium equations which are nonlinear in $\mathbf{v}$

$$
\mathbf{G}\,(\mathbf{v},\lambda) = \mathbf{R}\,(\mathbf{v}) - \lambda\,\mathbf{P} = \mathbf{0}. \tag{10}
$$

The vector $\mathbf{P}$ contains all loading terms, $\mathbf{R}$ denotes the stress divergence and the load parameter λ is introduced to describe the process of loading.

The linearization of (10) at $\bar{\mathbf{v}}$ leads to the definition of the tangent stiffness matrix of a finite element. With eq. (3) and (7) it remains

$$D\,G_e^h\,(\bar{\mathbf{v}}_e,\eta_{1e})\,\eta_{2e} = \mathbf{q}_{1e}^T \int_{\Omega_e} [\,\mathbf{B}^T(\bar{\mathbf{v}}_e)\,\mathbf{D}\,\mathbf{B}(\bar{\mathbf{v}}_e) + \mathbf{G}^T\hat{\mathbf{S}}(\bar{\mathbf{v}}_e)\mathbf{G}\,]\,d\Omega\,\mathbf{q}_{2e} = \mathbf{q}_{1e}\mathbf{K}_{Te}\mathbf{q}_{2e}\,. \tag{11}$$

The matrix $\hat{\mathbf{S}}$ contains values of the stress vector $\mathbf{S} = \mathbf{D}\,\tilde{\mathbf{B}}(\bar{\mathbf{v}}_e)\,\bar{\mathbf{v}}_e$. Standard assembly procedures denoted by the operator $\cup$ for the element stiffness matrices (11) lead to the global tangent stiffness matrix $\mathbf{K}_T$

$$\mathbf{q}_1^T\,\mathbf{K}_T\,\mathbf{q}_2 = \bigcup_{e=1}^{n_e} \mathbf{q}_{1e}\,\mathbf{K}_{Te}\,\mathbf{q}_{2e}\,. \tag{12}$$

The discrete form of the derivation of the tangent stiffness matrix is based on eq. (5). Using finite element approximations (7) of strains and virtual strains we arrive on the element level at

$$\begin{aligned} D\,[\,D\,G_e\,\eta_{2e}\,]\,\eta_{3e} &= \mathbf{q}_{1e}^T[\mathbf{K}_{Te}(\bar{\mathbf{v}}_e,\bar{\mathbf{v}}_e)\mathbf{q}_{2e}]_{,v}\,\mathbf{q}_{3e} \\ &= \mathbf{q}_{1e}^T\mathbf{h}_e(\bar{\mathbf{v}}_e,\mathbf{q}_{3e},\mathbf{q}_{2e})\,, \end{aligned} \tag{13}$$

with the vector

$$\mathbf{h}_e = \int_{\Omega_e} [\mathbf{B}^T(\bar{\mathbf{v}}_e)\mathbf{D}\mathbf{B}^{nl}(\mathbf{q}_{3e}) + \mathbf{B}^{nlT}(\mathbf{q}_{3e})\mathbf{D}\mathbf{B}(\bar{\mathbf{v}}_e) + \mathbf{G}^T\hat{\mathbf{S}}_1(\bar{\mathbf{v}}_e,\mathbf{q}_{3e})\mathbf{G}]\,d\Omega\,\mathbf{q}_{2e}\,. \tag{14}$$

Abbreviation $(...),v$ indicates the directional derivative with respect to the displacements. The matrix $\hat{\mathbf{S}}_1$ contains the components of the stress vector given by $\mathbf{S}_1(\bar{\mathbf{v}}_e,\mathbf{q}_{3e}) = \mathbf{D}\mathbf{B}(\bar{\mathbf{v}}_e)\mathbf{q}_{3e}$.

The application of similar formulations to the direct computation of stability points has been discussed in Wriggers, Wagner, Miehe [20] and Wriggers, Simo [21].

3 Definition of Singular Points

Singular points which are related to the formulation in the next chapters are defined in this section to provide the necessary background. Thus before studying the two parameter problem we discuss some results on singular points for the one parameter problem $\mathbf{G}(\mathbf{v},\lambda)$.

A detailed description of singular points is possible when a parametrization of the nonlinear solution curve $\mathbf{G}(\mathbf{v},\lambda) = \mathbf{G}(\mathbf{v}(s),\lambda(s))$ in terms of the path parameter s is introduced. The different types of singular points can then be examined by analyzing the path derivatives $\dot{\mathbf{G}}$ and $\ddot{\mathbf{G}}$, where $\dot{(...)}$ denotes the derivative with respect to the path parameter s. In this case $\dot{\mathbf{G}}$ is the tangent and $\ddot{\mathbf{G}}$ the curvature of the equilibrium curve $\mathbf{G} = \mathbf{0}$. For simplification we assume within the following analysis that only simple bifurcation points occur. The tangent to $\mathbf{G}$ is given by

$$\dot{\mathbf{G}} = \mathbf{G}_v \dot{\mathbf{v}} + \mathbf{G}_\lambda \dot{\lambda} = \mathbf{0} . \tag{15}$$

To discuss singular points we introduce the eigenvalue problem

$$(\mathbf{G}_v - \omega \mathbf{1})\boldsymbol{\phi} = 0 , \quad (\mathbf{G}_v^T - \omega \mathbf{1})\boldsymbol{\psi} = 0 \tag{16}$$

with the left eigenvector $\boldsymbol{\psi}$ and the right eigenvector $\boldsymbol{\phi}$. At singular points the eigenvalue ω is zero and eq. (16) can be transformed to

$$\mathbf{G}_v \boldsymbol{\phi}_0 = 0 , \quad \boldsymbol{\psi}_0^T \mathbf{G}_v = 0 . \tag{17}$$

A determination of singular points bases on the product of the left eigenvector and the tangent, eq. (15)

$$\boldsymbol{\psi}_0^T [\mathbf{G}_v \dot{\mathbf{v}} + \mathbf{G}_\lambda \dot{\lambda}] = 0 , \tag{18}$$

which leads with (15) to the equation

$$\boldsymbol{\psi}_0^T \mathbf{G}_\lambda \dot{\lambda} = 0 . \tag{19}$$

The discussion of (19) furnishes the well known condition for turning and bifurcation points

$$\boldsymbol{\psi}_0^T \mathbf{G}_\lambda = \begin{cases} = 0 \ldots\ldots \text{ simple bifurcation point} \\ \neq 0 \ldots\ldots \text{ turning point} \quad . \end{cases} \tag{20}$$

To solve the nonlinear system of equations (15) defined by the tangent, $\mathbf{G}_v \dot{\mathbf{v}} = -\dot{\lambda} \mathbf{G}_\lambda$ we introduce an orthogonal projector, see e.g. Chan [2],

$$\mathbf{P}_\perp = \mathbf{1} - \boldsymbol{\psi}_0 \boldsymbol{\psi}_0^T . \tag{21}$$

Making use of condition $\boldsymbol{\psi}_0^T \mathbf{G}_v = 0$, see eq. (17), the following system of equations has to be solved

$$\mathbf{G}_v \dot{\mathbf{v}} = -\mathbf{P}_\perp \mathbf{G}_\lambda \dot{\lambda} = -\mathbf{G}_\lambda \dot{\lambda} + (\boldsymbol{\psi}_0^T \mathbf{G}_\lambda) \dot{\lambda} \boldsymbol{\psi}_0 . \tag{22}$$

Based on the assumption

$$\dot{\mathbf{v}} = \dot{\lambda} \mathbf{w}_0 + \alpha \boldsymbol{\phi}_0 \tag{23}$$

two situations are possible: $\boldsymbol{\psi}_0^T \mathbf{G}_\lambda = 0$ and $\dot{\lambda} = 0$, see eq. (19). In case of $\boldsymbol{\psi}_0^T \mathbf{G}_\lambda = 0$ (bifurcation point) the result is

$$\left. \begin{aligned} \mathbf{G}_v \dot{\lambda} \mathbf{w}_0 &= -\dot{\lambda} \mathbf{G}_\lambda \\ \mathbf{G}_v \alpha \boldsymbol{\phi}_0 &= \mathbf{0} \end{aligned} \right\} \Longrightarrow \quad \dot{\mathbf{v}} = \dot{\lambda} \mathbf{w}_0 + \alpha \boldsymbol{\phi}_0 , \tag{24}$$

whereas $\dot{\lambda} = 0$ (turning point) leads to

$$\mathbf{G}_v \alpha \boldsymbol{\phi}_0 = \mathbf{0} \quad \} \Longrightarrow \quad \dot{\mathbf{v}} = \alpha \boldsymbol{\phi}_0 . \tag{25}$$

Due to the singularity (rank deficiency of 1) of the system $\mathbf{G}_v \mathbf{w}_0 = -\mathbf{G}_\lambda$ we have to add one constraint equation. Normally the orthogonality condition $\boldsymbol{\psi}_0^T \mathbf{w}_0 = 0$ is used.

Bifurcation Points. In a next step we define different types of bifurcation points. For this purpose we examine the second derivative of $\mathbf{G}$ which is given by

$$\ddot{\mathbf{G}} = \mathbf{G}_{vv}\dot{\mathbf{v}}\dot{\mathbf{v}} + 2\mathbf{G}_{v\lambda}\dot{\mathbf{v}}\dot{\lambda} + \mathbf{G}_v\ddot{\mathbf{v}} + \mathbf{G}_{\lambda\lambda}\dot{\lambda}^2 + \mathbf{G}_\lambda\ddot{\lambda} = \mathbf{0}. \tag{26}$$

Multiplication of $\ddot{\mathbf{G}}$ by the left eigenvector $\boldsymbol{\psi}_0$ yields under the conditions for a bifurcation point $\boldsymbol{\psi}_0^T \mathbf{G}_\lambda = 0$ and $\boldsymbol{\psi}_0^T \mathbf{G}_v = 0$

$$\boldsymbol{\psi}_0^T \mathbf{G}_{vv}\dot{\mathbf{v}}\dot{\mathbf{v}} + 2\boldsymbol{\psi}_0^T \mathbf{G}_{v\lambda}\dot{\mathbf{v}}\dot{\lambda} + \boldsymbol{\psi}_0^T \mathbf{G}_{\lambda\lambda}\dot{\lambda}^2 = 0. \tag{27}$$

With eq. (23) we gain from (27) the so called bifurcation equation

$$a_g \alpha^2 + 2b_g \alpha\dot{\lambda} + c_g \dot{\lambda}^2 = 0\,, \tag{28}$$

with the constants

$$\begin{aligned} a_g &= \boldsymbol{\psi}_0^T[\mathbf{G}_{vv}\boldsymbol{\phi}_0\boldsymbol{\phi}_0] \\ b_g &= \boldsymbol{\psi}_0^T[\mathbf{G}_{vv}\boldsymbol{\phi}_0\mathbf{w}_0 + \mathbf{G}_{v\lambda}\boldsymbol{\phi}_0] \\ c_g &= \boldsymbol{\psi}_0^T[\mathbf{G}_{vv}\mathbf{w}_0\mathbf{w}_0 + 2\mathbf{G}_{v\lambda}\mathbf{w}_0 + \mathbf{G}_{\lambda\lambda}] \\ D_g &= b_g^2 - a_g\, c_g\,. \end{aligned} \tag{29}$$

and the solution for the load direction

$$\dot{\lambda}_{1,2} = -\frac{b_g \alpha}{c_g} \pm \frac{\alpha_g}{c_g}\sqrt{b_g^2 - a_g\, c_g} \quad . \tag{30}$$

For $D_g > 0$ a real square root exists which leads to two distinct solution curves in the bifurcation point. The singular point associated with $a_g \neq 0$ is called in the engineering literature asymmetric bifurcation point, see e.g. Koiter [9], whereas in the mathematical literature it is entitled simple transcritical bifurcation point, see e.g. Jepson, Spence [5].

For the special case $a_g = \boldsymbol{\psi}_0^T[\mathbf{G}_{vv}\boldsymbol{\phi}_0\boldsymbol{\phi}_0] = 0$ the solution of (28) yields

$$\dot{\lambda}_1 = -\frac{2\,b_g}{c_g}\,\alpha \quad \dot{\lambda}_2 = 0\,. \tag{31}$$

This point is called a symmetric or simple pitchfork bifurcation point. The evaluation of $\ddot{\lambda}_2$ indicates whether the symmetric bifurcation point is stable or unstable.

In addition we have for $\boldsymbol{\psi}_0^T \mathbf{G}_\lambda = 0, D_g = 0$ a cusp point and for $\boldsymbol{\psi}_0^T \mathbf{G}_\lambda = 0, D_g < 0$ a simple isola formation point.

Limit Points. Furthermore we can discuss singular points obeying the condition $\dot{\lambda} = 0$. Since a_g denotes the curvature of the solution path $\mathbf{G} = \mathbf{0}$ the condition $a_g \neq 0$ is associated with a limit point or simple quadratic turning point. Thus the type of limit point is determined by the sign of a_g. A special turning point can be found for $a_g = 0$ and is called a simple cubic turning point. For a further discussion see e.g. Wagner [19].

4 Computation of Fold Curves

Fold curves are defined by two–parameter systems which have the form

$$\mathbf{F}(\mathbf{v},\,\lambda,\,\mu) = \mathbf{0} \tag{32}$$

where λ is the load parameter. The additional parameter μ describes a perturbation of the equation $\mathbf{G}(\mathbf{v},\lambda) = \mathbf{0}$. In our example this will be the perturbation of a structural system. For a constant parameter $\mu = \bar{\mu}$ the relation to the one parameter problem is given by

$$\mathbf{F}(\mathbf{v},\,\lambda,\,\bar{\mu}) = \mathbf{G}(\mathbf{v},\,\lambda)|_{\mu=\bar{\mu}}\;. \tag{33}$$

Singular points of the two parameter problem $\mathbf{F}$ can be computed for a constant parameter $\mu = \bar{\mu}$ with an extended system which has been introduced in Keener, Keller [7] and was since then applied to different problem classes, see Seydel [14], Moore, Spence [10] or for structural problems in Wriggers, Wagner, Miehe [20], Wriggers, Simo [21]. Using this approach we have to solve the following extended system

$$\bar{\mathbf{F}}(\mathbf{v},\boldsymbol{\phi},\lambda,\mu) = \left\{ \begin{array}{c} \mathbf{F}(\mathbf{v},\lambda,\mu) \\ \mathbf{F}_v(\mathbf{v},\lambda,\mu)\,\boldsymbol{\phi} \\ l(\boldsymbol{\phi}) \end{array} \right\} = \mathbf{0}\,. \tag{34}$$

To shorten the notation we introduce the vector $\mathbf{x} = (\mathbf{v},\boldsymbol{\phi},\lambda)$. Different values of μ lead now to different solution points $(\mathbf{x}_0,\bar{\mu})$ of the so called fold curve. Remember that each point of this fold curve is a singular point of $\mathbf{F}$.

Now we want to discuss how to find regular points of the fold curve. For this purpose we derive the linearization of $\bar{\mathbf{F}}$ with respect to $\mathbf{x}$ for a fixed $\mu = \bar{\mu}$

$$\bar{\mathbf{F}}_x\,\dot{\mathbf{x}} = \begin{bmatrix} \mathbf{F}_v & \mathbf{0} & \mathbf{F}_\lambda \\ \mathbf{F}_{vv}\,\boldsymbol{\phi}_0 & \mathbf{F}_v & \mathbf{F}_{v\lambda}\,\boldsymbol{\phi}_0 \\ \mathbf{0}^T & \mathbf{1}_\phi^T & 0 \end{bmatrix} \left\{ \begin{array}{c} \dot{\mathbf{v}} \\ \dot{\boldsymbol{\phi}} \\ \dot{\lambda} \end{array} \right\} = \mathbf{0}\,. \tag{35}$$

With eq. (23) $\dot{\mathbf{v}} = \dot{\lambda}\,\mathbf{w}_0 + \alpha\,\boldsymbol{\phi}_0$ which can be chosen with respect to $\mathbf{F}_v = \mathbf{G}_v$ and $\mathbf{F}_\lambda = \mathbf{G}_\lambda$ the second and third row in (35) lead to

$$\begin{aligned} \mathbf{F}_v\dot{\boldsymbol{\phi}} &= -\dot{\lambda}\,[\,\mathbf{F}_{vv}\boldsymbol{\phi}_0\mathbf{w}_0 + \mathbf{F}_{v\lambda}\boldsymbol{\phi}_0\,] - \alpha\mathbf{F}_{vv}\boldsymbol{\phi}_0\boldsymbol{\phi}_0\,, \\ \mathbf{1}_\phi^T\,\dot{\boldsymbol{\phi}} &= 0\,. \end{aligned} \tag{36}$$

Multiplying then the first and second equation of (35) with $\boldsymbol{\psi}_0^T$ yields the following 2×2 system of equations for the unknown parameters $\dot{\lambda}$ and α

$$\begin{bmatrix} \boldsymbol{\psi}_0^T\mathbf{F}_\lambda & 0 \\ b_f & a_f \end{bmatrix} \left\{ \begin{array}{c} \dot{\lambda} \\ \alpha \end{array} \right\} = \mathbf{0}\,, \tag{37}$$

with the constants

$$\begin{aligned} a_f &= \boldsymbol{\psi}_0^T[\mathbf{F}_{vv}\boldsymbol{\phi}_0\boldsymbol{\phi}_0] \\ b_f &= \boldsymbol{\psi}_0^T[\mathbf{F}_{vv}\boldsymbol{\phi}_0\mathbf{w}_0 + \mathbf{F}_{v\lambda}\boldsymbol{\phi}_0]\,. \end{aligned} \tag{38}$$

The regularity of $\tilde{\mathbf{F}}_x$ can be observed by investigating the coefficient matrix of the homogeneous system in eq. (38) which yields the result that $\tilde{\mathbf{F}}_x$ is regular only for $\boldsymbol{\psi}_0^T \mathbf{F}_\lambda \, a_f \neq 0$. For the different types of singular points described in section 2, this condition can only be satisfied at limit points or simple quadratic turning points ($\boldsymbol{\psi}_0^T \mathbf{F}_\lambda \neq 0 \, , a_f \neq 0$).

This means that a simple quadratic turning point of $\mathbf{F}$ with respect to λ given by $(\mathbf{v}_0, \lambda_0, \bar{\mu})$ is a regular point $(\mathbf{x}_0, \bar{\mu})$ on the fold curve. Using the implicit function theorem, see Hildebrandt, Graves [4], one can show that the fold curve exists in a certain neighbourhood of the given point $(\mathbf{x}_0, \bar{\mu})$, see e.g. Jepson, Spence [5].

Often the projection of the fold curve onto the $\lambda - \mu$ plane will be of interest which we will discuss later on in detail.

5 Computation of simple turning points of Fold Curves

In the last section we have seen that a regular point of the fold curve is only possible for a simple quadratic turning point of $\mathbf{F}$. Now we want to characterize some special singular points of the extended system $\tilde{\mathbf{F}}$, see eq. (34).

For this purpose we formulate again a linearization of $\tilde{\mathbf{F}}$ with respect to $\mathbf{x}$ but also for μ

$$\dot{\tilde{\mathbf{F}}} = \tilde{\mathbf{F}}_x \, \dot{\mathbf{x}} + \tilde{\mathbf{F}}_\mu \, \dot{\mu} = \begin{bmatrix} \mathbf{F}_v & \mathbf{0} & \mathbf{F}_\lambda & \mathbf{F}_\mu \\ \mathbf{F}_{vv} \boldsymbol{\phi}_0 & \mathbf{F}_v & \mathbf{F}_{v\lambda} \boldsymbol{\phi}_0 & \mathbf{F}_{v\mu} \boldsymbol{\phi}_0 \\ \mathbf{0}^T & \mathbf{1}_\phi^T & 0 & 0 \end{bmatrix} \begin{Bmatrix} \dot{\mathbf{v}} \\ \dot{\boldsymbol{\phi}} \\ \dot{\lambda} \\ \dot{\mu} \end{Bmatrix} = \mathbf{0} \, . \tag{39}$$

From the first equations we obtain the solution

$$\dot{\mathbf{v}} = \dot{\lambda} \, \mathbf{w}_0 + \dot{\mu} \, \mathbf{z}_0 + \alpha \, \boldsymbol{\phi}_0 \tag{40}$$

with

$$\begin{aligned} \mathbf{F}_v \, \mathbf{w}_0 &= -\mathbf{F}_\lambda \, , \qquad & \boldsymbol{\psi}_0^T \, \mathbf{w}_0 &= 0 \, , \\ \mathbf{F}_v \, \mathbf{z}_0 &= -\mathbf{F}_\mu \, , \qquad & \boldsymbol{\psi}_0^T \, \mathbf{z}_0 &= 0 \, . \end{aligned} \tag{41}$$

Using eq. (41) and multiplying the first and second equation of (39) with $\boldsymbol{\psi}_0^T$ yields the following 2 * 2 system of equations for the unknown parameters $\dot{\lambda}$, α and μ

$$\begin{bmatrix} \boldsymbol{\psi}_0^T \mathbf{F}_\lambda & 0 \\ b_f & a_f \end{bmatrix} \begin{Bmatrix} \dot{\lambda} \\ \alpha \end{Bmatrix} + \begin{Bmatrix} \boldsymbol{\psi}_0^T \mathbf{F}_\mu \\ d_f \end{Bmatrix} \dot{\mu} = \mathbf{0} \, , \tag{42}$$

with the the additional constant $d_f = \boldsymbol{\psi}_0^T [\mathbf{F}_{vv} \boldsymbol{\phi}_0 \mathbf{z}_0 + \mathbf{F}_{v\mu} \boldsymbol{\phi}_0]$.

Analogous to the results based on (37) singular points of the extended system $\tilde{\mathbf{F}}$ are possible for

$$\begin{aligned} \boldsymbol{\psi}_0^T \mathbf{F}_\lambda &= 0 \, , \qquad & \boldsymbol{\psi}_0^T \mathbf{F}_\lambda &\neq 0 \, , \\ a_f &\neq 0 \, , \qquad & a_f &= 0 \, . \end{aligned} \tag{43}$$

In the first case we have an asymmetric bifurcation point of $\mathbf{F}$ whereas in the second case a simple cubic turning point of $\mathbf{F}$ is characterized, see section 3. The special case of a symmetric bifurcation point ($\boldsymbol{\psi}_0^T \mathbf{F} = 0$ and $a_f = 0$) will not be studied in this paper.

Because of the additional variable $\dot{\mu}$ further investigations have to be performed. Especially we want to find simple turning points of the fold curve with respect to μ. Thus we can assume in eq. (42) $\dot{\mu} = 0$ and the complete conditions for the calculation of simple turning points of $\bar{\mathbf{F}}$ are either

$$\begin{aligned} \boldsymbol{\psi}_0^T \mathbf{F}_\lambda &= 0, \\ a_f &\neq 0, \\ \boldsymbol{\psi}_0^T \mathbf{F}_\mu &\neq 0, \end{aligned} \tag{44}$$

for an asymmetric bifurcation point of $\mathbf{F}$ or

$$\begin{aligned} \boldsymbol{\psi}_0^T \mathbf{F}_\lambda &\neq 0, \\ a_f &= 0, \\ \boldsymbol{\psi}_0^T \mathbf{F}_\mu &\neq 0, \\ d_f &= 0, \end{aligned} \tag{45}$$

for a simple cubic turning point of $\mathbf{F}$. It can be shown that the simple turning point of $\bar{\mathbf{F}}$ is a quadratic turning point when the conditions in (44) are fulfilled, see Jepson, Spence [5].

These results enable us to compute asymmetric bifurcation points of two parameter systems. We obtain an asymmetric bifurcation point of $\mathbf{G}$ when the fold curve $\bar{\mathbf{F}}$ exhibits a quadratic turning point at $\mu = \mu_0$. This quadratic turning point can be calculated from $\bar{\mathbf{F}}$ using standard path following techniques.

For this purpose let use define the following set of equations

$$\begin{aligned} \bar{\mathbf{F}}(\mathbf{x}, \mu) &= \mathbf{0} \\ \bar{f}(\mathbf{x}, \mu) &= 0 \end{aligned} \tag{46}$$

where $\bar{f}$ denotes a constraint equation which can be chosen in different ways, see e.g. Riks [12], Keller [8], Ramm [11], Crisfield [3], Wagner, Wriggers [17] and Stein et. al [15]. The linearization of this equation yields

$$\begin{bmatrix} \bar{\mathbf{F}}(\mathbf{x}, \mu)_{,x} & \bar{\mathbf{F}}(\mathbf{x}, \mu)_{,\mu} \\ \bar{f}(\mathbf{x}, \mu)_{,x} & \bar{f}(\mathbf{x}, \mu)_{,\mu} \end{bmatrix} \begin{Bmatrix} \dot{\mathbf{x}} \\ \dot{\mu} \end{Bmatrix} = - \begin{Bmatrix} \bar{\mathbf{F}} \\ \bar{f} \end{Bmatrix} . \tag{47}$$

$\bar{\mathbf{F}}_{,x}$ is nonsingular for quadratic turning points of $\mathbf{F}$ with respect to λ. However efficient standard block elimination techniques for (47) have to be used with caution. Note that the solution of (47) involves the solution of an extended system as given in (34) within each increment which leads to a singularity of $\mathbf{F}_{,v}$.

$$\begin{bmatrix} \mathbf{F}_v & \mathbf{0} & \mathbf{F}_\lambda & \mathbf{F}_\mu \\ \mathbf{F}_{vv}\boldsymbol{\phi}_0 & \mathbf{F}_v & \mathbf{F}_{v\lambda}\boldsymbol{\phi}_0 & \mathbf{F}_{v\mu}\boldsymbol{\phi}_0 \\ \mathbf{0}^T & \mathbf{1}_\phi^T & 0 & 0 \\ \bar{f}_v & \bar{f}_\phi & \bar{f}_\lambda & \bar{f}_\mu \end{bmatrix} \begin{Bmatrix} \dot{\mathbf{v}} \\ \dot{\boldsymbol{\phi}} \\ \dot{\lambda} \\ \dot{\mu} \end{Bmatrix} = - \begin{Bmatrix} \mathbf{F}(\mathbf{v}, \lambda, \mu) \\ \mathbf{F}_v(\mathbf{v}, \lambda, \mu)\boldsymbol{\phi} \\ l(\boldsymbol{\phi}) \\ \bar{f}(\mathbf{v}, \boldsymbol{\phi}, \lambda, \mu) \end{Bmatrix} . \tag{48}$$

An efficient formulation using block elimination can be found in Moore and Spence [10], Wriggers, Wagner, Miehe [20] or Wriggers, Simo [21].

6 Finite Element Formulation

In this section we state the finite element discretization of the equations given in the previous sections. For this purpose we recall the notation of section 2.

For a given $\bar{\mu}$ $\mathbf{F}$ can be expressed in terms of the residual $\mathbf{G}$ see eq. (10). $\mathbf{F}_v$ is represented by the tangent matrix $\mathbf{K}_T$ in (12). With (10) we obtain for $\mathbf{F}_\lambda = -\mathbf{P}$. In our examples we assume that only the load vector depends on μ thus $\mathbf{F}_\mu = -\mathbf{P}_\mu$. Furthermore $\mathbf{P}$ should not depend on the displacement $\mathbf{v}$. Thus the terms $\mathbf{F}_{v\lambda}$ and $\mathbf{F}_{v\mu}$ are zero. The term $\mathbf{F}_{vv}\,\boldsymbol{\phi}_0\dot{\mathbf{v}}$ is defined by (13). The normalization function $l(\boldsymbol{\phi})$ appearing in (34) is given by $l(\boldsymbol{\phi}) = \|\boldsymbol{\phi}\| - 1 = 0$. This yields $\mathbf{l}_\phi = \boldsymbol{\phi}/\|\boldsymbol{\phi}\|$.

For the iteration procedure we define the increments $\Delta\mathbf{v} = \dot{\mathbf{v}}\,\Delta s$, $\Delta\lambda = \dot{\lambda}\,\Delta s$ and $\Delta\mu = \dot{\mu}\,\Delta s$. Since our formulation starts from a potential the tangent stifness matrix is symmetrical and we do not have to differentiate between left and right eigenvectors. Hence we obtain $\boldsymbol{\varphi} = \boldsymbol{\phi} = \boldsymbol{\psi}$ and can define its increment $\Delta\boldsymbol{\varphi} = \dot{\boldsymbol{\varphi}}\,\Delta s$.

Now we introduce the classical constant arc–length constraint used in the formulation

$$\bar{f}(\mathbf{v},\boldsymbol{\varphi},\lambda,\mu) = \mathbf{t}_v^T(\mathbf{v}-\bar{\mathbf{v}}) + \mathbf{t}_\varphi^T(\boldsymbol{\varphi}-\bar{\boldsymbol{\varphi}}) + t_\lambda(\lambda-\bar{\lambda}) + t_\mu(\mu-\bar{\mu}) = 0 \qquad (49)$$

where the quantities $\mathbf{t}$ denote an approximation to the tangent of $\bar{\mathbf{F}} = \mathbf{0}$ at the last equilibrium point. The quantities $(\bar{..})$ are commonly associated with the last iteration step.

With these preliminaries we are able to state the discretized form of (48)

$$\begin{bmatrix} \mathbf{K}_T & \mathbf{0} & -\mathbf{P} & -\mathbf{P}_{,\mu} \\ (\mathbf{K}_T\boldsymbol{\varphi})_{,v} & \mathbf{K}_T & \mathbf{0} & \mathbf{0} \\ \mathbf{0}^T & \frac{\boldsymbol{\varphi}^T}{\|\boldsymbol{\varphi}\|} & 0 & 0 \\ \mathbf{t}_v^T & \mathbf{t}_\varphi^T & t_\lambda & t_\mu \end{bmatrix} \begin{Bmatrix} \Delta\mathbf{v} \\ \Delta\boldsymbol{\varphi} \\ \Delta\lambda \\ \Delta\mu \end{Bmatrix} = - \begin{Bmatrix} \mathbf{G}(\mathbf{v},\lambda,\mu) \\ \mathbf{K}_T(\mathbf{v},\lambda)\,\boldsymbol{\varphi} \\ l\,(\boldsymbol{\varphi}) \\ \bar{f}(\mathbf{v},\boldsymbol{\varphi},\lambda,\mu) \end{Bmatrix}. \qquad (50)$$

7 Numerical example

In this section we like to present one numerical example which illustrates the use of the two parameter system (32) for the computation of asymmetric bifurcation points. The algorithm has been implemented into the finite element code F E A P, developed by R. L. Taylor and described in Zienkiewicz, Taylor [22].

We discuss the stability behaviour of the so called Roorda frame which can be seen in Fig. 1. Roorda [13] studied this problem experimentally. As a result he obtained a relationship between the load and the eccentricity of the load in terms of F/F_c versus μ/L. The parameters and material constants for the problem are given in Fig. 1, and have been chosen to compare our calculations with that of Jürcke [6].

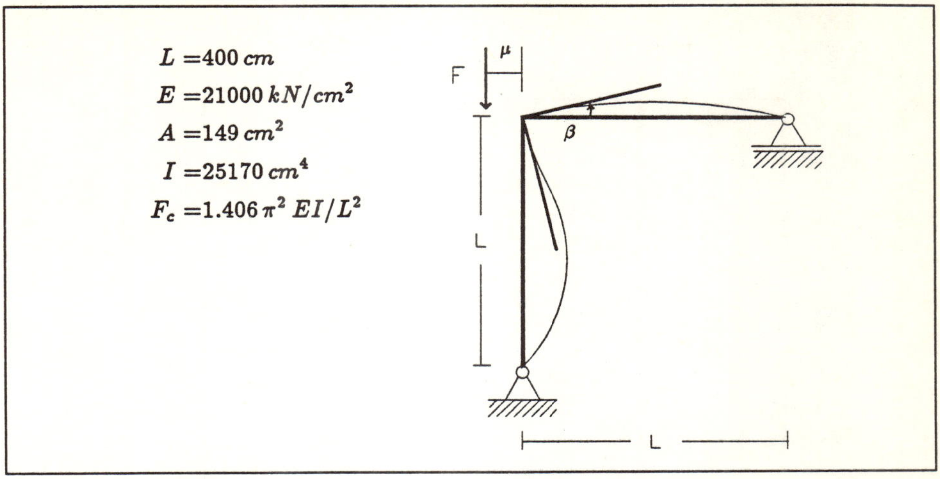

Fig. 1: Roorda–Frame, System and Material Data

The analytical solution of this problem is given for an inextensible cord length of the frame members, see e.g. Brush, Almroth [1]. This leads to an asymmetrical bifurcation point for the eccentricity $\mu = 0$ as shown in Fig. 2.

However, in real structures this constraint is not fulfilled. Using a standard finite element formulation which includes the extensibility of the frame members one does not obtain a bifurcation point. The load deflection curve becomes then the curve denoted by the triangles, see Fig. 2. If one tries to obtain the bifurcation point of the inextensible formulation as a limit case one has to increase membrane stiffness artificially. However this leads due to the ill–conditioning of the associated system of equations often to divergence within the solution procedure.

For a value of $\mu > \mu_0$ we obtain a load deflection curve with positive rotation angle β which exhibits a limit point. For $\mu < \mu_0$ (negative rotation angle β) a pure bending behaviour occurs. As an example Fig. 2. contains the load deflection curves for $\mu = \pm 2$ cm. Here μ_0 defines an eccentricity for which bifurcation occurs.

The load deflection curve is shown in Fig. 3 for the value of $\mu = \pm 2$ cm. It depicts the global stability diagram for the first and second singular point. Observe that we obtain for the parameter $\mu = +2$ cm two solution paths. By changing the sign of μ we get two other curves. These solutions close in on the asymmtrical bifurcation path of the inextensible theory. Due to presence of large deflections the asymmetrical bifurcation path is no longer linear and the load deflection curves exhibit a decrease of load for large values of β. Thus we employed for the finite element formulation a geometrically exact beam element derived from an axisymmetrical shell element, see Wagner [18]. In total 20 finite elements were used to discretize the Roorda frame.

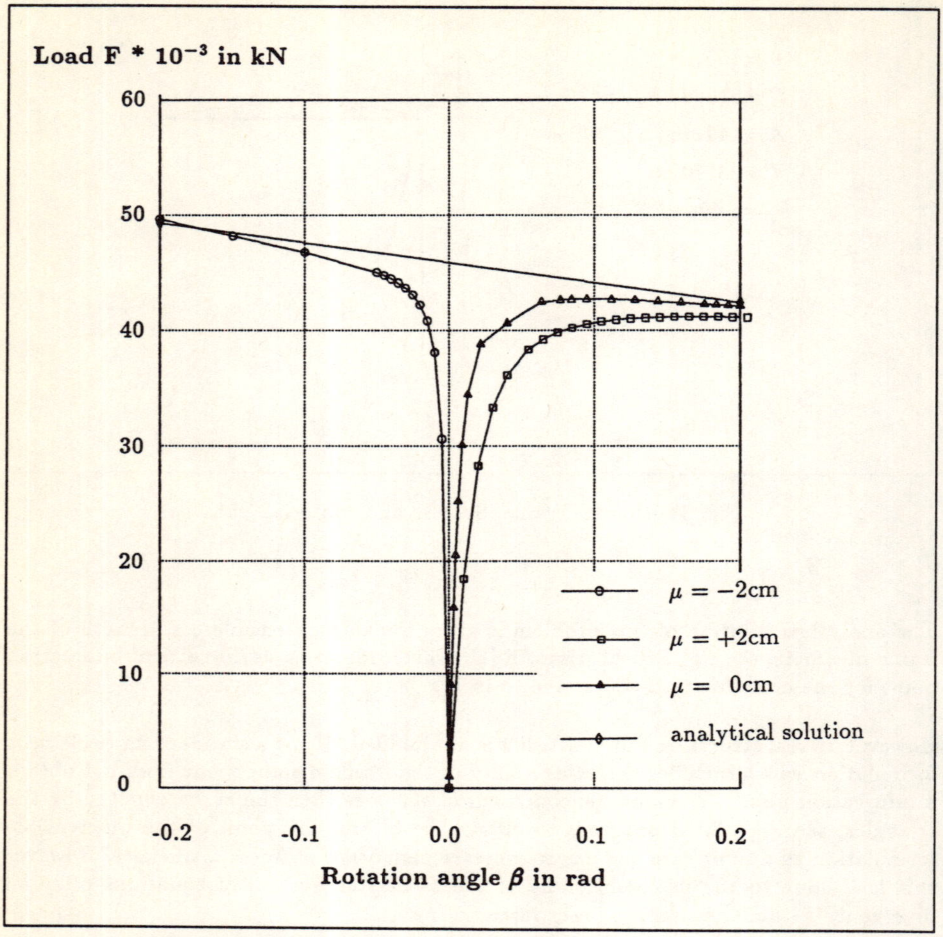

Fig. 2: Load deflection curves for different values of μ

It is now interesting to know whether also the extensible formulation of the Roorda–frame exhibits an asymmetrical bifurcation point. This question can be answered approximately by changing the parameter μ. One obtains for $\mu > \mu_0$ a limit point which can be viewed as an approximation of the bifurcation point. Since the value of μ_0 is not known the asymmetric bifurcation point cannot be found exactly within this strategy.

With the above derived algorithms we are able to detect the asymmetric bifurcation point as a simple quadratic turning point of the fold curve $\tilde{\mathbf{F}}$. The fold curve can be calculated in a simple way, if we introduce a procedure similar to the load control algorithm in terms of the eccentricity μ. This yields for a given $\mu = \bar{\mu}$ a limit point $(\mathbf{v}_0, \varphi_0, \lambda_0)$ of $\mathbf{F}$ using the extended system (34).

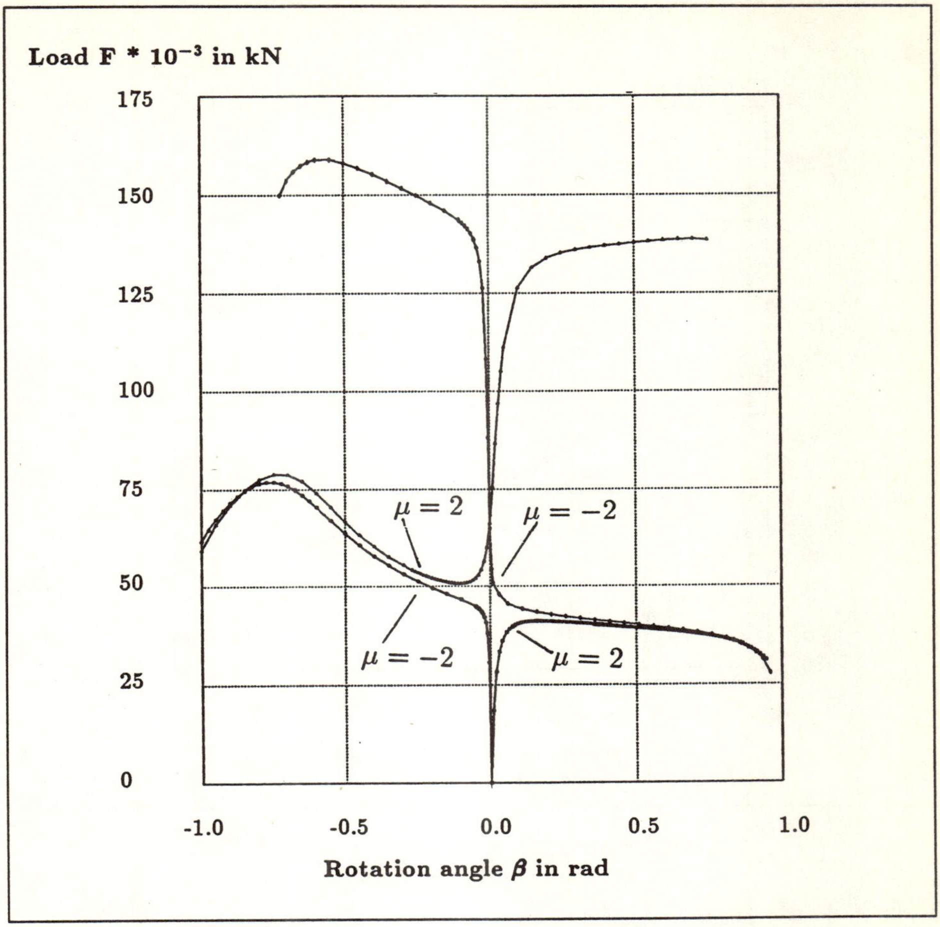

Fig. 3: Global stability behaviour for $\mu = \pm 2$ cm

With this procedure we were able to find the fold curve denoted in Fig. 4 which contains all limit points of the load deflection curves for different values of μ. The projection of the fold curve on the μ–λ plane yields Fig. 5. Here we clearly see the quadratic turning point associated with the asymmtrical bifurcation point of the extensible theory. The value of μ at the turning point is $\mu_0 = -1.31$ cm. Furthermore the load factor is given by $\lambda_0 = 45495$ kN. The left part of the projected fold curve is associated with limit points of **F** which are smaller than λ_0 whereas the right part of is given by limit points which are larger than λ_0.

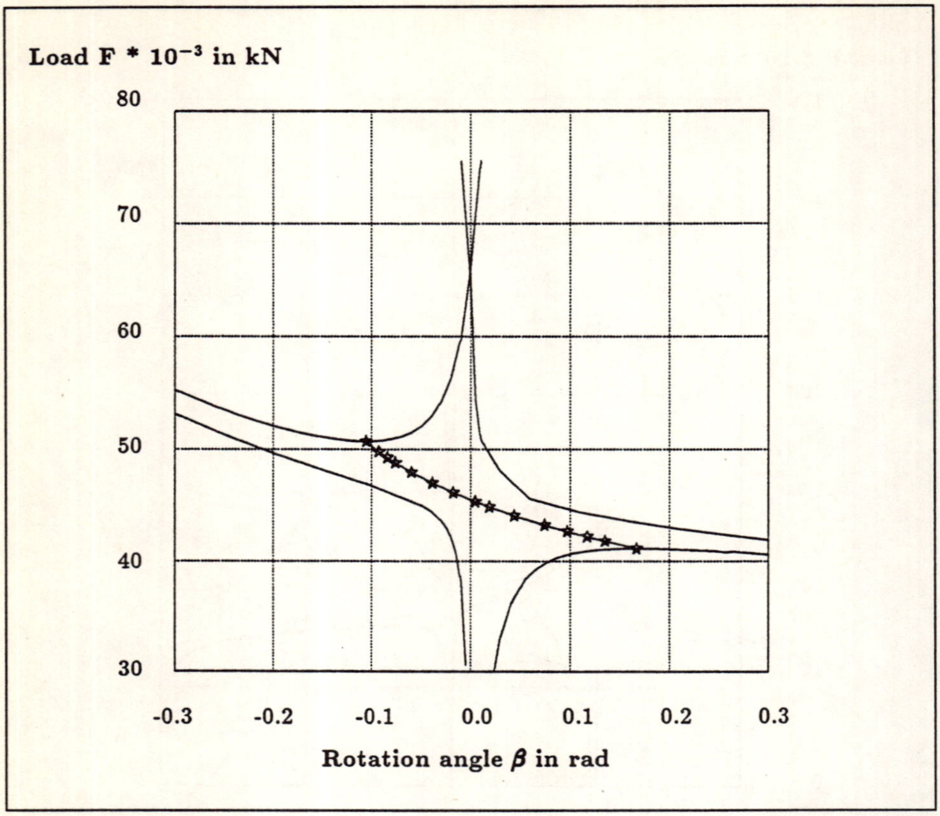

Fig. 4: Load deflection curves for $\mu = \pm 2cm$ and Fold Curve

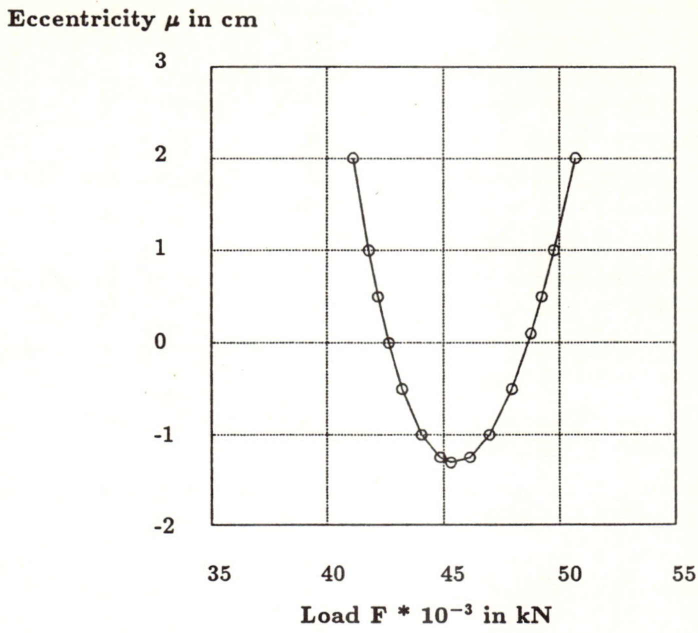

Fig. 5: Projection of Fold Curve on $\mu - \lambda$–plane

8 References

1. Brush, D. O., Almroth, B. O.: *Buckling of Bars, Plates and Shells* , Mc Graw Hill, London 1975.
2. Chan, T. F.: Deflation Techniques and Block–Elimination Algorithms for Solving Bordered Singular Systems, SIAM, J. Sci. Stat. Comput. **5** (1984-1), 121–134.
3. Crisfield, M.A.: A Fast Incremental/Iterative Solution Procedure that Handles Snap Through, Computers & Structures, **13** (1981), 55-62.
4. Hildebrandt, T. H., Graves, L. M.: Implicit Functions and their Differentials in General Analysis. A.M.S. Transactions **29** (1927), 127–153.
5. Jepson A. D., Spence, A.: Folds in Solutions of two Parameter Systems and their Calculation, SIAM, J. Numer. Anal. **22** (1985), 347–369.
6. Jürcke, R. K.: Zur Stabilität und Imperfektionsempfindlichkeit elastischer Schalentragwerke – Finite Element Formulierung der Anfangs–Nachbeul–Theorie, Tech. Rep. No. 85-5, Inst. KIB, Ruhr Universität Bochum (1985).
7. Keener, J. P., Keller, H. B.: Perturbed Bifurcation theory, Arch. Rational Mech. Anal. **50** (1973), 159–175.

8. Keller, H. B.: Numerical Solution of Bifurcation and Nonlinear Eigenvalue Problems. In: Rabinowitz, P. (ed.): Application of Bifurcation Theory, Academic Press, New York (1977), 359–384.

9. Koiter W. T.: On the Stability of Elastic Equilibrium, Translation of 'Over de Stabiliteit von het Elastisch Evenwicht', Polytechnic Institute Delft, H. J. Paris Publisher Amsterdam 1945, NASA TT F-10,833, 1967.

10. Moore, G., Spence, A.: The Calculation of Turning Points of Nonlinear Equations, SIAM, J. Numer. Anal. Comput. **17** (1980), 567–575.

11. Ramm, E.: Strategies for Tracing the Nonlinear Response Near Limit Points, in: Nonlinear Finite Element Analysis in Structural Mechanics, eds. W. Wunderlich, E. Stein, K.-J. Bathe, Springer, Berlin-Heidelberg-New-York (1981).

12. Riks, E.: The Application of Newtons Method to the Problem of Elastic Stability, J. Appl. Mech., **39** (1972), 1060-1066.

13. Roorda, J.: Stability of Structures with Small Imperfections, J. Engng. Mech. Div. Am. Soc. civ.Engrs. , **91** (1965), 87.

14. Seydel, R.: Numerical Computation of Branch Points in Nonlinear Equations, Numer. Math. **33** (1979), 339–352.

15. Stein, E., Wagner, W., Wriggers, P.: Nonlinear Stability Analysis of Shell– and Contact–Problems including Branch–Switching, Computational Mechanics, **5** (1990), 428–446.

16. Thompson, J. M. T, Hunt, G. W.: *A General Theory of Elastic Stability*, John Wiley & Sons Ltd., London 1973.

17. Wagner, W., Wriggers, P.: A Simple Method for the Calculation of Secondary branches, Engineering Computations, **5** (1988), 103–109.

18. Wagner, W.: A Finite Element Model for Nonlinear Shells of Revolution with Finite Rotations, Int. J. Num. Meth. Engng. **29** (1990), 1455–1471.

19. Wagner, W.: Zur Behandlung von Stabilitätsproblemen der Elastostatik mit der Methode der Finiten Elemente, Forschungs- und Seminarberichte aus dem Bereich der Mechanik der Universität Hannover, F91/1 (1991).

20. Wriggers, P., Wagner, W., Miehe, C.: A Quadratically Convergent Procedure for the Calculation of Stability Points in Finite Element Analysis, Comp. Meth. Appl. Mech. Engng. **70** (1988), pp 329–347.

21. Wriggers, P., Simo, J.C.: A General Procedure for the Direct Computation of Turning and Bifurcation Points, Int. J. Num. Meth. Engng. **30** (1990), 155–176.

22. Zienkiewicz, O. C.,Taylor, R. L.:*The Finite Element Method*, 4th edn., Mc Graw–Hill, London, 1989.

On Stability Conditions for Nonlinear Static and Dynamic Buckling Responses of Arbitrary Structures

W.B. Krätzig, Y.-y. Qian*

Institute of Statics and Dynamics, Ruhr-University Bochum, F.R.Germany; *on leave from Shanghai University of Engineering Science, P.R. China

Summary

Nonlinear static and dynamic structural responses may be susceptible to instability phenomena of various nature. In the framework of finite element analyses such responses in general are constructed by use of path-following strategies operating in tangential subspaces to the real vector space of loads and displacements. By aid of Ljapunov's direct method the paper derives and carefully discusses very general stability criteria, also based on tangential properties of the real response. These criteria hold for the static as well as the dynamic case, and are easily expansible to physically nonlinear problems.

1. Introduction

Instabilities are inherent to nonlinear structural responses, so buckling and bifurcation phenomena have been researched since the beginning of engineering mechanics. But until now, instability effects are difficult to detect and often tough to understand.

If only geometrical nonlinearities are considered, difficulties can be reduced to some degree. In this case and for simple structures, even dynamic buckling problems have been treated also in an early stage [1]. For more complicated structural models, such as thin shells, only theoretical foundations could be laid then [2,3,4].

This situation changed dramatically with the penetration of computers into engineering. Numerical simulations stimulated the development of new theories, algorithms and criteria both for static [5,6,7,8,9] and dynamic [10,11,12] unstable systems. More generalized theories even enable the integration of material nonlinearities into unified stability algorithms [13,14].

In this paper we start repeating the derivation of the fundamental equation for the finite element analysis of arbitrary nonlinear structural responses: the tangential equation of motion [11,12]. From this discretized 2nd order differential equation we will derive the most general stability criteria for time-dependent paths, from which the static case will be derived as a special one. Within this context, also material nonlinearities can be considered without fundamental difficulties.

2. Nonlinear Field Equations

During the following derivations we have thin shell structures in mind [15], but the transfer to other continuum models obviously is uncomplicated. Let us consider an arbitrary shell, imbedded into the E3, in its initial state with middle surface domain F^0 and boundary C^0. General convective curvilinear coordinates Θ^α: Θ^1, Θ^2 determine points on F^0, and the following fields of variables are defined on F^0, respectively along C^0:

$\mathbf{p}=\mathbf{p}(\Theta^\alpha, \mathbf{u}, t)$	prescribed loads on F^0, related to dF^0;
$\mathbf{u}=\mathbf{u}(\Theta^\alpha, t)$	displacement variables on F^0;
$\mathbf{f}=\mathbf{f}(\mathring{\rho}, \mathbf{u}, t)$	d'Alembert's inertia forces with $\mathring{\rho}$ the mass density function of F^0;
$\mathbf{t}=\mathbf{t}(\Theta^\alpha, \mathbf{u}, t)$	boundary force variables, related to dC^0;
$\mathbf{r}=\mathbf{r}(\Theta^\alpha, t)$	boundary displacement variables along C^0;
$\boldsymbol{\sigma}=\boldsymbol{\sigma}(\Theta^\alpha, t)$	Piola-Kirchhoff stress variables of 2nd kind;
$\boldsymbol{\varepsilon}=\boldsymbol{\varepsilon}(\Theta^\alpha, t)$	Green-Lagrange strain variables;
$\dot{\boldsymbol{\varepsilon}}=\dot{\boldsymbol{\varepsilon}}(\Theta^\alpha, t)$	corresponding strain velocities.

These column variables consisting of tensor components permit the formulation of arbitrary nonlinear shell theories [14] with the following

field equations $\forall \{\mathbf{p}, \mathbf{u}, \mathbf{f}, \boldsymbol{\sigma}, \boldsymbol{\varepsilon}\} \in \mathring{F}$

$$-(\mathbf{p}+\mathbf{f}) = \mathbf{D}_e \cdot \boldsymbol{\sigma} = (\mathbf{D}_{eL} + \mathbf{D}_{eN}(\mathbf{u}))\cdot\boldsymbol{\sigma}, \tag{1}$$

$$\boldsymbol{\varepsilon} = \mathbf{D}_k \cdot \mathbf{u} = (\mathbf{D}_{kL} + 1/2\, \mathbf{D}_{kN}(\mathbf{u}))\cdot \mathbf{u}, \tag{2}$$

boundary conditions: $\forall \{\mathbf{u}, \boldsymbol{\sigma}\} \in \mathring{F}, \quad \{\mathbf{r}, \mathbf{t}\} \in \mathring{C}$

$$\mathbf{t} = \mathring{\mathbf{t}} = \mathbf{R}_t\, \boldsymbol{\sigma} = (\mathbf{R}_{tL} + \mathbf{R}_{tN}(\mathbf{u}))\, \boldsymbol{\sigma} \quad \text{along } \mathring{C}_t, \tag{3}$$

$$\mathbf{r} = \mathring{\mathbf{r}} = \mathbf{R}_r \cdot \mathbf{u} \quad \text{along } \mathring{C}_r. \tag{4}$$

Herein $\mathbf{D}_e$ and $\mathbf{D}_k$ represent the dynamic and kinematic field operators. Both consist of linear differential operators $\mathbf{D}_{eL}$, $\mathbf{D}_{kL}$ and nonlinear parts $\mathbf{D}_{eN}(\mathbf{u})$, $\mathbf{D}_{kN}(\mathbf{u})$, which are linear functionals of elements of $\mathbf{u}$ or their covariant derivatives. An indispensable property for the future discretization is the symmetry of $\mathbf{D}_{kN}(\mathbf{u})$:

$$\mathbf{D}_{kN}(\mathbf{u})\cdot \mathbf{v} = \mathbf{D}_{kN}(\mathbf{v})\cdot \mathbf{u}\,. \tag{5}$$

$\mathbf{R}_t$ and $\mathbf{R}_r$ are recognized as dynamic and kinematic trace operators.

For an arbitrary dynamic equilibrium state the following principle of virtual displacements

$$\iint_{\mathring{F}} \boldsymbol{\sigma}^T \cdot \delta\boldsymbol{\varepsilon} \cdot d\mathring{F} = \iint_{\mathring{F}} \mathbf{p}^T \cdot \delta\mathbf{u} \cdot d\mathring{F} + \iint_{\mathring{F}} \mathbf{f}^T \cdot \delta\mathbf{u} \cdot d\mathring{F} + \int_{\mathring{C}_t} \mathbf{t}^T \cdot \delta\mathbf{r} \cdot d\mathring{S} \tag{6}$$

$$\forall \{\mathring{\mathbf{p}}, \mathbf{u}, \boldsymbol{\sigma}, \boldsymbol{\varepsilon}\} \in \mathring{F}, \quad \{\mathring{\mathbf{t}}, \mathbf{r}\} \in \mathring{C}_t$$

describes a week formulation of this state of equilibrium (1,3) under the constraints (2,4)

$$\delta\boldsymbol{\varepsilon} = (\mathbf{D}_{kL} + \mathbf{D}_{kN}(\mathbf{u}))\cdot\delta\mathbf{u} \text{ on } \mathring{F}, \quad \delta\mathbf{r} = \mathbf{R}_r\cdot\delta\mathbf{u} = \mathbf{0} \quad \text{along } \mathring{C}_r, \tag{7}$$

independent from the kinds of nonlinearities in the process.

3. Nonlinear Incremental Principle of Virtual Displacements

In the context of stability investigations due to Fig. 1 the independent displacement vector of a shell's middle surface is incremented into three parts: these are the initial displacements $\mathring{\mathbf{u}}$ in the initial state, the fundamental displacements $\bar{\mathbf{u}}$ under fundamental dynamic equilibrium loads as prescribed, and the neighbouring motions $\overset{+}{\mathbf{u}}$ as perturbations to the $\bar{\mathbf{u}}$ [10, 11, 12]:

$$\begin{aligned} \mathbf{u} &= \mathring{\mathbf{u}} + \bar{\mathbf{u}} + \overset{+}{\mathbf{u}} = \mathring{\mathbf{u}} + \overset{*}{\mathbf{u}}, \quad \overset{*}{\mathbf{u}} = \bar{\mathbf{u}} + \overset{+}{\mathbf{u}}, \\ \dot{\mathbf{u}} &= \dot{\bar{\mathbf{u}}} + \dot{\overset{+}{\mathbf{u}}}, \\ \mathbf{r} &= \mathring{\mathbf{r}} + \bar{\mathbf{r}} + \overset{+}{\mathbf{r}}. \end{aligned} \tag{8}$$

The strain due to (8) are assumed to satisfy the constraints

$$\boldsymbol{\varepsilon} = [\mathbf{D}_{kL} + 1/2\, \mathbf{D}_{kN}(2\mathring{\mathbf{u}} + \overset{*}{\mathbf{u}})]\cdot\mathbf{u}, \tag{9}$$

$$\mathbf{r} = \mathring{\mathbf{r}} = \mathbf{R}_r\overset{*}{\mathbf{u}} \quad \text{along } \mathring{C}_r, \tag{10}$$

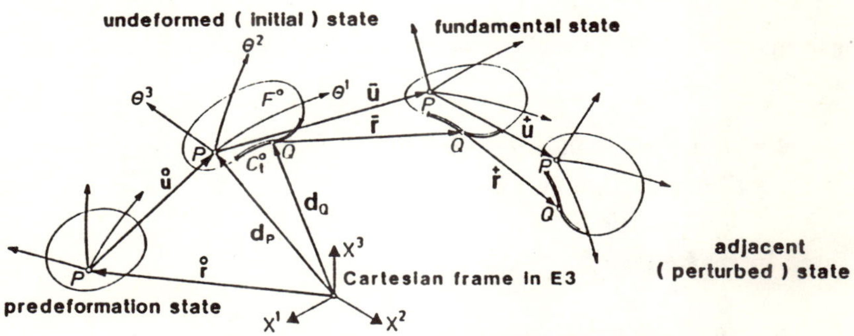

Fig. 1. Different States of Shell Deformation

where for the operators $\mathbf{D}_{kL}$, $\mathbf{D}_{kN}$ and $\mathbf{R}_r$ exist different forms for different structural models as well as model approximations [2, 15].

If material nonlinearities are taken into consideration, the stresses of the present state can no longer be found explicitly from the strains. An integral is needed for this evaluation

$$\boldsymbol{\sigma} = \boldsymbol{\sigma}_0 + \int_{\sigma} d\boldsymbol{\sigma} = \boldsymbol{\sigma}_0 + \iint_{(0,0)}^{(\varepsilon,\dot{\varepsilon})} (\mathbf{E}\,d\boldsymbol{\varepsilon} + \mathbf{D}\,d\dot{\boldsymbol{\varepsilon}}) \tag{11}$$

with
$$d\boldsymbol{\sigma} = \mathbf{E}d\boldsymbol{\varepsilon} + \mathbf{D}d\dot{\boldsymbol{\varepsilon}}, \quad \mathbf{E} = \mathbf{E}(\boldsymbol{\varepsilon},\dot{\boldsymbol{\varepsilon}},\boldsymbol{\sigma}), \quad \mathbf{D} = \mathbf{D}(\boldsymbol{\varepsilon},\dot{\boldsymbol{\varepsilon}},\boldsymbol{\sigma}) \tag{12}$$

in which **E** is the tangential elasticity tensor for the present state, and **D** is the corresponding tangential viscosity tensor. Both **E** and **D** are state functions. The tensors **E** and **D** are constant tensors only in linear problems.

Equations (11, 12) represent a very general description of nonlinear material behavior. In the case of elasto-plastic materials with hardening properties for instance, we find for

$$\mathrm{E} = \mathrm{E}(\boldsymbol{\varepsilon},\boldsymbol{\sigma},\mathrm{F}), \quad \mathrm{D} = 0. \tag{13}$$

(12) then transforms into the well-known elasto-plastic constitutive law

$$d\boldsymbol{\sigma} = \mathrm{E}\, d\boldsymbol{\varepsilon}, \tag{14}$$

where **E**, combining elastic and flow rule components, depends on the sign of the yield criterion:

$$\begin{aligned} \mathrm{F}(\boldsymbol{\sigma},\sigma_F,\mathbf{k}_F) &< 0\text{: } \mathbf{E} \text{ describes elastic actions,} \\ &\geq 0\text{: } \mathbf{E} \text{ describes elasto-plastic actions.} \end{aligned} \tag{15}$$

In (15) σ_F describes the yield stress and $\mathbf{k}_F$ a set of hardening evolution functions. The following derivations will be continued with the general law (12), since the additional transformation and discretization of supplementary inequalities like (15) add no principle difficulties into the deductions.

According to equations (8, 9), the strains under present displacement can be calculated as follows:

$$\begin{aligned} \boldsymbol{\varepsilon} &= [\,D_{kL} + 1/2\; D_{kN}(2\mathring{u} + \bar{u} + \overset{+}{u})]\cdot(\bar{u} + \overset{+}{u}) \\ &= [D_{kL} + 1/2\; D_{kN}(2\mathring{u} + \bar{u})]\cdot\bar{u} + [\,D_{kL} + D_{kN}(\mathring{u} + \bar{u})]\cdot\overset{+}{u} \\ &\quad + 1/2\; D_{kN}(\overset{+}{u})\cdot\overset{+}{u} \\ &= \bar{\boldsymbol{\varepsilon}} + \overset{+}{\boldsymbol{\varepsilon}} + 1/2\,\overset{++}{\boldsymbol{\varepsilon}}, \end{aligned} \tag{16}$$

$$\begin{aligned} \dot{\boldsymbol{\varepsilon}} &= [D_{kL} + D_{kN}(\mathring{u} + \bar{u})]\cdot\dot{\bar{u}} + [\,D_{kL} + \; D_{kN}(\mathring{u} + \bar{u})]\cdot\dot{\overset{+}{u}} \\ &\quad + D_{kN}(\dot{\bar{u}})\cdot(\overset{+}{u}) + D_{kN}(\overset{+}{u})\cdot\dot{\overset{+}{u}} \\ &= \dot{\bar{\boldsymbol{\varepsilon}}} + \dot{\overset{+}{\boldsymbol{\varepsilon}}} + 1/2\,\dot{\overset{++}{\boldsymbol{\varepsilon}}}. \end{aligned} \tag{17}$$

Consequently, we find

$$\begin{aligned} \boldsymbol{\sigma} &= \boldsymbol{\sigma}_0 + \int_{(0,0)}^{(\bar{\varepsilon},\dot{\bar{\varepsilon}})} \mathbf{E}d\boldsymbol{\varepsilon} + \mathbf{D}d\dot{\boldsymbol{\varepsilon}}) + \int_{(\bar{\varepsilon},\dot{\bar{\varepsilon}})}^{(\varepsilon,\dot{\varepsilon})} \mathbf{E}d\boldsymbol{\varepsilon} + \mathbf{D}d\dot{\boldsymbol{\varepsilon}}) \\ &= \boldsymbol{\sigma}_0 + \bar{\boldsymbol{\sigma}} + (\mathbf{E}\overset{+}{\boldsymbol{\varepsilon}} + \mathbf{D}\dot{\overset{+}{\boldsymbol{\varepsilon}}}) + (1/2\mathbf{E}\overset{++}{\boldsymbol{\varepsilon}} + 1/2\mathbf{D}\dot{\overset{++}{\boldsymbol{\varepsilon}}}) \\ &= \boldsymbol{\sigma}_0 + \bar{\boldsymbol{\sigma}} + \overset{+}{\boldsymbol{\sigma}} + \overset{++}{\boldsymbol{\sigma}}. \end{aligned} \tag{18}$$

The superscript + + denotes quadratic contributions of $\overset{+}{\mathbf{u}}$.

Then the variation of the displacements produces the following equation:

$$\delta \mathbf{u} = \delta \overset{+}{\mathbf{u}}, \quad \delta \mathbf{r} = \delta \overset{+}{\mathbf{r}} . \tag{19}$$

Since only $\overset{+}{\mathbf{u}}$ is the neighbouring perturbation motion around the fundamental equilibrium state, the variational symbol δ operates only on the perturbation displacements $\overset{+}{\mathbf{u}}$. So that

$$\begin{aligned} \delta \boldsymbol{\varepsilon} &= [\, \mathbf{D}_{kL} + \mathbf{D}_{kN}(\mathring{\mathbf{u}} + \bar{\mathbf{u}})] \cdot \delta \overset{+}{\mathbf{u}} + \mathbf{D}_{kN}(\overset{+}{\mathbf{u}}) \cdot \delta \overset{+}{\mathbf{u}} \\ &= \delta \overset{+}{\boldsymbol{\varepsilon}} + \delta \overset{++}{\boldsymbol{\varepsilon}} , \end{aligned} \tag{20}$$

$$\begin{aligned} \delta \dot{\boldsymbol{\varepsilon}} &= [\, \mathbf{D}_{kL} + \mathbf{D}_{kN}(\mathring{\mathbf{u}} + \bar{\mathbf{u}})] \cdot \delta \dot{\overset{+}{\mathbf{u}}} + \mathbf{D}_{kN}(\dot{\bar{\mathbf{u}}}) \cdot \delta \overset{+}{\mathbf{u}} \\ &\quad + 2\mathbf{D}_{kN}(\overset{+}{\mathbf{u}}) \cdot \delta \dot{\overset{+}{\mathbf{u}}} \\ &= \delta \overset{+}{\boldsymbol{\varepsilon}} + \delta \dot{\overset{++}{\boldsymbol{\varepsilon}}} . \end{aligned} \tag{21}$$

Corresponding to this variation of the displacement, the applied forces will also be incremented, denoted by the superscript (+) as

$$\mathbf{p} = \bar{\mathbf{p}} + \overset{+}{\mathbf{p}} , \quad \mathbf{f} = \bar{\mathbf{f}} + \overset{+}{\mathbf{f}} = -\mathring{\rho}(\ddot{\bar{\mathbf{u}}} + \ddot{\overset{+}{\mathbf{u}}}), \quad \mathbf{t} = \bar{\mathbf{t}} + \overset{+}{\mathbf{t}} . \tag{22}$$

We now can substitute equations (18-22) into principle (6), and obtain

$$\begin{aligned} &\iint_{\mathring{F}} \mathring{\rho} \overset{+}{\mathbf{u}} \cdot \delta \overset{+}{\mathbf{u}} \cdot d\mathring{F} + \iint_{\mathring{F}} [(\boldsymbol{\sigma}_0 + \bar{\boldsymbol{\sigma}})^T \cdot \delta \overset{++}{\boldsymbol{\varepsilon}} + \overset{+}{\boldsymbol{\sigma}}{}^T \cdot \delta \overset{++}{\boldsymbol{\varepsilon}} \,] \cdot d\mathring{F} \\ &= \iint_{\mathring{F}} \overset{+}{\mathbf{p}}{}^T \cdot \delta \overset{+}{\mathbf{u}} \cdot d\mathring{F} + \int_{\mathring{C}_t} \overset{+}{\mathbf{t}}{}^T \cdot \delta \overset{+}{\mathbf{r}} \cdot d\mathring{S} \\ &- [\iint_{\mathring{F}} \mathring{\rho} \bar{\mathbf{u}} \cdot \delta \overset{+}{\mathbf{u}} \cdot d\mathring{F} + \iint_{\mathring{F}} (\boldsymbol{\sigma}_0 + \bar{\boldsymbol{\sigma}}) \cdot \delta \overset{+}{\boldsymbol{\varepsilon}} \cdot d\mathring{F} - \iint_{\mathring{F}} \bar{\mathbf{p}}^T \cdot \delta \overset{+}{\mathbf{u}} \cdot d\mathring{F} \\ &- \int_{\mathring{C}_t} \bar{\mathbf{t}}^T \cdot \delta \overset{+}{\mathbf{r}} \cdot d\mathring{S} \,] \end{aligned} \tag{23}$$

by neglecting terms of higher than quadratic power of $\overset{+}{\mathbf{u}}$.

Equation (23) is the incremental principle of virtual displacements. Two nonlinear algorithms can be based on it. One is the algorithm for the evaluation of arbitrary points along nonlinear load-displacement paths. The other is the algorithm for the evaluation of critical points of instability phenomena. The first one uses the last term of equation (23) as virtual works of external and internal forces for fundamental motions. For the analysis of buckling or bifurcation, these virtual works vanish because of the equilibrium of the fundamental state.

4. Finite Element Discretization

Interpolation of the displacement fields is adopted here. Hence we introduce the nodal displacements $\mathbf{v}^p$ of an arbitrary finite element $F^{0p} \in F^o$, and the interpolation functions $\mathbf{\Omega}^p$ to evaluate the displacement field within a finite element leading to

$$\mathbf{u}^p = \mathbf{\Omega}^p \cdot \mathbf{v}^p . \tag{24}$$

Then

$$\begin{aligned} \mathbf{D}_{kL}\mathbf{u}^p &= \mathbf{D}^p_{kL}\mathbf{v}^p, & \mathbf{D}^p_{kL} &= \mathbf{D}_{kL}\,\mathbf{\Omega}^p , \\ \mathbf{D}_{kN}(\mathbf{u}^p)\cdot\mathbf{u}^p &= \mathbf{D}^p_{kN}(\mathbf{v}^p)\cdot\mathbf{v}^p, & \mathbf{D}^p_{kN}(\mathbf{v}^p) &= \mathbf{D}_{kN}(\mathbf{\Omega}^p\cdot\mathbf{v}^p)\cdot\mathbf{\Omega}^p . \end{aligned} \tag{25}$$

Herein $\mathbf{D}_{kL}{}^p$ and $\mathbf{D}_{kN}{}^p(\mathbf{v}^p)$ are matrix functions, descending from operators $\mathbf{D}_{kL}$ and $\mathbf{D}_{kN}(\mathbf{u})$. Consequently for all mentioned boundary displacements we find

$$\mathbf{r} = \mathbf{R}^p_r \cdot \mathbf{u}^p \tag{26}$$

with $\mathbf{R}_r{}^p$ an elementary transformation matrix from internal displacements to its boundary counterparts.

Through the application of equations (24-26) in principle (23), a discrete incremental principle of virtual displacements is obtained as follows. It also expresses the virtual work of the forces in the present state by simply assembling all element contributions:

$$\begin{aligned} \sum_p \delta\overset{+}{\mathbf{v}}{}^{pT} [\mathbf{m}^p \ddot{\overset{+}{\mathbf{v}}}{}^p + (\mathbf{C}^p_v + \mathbf{C}^p_{uL} + \mathbf{C}^p_{uN})\cdot\dot{\overset{+}{\mathbf{v}}}{}^p \; + (\mathbf{k}^p_e + \mathbf{k}^p_{\sigma_o} \\ + \mathbf{k}^p_{\sigma L} + \mathbf{k}^p_{\sigma N} + \mathbf{k}^p_{\dot{\sigma} L} + \mathbf{k}^p_{\dot{\sigma} N} + \mathbf{k}^p_{uL} + \mathbf{k}^p_{uN} + \mathbf{k}^p_{\dot{u}L} + \mathbf{k}^p_{\dot{u}N})\,\overset{+}{\mathbf{v}}{}^p \\ - \bar{\mathbf{p}}^p - \overset{+}{\mathbf{p}}{}^p + \mathbf{f}^p_{\sigma_o} + \mathbf{f}^p_{\sigma} + \mathbf{f}^p_{\dot{\sigma}} + \mathbf{f}^p_m] = 0 , \end{aligned} \tag{27}$$

in which the lower indices uL and uN denote linearity in $\bar{\mathbf{v}}^p(\bar{\mathbf{u}})$ and nonlinearity in $\bar{\mathbf{v}}^p(\bar{\mathbf{u}})$ respectively. All element matrices are defined in Table 1. These are: the element mass matrix $\mathbf{m}^p$ (symmetrical and constant); the element viscous damping matrix $\mathbf{c}_v{}^p$; the element damping matrices due to initial deformations $\mathbf{c}_{uL}{}^p$ and $\mathbf{c}_{uN}{}^p$; the element tangential stiffness matrix $\mathbf{k}_e{}^p$; the element geometric stiffness matrices $\mathbf{k}_{\sigma 0}{}^p$ and $\mathbf{k}_{\dot{\sigma}}{}^p$ which correspond to stresses; the element geometric stiffness matrices $\mathbf{k}_u{}^p$ and $\mathbf{k}_{\dot{u}}{}^p$ which correspond to deformations; the element external force matrices $\bar{\mathbf{p}}^p$ and $\overset{+}{\mathbf{p}}{}^p$; and the element internal force matrices $\mathbf{f}_m{}^p$ and $\mathbf{f}_{\sigma}{}^p$.

From the definitions in Table 1, we find that all damping matrices will be symmetrical matrices if and only if the tangential viscosity tensor $\mathbf{D}$ is symmetrical. Obviously this differs from material linear problems such that stiffness submatrices here generally can lose their symmetry because of the introduction of material nonlinearities. Only if the tangential elasticity tensor $\mathbf{E}$ is symmetrical, the stiffness submatrices $\mathbf{k}_e$, $\mathbf{k}_u$ retain their symmetry. Another difference compared to linear material problems is clearly that the stiffness here are functions of the present state. Hence no definiteness statements are possible even if $\mathbf{E}$ and $\mathbf{D}$ remain symmetric.

We finally assemble all element nodal parameters v^p to form the global degrees of freedom V_j. This forms a standard procedure to assemble all mentioned submatrices in equation (27) in global matrices. By this procedure we form a global dynamic equilibrium equation from (27), the tangential equation of motion:

$$
\begin{aligned}
&\mathbf{M}\cdot\overset{+}{\ddot{\mathbf{V}}} + (\mathbf{C}_v + \mathbf{C}_{uL} + \mathbf{C}_{uN})\cdot\overset{+}{\dot{\mathbf{V}}} + (\mathbf{K}_e + \mathbf{K}_{\sigma_0} + \mathbf{K}_{\sigma L} + \mathbf{K}_{\sigma N} \\
&+ \mathbf{K}_{\dot{\sigma} L} + \mathbf{K}_{\dot{\sigma} N} + \mathbf{K}_{uL} + \mathbf{K}_{uN} + \mathbf{K}_{\dot{u} L} + \mathbf{K}_{\dot{u} N})\cdot\overset{+}{\mathbf{V}} \\
&= \mathbf{P} + \mathbf{F}_{\sigma_0} + \mathbf{F}_{\sigma} + \mathbf{F}_{\dot{\sigma}} + \mathbf{F}_m \quad \text{with} \quad \mathbf{P} = \bar{\mathbf{P}} + \overset{+}{\mathbf{P}} .
\end{aligned} \tag{28}
$$

We abbreviate this equation by

$$
\mathbf{M}\,\overset{+}{\ddot{\mathbf{V}}} + \mathbf{C}_T\,\overset{+}{\dot{\mathbf{V}}} + \mathbf{K}_T\,\overset{+}{\mathbf{V}} = \mathbf{P} - \mathbf{F}_I, \tag{29}
$$

where $\mathbf{C}_T$ descrides the tangential damping; $\mathbf{K}_T$ the tangential stiffness properties of the process, $\mathbf{F}_I$ is called the vector of internal nodal forces. As discussed earier: if the fundamental motion is in equilibrium, $\mathbf{F}_I = \mathbf{0}$.

5. Stability Criteria

In order to execute the stability analysis of a certain reached point of equilibrium of a given deformation path we transform (29) into the following phase form:

$$
\begin{bmatrix} \overset{+}{\dot{\mathbf{V}}} \\ \hline \overset{+}{\mathbf{V}} \end{bmatrix}_{,t} = \left[\begin{array}{c|c} -\mathbf{M}^{-1}\mathbf{C}_T & -\mathbf{M}^{-1}\mathbf{K}_T \\ \hline \mathbf{I} & \mathbf{0} \end{array}\right] \cdot \begin{bmatrix} \overset{+}{\dot{\mathbf{V}}} \\ \hline \overset{+}{\mathbf{V}} \end{bmatrix} + \begin{bmatrix} \mathbf{M}^{-1}(\mathbf{P} - \mathbf{F}_I) \\ \hline \mathbf{0} \end{bmatrix} \tag{30}
$$

$$
\overset{+}{\dot{\mathbf{X}}} = \mathbf{A}\,\overset{+}{\mathbf{X}} + \mathbf{B} .
$$

Here the assumption $\det\mathbf{M} \neq 0$, $\mathbf{M} > 0$ is introduced, which in general is satisfied for all allowable deformations of real structures. Because of equilibrium $\mathbf{F}_I = \mathbf{0} \rightarrow \mathbf{B} = \mathbf{0}$ the stability analysis will be based on

$$
\overset{+}{\dot{\mathbf{X}}} = \mathbf{A}\cdot\overset{+}{\mathbf{X}} \tag{31}
$$

where $\mathbf{A}$ is a nonlinear matrix of the fundamental motion $\bar{\mathbf{V}}$.

The stability of (27) with $\mathbf{P} - \mathbf{F}_I = \mathbf{0}$ respectively of (31) is discussed in large generality in [16]. To adopt the argumentation there we find as a **first statement**:

- . sufficient for stability is the negative semi-definiteness,
- . sufficient for asympototic stability is the negative definiteness,
- . necessary for instability is the indefiniteness or positive definiteness

of $\mathbf{A} + \mathbf{A}^T$. But this stability condition seems too strictly to be a criterion. In the next section we will detail this point by an example.

Here we remind our reader that for an arbitrary symmetric matrix the positive definiteness, positive semi-definiteness, negative definiteness, negative semi-definiteness and indefiniteness are equivalent to requirements that the eigenvalues of this matrix are all positive, non-negative, all negative, non-positive and indefinite respectively.

There exist a useful, more detailed sufficient condition both for stability and instability [16]. That is the system governed by (31) is

. asympototically stable, if all the eigenvalues of **A** have negative real parts;

. critically stably, if none of the eigenvalues of **A** has a positive real parts and at least one eigenvalue has a zero real part with a corresponding Jordan block of order 1;

. unstable, if **A** has at least one eigenvalue with either positive real part or zero real part whose order of the corresponding Jordan block is larger than 1.

As **second statement**, no symmetry of the matrix **A** is required in the above criterion. This means no requirement concerning the symmetry of matrices $\mathbf{C_T}$ and $\mathbf{K_T}$ has to be set up. The direct application of this criterion will further on be expanded to both geometrically and physically nonlinear problems.

To go into more detail we now assume in addition to $\mathbf{M} > \mathbf{0}$ at least positive semi-definiteness of the symmetric part of $\mathbf{C_T}$ and positive definiteness of the symmetric part of $\mathbf{K_T}$:

$$(\mathbf{C_T} + \mathbf{C_T^T}) \geq 0, \qquad (\mathbf{K_T} + \mathbf{K_T^T}) > 0\,. \tag{32}$$

If we apply the Ljapunov-Function

$$L = \mathbf{V^T} \cdot \begin{pmatrix} \mathbf{M} & \mathbf{O} \\ \mathbf{O} & \mathbf{K_T} + \mathbf{K_T^T} \end{pmatrix} \cdot \mathbf{V}\,, \tag{33}$$

we are able to detail the first given statement to the following **third statement**:

. a sufficient condition for stability is

$$\begin{pmatrix} -(\mathbf{C_T} + \mathbf{C_T^T}) & \mathbf{K^T} \\ \mathbf{K} & \mathbf{O} \end{pmatrix} \leq 0\,; \tag{34}$$

. a sufficient condition for asympototic stability is

$$\begin{pmatrix} -(\mathbf{C_T} + \mathbf{C_T^T}) & \mathbf{K^T} \\ \mathbf{K} & \mathbf{O} \end{pmatrix} < 0\,. \tag{35}$$

In all other cases one has to account for instability.

If all the matrices **M**, C_T, K_T are symmetric, the stability condition will be more simple. We then receive the **fourth statement** [16]:

. a sufficient condition for stability is

$$K_T \geq 0 \quad \text{and} \quad C_T \geq 0 \quad \text{and} \quad (C_T + K_T) > 0, \tag{36}$$

. a sufficient condition for asympototic stability is

$$C_T \geq 0 \quad \text{and} \quad K_T > 0\,, \tag{37}$$

. a sufficient condition for instability is

$$C_T \geq 0 \quad \text{and} \quad K_T \not\geq 0\,. \tag{38}$$

We have discussed much on dynamic stability conditions. If the problem is time-invariant, equation (31) will reduce to

$$K_T \cdot \overset{+}{V} = 0\,. \tag{39}$$

For this static system, the **fifth statement** is found as sufficient condition [12] for

. stability, if all eigenvalues of K_T are positive,

. instability, if at least one eigenvalue of K_T is negative,

. indifference, if $\det K_T = 0$.

6. Numerical Algorithms and Applications

The remaining question is how to use the stability conditions in an arbitrary response analysis. Since all the criteria mentioned based on the positivity or negativity of the eigenvalues, the direct application is to determine whether all the eigenvalues of the corresponding matrix are positive (negative) or not. The calculation procedure maybe complicated in some cases since the considered matrices are no longer symmetrical.

In numerical applications, first let's consider the following vibration problem:

$$\begin{pmatrix} 1 & 0 \\ 0 & 1 \end{pmatrix} \frac{d^2}{dt^2} \begin{pmatrix} V_1 \\ V_2 \end{pmatrix} + \begin{pmatrix} k_1 & 0 \\ l & k_2 \end{pmatrix} \begin{pmatrix} V_1 \\ V_2 \end{pmatrix} = 0\,, \tag{40}$$

where k_1,k_2 and l are arbitrary real numbers. Obviously, the sufficient conditions for stability in both mentioned first and third statements are not satisfied here. And just because only the sufficient conditions for stability have been identified in both the first and third statements, stability or instability can not be decided unless these sufficient conditions for stability are satisfied.

Eq. (40) shows a free vibration with a non-symmetric stiffness matrix. By use of the criterion from the second statement, it is found that the motion will be stable (or critically stable) if

$$k_1 \neq k_2 \quad \text{and} \quad k_1 > 0 \quad \text{and} \quad k_2 > 0. \tag{41}$$

Meanwhile, the motion will be unstable if

$$k_1 = k_2 \quad \text{or} \quad k_1 \leq 0 \quad \text{or} \quad k_2 \leq 0. \tag{42}$$

This instability is also determined by the second statement. Here, the negative values of both k_1 and k_2 mean that the eigenvalues of **A** have positive real parts; equality between k_1 and k_2 means that at least one Jordan block of **A** has an order larger than 1, and so on.

Another application is to determine the unstable critical time of a 3D toggle under sudden applied load. This example showed by Fig. 2 and 3 demonstrates the dynamic stability of this nonlinear problem. This similar truss has been used by Bergan [5] to illustrate the nonlinear behaviour in the static case. Now in our case, it is found that the fundamental motion should be

$$\overline{V}_1 \neq 0, \qquad \overline{V}_2 = 0. \tag{43}$$

$E = 2.1\times10^7 (kN/m^2)$, $K = 40.0 (kN/m)$, $A = 6.0\times10^{-4} (m^2)$,
$h = 1.0 (m)$, $L = 2.0 (m)$, $m = 1.0 (kg)$.

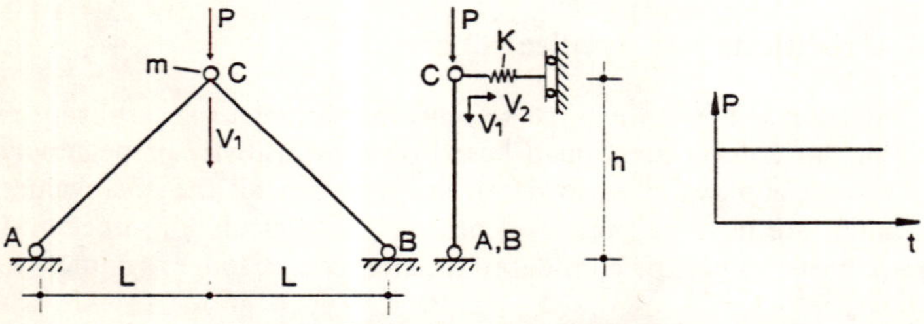

Fig. 2 3D Toggle under Sudden Step Load

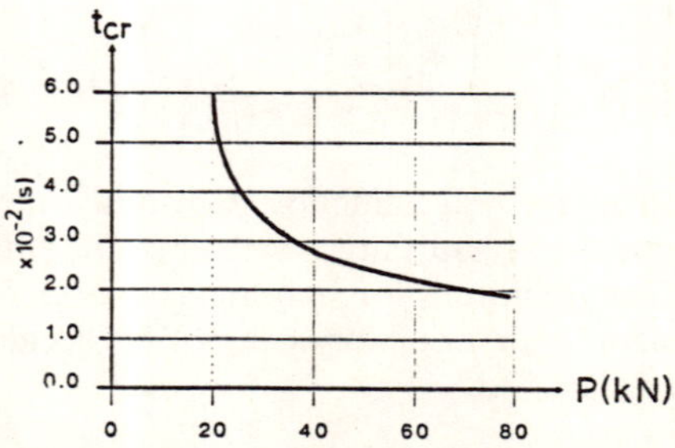

Fig. 3 Critical Time vs the Magnitute of Step Load

On this equilibrium state, the tangential stiffness matrix for the perturbed motion $\overset{+}{V}_i$ is symmetric. By the fourth statement for stability, the critical state is defined as

$$\det(\mathbf{K}_T) = 0 \ . \tag{44}$$

At this point, the tangential stiffness matrix has a non-zero eigenvalue of {0,a} which corressponds to the zero eigenvalue. We denote the corresponding time as critical time. And the truss begins its quasi-birfurcated motion in the direction of V_2 (dynamic quasi-bifurcation).

7. Conclusions

Incremental-iterative algorithms of finite element methods enable the numerical evaluation of arbitrary nonlinear structural responses. Both static and dynamic problems can be treated on this basis. They contain simultaneously all the necessary information concerning possible instabilities of the mentioned responses. As demonstrated, the generalized and more detailed conditions for stability/instability are carefully introduced and discussed. The advantages of these conditions appear in both static and dynamic, geometrically and physically nonlinear problems.

Achnowledgement This paper was gained within project B3 of the SFB 151-Special Research Center on Structural Dynamics at the Ruhr-University Bochum. Financial support of the DFG, the German Science Foundation, and the AvH, the Alexander-von-Humboldt-Foundation, is gratefully acknowledged. With gratitude we register valuable comments from Professor P.C. Müller, Wuppertal-University, which improved the results.

References

[1] Budiansky. B., Dynamic Buckling of Elastic Structures: Criteria and Estimates, in Herrmann, G. (ed.): Dynamic Stability of Structures, Pergamon Press, Oxford 1967, 83-106.

[2] Koiter, W. T., On the Nonlinear Theory of Thin Shells, Proc. Kon. Ned. Akad. Wet. (1966), ser. B, Val. 66, 0.1, 1-54.

[3] Koiter, W. T., General Equations of Elastic Stability for Thin Shells Proc. Symp. of the Theory of Thin Shells, University of Houston, Texas 1967, 187-203.

[4] Krätzig, W.B., Beitrag zu einer linearen Approximation der Stabilitätstheorie elastischer Flachentragwerke, Dr.-Ing. habil. Thesis, T.U. Hannover, 1968.

[5] Bergan, P. G., Solution Algorithms for Nonlinear Structural Problems, Comp. Struct., Vol. 12, 1980, 497-509.

[6] Ramm, E., Strategies for Tracing the Nonlinear Response near Limit Points, in Wunderlich, W. et al. (eds.): Nonlinear Finite Element Analysis in Structural Mechanics, Springer-Verlag, Berlin 1981.

[7] Eckstein, U., Nichtlineare Stabilitätsberechnung elastischer Schalentragwerke, Institut für Konstruktiven Ingenieurbau, Ruhr-Universität Bochum, 1983, TWM Nr. 83-3.

[8] Krätzig, W. B., Wittek, U. and Basar, Y., Buckling of General Shells, Theory and Numerical Analysis, in Thompson, M., Hunt, G. (eds.): Collapse: the buckling of Structures in Theory and Practice, ITUAM Symposium, Cambridge University Press, Cambridge 1983, 203-213.

[9] Stein, E., Wagner, W., Wriggers, P., Nonlinear Stability Analysis of Shell and Contact Problems Including Branch-Switching, Comput. Mechanics, **5**(1990), 428-446.

[10] Eller, C., Lineare und Nichtlineare Stabilitätsanalyse periodisch erregter viskoelastischer Strukturen, Institut für Konstruktiven Ingenieurbau, Ruhr-Universität Bochum, 1988, TWM Nr. 88-2.

[11] Krätzig, W. B., Eine einheitliche statische und dynamische Stabilitätstheorie für Pfadverfolgungsalgorithmen in der numerischen Festkörpermechanik, ZAMM, **69**(1989), 203-213.

[12] Krätzig, W. B., Fundamentals of Numerical Algorithms for Static and Dynamic Instability Phenomena of Thin Shells, in: Krätzig, W. B., Onate, E. (eds.): Computational Mechanics of Nonlinear Response of Shells, Springer-Verlag, Berlin Heidelberg New York, 1990, 102-124.

[13] Krätzig, W. B. and Zahlten, W., The Application of Plastic-Fracturing Theory to the Finite Element Analysis of General Reinforced Shells, Proc. of the 10th Int. Conf. on Struct. Mech. in Reactor Technology, SMIRT, 1989, Vol. B, 311-322.

[14] Quante, R., Stabilitätsanalyse elasto-visco-plastischer Flächentragwerke unter zeitvarianten Einwirkungen, Dr.-Ing.-Thesis, Ruhr-Universität Bochum 1991.

[15] Basar, Y., Krätzig, W.B., Mechanik der Flächentragwerk, Braunschweig/Wiesbaden: Vieweg & Sohn, 1985.

[16] Müller, P. C., Stabilität und Matrizen, Springer-Verlag, Berlin Heidelberg New York, 1977.

Table 1. Definition of Element Matrices

$$m^p = \iint_{\mathring{F}^p} \mathring{\rho}\, \Omega^{pT} \cdot \Omega^p \cdot d\mathring{F}$$

$$c_v^p = \iint_{\mathring{F}^p} D_{kL}^{pT} \cdot D_{kL}^p \cdot d\mathring{F}$$

$$c_{uL}^p = \iint_{\mathring{F}^p} [D_{kL}^{pT} \cdot D \cdot D_{kN}^p(\mathring{v}^p + \bar{v}^p) + D_{kN}^{pT}(\mathring{v}^p + \bar{v}^p) \cdot D \cdot D_{kL}^p] \cdot d\mathring{F}$$

$$c_{uN}^p = \iint_{\mathring{F}^p} [D_{kL}^{pT}(\mathring{v}^p + \bar{v}^p) \cdot D \cdot D_{kL}^p(\mathring{v}^p + \bar{v}^p)] \cdot d\mathring{F}$$

$$k_e^p = \iint_{\mathring{F}^p} D_{kL}^{pT} \cdot E \cdot D_{kL}^p \cdot d\mathring{F}$$

$$k_{\sigma_0}^p = \iint_{\mathring{F}^p} D_{kN}^{pT}(\overset{+}{v}^p) \cdot \sigma_0 \cdot d\mathring{F} \,/,_{\overset{+}{v}^p}$$

$$k_{\sigma L}^p = (\iint_{\mathring{F}^p} D_{kN}^{pT}(\overset{+}{v}^p) \cdot E \cdot (D_{kL}^p + D_{kN}^p(\mathring{v}^p)) \cdot \bar{v}^p \cdot d\mathring{F}^p) \,/,_{\overset{+}{v}^p}$$

$$k_{\sigma N}^p = (\tfrac{1}{2}\iint_{\mathring{F}^p} D_{kN}^{pT}(\overset{+}{v}^p) \cdot E \cdot D_{kN}^p(\bar{v}^p) \cdot \bar{v}^p \cdot d\mathring{F}^p) \,/,_{\overset{+}{v}^p}$$

$$k_{\dot{\sigma} L}^p = (\iint_{\mathring{F}^p} D_{kN}^{pT}(\overset{+}{v}^p) \cdot D \cdot D_{kL}^p \cdot \dot{\bar{v}}^p \cdot d\mathring{F}^p) \,/,_{\overset{+}{v}^p}$$

$$k_{\dot{\sigma} N}^p = (\iint_{\mathring{F}^p} D_{kN}^{pT}(\overset{+}{v}^p) \cdot D \cdot D_{kN}^p(\mathring{v}^p + \bar{v}^p) \cdot \dot{\bar{v}}^p \cdot d\mathring{F}^p) \,/,_{\overset{+}{v}^p}$$

$$k_{uL}^p = \iint_{\mathring{F}^p} [D_{kL}^{pT} \cdot E \cdot D_{kN}^p(\mathring{v}^p + \bar{v}^p) + D_{kN}^{pT}(\mathring{v}^p + \bar{v}^p) \cdot E \cdot D_{kL}^p] \cdot d\mathring{F}$$

$$k_{uN}^p = \iint_{\mathring{F}^p} [D_{kN}^{pT}(\mathring{v}^p + \bar{v}^p) \cdot E \cdot D_{kN}^p(\mathring{v}^p + \bar{v}^p)] \cdot d\mathring{F}$$

$$k_{\dot{u}L}^p = \iint_{\mathring{F}^p} D_{kL}^{pT} \cdot D \cdot D_{kN}^p(\bar{v}^p) \cdot d\mathring{F}$$

$$k_{\dot{u}N}^p = \iint_{\mathring{F}^p} D_{kN}^{pT}(\mathring{v}^p + \bar{v}^p) \cdot D \cdot D_{kN}^p(\bar{v}^p) \cdot d\mathring{F}$$

$$\bar{p}^p = \iint_{\mathring{F}^p} \Omega^{pT} \cdot \bar{p} \cdot d\mathring{F} + \int_{\mathring{C}_t^p} R_r^{pT} \cdot \bar{t} \cdot d\mathring{S}$$

$$\overset{+}{p}^p = \iint_{\mathring{F}^p} \Omega^{pT} \cdot \overset{+}{p} \cdot d\mathring{F} + \int_{\mathring{C}_t^p} R_r^{pT} \cdot \overset{+}{t} \cdot d\mathring{S}$$

$$f_{\sigma_0}^p = \iint_{\mathring{F}^p} [D_{kL}^p + D_{kN}^p(\mathring{v}^p + \bar{v}^p)]^T \cdot \sigma_0 \cdot d\mathring{F}$$

$$f_\sigma^p = \iint_{\mathring{F}^p} (D_{kL}^{pT} + D_{kN}^{pT}(\mathring{v}^p + \bar{v}^p)) \cdot E \cdot (D_{kL}^p + 1/2 D_{kN}^p(\bar{v}^p) + D_{kN}^p(\mathring{v}^p)) \cdot \bar{v}^p \cdot d\mathring{F}^p$$

$$f_{\dot{\sigma}}^p = \iint_{\mathring{F}^p} (D_{kL}^{pT} + D_{kN}^{pT}(\mathring{v}^p + \bar{v}^p)) \cdot D \cdot (D_{kL}^p + D_{kN}^p(\mathring{v}^p + \bar{v}^p)) \cdot \dot{\bar{v}}^p \cdot d\mathring{F}^p$$

An Energy-Based Concept of Structural Stability

B.Kröplin and D.Dinkler
Institut für Statik und Dynamik - Universität Stuttgart

Fundamental investigations of system stability upon perturbation of the initial conditions can be accredited to Liapunov. Malkin [1] gives an overview of this and other works. A motion can be termed *stable* if its reaction upon a disturbance of the initial conditions by a measure $\delta_1(\epsilon)$ differs by a maximal deviation of ϵ from the undisturbed response (see also Figure 1 and equation 1). Motions not following this condition are called *unstable*.

$$\begin{array}{lcrcl} |v_{t_0,a_S} & - & v_{t_0,a_0}| & \leq & \delta_1(\epsilon) \quad , \quad t = 0, \\ |v_{t,a_S} & - & v_{t,a_0}| & \leq & \delta_1\epsilon \quad\;\; , \quad t > t_0. \end{array} \tag{1}$$

If the following condition (2) holds, a motion is termed *asymptotically stable*

$$\lim_{x \to \infty} |v_{t,a_S} - v_{t,a_0}| = 0. \tag{2}$$

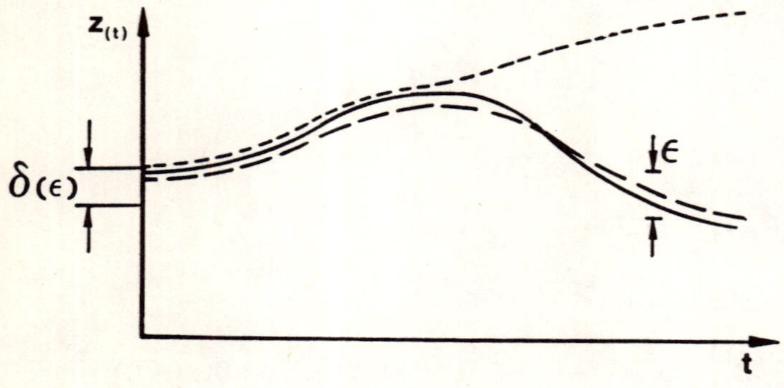

Figure 1: Definition of stability

Instability for structures with a linear load-deformation behavior is only possible if the occuring forces (i.e. in a self-exciting system) result in an unsteady amplification of the deformation amplitudes. Instability is also possible without continuous excitation forces if nonlinearities are taken into account. An example for this is when high membrane stresses on thin-walled shell structures lead to a sudden increase of bending deformation and even to a change of the buckling shape. Because of the strong nonlinearity in the deformation behavior of shell structures, investigations of the occuring phenomena must be made. In the following, only the stability of structures under time-independent loading is taken into account. The development of the main ideas of the theory can be found in [2].

In the case of time independent effects the deformation state is also constant in time. A disturbance of the initial condition leads to an infinitesimal initial deflection, which produces a motion around the state of equilibrium. The state of equilibrium usually consists of the equilibrium of a fundamental state and a finite disturbance of that position . This disturbance could be brought on by additional time-invariant and spatially arbitrary effects. If the fundamental state is to be investigated, the initial deflection is superposed upon that state, otherwise upon the total state.

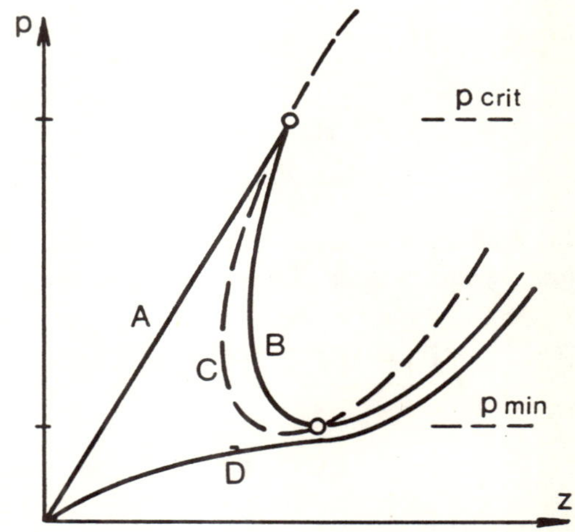

Figure 2: Regions of stability

The prebuckling region A is generally stable, since additional deformations ϵ caused by initial deformation are small. Upon reaching critical states, that are characterized by the vanishing of stiffness in direction of specific buckling patterns, the limitation of deformations brought on even by infinitesimal initial displacements is no longer possible. In this case the equilibrium position is *indifferent* or *unstable* if the stiffness in the vicinity are negative. Deformations brought on by finite disturbances of the fundamental state stay bounded only for effects p that are lower than the postbuckling minimum p_{min} . If the disturbance of the fundamental state in the pre-buckling region (brought on by effects larger than p_{min}) lead to a transgression of the curve C in figure 2, then a disturbance of the initial conditions of the fundamental state will initiate increasing of the deformations into the post-buckling region.

In all cases instability, i.e. the unbounded growth of deformations, is characterized by leaving the region of states of equilibrium in the pre-buckling range A.

Therefore, instabilities are possible if the solution is not unique. If the solution is unique as in case D , the perturbation of the fundamental state may be arbitrarily large without leaving the region of the fundamental solution. This implies that the structure is globally stable even in the case of large deformations, otherwise the structure is locally stable.

The condition for all trivial and nontrivial states of equilibrium is the vanishing of the first variation of the potential of effects and the strain energy.

$$\begin{aligned} \Pi &= \Pi^i + \Pi^e \\ \Pi &= \Pi(z) \\ \Pi^i &: \text{ strain energy} \\ \Pi^e &: \text{ external energy} \end{aligned}$$

Starting with the fundamental state $\Pi = \Pi(z_0)$ nontrivial solutions are possible, if the first variation of the increment of the potential vanishes:

$$\Delta\Pi = \Pi_2 - \Pi_0 = \delta\not{\Pi}_0^0 + \frac{1}{2!}\delta^2\Pi_0 + \frac{1}{3!}\delta^3\Pi_0 + \ldots \tag{3}$$

it follows that

$$\delta\Delta\Pi = 0. \tag{4}$$

Especially in case of constant loads, equation 4 is valid for the external energy of the increment

$$\Delta\Pi^e = 0\ , \tag{5}$$

and therefore

$$\delta\Delta\Pi^i(\Delta z_2) = 0. \tag{6}$$

This equation is nonlinear in the state vector Δz_2 , because of the higher variation of Π_0, and must therefore be solved iteratively.

Clearly this is the condition for the vanishing of the secant stiffness and for maximum and minimum values of the strain energy in the vicinity (see figure 3).

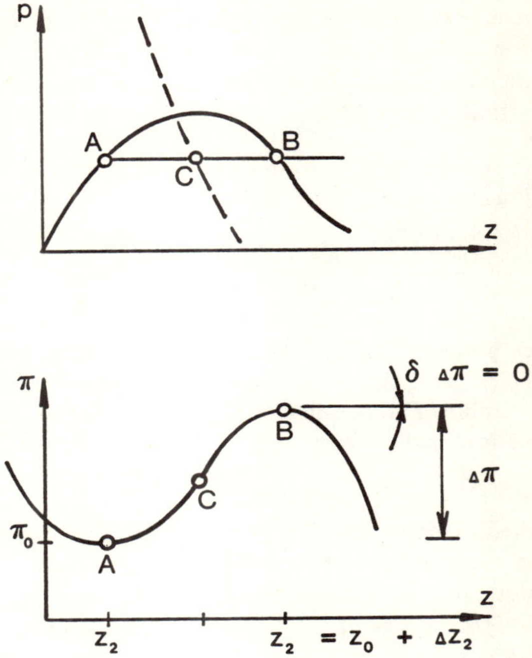

Figure 3: Properties of nontrivial solutions

If the fundamental state A is disturbed, additional energy is induced into the structure. The total state and its corresponding nontrivial solution moves toward the point C in figure 1 . If the disturbance of the structure is such that the point C is just reached, points A, B and C are coincident. This means that the total state is indifferent and an infinitesimal disturbance leads to a steady increase of the deformations. If Π_2 describes the total potential (fundamental state and disturbance), then the requirement for the critical state C is given by equation 7 :

$$\delta \Delta \Pi^i = 0. \tag{7}$$

The nontrivial solution gives the state of deformation that the structure will take on without further increase of its strain energy. This also means that

$$\delta^2 \Delta \Pi^i = 0 \tag{8}$$

must hold.

In order to judge the stability of the fundamental state Π_0 against arbitrary finite disturbances it is neccesary to compare different disturbances with each other as well as with the critical state. The strain energy $\Delta\Pi^e$ that is induced to the structure as a result of the disturbance is suitable for such a comparison, because it is a scalar measure. This comparison can be made even in case of spatially arbitrarily distributed effects. The comparative measure that indicates if the critical state has been reached or not, can be defined as the strain energy $\Delta\Pi_{crit}$ corresponding to that critical state.

The stability of the fundamental state against a finite disturbance can now be described by using the *degree of stability* :

$$\eta = \frac{\Delta\Pi_{crit} - \Delta\Pi^e}{\Delta\Pi_{crit}} \quad \begin{matrix} > \\ = \\ < \end{matrix} \quad 0 \quad \begin{matrix} \text{stable} \\ \text{indifferent} \\ \text{unstable} \end{matrix} . \tag{9}$$

The disturbance potential $\Delta\Pi^e(\Delta z^e)$ is calculated from the given disturbance Δp^e. In calculating the critical disturbance potential $\Delta\Pi_{crit}(\Delta z_{crit})$, the most unfavourable of all possible critical states on the curve C must be used, so that the minimality condition

$$\delta\Delta\Pi_{crit} = 0 \tag{10}$$

has to be satisfied. This means that the unfavourable direction of a disturbance always points straight to a neighboring state of equilibrium Δz_2. Because the minimality condition (eq. 10) can be used to calculate the nontrivial neighboring states of equilibrium Δz_2 , the critical state Δz_{crit} has the same shape as Δz_2 and differs only in amplitude. This amplitiude can be determined by the requirement

$$\delta^2\Delta\Pi^i(z_0 + \lambda\Delta z_2) = 0 \tag{11}$$

because the critical condition is defined for a specific value of λ.

References

[1] Malkin, J.G.: "Theorie der Stabilität einer Bewegung ", München : R.Oldenbourg 1959

[2] Duddeck,H.; Kröplin,B.; Dinkler,D.; Hillmann,J.; Wagenhuber,W.: "Berechnung des nichtlinearen Tragverhaltens dünner Schalen im Vor- und Nachbeulbereich." in: Stein,E. Edt. : Nichtlineare Berechnungen im Konstruktiven Ingenieurbau, Springer-Verlag, Berlin 1989

[3] Dinkler,D.; Kröplin,B.: "Dynamic stability of buckling elastic structures." in: C.L.Kirk and J.L.Junkins Edts. : Dynamics of flexible structures in space, Comp. Mech. Publ., Springer-Verlag, Berlin 1990

Two and Three Phase Behaviour in Soil Dynamics

O.C. Zienkiewicz*, B.A. Schrefler**, L. Simoni**, Y.M. Xie* and X.Y. Zhan**

* The Institute for Numerical Methods in Engineering, University College of Swansea, U.K.

** Istituto di Scienza e Tecnica delle Costruzioni University of Padova, Italy

Summary

Negative pore pressures developing in a geomaterial cause release of dissolved air from the pore fluid and the entrance of air in the mixture through the boundaries. In this semi-saturated state, the negative pore pressures are maintained through the mechanism of capillary forces. A three-phase model, which takes into account such capillary effects, is developed to solve most static and dynamic semi-saturated problems. This model contains, as limiting case, two-phase semi-saturated geomaterials, where the air phase is considered as static. A numerical solution of the governing equations is presented and two examples of computations are shown.

1. INTRODUCTION

Many fascinating problems in the field of the mechanics of geomaterials can be solved today by elaborate numerical computations. A full account of the development of such numerical procedures and of the state of the art of our predictive capacities was given by Zienkiewicz et al. [1,2]. The field of applications presented ranges from civil engineering problems to oil reservoir engineering, hydrology and, with appropriate constitutive models, even to bio-engineering. Both fully saturated and semi-saturated geomaterials were considered in Reference [1,2]. However, for the analysis of semi-saturated geomaterials the simplifying hypothesis of static air phase was introduced, allowing thus the fully saturated model to comprise also such a particular semi-saturated material. This assumption is acceptable in many soil mechanics situations, but prevents the solution of others, where the flow of air (or of a different third fluid phase) is of importance.

In this chapter the assumption of static air phase is removed and a full three-phase model is developed. The governing equations were already derived for slow phenomena, such as consolidation, in Reference [3] and are extended here to comprise the full dynamic behaviour. As in Reference [3] the equations are obtained within the framework of the continuum theory of mixtures, using a spatially averaged approach. The reader is considered familiar with the averaging principles used in this context. However a short summary of these principles may be found in Reference [3], while a full account of the techniques is reported in [4,5,6].

2. GENERAL HYPOTHESES

The partially saturated porous medium sample is treated as a multiphase system undergoing mechanical deformation of the solid and with flux of the fluid phases (water and gas). The constituents are assumed to be immiscible and chemically non-reacting. All constituents are at the same temperature, which is assumed constant in time and space.

For the macroscopic behaviour of the single constituents, the following hypotheses are made:

solid phase:
- infinitesimal strain is assumed;
- deformation of the solid depends only on the effective stress;
- the stress-strain relationship is non-linear (i.e. non-linear elastic or elastoplastic);
- creep effects and strains not directly associated with stress changes are accounted for;
- stress is defined as tension positive.

fluid phases:
- both fluid phases are barotropic and in contact with the solid phase;
- pore pressure is defined as compressive positive;
- at macroscopic level, shear stresses are neglected in both phases; of course they exist at microscopic level and their effect is accounted for by viscous drag forces exerted on the solid

phase. These forces are assumed to be dependent on the relative velocities of the phases;

- the saturation S_π of the π phase is a known function of the capillary pressure p_c:

$$S_\pi = S_\pi(p_c) \quad \text{or} \quad p_c = p_c(S_\pi) \tag{1}$$

where

$$p_c = p_g - p_w \tag{2}$$

and the subscripts g and w refer to gas (air) and water respectively;

- the transport law of both water and gas is obtained from general mechanical principles;
- the permeability of the medium to each fluid is defined by the product of the intrinsic permeability times the relative permeability $k_{r\pi}$ of the π phase;
- the relative permeabilities are known functions of the degree of saturation:

$$k_{r\pi} = k_{r\pi}(S_\pi). \tag{3}$$

Under these hypotheses the conservative equations of mechanical quantities may be stated using the principles of continuum mixture theory. This step is performed using averagedvalues of the field variables. This fact allows for the constitutive model to be easily related to experimental data.

In a dynamical environment, some further assumptions can be introduced, which are useful at a later stage:

- the convective terms, which are related to the gradients of velocity of each phase, are assumed to be negligible;
- the momentum exchange term between solid and fluids depends only on the relative velocities, as in the quasi-static case;
- the local acceleration of fluid velocities is neglected;
- the seepage flow is not influenced by the gradient of the fluid densities.

3. THE STRESS AND PRESSURE FIELDS

An important question in mechanical problems concerning partially saturated porous media is the definition of the effective stress, i.e. the stress controlling both changes in volume and strength of the medium [7]. For this purpose, we apply the definition of averaged quantities [3] to the total stress vector $\underline{\sigma}$ in a multiphase system with a solid phase, a liquid (water) and a gaseous phase (air).

$$
\begin{aligned}
< \underline{\sigma} > &= \frac{1}{V}\int_V \underline{\sigma} dV = \frac{1}{V}\left[\int_{V_s} \underline{\sigma} dV + \int_{V_f} \underline{\sigma} dV \right] \\
&= \frac{V_s}{V} < \underline{\sigma}_s >^s + \frac{V_f}{V}\left[\frac{V_w}{V_f} < \underline{\sigma}_w >^w + \frac{V_g}{V_f} < \underline{\sigma}_g >^g \right] \qquad (4) \\
&= (1-n) < \underline{\sigma}_s >^s + n\left[S_w < \underline{\sigma}_w >^w + S_g < \underline{\sigma}_g >^g \right]
\end{aligned}
$$

where V and V_π are respectively the total volume of the Representative Volume Element (R.E.V.) and the volume of the π-phase in the R.E.V., n is the porosity, $\underline{\sigma}_\pi$ the stress of the π-phase and S_π its saturation. Moreover the following notation is used in the following:

$< * >$ = averaged value of * over the R.E.V.

$< * >^\pi$ = value of * averaged over the volume V_π

For the fluid phases, the intrinsic stress vector is given by

$$
< \underline{\sigma}_\pi >^\pi = < \underline{\tau}_\pi >^\pi - \underline{m} < p_\pi >^\pi \qquad (5)
$$

where p_π is the pressure of the π-phase, $\underline{\tau}_\pi$ is the shear stress and $\underline{m} = [1,1,1,0,0,0]^T$. Under the assumption that the shear stress $\underline{\tau}_\pi$ is negligible, we obtain

$$
< \underline{\sigma} > = (1-n) < \underline{\sigma}_s >^s - n\, \underline{m} \left[S_g < p_g >^g + S_w < p_w >^w \right] \qquad (6)
$$

The term in square brackets of eq. (6) represents the intrinsically averaged (or macroscopic) pressure $\bar{p}$ of the fluid phases

$$\bar{p} = < \bar{p} >^f = \left[S_g < p_g >^g + S_w < p_w >^w \right]. \tag{7}$$

Equation (6) can be modified, also accounting for the definition of the mean pressure $\bar{p}$, as

$$\begin{aligned} < \underline{\sigma} > &= (1-n) < \underline{\sigma}_s >^s - n \, \underline{m} < \bar{p} >^f \\ &= (1-n)(< \underline{\sigma}_s >^s + \underline{m} < \bar{p} >^f) - (1-n)\underline{m} < \bar{p} >^f - n\underline{m} < \bar{p} >^f \\ &= \underline{\sigma}' - \underline{m} \; \bar{p} \end{aligned} \tag{8}$$

The total stress tensor is split into two components, the pore pressure effect and the part exerted on the solid skeleton only.

For the moment, we assume that the grains do not undergo any deformation due to the mean pressure $\bar{p}$ exerted by the fluids [8]. The deformation of the solid skeleton consists of the rearrangement of the grains and depends only on $\underline{\sigma}'$, which is the effective stress. This latter is given by

$$\underline{\sigma}' = < \sigma' > = (1-n) \left(< \underline{\sigma}_s >^s + \underline{m} < \bar{p} >^f \right) \tag{9}$$

Under more general hypotheses, a corrective term, known as Biot's constant [9], must be introduced to account for the deformability of the grains. On the basis of experimental results and by omitting the averaging symbol for the total stress vector, we assume the more general expression of the effective stress as [10]

$$\underline{\sigma}' = \underline{\sigma} + \underline{m} \; \bar{p} \; \alpha \tag{10}$$

where the corrective coefficient α will be determined in the following, after definition of the constitutive relationships.

4. CONSTITUTIVE EQUATIONS

It is worth remarking that all constitutive equations are written in terms of macroscopic quantities.

4.1 Solid phase

The intrinsically averaged pore pressure $\bar{p}$ of the fluids occupying the void space introduces a hydrostatic stress distribution in the solid phase. The ensuing deformation is the purely volumetric strain

$$\varepsilon_v^P = - \frac{\bar{p}}{K_s} \tag{11}$$

or, in incremental form

$$d\underline{\varepsilon}^P = - \underline{m} \frac{d\bar{p}}{3K_s} \tag{12}$$

where K_s is the averaged bulk modulus of the solid grains. As indicated previously, the effective stress causes all relevant deformation of the solid skeleton. The constitutive relationship may be written as

$$d\underline{\sigma}' = \underline{D}_T \left(d\underline{\varepsilon} - d\underline{\varepsilon}^c - d\underline{\varepsilon}^p - d\underline{\varepsilon}^o \right) \tag{13}$$

where:
- $d\underline{\sigma}'$ represents the stress responsible for all deformations, except for grain compressibility;
- $d\underline{\varepsilon}$ represents the total strain of the solid;
- $\underline{D}_T$ is the tangent constitutive matrix:

$$\underline{D}_T = \underline{D}_T \, (\underline{\sigma}', \underline{\varepsilon} , \dot{\underline{\varepsilon}}) \tag{14}$$

- $d\underline{\varepsilon}^c = g(\underline{\sigma}')dt$ accounts for creep strain;
- $d\underline{\varepsilon}^c$ represents all other strains in the solid skeleton not directly dependent on effective stress.

We introduce now a modified effective stress $\underline{\sigma}''$ to account also for grain compression. This allows us to obtain Biot's constant α. By substituting equation (12) in (13) and by omitting, for brevity, $d\underline{\varepsilon}^o$ and $d\underline{\varepsilon}^c$ we obtain

$$d\underline{\sigma}' = \underline{D}_T \, d\underline{\varepsilon} + \underline{D}_T \, \underline{m} \, \frac{d\bar{p}}{3K_s} \tag{15}$$

By using the effective stress equation (8), in incremental form, we have

$$d\underline{\sigma} = d\underline{\sigma}'' - \left(\underline{m} - \underline{D}_T \, \underline{m} \, \frac{1}{3K_s} \right) d\bar{p} \tag{16}$$

where $\underline{\sigma}''$ represents the stress responsible for all deformations in the solid. Noting that

$$\frac{\underline{m}^T \underline{m}}{3} = 1 \tag{17}$$

we can rewrite the above equation as

$$d\underline{\sigma} = d\underline{\sigma}'' - \frac{1}{3}\left(\underline{m}^T\underline{m} - \underline{m}^T\underline{D}_T\underline{m} \frac{1}{3K_s}\right) \underline{m}\, d\bar{p}$$

$$= d\underline{\sigma}'' - \left(1 - \frac{\underline{m}^T\underline{D}_T\underline{m}}{9\,K_s}\right) \underline{m}\, d\bar{p} = d\underline{\sigma}'' - \alpha\, \underline{m}\, d\bar{p} \tag{18}$$

where, in the case of an isotropic material [9]

$$\alpha = 1 - \frac{\underline{m}^T\underline{D}_T\underline{m}}{9K_s} = 1 - \frac{K_T}{K_s} \tag{19}$$

and K_T is the bulk modulus of the overall skeleton.
The resulting expression of the constitutive relationship is therefore

$$d\underline{\sigma}'' = \underline{D}_T \left(d\underline{\varepsilon} - d\underline{\varepsilon}^c - d\underline{\varepsilon}^o \right) \tag{20}$$

4.2 Liquid phases

The intrinsic pore pressure p_π of the π fluid phase causes a purely volumetric strain in the phase , which is given by

$$\varepsilon_v^\pi = - \frac{p_\pi}{K_\pi} \tag{21}$$

where K_π is the averaged bulk modulus of the phase.

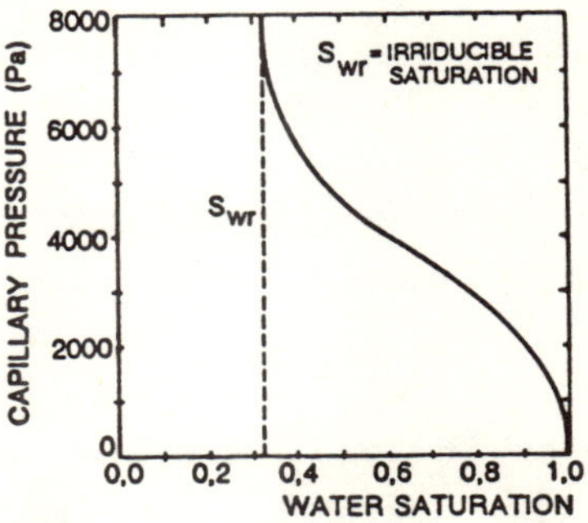

Fig. 1- Typical curve for capillary pressure versus saturation of water.

In addition to the constraint on the saturation of the two fluid phases, that their sum must be equal to 1, the pressures of air and water are subjected to the constraint that the capillary pressure

$$p_c = p_g - p_w \tag{22}$$

is a function of S_w. The relationship $p_c(S_w)$ is known from laboratory experiments. Figure 1 represents a characteristic $p_c(S_w)$ curve.

5. BALANCE EQUATIONS

To obtain the governing equations of the mechanics of partially saturated porous media, we use the continuum theory of mixtures, i.e. we suppose that all constituents of the porous medium occupy at the same time every point of the domain. Due to the assumed immiscibility of the constituents [3]:

- in some local sense each constituent will obey the constitutive relation for that constituent alone;
- the constituents intrinsically have a microstructure, i.e. interfaces are present which separate the constituents.

As a consequence, the mechanical behaviour of the mixture is described by the usual thermodynamical variables, as if the components occupied at the same time all of the domain, but in addition the volume fraction of the constituents is introduced to account for the discontinuous distribution. The volume fraction is defined as the volume of each constituent per unit volume of the material.

In the context of the continuum approach used, two types of theories have evolved:

- postulated theories, in which the thermodynamical balance equations are written in terms of postulated mean values of the variables;
- averaged theories, in which the thermodynamical balance equations are written in local form and are then integrated (averaged) in space, time, or space and time [4].

Spatially averaged theories will be used in the sequel.

5.1 Mass balance equations

5.1.1 Solid phase

The mass of the soil skeleton in a representative elementary volume may be written as

$$M_S = \int_V \rho^s \, dV \tag{23}$$

where ρ^s is the local value of the density of the solid. We express the same quantity using intrinsic averaged density ρ_s and the volume fraction of the solid as

$$M_s = \int_V (1-n) \, \rho_s dV \tag{24}$$

where n is the porosity of the medium, which may vary in the time span investigated.

The mass conservation requires that (using summation over repeated indices)

$$\begin{aligned} \frac{D}{Dt}\int_V (1-n)\rho_s dV &= \int_V \left(\frac{D}{Dt} [(1-n)\,\rho_s] + (1-n)\,\rho_s \frac{\partial \dot{u}_{s,i}}{\partial x_i} \right) dV \\ &= \int_V \left[\frac{\partial (1-n)\rho_s}{\partial t} + \frac{\partial (1-n)\rho_s \, \dot{u}_{s,i}}{\partial x_i} \right] dV = 0 \end{aligned} \tag{25}$$

The term $\dot{u}_{\pi,i}$ represents the i-th component of the absolute velocity of the π-phase.

Then, by localization, we obtain the equation of continuity of the solid which is expressed in terms of intrinsic averaged quantities:

$$\frac{\partial (1-n)\rho_s}{\partial t} + \frac{\partial (1-n)\rho_s \, \dot{u}_{s,i}}{\partial x_i} = 0 \tag{26}$$

5.1.2 Fluid phases (water and air)

The mass of water in a R.E.V., in partially saturated conditions, may be expressed in terms of intrinsic averaged quantities using the pertinent volume fraction as

$$M_w = \int_V n\, S_w\, \rho_w\, dV \tag{27}$$

where ρ_w is the intrinsic averaged density of water. External supply is not accounted for in the following.

The water mass conservation requires the material time derivative in respect to the flowing water to be zero:

$$\begin{aligned}\frac{DM_w}{Dt} &= \frac{D}{Dt}\int_V n\, S_w\, \rho_w\, dV \\ &= \int_V \left[\frac{\partial(n\, S_w\, \rho_w)}{\partial t} + \frac{\partial(n\, S_w\, \rho_w\, \dot{u}_{w,i})}{\partial x_i}\right] dV = 0\end{aligned} \tag{28}$$

Then, by localization, we obtain the continuity equation for water:

$$\frac{\partial(nS_w\rho_w)}{\partial t} + \frac{\partial(nS_w\rho_w\dot{u}_{w,i})}{\partial x_i} = 0 \tag{29}$$

For the gaseous phase we have in a similar way:

$$\frac{\partial(nS_g\rho_g)}{\partial t} + \frac{\partial(nS_g\rho_g\dot{u}_{g,i})}{\partial x_i} = 0 \tag{30}$$

It is worth remarking that all equations are expressed in terms of intrinsic averaged quantities and contain the volume fraction as variable. It is costumary to describe the flow of the fluid phases through the porous medium using Darcy type relationships. These are expressed in terms of volume averaged velocities relative to the moving solid. To define the necessary quantities $\dot{\kappa}_i$, in the R.E.V. we apply the averaging process to the previously defined intrinsic phase averaged quantities:

$$\begin{aligned}\dot{\kappa}_i &= \frac{1}{V}\int_V \left(\dot{u}_{\pi,i} - \dot{u}_{s,i}\right) dV = \frac{1}{V}\int_{V_\pi} \left(\dot{u}_{\pi,i} - \dot{u}_{s,i}\right) dV \\ &= \frac{V_\pi}{V}\left(\dot{u}_{\pi,i} - \dot{u}_{s,i}\right)\end{aligned} \tag{31}$$

In this way we obtain the water ($\dot{w}_i$) and gas ($\dot{v}_i$) velocities relative to the moving solid:

$$\dot{w}_i = n\, S_w \left(\dot{u}_{w,i} - \dot{u}_{s,i} \right)$$

$$\dot{v}_i = n\, S_g \left(\dot{u}_{g,i} - \dot{u}_{s,i} \right) \tag{32}$$

From a practical point of view, it is useful to express the balance equations in terms of the solid skeleton velocity ($\dot{u}_i$), where the subscript s has been omitted, and in terms of fluid phases velocity relative to the moving solid. This yields

$$\frac{(1-n)}{\rho_s}\frac{D\rho_s}{Dt} - \frac{Dn}{Dt} + (1-n)\,\dot{u}_{i,i} = 0 \tag{33}$$

$$\frac{n}{S_w\rho_w}\frac{D(S_w\rho_w)}{Dt} + \frac{Dn}{Dt} + n\dot{u}_{i,i} + \frac{1}{S_w}\dot{w}_{i,i} + \frac{\dot{w}_i}{S_w\rho_w}\frac{\partial\rho_w}{\partial x_i} = 0 \tag{34}$$

$$\frac{n}{S_g\rho_g}\frac{D(S_g\rho_g)}{Dt} + \frac{Dn}{Dt} + n\dot{u}_{i,i} + \frac{1}{S_g}\dot{v}_{i,i} + \frac{\dot{v}_i}{S_g\rho_g}\frac{\partial\rho_g}{\partial x_i} = 0 \tag{35}$$

respectively for solid, water and air, where the notation

$$(*)_{i,i} = \frac{\partial(*)_i}{\partial x_i} = \text{div}\ (*) \tag{36}$$

is used, with summation over repeated indices. All material time derivatives are performed with respect to the moving solid, i.e.

$$\frac{D\rho_s}{Dt} = \frac{\partial\rho_s}{\partial t} + \frac{\partial\rho_s}{\partial x_i}\dot{u}_i \tag{37}$$

$$\frac{Dn}{Dt} = \frac{\partial n}{\partial t} + \frac{\partial n}{\partial x_i}\dot{u}_i \tag{38}$$

$$\frac{D(S_w\,\rho_w)}{Dt} = \frac{\partial(S_w\,\rho_w)}{\partial t} + \frac{\partial(S_w\,\rho_w)}{\partial x_i}\,\dot{u}_i \tag{39}$$

$$\frac{D(S_g\,\rho_g)}{Dt} = \frac{\partial(S_g\,\rho_g)}{\partial t} + \frac{\partial(S_g\,\rho_g)}{\partial x_i}\,\dot{u}_i \tag{40}$$

A further simplification is obtained by summing the mass balance equation of each fluid phase with that of the solid, obtaining respectively for water and air

$$\frac{(1-n)}{\rho_s}\frac{D\rho_s}{Dt} + \frac{n}{\rho_w}\frac{D\rho_w}{Dt} + \frac{n}{S_w}\frac{DS_w}{Dt} + \dot{u}_{i,i} + \frac{1}{S_w}\,\dot{w}_{i,i} + \frac{\dot{w}_i}{S_w\,\rho_w}\,\frac{\partial\rho_w}{\partial x_i} = 0 \tag{41}$$

$$\frac{(1-n)}{\rho_s}\frac{D\rho_s}{Dt} + \frac{n}{\rho_g}\frac{D\rho_g}{Dt} + \frac{n}{S_g}\frac{DS_g}{Dt} + \dot{u}_{i,i} + \frac{1}{S_g}\,\dot{v}_{i,i} + \frac{\dot{v}_i}{S_g\,\rho_g}\,\frac{\partial\rho_g}{\partial x_i} = 0 \tag{42}$$

The material derivatives of the intrinsic phase averaged densities, which appear in equations (41) and (42) are obtained next. For the fluid phases we recall that these are assumed barotropic. Consequently the equations of state may be directly obtained from the constitutive equations. From the definitions of the bulk modulus of water K_w and of air K_g

$$\frac{1}{\rho_\pi}\,\frac{\partial\rho_\pi}{\partial p_\pi} = \frac{1}{K_\pi} \tag{43}$$

we obtain

$$\frac{1}{\rho_w}\,\frac{D\rho_w}{Dt} = \frac{1}{K_w}\,\frac{Dp_w}{Dt} \tag{44}$$

$$\frac{1}{\rho_g}\,\frac{D\rho_g}{Dt} = \frac{1}{K_g}\,\frac{Dp_g}{Dt} \tag{45}$$

As far as the solid skeleton is concerned, this is usually assumed as microscopically incompressible, i.e. grains do not undergo any deformation. As a consequence

$$\frac{D\rho_s}{Dt} = 0 \tag{46}$$

However to account for the effect of grain deformation on solid density we use the solid mass conservation equation in differential form:

$$\frac{D(\rho_s V_s)}{Dt} = 0 \tag{47}$$

which yields

$$\frac{1}{\rho_s}\frac{D\rho_s}{Dt} = -\frac{1}{V_s}\frac{DV_s}{Dt} \tag{48}$$

The change of volume of solid V_s depends on the intrinsically averaged fluid pressure and the mean effective stress σ_0', which is a volume averaged quantity:

$$-\frac{1}{V_s}\frac{DV_s}{Dt} = -\frac{1}{V_s}\left[-\frac{V_s}{K_s}\frac{D\bar{p}}{Dt} + \frac{1}{K_s}\frac{V_s}{V_s}\frac{D\sigma_0'}{Dt}\right]$$

$$= -\frac{1}{V_s}\left[-\frac{V_s}{K_s}\frac{D\bar{p}}{Dt} + \frac{1}{K_s}\left(K_T\frac{D\varepsilon_v}{Dt} + \frac{K_T}{K_s}\frac{D\bar{p}}{Dt}\right)\right]$$

$$= \frac{1}{1-n}\left[\frac{\alpha - n}{K_s}\frac{D\bar{p}}{Dt} - (1-\alpha)\frac{D\varepsilon_v}{Dt}\right] \tag{49}$$

where ε_v is the volumetric strain and α is defined in eq. (19). Substituting equations (44), (45), (48) and (49) in (41) and (42), we obtain the mass balance equation of the mixture 'solid + water'

$$\left[\frac{\alpha-n}{K_s}\frac{D\bar{p}}{Dt} - (1-\alpha)\,\dot{u}_{i,i}\right] + \frac{n}{K_w}\frac{Dp_w}{Dt}$$
$$+ \frac{n}{S_w}\frac{DS_w}{Dt} + \dot{u}_{i,i} + \frac{1}{S_w}\dot{w}_{i,i} + \frac{\dot{w}_i}{S_w\rho_w}\frac{\partial\rho_w}{\partial x_i} = 0 \tag{50}$$

and of the mixture 'solid + air'

$$\left[\frac{\alpha-n}{K_s}\frac{D\bar{p}}{Dt}-(1-\alpha)\,\dot{u}_{i,i}\right]+\frac{n}{K_g}\frac{Dp_g}{Dt}$$

$$+\frac{n}{S_g}\frac{DS_g}{Dt}+\dot{u}_{i,i}+\frac{1}{S_g}\dot{v}_{i,i}+\frac{\dot{v}_i}{S_g\rho_g}\frac{\partial\rho_g}{\partial x_i}=0 \tag{51}$$

We recall the definition of the intrinsic averaged fluid pressure

$$\bar{p}=S_w\,p_w+S_g\,p_g \tag{52}$$

and the constraint relative to the degrees of saturation:

$$S_w+S_g=1 \tag{53}$$

From the definition of the time material derivative, it follows

$$\frac{DS_g}{Dt}=-\frac{DS_w}{Dt} \tag{54}$$

and the mass balance equations result in

$$\left[\frac{\alpha-n}{K_s}S_w^2+\frac{nS_w}{K_w}\right]\frac{Dp_w}{Dt}+\frac{\alpha-n}{K_s}S_gS_w\frac{Dp_g}{Dt}+\alpha S_w\dot{u}_{i,i}$$

$$+\left[\frac{\alpha-n}{K_s}S_w(p_w-p_g)+n\right]\frac{DS_w}{Dt}+\dot{w}_{i,i}+\frac{\dot{w}_i}{\rho_w}\frac{\partial\rho_w}{\partial x_i}=0 \tag{55}$$

$$\left[\frac{\alpha-n}{K_s}S_g^2+\frac{nS_g}{K_g}\right]\frac{Dp_g}{Dt}+\frac{\alpha-n}{K_s}S_wS_g\frac{Dp_w}{Dt}+\alpha S_g\dot{u}_{i,i}$$

$$+\left[\frac{\alpha-n}{K_s}S_g(p_g-p_w)+n\right]\frac{DS_g}{Dt}+\dot{v}_{i,i}+\frac{\dot{v}_i}{\rho_g}\frac{\partial\rho_g}{\partial x_i}=0 \tag{56}$$

5.2 Linear momentum balance equations

The microscopic (or local) momentum balance equation for the π-phase

$$\frac{D}{Dt}\left(\rho_\pi\,\underline{v}_\pi\right)-\underline{\nabla}\,\underline{\sigma}_\pi-\rho_\pi\,\underline{g}_\pi=0 \tag{57}$$

has to be integrated over the region occupied by the phase and then expressed in terms of averaged quantities. In doing so,

attention has to be paid to the fact that the limiting surface of the volume V_π is different from that of the R.E.V. and mechanical interactions arise at the interfaces. We assume that the external momentum supply is $\underline{g}$, no phase change takes place and the temperature is stationary in space and time.
To account for these conditions, we use the averaging theorem by Hassanizadeh and Gray [11], by which we obtain the following form of momentum balance equation (in indicial notation):

$$\varepsilon_\pi \rho_\pi \frac{D^\pi v_l^\pi}{Dt} - \varepsilon_\pi \sigma_{kl,k}^\pi - \varepsilon_\pi \rho_\pi g_l^\pi - \varepsilon_\pi \rho_\pi T_l^\pi = 0 \tag{58}$$

where ε_π represents the volume fraction for the π phase. The material derivative is performed with respect to the moving phase. $\underline{\sigma}^\pi_{,k}$ represents the divergence of the intrinsic stress vector for the π phase and $\underline{T}^\pi$ accounts for the exchange of momentum due to mechanical interactions with the other phases.

The constitutive relationship has to be introduced in eq. (58) for the momentum exchange term. As previously stated, a simplified constitutive equation may be assumed as

$$\varepsilon_\pi \rho_\pi \underline{T}^\pi = -\underline{R}^\pi \underline{v}^{d\pi} \tag{59}$$

This term represents the dissipative part of the fluid-solid exchange of momentum and is related to the velocity $\underline{v}^{d\pi}$ of the the π-fluid relative to the solid. In this way we account for local effects, depending on the viscosity of the fluid, assuming the fluid to be macroscopically inviscid. This assumption leads, in a static environment, to Darcy's law [3]. In such a situation, assuming the resistivity matrix $\underline{R}^\pi$ to be invertible we define

$$\frac{\underline{k}_\pi}{\mu_\pi} = \varepsilon_\pi (\underline{R}^\pi)^{-1} = nS_\pi (\underline{R}^\pi)^{-1} \tag{60}$$

with $\underline{k}_\pi$ being the permeability matrix of the medium to the π fluid and μ_π its dynamic viscosity. In a dynamic environment, it is assumed that equations (59) and (60) still hold, because of the different time scale of the consolidation process and the dynamic phenomenon (e.g. an earthquake) considered.

When several fluid phases are present, dissipative terms arise at the fluid-solid and fluid-fluid interfaces. To express the constitutive equations for these terms, we assume the permeability of the fully saturated case (intrinsic permeability), which depends on the geometry of the solid skeleton, modified by the presence of other fluid phases. This modification depends on the volume fraction of the other phases. As a consequence we assume the resulting permeability as the product of the intrinsic permeability times a coefficient $k_{r\pi}$ (relative permeability), which is a function of the degree of saturation S_π. In isotropic conditions we have $\underline{k}_\pi$ evaluated by the following expression:

$$\underline{k}_\pi = \underline{k}_{a\pi} k_{r\pi} \left(S_\pi \right) \tag{61}$$

In this equation $\underline{k}_{a\pi}$ represents the absolute permeability of the medium.

5.2.1 Fluid phases

A more suitable form for the linear momentum balance equation of the π-phase can be obtained by recalling the assumption for the stress vector in a fluid phase, by introducing eq. (59) in eq. (58) and by expressing the volume fraction as a function of the porosity and saturation of the π-phase. Furthermore we express the intrinsic velocity of the fluid by means of eq. (32). This yields

$$nS_\pi \rho_\pi \left[\frac{D^s v_l^s}{Dt} + \frac{D^s}{Dt} \left(\frac{v_l^{d\pi}}{n\, S_\pi} \right) + \frac{v_i^{d\pi}}{n\, S_\pi} \; \frac{\partial}{\partial x_i} \dot{v}_l^\pi \right]$$

$$+ nS_\pi p_{\pi,k} - nS_\pi \rho_\pi g_l^\pi + R_l^\pi \, v_l^{d\pi} = 0 \tag{62}$$

where all material derivatives are performed with respect to the intrinsic velocity of the solid. If we assume that the terms dependent on the gradient of the velocity vector components for each fluid are negligible and by omitting the relative acceleration term we obtain the relative velocity of the π fluid as

$$\underline{v}^{d\pi} = nS_{\pi}\left(\underline{\underline{R}}^{\pi} \right)^{-1} \left[-p_{\pi,k} + \rho_{\pi}\underline{g}^{\pi} - \rho_{\pi}\left(\frac{D^s\underline{v}^s}{Dt} \right)\right]$$

$$= \underline{\underline{k}}_{a\pi}\frac{k_{r\pi}}{\mu_{\pi}} \left[-p_{\pi,k} + \rho_{\pi}\underline{g}^{\pi} - \rho_{\pi}\left(\frac{D^s\underline{v}^s}{Dt} \right)\right] \tag{63}$$

This relationship has to be written for the water relative velocity $\underline{\dot{w}}$ and for the gas relative velocity $\underline{\dot{v}}$.

5.2.2 Solid phase

By operating in the same way as in the previous case, we obtain the momentum balance equation for the solid as

$$\left(1-n\right)\rho_s\frac{D^s v^s_l}{Dt} = \left(1-n\right)\sigma^s_{kl,k} + \left(1-n\right)\rho_s g^s_l + R^f_{kl}v^d_l \tag{64}$$

5.2.3 Whole mixture

By summing the momentum balance equation of each component, we obtain the equilibrium condition for the whole mixture. We remark however that:

- the exchanges of momentum due to mechanical interactions are subjected to the following condition:

$$\Sigma_{\pi}\ \varepsilon_{\pi}\rho_{\pi}T^{\pi}_{l} = 0 \tag{65}$$

- the external supply for each phase comes from gravity effects, consequently

$$\underline{g}^s = \underline{g}^f = \underline{g}^g = \underline{g} \tag{66}$$

- the averaged density of the mixture is obtained by an averaging process as

$$\rho = (1-n)\rho_s + nS_g\rho_g + nS_w\rho_w \tag{67}$$

- the averaged total stress for the mixture is given by eq.

(10). This results in the following expression of the linear momentum balance equation for the whole mixture:

$$-\left(1-n\right)\rho_s \frac{D^s v_l^s}{Dt} - nS_w\rho_w \frac{D^w v_l^w}{Dt} - nS_g\rho_g \frac{D^g v_l^g}{Dt} + \rho g_l + \sigma_{kl,k} = 0 \tag{68}$$

where the material time derivatives are performed with respect to the velocity of each phase. This equilibrium equation can be further transformed by introducing the velocities of the fluids relative to the solid (by using eq. (32)). This yields, by accounting for (67),

$$-\rho \frac{D^s v_l^s}{Dt} - nS_w\rho_w \frac{D^s(\dot{w}_l/nS_w)}{Dt} - nS_g\rho_g \frac{D^s(\dot{v}_l/nS_g)}{Dt} - \rho_w \dot{w}_i \frac{\partial v_l^w}{\partial x_i}$$

$$-\rho_g \dot{v}_i \frac{\partial v_l^g}{\partial x_i} + \rho g_l + \sigma_{kl,k} = 0 \tag{69}$$

A more convenient form of the Eulerian equation of motion for the mixture may be obtained in the following way:

- multiply the momentum balance equation (in vectorial form) by a vector function $\delta \underline{u}$ which has continuous first partial derivatives with respect to the coordinates x_i and satisfies displacement boundary conditions, where prescribed, (virtual displacement). Integrate over the body volume:

$$-\int_\Omega \rho \delta \underline{u}^T \frac{D^s \underline{v}^s}{Dt} d\Omega - \int_\Omega nS_w\rho_w \delta \underline{u}^T \frac{D^s(\dot{\underline{w}}/nS_w)}{Dt} d\Omega - \int_\Omega nS_g\rho_g \delta \underline{u}^T \frac{D^s(\dot{\underline{v}}/nS_g)}{Dt} d\Omega$$

$$-\int_\Omega \delta \underline{u}^T \left(\rho_w \nabla \underline{v}^w \;\dot{\underline{w}} + \rho_g \nabla \underline{v}^g \;\dot{\underline{v}}\right) d\Omega + \int_\Omega \rho \delta \underline{u}^T \underline{g} d\Omega + \int_\Omega \delta \underline{u}^T \underline{\sigma} d\Omega = 0 \tag{70}$$

- by integrating the last term by parts and owing to Gauss' theorem, we obtain

$$-\int_\Omega \rho \delta \underline{u}^T \frac{D^s \underline{v}^s}{Dt} d\Omega - \int_\Omega nS_w\rho_w \delta \underline{u}^T \frac{D^s(\dot{\underline{w}}/nS_w)}{Dt} d\Omega - \int_\Omega nS_g\rho_g \delta \underline{u}^T \frac{D^s(\dot{\underline{v}}/nS_g)}{Dt} d\Omega$$

$$-\int_{\Omega} \delta\underline{u}^T \left(\rho_w \nabla \underline{v}^w \; \dot{\underline{w}} + \rho_g \nabla \underline{v}^g \; \dot{\underline{v}}\right) d\Omega + \int_{\Omega} \rho \delta\underline{u}^T \underline{g} d\Omega$$

$$-\int_{\Omega} \delta\underline{\varepsilon}^T \underline{\sigma} d\Omega + \int_{\Gamma} \delta\underline{u}^T \hat{\underline{t}} d\Gamma = 0 \tag{71}$$

where $\delta\varepsilon^T$ is the strain relative to the virtual displacement and $\hat{\underline{t}}$ are the boundary tractions. Equation (71) is the weak form of the differential equation of motion, i.e. the principle of virtual work.

6 SUMMARY OF THE GOVERNING EQUATIONS

For a more compact form of the governing equations, we define the following coefficients :

$$c_{ww} = \frac{S_w^2(\alpha-n)}{K_s} + \frac{S_w n}{K_w}$$

$$c_{gg} = \frac{S_g^2(\alpha-n)}{K_s} + \frac{S_g n}{K_g} \tag{72}$$

$$c_{wg} = c_{gw} = \frac{(\alpha-n) S_w S_g}{K_s}$$

and use a short notation for the material derivative

$$\dot{F} = \frac{DF}{Dt} \tag{73}$$

Furthermore the relative velocities of equation (63) are substituted in the mass balance equations (55) and (56); the total stress of the mixture expressed by eq. (18) and the mean pore pressure $\bar{p}$ expressed by eq. (7) are substituted in the equilibrium statement of eq. (69). Hence the final form of the governing equations is the following:

- equilibrium equation for the mixture:

$$-\int_{\Omega}\rho\delta\underline{u}^{T}\frac{D^{s}\underline{v}^{s}}{Dt}d\Omega-\int_{\Omega}nS_{w}\rho_{w}\delta\underline{u}^{T}\frac{D^{s}(\dot{\underline{w}}/nS_{w})}{Dt}d\Omega-\int_{\Omega}n_{g}S_{g}\rho\ \delta\underline{u}^{T}\frac{D^{s}(\dot{\underline{v}}/nS_{g})}{Dt}d\Omega$$

$$-\int_{\Omega}\delta\underline{u}^{T}\left(\rho_{w}\nabla\underline{v}^{w}\ \dot{\underline{w}}+\rho_{g}\nabla\underline{v}^{a}\ \dot{\underline{v}}\right)d\Omega+\int_{\Omega}\rho\delta\underline{u}^{T}\underline{g}d\Omega-\int_{\Omega}\delta\underline{\varepsilon}^{T}\underline{\sigma}''d\Omega$$

$$+\int_{\Omega}\delta\underline{\varepsilon}^{T}\alpha\underline{m}S_{w}p_{w}d\Omega+\int_{\Omega}\delta\varepsilon^{T}\alpha\underline{m}S_{g}p_{g}d\Omega+\int_{\Gamma}\delta u^{T}\hat{\underline{t}}\ \underline{d}\Gamma=0 \tag{74}$$

- continuity equation for water:

$$C_{ww}\dot{p}_{w}+C_{wg}\dot{p}_{g}+\left[\frac{C_{wg}}{S_{g}}(p_{w}-p_{g})+n\right]\dot{S}_{w}+\alpha S_{w}\dot{u}_{i,i}$$

$$+\left(k_{w_{ij}}\left(-p_{w,i}+\rho_{w}(b_{i}-\dot{v}_{i}^{s})\right)\right)_{,i}+\frac{1}{\rho_{w}}\left(k_{w_{ij}}\left(-p_{w,i}+\rho_{w}(b_{i}-\dot{v}_{i}^{s})\right)\right)\frac{\partial\rho_{w}}{\partial x_{i}}=0 \tag{75}$$

- continuity equation for air:

$$C_{gg}\dot{p}_{g}+C_{gw}\dot{p}_{w}+\left[-\frac{C_{wg}}{S}(p_{w}-p_{g})+n\right]\dot{S}_{g}+\alpha S_{g}\dot{u}_{i,i}$$

$$+\left(k_{g_{ij}}\left(-p_{g,i}+\rho_{g}(b_{i}-\dot{v}_{i}^{s})\right)\right)_{,i}+\frac{1}{\rho_{g}}\left(k_{g_{ij}}\left(-p_{g,i}+\rho_{g}(b_{i}-\dot{v}_{i}^{s})\right)\right)\frac{\partial\rho_{g}}{\partial x_{i}}=0 \tag{76}$$

The boundary value problem for the mechanical behaviour of a partially saturated sample requires that equations (74)-(76) be satisfied within all points of the domain Ω and that the boundary conditions are satisfied on the boundary Γ of the domain. Such type of conditions are imposed displacements or boundary tractions for the displacement field, imposed pressures or fluxes for the pressure fields. Moreover the initial conditions has to be specified to completely define the initial value problem.

6.1 The case $p_g = 0$

Sometimes phenomena regarding desaturating porous media occur with the air or gas at atmospheric pressure, which represents the reference pressure. Consequently $p_g=0$, $p_c=-p_w$ and the averaged pressure of the fluid phase is simply

$$\bar{p} = S_w \, p_w \tag{77}$$

In this case, the constitutive relationships for the hydraulic parameters are assumed to be as

$$S_w = S_w(p_w) \qquad \text{and} \qquad k_{rw} = k_{rw}(p_w)$$

We define the specific moisture content C_s as

$$C_s = n \frac{dS_w}{dp_w} = -n \frac{dS_w}{dp_c} \tag{78}$$

By accounting for this definition and by neglecting the convective terms and the gradient of the fluid density, the mass balance equation for water (75) results in

$$\left[C_{ww} + \frac{C_{wg}}{S_g} \; \frac{C_s}{n} p_w + C_s\right] \frac{\partial p_w}{\partial t} + \alpha S_w \dot{u}_{i,i} + \left(k_{w_{ij}} \left(-p_{w,i} + \rho_w (b_i - \dot{v}_i^s)\right)\right)_{,i} = 0 \tag{79}$$

If we do not want to control the flow of air in the sample, the gas mass balance equation may be neglected and the only mass balance equation is that for the mixture 'solid plus water'.

The equilibrium statement is the same as in eq. (74), provided that the condition $p_g=0$ is accounted for. Then we have

$$-\int_\Omega \rho \delta \underline{u}^T \frac{D^s \underline{v}^s}{Dt} d\Omega - \int_\Omega n S_w \rho_w \delta \underline{u}^T \frac{D^s(\underline{\dot{w}}/nS_w)}{Dt} d\Omega - \int_\Omega n S_g \rho_g \delta \underline{u}^T \frac{D^s(\underline{\dot{v}}/nS_g)}{Dt} d\Omega$$
$$-\int_\Omega \delta \underline{u}^T \left(\rho_w \nabla \underline{v}^w \; \underline{\dot{w}} + \rho_g \nabla \underline{v}^a \; \underline{\dot{v}}\right) d\Omega + \int_\Omega \rho \delta \underline{u}^T \underline{g} d\Omega - \int_\Omega \delta \underline{\varepsilon}^T \underline{\sigma}'' d\Omega$$
$$+\int_\Omega \delta \underline{\varepsilon}^T \alpha \underline{m} S_w p_w d\Omega + \int_\Gamma \delta u^T \hat{\underline{t}} \; d\Gamma = 0 \tag{80}$$

7. FINITE ELEMENT DISCRETIZATION IN SPACE

The solution of the boundary value problem represented by equations (74) - (76) becames easier if a weak form for all equations is obtained. In this way in fact second derivatives in space do not appear and natural boundary conditions are automatically satisfied. The weak form of eqs. (75) and (76) is easily obtained using the same technique as for the equilibrium equation:

- each term of the balance equation is multiplied by a vector function $\delta \underline{p}_\pi$ which has continuous first partial derivatives with respect to the coordinates x_i and satisfies the boundary conditions where prescribed (virtual pressure). Integration is then performed over the domain Ω.
- terms involving second spatial derivatives are transformed according to the Gauss' theorem. q_π represents the flux of the π phase at the boundary Γ of the domain.

Consequently the continuity equations assume the following form for the system 'water plus solid'

$$\int_\Omega \delta \underline{p}_w^T \left\{ C_{ww}\dot{p}_w + C_{wg}\dot{p}_g + \left(\frac{C_{wg}}{S_g}(p_w - p_g) + n \right) \dot{S}_w + \alpha S_w \dot{u}_{i,i} + \frac{\underline{k}_w}{\rho_w} \left(-p_{w,i} + \rho_w (g_i - \dot{v}_i^s) \right) \frac{\partial \rho_w}{\partial x_i} \right\} d\Omega - \int_\Omega \nabla \left(\delta \underline{p}_w \right)^T \left(\underline{k}_w \left(-p_{w,i} + \rho_w (g_i - \dot{v}_i^s) \right) \right) d\Omega + \int_\Gamma \delta \underline{p}_w^T q_w \, d\Gamma = 0 \tag{81}$$

and for the mixture 'gas plus solid'

$$\int_\Omega \delta \underline{p}_c^T \left\{ C_{gw}\dot{p}_w + C_{gg}\dot{p}_w + \left(- \frac{C_{wg}}{S_w}(p_w - p_g) + n \right) \dot{S}_g + \alpha S_g \dot{u}_{i,i} + \frac{\underline{k}_g}{\rho_g} \left(-p_{g,i} + \rho_g (g_i - \dot{v}_i^s) \right) \frac{\partial \rho_g}{\partial x_i} \right\} d\Omega - \int_\Omega \nabla \left(\delta \underline{p}_c \right)^T \left(\underline{k}_g \left(-p_{g,i} + \rho_g (g_i - \dot{v}_i^s) \right) \right) d\Omega + \int_\Gamma \delta \underline{p}_w^T q_g \, d\Gamma = 0 \tag{82}$$

In solving the boundary value problem represented by eqs. (74), (81) and (82), three formulations are possible, depending on the choice of the primary field variables. Whereas the displacement vector of the solid phase and the water pressure are natural choices, the third field variable may be:

a) the capillary pressure p_c;
b) the gas pressure p_g;
c) the water degree of saturation S_w.

In each case, all the other variables are expressed by means of the constitutive relationships.

The finite element discretization in space is now carried out and $\underline{u}$, p_w and s are chosen as basic variables, with s representing one of the possible choices for the third variable. Obviously each of the three formulations require a particular form of the constitutive relationships, resulting in three different forms of the governing equations.

The field variables are expressed in terms of their nodal values $\bar{\underline{u}}$, $\underline{p}$ and $\underline{s}$ by means of the shape functions $\underline{N}^d$ and $\underline{N}^p$ as

$$\underline{u} = \underline{N}^d \bar{\underline{u}}, \quad p_w = \underline{N}^p \underline{p}_w, \quad s = \underline{N}^p \underline{s} \tag{83}$$

The numbers and locations of the nodes are not necessarily the same for all variables [9].

As initially specified, we assume some terms in the governing equations to be negligible. In particular

- in the equilibrium equation, the terms related to the gradient of the fluid velocity vectors and those involving the material derivatives of the relative velocities;

- in the continuity equations, the terms related to the gradient of the density of the fluids.

By substituting the piecewise interpolation functions (83) into the governing equations, and taking into account the arbitrary choice of $\delta\underline{u}$, $\delta\underline{p}_w$ and $\delta\underline{s}$, we may obtain the semidiscretized non-linear system of differential equations in time as follows

$$\underline{M}\,\ddot{\bar{\underline{u}}} + \int_\Omega \underline{B}^T\underline{\sigma}''d\Omega + \underline{C}_{sw}\underline{p}_w + \underline{C}_{sc}\underline{s} + \underline{F}_s = 0$$

$$\underline{M}_p\ddot{\bar{\underline{u}}} + \underline{C}_{ws}\dot{\underline{u}} + \underline{P}_{ww}\dot{\underline{p}}_w + \underline{C}_{wc}\dot{\underline{s}} + \underline{H}_{pp}\underline{p}_w + \dot{\underline{F}}_w = 0 \tag{84}$$

$$\underline{M}_c\ddot{\bar{\underline{u}}} + \underline{C}_{cs}\dot{\underline{u}} + \underline{C}_{cw}\dot{\underline{p}}_w + \underline{P}_{cc}\dot{\underline{s}} + \underline{H}_{cc}\underline{s} + \underline{H}_{cw}\underline{p}_w + \dot{\underline{F}}_c = 0$$

The matrices of eq. (84) are listed in appendix I, where the coefficients that distinguish the different formulations are also specified. In all cases the system (84) is non-linear, non-symmetric and coupled, i.e. we can not solve any field separately from the others nor can any set of variables be explicitly eliminated. The complete solution may now be obtained by means of appropriate time stepping algorithms.

8. SOLUTION IN THE TIME DOMAIN

The set of spatially discretized eq. (84), results to be a system of differential equations in time with second order derivatives. Among many available schemes for the numerical time integration [12], we use a Newmark type method in a direct solution procedure, i.e. the time stepping process is carried out simultaneously for the whole equation system (84).
To express the value of the variables in two successive time stations n and n+1, corresponding to the times t and t+Δt, we assume the following truncated series expansions [12]:

- for the displacement field

$$\bar{\underline{u}}_{n+1} = \bar{\underline{u}}_n + \Delta t\,\dot{\bar{\underline{u}}}_n + \frac{1}{2}\Delta t^2\,\ddot{\bar{\underline{u}}}_n + \frac{1}{2}\beta_2\,\Delta t^2\,\Delta\ddot{\bar{\underline{u}}}_n \tag{85}$$

$$\dot{\bar{\underline{u}}}_{n+1} = \dot{\bar{\underline{u}}}_n + \Delta t\,\ddot{\bar{\underline{u}}}_n + \frac{1}{2}\beta_1\,\Delta t\,\Delta\ddot{\bar{\underline{u}}}_n \tag{86}$$

where

$$\Delta\ddot{\bar{\underline{u}}}_n = \ddot{\bar{\underline{u}}}_{n+1} - \ddot{\bar{\underline{u}}}_n \tag{87}$$

- for p_w and s, simbolically represented through κ

$$\underline{\kappa}_{n+1} = \underline{\kappa}_n + \Delta t\, \dot{\underline{\kappa}}_n + \vartheta\, \Delta t\, \Delta\dot{\underline{\kappa}}_n \tag{88}$$

where

$$\Delta\dot{\underline{\kappa}}_n = \dot{\underline{\kappa}}_{n+1} - \dot{\underline{\kappa}}_n \tag{89}$$

$\Delta\ddot{\underline{u}}_n$, $\Delta\dot{\underline{p}}_n$ and $\Delta\dot{\underline{s}}_n$ are, for the moment, undetermined quantities and β_1, β_2 and ϑ are parameters in the range 0 - 1.
For unconditional stability of the solution procedure, in the linear case, it is required that [12]

$$1 \geq \beta_2 \geq \beta_1 \geq 0.5 \qquad \text{and} \qquad \vartheta \geq 0.5 \tag{90}$$

By inserting eqs.(85)-(89) in (84), we obtain the following non-linear equation set

$$\begin{aligned}
&\underline{M}\, \Delta\ddot{\underline{\bar{u}}}_n + \left[\int_{\Omega} \underline{\underline{B}}^T \underline{\sigma}''\, d\Omega\right]_{n+1} + \underline{C}_{sw}\vartheta\Delta t\Delta\dot{\underline{p}}_{w,n} + \underline{C}_{sc}\vartheta\Delta t\Delta\dot{\underline{s}}_n = -\underline{F}^*_{s,n+1} \\
&\underline{M}^*_p\Delta\ddot{\underline{\bar{u}}}_n + \left(\underline{P}_{ww} + \vartheta\Delta t\, \underline{H}_{pp}\right) \Delta\dot{\underline{p}}_{w,n} + \underline{C}_{wc}\Delta\dot{\underline{s}}_n = -\underline{F}^*_{w,n+1} \\
&\underline{M}^*_c\Delta\ddot{\underline{\bar{u}}}_n + \left(\underline{C}_{cw} + \vartheta\Delta t\, \underline{H}_{cw}\right)\Delta\dot{\underline{p}}_{w,n} + \left(\underline{P}_{cc} + \vartheta\Delta t\, \underline{H}_{cc}\right)\Delta\dot{\underline{s}}_n = -\underline{F}^*_{c,n+1}
\end{aligned} \tag{91}$$

where

$$\underline{M}^*_p = \left(\underline{M}_p + \frac{1}{2}\beta_1\Delta t\underline{C}_{ws}\right) \tag{92}$$

$$\underline{M}^*_c = \left(\underline{M}_c + \frac{1}{2}\beta_1\Delta t\underline{C}_{cs}\right) \tag{93}$$

$$\begin{aligned}
\underline{F}^*_{s,n+1} = &-\underline{F}_{s,n+1} - \underline{M}_p\ddot{\underline{\bar{u}}}_n - \underline{C}_{ws}\left(\dot{\underline{u}}_n + \Delta t\ddot{\underline{\bar{u}}}_n\right) - \underline{H}_{pp}\left(\underline{p}_{w,n} + \Delta t\dot{\underline{p}}_{w,n}\right) \\
&-\underline{P}_{ww}\dot{\underline{p}}_{w,n} - \underline{C}_{wc}\dot{\underline{s}}_n
\end{aligned} \tag{94}$$

$$\begin{aligned}
\underline{F}^*_{c,n+1} = &-\underline{F}_{c,n+1} - \underline{M}_c\ddot{\underline{\bar{u}}}_n - \underline{C}_{cs}\left(\dot{\underline{u}}_n + \Delta t\ddot{\underline{\bar{u}}}_n\right) - \underline{H}_{cw}\left(\underline{p}_{w,n} + \Delta t\dot{\underline{p}}_{w,n}\right) \\
&-\underline{H}_{cc}\left(\underline{s}_n + \Delta t\dot{\underline{p}}_n\right) - \underline{C}_{cw}\dot{\underline{p}}_{w,n} - \underline{P}_{cc}\dot{\underline{s}}_n
\end{aligned} \tag{95}$$

which has to be solved for $\Delta\ddot{\underline{u}}_n$, $\Delta\dot{\underline{p}}_n$ and $\Delta\dot{\underline{s}}_n$.
However it must be noted that terms such as $\underline{M}_p\Delta\ddot{\underline{u}}_n$, $\underline{M}_s\Delta\ddot{\underline{u}}_n$ are small in value for most slow response problems and can be neglected.
Attention must be paid when dealing with the integral term in eq. (91). The interested reader is referred to [13] for this problem and to [14] for problems concerning the convergence and stability analyses of the numerical procedure.

9. APPLICATIONS

The first application deals with the numerical simulation of the earthquake response of the San Fernando earth dam [2].

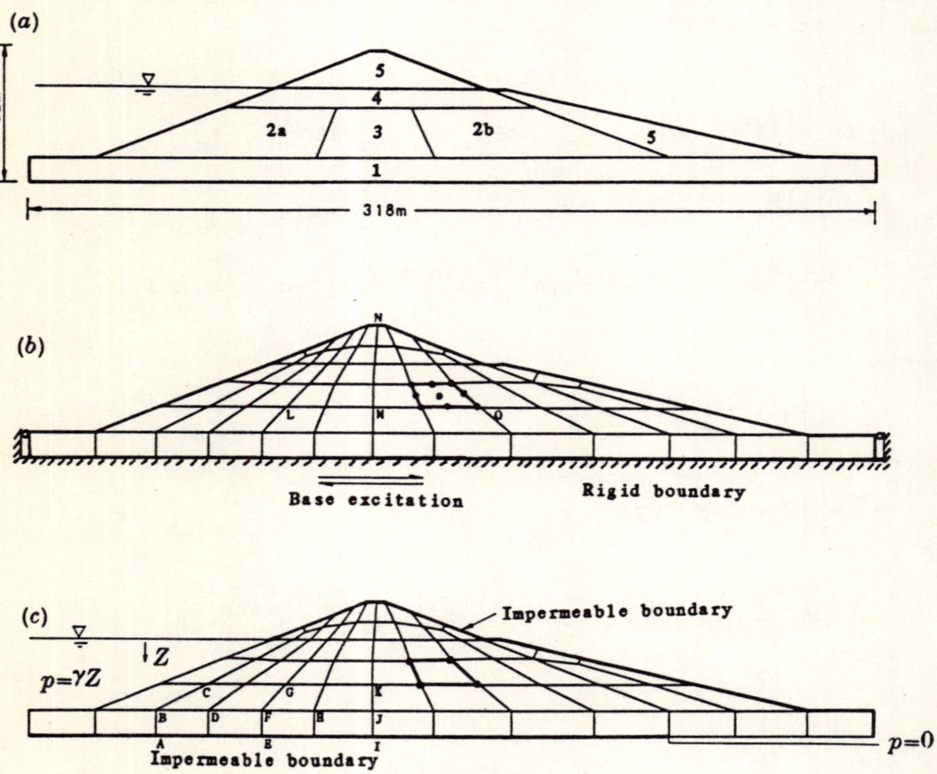

Fig. 2- Idealization of S. Fernando dam for analysis.
(a) material zones; spatial discretization and boundary conditions for the displacement field (a) and for the pressure field (b).

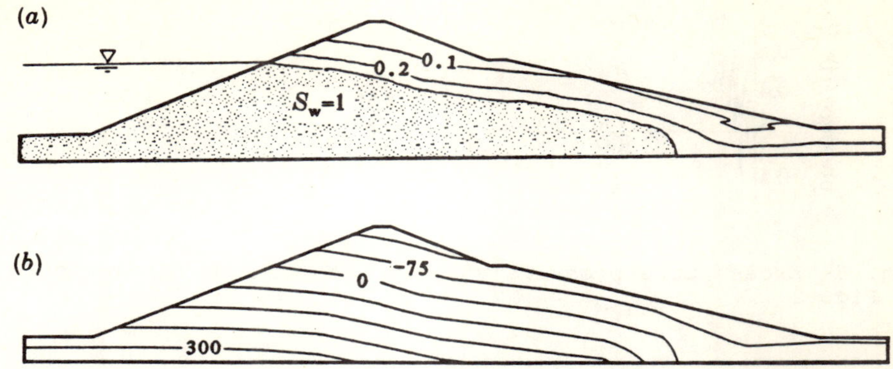

Fig. 3- Initial steady state solution: saturation (a) and water pressure contours (b).

This example illustrates the computational process applied to a two phase model where the gas pressure has been assumed constant at the reference value. Comparison with the actual occurrence is impossible, due to the lack of full records.

Figure 2 represents the cross section of the dam, the discretization and the boundary conditions for the displacement and water pressure fields. For a detailed presentation of the material properties and the acceleration time history of the earthquake we refer to [2]. For the soil behaviour a Pastor-Zienkiewicz model has been assumed [15]. The initial conditions for effective stress and pressure distribution (Fig. 3) have been obtained by means of a static analysis of the dam subjected to gravity forces and flow due to the pressure head. Starting with these initial conditions, the computation has been carried out for the full period of the earthquake and

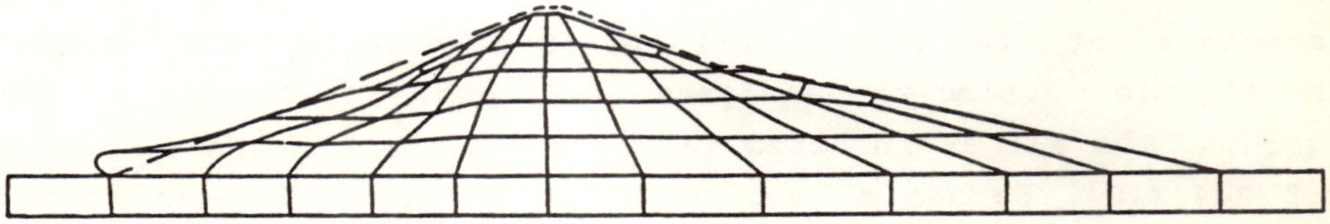

Fig. 4- Deformed shape of the dam after 30 sec.

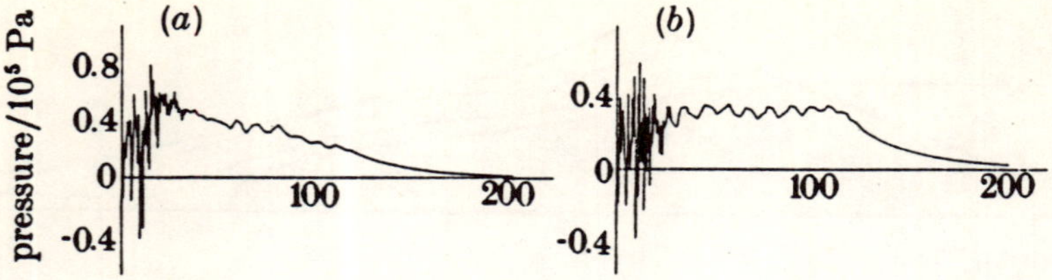

Fig. 5- Excess pore pressure vs. time at point B (a) and C (b) of Fig. 2.

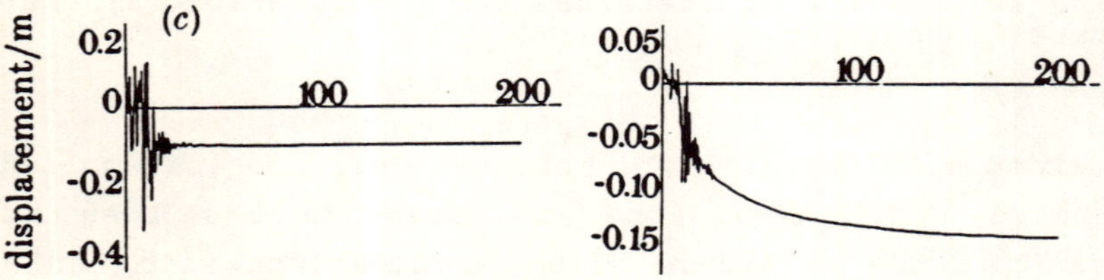

Fig. 6- Horizontal (left) and vertical displacements vs. time at point M of Fig. 2.

continued for a further time of 200 seconds. Some selected results are presented in Fig.4, which shows the deformed shape of the dam at 30 sec, and in Figs. 5 and 6, where excess pore pressure and displacements histories are shown.

In the second example the full three-phase behaviour is taken into account. It deals with the analysis of an aquifer used as storage space for a compressed air energy storage system. The reservoir is assumed as a column of a porous medium, which is initially fully saturated with reference permeability of $5*10^{-13}m^2$, porosity of 0.2 and initial aquifer pressure of 5066 kPa (or 50 atm). Temperature is assumed constant at 147°C and all dynamic effects are assumed negligible during the analyzed time span. At the top of the column, the air is injected into the system with the mass flow of $2.44*10^{-4}$ Kg/sec m^2 and no water is assumed to cross the

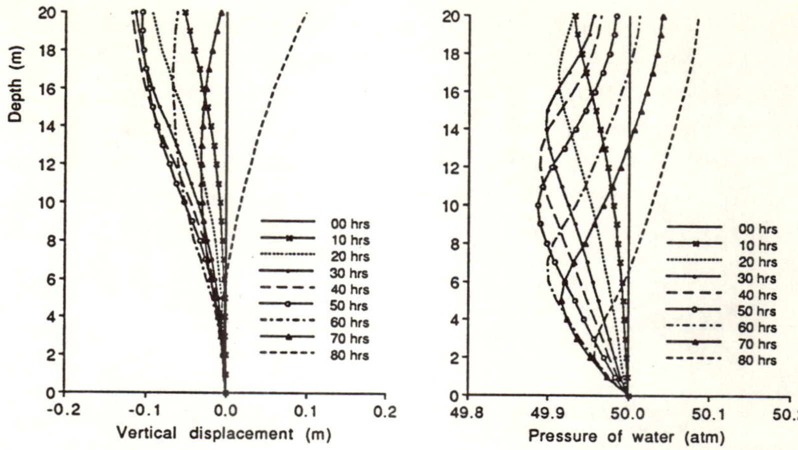

Fig. 7- Air storage in an aquifer: vertical displacements and pore water pressure vs. depth at different times.

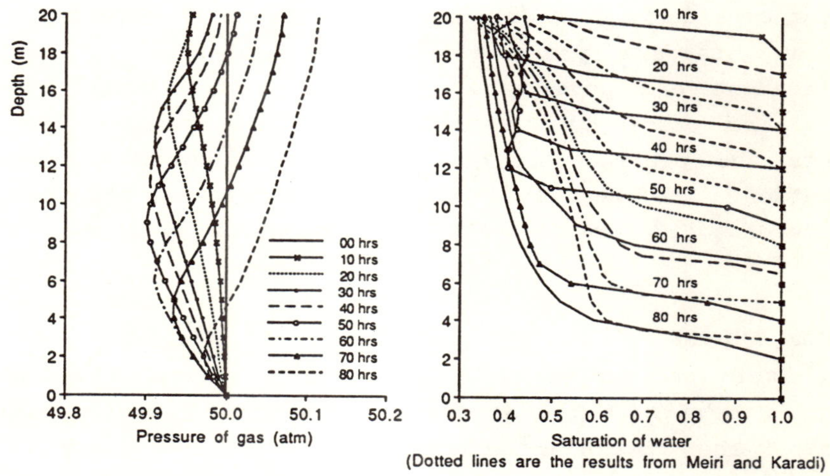

Fig. 8- Air storage in an aquifer: gas pressure and water saturation distribution vs. depth at different times.

upper boundary. At the bottom of the column, no flow of air is assumed and the water pressure is maintained constant at 5066 kPa. The water and air viscosities are selected as 0.3 cp and 0.024 cp respectively. The water density at standard condition is 1000 Kg/m^3 and the air density 1.22 Kg/m^3. The following relationships for the hydraulic parameters are employed in the analysis [16]:

$$k_{rw} = Se^{(2+3p)/p}$$
$$k_{rg} = (1-Se^2)(1-Se^{(2+p)/2})$$
$$p_c = p_b \ / \ (Se)^{1/p}$$

where $Se = (S_w - S_{wc})/(1-S_{wc})$ is the effective saturation, S_{wc} is the irriducible saturation, p means the pore size distribution index and p_b the bubbling pressure. The values for the parameters S_{wc}, p and p_b are given as 0.2, 3.0 and 1.68 kPa respectively, corresponding to a sand with the previously specified permeability. The same problem was simulated also by Meiri and Karadi [17], who used a 1-D finite element model, without considering the deformation of the solid skeleton.
Some results are presented here. In Fig. 7, vertical displacement and pore water pressure distributions in the sample are presented for different times, whereas Fig. 8 depicts the pressure of gas and the values of the water saturation. The results of the water saturation obtained in in Reference [17] are indicated for comparison purposes.

10. CONCLUSIONS

In this chapter we have developed a general three-phase model for the static and dynamic analysis of semi-saturated soils. The governing equations were derived within the framework of spatially averaging theories and then solved by a discrete approximation procedure. Examples referring to the seismic behaviour of an earth dam (two-phase problem) and to a compressed air energy storage system in an aquifer (three-phase problem) show how problems involving semi-saturation can be solved.

REFERENCES

1. Zienkiewicz, O.C., Chan, A.H.C., Pastor, M., Paul, D.K. and Shiomi, T.: Static and Dynamic behaviour of soils: a rational approach to quantitative solutions. I. Fully saturated problems. Proc. R. Soc. Lond. A 429, 285-309, 1990.

2. Zienkiewicz, O.C., Xie, Y.M., Schrefler, B.A., Ledesma, A. and Bicanic, N.: Static and Dynamic behaviour of soils: a rational approach to quantitative solutions. II. Semi-saturated problems. Proc. R. Soc. Lond. A 429, 311-321, 1990.

3. Schrefler, B.A., Simoni, L., Li Xikui and Zienkiewicz,O.C.: Mechanics of partially saturated porous media. in Numerical Methods and Constitutive Modelling in Geomechanics. CISM Lecture notes, eds. Desai C.S. and Gioda G., 169-209, Springer Verlag, Wien, 1990.

4. Bedford, A. and Drumheller, D.S.: Theories of immiscible and structured mixture, Int. J. Engng. Sci., 21 (1983), 863-960.

5 Bear, J. and Bachmat, Y.: Transport phenomena in porous media - Basic equations, in: Fundamentals of transport phenomena in porous media (Bear, J. and Corapcioglu, M.Y. eds.), Nato ASI Series, Nijhoff, Dordrecht 1984, 5-61.

6. Hassanizadeh, M. and Gray W.G.: General conservation equations for multiphase systems: 1. Averaging procedure, Adv. Water Resources, 2 (1979), 131-144.

7. Lewis, R.W. and Schrefler, B.A.: The finite element method in the deformation and consolidation of porous media, Wiley, Chichester 1987.

8. de Boer, R. and Ehlers, W.: The development of the concept of effective stress, Acta Mechanica, 83 (1990), 77-92.

9. Zienkiewicz, O.C. and Shiomi, T.: Dynamic behaviour of saturated porous media: the general Biot's formulation and its numerical solution, Int. J. Num. Anal. Meth. Geom., 8 (1985), 71-96.

10. Schrefler, B.A. and Simoni, L.: A unified approach to the analysis of saturated - unsaturated elastoplastic porous media, in Numerical Methods in Geomechanics. Innsbruck 1988. (ed. G. Swoboda), Balkema, Rotterdam 1988, 205-212.

11. Hassanizadeh, M. and Gray W.G.: General conservation equations for multiphase systems: 2. Mass, momenta, energy and entropy equations, Adv. Water Resources, 2 (1979), 191-203.

12. Zienkiewicz, O.C. and Taylor, R.L.: Coupled problems - A simple time stepping procedure, Comm. Appl. Num. Meth., 1 (1985), 233-239.

13. Simo, J.C.: A framework for finite strain elastoplasticity based on maximum plastic dissipation and the multiplicative decomposition. Part II: Computational aspects. Comp. Meths. Appl. Mech. Eng. 68 (1988), 1-31.

14. Schrefler, B.A. and Simoni, L.:Comparison between different finite element solutions for immiscible two phase flow in deforming porous media, to appear.

15. Pastor, M., Zienkiewicz, O.C., and Chan, A.H.C.: Generalized plasticity and the modelling of soil behaviour. Int. J. Num. Anal. Methods Geomech., 14, 151-190, 1990.

16. Brooks, R.N. and Corey, A.T.: Properties of porous media affecting fluid flow. J. Irrigation and Drainage Div., Proc. Am. Soc. Civ. Eng., 92, No. IR2, 61-68, 1966.

17. Meiri, D. and Karadi, G.M.: Simulation of air storage aquifer by finite element model. Int. J. Num. Anal. Meth. in Geomech., 6, 339-351, 1982.

APPENDIX I

List of the matrices of equation (84):

$$\underline{M} = \int \rho \, \underline{N}^{d^T} \underline{N}^d d\Omega$$

$$\underline{C}_{sw} = -\int_\Omega \underline{B}^T \alpha \, k_{12} \underline{m} \, \underline{N}^P d\Omega$$

$$\underline{C}_{sc} = -\int_\Omega \underline{B}^T \alpha \, k_{13} \, \underline{m} \, \underline{N}^P d\Omega$$

$$\underline{M}_p = \int_\Omega \left(\nabla \underline{N}^P\right)^T \underline{k}_w \rho_w d\Omega$$

$$\underline{C}_{ws} = \int_\Omega \underline{N}^{P^T} \alpha \, S_w \, \underline{m}^T \, \underline{B} \, d\Omega$$

$$\underline{P}_{ww} = \int_\Omega \underline{N}^{P^T} k_{22} \, \underline{N}^P \, d\Omega$$

$$\underline{C}_{wc} = \int_\Omega \underline{N}^{P^T} k_{23} \, \underline{N}^P \, d\Omega$$

$$\underline{H}_{ww} = \int_\Omega \left(\nabla \underline{N}^P\right)^T \underline{k}_w \nabla \underline{N}^P d\Omega - \int_\Omega \underline{N}^{P^T} \frac{\underline{k}_w}{\rho_w} \nabla \underline{N}^P \, \underline{\nabla}^T \rho_w d\Omega$$

$$\underline{M}_c = \int_\Omega \left(\nabla \underline{N}^P\right)^T \underline{k}_g \rho_g d\Omega$$

$$\underline{C}_{cs}=\int_{\Omega} \underline{N}^{P^T} \alpha\ \underline{m}^T\ \underline{B}\ d\Omega$$

$$\underline{C}_{cw}=\int_{\Omega} \underline{N}^{P^T} k_{32}\ \underline{N}^P\ d\Omega$$

$$\underline{P}_{cc}=\int_{\Omega} \underline{N}^{P^T} k_{33}\ \underline{N}^P\ d\Omega$$

$$\underline{H}_{cw}=\int_{\Omega}\left(\nabla \underline{N}^P\right)^T \underline{k}_g k_{34} \nabla \underline{N}^P d\Omega-\int_{\Omega} \underline{N}^{P^T} \frac{\underline{k}_g}{\rho_g} \nabla \underline{N}^P \nabla^T \rho_g d\Omega$$

$$\underline{H}_{cc}=\int_{\Omega}\left(\nabla \underline{N}^P\right)^T \underline{k}_g k_{35} \nabla \underline{N}^P d\Omega-\int_{\Omega} \underline{N}^{P^T} \frac{\underline{k}_g}{\rho_g} \nabla \underline{N}^P \nabla^T \rho_g d\Omega$$

$$\underline{F}_s=-\int_{\Omega} \underline{N}^{d^T} \rho \underline{g} d\Omega - \int_{\Gamma} \underline{N}^{d^T} \hat{\underline{t}} d\Gamma\ -\int_{\Omega} \underline{B}^T \alpha\ k_{14} \underline{m}\ p_g d\Omega$$

$$\dot{\underline{F}}_w=\int_{\Omega}\left[\underline{N}^{P^T} \frac{\underline{k}_w}{\rho_w} \underline{b}_w \nabla^T \rho_w-\left(\nabla \underline{N}^P\right)^T \underline{k}_w \underline{b}\ \right] d\Omega+\int_{\Gamma} \underline{N}^{P^T} q_w d\Gamma$$

$$\dot{\underline{F}}_c=\int_{\Omega}\left[\underline{N}^{P^T} \frac{\underline{k}_g}{\rho_g} \underline{b}_g \nabla^T \rho_g-\left(\nabla \underline{N}^P\right)^T \underline{k}_g \underline{b}_g\right] d\Omega+\int_{\Gamma} \underline{N}^{P^T} q_g d\Gamma$$

Where, for the approaches a, b and c of paragraph 7, we have respectively:

$k_{12}= 1$	$k_{13}= S_g$	$k_{14}= 0$
$k_{12}= S_w$	$k_{13}= S_g$	$k_{14}= 0$
$k_{12}= S_w$	$k_{13}= -p_g$	$k_{14}= 1$

$$k_{22}= \left(\frac{\alpha-n}{K_s} + \frac{n}{K_w}\right) S_w$$

$$k_{22}=\frac{\alpha-n}{K_s} S_w^2+ \frac{nS_w}{K_w} -\left[\frac{\alpha-n}{K_s} S_w\left(p_w-p_g\right)+n\right] \frac{dS_w}{dp_c}$$

$$k_{22}= \left(\frac{\alpha-n}{K_s} + \frac{n}{K_w}\right) S_w$$

$$k_{23}= n \frac{dS_w}{dp_c} + \frac{\alpha-n}{K_s} S_w\left(S_g - \frac{dS_w}{dp_c} p_c\right)$$

$$k_{23}= \frac{\alpha-n}{K_s} S_g S_w+\left[\frac{\alpha-n}{K_s} S_w\left(p_g-p_w\right)+n\right] \frac{dS_w}{dp_c}$$

$$k_{23}= \frac{\alpha-n}{K_s} \left[\frac{dp_c}{dS_w}\left(1-S_w\right) - p_c\right]S_w+ n$$

$$k_{32}= \frac{n\ S_g}{K_g} + \frac{\alpha-n}{K_s}\ S_g\left(\ S_w+\ S_g\right)$$

$$k_{32}=\frac{\alpha-n}{K_s}\ S_wS_g+\left[\frac{\alpha-n}{K_s}\ S_g\left(p_g-p_w\right)+n\right]\frac{dS_w}{dp_c}$$

$$k_{32}= \left[\frac{n}{K_g} + \frac{\alpha-n}{K_s}\right] S_g$$

$$k_{33}=\frac{S_g n}{K_g} - \frac{dS_w}{dp_c}\ n+\frac{\alpha-n}{K_s}\ S_g\left(S_g- \frac{dS_w}{dp_c}\ p_c\right)$$

$$k_{33}=\frac{n\ S_g}{K_g} +\frac{\alpha-n}{K_s} -\left[\frac{\alpha-n}{K_s}\ S_g\left(p_g-p_w\right)+n\right]\frac{dS_w}{dp_c}$$

$$k_{33}=\frac{dp_c}{dS_w}\ \left(\frac{n}{K_g} +\frac{\alpha-n}{K_s}\ S_g\right)S_g-\frac{\alpha-n}{K_s}\ S_gp_c-n$$

$k_{34}= 1$ $\qquad$ $k_{35}= 1$

$k_{34}= 0$ $\qquad$ $k_{35}= 1$

$k_{34}= 1$ $\qquad$ $k_{35}= \frac{dp_c}{dS_w}$

Adaptive Finite Element Analysis of Heat Transfer and Flow Problems

R.W.Lewis A.S.Usmani
J.T.Cross
Dept. of Civil Engineering, University College of Swansea

25 April 1991

Abstract

An error estimation technique for adaptive finite element analysis of heat conduction problems is described. The performance of this technique is demonstrated using a two-dimensional steady-state linear benchmark problem with a known analytical solution. The technique is then extended to solve steady and transient flow problems with examples given to demonstrate the utility of each. An improved version of the adaptive timestepping procedure due to Gresho is used in the solution of the transient flow problem.

1 Introduction

The solution of a continuum mechanics problem by the finite element method requires that the domain of interest is first subdivided into a number of discrete subdomains or elements. Variational, Galerkin or other techniques can be used to obtain an integral formulation; after evaluation of the integrals a set of algebraic equations, in matrix form, is obtained for each element. Once these equations are obtained for every element, the global matricies are assembled and the final solution is obtained.

The size and placement of these elements largely determines the accuracy with which the problem can be solved. Reducing the element sizes and thereby increasing the number of nodal points, usually yields a more accurate solution but at the cost of an increased computer time and memory requirement. The key to the efficient and economic solution of problems is not merely the *number* of nodal points and elements but also their *placement*. An accurate numerical analysis will invariably require judgement on behalf of the user as to the placement of mesh density. Regions with large gradients (e.g. a discontinuity, stress concentration or region of high heat flux), will need a high mesh density with quiescent regions requiring a comparatively coarser mesh. In many real engineering situations, attempts will be made either to obtain the most accurate solution possible within an upper problem size limit, or to try to model a numerically awkward feature without using an excessive number of elements.

The premise of adaptive procedures is that, by making use of the mathematics of error analysis, a finite element solution can indicate which regions need refinement. This information can then be used to automatically adapt the mesh to suit the problem. The error is calculated in each element and compared with a predefined limit. Any element with an error above, and in some cases below, the limit must be adjusted in size. The process is repeated, if necessary, with the ultimate aim that every element contains the same predefined, allowable error, thus yielding an optimal mesh density.

There are two distinct phases of the adaptive solution process, error analysis and mesh refinement. Error analysis allows us to calculate for each element, in the case of error estimation, an absolute value for the error, and in the case of error indication, an element value relative to other elements in the mesh. Much of the early mathematical work in error analysis was due to Babuska [1, 2] but was drawn together, in the form of an error estimate, in a notable paper by Kelly et al. [13]. A companion paper by Gago et al. [5] was later published which offered several strategies for using an error estimator to refine a mesh.

Although its potential was understood, the theory described in reference [13] was difficult for many engineers to follow, and little progress was made until a paper appeared by Zienkiewicz and Zhu [17] in 1987. In this paper, a simplified error estimate was outlined which, in particular, did not require the calculation of interelement traction jumps, and which was shown to give a satisfactory approximation to that of the error estimate reported in reference [13].

Once an estimate of the error in each element has been obtained, there are several ways of progressing to the next phase of the adaptive process i.e. improvement of the mesh. The two basic approaches are *p-refinement* and *h-refinement*. The first technique involves increasing the order of the polynomial approximation within an element while h-refinement simply means reducing or increasing the subdivision size.. P-refinement, especially when combined with a hierarchical formulation has several advantages — it is more efficient, converges faster and appeals to the purist by virtue of its mathematical elegance. However, incorporating p-refinement generally means restructuring if not rewriting an existing code. H-refinement on the other hand is simpler and works in a way that engineers intuitively understand. It is also easier to combine with an existing code and for these reasons h-refinement is more universally accepted and has been employed in the content of this paper.

Having decided on adopting the h-refinement method there is another choice to be made, *element subdivision* or *mesh regeneration*. With the element subdivion approach, every element that exceeds the allowable error threshold is subdivided into smaller elements. This is most effective when using four-node elements as it is otherwise very difficult to achieve the desired density distribution. However, constrained nodes are introduced which must be dealt with, and the method allows only one level of subdivision at a time.

The remeshing approach involves completely regenerating the mesh, either in regions of high error only, or over the entire domain. The advantage of regenerating the entire mesh is that areas can be coarsened if the calculated error is below the allowable error. It is this that allows the generation of a truly optimal mesh in which every element has approximately the same, predefined level of error, and it was this fact that led the authors to use the mesh regeneration approach reported in this paper. One of the disadvantages of mesh regeneration is that a high degree of spatial flexibility is required if use is to be made of the abundant information provided by an error estimation procedure. Obviously, triangles are the optimal shapes in this respect and the advancing front method [19, 10] has been a popular and successful method of generating such elements ([17, 18]. However, new work has laid open the possibility of using the advancing front method to generate unstructured quadrilateral grids [12], albeit with a slightly reduced degree of flexibility, and also of converting unstructured triangular meshes to quadrilateral ones automatically [9].

This work builds upon earlier work presented in references [6, 23], where an error estimation and adaptive technique was developed for linear steady state heat conduction and steady incompressible flow problems, which is now extended to the solution of transient incompressible flow problems. The validity of the error estimate is demonstrated by solving first a benchmark problem, for which an analytical solution exists, subsequently, an error indicator suitable for steady incompressible flow problems governed by the Navier-Stokes equations is presented, which is based upon the same technique as employed for the heat conduction problems. The validity of this indicator is demonstrated by solving the example of steady incompressible flow past a cylinder. Finally the performance of the adaptive scheme is tested by solving the transient flow problem of the 'Karman vortex street'.

2 Heat diffusion

For transient heat conduction problems the governing field equation can be written as follows:

$$\rho c\dot{T} = \nabla\cdot \mathrm{k}\,\nabla T \tag{1}$$

Where, ρ and c and k are density, specific heat and thermal conductivity and T is the temperature field. The boundary conditions used are given by,

$$\begin{aligned} T &= f(x,y,z,t) \qquad \text{on } \Gamma_T \\ k\frac{\partial T}{\partial n} + q + h(T - T_a) &= 0 \qquad \text{on } \Gamma_q \end{aligned} \tag{2}$$

Where q, h and T_a are specified boundary heat flux, convective heat tranfer coefficient and the ambient temperature respectively and n is the direction of the unit normal vector to Γ_q. Γ_T and Γ_q are portions of the boundary Γ of the computational domain Ω so that,

$$\Gamma_T \cup \Gamma_q = \Gamma$$

and,

$$\Gamma_T \cap \Gamma_q = \phi$$

Discretizing in space by the finite element method yields a system of first order ordinary differential equations in time. Finite differences are generally used for temporal discretization, yielding a fully discretized algebraic system of equations which, when solved, give us the temperature field. The details of the finite element solution procedure can be obtained from standard texts, such as [16].

2.1 Error estimation for heat diffusion

The temperature field obtained from a finite element analysis, denoted by $\hat{T}$, is an approximation of the exact temperature field T. Therefore, the error in the finite element solution can be expressed as,

$$\epsilon_T = T - \hat{T} \tag{3}$$

However, it is more suitable to define the error in terms of temperature gradients $\frac{\partial T}{\partial x}$, $\frac{\partial T}{\partial y}$ etc. The heat flux q is a quantity based on the temperature gradients, therefore the error can be defined as,

$$e_q = q - \hat{q} \tag{4}$$

where q is the exact heat flux, while $\hat{q}$ is the solution obtained from a finite element analysis. It is apparent that the errors, as defined above, are based on pointwise differences between the approximate and exact values. Such a definition is difficult to implement and scalar measures or norms are used instead. The error norm defined below has been found to be effective in heat diffusion problems.

$$\|e\| = \left(\int_\Omega \left(\nabla \mathrm{T} - \nabla \hat{\mathrm{T}}\right)^T \mathrm{k} \left(\nabla \mathrm{T} - \nabla \hat{\mathrm{T}}\right) d\Omega\right)^{\frac{1}{2}} \tag{5}$$

Along the lines of Zienkiewicz *et al.* [18] it can be shown that,

$$\int_\Omega (\nabla \mathrm{T})^T \mathrm{k}(\nabla \hat{\mathrm{T}}) = \int_\Omega (\nabla \hat{\mathrm{T}})^T \mathrm{k}(\nabla \mathrm{T}) = \int_\Omega (\nabla \hat{\mathrm{T}})^T \mathrm{k}(\nabla \hat{\mathrm{T}}) \tag{6}$$

Using this relation we can rewrite equation(5) as,

$$\|e\|^2 = \int_\Omega (\nabla \mathrm{T})^T \mathrm{k}(\nabla \mathrm{T}) - \int_\Omega (\nabla \hat{\mathrm{T}})^T \mathrm{k}(\nabla \hat{\mathrm{T}}) \tag{7}$$

This equation is now written as,

$$\|e\|^2 = \|q\|^2 - \|\hat{q}\|^2 \tag{8}$$

Here, the quantity $||q||$ can be considered to be a measure of the total heat dissipation in the whole domain. Such a definition allows a practical representation of the error norm in terms of a percentage error, i.e.

$$\eta = \frac{||e||}{||q||} \times 100\% \tag{9}$$

The significance of the term $\frac{||e||}{||q||}$ can be realised from the following analysis. Consider a boundary condition of zero flux (i.e. $\frac{\partial T}{\partial n} = 0$) on the domain boundary Γ. Using Green's theorem we can write

$$||q||^2 = \int_\Omega (\nabla \mathrm{T})^T \mathrm{k}(\nabla \mathrm{T}) d\Omega = -\int_\Omega \mathrm{T}\, \nabla(\mathrm{k}(\nabla \mathrm{T})) d\Omega \tag{10}$$

and similarly for $||\hat{q}||^2$

$$||\hat{q}||^2 = \int_\Omega (\nabla \hat{\mathrm{T}})^T \mathrm{k}(\nabla \hat{\mathrm{T}}) d\Omega = \int_\Omega (\nabla \hat{\mathrm{T}})^T \mathrm{k}(\nabla \mathrm{T}) d\Omega = -\int_\Omega \hat{\mathrm{T}}\, \nabla(\mathrm{k}(\nabla \mathrm{T})) d\Omega \tag{11}$$

From equations equations (10) and (11) we have,

$$||e||^2 = ||q||^2 - ||\hat{q}||^2 = -\int_\Omega (\mathrm{T} - \hat{\mathrm{T}})\, \nabla(\mathrm{k}(\nabla \mathrm{T})) d\Omega \tag{12}$$

Examining equations (12) and (10) we can write,

$$\frac{||e||^2}{||q||^2} = \frac{(\mathrm{T} - \hat{\mathrm{T}})}{\mathrm{T}} \tag{13}$$

Equation (13) establishes, in an average sense, the relation of the error in temperature to the error in the temperature gradient. It is evident that the temperature errors will be much lower than those in the gradient — for instance an error of 20% in the gradients should roughly correspond to an error of 4% in the temperatures.

Equation (9) suggests a strategy for implementing an adaptive finite element procedure for heat transfer problems, similar to that used in references [17] and [18] for elasticity and forming problems. Before the implementation is discussed, the problem of calculating the error norm $||e||$ must be addressed as it requires 'exact' values of the temperature gradients, which are obviously not available. A satisfactory answer to this problem was found by Zienkiewicz and Zhu [17], who showed that globally smoothed values of stresses, which represented a higher order approximation to the original stresses obtained from a finite element analysis, could be used instead of the exact value to calculate the error norm. This idea has been used here and a globally smoothed value of the temperature gradients is calculated. The smoothing is carried out in a least squares sense in the manner described by Hinton and Campbell [4], however basis functions of the same order have been used here for smoothing the main variable, T, unlike the lower order basis functions used in reference [4]. The smoothing procedure involves representing the smoothed gradients $\tilde{G}$ in the element in terms of their unknown nodal values $\tilde{G}_i$, i.e.

$$\tilde{G} = \sum_{i=1}^{n} N_i \tilde{G}_i \tag{14}$$

where n is the number of nodes per element. This is followed by imposing in a mean sense the constraint,

$$\int_\Omega (\tilde{G} - \hat{G}) d\Omega = 0 \tag{15}$$

where, $\hat{G}$ represents the non-smooth temperature gradients obtained from the original finite element analysis. This yields a familiar system of equations for each element of the form,

$$\left(\int_{\Omega_e} N_i N_j d\Omega_e\right)(\tilde{G}_i) = \left(\int_{\Omega_e} N_i \hat{G}_i d\Omega_e\right) \tag{16}$$

which can be assembled over the whole domain and solved in the usual manner to obtain the globally smoothed values of the temperature gradients. These values can now be used in the expression for the error norm to obtain an approximation.

2.2 Implementation of the adaptive procedure

The general procedure for implementing an adaptive analysis for problems considered here follows closely the strategy given by Zienkiewicz and Zhu [17]. The first step is to specify a maximum allowable error $\bar{\eta}$ that is to be achieved by the end of the analysis. The requirement for a near optimum analysis is that all the elements of the final mesh must contain an approximately equal error. After the first analysis, from a preliminary (coarse) mesh, the quantity $\|q\|$ is calculated by summing over all elements as follows;

$$\|q\|^2 \simeq \|\hat{q}\|^2 = \sum_{e=1}^{m} \int_{\Omega_e} (\nabla \hat{\mathbf{T}}) \mathbf{k} (\nabla \hat{\mathbf{T}}) d\Omega_e \tag{17}$$

or more explicitly, for two dimensions

$$\|q\|^2 \simeq \|\hat{q}\|^2 = \sum_{e=1}^{m} \int_{\Omega_e} \left(k_x \left(\frac{\partial \hat{T}}{\partial x}\right)^2 + k_y \left(\frac{\partial \hat{T}}{\partial y}\right)^2 \right) dx dy$$

where m is the total number of elements. Now, the maximum permissible error for each element can be calculated by distributing $\|q\|^2$ equally over all the elements i.e.,

$$\|\bar{\epsilon}\|_e \leq \bar{\eta} \left(\frac{\|q\|^2}{m}\right)^{\frac{1}{2}} \tag{18}$$

The approximate error in each element after the first analysis can be calculated according to equation (5) using, of course, the smoothed values of the temperature gradients instead of the exact values. This can be written for a two-dimensional element e as,

$$\|\epsilon\|_e = \left(\int_{\Omega_e} \left(k_x \left(\frac{\partial \tilde{T}}{\partial x} - \frac{\partial \hat{T}}{\partial x} \right)^2 + k_y \left(\frac{\partial \tilde{T}}{\partial y} - \frac{\partial \hat{T}}{\partial y} \right)^2 \right) dxdy \right)^{\frac{1}{2}}$$

Now this error is compared, for all elements, with the maximum permissible error in an element as calculated above and used to modify the mesh for a second analysis. A variable ξ is defined as in [17], and is written for element e as,

$$\xi_e = \frac{\|e\|_e}{\|\bar{e}\|_e} \tag{19}$$

If $\xi_e > 1$ the size of element e must be reduced and vice versa. If the existing element size is h_e the new element size $\bar{h}_e$ is given by,

$$\bar{h}_e = \frac{h_e}{\xi_e^{\frac{1}{p}}} \tag{20}$$

where, p is generally taken as the order of the shape functions. When the information regarding all the element sizes is available, a new mesh is generated on which another analysis may be carried out. This procedure is continued until the calculated error falls below the specified maximum.

2.3 Benchmark example

An example to test the adaptive procedure, outlined above, was obtained from a list of selected benchmarks published by NAFEMS [8], the details of which are shown in Figure 1. This is a linear steady state 2-D heat conduction problem, where the severe discontinuity in boundary conditions at point B gives rise to a quasi-singular type behaviour at this location. This fact, combined with the relatively coarse meshes specified in reference [8], made it almost impossible to obtain a good comparison to the target value of temperature at point E obtained from an analytical solution to this problem. This analytical solution is shown for the whole domain in Figure 2, from which it is clearly seen that the steep temperature gradient between points B and E requires the use of a relatively fine mesh in this area. In many practical problems it is not immediately obvious where a fine mesh may be required and this is a major advantage of the adaptive finite element method.

The problem as defined Figure 1 was solved using the adaptive procedure described earlier and is also reported in reference [6]. A set of three meshes of 6-noded triangular elements is shown in Figure 3, beginning with the uniform starting mesh shown in Figure 3(a) which gives a norm error of 15%. The two subsequent meshes in Figures 3(b) and 3(c) were generated according to the adaptive procedure which lead to a norm error of 3.7% in the final mesh of Figure 3(c). This corresponds to a temperature error of approximately 0.1%. The solution obtained from this mesh is virtually indistinguishable

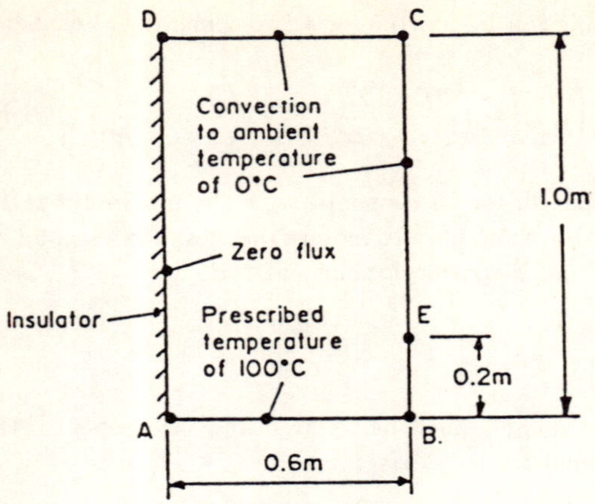

Figure 1: Problem definition for the benchmark example.

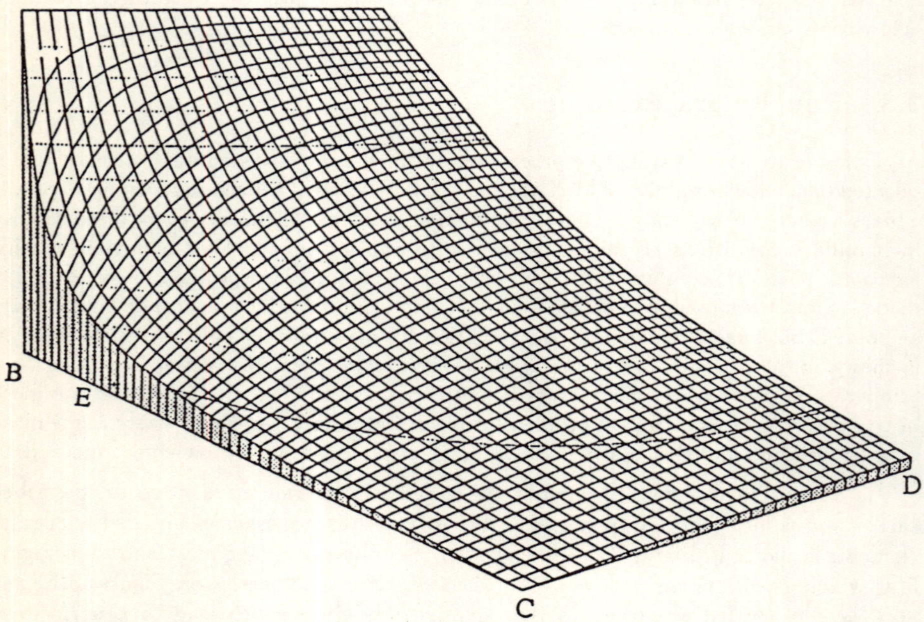

Figure 2: Analytical solution for the benchmark example.

(a) (b) (c)

Figure 3: Set of meshes based on the error estimator.

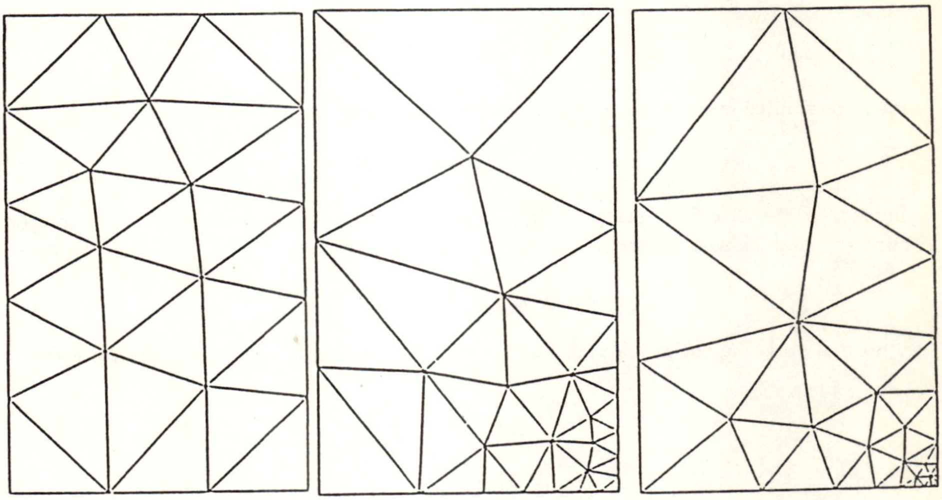

(a) (b) (c)

Figure 4: Set of meshes based on the exact solution.

from the analytical solution. The mesh can be seen to be finer in the area of points B and E, as expected, and much coarser elsewhere. Incidentally, the number of elements in the final mesh (26) is lower than the number of elements in the first mesh (28) which demonstrates the fact that adaptive remeshing doesn't necessarily mean a global increase in the number of elements. Figures 4(a), (b) and (c) shows a similar set of meshes obtained by calculating the error based on the exact solution (which in this case, is available) and the same target error. The meshes of Figures 3 and 4 follow a very similar pattern indicating the validity of the error estimator. This procedure has been found to be of significant benefit when applied to transient heat transfer problems involving change of phase [21, 22, 23].

3 Adaptivity for incompressible flow problems

An adaptive procedure applicable to steady state incompressible fluid flow problems using the same principles as outlined previously was developed and reported in reference [23]. This procedure will be applied to a transient flow problem governed by the Navier-Stokes equations for incompressible flow, viz:

$$\frac{\partial \mathbf{v}}{\partial t} + \rho(\mathbf{v} \cdot \nabla)\mathbf{v} = -\nabla \tau \tag{21}$$

where,

$$\tau = -P\delta_{ij} + \mu \left(\frac{\partial v_i}{\partial x_j} + \frac{\partial v_j}{\partial x_i} \right)$$

Incompressibility is enforced by the continuity equation

$$\nabla \cdot \mathbf{v} = 0 \tag{22}$$

where, $\mathbf{v}$ is the velocity vector, P is the pressure and μ is the viscosity. The relevent boundary conditions are specified velocities on the boundary,

$$\mathbf{v} = f(x, y, z) \qquad \text{on } \Gamma_1 \tag{23}$$

or, normal and tangential traction forces may be specified as below, on boundary Γ_2,

$$f_n = -P + 2\mu \frac{\partial u_n}{\partial n} \tag{24}$$

$$f_\tau = \mu \left(\frac{\partial u_n}{\partial \tau} + \frac{\partial u_\tau}{\partial n} \right) \tag{25}$$

n and τ are the unit normal and tangent vectors respectively. Here again Γ_1 and Γ_2 are portions of the boundary Γ of the computational domain Ω so that,

$$\Gamma_1 \cup \Gamma_2 = \Gamma$$

and,

$$\Gamma_1 \cap \Gamma_2 = \phi$$

The equations and boundary conditions are discretised by the Galerkin finite element method in the standard manner using a mixed interpolation for pressure and velocities. A spatially discretised system of first order differential equations in time is obtained in the familiar form, i.e.,

$$\mathbf{M}\dot{\boldsymbol{\theta}} + \mathbf{K}\boldsymbol{\theta} = \mathbf{F} \tag{26}$$

where $\boldsymbol{\theta}$ represents all the unknowns (velocities, pressure) and the matrices **M**,**K** and **F** have the usual meanings.

3.1 Error estimation and adaptive procedure for steady flow problems

Wu *et. al.* [11] recently suggested an adaptive procedure for incompressible flow problems governed by the Navier-Stokes equations. This procedure is based on an 'energy norm' which is shown to be an effective measure for guiding the refinement of meshes. Here, a simpler error norm is defined similar to that presented earlier for the heat diffusion problem, however in this case velocity gradients are used instead of temperature gradients. We can write this norm for a two-dimensional element e as,

$$\|e\|_e = \left(\int_{\Omega_e}\left(\left(\frac{\partial u}{\partial x}-\frac{\partial \hat{u}}{\partial x}\right)^2+\left(\frac{\partial u}{\partial y}-\frac{\partial \hat{u}}{\partial y}\right)^2+\left(\frac{\partial v}{\partial x}-\frac{\partial \hat{v}}{\partial x}\right)^2+\left(\frac{\partial v}{\partial y}-\frac{\partial \hat{v}}{\partial y}\right)^2\right)dxdy\right)^{\frac{1}{2}} \tag{27}$$

Where, u and v are the velocity components in the x and y directions respectively. The exact values of the gradients must of course be approximated by smoothed values. Pressure gradients are not included in this particular norm because pressure is an intrinsic variable in equation (21) and must propogate with infinite speed in order to satisfy the incompressibility condition.

The adaptive procedure based on this norm can also be implemented in exactly the same way as discussed earlier for the heat diffusion problem. The quantity $\|q\|$ in this context is written as,

$$\|q\|^2 \simeq \|\hat{q}\|^2 = \sum_{e=1}^{m}\int_{\Omega_e}\left(\left(\frac{\partial \hat{u}}{\partial x}\right)^2+\left(\frac{\partial \hat{u}}{\partial y}\right)^2+\left(\frac{\partial \hat{v}}{\partial x}\right)^2+\left(\frac{\partial \hat{v}}{\partial y}\right)^2\right)dxdy \tag{28}$$

The rest of the procedure follows exactly the same pattern as before.

3.2 Steady flow example

The problem of steady flow past a circular cylinder [7] was solved using the adaptive procedure discussed above. The problem is symmetric and so only half the domain was

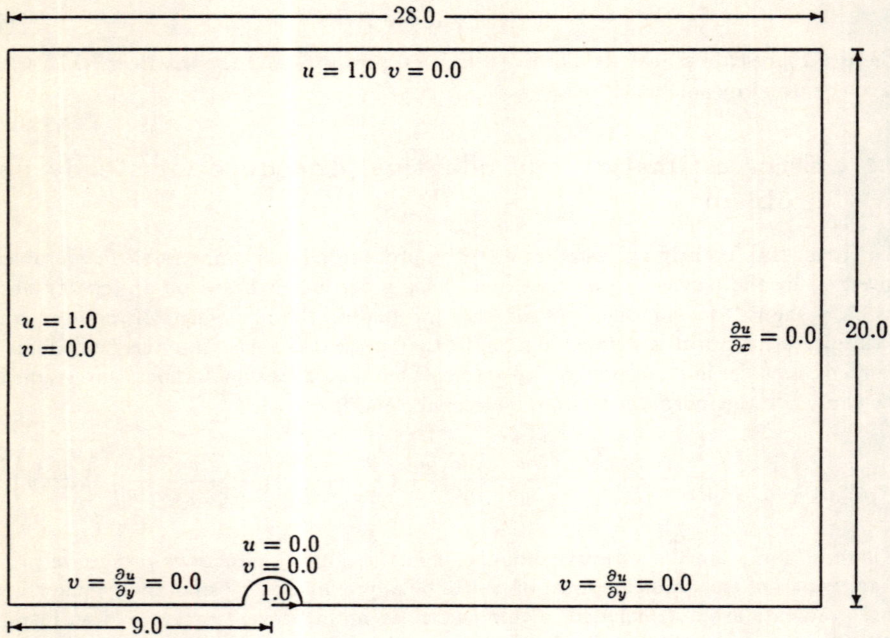

Figure 5: Problem definition for the steady flow problem.

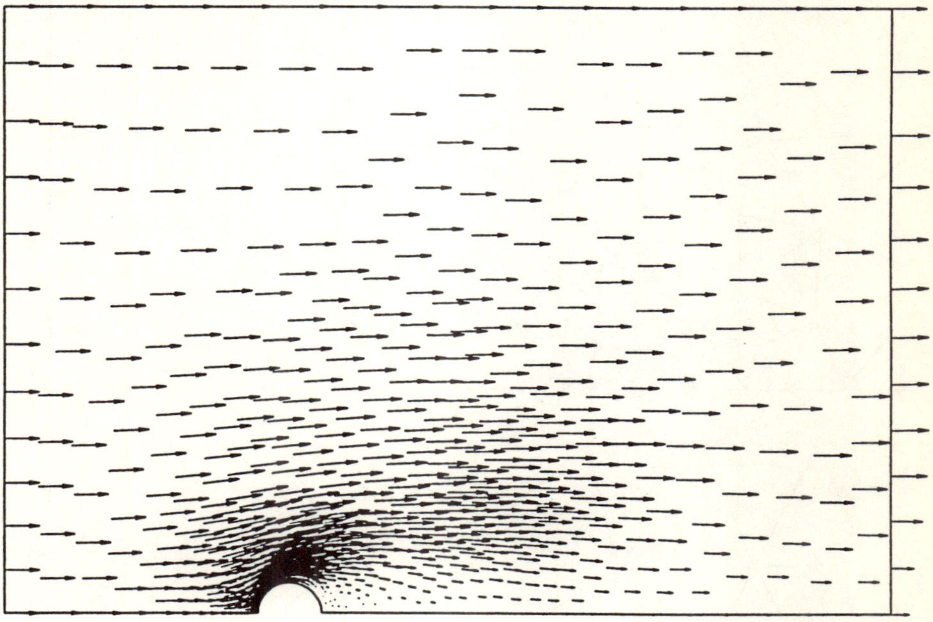

Figure 6: The velocity vectors for steady flow over a cylinder.

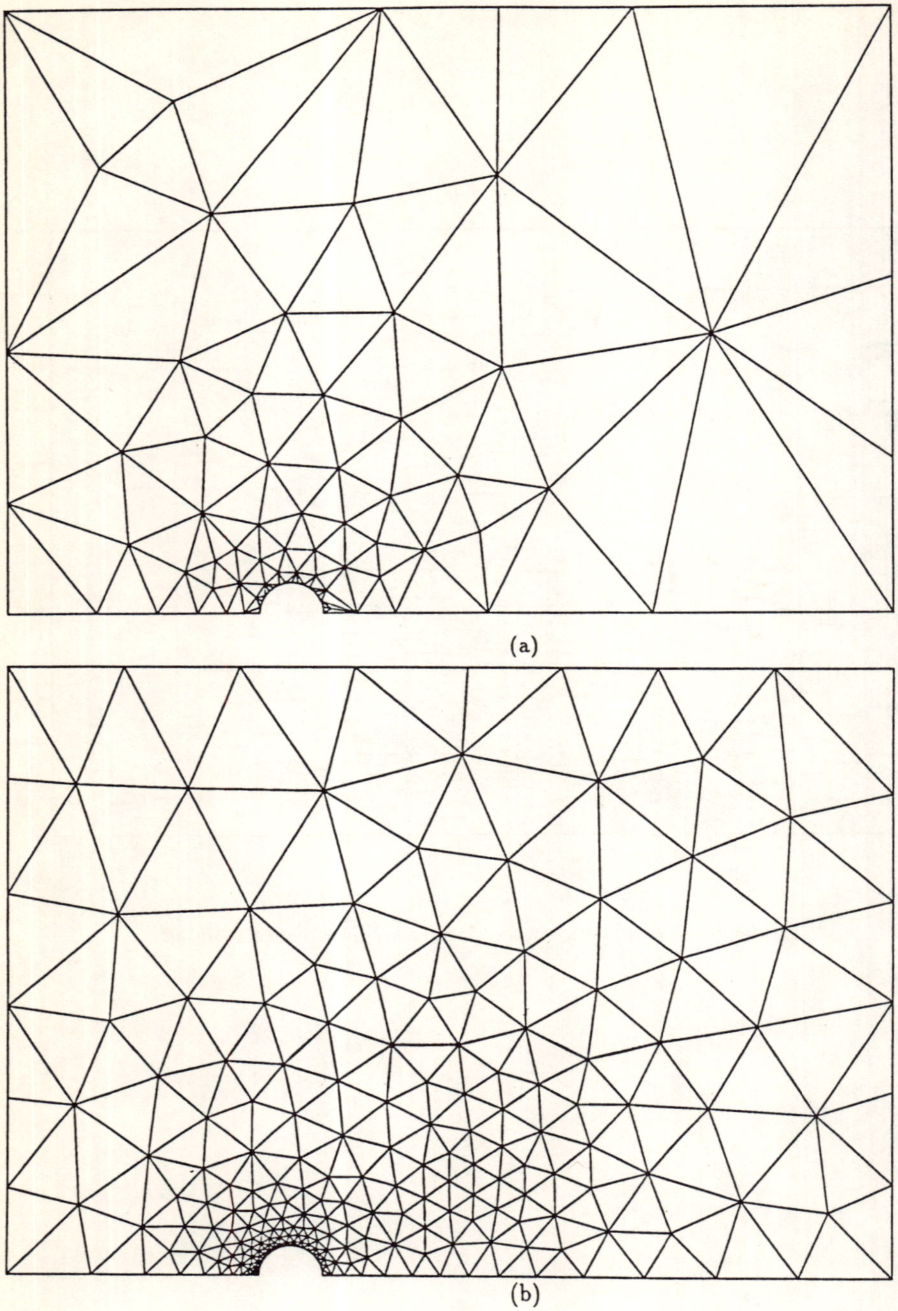

(a)

(b)

Figure 7: The meshes used for steady flow over a cylinder problem.

analysed, as shown in Figure 5 along with the appropriate boundary conditions and dimensions, for a Reynolds Number of 20. The target norm error was fixed at 6.0%, the solution as shown in Figure 6 was arrived at after only one remeshing with an error of 4.0%. The two meshes used are shown in Figure 7, where the mesh of Figure 7(a) was used as the initial mesh. The first refinement, through the adaptive procedure, resulted in the mesh shown in Figure 7(b) from which it may be noted that the highest element density (refinement) occurs, as expected, in the region near the cylinder where the highest velocity gradients exist, while the mesh is largely uniform over the rest of the domain.

3.3 Temporal Discretization for time dependant flow

It is advantageous in most transient problems to adapt the timestep size to the temporal gradients of solution in order to reduce running costs. This may be done by simply reducing, or increasing, the timestep size depending upon the number of iterations required for convergence in the previous step. A better technique was presented by Gresho et. al. [20] based on the comparisons of local time truncation error between a predictor i.e., AB (Adams- Bashforth) formula and a corrector TR (Trapezoidal rule). This technique is briefly described in the following lines. The AB formula gives for an equation, $y' = f(y,t)$, (where y' represents the first derivative of y with respect to time, and so on).

$$y^p_{n+1} = y_n + \frac{\Delta t_n}{2}\left[\left(2 + \frac{\Delta t_n}{\Delta t_{n-1}}\right) y'_n - \left(\frac{\Delta t_n}{\Delta t_{n-1}}\right) y'_{n-1}\right] \tag{29}$$

where p stands for predictor and n is the time level. Subtracting the Taylors series expansion of y_{n+1} from (29) we get the time truncation error for the AB formula, as,

$$y^p_{n+1} - y(t_{n+1}) = -\frac{1}{12}\left(2 + 3\frac{\Delta t_{n-1}}{\Delta t_n}\right) \Delta t_n^3 y'''_n + O(\Delta t_n)^4 \tag{30}$$

We have from the trapezoidal rule,

$$\begin{aligned} y_{n+1} &= y_n + \frac{\Delta t_n}{2}\left(f_{n+1} + f_n\right) \\ &= y_n + \frac{\Delta t_n}{2}\left(y'_{n+1} + y'_n\right) \end{aligned} \tag{31}$$

Subtracting the Taylors series expansion from (31) we get the time truncation error for the TR, as,

$$y_{n+1} - y(t_{n+1}) = \frac{\Delta t_n^3}{12} y'''_n + O(\Delta t_n)^4 \tag{32}$$

Subtracting (30) from (32) and eliminating the unknowns i.e., y'''_n and $y(t_{n+1})$, we obtain,

$$y_{n+1} - y^p_{n+1} = 3\left(\frac{\Delta t_n^3}{12} y'''_n\right)\left(1 + \frac{\Delta t_{n-1}}{\Delta t_n}\right) + O(\Delta t_n)^4 \tag{33}$$

We now define the diffrence in equation (32) as $d(y_{n+1})$ and substitute its value from (32) in (33), to obtain,

$$d(y_{n+1}) = \frac{y_{n+1} - y^p_{n+1}}{3\left(1 + \frac{\Delta t_{n-1}}{\Delta t_n}\right)} + O(\Delta t_n)^4 \tag{34}$$

Writing an expression for $d(y_{n+2})$ from (32), and comparing it with $d(y_{n+1})$, and specifying an acceptable error ϵ for $\mid d(y_{n+2}) \mid$ leads to the formula for the next timestep Δt_{n+1}, as,

$$\Delta t_{n+1} = \Delta t_n \left(\frac{\epsilon}{\mid d(y_{n+1}) \mid}\right)^{\frac{1}{3}} \tag{35}$$

The norm suggested by Gresho et. al. [20] is,

$$\mid d(y_{n+1}) \mid = \sqrt{\frac{1}{N}\left[\frac{1}{\theta^2_{max}} \sum_{i=1}^{N} d_i^2(\boldsymbol{\theta}_{n+1})\right]} \tag{36}$$

Where N is the total number of nodal variables $\boldsymbol{\theta}$ and neither includes the pressure term **P**.

Bixler [15] suggests some improvements in the scheme as described above. He suggests the use of the mid point rule instead of the trapezoidal rule as corrector while keeping the AB formula as predictor. The midpoint rule for the equation $y' = f(y,t)$ can be written as,

$$\begin{aligned} y_{n+1} &= y_n + \Delta t_n \left(f_{n+\frac{1}{2}}\right) \\ &= y_n + \Delta t_n \left(y'_{n+\frac{1}{2}}\right) \end{aligned} \tag{37}$$

Bixler calculates the time truncation error for the midpoint rule by subtracting the Taylor series expansion from (37), as,

$$y_{n+1} - y(t_{n+1}) = -\frac{\Delta t_n^3}{24}\left(y'''_n - \frac{3\left(y''_n\right)^2}{y'_n}\right) + O(\Delta t_n)^4 \tag{38}$$

In calculating the time derivative y' for use in the predictor, Bixler [15] suggests a formula based on the values of y at different time levels, to improve the stability of the predictor, i.e.,

$$y'_n = \frac{2}{\Delta t_{n-1}}(y_n - y_{n-1}) - y'_{n-1} \tag{39}$$

where,

$$y'_{n-1} = \frac{\Delta t_{n-2}}{\Delta t_{n-1} - \Delta t_{n-2}}\left(\frac{y_n - y_{n-1}}{\Delta t_{n-1}}\right) + \frac{\Delta t_{n-1}}{\Delta t_{n-1} - \Delta t_{n-2}}\left(\frac{y_{n-1} - y_{n-2}}{\Delta t_{n-2}}\right) \tag{40}$$

The formula for the succeeding timestep remains the same except, $d(y_{n+1})$ is redefined for the new corrector, i.e.,

$$d(y_{n+1}) = \frac{\gamma}{2 + \gamma + 3\frac{\Delta t_{n-1}}{\Delta t_n}}(y_{n+1} - y^p_{n+1}) \tag{41}$$

where, $0.25 \leq \gamma \leq 1.0$. with $\gamma = 1$, the original formula is obtained. This scheme helps to increase the timestep much more rapidly when steady state impends.

For both the above time integration methods the AB formula is used as predictor which requires the time derivative, $\dot{\boldsymbol{\theta}}_o$ at the first step, which can be obtained by solving (26) as,

$$\mathbf{M}_o \dot{\boldsymbol{\theta}}_o = \mathbf{F}_o - \mathbf{K}_o \boldsymbol{\theta}_o \tag{42}$$

The generalized mid-point family of methods [24, page 145] is written as,

$$\mathbf{M}(\boldsymbol{\theta}_{n+\alpha}, t_{n+\alpha})\dot{\boldsymbol{\theta}}_{n+\alpha} + \mathbf{K}(\boldsymbol{\theta}_{n+\alpha}, t_{n+\alpha})\boldsymbol{\theta}_{n+\alpha} = \mathbf{F}(\boldsymbol{\theta}_{n+\alpha}, t_{n+\alpha}) \tag{43}$$

where,

$$\begin{aligned} \boldsymbol{\theta}_{n+\alpha} &= (1-\alpha)\boldsymbol{\theta}_n + \alpha\boldsymbol{\theta}_{n+1} \\ \dot{\boldsymbol{\theta}}_{n+\alpha} &= \frac{\boldsymbol{\theta}_{n+1} - \boldsymbol{\theta}_n}{\Delta t} \\ t_{n+\alpha} &= t_n + \alpha\Delta t \end{aligned} \tag{44}$$

Substituting (44) into (43) we obtain,

$$[\mathbf{M}_{n+\alpha} + \alpha\Delta t\mathbf{K}_{n+\alpha}](\boldsymbol{\theta}_{n+1}) = [\mathbf{M}_{n+\alpha} - (1-\alpha)\Delta t\mathbf{K}_{n+\alpha}](\boldsymbol{\theta}_n) + (\Delta t\mathbf{F}_{n+\alpha}) \tag{45}$$

No calculation of derivatives is necessary for this method. By changing the values of α from 0 to 1, different members of this family of methods are identified, i.e.,

$\alpha = 0$ -Forward Difference or Forward Euler.
$\alpha = \frac{1}{2}$ -Midpoint rule or Crank Nicolson.
$\alpha = \frac{2}{3}$ -Galerkin.
$\alpha = 1$ -Backward Difference or Backward Euler.

All, except the first (forward Euler), of the above schemes are implicit.

In solving the transient flow problem of section 3.5 the generalized midpoint rule was used with adaptive timestepping. Following [20], only one Newton Raphson iteration was used for each timestep after which the next timestep size was predicted as described earlier.

3.4 Adaptive strategy for transient problems

For a transient problem the adaptive meshing procedure needs to be varied slightly. The approach preferred is to check the error after a fixed number of timesteps and modify the mesh accordingly. The additional work involved here is the need to interpolate the nodal information from the previous mesh to the new mesh in order to continue the time integration process.

3.5 Transient flow example

The problem of flow around a circular cylinder is again chosen to test the adaptive procedure for transient analysis. The main advantage of using this problem in testing a transient Navier-Stokes equation solving code is due to the fact that beyond a certain value of Re the two symmetrically rotating eddies on either side of the cylinder become unstable for flows of lower Re ($Re \leq 40$) and begin to elongate and oscillate laterally. For higher Re ($Re \geq 70$) the eddies start to break off alternately and are transported downstream. This phenomenon is known as 'Karman vortex street' and greater detail is given by Slichting [7]. The numerical modelling of this phenomenon offers an excellent opportunity for verifying the method and the code. Many solutions of such modelling using finite elements have been published, for instance [20, 14, 3]. The first mentioned reference of Gresho *et. al.* provides a first and comprehensive solution to this problem in a finite element context. The transient example solved here has mainly followed this reference for the problem details etc. The problem definition for the transient example is given in Figure 8. The inflow condition is a variable normal traction obtained from the pressure values of the steady state analysis done for $Re = 100$ according to Gresho's procedure, and the outflow condition is of zero traction. The fluid is assumed to start from rest for the initial conditions, except for the non-zero velocities specified at the side walls. The initial mesh of six noded triangular elements is shown in Figure 9.

The transient analysis was continued until a steady cycle of vortex shedding was reached. The results from three consecutive timesteps which show the shedding of vortices are shown in Figure 10, as velocity vectors, and in Figure 11, as pressure contours. Figures 12(a), 12(b) and 12(c) show a succession of three adaptively generated meshes after the commencement of the vortex shedding for the times of 187.812, 188.410 and 188.703 respectively, using a norm error of 20%. The refinement in all the meshes can be seen to simulate the pattern of the vortices due to the existence of high and low velocity gradient zones in the vortices. The same kind of refinement appears in the analysis of the same problem by Wu *et.al.* [11].

Lastly, Figure 13 shows the timestep size variation for the analysis. The timestep size increases in the beginning as the flow develops, reaches a peak and then suddenly begins to drop as instabilities begin to take effect on the symmetric recirculating eddies causing them to oscillate. The timestep size goes on decreasing as the oscillations lead to the shedding of vortices. The decrease in timestep size levels off at the end as a pseudo-steady state of a steady cycle of vortex shedding impends. The whole analysis was completed in 95 timesteps.

4 Conclusions

The adaptive procedures presented herein are shown to give satisfactory results for nonlinear transient heat diffusion [23] and, from the examples solved in this paper, for steady

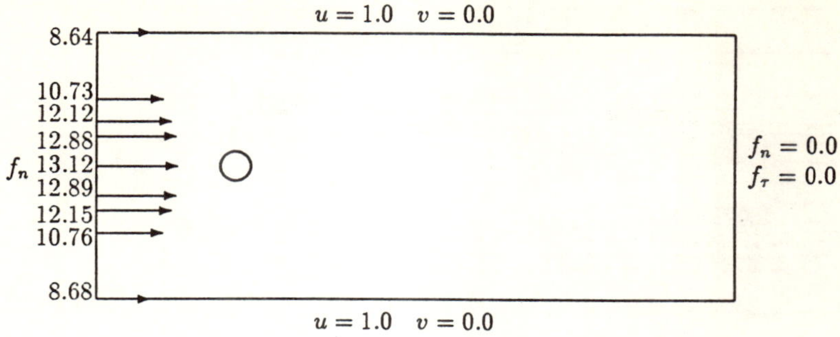

Figure 8: Problem definition for transient flow around a cylinder.

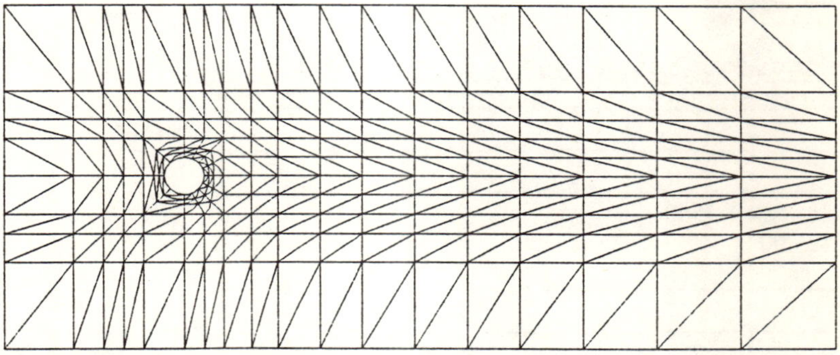

No. of Nodes: 850 No. of Elements: 392
No. of Nodes/Element: 6

Figure 9: The initial mesh of 6-noded triangular elements.

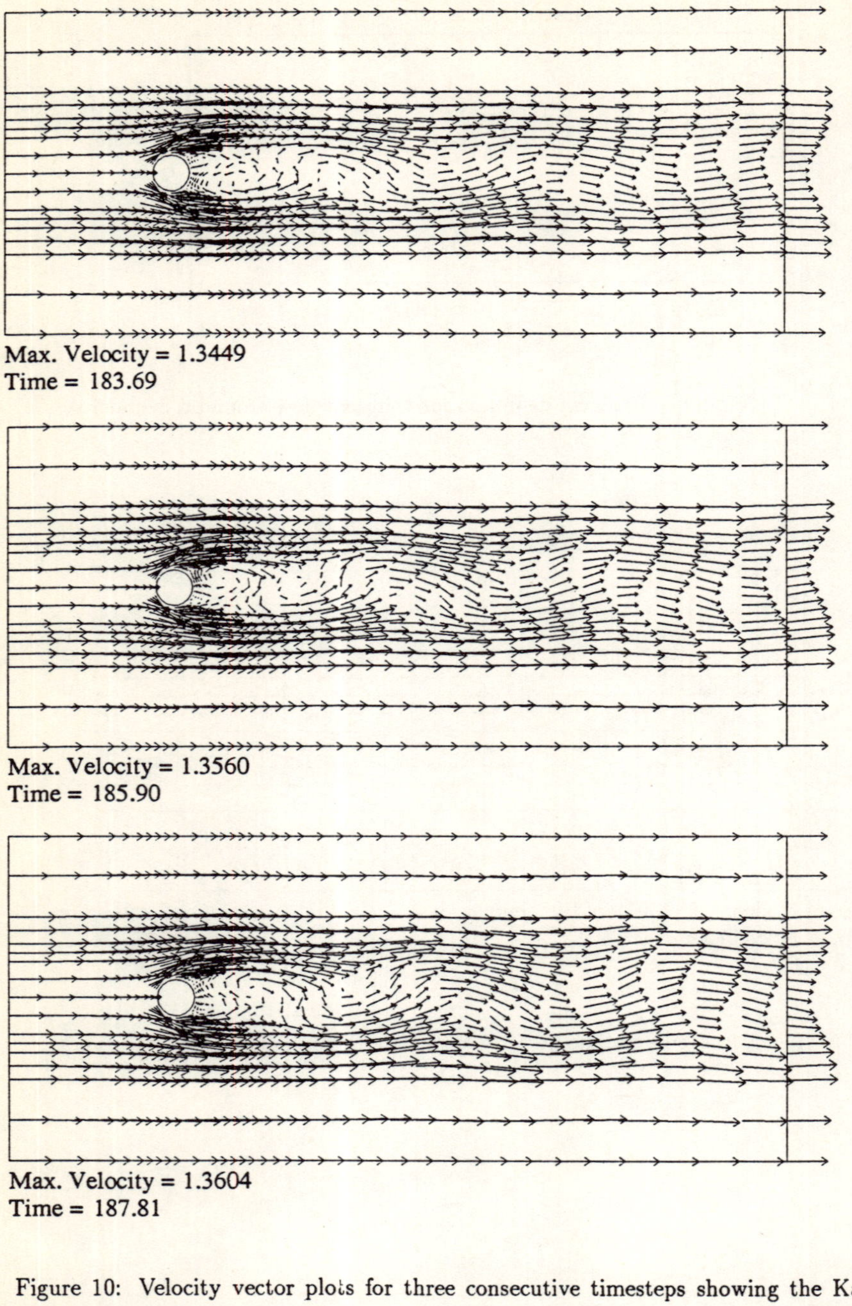

Figure 10: Velocity vector plots for three consecutive timesteps showing the Karman vortex street.

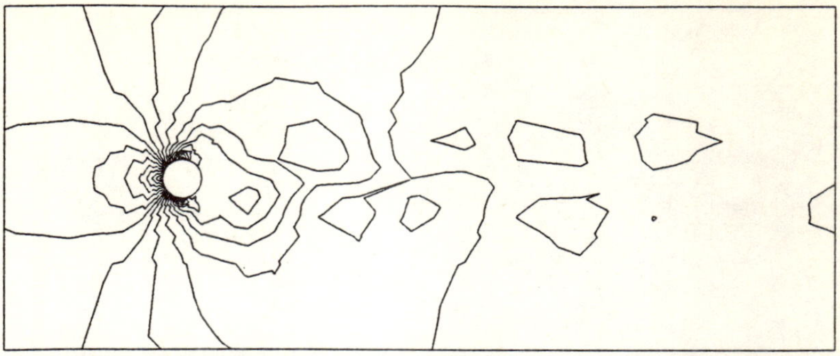

Min. value = -118.8886 Interval = 11.0334
Max. value = 112.8125 Time = 183.69

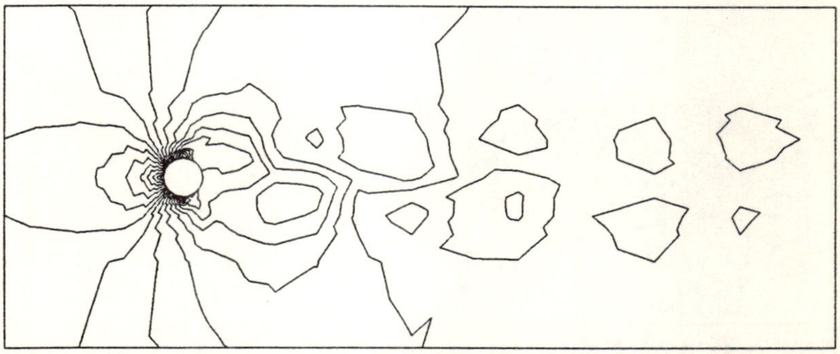

Min. value = -116.5805 Interval = 10.9312
Max. value = 112.9749 Time = 185.90

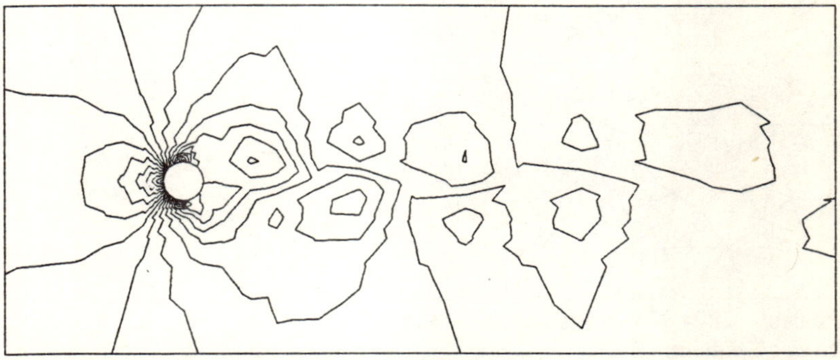

Min. value = -129.9460 Interval = 11.5666
Max. value = 112.9524 Time = 187.81

Figure 11: Pressure plots for the same three consecutive timesteps.

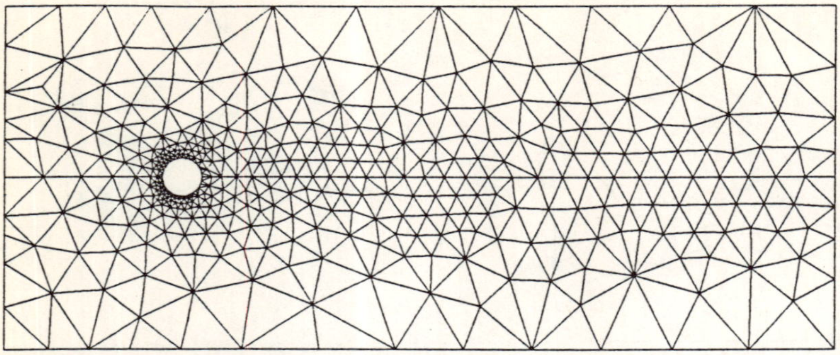

No. of Nodes: 1625 No. of Elements: 773
No. of Nodes/Element: 6

(a)

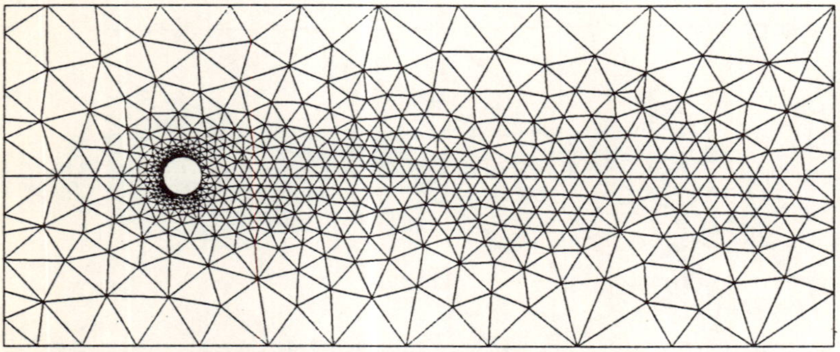

No. of Nodes: 2479 No. of Elements: 1189
No. of Nodes/Element: 6

(b)

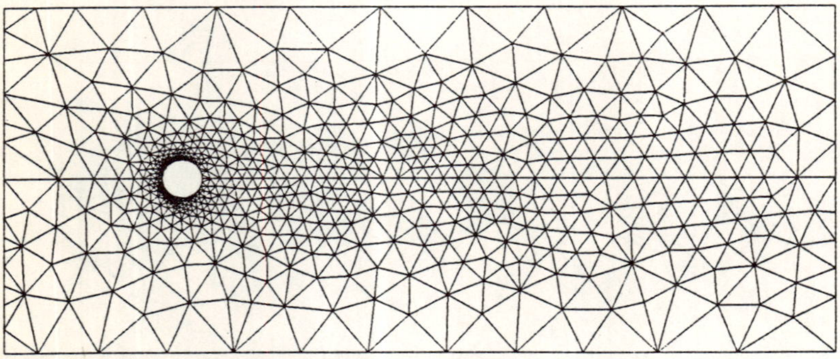

No. of Nodes: 2824 No. of Elements: 1356
No. of Nodes/Element: 6

(b)

Figure 12: The adaptively generated meshes of 6-noded triangular elements after the shedding vortices began.

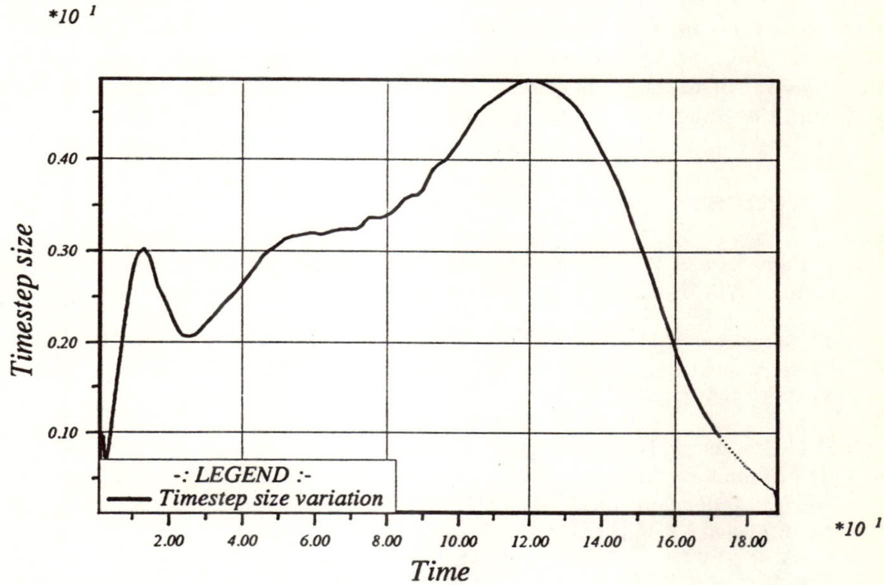

Figure 13: Curve showing the timestep size variation for the midpoint-rule based adaptive timestepping scheme versus the time of analysis.

and transient incompressible flow problems. While the error norm used for the heat diffusion problem has a theoretical basis and can provide an *estimate* of the error, the error norm used for the incompressible flow problem is less well founded and can only provide an *indication* of the magnitude of the error and where it occurs. A rigorous theoretical analysis will be required to prove that the adaptively refined meshes do indeed converge towards an exact solution before this or any other norm is selected for incompressible flow problems. However, the results from this velocity gradient based error indicator appear to be equivalent to the results from the energy norm based error estimator of Wu *et.al.* [11], with the added bonus of being simpler and more economical.

References

[1] I. Babuska and W. Rheinboldt. Error estimates for adaptive finite element computations. *SIAM Journal of Numerical Analysis*, 15, 1978.

[2] I. Babuska and W. Rheinboldt. Adaptive approaches and reliability estimates in finite element analysis. *Computer Methods in Applied Mechanics and Engineering*, 17/18:519–514, 1979.

[3] D.M.Hawken, P.Townsend, and M.F.Webster. A finite element simulation of viscous flow around a cylinder. In *Proceedings of the Third International Conference on Numerical Methods in Engineering: Theory and Applications*, Swansea, U.K., January 1990. Elsevier Applied Science.

[4] E.Hinton and J.S.Campbell. Local and global smoothing of discontinuous finite element functions using a least squares method. *International Journal for Numerical Methods in Engineering*, 8:461–480, 1974.

[5] J. D. S. Gago, D. Kelly, O. Zienkienwicz, and I. Babuska. A posteriori error analysis and adaptive processes in the finite element method: part 2 — adaptive mesh refinement. *International Journal for Numerical Methods in Engineering*, 19:1621–1656, 1983.

[6] H.C.Huang and R.W.Lewis. Adaptive analysis for heat flow problems using error estimation techniques. In *Sixth International Conference for Numerical Methods in Thermal Problems*, Swansea, U.K., July 1989. Pineridge Press, Swansea.

[7] H.Schlichting. *Boundary-Layer Theory.* McGraw-Hill Book Company, 1968.

[8] J.Barlow and G.A.O.Davies. Selected FE benchmarks in structural and thermal analysis. Technical Report FEBSTA REV 1, NAFEMS, 1986.

[9] B. Johnston, J. Sullivan, and A. Kwasnik. Automatic conversion of triangular finite element meshes to quadrilateral elements. *International Journal for Numerical Methods in Engineering*, 31:67–84, 1991.

[10] J.T.Cross. A two-dimensional triangular mesh generator using the advancing front method. Technical Report CR/662/91, Department of Civil Engineering, University College Swansea, 1990.

[11] J.Wu, J.Z.Zhu, J.Szmelter, and O.C.Zienkiewicz. Error estimation and adaptivity in Navier-Stokes incompressible flows. *Computational Mechanics*, 6:259–270, 1990.

[12] J.Z.Zhu and O.C.Zienkiewicz. A new approach to the development of automatic quadrilateral mesh generation. *International Journal for Numerical Methods in Engineering*, 1991. In this issue.

[13] D. Kelly, J. D. S. Gago, O. Zienkienwicz, and I. Babuska. A posteriori error analysis and adaptive processes in the finite element method: part 1 — error analysis. *International Journal for Numerical Methods in Engineering*, 19:1596–1619, 1983.

[14] M.Kawahara, H.Hirano, and T.Kodama. Two-step explicit finite element method for high Reynolds number flow passed through oscillating body. In R.H.Gallagher, J.T.Oden, O.C.Zienkiewicz, T.Kawai, and M.Kawahara, editors, *Finite Elements in Fluids - Volume 5*. John Wiley and Sons, 1984.

[15] N.E.Bixler. An improved time integrator for finite element analysis. *Commnunications in Applied Numerical Methods*, 5:69–78, 1989.

[16] O.C.Zienkiewicz. *The Finite Element Method.* McGraw-Hill Book Company (UK) Limited, London, 1977.

[17] O.C.Zienkiewicz and J.Z.Zhu. A simple error estimator and adaptive procedure for practical engineering analysis. *International Journal for Numerical Methods in Engineering*, 24:337–357, 1987.

[18] O.C.Zienkiewicz, Y.C.Liu, and G.C.Huang. Error estimation and adaptivity in flow formulation for forming problems. *International Journal for Numerical Methods in Engineering*, 25:23–42, 1988.

[19] J. Peraire, M. Vahdati, K. Morgan, and O. Zienkiewicz. Adaptive remeshing for compressible flow computations. *Journal of Computational Physics*, 72:449–466, 1987.

[20] P.M.Gresho, R.L.Lee, and R.L.Sani. On the time-dependent solution of the incompressible Navier-Stokes equations in two and three dimensions. In *Recent Advances in Numerical Methods in Fluids*, volume 1. Pineridge Press Limited, Swansea, 1980.

[21] R.W.Lewis and H.C.Huang. A finite element analysis of fire resisting cabinets using an adaptive remeshing technique. *Applied Mathematical Modelling*, (to be published in 1991).

[22] R.W.Lewis, H.C.Huang, and A.S.Usmani. Remeshing approaches to finite element simulation of casting problems. In *Eurotherm Seminar No.6, Heat Transfer in Phase Change Problems*, Delft, The Netherlands, October 1988.

[23] R.W.Lewis, H.C.Huang, A.S.Usmani, and J.T.Cross. Finite element analysis of heat transfer and flow problems using adaptive remeshing including application to solidification. *International Journal for Numerical Methods in Engineering*, 1991. To be published in a special issue on adaptivity and meshing.

[24] T.J.R.Hughes. Analysis of transient algorithms with particular reference to stability behaviour. In *Computaional methods for transient analysis*. Elsevier Science Publishers, 1983.

Strategies in Shape Optimization of Free Form Shells

E. Ramm, K.-U. Bletzinger and S. Kimmich
Institut für Baustatik
Universität Stuttgart

0 Summary

Shape optimization of general shells requires the synthesis of structural analysis (FEM), mathematical programming schemes, behavior sensitivity analyses and computer aided geometrical design (CAGD) concepts. Plane stress, plate bending problems or axisymmetric shells may be looked upon as special cases. The paper focuses on two aspects: mathematical solution methods with special regard to an extension of the method of moving asymptotes and geometrical parametrization of arbitrary surfaces in space. The design element concept is introduced in which a few characteristic variables, driven by the optimizer, are the controlling parameters. The underlying program concept, the used shell element, and the related discrete sensitivity analysis (DSA) are only briefly described. Several examples are given applying different objectives like weight or strain energy.

1 Introduction

1.1 General Remarks

Typical design objectives for structures in general and shells in particular are high stiffness, a low almost homogeneous stress state, no stress concentrations, little bending, minimum material, a certain frequency range etc. This is the typical area of application for shape optimization in which geometrical design aspects are combined with structural analysis and mathematical programming schemes [37]. The efficiency directly depends on the number of independent design variables. Therefore it became clear that in shape optimization the choice of all nodal geometrical parameters of the finite element mesh as variables is prohibitive. Traditionally, the engineer has always tried to concentrate on a few key parameters to get a feasible, economic, and aesthetically pleasing structure. It is therefore obvious that the choice of these "natural" design variables becomes a key factor for a realistic problem definition. The design element concept introduced in this paper shows that a few geometrical parameters offer an almost infinite number of different shapes for shell surfaces in space.

The different levels of structural optimization have an increasing degree of complexity: optimization of material properties, cross sections (sizing), structural shape and finally topology. Up to now relatively little has been achieved in topology optimization, see for example the micro hole approach [3]. The bulk of investigations is still devoted towards sizing problems. In shape optimization [4,13,20] most work addresses two dimensional systems, either truss and frame structures [25] or plane stress [11] and axisymmetric problems. Based on a second order plastic hinge theory two- and three-dimensional frame structures are investigated by Stein and co-workers [2,6,38] with respect to weight and other objectives. Shell problems are usually also restricted to the axisymmetric case [31]; for folded plates see [10]. Relatively little has been described for general shells up to now [28].

1.2 Scope of Study

The present study addresses the shape optimization of free form shell structures including thickness variations. Several objectives like weight, strain energy (stiffness), and stress levelling (strength) are introduced. Although the task involves several disciplines with almost equal importance, like structural analysis, sensitivity analyses, optimization strategies, geometrical design concepts, the present paper focuses on the last two subjects; the other areas are only briefly described. The underlying computer program – the in house research tool CARAT [26] - includes several optimization schemes; for example an SQP algorithm [35] and the method of the moving asymptotes [39]. This method has been supplemented by several modifications [7]. The second main part is devoted to the design element concept. Different geometrical parametrization schemes of arbitrary surfaces in the three-dimensional space (Bézier, B-spline, Coons etc.) supplemented by continuity constraints and linking rules are discussed.

Although the computer program CARAT is especially designed for nonlinear structural analysis the present study is restricted to linear mechanical response.

2 Concept

As already mentioned a major task in structural optimization is the integration of several individual disciplines which in themselves are to a large extend developed. The interaction reflects the typical loop of each design process (fig. 1).

Supposedly that a few characteristic variables have been selected as optimization variables. These may be either typical design variables like certain selected geometrical parameters defining the overall shape of the structure or structural parameters like stresses or displacements. They, in turn, are subsets of the entire group of either design or structural parameters; some of them may already be automatically generated from previous variables. The geometrical part of the problem description defines the design model and is primarily based on CAGD concepts. The mechanical part is as usual integrated in the analysis model processed for example by the finite element method and supplying the structural response. At this point it is important to realize that now the standard procedure has to be supplemented by an efficient sensitivity analysis. Although independent from the optimization model, it supplies important gradient information for the optimizer. Structural data are transformed satisfying the requirements for optimization, i.e. objective, constraints and the related derivatives are calculated. This is the starting point for the mathematical optimizer which finally delivers a new proposal for the structure.

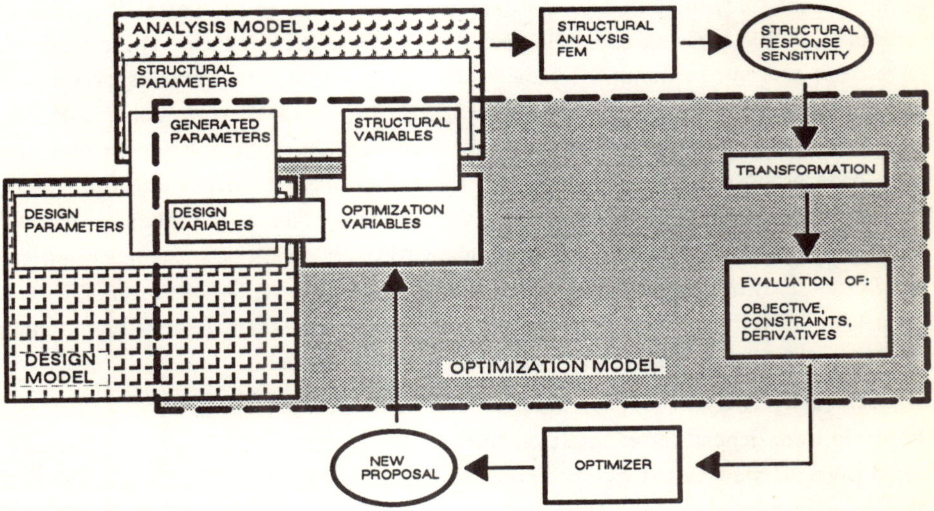

Fig. 1. Interaction of optimization, design and analysis models

Figure 1 almost suggests that optimization requires simply to join traditional independent programs together like a toy kit. Although this fits into the common interest of a usual engineering environment where CAD modules, FEM programs and mathematical programming libraries are available, it is not as easy as it looks like. First of all, there are quite a few rather complex dependencies and linking rules which have to be created and maintained during the optimization progress to control parameter variations and to guide the transformation of structural response. Secondly, the program and data flow between independent modules, having probably their own data bases, needs special interfaces controlled by the operating system. This can lead to an inefficient and inflexible optimization model. A rather opposite approach is to develop different modules which are coordinated from the root of development and share one common incore data base. This integrated approach is realized in the program CARAT (Computer Aided Research and Analysis Tool) [26] used herein. It allows a strictly modular, flexible and expandable program structure with efficient data exchanges during the entire process of optimization and includes all the necessary software from preprocessing via structural analysis to post-processing. Of course, existing outside codes (e.g. PATRAN) can be linked also here in the conventional way.

3 Structural and Sensitivity Analysis

In the present study the well-known isoparametric degenerated solid elements are used for shell analysis applying an explicit integration across the thickness. These elements have been proven to contain the same mechanical content than elements based on a corresponding shear flexible shell theory [12]. The different locking phenomena of fully integrated displacement based elements are very well understood, a variety of remedies have been proposed to avoid locking and to improve their efficiency and robustness, among these are the reduced integration, partially with stabilization schemes to avoid zero-energy modes, or the assumed and enhanced strain concepts; for an overview see [1]. In this study the former scheme is applied, in particular the 8-node serendipity element with 2 x 2 integration is used. Since structural optimization requires high quality stresses smoothing procedures for stress recovery are essential.

The sensitivity analysis supplies gradient information on objective(s) and constraints which, in turn, depend for example on stresses and displacements. From the computational point of view this is a very important aspect since it may require a great deal of the entire computer time. Here the discrete design sensitivity analysis (DDSA) is applied in which - opposite to the variational design sensitivity analysis (VDSA) [23] - the gradients are obtained after discretization. Program CARAT includes three different versions of DDSA: the numerical, semi-analytical and analytical approach. Opposite to the two other variants the analytical method requires analytical derivatives of stiffness matrix, load vector and the stresses with respect to the design parameters. A selection of static response sensitivity coefficients and some characteristic objective and constraint functions are given in fig. 2 for the isoparametric shell element.

For an extensive discussion and evaluation of the different versions see [27]. In [27] it is also shown that locking is equally transferred to the sensitivity coefficients if it is not à priori avoided.

Finite element operations

Equilibrium $$0 = \mathbf{K}\,\mathbf{u} - \mathbf{R}$$

Sensitivity of displacements $$\mathbf{u}_{,x} = \mathbf{K}^{-1}\left[\mathbf{R}_{,x} - \mathbf{K}_{l,x}\,\mathbf{u}\right]$$

assembly $$\mathbf{K} = \sum_{e} \mathbf{k}_e \quad \mathbf{R}_{,x} = \sum_{e} \mathbf{r}_{e,x} \quad \mathbf{K}_{,x} = \sum_{e} \mathbf{k}_{e,x}$$

Sensitivity of weight objective $$W_{,x} = \sum_{e} w_{e,x}$$

Sensitivity of stress constraint $$g_{\sigma,x} = -\frac{\sigma_{,x}}{\sigma_{all}}$$

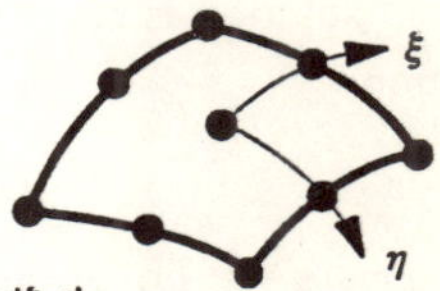

Isoparametric degenerated shell element

Weight objective $$w_e = \varrho \int |\mathbf{J}|\, d\xi\, d\eta \qquad w_{e,x} = \varrho \int |\mathbf{J}_{,x}|\, d\xi\, d\eta$$

Stiffness matrix $$\mathbf{k}_e = \int \mathbf{B}^T \mathbf{C}\, \mathbf{B}\, |\mathbf{J}|\, d\xi\, d\eta$$

$$\mathbf{k}_{e,x} = \int \mathbf{B}_{,x}^T \mathbf{C}\, \mathbf{B}\, |\mathbf{J}|\, d\xi\, d\eta + \int \mathbf{B}^T \mathbf{C}_{,x}\, \mathbf{B}\, |\mathbf{J}|\, d\xi\, d\eta$$
$$+ \int \mathbf{B}^T \mathbf{C}\, \mathbf{B}_{,x}\, |\mathbf{J}|\, d\xi\, d\eta + \int \mathbf{B}^T \mathbf{C}\, \mathbf{B}\, |\mathbf{J}_{,x}|\, d\xi\, d\eta$$

Dead load vector $$\mathbf{r}_e = \varrho \int \mathbf{N}\, |\mathbf{J}|\, d\xi\, d\eta \quad \mathbf{r}_{e,x} = \varrho \int \mathbf{N}\, |\mathbf{J}_{,x}|\, d\xi\, d\eta$$

Fig. 2. Isoparametric shell element, response sensitivity coefficients

4 Optimization Strategies

4.1 Problem Types in Structural Optimization

Typical problems of structural optimization are characterized by few objectives and many constraints which are nonlinear functions of the optimization variables. The functions can be discontinuous, non–convex, and are often only implicitly known. Furthermore detailed formulations of practical optimization problems vary a lot with every application demanding for powerful but flexible solution methods. The methods of mathematical programming (MP) meet all these requirements [21,41]. In contrast to the alternative methods of optimality criteria [5] which are very efficient for the solution of special problems MP methods are generally applicable and fit perfectly to numerical analysis methods like FEM to yield practicable design tools. In the context of structural optimization MP methods have been used successfully since the beginning of the sixties and reached now acceptance in research and industry [14,21]. Without loss of generality most of the problems are stated to be continuous in gradients and curvature of the problem functions and are solved efficiently by the well known methods of nonlinear programming (NLP) [18,29].

4.2 Standard Formulation of Nonlinear Programming

A nonlinear program is stated as the optimization problem:

minimize the objective function f(x) subject to the equality constraints $h_j(x)$ (j = 1,...,m_e) and non-equality constraints $g_j(x)$ (j = m_e+1,...,m).

All functions are nonlinear functions of the optimization variables **x**. They can be chosen in R^n but have to stay within their lower and upper bounds $\underline{\mathbf{x}}$ and $\overline{\mathbf{x}}$ (side constraints). A local solution is characterized as a stationary point of the corresponding Lagrangian function:

$$L(\mathbf{x}, \mathbf{u}, \mathbf{v}) = f(\mathbf{x}) + \mathbf{u}^T \mathbf{h}(\mathbf{x}) + \mathbf{v}^T \mathbf{g}(\mathbf{x}) \tag{1}$$

u and **v** are the Langrangian multipliers or dual variables. The necessary condition for the stationarity of L or the corresponding constrained minimum of f(x) is defined by the Kuhn–Tucker conditions. They give a set of nonlinear equations to determine the optimal solution $\mathbf{x}^*$, $\mathbf{u}^*$, $\mathbf{v}^*$:

$$\frac{\partial L^*}{\partial \mathbf{x}} = \frac{\partial f^*}{\partial \mathbf{x}} + \left(\frac{\partial \mathbf{h}^*}{\partial \mathbf{x}}\right)^T \mathbf{u}^* + \left(\frac{\partial \mathbf{g}^*}{\partial \mathbf{x}}\right)^T \mathbf{v}^* = 0 \tag{2}$$

$$\frac{\partial L^*}{\partial u_j} = h_j^* = 0 \quad ; j = 1,\ldots, m_e$$

$$\frac{\partial L^*}{\partial v_j} v_j^* = g_j^* v_j^* = 0 \quad ; j = m_e + 1,\ldots, m$$

$$v_j^* \geq 0$$

Where $\mathbf{f}^*$, $\mathbf{h}^*$, $\mathbf{g}^*$ and L^* are the function values at the optimal solution. Constraints which are limiting the optimal solution are called active.

4.3 Methods of Nonlinear Programming

Due to Luenberger [29] NLP methods can be divided into (i) primal methods, (ii) penalty and barrier methods, (iii) dual methods, and (iv) Lagrange methods. They are distinguished by the type and number of independent variables they use.

By now Lagrange methods are the most sophisticated optimization techniques and are applicable for all kinds of constrained problems. They are designed to solve the Kuhn–Tucker conditions (2) directly and are operating in the full space of primal and dual variables. Iterative solution of eqns. (2) by subsequent linearization leads to a natural extension of the classical Newton/Raphson procedure which became known as RQP–(recursive quadratic programming) or SQP–method (sequential quadratic programming). Han [22] and Powell [32] introduced the quasi Newton idea based on the BFGS update. This method underwent further refinements by several researchers. The version by Schittkowski [35] was used in this work. In the k-th iteration step the corresponding quadratic subproblem states as:

$$\begin{aligned} &\text{minimize} && \frac{1}{2}\mathbf{d}^{(k)T}\mathbf{B}^{(k)}\mathbf{d}^{(k)} + \left(\frac{\partial f(\mathbf{x}^{(k)})}{\partial \mathbf{x}}\right)^T \mathbf{d}^{(k)} \\ &\text{subject to} && \frac{\partial \mathbf{h}(\mathbf{x}^{(k)})}{\partial \mathbf{x}}\mathbf{d}^{(k)} + \mathbf{h}(\mathbf{x}^{(k)}) = \mathbf{0} \\ &&& \frac{\partial \mathbf{g}(\mathbf{x}^{(k)})}{\partial \mathbf{x}}\mathbf{d}^{(k)} + \mathbf{g}(\mathbf{x}^{(k)}) \leq \mathbf{0} \\ &&& \left.\begin{aligned} \mathbf{d}^{(k)} &\leq \bar{\mathbf{x}} - \mathbf{x}^{(k)} \\ -\mathbf{d}^{(k)} &\leq \mathbf{x}^{(k)} - \underline{\mathbf{x}} \end{aligned}\right. \quad ; \quad \mathbf{d}^{(k)} = \mathbf{x}^{(k+1)} - \mathbf{x}^{(k)} \end{aligned} \tag{3}$$

$\mathbf{B}^{(k)}$ ... current approximation of Lagrangian second derivatives

Up to now SQP-methods have been used rather infrequently within structural optimization. But with the increasing complexity of problems like shape optimization these methods get more and more accepted [40].

4.4 Approximation Techniques

Direct application of NLP methods may turn out to be troublesome and time consuming because of two facts: (i) the internal approximation scheme of the chosen method (usually linear or quadratic) does not reflect to the physical properties of the problem and causes many iteration steps with a lot of expensive evaluations of functions and gradients, and (ii) most of the algorithms need line searches to ensure global convergence with again many evaluations of functions or even gradients according to the line search method (fig. 3a). Performance can be improved tremendously if in every iteration step objective function and constraints are replaced by proper approximations (fig. 3b). Usually, approximations are derived from first order linearizations with respect to generalized variables $\mathbf{y}$ as functions of $\mathbf{x}$ [36]:

$$\begin{aligned} \tilde{f}(\mathbf{y})^{(k)} &= f(\mathbf{y}^{(k)}) + \left(\frac{\partial f(\mathbf{y}^{(k)})}{\partial \mathbf{y}}\right)^T \left(\mathbf{y}^{(k+1)} - \mathbf{y}^{(k)}\right) \\ \tilde{f}(\mathbf{x})^{(k)} &= f(\mathbf{x}^{(k)}) + \left(\frac{\partial f(\mathbf{x}^{(k)})}{\partial \mathbf{x}}\right)^T \frac{\partial \mathbf{x}^{(k)}}{\partial \mathbf{y}} \left(\mathbf{y}(\mathbf{x}^{(k+1)}) - \mathbf{y}(\mathbf{x}^{(k)})\right) \end{aligned} \tag{4}$$

where $\quad \mathbf{y} = \mathbf{y}(\mathbf{x})$

The resulting approximated sub-problem still remains nonlinear in terms of the original variables $\mathbf{x}$ and has to be solved by suitable NLP methods. Functions and gradients can be evaluated very efficiently since the substituted functions are known explicitly. Because of the linearization technique sub-problems show special properties like separability which can be exploited successfully e.g. by use of dual solution methods [17]. Originally the described approximation technique was designed to generate a series of approximated sub-problems without an additional line search in the original problems to further improve efficiency. This may lead to a lack of the global convergence property of the overall procedure as was indicated by Mahnken and Stein [30].

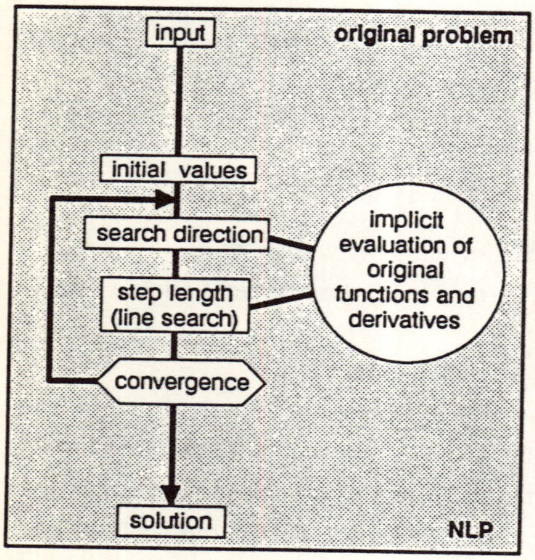

Fig. 3a. Direct use of NLP methods

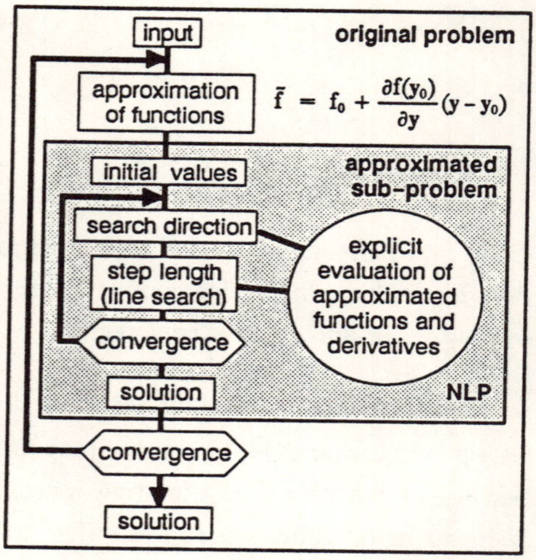

Fig. 3b. NLP to solve appr. sub-problem

An often used approximation scheme which became known as "hybrid approximation" is based on selective linearization of functions with respect to the original variables or their inverse depending on the sign of partial derivatives. It is defined as:

$$\tilde{t}_j(\mathbf{x}) = t_j(\mathbf{x}^{(k)}) + \begin{cases} \sum_{i=1}^{n} \frac{\partial t_j(\mathbf{x}^{(k)})}{\partial x_i} \left(x_i^{(k+1)} - x_i^{(k)} \right) & ; \frac{\partial t_j(\mathbf{x}^{(k)})}{\partial x_i} \geq 0 \\ \sum_{i=1}^{n} - \frac{\partial t_j(\mathbf{x}^{(k)})}{\partial x_i} \left(x_i^{(k)} \right)^2 \left(\frac{1}{x_i^{(k+1)}} - \frac{1}{x_i^{(k)}} \right) & ; \frac{\partial t_j(\mathbf{x}^{(k)})}{\partial x_i} < 0 \end{cases} \tag{5}$$

$$j = 0, \ldots, m$$

$$t_0(\mathbf{x}) = f(\mathbf{x})$$

$$t_j(\mathbf{x}) = h_j(\mathbf{x}) \quad ; \; j = 1, \ldots, m_e$$

$$t_j(\mathbf{x}) = g_j(\mathbf{x}) \quad ; \; j = m_e + 1, \ldots, m$$

This technique is exact in the case of sizing of statically determinate linear trusses. Although there exists no physical relevance this approximation appears to be very efficient for other problems of structural optimization also, including shape optimization. The method was generalized and gained wide-spread recognition as versions called CONLIN (convex linearization [17]) and MMA (method of moving asymptotes [39]). The latter will be discussed more in detail in the next chapters.

4.5 The Method of Moving Asymptotes and some Extensions

1987 Svanberg [39] introduced an approximation technique which he derived from a problem linearization with respect to generalized variables:

$$y_i = \begin{cases} \dfrac{1}{U_i - x_i} & ; \quad \dfrac{\partial t_j(\mathbf{x}^{(k)})}{\partial x_i} \geq 0 \\[2ex] \dfrac{1}{x_i - L_i} & ; \quad \dfrac{\partial t_j(\mathbf{x}^{(k)})}{\partial x_i} < 0 \end{cases} \tag{6}$$

U_i and L_i are upper and lower asymptotes of the variables which are updated in every iteration step to adjust the curvature of the problem (moving asymptotes). This choice of linearization leads to convex and separable subproblems which can be solved efficiently by dual methods since minimization with respect to the primal variables can be done explicitly. After the method was published a lot of applications showed its effectiveness for many structural optimization problems.

4.5.1 Bounded Approximation by Use of Second Order Informations

The proposed linearization scheme (6) may generate unbounded approximated sub-problems in the case of little or even unconstrained original problems as they often appear in shape optimization. In those circumstances the method oscillates and convergence can be ensured only by a clever use of move limits. This drawback is overcome by a modified approximation of the objective function which guarantees that the approximated objective always has a minimum [7]. Now the modified problem approximation states as:

$$\bar{f}(\mathbf{x}) = r_0 + \sum_{i=1}^{n} \left(\frac{p_{0i}}{(U_i - x_i)} + \frac{q_{0i}}{(x_i - L_i)} \right) \tag{3}$$

where the coefficients are:

$$p_{0i} = c_i \left(U_i - x_i^{(k)}\right)^2 \left[\frac{f'^{(k)}_i}{x_i^{(k)} - L_i} + \frac{1}{2} f''^{(k)}_i\right]$$

$$q_{0i} = c_i \left(x_i^{(k)} - L_i\right)^2 \left[-\frac{f'^{(k)}_i}{U_i - x_i^{(k)}} + \frac{1}{2} f''^{(k)}_i\right]$$

$$r_0 = f(\mathbf{x}^{(k)}) - \sum_{i=1}^{n} \left(\frac{p_{0i}}{(U_i - x_i^{(k)})} + \frac{q_{0i}}{(x_i^{(k)} - L_i)} \right)$$

and $$c_i = \frac{(x_i^{(k)} - L_i)\,(U_i - x_i^{(k)})}{(x_i^{(k)} - L_i) + (U_i - x_i^{(k)})}$$

$$f'^{(k)}_i = \frac{\partial f(\mathbf{x}^{(k)})}{\partial x_i} \quad ; \quad f''^{(k)}_i = \max \left\{ \begin{array}{l} \dfrac{\partial^2 f(\mathbf{x}^{(k)})}{\partial x_i^2} \\ -\dfrac{2}{x_i^{(k)} - L_i} f'^{(k)}_i (1 + c_\kappa) \\ \dfrac{2}{U_i - x_i^{(k)}} f'^{(k)}_i (1 + c_\kappa) \end{array} \right\}$$

$\mathbf{x}^{(k)}$ ... point of expansion

U_i, L_i ... asymptotes of variable x_i

c_K ... curvature correction factor ($c_K > 0$) to ensure convexity

This kind of approximation makes additional use of some second order information which is estimated by forward differences according to the BFGS or DFP recursive formulae. It still fits into Svanberg's frame: the subproblem remains convex and separable, and one dimensional minimization with respect to the primal variables can be done explicitly which makes dual solvers efficient.

The following academic but typical problem shows the potential of the proposed extension. Shape and cross sections of the three bar truss (fig. 4) are to be determined. The objective to be minimized is structural weight. Variables are the location x_2, y_2 of the center node and the common cross section A of all bars. The truss is loaded by a vertical force where the tensile stresses are limited to 1.0 kN/cm^2. At the optimum the variables are uncoupled. This means that the original method converges only if the asymptotes are adapted, eventually leading to failure. The adaptation rule by Svanberg [39] with an adaptation factor of s = 0.3 was used. The extended method appears to be substantially more stable as can be seen in fig. 5. This behavior can be improved further by an additional line search [30].

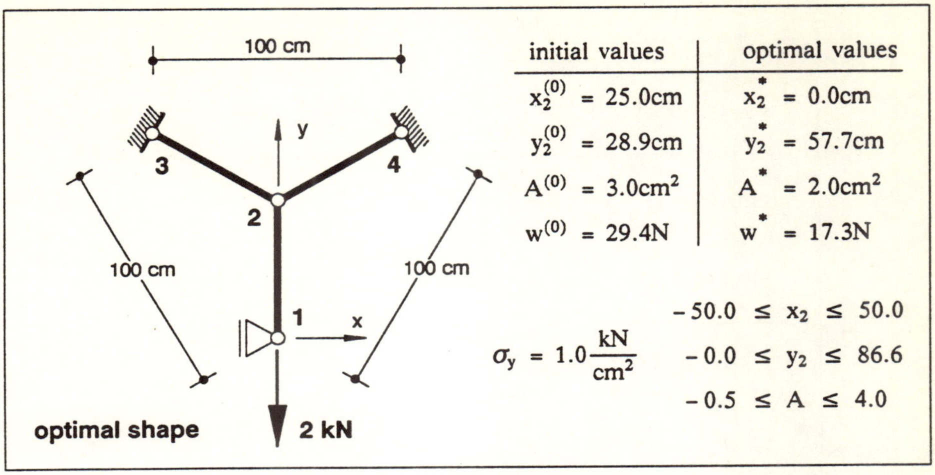

Fig. 4. Another three bar truss problem

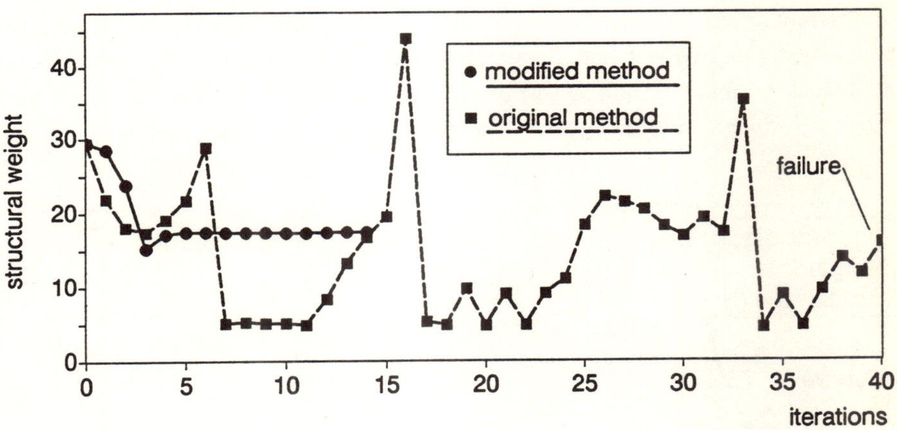

Fig. 5. Three bar truss: iteration histories

4.5.2 General Power Approximation

The ideas of general power approximation [33] are introduced which make the whole family of hyperbolic and parabolic convex and concave approximations (fig. 6) available to the method of moving asymptotes. To get convex sub-problems the problem functions have to be linearized with respect to the following generalized variables [7]:

$$
y_i = \begin{cases}
-\frac{1}{b}\,(U_i - x_i)^b \;;\; \frac{\partial t_j(\mathbf{x}^{(k)})}{\partial x_i} > 0 & ;\; b < 0 \text{ and } 0 < b < 1 \\
-\frac{1}{b}\,(x_i - L_i)^b \;;\; \frac{\partial t_j(\mathbf{x}^{(k)})}{\partial x_i} < 0 & \text{(hyperbola)} \quad \text{(root)} \\
\hline
-\ln\,(U_i - x_i) \;;\; \frac{\partial t_j(\mathbf{x}^{(k)})}{\partial x_i} > 0 & ;\; b = 0 \\
-\ln\,(x_i - L_i) \;;\; \frac{\partial t_j(\mathbf{x}^{(k)})}{\partial x_i} < 0 & \text{(nat. logarithm)} \\
\hline
-\frac{1}{b}\,(x_i - L_i)^b \;;\; \frac{\partial t_j(\mathbf{x}^{(k)})}{\partial x_i} > 0 & ;\; b \geq 0 \\
-\frac{1}{b}\,(U_i - x_i)^b \;;\; \frac{\partial t_j(\mathbf{x}^{(k)})}{\partial x_i} < 0 & \text{(parabola)}
\end{cases} \tag{7}
$$

The exponent b can be chosen with respect to the physical behavior of the problem. Alternatively it can be determined using second order informations. In the latter case or if different approximation schemes for every problem function are used the subproblem is solved by a SQP-method. This appears to be superior to the dual method since one dimensional minimization as part of the dual approach would have now to be done numerically which comes along with a lack of efficiency and numerical stability.

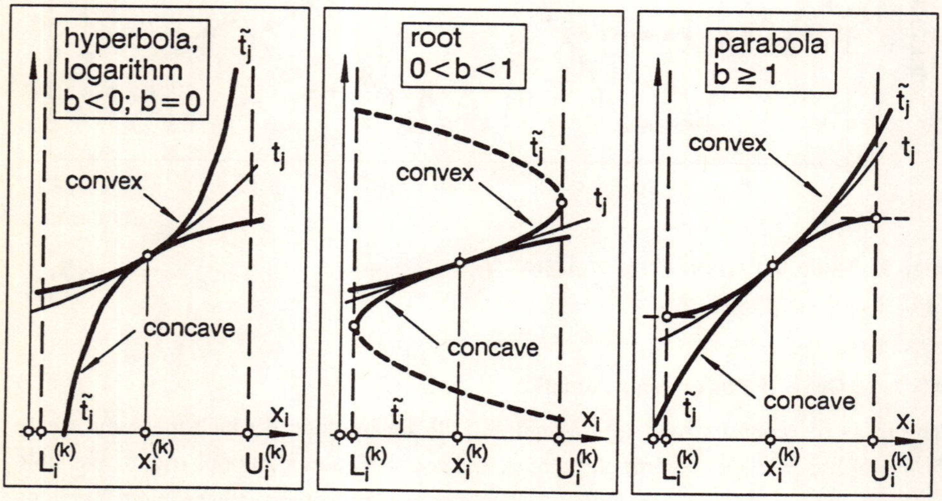

Fig. 6. Convex and concave approximations of problem function

5 Design Model

5.1 Concept

Nowadays it is common practice to use pre–processors to generate finite element data. The idea is to describe the geometrical layout of the structure in terms of as few as possible characteristic or "natural" parameters which are part of the "design model". Besides very specialized and effective procedures for the design of distinct structures the methods of computer aided geometrical design (CAGD) [8,16] have been adopted to yield general and flexible tools for the design of geometries in two and three dimensions. With these methods even complex geometries can be described piecewise by the use of few "design elements" which are orientated in size and shape at the desired geometrical properties of the structures. Within each design element the resulting shape **r** is parametrized in terms of shape functions Φ_1, element parameters u,v,w and design nodes $\mathbf{r}_i$ which describe the elements location in space:

$$\mathbf{r}(u,v,w) = \sum_{i=1}^{n} \Phi_i(u,v,w)\ \mathbf{r}_i \tag{8}$$

This concept has the following main advantages which are attractive for effective data generation as well as in structural optimization when the geometrical entities are used as optimization variables:

- clear distinction of geometry (CAGD) and mechanics (FEM)
- few but characteristic degrees of freedom
- shape approximation as accurate as desired, by
  - refinement of design element mesh
  - different shape functions
- combination of different shape functions and design elements within one model
- little effort of input data preparation, therefore
- comfortable and non–defective generation of smooth FE meshes
- interactive shape modifications by affecting design node coordinates

Imam [24] was one of the first who used the design element concept in shape optimization for the design of three dimensional structures. An early application for the design of plates and shells is known by Botkin [10]. A lot of other contributions followed in which the terminology and specific capabilities of CAGD were used e.g. by Braibant and Fleury [11] and Esping [15].

5.2 Consequences for the Overall Design Process

An essential feature of pre–processors used in interactive or automatic shape optimization procedures is the feed back to FE–analysis results. As a consequence those pre–processors should be able to generate sensitivity data also. This means that FE pre-processing now becomes an inherent part of the overall design process. As already mentioned design modeling and FE pre-processing are perfectly integrated into the CARAT design concept. All modules share the same incore data base. This leads to further improvements of effectiveness and accuracy of the methods [27].

5.3 Element Catalogue and Hierarchy

In CARAT-Design [7] as well as in other CAD systems design elements are ordered hierarchically. Higher levels are defined by the previous lower level:

- 1-D: design *edges* by design nodes
- 2-D: design *patches* by design edges
- 3-D: design *bricks* by design patches

Additional internal nodes are necessary to define 9-, 10-, and 16-node patches. The hierarchical approach ensures that patches at common edges or bricks at common surfaces will be compatible. The entire catalogue of design elements available in CARAT is given in fig 7.

5.4 Bézier- and B-Splines

An obvious characteristic of Bézier- and B-splines is that only the end control nodes are interpolated by the resulting polynomial curve while the inner nodes are approximating the shape (fig. 7). (In the most general formulation of B-splines even the end nodes are not interpolated.) As a consequence of this definitions the end tangents of the approximated curves are determined by the difference vectors of the first two or last two control nodes respectively. This is very useful in interactive applications (CAD) or in structural optimization since continuity conditions can be defined and maintained in an obvious manner as demonstrated in the next chapter.

Bézier-splines are defined by the Bernstein polynomials of degree n [16]. The cubic curve (i.e. n=3) is commonly used. It is defined by four control nodes and a parameter u running from 0 to 1. The approximated curve is defined by:

$$\mathbf{r}(u) = (1-u)^3\mathbf{r}_0 + 3u(1-u)^2\mathbf{r}_1 + 3u^2(1-u)\mathbf{r}_2 + u^3\mathbf{r}_3 \qquad (9)$$

where $\mathbf{r}_0$, $\mathbf{r}_1$, $\mathbf{r}_2$, $\mathbf{r}_3$ are the position vectors of the control nodes.

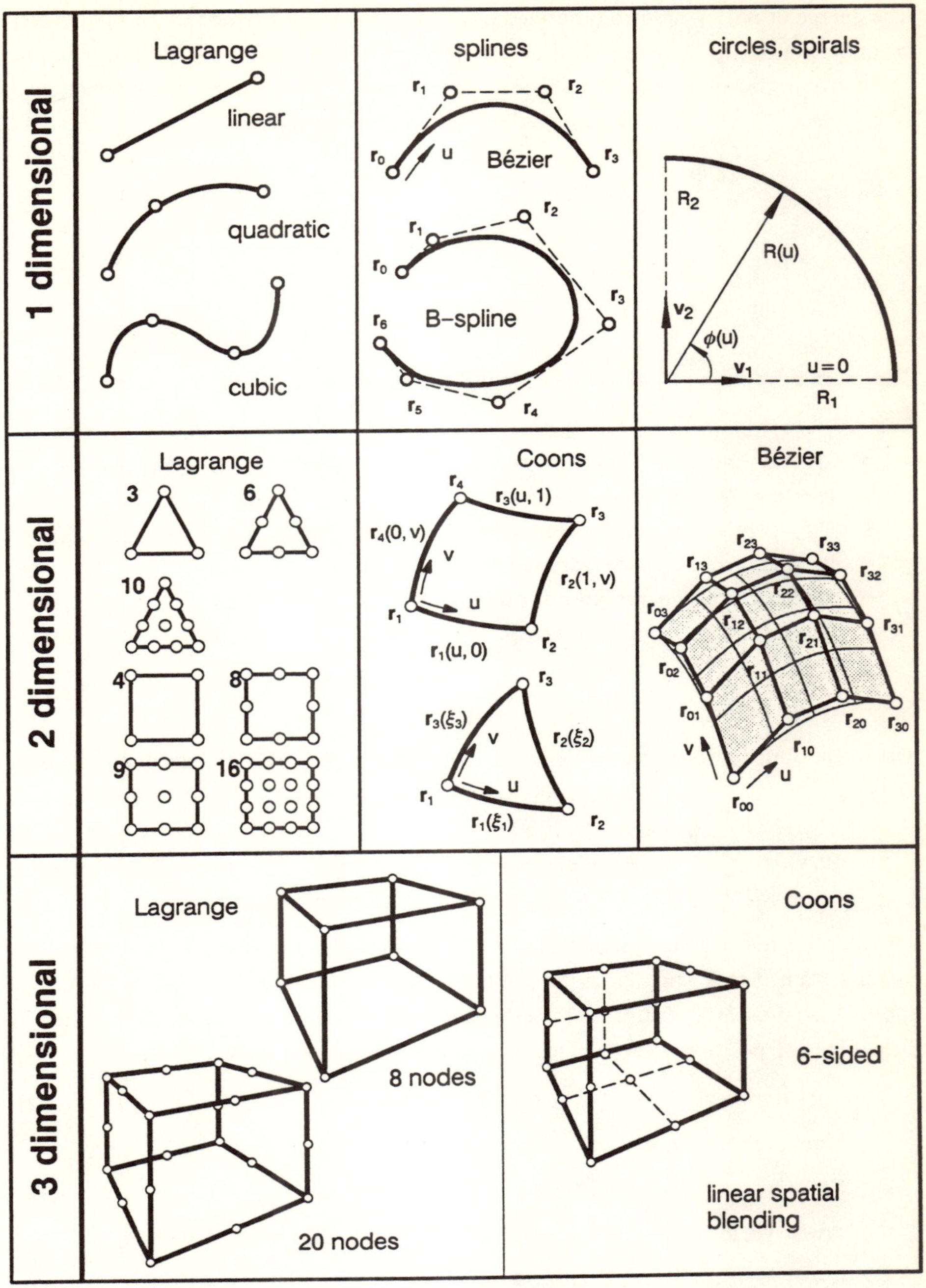

Fig. 7. Design elements available in CARAT design concept

The degree of Bézier-splines is defined by the number of control nodes. This is different from B-splines (fig. 7). A B-spline of degree n can be controlled by an arbitrary number of control nodes m > n+1. The shape functions are defined piecewise by polynomials of degree n. They can be determined by a recursive formula [9]:

$$\Phi_i^r(u) = (u - u_i)\,\frac{\Phi_i^{r-1}(u)}{u_{i+r} - u_i} + (u_{i+r+1} - u)\,\frac{\Phi_{i+1}^{r-1}(u)}{u_{r+i+1} - u_{i+1}} \quad ; \quad r = 1,\dots,n \tag{10}$$

$$\Phi_i^0 = \begin{cases} 1 \ ; \ u \in [u_i, u_{i+1}) \\ 0 \ ; \ u \notin [u_i, u_{i+1}) \end{cases}$$

The main advantages of B-splines for the use in optimization of free formed shapes are:

- arbitrary number of control nodes with given degree of polynomial (preferably n=3, cubic)
- C_{n-1}-continuity; i.e. C_2-continuity of cubic splines
- local control of shape variations
- approximated curves do not oscillate
- convex hull of control nodes [8,16]

5.5 Bézier Patch

The shape function of cubic Bézier patches is derived by a two dimensional extension of (9):

$$\mathbf{r}(u,v) = \sum_{i=0}^{3}\sum_{j=0}^{3} \mathbf{r}_{ij}\,\frac{3!}{(3-i)!i!}\,\frac{3!}{(3-j)!j!}\; u^i\,(1-u)^{3-i}\,v^j\,(1-v)^{3-j} \tag{11}$$

where $\mathbf{r}_{ij}$ are the position vectors of the 16 control nodes and u and v are parameters running from 0 to 1 (fig. 7). Again, only the four vertices $\mathbf{r}_{00}$, $\mathbf{r}_{30}$, $\mathbf{r}_{03}$, $\mathbf{r}_{33}$ lie on the approximated surface. Tangential and twist vectors at any vertex can be determined by differences of position vectors. This is shown at vertex u=0, v=0:

position vector:	$\mathbf{r}(0,0)$	$= \mathbf{r}_{00}$	(12)
tangential vectors:	$\mathbf{r}_{,u}(0,0)$	$= 3\,(\mathbf{r}_{10} - \mathbf{r}_{00})$	(13)
	$\mathbf{r}_{,v}(0,0)$	$= 3\,(\mathbf{r}_{01} - \mathbf{r}_{00})$	(14)
twist vector:	$\mathbf{r}_{,uv}(0,0)$	$= 9\,(\mathbf{r}_{00} - \mathbf{r}_{01} - \mathbf{r}_{10} + \mathbf{r}_{11})$	(15)

where partial derivatives of the surface with respect to the parameters are indicated by $(\;)_{,u}$, $(\;)_{,v}$ and $(\;)_{,uv}$ respectively.

5.6 Coons Patches

These patches are defined by a very flexible interpolation scheme which is also called "transfinite interpolation" [19]. It allows to combine totally different edge types within one e.g. quadrilateral design element:

$$\begin{aligned} \mathbf{r}(u,v) = &\left[L_0^1(u)\ L_1^1(u)\right] \begin{Bmatrix} \mathbf{r}_4(0,v) \\ \mathbf{r}_2(1,v) \end{Bmatrix} + \left[\mathbf{r}_1(u,0)\ \mathbf{r}_3(u,1)\right] \begin{Bmatrix} L_0^1(v) \\ L_1^1(v) \end{Bmatrix} \\ &- \left[L_0^1(u)\ L_1^1(u)\right] \begin{bmatrix} \mathbf{r}_1 & \mathbf{r}_4 \\ \mathbf{r}_2 & \mathbf{r}_3 \end{bmatrix} \begin{Bmatrix} L_0^1(v) \\ L_1^1(v) \end{Bmatrix} \end{aligned} \tag{16}$$

with the linear blending functions:

$$\begin{aligned} L_0^1(u) &= 1-u & \qquad L_1^1(u) &= u \\ L_0^1(v) &= 1-v & \qquad L_1^1(v) &= v \end{aligned} \tag{17}$$

and the edge functions $\mathbf{r}_1(u,0)$, $\mathbf{r}_2(1,v)$, $\mathbf{r}_3(u,1)$, $\mathbf{r}_4(0,v)$ (fig. 7). Equivalent formulations exist for triangular patches [8]. Because of the linear blending functions Coons patches are C_0–continuous. Therefore they are prefered for plane or axisymmetric shapes. With the use of cubic blending functions C_1–continuous patches can be constructed [16].

5.7 Continuity Conditions

Shapes of realistic structures are usually modelled by more than one design element. Especially in the case of free formed surfaces it is important to consider continuity requirements at design element interfaces. To get sufficiently smooth shapes it is necessary to force at least continuous slopes whereas continuous curvatures improve the impression of smoothness even more. Since continuity requirements apply for the generated (i.e. visual) shape and have to be independent of curve or surface parametrizations they are called "visual" or "geometrical" continuity requirements [8]. Because of their special construction C_1–geometrical or G_1–continuity between Bézier– and B–splines can be generated in a very elegant manner. It is sufficient to force the common node and the next control nodes on either side to stay on a common line (fig. 8). This in terms of structural optimization additional constraint is expressed by the relation:

$$3\,\delta\left(\mathbf{r}_n^{(1)} - \mathbf{r}_{n-1}^{(1)}\right) = 3\left(\mathbf{r}_1^{(2)} - \mathbf{r}_0^{(2)}\right) \tag{18}$$

$$\text{with} \qquad \mathbf{r}_n^{(1)} = \mathbf{r}_0^{(2)}$$

where $\mathbf{r}_i^{(1)}$ and $\mathbf{r}_i^{(2)}$ are the position vectors of control nodes of the two adjacent splines. δ is a free parameter which will be called tangent factor [7].

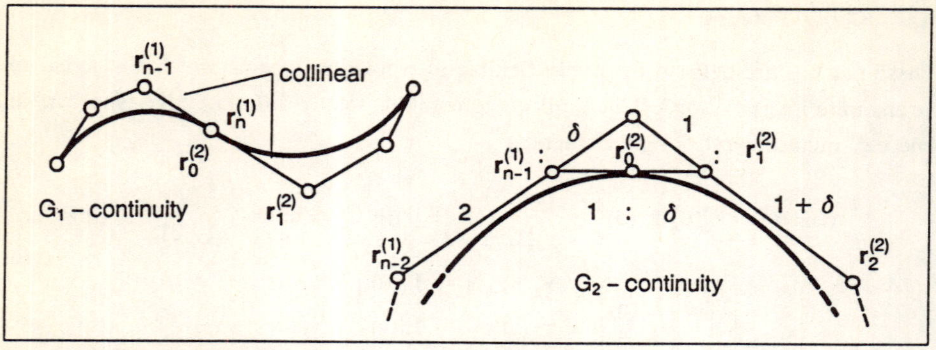

Fig. 8. Geometrical continuity of Bézier- and B-splines

It is convenient to fulfill eqn. (18) explicitly. This will reduce the number of optimization variables and will speed up and stabilize the optimization procedure. $\mathbf{r}_1^{(2)}$ can be expressed in terms of the two other nodes and δ (fig. 8):

$$\mathbf{r}_1^{(2)} = \mathbf{r}_0^{(2)} + \delta\,(\mathbf{r}_n^{(1)} - \mathbf{r}_{n-1}^{(1)}) \tag{19}$$

An equivalent rule which ensures continuous curvature between two B-splines affects in total five control nodes: the common node and the two next nodes of each connected curve.

The relations state as (compare with fig. 8):

$$\begin{aligned} \mathbf{r}_{n-1}^{(1)} &= \frac{1}{(1+3\delta)(2+\delta)}\left[\delta\,(1+\delta)\,\mathbf{r}_{n-2}^{(1)} + 2\,(1+\delta)\,(2+\delta)\,\mathbf{r}_n^{(1)} - 2\,\mathbf{r}_2^{(2)}\right] \\ \mathbf{r}_1^{(2)} &= \frac{1}{(1+3\delta)(2+\delta)}\left[-\delta^2(1+\delta)\,\mathbf{r}_{n-2}^{(1)} + (1+\delta)^2\,(2+\delta)\,\mathbf{r}_0^{(2)} + 2\,\delta\,\mathbf{r}_2^{(2)}\right] \end{aligned} \tag{20}$$

G_1-continuity of Bézier patches is enforced by linear relations between at most nine control nodes. The location of five dependent nodes can be described in terms of four remaining free nodes and two so called "line factors" δ_1 and δ_2 [7].

All these topological dependencies are preserved by superimposed "tangent elements" (1-D case) and "continuity patches" (2-D case) [7]. On request they are generated automatically by CARAT and are preserved during interactive user manipulations and shape optimization.

5.8 Linking of Design Node Coordinates

In shape optimization it is important to control the change of variables during the procedure very carefully. This holds especially when contours of plane structures have to be optimized but also in 3D applications. In those cases additional linking rules between design parameters help to fix typical and problem oriented degrees of freedom [11]. The general type of linking rules is given by:

$$\mathbf{z} = \mathbf{z}^0 + \mathbf{f}_{zy}\left(\mathbf{y}^0 + \mathbf{L}_{yx}\mathbf{x}\right) \tag{21}$$

$\mathbf{x}$ ... independent optimization variables
$\mathbf{y}, \mathbf{z}$ ... dependent variables
$\mathbf{y}^0, \mathbf{z}^0$ ... constant part of dependent variables
$\mathbf{L}_{yz}$ ... linear linking matrix
$\mathbf{f}_{zy}$ ... nonlinear linking function

in the special case of linking between generated FE coordinates and design nodes the rules state as:

$$\mathbf{r} = \Phi\left(\mathbf{r}_D^0 + \mathbf{L}_{rx}\,\mathbf{x}\right) \tag{22}$$

$\mathbf{r}$ ... FE coordinates
$\mathbf{x}$ ... independent optimization variables
$\mathbf{r}_D$ ... constant part of design node coordinates
$\mathbf{L}_{rx}$ ... linear linking matrix
Φ ... shape functions

Experience shows that it is impossible to state a good optimization problem in the first guess. It is therefore very important that linking rules can be easily defined by simple input statements. CARAT supplies with a comfortable input format which allows even complex links by a few easy readable and mnemonic designed statements [7].

Several linear linking relations between design nodes are available which are useful in practical applications:

- prescribed move directions
- linear combinations
- change of basis
- symmetry due to any plane in space
- projection of movement

6 Problem Formulation

The current version of the optimization module in program CARAT includes a three dimensional truss element, an isoparametric two-dimensional element, and a quadrilateral shell element. Possible optimization variables **x**, objective functions f(**x**) and constraints h(**x**), g(**x**) are listed in fig. 9.

The typical weight optimization is supplemented by a minimization of the strain energy rendering a membrane oriented structure with optimal stiffness, and by a stress levelling function eventually leading to an optimal strength in the structures [27]. The usual constraints, e.g. in displacements and stresses, are added to limit the structural response to reasonable values. Shape and thickness optimization may be combined. A particular appealing procedure for shell optimization is that several load cases can be investigated; for example the compromise solution follows from the sum of all individual strain energies.

Design variables **x**:

$\mathbf{r}_D$, $\mathbf{r}$	Coordinates of design and structural nodes
$\mathbf{t}_D$, $\mathbf{t}$	Thicknesses at design and structural nodes
$\mathbf{A}$	Element areas of truss elements

Objective functions f(**x**):

$W = \int \varrho dv$	Structural weight or volume
$E = \frac{1}{2}\int \sigma\epsilon dv$	Strain energy
$S = \int (\sigma - \sigma_{avg})^2 da$	Stress leveling function
$C = \int \varrho r^2 dv$	Rotational mass

Constraint functions h(**x**), g(**x**):

$h_W = \frac{W}{W_{all.}} - 1 = 0 \quad h_V = \frac{V}{V_{all.}} - 1 = 0$	Weight or volume constraint
$h_C = \frac{C}{C_{all.}} - 1 = 0$	Rotational mass constraint
$g_u = \frac{u}{u_{all,}} - 1 \leq 0$	Displacement constraint
$g_\sigma = \frac{\sigma}{\sigma_{all.}} - 1 \leq 0$	Stress constraints

Fig. 9. Details of optimization model

7 Applications

7.1 Shells of Revolution

The example is used as benchmark demonstrating the principle capabilities of shape optimization in a qualitative way. It also shows the performance of the SQP and MMA solution schemes. The objective is either weight or strain energy. Starting from a spherical shell of uniform thickness (radius 10.0 m) with an apex hole of radius 2.5m the optimal shape is obtained for four different load cases: concentrated ring load c = 255 kN/m on the upper free edge, dead load (γ = 78.5 kN/m^3), uniform vertical load of 0.75 kN/m^2 (snow) and a uniform horizontal load of 0.72 kN/m^2 with pressure on one side and suction on the other side roughly simulating wind. Either a fixed hinged or a horizontal roller support is assumed. The material properties of steel are chosen: E = 2.1 $\cdot$ 10^5 N/mm^2; υ = 0.3. Stress constraints with a v. Mises stress limit of 160 N/mm^2 are considered.

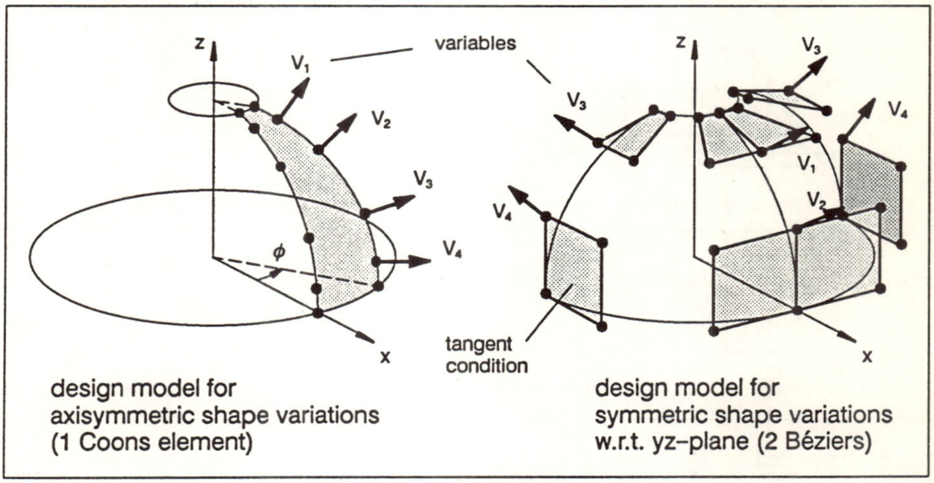

Fig. 10. Design elements for shell of revolution

The surface can be defined by one single Coons element described in cylindrical coordinates using a 6-node B-spline and a 2-node Lagrange side in meridional and hoop direction, respectively (fig. 10, left). Since the optimization for the horizontal load leads to a non-axisymmetric shell two Bézier elements are applied for one half of the structure (fig. 10, right). In both cases only 4 geometrical design variables V_i are introduced. In case of weight minimization the uniform thickness is taken as extra variable.

Figure 11 shows the optimal shapes for different loads and boundary conditions. Except for the "wind load" the final shape allows the shell to carry the load by almost pure membrane action. This holds for weight (concentrated and snow load) as well as strain energy (dead load) minimization. Table 1 summarizes the main results of the parametric study.

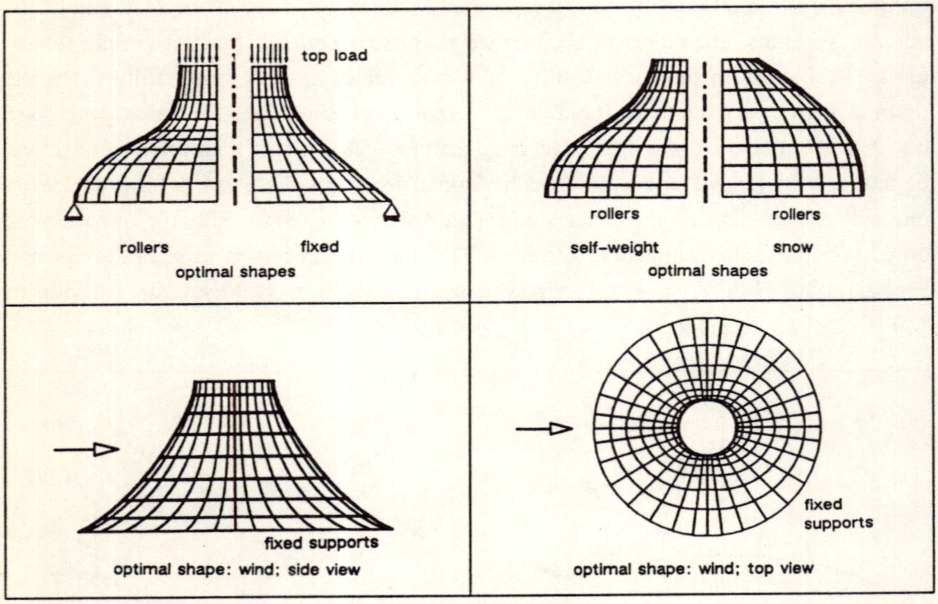

Fig. 11. Shape optimization of shell of revolution

load	support	objective	improve-ment [%]	optimizer	no. of iterations
concentrated load	roller	weight	86	MMA	10
concentrated load	fixed	weight	75	MMA	12
dead load	roller	strain energy	3.5 *	SQP	3 *
snow	roller	weight	63	MMA	15
"wind"	fixed	strain energy	88	SQP	9

Table 1. Parametric study for shell of revolution; (* pre-optimized initial design)

7.2 Non-Regular Shell

Two concrete shells with a uniform thickness t ($E = 3.4 \cdot 10^7$ kN/m², $\upsilon = 0.0$, $\gamma = 25$ kN/m²) have been investigated with respect to strain energy minimization. One load case of dead load and uniform life load $p = 5$ kN/m² is considered. The initial geometry of both shells with a thickness of $t = 0.1$ m is defined in fig. 12 in which the height is plotted in a different scale.

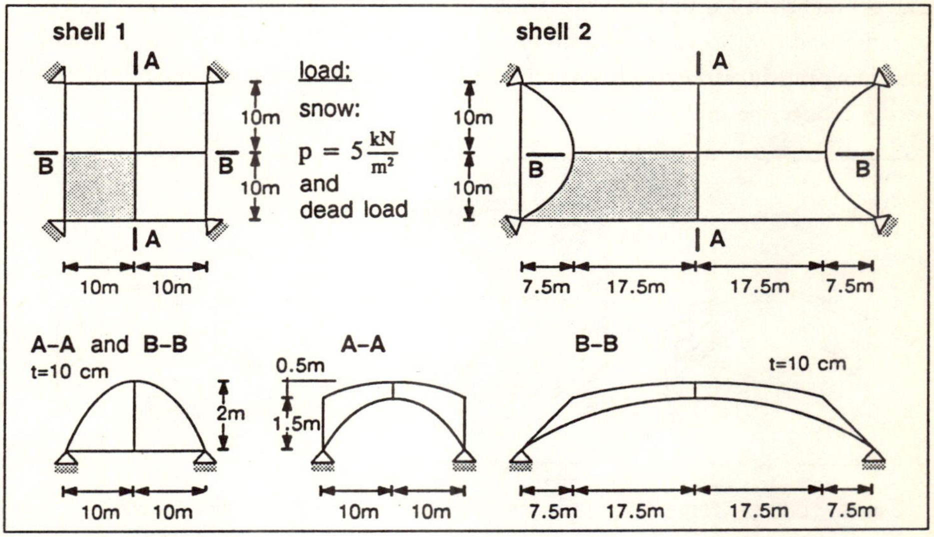

Fig. 12. Initial geometries of 4-point supported shells

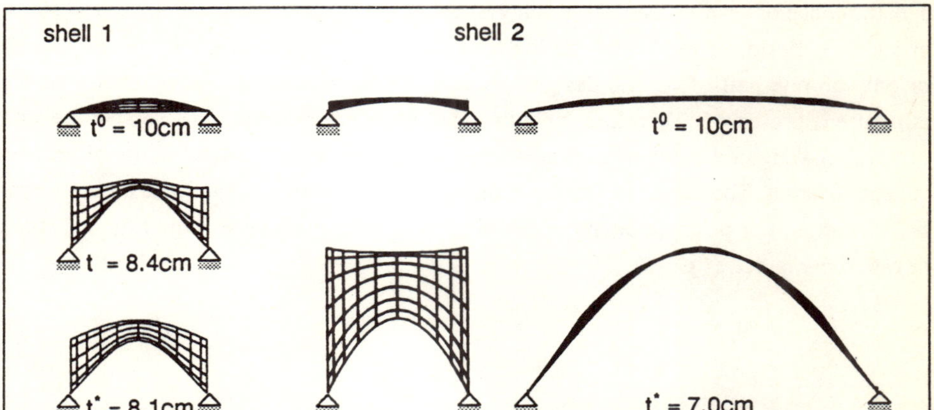

Fig. 13. Initial and optimal shape

The curved free edges are quadratic parabolas. Except that the total material volume is kept constant during shape optimization no further constraints are introduced. Fig. 13 shows the initial shapes in real scale and their optimal counterparts. Only shell 2 has a slight negative curvature at the longitudinal edges known from experiments with hanging fabric. The intermediate solution for shell 1 just before the optimum also indicates this phenomenon. Presumedly several local minima are close together.

The optimal shell 1 is further processed by thickness optimization based on weight minimization and a v. Mises stress limit of 2.1 N/mm^2. The thickness variation, defined by four independent variables, is given in fig. 14. A subsequent shape optimization does hardly change the initial shape.

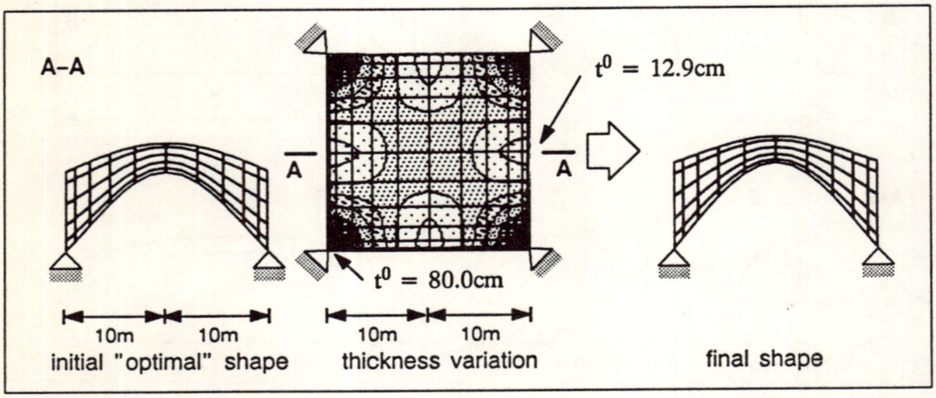

Fig. 14. Thickness and shape optimization

The influence of several load cases is investigated next, again using the objective "strain energy" f_E. Besides dead load g the life load p = 5 kN/m^2 acts either on the entire shell or only on one half of it. The design variables are linked so that the symmetry of the structure is preserved. Therefore, only two load cases are relevant; as objective the sum of the non-weighted individual strain energies is considered. The total material volume is kept constant. The initial and optimal shapes are plotted in fig. 15. Since a pure membrane state is not possible anymore the shell needs a boundary stiffened by a distinct negative curvature (fig. 15).

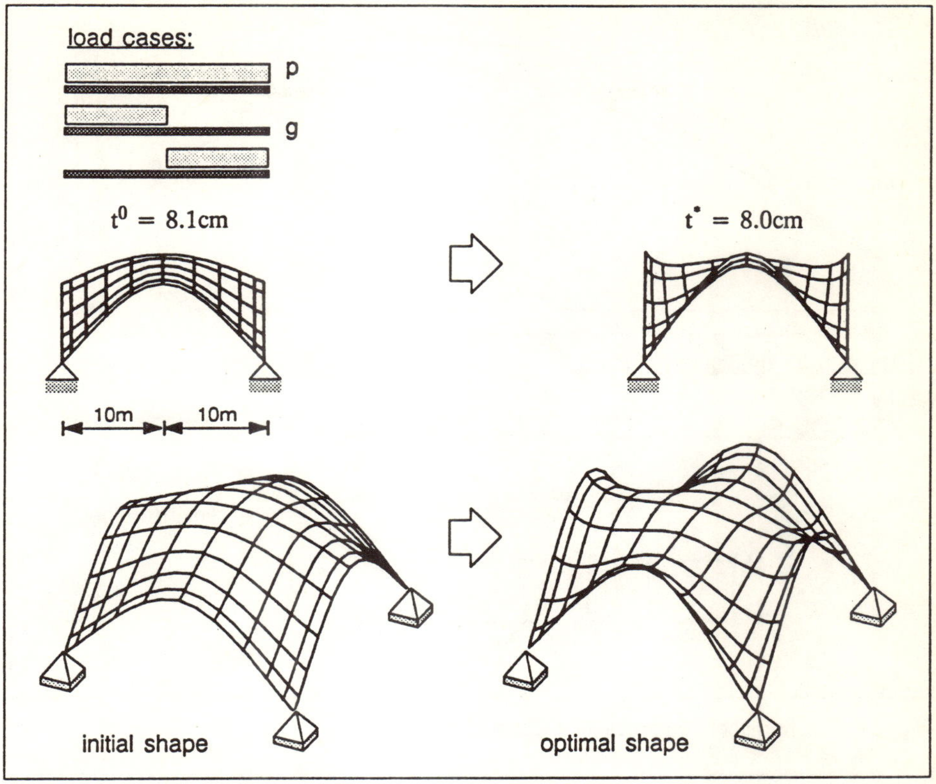

Fig. 15. Shape optimization with different load cases

7.3 Free Form Concrete Shell

The shell in fig. 16 is related to the reinforced concrete tennis hall designed by H. Isler [34]. The material data are $E = 3.0 \cdot 10^7$ kN/m², $\nu = 0.2$, $\gamma = 25$ kN/m³. One load case of dead load plus uniform life load 5 kN/m² is considered. One quarter of the shell is idealized by 4 Bézier elements linked by continuity patches and with 23 vertical nodal coordinates S_i as design variables. 126 reduced integrated 8-node shell elements are used for a linear structural analysis. Again the strain energy is minimized. Since the material volume is kept constant the uniform initial thickness $t^o = 0.10$ m slightly varies during optimization. It could be recognized that the shape near the free edges is sensitive and might even result in a sharp local curvature representing a kind of edge beam. If this is not tolerated by prescribed geometrical constraints the optimal shape again shows a clear negative curvature near the free boundary (fig. 17). For a comparison of different sensitivity analyses see [27].

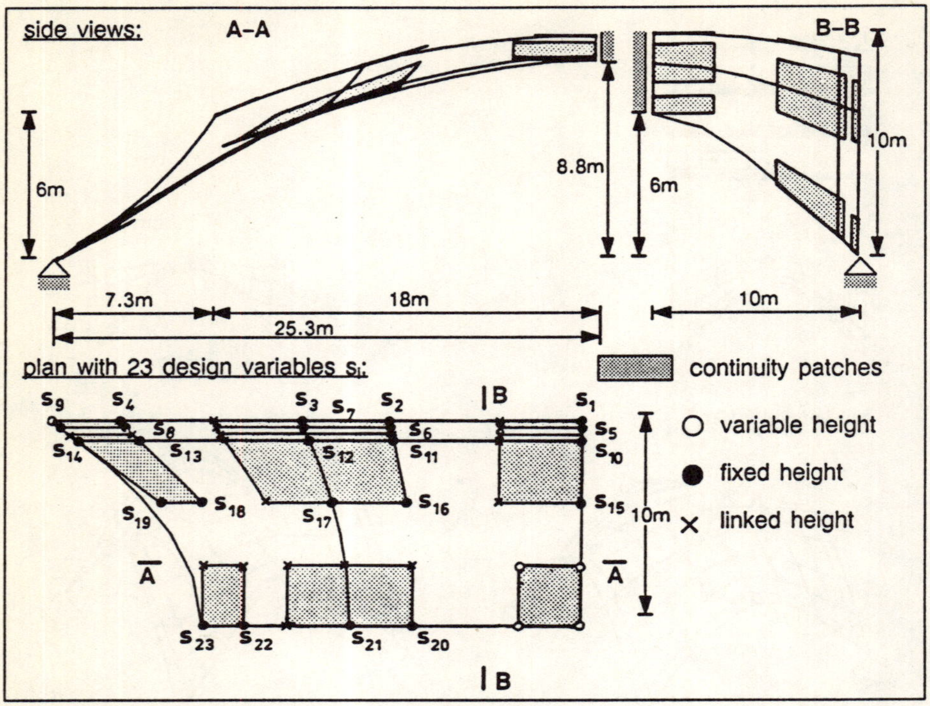

Fig. 16. Geometrical model of free form shell

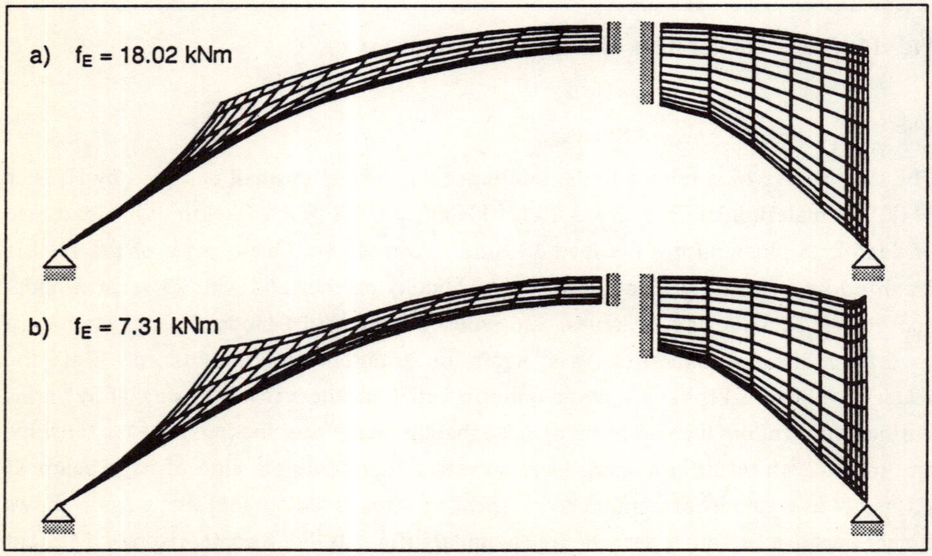

Fig. 17. a) Initial and b) optimal shape of free form shell

8 Conclusions

The following conclusions may be drawn:

- In structural optimization a key for a realistic approach for the design is the reduction of the number of design variables to a reasonable value; in shape optimization this is achieved by the design (macro) element concept. The typical CAGD concepts like B-spline or Bézier interpolation schemes and related linking rules have been proven to be an ideal tool for this purpose.
- The minimum weight design traditionally applied can be supplemented by further interesting objectives; in particular, the strain energy is an appealing alternative in shell design. For stress homogenization see [27].
- Although not detailed herein the sensitivity analysis is a major part in structural optimization; for details see [27].

Further studies are necessary to further improve the efficiency of all solution schemes and to include nonlinear and time dependent structural behavior.

9 Acknowledgements

This work is part of the research project SFB 230 "Natural Structures - Light Weight Structures in Architecture and Nature" supported by the German Research Foundation (DFG) at the University of Stuttgart. The support is gratefully acknowledged.

10 References

1. Andelfinger, U.: Untersuchungen zur Zuverlässigkeit hybrid-gemischter finiter Elemente für Flächentragwerke. Ph.D.-Dissertation, Institut für Baustatik, Universität Stuttgart 1991.
2. Becker, A., Berkhahn, V., Stein, E.: Weight optimization of steel frames with consideration of geometrical and material nonlinearities. Proc. GAMM-Seminar, Oct. 5- 7, 1988, Siegen, on Discretization Methods and Structural Optimization - Procedures and Applications (H.A. Eschenauer, G. Thierauf, eds.). Lecture Notes in Engineering 42. Heidelberg: Springer (1989) 32-39.
3. Bendsøe, M.P.: Optimal shape design as a material distribution problem. Structural Optimization 1 (1989) 193-202.
4. Bennett, J.A.; Botkin, M.E. (eds.): The Optimum Shape - Automated Structural Design. New York, London: Plenum Press 1986.
5. Berke, L.; Khot, N.S.: Structural optimization using optimality criteria. Proceedings of NATO ASI on Computer Aided Optimal Design - Structural and Mechanical Systems (Mota Soares, C.A. ed.); NATO ASI Series F: Computer and Systems Siences 27. Heidelberg, NewYork: Springer (1987) 271-311.

6. Berkhahn, V.: Optimale Formgebung von Stabtragwerken mit Nichtlinearitäten in der Zielfunktion und in den Restriktionen unter Verwendung der Finite - Element - Methode. Ph.D.-Dissertation, Bericht Nr. F 88/2, Institut für Baumechanik und Numerische Mechanik, Universität Hannover 1988.

7. Bletzinger, K.-U.: Formoptimierung von Flächentragwerken. Ph.D. Dissertation, Institut für Baustatik, Universität Stuttgart 1990.

8. Böhm, W.; Farin, G.; Kahmann, J.: A survey of curve and surface methods in CAGD. Comp. Aided Geom. Des. 1 (1984) 1-60.

9. de Boor, C.: A Practical Guide to Splines. Heidelberg, NewYork: Springer 1978.

10. Botkin, M.E.; Bennett, J.A.: Shape optimization of three-dimensional folded plate structures. AIAA Journal 23 (1985) 1804-1810.

11. Braibant, V.; Fleury, C.: Shape optimal design and CAD oriented formulation. Engng. with Comp. 1 (1986) 193-204.

12. Büchter, N.; Ramm, E.: Shell theory versus degeneration - a comparison in large rotation finite element analysis. submitted to Int. J. Num. Meth. Engng. 1990.

13. Ding, Y.: Shape optimization of structures: a literature survey. Comp. & Struct. 24 (1986) 985-1004.

14. Eschenauer, H.; Post, U.; Bremicker, M.: Einsatz der Optimierungsprozedur SAPOP zur Auslegung von Bauteilkomponenten. Bauingenieur 63 (1988) 515-526.

15. Esping, B.J.D.: A CAD approach to the minimum weight design problem. Int. J. Num. Meth. Engng. 21 (1985) 1049-1066.

16. Faux, I.D.; Pratt, M.J.: Computational Geometry for Design and Manufacture. Chichester: Ellis Horwood Publishers 1979.

17. Fleury, C.: Sructural weight optimization by dual methods of convex programming. Int. J. Num. Meth. Engng. 14 (1979) 1761-1783.

18. Gill, P.E.; Murray, W.; Wright, M.H.: Practical Optimization. London: Academic Press 1981.

19. Haber, R.; Shephard, M.S.; Abel, J.F.; Gallagher, R.H.; Greenberg, D.P.: A general two-dimensional, graphical finite element preprocessor utilizing discrete transfinite mapping. Int. J. Num. Meth. Engng. 17 (1981) 1015-1044.

20. Haftka, R.T.; Grandhi, R.V.: Structural shape optimization - a survey. Comp. Meth. Appl. Mech. Engng. 57 (1986) 91-106.

21. Haftka, R.T.; Gürdal, Z.; Kamat, M.P.: Elements of Structural Optimization. 2nd edition. Dordrecht, Boston, London: Kluwer 1990.

22. Han, S.-P.: A globally convergent method for nonlinear programming. J. Optimization Theory Appl. 22 (1977) 297–309.

23. Haug, E.J.; Choi, K.K.; Komkov, V.: Design Sensitivity Analysis of Structural Systems. Orlando, New York, London: Academic Press 1986.

24. Imam, M.H.: Three-dimensional shape optimization. Int. J. Num. Meth. Engng. 18 (1982) 661–673.

25. Karihaloo, B.L.; Kanagasundaram, S.: Minimum-weight design of structural frames. Comp. & Struct. 31 (1989) 647–655.

26. Kimmich, S.; Ramm, E.: Structural optimization and analysis with program system CARAT. Proc. GAMM-Seminar, Oct. 5– 7, 1988, Siegen, on Discretization Methods and Structural Optimization - Procedures and Applications (H.A. Eschenauer, G. Thierauf, eds.). Lecture Notes in Engineering 42. Heidelberg: Springer (1989) 186–193.

27. Kimmich, S.: Strukturoptimierung und Sensibilitätsanalyse mit finiten Elementen. Ph.D.-Dissertation, Institut für Baustatik, Universität Stuttgart 1990.

28. Kruzelecki, J.; Zyczkowski, M.: Optimal structural design of shells - a survey. SM Archives 10 (1985) 101–170.

29. Luenberger, D.G.: Linear and Nonlinear Programming. Reading: Addison-Wesley 1984.

30. Mahnken, R.; Stein, E.: A globally convergence criterion for first order approximation strategies in structural optimization. submitted to Int. J. Num. Meth. Engng. 1990.

31. Plaut, R.H.; Johnson, L.W.; Parbery, R.: Optimal forms of shallow shells with circular boundary; Part 1: maximum fundamental frequency; Part 2: maximum buckling load; Part 3: maximum enclosed volume. J. Appl. Mech. 51 (1984) 526–539.

32. Powell, M.J.D.: A fast algorithm for nonlinearly constrained optimization calculations. Proceedings of the Biennial Conference held at Dundee, June 1977; Lecture Notes in Mathematics 630. Berlin, Heidelberg, New York: Springer 1978.

33. Prasad, B.: Explicit constraint approximation forms in structural optimization, part 1: analyses and projections. Comp. Meth. Appl. Mech. Engng. 40 (1983) 1–26.

34. Ramm, E.; Schunck, E.: Heinz Isler - Schalen. Stuttgart: Krämer 1986.

35. Schittkowski, K.: The nonlinear programming method of Wilson, Han and Powell with an augmented Lagrangian type line search function. Numerische Mathematik 38 (1981) 83–114.

36. Schmit, L.A.; Farshi, B.: Some approximation concepts for Structural Synthesis. AIAA Journal 12 (1974) 692–699.

37. Schmit, L.A.: Structural synthesis – its genesis and development. AIAA Journal 19 (1981) 1249–1263.

38. Stein, E., Becker, A., Barthold, F.J.: Structural optimization of beam structures including imperfections. Proc. Int. Conference on Numerical Methods in Engineering, NUMETA 90, Swansea, Jan. 7–11, 1990 (G.N. Pande, J. Middleton, eds.). London, New York: Elsevier Applied Science (1990) 454–463.

39. Svanberg, K.: The method of moving asymptotes – a new method for structural optimization. Int. J. Num. Meth. Engng. 24 (1987) 359–373.

40. Thanedar, P.B.; Arora, J.S.; Tseng, C.H. et al.: Performance of some SQP algorithms on structural optimization problems. J. Num. Meth. Engng. 23 (1986) 2187–2203.

41. Vanderplaats, G.N.: Numerical Optimization Techniques for Engineering Design: With Applications. New York: McGraw–Hill 1984.

Unilateral Contact with Friction by a Time Domain BEM

H. ANTES and B. STEINFELD

Institut für Angewandte Mechanik
Universität Carolo Wilhelmina, D-3300 Braunschweig, Germany

SFB 151, Dept. Civil Engineering
Ruhr-Universität, D-4300 Bochum, Germany

Summary

The subject of this paper is to analyze plane elastic, time dependent contact problems under frictional conditions. The initial state of contact is assumed to be known by geometric features, but its extent changes when the contacting elastic structures are deformed by dynamic loadings or other sources of stresses, e.g. incident waves. Thus, the problem is non-linear by virtue of this variable area of contact. In order to determine the changes of the contact extent and of the sliding area, an iterative technique, based on a time domain direct boundary element approach, has been developed. Some examples concerning massive surface foundations are presented.

Introduction

The analysis of elastic bodies in unilateral contact is a common problem for engineers. Contact components being in or coming into contact may be subjected to quasi-static loadings, i.e. to progressively applied loads, to transient dynamic excitations, or to incident waves. In any case, the study of the load transfer is important: either because high peak stresses and strains can arise at the contact locations and the associated large fatigue can remarkably shorten the structural lifetime, or because partial uplift and sliding can happen for small time durations and this behavior can be responsible for the good performance of a seemingly unstable structure.

All existing analytical solutions of contact problems are valid for idealized model configurations, the engineering application of which is rather limited. Therefore, the need for accurate numerical methods is apparent.

Currently, the most popular numerical technique used to perform the analysis of contacting bodies is the Finite Element method. Ignoring the effects of friction, Wilson and Parson [1] were the first to use the FEM for determining in a static analysis the surface contact stresses between two plane bodies. Since then, the FE methods were extended rapidly: friction was introduced in the formulation through the Coulomb law (Ohte [2]), through various other frictional conditions (Okamoto and Nakaza [3]) and, for a more realistic modeling, through laws of 'non-local' and 'non-linear' type (e.g. Oden and Pires [4], Gallego and Anza [5]). Moreover, also time dependent excitations and their influences on contacting bodies have been analyzed by FEM, for two-dimensional as well as for three-dimensional problems (Zolti [6], Chaudhary and Bathe [7]). These FE contact algorithms are iterative, incremental procedures of the type 'trial and error' which are not necessarily convergent to the correct solution, or require, based on variational principles, the use of optimization techniques with restrictions by means of Lagrange multipliers [7,8] or penalty [3] procedures. Based on some theorems of quadratic programming, Panagiotopoulos and Talaslidis [9] have developed algorithms for contact problems, combining the advantages of the "trial and error" methods (use of general purpose programs which effectively treat large problems) and those of optimization procedures (systematic choice of the iteration steps and convergence)

The Boundary Element Method (BEM) has already proved to be an extremely versatile and powerful tool on various fields of applications (for a rather recent review in dynamic analysis see Beskos [10]). Obviously, it is particularly well suited for solving contact problems since the variables of interest of both the problem and the solution procedure are on the boundaries. Moreover, the boundary tractions are calculated with the same accuracy as the boundary displacements, and the normal and tangential tractions can be coupled directly in the equation system.

Anderson et al. [11] were the first to apply the direct BEM to two-dimensional frictionless static problems using constant elements with successful results, and later extended the formulation and application to frictional contact [12] and linear and parabolic elements [13]. Also, discontinuous linear elements have been implemented to model typical friction contact problems [14], and, moreover, a non-classical friction procedures were used to solve several mechanical problems in elastostatics (e.g. the effects of inclusions embedded in an elastic medium [16] or contact between plates [17]) and for thermoelastic analysis [18].

As for the FE approach, Panagiotopoulos developed also for the BEM - for both the Direct and the Indirect method [18,19] and for friction problems [20] - suitable boundary variational principles that allow to use very effective quadratic programming tools. Some of these boundary variational statements have been derived independently by Tralli et al. [21,22] by means of standard direct BE procedures.

But, although most of the real contacting phenomena are time history dependent, adequate BE formulations acting directly in the time domain are rare. Besides some investigations pertinent to foundations (for a survey see Spyrakos [23]) and other frictionless uplifting problems (Antes and Steinfeld [24,25]), only recently, almost at the same time, but independently Dominguez et al. [26] and Antes et al. [27,28] developed iterative time-stepping direct BE techniques to solve the more realistic, frictional contact problem, i.e. to determine which part of an interface remains sticking, changes to sliding, looses the contact by uplifting, returns to sliding a.s.o.

In this paper, these time-stepping BE algorithms for solving unilateral dynamic contact under Coulomb's frictional conditions are given. Considering impulsive transient excitations the dynamic interaction of a flexible massive foundation with the supporting elastic soil is analyzed. Thereby, the influence of the frictional resistance on the dynamic behavior is especially investigated.

Time-Stepping BE-Formulation of Two-Dimensional Contact Problems with Friction

Two structures or domains A and B may have contact along the interface Γ_I. A convenient boundary element mesh is defined for the boundary Γ^A of A and Γ^B of B, where each pair of boundary nodes of the interface zone is initially at the some geometric location. Besides, the following assumptions are made:

1. The materials are linearly elastic and homogeneous.

2. The motions are limited to small amplitudes, and the strains are infinitesimal.
3. The gravity effect, i.e. the dead weight of a structure is taken into account, while that of the supporting soil is ignored.
4. The frictional forces acting at the contact interface follow the Coulomb type criterion for dry friction, i.e. during sliding the Coulomb coefficient μ determines the ratio between the tangential shear and normal compressive stress along the interface.

The contact region between two bodies transmits forces between them by means of normal compressive stresses, and shear stress if friction is present. Frictional contact may be of either the sticking (adhesive) or sliding type, depending on whether or not relative tangential displacements occur. Indeed, a particular contact region may contain zones of each type, the extents of which are not known in advance and, moreover, which are time-dependent in the case of dynamic loadings.

Contact boundary conditions can be expressed in terms of displacements **u** and tractions **T**, in the local outward normal and tangential directions relative to the contact surface, indicated by subscripts n and s, respectively. Also, indices A and B are used to distinguish between corresponding values of nodal points (a nodepair) on the boundaries of the two contacting bodies. For a prescribed dynamic loading on the system, a node pair either enters or remains in one of the following four modes of behaviour:

(i) Stick Mode:

As long as no tensile stresses occur and the tangential stresses do not reach the Coulomb limit, stress equilibrium and displacement compatibility require ($n^I = n^A = -n^B$, $s^I = s^A = -s^B$):

$$T_s^A = T_s^B = T_s^I \qquad \text{or} \qquad T_s^A - T_s^B = 0 \tag{1}$$

$$T_n^A = T_n^B = T_n^I \qquad \text{or} \qquad T_s^A - T_s^B = 0 \tag{2}$$

$$u_s^A = -u_s^B = u_s^I. \qquad \text{or} \qquad u_s^A + u_s^B = 0 \tag{3}$$

$$u_n^A = -u_n^B = u_n^I. \qquad \text{or} \qquad u_n^A + u_n^B = 0 \tag{4}$$

(ii) Coulomb Slip Mode:

If no tensile normal stresses occur, but the level of tangential stress reaches the limit prescribed by Coulomb's law (the dependency of the friction coefficient μ from the total effective slip [12] will be neglected), slipping will occur with a final rupture of the adhesion, i.e. relative tangential displacements $(u_s^A - u_s^B) \neq 0$ are permitted; by the condition of energy dissipation, the direction of the "friction force" T_s^I has to be opposite to the direction of the relative movements ("•" denotes the time derivative):

$$| T_s^I | = \mu \, | T_n^I | \qquad \text{with} \qquad T_s^I \, (\dot{u}_s^A + \dot{u}_s^B) < 0 \tag{5}$$

In the normal direction, equilibrium and compatibility still require:

$$T_n^A = T_n^B = T_n^I \qquad (< 0) \tag{6}$$

$$u_n^A = - u_n^B = u_n^I \, . \tag{7}$$

(iii) Separation Mode:

If tensile normal stresses occur in some region of the interface, such parts debond and open a gap. There, both boundaries are free of stresses, i.e.

$$T_n^A = T_s^A = 0 \tag{8}$$

$$T_n^B = T_s^B = 0 \, , \tag{9}$$

and they can move independently, but under certain constraints, i.e. their relative displacements have to be compatible. Neglecting relative rotations, this gap is given by

$$\Delta u_n^I : = - u_n^A - u_n^B > 0. \tag{10}$$

If an opened gap Δu_n^I tends to zero again, either sliding or sticking of the two contacting structures can take place, i.e either the Coulomb slip mode conditions (ii) hold again, or those of the following

(iiii) Rebonding mode:

If separation or sliding contact returns to stick contact, at the n-th time step, for instance, the condition of tangential compatibility (3) must be changed to be

$$\overset{(n)}{u_s^A} + \overset{(n)}{u_s^B} =: \overset{(n)}{g_s} = \overset{(n-1)}{g_s} = \overset{(n-1)}{u_s^A} + \overset{(n-1)}{u_s^B} \, , \tag{11}$$

where $\overset{(n-1)}{g_s}$ is the non-reversible tangential displacement at time $(n\text{-}1)\Delta t$, i.e. the total effective slip which has occured during the separation - sliding process till the node returns to welded contact; it remains unchanged during the whole welded contact period. Then, by introducing the "new" tangential displacements

$$\overset{(n)}{u}{}^{A}_{sr} := \overset{(n)}{u}{}^{A}_{s} - \alpha \overset{(n)}{g}_{s} \qquad \text{and} \qquad \overset{(n)}{u}{}^{B}_{sr} := \overset{(n)}{u}{}^{B}_{s} - (1-\alpha) \overset{(n)}{g}_{s} \tag{12}$$

the generalized tangential compatibility condition

$$\overset{(n)}{u}{}^{A}_{sr} + \overset{(n)}{u}{}^{B}_{sr} = 0 = \overset{(n)}{u}{}^{A}_{s} + \overset{(n)}{u}{}^{B}_{s} - \overset{(n)}{g}_{s} \tag{13}$$

is formulated and substitutes the condition (3). All other conditions of the stick mode, i.e. (1), (2), and (4), hold unchanged.

A time dependent integral representation of the dynamical initial-boundary value problem describing time-dependent displacements in elastic media can be deduced by employing a reciprocal theorem and a subsequent integral transformation [29]. Finally, as a general analytical solution is impossible, a numerical evaluation is needed and leads to a system of algebraic equations for each domain (or for each part of a domain with equal material properties):

$$[\, 0.5\, \mathbf{I} + \overset{(1)}{\mathbf{T}}\,] \cdot \overset{(n)}{\mathbf{u}} + \sum_{m=1}^{n-1} \overset{(n-m+1)}{\mathbf{T}} \cdot \overset{(m)}{\mathbf{u}} = \sum_{m=1}^{n} \overset{(n-m+1)}{\mathbf{U}} \cdot \overset{(m)}{\mathbf{t}} + \overset{(n)}{\mathbf{r}}. \tag{14}$$

This system is generated based on a discretization of both, geometry and time, and by collocation:

(i) The boundary is subdivided into a number L of lines over which displacements and tractions are assumed to be constant

(ii) The whole observation period is discretized by equal time increments Δt, and linear and constant shape functions are used to approximate the transient behavior of the displacements and tractions, respectively

(iii) Collocation is taken at each boundary node ξ and at all time steps $n\Delta t$.

$\overset{(\mu)}{\mathbf{T}}$ and $\overset{(\mu)}{\mathbf{U}}$ represent the influence matrices whose elements are obtained by integrations over each boundary element (numerically) and over each time steps (analytically). The index μ=n-m indicates the difference between the observation time $n\Delta t$ and the impulse time $m\Delta t$. The vectors $\overset{(m)}{\mathbf{u}}$ and $\overset{(m)}{\mathbf{t}}$ contain all nodal displacements and tractions, respectively, of time step m, while the vector $\overset{(n)}{\mathbf{r}}$ gives the influence of the non-zero body forces.

The dynamic interaction analyses of elastic structures, e.g. Ω^A and Ω^B in unilateral frictional contact, require the combination of these systems of algebraic equations, generated for each structure, by imposing adequate, i.e. one of the above listed interface conditions. For any coupling, the algebraic equations, relating at each time step n boundary values of the heavy, elastic structure Ω^A, e.g. an elastic massive strip foundation, and those of the domain Ω^B, e.g. the supporting soil, have to be subdivided and rearranged corresponding to the position of the collocation and observation points at the contact and "outer" boundary, respectively.

Coupling via the stick contact conditions, and rearranging according to the unknown values, to the prescribed ones, and to values already known or found in the previous time steps, yields a system of equations which can be cast into the form [12,13]

$$\overset{(1)}{[\mathbf{Q}]}\cdot\begin{Bmatrix}\overset{(n)}{\mathbf{u}}{}^A_o\\ \overset{(n)}{\mathbf{u}}{}^A_c\\ \overset{(n)}{\mathbf{t}}{}^A_c\\ \overset{(n)}{\mathbf{u}}{}^B_o\end{Bmatrix} = \overset{(1)}{[\mathbf{P}]}\cdot\{\overset{(n)}{\mathbf{t}}{}^A_o\} + \begin{Bmatrix}\overset{(n)}{\mathbf{R}}{}^A_o\\ \overset{(n)}{\mathbf{R}}{}^A_c\\ \overset{(n)}{\mathbf{R}}{}^B_c\\ \overset{(n)}{\mathbf{R}}{}^B_o\end{Bmatrix}, \qquad (15)$$

where the matrices $\overset{(1)}{\mathbf{Q}}$ and $\overset{(1)}{\mathbf{P}}$ are explicitly ($\overset{*}{\mathbf{T}} := \frac{1}{2}\mathbf{I} + \mathbf{T}$)

$\overset{(1)}{\mathbf{C}}$

$$\overset{(1)}{\mathbf{Q}} := \begin{bmatrix} \overset{(1)}{\overset{*}{\mathbf{T}}}{}^A_{oo} & \overset{(1)}{\mathbf{T}}{}^A_{oc} & -\overset{(1)}{\mathbf{U}}{}^A_{oc} & 0\\ \overset{(1)}{\mathbf{T}}{}^A_{co} & \overset{(1)}{\overset{*}{\mathbf{T}}}{}^A_{cc} & -\overset{(1)}{\mathbf{U}}{}^A_{cc} & 0\\ 0 & -\overset{(1)}{\overset{*}{\mathbf{T}}}{}^B_{cc} & -\overset{(1)}{\mathbf{U}}{}^B_{cc} & \overset{(1)}{\mathbf{T}}{}^B_{co}\\ 0 & -\overset{(1)}{\mathbf{T}}{}^B_{oc} & -\overset{(1)}{\mathbf{U}}{}^B_{oc} & \overset{(1)}{\overset{*}{\mathbf{T}}}{}^B_{oo} \end{bmatrix}, \quad \overset{(1)}{\mathbf{P}} := \begin{bmatrix} \overset{(1)}{\mathbf{U}}{}^A_{oo}\\ \overset{(1)}{\mathbf{U}}{}^A_{co}\\ 0\\ 0 \end{bmatrix}. \qquad (16)$$

The subscripts "c" and "o" indicate the position of the collocation and nodal points on the "contact" boundary, i.e. the interface, and on the "outer" boundaries. The terms $\overset{(n)}{R}{}^A$ and $\overset{(n)}{R}{}^B$ represent the influence of all the previous time steps $(m<n)$ and, if it is taken into account, of the dead weight of a heavy structure, e.g.

$$\{\overset{(n)}{R}{}^A_c\} = \{\overset{(n)}{R}{}^A_c(\overset{(m)}{t}{}^A, \overset{(m)}{u}{}^A)\} = \sum_{m=1}^{n-1} [\, \overset{(n-m+1)}{U}{}^A_c \bullet \overset{(m)}{t}{}^A - \overset{(n-m+1)}{T}{}^A_c \bullet \overset{(m)}{u}{}^A \,] + \overset{(n)}{r}{}^A. \qquad (17)$$

Formulating the Equ. (15), it has been assumed that the non-contacting surface of the soil Ω^B is free of stresses $(t^B_o=0)$, and that dynamic loadings t^A_o are prescribed along the outer boundary of the structure Ω^A.

For stepping from time $(n-1)\Delta t$ to time $n\Delta t$, it is only necessary to change the current prescribed boundary values, and to update the terms $\{\overset{(n)}{R}{}^A\}$ and $\{\overset{(n)}{R}{}^B\}$. Moreover, in order to verify the sliding and the debonding on certain interface elements, nothing but the part $\overset{(1)}{C}$ of the matrix $\overset{(1)}{Q}$ has to be changed, while for the rebonding mode "non-reversible" relative tangential displacements, the relative slips have to be taken into account additionally at the right hand side of Eq. (15).

When at a certain time step n sliding takes place on some interface elements, the tangential and the normal components of those nodal values behave different: while in the normal direction equilibrium and compatibility are still preserved, tangentially relative displacements are permitted and the "friction force" is determined by the normal compressive stress via the relation (5). To find the right sign of the friction force for each sliding node, it is assumed, that its direction is the same as under sticking condition just before:

$$\overset{(n)}{T}{}^I_s = \mathrm{sign}(\overset{(n-1)}{T}{}^{I\,stick}_s / \overset{(n-1)}{T}{}^{I\,stick}_n)\, \mu\, \overset{(n)}{T}{}^I_n \qquad (18)$$

But in some cases the direction might change. Therefore it must be checked whether the relation (5) is really satisfied:

$$\begin{aligned} \mathrm{sign}(\overset{(n)}{T}{}^I_s) &= -\,\mathrm{sign}(\overset{(n)}{\dot{u}}{}^A_s + \overset{(n)}{\dot{u}}{}^B_s) \\ &\cong \mathrm{sign}(\overset{(n-1)}{u}{}^A_s + \overset{(n-1)}{u}{}^B_s) - \mathrm{sign}(\overset{(n)}{u}{}^A_s + \overset{(n)}{u}{}^B_s). \end{aligned} \qquad (19)$$

If not, sticking is again assumed for this time step and this element, and the iteration cycle starts again.

When separation takes place because the normal traction in certain interface elements becomes tensile, both traction components disappear in the debonded elements and both faces move separately. But, such debonded elements move under the rebonding constraint $\Delta u_n \geqq 0$, numerically approximated by $|\Delta u_n| \leqq |\varepsilon_n|$, where ε_n represents a small "admissible" incompatibility, either a tiny interpenetration or a little gap.

Thus, these types of contact modes causes typical changes in the contact related part $\overset{(1)}{\mathbf{C}}$ of the matrix $\overset{(1)}{\mathbf{Q}}$ (see definition (16)):

for sliding elements

$$\overset{(1)}{\mathbf{C}} := \begin{bmatrix} \overset{(1)}{\mathbf{T}}{}^{A}_{noc} & \overset{(1)}{\mathbf{T}}{}^{A}_{soc} & 0 & -\overset{(1)}{\mathbf{U}}{}^{A}_{noc} + \mu \operatorname{sign}(\overset{(n)}{\mathbf{T}}{}^{I}_{s})\overset{(1)}{\mathbf{U}}{}^{A}_{soc} \\ \overset{(*)}{\mathbf{T}}{}^{A}_{ncc} & \overset{(*)}{\mathbf{T}}{}^{A}_{scc} & 0 & -\overset{(1)}{\mathbf{U}}{}^{A}_{ncc} + \mu \operatorname{sign}(\overset{(n)}{\mathbf{T}}{}^{I}_{s})\overset{(1)}{\mathbf{U}}{}^{A}_{scc} \\ -\overset{(*)}{\mathbf{T}}{}^{B}_{ncc} & 0 & \overset{(*)}{\mathbf{T}}{}^{B}_{scc} & -\overset{(1)}{\mathbf{U}}{}^{B}_{ncc} + \mu \operatorname{sign}(\overset{(n)}{\mathbf{T}}{}^{I}_{s})\overset{(1)}{\mathbf{U}}{}^{B}_{scc} \\ -\overset{(1)}{\mathbf{T}}{}^{B}_{noc} & 0 & \overset{(1)}{\mathbf{T}}{}^{B}_{soc} & -\overset{(1)}{\mathbf{U}}{}^{B}_{noc} + \mu \operatorname{sign}(\overset{(n)}{\mathbf{T}}{}^{I}_{s})\overset{(1)}{\mathbf{U}}{}^{B}_{soc} \end{bmatrix} \tag{19}$$

for debonded elements

$$\overset{(1)}{\mathbf{C}} := \begin{bmatrix} \overset{(1)}{\mathbf{T}}{}^{A}_{noc} & 0 & \overset{(1)}{\mathbf{T}}{}^{A}_{soc} & 0 \\ \overset{(*)}{\mathbf{T}}{}^{A}_{ncc} & 0 & \overset{(*)}{\mathbf{T}}{}^{A}_{scc} & 0 \\ 0 & \overset{(*)}{\mathbf{T}}{}^{B}_{ncc} & 0 & \overset{(*)}{\mathbf{T}}{}^{B}_{scc} \\ 0 & \overset{(1)}{\mathbf{T}}{}^{B}_{noc} & 0 & \overset{(1)}{\mathbf{T}}{}^{B}_{soc} \end{bmatrix} \tag{20}$$

for bonded or rebonded elements (see definition (16) also)

$$\overset{(1)}{\mathbf{C}} := \begin{bmatrix} \overset{(1)}{\mathbf{T}}{}^{A}_{noc} & \overset{(1)}{\mathbf{T}}{}^{A}_{soc} & -\overset{(1)}{\mathbf{U}}{}^{A}_{noc} & -\overset{(1)}{\mathbf{U}}{}^{A}_{soc} \\ \overset{(*)}{\mathbf{T}}{}^{A}_{ncc} & \overset{(*)}{\mathbf{T}}{}^{A}_{scc} & -\overset{(1)}{\mathbf{U}}{}^{A}_{ncc} & -\overset{(1)}{\mathbf{U}}{}^{A}_{scc} \\ -\overset{(*)}{\mathbf{T}}{}^{B}_{ncc} & -\overset{(*)}{\mathbf{T}}{}^{B}_{scc} & -\overset{(1)}{\mathbf{U}}{}^{B}_{ncc} & -\overset{(1)}{\mathbf{U}}{}^{B}_{scc} \\ -\overset{(1)}{\mathbf{T}}{}^{B}_{noc} & -\overset{(1)}{\mathbf{T}}{}^{B}_{soc} & -\overset{(1)}{\mathbf{U}}{}^{B}_{noc} & -\overset{(1)}{\mathbf{U}}{}^{B}_{soc} \end{bmatrix} \tag{21}$$

where the respective unknown node values are

$$[\overset{(n)}{u}{}^A_n,\ \overset{(n)}{u}{}^A_s,\ \overset{(n)}{u}{}^B_s,\ \overset{(n)}{t}{}^A_n]\quad \text{in the sliding elements} \tag{22}$$

$$[\overset{(n)}{u}{}^A_n,\ \overset{(n)}{u}{}^B_n,\ \overset{(n)}{u}{}^A_s,\ \overset{(n)}{u}{}^B_s]\quad \text{in the debonded elements} \tag{23}$$

$$[\overset{(n)}{u}{}^A_n,\ \overset{(n)}{u}{}^A_{sr},\ \overset{(n)}{t}{}^A_n,\ \overset{(n)}{t}{}^A_s]\quad \text{in the rebonded elements} \tag{24}$$

If an interface node returns from sliding or separation to the stick mode, the relative slip $\{\overset{(n)}{g}\}$ of the interface elements has to be taken into account, i.e. it has to be added at the right hand side of Eq. (15):

$$[\overset{(1)}{Q}]\cdot\begin{Bmatrix}\overset{(n)}{u}{}^A_o\\ \overset{(n)}{u}{}^A_c\\ \overset{(n)}{t}{}^A_c\\ \overset{(n)}{u}{}^B_o\end{Bmatrix} = [\overset{(1)}{P}]\cdot\{\overset{(n)}{\bar{t}}{}^A_o\} - \begin{bmatrix}\alpha\,\overset{(1)}{T}{}^A_{soc}\\ \alpha\,\overset{(*)}{T}{}^A_{scc}\\ (1-\alpha)\overset{(*)}{T}{}^B_{scc}\\ (1-\alpha)\overset{(1)}{T}{}^B_{soc}\end{bmatrix}\cdot\{\overset{(n)}{g}_s\} + \begin{Bmatrix}\overset{(n)}{R}{}^A_o\\ \overset{(n)}{R}{}^A_c\\ \overset{(n)}{R}{}^B_c\\ \overset{(n)}{R}{}^B_o\end{Bmatrix}, \tag{25}$$

The value of the factor α can be chosen arbitrarily in the range 0 to 1. This statement (see also Anderson [12]) has been checked in running some parametric studies, e.g. a heavy surface foundation. Trial values α=0.0, 0.5, and 1.0 gave almost identical interface displacements and tractions. The examples in this paper are calculated with α = 0.0, i.e. $u^A_{sr} = u^A_s$ and $u^B_{sr} = u^B_s - g_s$.

Finally, after such modifications in the matrix $\overset{(1)}{Q}$ corresponding to the current situation along the interface, the Eqs. (15) or (25) have to be solved a second time for providing the unknown of the new, i.e. partially sliding, debonded or rebonded status.

Numerical Studies

Consider a rectangular shaped, massive, and elastic strip foundation (see Fig. 1) of height H = 1.0 m and width 2B = 4.0 m resting on the free surface of a supporting homogeneous soil medium. Adopting plane strain assumptions, the soil material is assumed to have a mass density ρ_s = 2000 kg/m^3, Poisson's ratio υ_s = 0.33, and Young's modulus $E_s = 4.788\cdot10^6$ KN/m^2, while the foundation material is taken to be more dense and stiffer with a mass density $\rho_f = 2500$ kg/m^3, Poisson's ratio $\upsilon_f = 0.20$, and Young's modulus $E_f = 1.968 \cdot 10^7$ KN/m^2. Hence, the velocities of the pressure wave and the shear wave are $c_{1s} = 1833.4$ m/s and $c_{2s} = 948.7$ m/s in the soil medium, and $c_{1f} = 2957.5$ m/s and $c_{2f} = 1811.1$ m/s in the foundation, respectively.

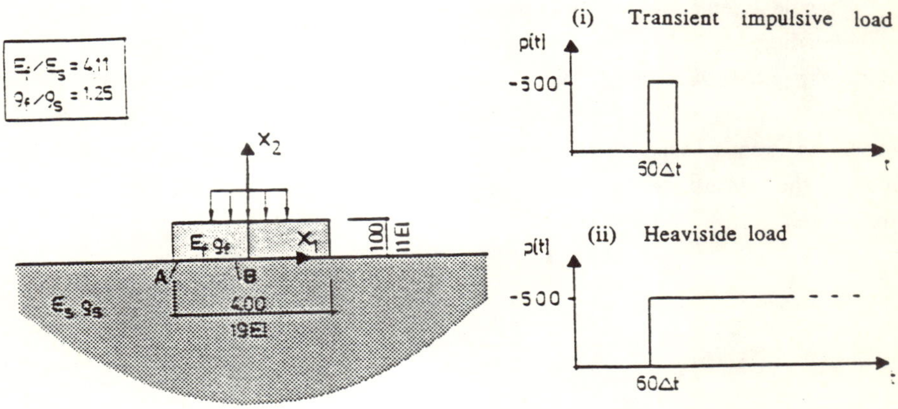

Fig. 1. Geometry, discretization, and excitations of a flexible massive strip foundation resting on the elastic halfspace

The foundation boundary is discretized by using 11 elements on both free side faces, 13 elements on the top surface, and 19 elements along the interface, i.e. by 54 elements in the whole. The length of these elements varies; it is chosen to decrease towards the expected stress concentration points, e.g. along the interface from 0.45 m of both the "central" elements to 0.01 m of the "corner" elements. In addition to the 19 interface elements, the free soil surface discretization with 11 unequal elements on each side (the first, next the foundation, is 0.055 m, the last is 3.0 m long) extends to a distance of 10.0 m. In order to evaluate the foundation dead weight integral numerically using Gaussian quadrature formula, the foundation domain has been subdivided into interior cells. The time increment used is $\Delta t = 2.258 \cdot 10^{-4}$ secs; the whole observation period is $177\Delta t = 0.04$ secs.

Now, the results of two studies are presented in order to demonstrate the applicability of the time-stepping BEM procedure, and to show the importance of considering unilateral frictional contact along the foundation-soil interface.

First, assuming a Coulomb friction coefficient $\mu=0.1$, the effects of two different impulsive loadings, acting along 2 meters on the central part of the foundation's upper surface (see Fig. 1), are examined:

(i) a transient vertical impulse with a duration of only two time steps, i.e. of $5.516 \cdot 10^{-4}$secs, and

(ii) a suddenly applying vertical load acting continously during the whole remaining observation period, i.e. a vertical "Heaviside" loading.

Both the excitations have the equal constant intensity $p(t)=p_0=-500$ KN/m, and start acting only after 0.01348 secs ($=60\ \Delta t$). This time is needed in order to reach the distribution of normal compressive stress resultants along the interface elements equilibrating the dead weight of the foundation.

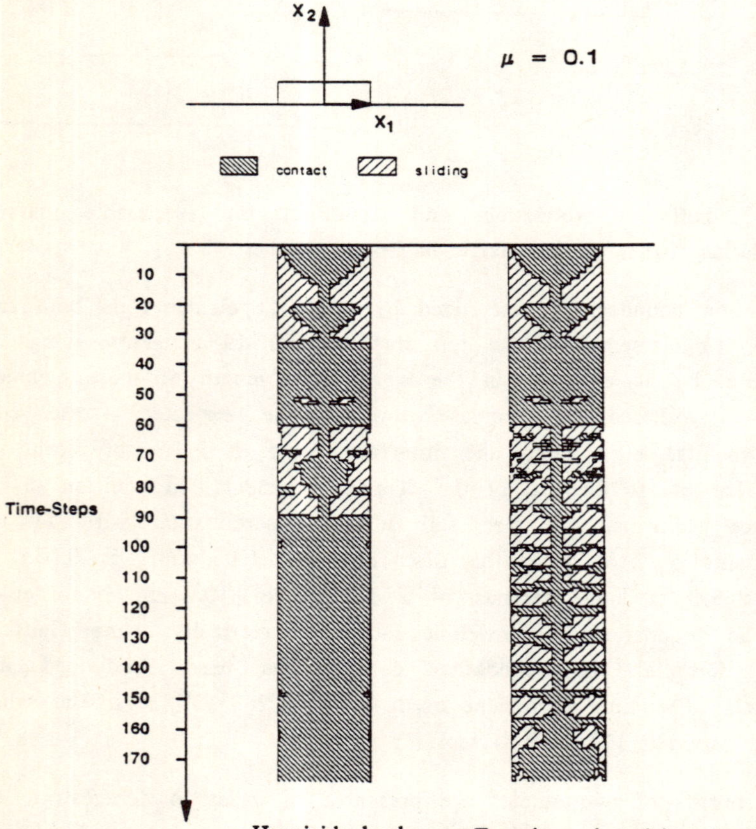

Fig. 2. Time-dependent distribution of the bonded, the sliding, and the uplifted interface areas: comparison of transient and continuous loading.

In Fig. 2, the time history of the unilateral contact behavior, i.e. the time-dependent distribution of the uplifted, the sliding, and the bonded regions in the interface is plotted for both the transient impulse and the Heaviside loading. Obviously, starting at the 60th time step, both excitations cause large areas of sliding, but, different from the short impulse which effects only slowly damped sliding cycles, the suddenly applied Heaviside load generates sliding only during the first 30 time steps.

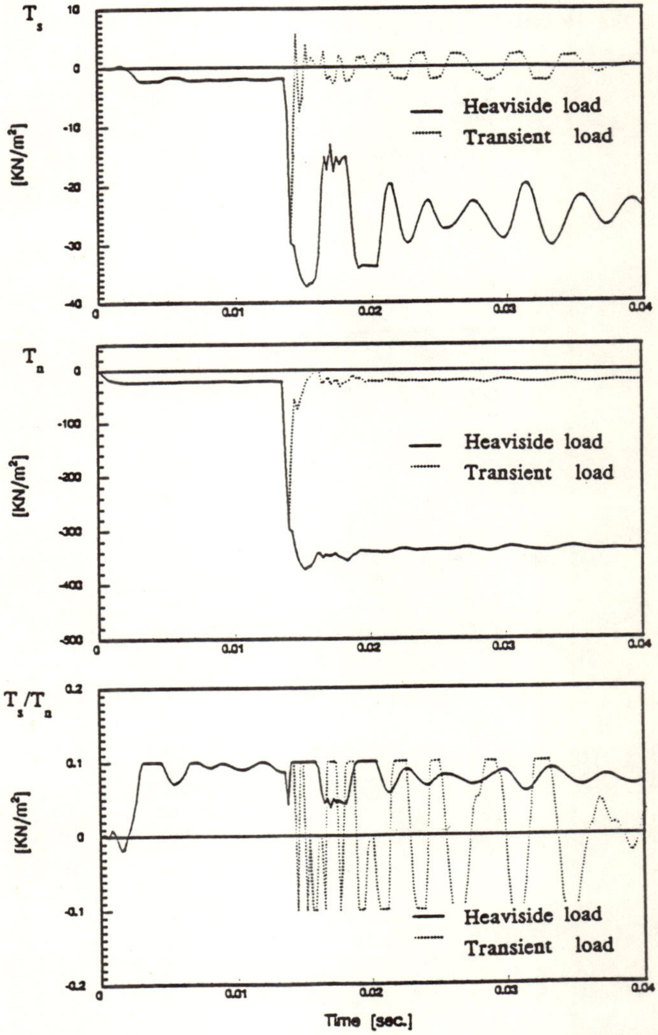

Fig. 3. Time history of the normal traction, the tangential traction, and the ratio of both at point B: comparison of transient and continuous load

Certainly, this is due to the fact that the continuously acting Heaviside load produces higher compressive stress resultants along the interface. This fact becomes apparent by comparing the time history of the tractions at a point on the interface, e.g. at point B located 0.45 m from the interface center. In Fig. 3, the normal traction T_n, the tangential traction T_s, and the ratio T_s/T_n are depicted for both the excitations. Since the Heaviside load increases the compressive normal stress, the consequently also raised friction force limit is reached only twice after the start of the Heaviside load action.

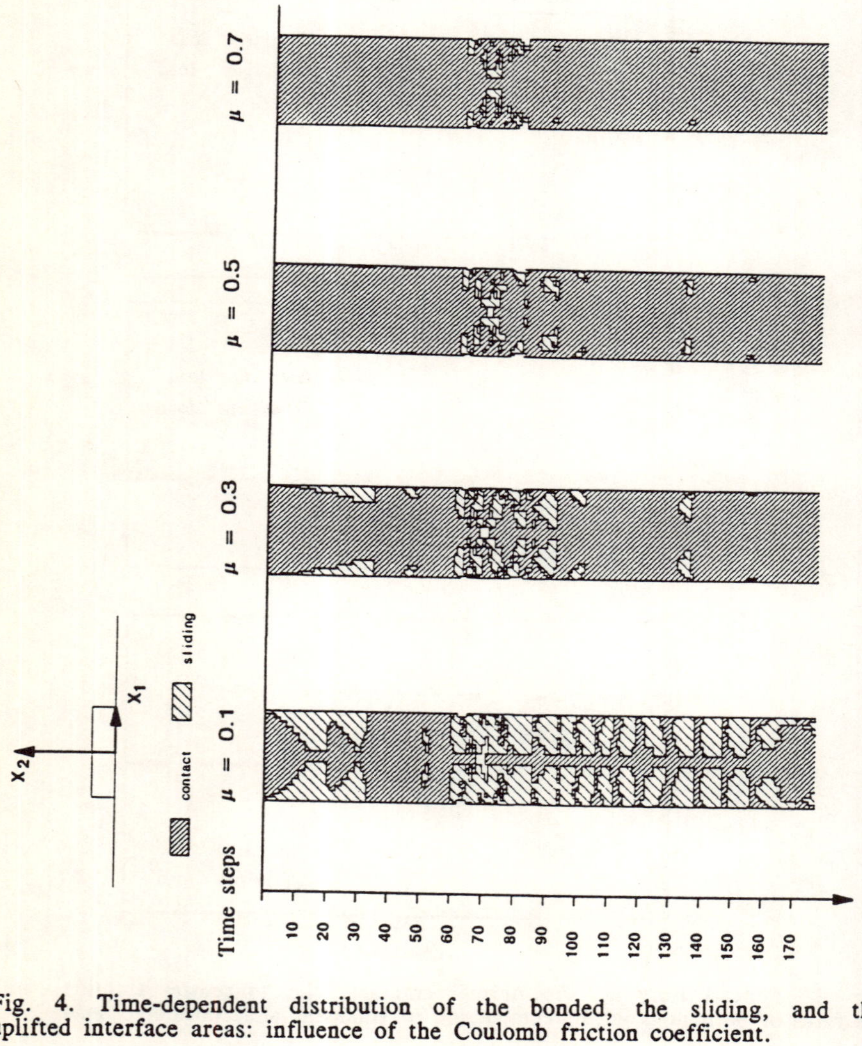

Fig. 4. Time-dependent distribution of the bonded, the sliding, and the uplifted interface areas: influence of the Coulomb friction coefficient.

In the following study considering the transient impulsive loading (i) again, the effects of raising the frictional resistance are analyzed. First, in Fig. 4, the great reduction of the sliding areas by taking higher Coulomb friction coefficients is very impressive.

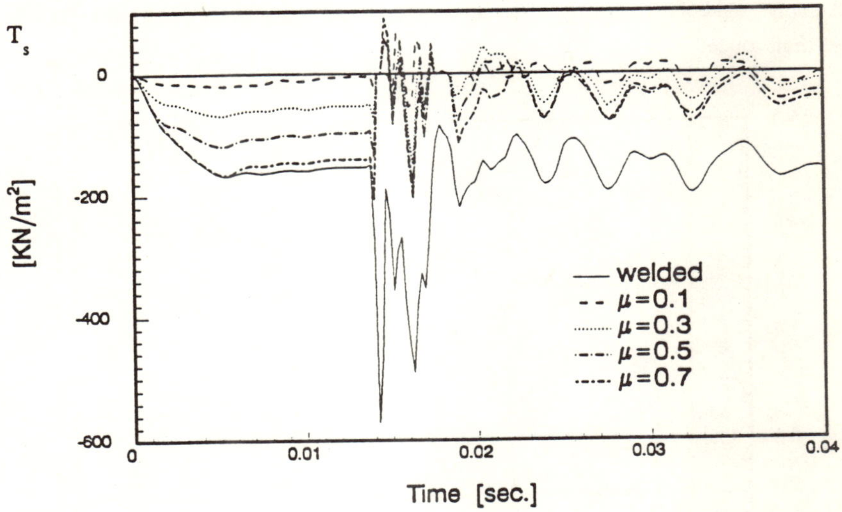

Fig. 5. Time history of the tangential traction at point A: influence of different Coulomb friction coefficients μ.

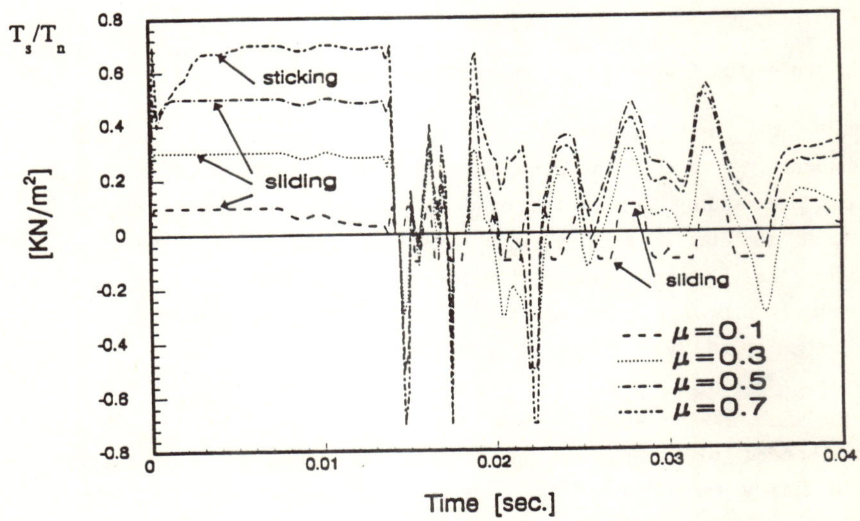

Fig. 6. Time history of the ratio between the tangential and the normal traction at point A: influence of the Coulomb friction coefficient μ.

Figs. 5 and 6, depicting the time history of the tangential traction, and again, the ratio between the tangential and the normal tractions at point A, 0.005 m from the left interface corner of the foundation, show clearly the expected effects. The higher the frictional resistance, the more the "oscillations" are damped. Thus, only for the small Coulomb friction coefficient $\mu=0.1$ both the friction limits indicating sliding are reached more than once.

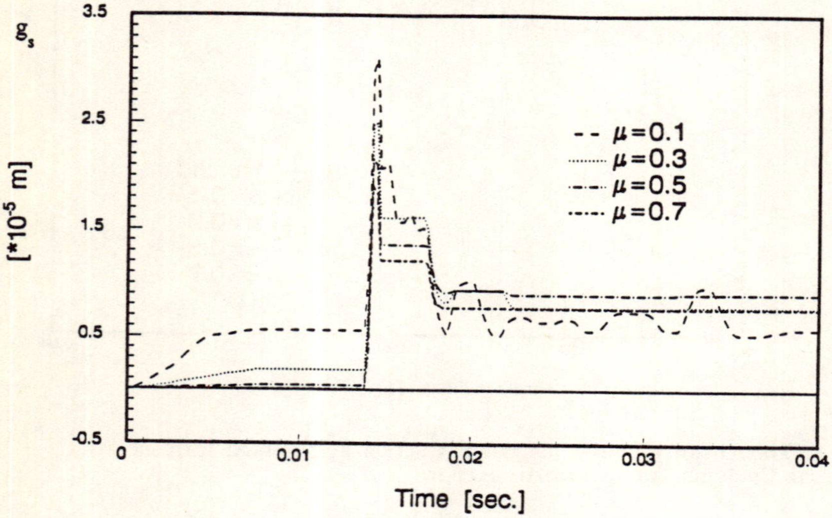

Fig. 7. Time history of the foundation's relative slip at point A: influence of the Coulomb friction coefficient μ.

Finally, the time history of the relative slip (relative tangential displacement) $g_s = u^f_s + u^s_s$ at point A is plotted in Fig. 7. Obviously, as expected, the sliding due to the dead weight (the first 0.0135 secs= $60\Delta t$) produces the larger relative shift (constant after some time steps), the smaller the Coulomb friction coefficient is assumed to be. The subsequent transient excitation, the short impulsive loading causes a different, partially unexpected behavior.

In the case of the small friction coefficient $\mu=0.1$, the frequent changes from bonded to sliding contact produce also frequently alternating slips which finally tend towards the "static" dead weight slip. The higher the frictional resistance, the lower is the number of slip changes after the impulse action. In the case $\mu=0.7$, for instance, only two different constant slip levels are reached. The final slip of the (μ=0.3)-case is smaller than that of the (μ=0.5)-solution, and, moreover, identical with the final value related to μ=0.7. The (μ=0.3)-solution reaches this rather unexpected value by the "last" sliding which does not happen in the higher frictional resistance case μ=0.5. This may be the reason for that ending at a smaller slip level, i.e. for ending more "relaxed".

Conclusions

Determining the sliding and nonsliding contact areas is the most challenging aspect of a frictional contact problem. Here, an iterative time-stepping Boundary Element procedure has been presented for analyzing time dependent contact with Coulomb friction. Applications to sample problems indicate the approach is efficient. The formulation can and shall be extended to three-dimensional contact problems.

Acknowledgements

The financial support by the Deutsche Forschungsgemeinschaft through project SFB 151/C1, and by the Spanish-German collaborative research grant 313-AI (Acciones Integradas) is gratefully acknowledged.

References

1. Wilson, E. A.; Parsons, B.: Finite element analysis of elastic contact problems using differential displacements. Int. J. Numer. Methods in Engineering 2 (1970) 387-395.
2. Ohte, B.: Analysis of elastic contact stress by using finite element method. Trans. Jap. Soc. Mech. Engng. 38 (1972) 2210-2216 (in Japanese).
3. Okamoto, N.; Nakazawa, M.: Finite element incremental contact analysis with various frictional conditions. Int. J. Numer. Methods in Engineering 14 (1979) 337-357.
4. Oden, J.T.; Pires, E.B.: Nonlocal and nonlinear friction laws and variational principles for contact problems in elasticity. J. Appl. Mechanics 50 (1983) 67-76.
5. Gallego, F.J.; Anza, J.J.: A mixed finite element method for the elastic contact problem. Int. J. Numer. Methods in Engineering 28 (1989) 1249-1264.
6. Zolti, E. A: Finite element procedure to time dependent contact analysis. Computers & Structures 17 (1983) 555-561.
7. Chaudhary, A.B.; Bathe, K.-J.: A solution method for static and dynamic analysis of three-dimensional contact problems with friction. Computers & Structures 24 (1986) 855-873.
8. Vu Van; T.: Zur Behandlung von Stoß- Kontaktproblemen mit Reibung unter Verwendung der Finite-Element-Methode. Bericht F90/2, Institut für Baumechanik und Numerische Mechanik, Universität Hannover 1990.

9. Talaslidis, D.; Panagiotopoulos, P.D.: A linear finite element approach to the solution of the variational inequalities arising in unilateral contact problems in structural dynamics. Int. J. Numer. Methods in Engineering 18 (1982) 1505-1520.
10. Beskos, D.E.: Boundary element methods in dynamic analysis. Applied Mech. Review 40 (1987) 1-23.
11. Andersson, T.; Fredriksson, B.; Allan-Persson, B.G.: The boundary element method applied to two-dimensional contact problems. In: Brebbia, C.A. (ed.): New Developments in Boundary Element Methods. Southampton:CML Publ 1980.
12. Andersson, T.: The boundary element method applied to two-dimensional contact problems with friction. 239-258 in: Brebbia, C.A. (ed.): Proc. Third Int. Seminar on Boundary Element Methods. Berlin: Springer-Verlag 1981.
13. Andersson, T.; Allan-Persson, B.G.: The boundary element method applied to two-dimensional contact problems. in: Brebbia, C.A. (ed.): Progress in Boundary Element Methods 2. London: Pentech Press 1983.
14. Paris, F.; Garrido, J.A.: On the use of discontinuous elements in two-dimensional contact problems. (13-17)-(13-39) in: Brebbia, C.A.; Maier, G. (eds.): Boundary Elements VII , CML Publ., Southampton, 1985.
15. Jin, H.; Runesson, K.; Samuelsson, A.: Application of the boundary element method to contact problems in elasticity with a nonclassical friction law. 397-415 in: Brebbia, C.A.; Wendland, W.L.; Kuhn, G. (eds.): Boundary Elements IX, Vol 2: Stress Analysis Applications. Berlin: Springer Verlag 1986.
16. Selvadurai, A.P.S.; Ap, M.C.: Response of inclusions with interface separation, friction and slip, (14-109)-(14-127) in: Brebbia, C.A.; Maier, G.: Boundary Elements VII. Berlin: Springer-Verlag 1985.
17. Bezine, G.; Fortune, D.: Contact between plates by a new direct boundary integral equation formulation. Int. J. Solids Structures 20 (1984) 739-746.
18. Panagiotopoulos, P.D.: Boundary integral "equation" methods for the Signorini-Fichera problem. (12-73)-(12-83) in: Brebbia C.A., Maier, G. (eds.): Boundary Elements VII. Southampton: CML Publ. 1985.
19. Lazaridis, P.P.; Panagiotopoulos, P.D.: Boundary variational principles for inequality structural analysis problems and numerical applications. Computers & Structures 25 (1987) 35-49.
20. Panagiotopoulos, P.D.: Boundary integral equation methods for the friction problem. Engineering Analysis 4 (1987) 100-105.
21. Tralli, A.; Alessandri, C.: On B.E.M. solutions of elastic frictionless contact problems. 337-352 in: Brebbia, C.A. (ed.): Boundary Elements X, Vol 3: Stress Analysis. Berlin: Springer Verlag 1989.
22. Alliney, S.; Tralli, A.; Alessandri, C.: Boundary variational formulations and numerical solution techniques for unilateral contact problems. Comput. Mechanics 6 (1990) 247-257.
23. Spyrakos, C.C.: Dynamic behaviour of foundations in bilateral and unilateral contact. The Shock and Vibration Digest 20 (1988) 3-12.
24. Antes, H.; Steinfeld, B.: Unilateral contact in dynamic soil-structure interaction by a time domain boundary element method. 45-58 in: Brebbia, C.A. (ed.): Boundary Elements X, Vol 4: Geomechanics, Wave Propagation and Vibrations. Berlin: Springer Verlag 1988.
25. Antes, H.; Steinfeld, B.: Seismic wave response of elastic structures in unilateral soil contact. Proc. 2nd National Cong. on Mechanics, Athens, Greece, 1989 (in press).
26. Dominguez, J.; Abascal, R.: Non-linear effects due to the contact

Brebbia, C.A.; Cheng, A. (eds.): Comp. Engng. with Boundary Elements 2: Solid and Computational Problems. Southampton: CML Publ. 1990.

27. Antes, H.; Steinfeld, B.: The influence of friction on unilateral dynamic soil-structure contact. 3-17 in: Brebbia, C.A.; Cheng, A. (eds.): Comp. Engng. with Boundary Elements 2: Solid and Computational Problems. Southampton: CML Publ. 1990.
28. Antes, H.; Steinfeld, B.: Dynamic response of elastic structures in unilateral soil contact with friction. Krätzig, W. (ed.): Proc. EURODYN. Rotterdam: Balkema 1990.
29. Antes, H.: A boundary element procedure for transient wave propagation in two-dimensional isotropic elastic media. Finite Elements in Analysis and Design 1 (1985) 313-322.

Boundary Element Methods for Problems Involving Unilateral Boundary Conditions

H. SCHMITZ, G. SCHNEIDER, W. WENDLAND
University of Stuttgart, Germany

Summary

The Poisson equation with mixed Dirichlet-Signorini boundary conditions is reduced to a boundary variational inequality. A penalty method leads to an approximation by a sequence of nonlinear boundary integral equations and corresponding boundary element methods. Stability and convergence of Galerkin's method and the collocation method are analyzed. A suitable balance of the meshsize h and the penalty parameter yields an asymptotic convergence of order $h^{\frac{1}{2}}$.

Introduction

The application of boundary element methods to boundary value problems with unilateral boundary conditions can already be found in many applications. For the Poisson equation as a simple model problem, the corresponding analysis was developed in [12] , [13] and [23]. The corresponding method for unilateral boundary value problems in elasticity was proposed in [17].

In particular, Spann [23] considered in his thesis the Laplacian and inhomogeneous mixed Dirichlet-Neumann-Signorini boundary conditions in a bounded domain $\Omega \subset \mathbb{R}^2$ whose boundary consists of finitely many boundary parts corresponding to the different boundary conditions. Applications for this mathematical model can be found in [1] and [2]. If the different boundary parts are separated then the solution is in the Sobolev space $H^2(\Omega)$ provided the given data are smooth enough (see [6] , [15]).

If the different boundary parts are connected or the boundary has corners then the solution develops singularities which are still not completely analyzed. Kawohl in [15] obtained regularity estimates in weighted spaces. However, these results are probably not sharp and, correspondingly, the performance of numerical methods cannot be analyzed. Here, we therefore consider the case with disjoint boundary parts where a thorough analysis of a penalty method becomes possible.

We first formulate a penalty method which corresponds to the Yosida approximation of the Signorini boundary condition. In [22] the convergence in the energy norm of this method is shown if the penalty parameter tends to zero.
To the penalty method we apply a boundary element Galerkin approximation and also a boundary element collocation method. For the Galerkin method, the analysis from [19] for similar nonlinear integral equations was extended in [22] to the case of the penalty parameter tending to zero and corresponding asymptotic error analysis.
Here we follow the ideas of [21] where the collocation method from [18] was also extended to the penalty approximation of the Dirichlet-Signorini problem. In particular, we are able to present a unified analysis of the Galerkin and the collocation boundary element schemes using arguments of the degree of mappings. We present asymptotic stability and error estimates by a suitable combination of the mesh-width and the penalty parameter. The numerical algorithm is based on Broyden's method for solving the nonlinear discrete equations. Numerical experiments at the end of the paper confirm the effectivity of the method.

The Dirichlet-Signorini problem for the Laplacian

Let $\Omega \subset \mathbb{R}^2$ be a bounded domain. The boundary Γ is assumed to be sufficiently smooth consisting of two disjoint parts Γ_D and Γ_S, i.e. with Dirichlet conditions on Γ_D and Signorini conditions on Γ_S. We consider the solution u of

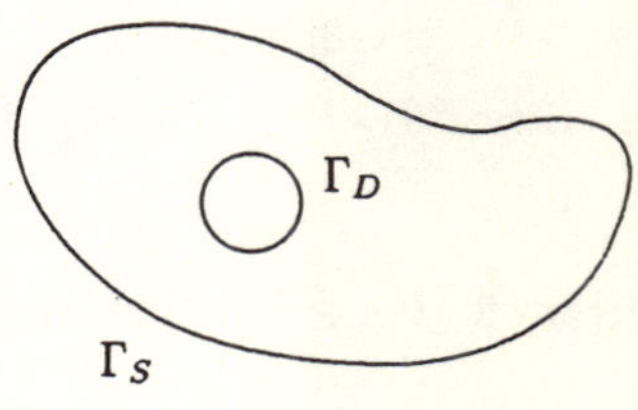

$$\begin{aligned} -\Delta u &= f && \text{in } \Omega, \\ u &= 0 && \text{on } \Gamma_D, \\ \frac{\partial u}{\partial n}\cdot u = 0\,,\ \frac{\partial u}{\partial n} &\geq 0\,,\ u \geq 0 && \text{on } \Gamma_S. \end{aligned} \tag{1}$$

The variational formulation of this problem is well known and reads as (see [10])

Problem P: Find $u \in K := \left\{ v \in H^1(\Omega) \mid v|_{\Gamma_D} = 0\,,\ v|_{\Gamma_S} \geq 0 \right\}$ with

$$\int_\Omega \nabla u \cdot \nabla(v-u)\, dx \geq \int_\Omega f\,(v-u)\, dx \quad \text{for all } v \in K.$$

This variational inequality is uniquely solvable due to the coerciveness of the Dirichlet bilinear form

$$a(u,v) := \int_\Omega \nabla u \cdot \nabla v \, dx \tag{2}$$

on

$$H_0^1(\Omega) := \left\{ v \in H^1(\Omega) \mid v|_{\Gamma_D} = 0 \right\} .$$

Problem P can be discretized by finite element schemes directly (see [9], [10]). Here, however, we want to use a boundary element method. For this purpose we have to remove the right-hand side f. Hence, in the following, let u_f be the solution of the Poisson problem

$$\begin{aligned} -\Delta u_f &= f \qquad && \text{in} \quad \Omega, \\ u_f &= 0 && \text{on} \quad \Gamma. \end{aligned} \tag{3}$$

For the desired boundary integral equation let us introduce the boundary integral operators

$$K\varphi(x) := \frac{1}{\pi} \int_\Gamma \varphi(y) \frac{\partial}{\partial n_y} \log|x-y| \, ds_y \,, \tag{4}$$

$$V\psi(x) := -\frac{1}{\pi} \int_\Gamma \psi(y) \log|x-y| \, ds_y \,. \tag{5}$$

Let $T : H^{\frac{1}{2}}(\Gamma) \to H^{-\frac{1}{2}}(\Gamma)$ denote the Steklov-Poincaré mapping which assigns to τ the normal derivative of the harmonic function v with $v|_\Gamma = \tau$. The operator $V : H^{-\frac{1}{2}}(\Gamma) \to H^{\frac{1}{2}}(\Gamma)$ is invertible provided the conformal radius of Γ is not equal to 1. This assumption is satisfied if the diameter of Ω is less than 1 which will be assumed throughout the paper. Hence, $T = V^{-1}(I-K)$. By Green's representation formula we get the following theorem.

Theorem 1. *If $u \in H^1(\Omega)$ solves Problem P then the boundary datum $u|_\Gamma$ solves*

Problem Q: *Find $t \in C := \left\{ \tau \in H^{\frac{1}{2}}(\Gamma) \mid \tau|_{\Gamma_D} = 0,\ \tau|_{\Gamma_S} \geq 0 \right\}$ such that*

$$(T\,t, \tau - t)_\Gamma \geq -\left(\frac{\partial u_f}{\partial n}, \tau - t \right)_\Gamma \qquad \textit{for all } \tau \in C.$$

If $t \in H^{\frac{1}{2}}(\Gamma)$ is a solution of Problem Q then the formula

$$\begin{aligned} u(x) = u_f(x) &+ \frac{1}{2\pi} \int_\Gamma \frac{\partial}{\partial n_y} \log|x-y| \, t(y) \, ds_y \\ &- \frac{1}{2\pi} \int_\Gamma \log|x-y| \, T\,t\,(y) \, ds_y \quad \textit{for } x \in \Omega \end{aligned} \tag{6}$$

represents the solution of Problem P.

For the proof see [22]. A discretization of the variational inequality of Problem Q can be found in [23]. Here we consider a different method, a penalty method. For this purpose we introduce $P\tau$ to be the L^2–Projection of τ onto the L^2–closure of C. The projection P satisfies the identity

$$P\tau(x) = \begin{cases} \tau^+(x) := \sup\{0, \tau(x)\} & \text{for } x \in \Gamma_S\,, \\ 0 & \text{for } x \in \Gamma_D\,. \end{cases} \tag{7}$$

With P define the nonlinear operator B by

$$B\tau := \tau - P\tau. \tag{8}$$

Then the penalty solutions t_ε for $\varepsilon > 0$ are defined by solving the following problem.

Problem Q_ε: Find $t_\varepsilon \in H^{\frac{1}{2}}(\Gamma)$ with

$$(T\,t_\varepsilon, \tau)_\Gamma + \frac{1}{\varepsilon}\,(B\,t_\varepsilon\,, \tau)_\Gamma = -\left(\frac{\partial u_f}{\partial n}, \tau\right)_\Gamma \qquad \text{for all } \tau \in H^{\frac{1}{2}}(\Gamma).$$

By the identity

$$\frac{1}{2}\,(I-K)\,u_f\,(x) - \frac{1}{2}\,V\,\frac{\partial u_f}{\partial n}\,(x) = \frac{1}{2\pi}\int_\Omega f(y)\,\log|x-y|\,dy \quad \text{for } x \in \Gamma\,, \tag{9}$$

Problem Q_ε corresponds to the nonlinear boundary integral equation

$$(I-K)\,t_\varepsilon\,(x) + \frac{1}{\varepsilon}\,V\,B\,t_\varepsilon\,(x) = \frac{1}{\pi}\int_\Omega f(y)\,\log|x-y|\,dy \tag{10}$$

for t_ε and the Yosida approximation of the variational inequality Problem Q.

Proposition 2. *Problem Q_ε is uniquely solvable and t_ε converges weakly to the solution t_0 of problem Q in $H^{\frac{1}{2}}(\Gamma)$ for $\varepsilon \to 0$. Furthermore, the functions t_ε are uniformly bounded in $H^{\frac{3}{2}}(\Gamma)$ and converge weakly to t_0 in this space.*

The proof of this proposition can be found in [15].

Theorem 3. *The penalty solutions t_ε of Problem Q_ε satisfy the asymptotic error estimate*

$$\|t_\varepsilon - t_0\|_{\frac{1}{2}} \le C\,\varepsilon \tag{11}$$

where t_0 denotes the solution of Problem Q and where C depends on u_f but is independent of ε.

Proof. The proof is based on Proposition 1.2, the fact that P maps $H^{\frac{1}{2}}(\Gamma)$ into $H^{\frac{1}{2}}(\Gamma)$ (see [16]) and a Gårding inequality for the nonlinear mapping T,

$$(T\tau, \tau)_\Gamma \geq \gamma \|\tau\|^2_{\Gamma_D} - c(\tau, \tau)_{\Gamma_D} \quad \text{for all } \tau \in H^{\frac{1}{2}}(\Gamma)$$

where $\gamma > 0$ and $c \geq 0$. The complete proof can be found in [22]. □

Boundary element approximations

Here we consider the two standard boundary element methods for Equation (10), namely the Galerkin method and collocation at the break points with piecewise linear splines for solving the nonlinear integral equation

$$At_\varepsilon(x) := (I-K)\,t_\varepsilon(x) + \frac{1}{\varepsilon} V B\, t_\varepsilon(x) = \frac{1}{\pi}\int_\Omega f(y)\,\log|x-y|\,dy \;=: g(x) \tag{12}$$

in the Banach space $H^{\frac{1}{2}}(\Gamma)$. For this purpose let Δ_n be a sequence of grids with meshsize $h \to 0$. Let $S(\Delta_n)$ be the corresponding space of piecewise linear continuous splines with respect to a given parametric representation of Γ. The break points of $S(\Delta_n)$ on Γ are denoted by $\{x_j\}_{j=1}^n$. As usual, for fixed n we associate with Δ_n the basis $\{\mu_j\}_{j=1}^n$ of $S(\Delta_n)$ defined by the usual hat functions with $\mu_j(x_k) = \delta_{jk}$. Then the approximate solution has the form

$$t_\varepsilon^n(x) = \sum_{j=1}^n \gamma_j \mu_j(x)\,. \tag{13}$$

The Galerkin method for (12) reads as to find $\{\gamma_j\}_{j=1}^k$ as the solution of the nonlinear system of equations

$$\sum_{j=1}^n ((I-K)\mu_j, \mu_k)_\Gamma\, \gamma_j + \frac{1}{\varepsilon}\left(VB\sum_{j=1}^n \gamma_j\mu_j, \mu_k\right)_\Gamma = (g, \mu_k)_\Gamma, \quad k = 1, \ldots, n. \tag{14}$$

With the help of the L_2-orthogonal projection $P_n : H^{\frac{1}{2}}(\Gamma) \to S(\Delta_n)$, the Galerkin equations (14) can also be written as

$$A_n t_\varepsilon^n := P_n A t_\varepsilon^n = P_n g. \tag{15}$$

The collocation method for (12) reads as to find $\{\gamma_j\}_{j=1}^n$ from the collocation equations

$$\sum_{j=1}^n (I-K)\mu_j(x_k)\gamma_j + \frac{1}{\varepsilon}\left(VB\sum_{j=1}^n \gamma_j\mu_j\right)(x_k) = g(x_k) \text{ for } k = 1, \ldots, n. \tag{16}$$

Clearly, for (16) we need at least $g \in C(\Gamma) \cap H^{\frac{1}{2}}(\Gamma)$. With the interpolation operator $I_n : C(\Gamma) \to S(\Delta_n)$ interpolating at the break points, (16) can be written in short as

$$A_n t_\varepsilon^n := I_n A t_\varepsilon^n = I_n g. \tag{17}$$

Lemma 4. *Let $A_n := P_n A|_{S(\Delta_n)}$ in* (15) *or $A_n := I_n A|_{S(\Delta_n)}$ in* (17). *Then, in both cases, $A_n : S(\Delta_n) \to S(\Delta_n)$ is continuous for every $n \in \mathbb{N}$ if $S(\Delta_n)$ is equipped with the $H^{\frac{1}{2}}(\Gamma)$-norm.*

Proof. Since $P_n(I-K)|_{S(\Delta_n)}$ and $I_n(I-K)|_{S(\Delta_n)}$ are continuous as linear operators, it remains to show the continuity of the nonlinear part. Since $H^{\frac{1}{2}}(\Gamma)$ is continuously imbedded into $L^2(\Gamma)$ and B is continuous on $L^2(\Gamma)$, the composition VB is a continuous mapping from $H^{\frac{1}{2}}(\Gamma)$ into $H^1(\Gamma)$. Here $P_n : H^1(\Gamma) \to H^{\frac{1}{2}}(\Gamma)$ as well as $I_n : H^1(\Gamma) \to H^{\frac{1}{2}}(\Gamma)$ are continuous. □

In the following, we shall consider the two limits $\varepsilon \to 0$ and $n \to \infty$, i.e. $h \to 0$ and show that the methods are stable and convergent. To this end we need the approximation properties of $S(\Delta_n)$, P_n and I_n. Let us suppose that the family Δ_n is given by smoothly graded meshes in the sense of [3]. Then

$$\|P_n v - v\|_t \le c h^{s-t} \|v\|_s \tag{18}$$

and

$$\|I_n v - v\|_t \le c h^{s-t} \|v\|_s \tag{19}$$

for $0 \le t < \frac{3}{2}$, $\frac{1}{2} < s \le 2$, $t \le s$ (see e.g. [4] , [5] , [8] , [20]).

Lemma 5. *For $\varepsilon \le 1$ the sequence A_n for the Galerkin method* (15) *and for the collocation method* (17), *correspondingly, are asymptotically inversely stable, i.e. there exist constants $c, d > 0$ such that*

$$\|A_n u - A_n v\|_{\frac{1}{2}} \ge c\ \|u - v\|_{\frac{1}{2}} \tag{20}$$

for all n with $h^{\frac{1}{2}} \le d\varepsilon$ and $u, v \in S(\Delta_n)$.

Proof. The proof rests on the strong V^{-1}-monotonicity of A

$$\left(Au - Av\,,\, V^{-1}(u-v)\right)_\Gamma \ge c\ \|u-v\|^2_{\frac{1}{2}}$$

which was shown in [22] for $\varepsilon \le 1$. We now proceed with the proof for the collocation. For Galerkin's method just replace I_n by P_n, the remaining proof is the same. We conclude

$$\left(A_n u - A_n v\,,\, V^{-1}(u-v)\right)_\Gamma \ge c_1\ \|u-v\|^2_{\frac{1}{2}} - \|(I - I_n)(Au - Av)\|_{\frac{1}{2}}\ \|u-v\|_{\frac{1}{2}}\,.$$

Using the approximation property of the interpolation operator I_n we get

$$\begin{aligned} \|(I - I_n)(Au - Av)\|_{\frac{1}{2}} &\le \|(I-I_n)K(u-v)\|_{\frac{1}{2}} + \varepsilon^{-1}\|(I-I_n)V(Bu - Bv)\|_{\frac{1}{2}} \\ &\le c_2 h^{\frac{3}{2}}\ \|u-v\|_{\frac{1}{2}} + c_3\varepsilon^{-1} h^{\frac{1}{2}}\ \|u-v\|_0 \\ &\le c_4\left(h^{\frac{3}{2}} + \frac{h^{\frac{1}{2}}}{\varepsilon}\right)\|u-v\|_{\frac{1}{2}} \end{aligned}$$

by using the mapping properties of K, V and B. Hence,

$$\|A_n u - A_n v\|_{\frac{1}{2}} \; \|u-v\|_{\frac{1}{2}} \geq C \; \|u-v\|^2_{\frac{1}{2}}$$

for $h^{\frac{3}{2}} + \frac{h^{\frac{1}{2}}}{\varepsilon} \leq \frac{c_1}{2c_4}$ with $C = \frac{c_1}{2} > 0$ independent of h and ε. □

Remark: The above proof follows closely [18] and [19].

Proposition 6. *There exists a constant $d > 0$ such that for $\varepsilon \leq 1$ the Galerkin equations (14) and the collocation equations (16) are uniquely solvable as long as $h^{\frac{1}{2}} \leq d\,\varepsilon$.*

Proof. In both cases A_n is continuous and satisfies the a-priori estimate (20). Let $\{y_m\} \subset S(\Delta_n)$ be a sequence of spline functions $y_m = A_n x_m$ converging to $y \in S(\Delta_n)$. By (20) x_m is a Cauchy sequence converging to $x \in S(\Delta_n)$. By the continuity of A_n we get $A_n x = y$ and, hence, the range of A_n is closed. The range of A_n is also open. Indeed, let $y_0 \in$ range A_n. Then the map $x \to A_n x - y_0$ has a critical value at x_0. Because of (20), the surface $\|x - x_0\|_{\frac{1}{2}} = 1$ is free of critical values for all mappings $x \to A_n x - y$ which are homotopic as long as $\|y - y_0\|$ is sufficiently small. Hence, $y \in$ range A_n, too, by mapping degree theory. Since range $A_n \neq \emptyset$ we have range $A_n = S(\Delta_n)$ and the solvability of (14) and of (16) follows. By (20) we also obtain uniqueness. □

Proposition 7. *The sequences t^n_ε converge strongly to t_ε and satisfy the asymptotic error estimate*

$$\|t^\varepsilon - t^\varepsilon_n\|_{\frac{1}{2}} \leq c\left(h^{\frac{3}{2}} + \frac{h^{\frac{3}{2}}}{\varepsilon}\right) \|t^\varepsilon\|_{\frac{3}{2}} .$$

Proof. By (20) we get for collocation

$$\begin{aligned}
\|I_n t^\varepsilon - t^\varepsilon_n\|_{\frac{1}{2}} &\leq c \, \|A_n I_n t^\varepsilon - A_n t^\varepsilon_n\|_{\frac{1}{2}} = c\,\|I_n A I_n t^\varepsilon - I_n A t^\varepsilon\|_{\frac{1}{2}} \\
&\leq c\,\|I_n K (I_n - I) t^\varepsilon\|_{\frac{1}{2}} + c\varepsilon^{-1} \|I_n (V B I_n t^\varepsilon - V B t^\varepsilon\|_{\frac{1}{2}} \\
&\leq c(1+\varepsilon^{-1}) \; h^{\frac{3}{2}} \; \|t_\varepsilon\|_{\frac{3}{2}} .
\end{aligned}$$

Replacing I_n by P_n, one gets the same result for Galerkin's method. □

Coupling the meshsize h and the penalty parameter ε we end up with

Theorem 8. *There is a constant $d > 0$ such that for $h^{\frac{1}{2}} = d\,\varepsilon$ and $\varepsilon \leq 1$ the collocation equations (15) are uniquely solvable. The solution satisfies the asymptotic error estimate*

$$\left\|t^\varepsilon_n - t^0\right\|_{\frac{1}{2}} \leq \|t^\varepsilon_n - t^\varepsilon\|_{\frac{1}{2}} + \left\|t^\varepsilon - t^0\right\|_{\frac{1}{2}} \leq c\,h^{\frac{1}{2}}.$$

Numerical results

To verify the theoretical results, we implemented a Galerkin procedure. For simplicity, Ω is chosen as a circle with radius $r < \frac{1}{2}$. For the boundary part in the upper half-plane we impose the Signorini boundary condition; and for the boundary part in the lower half-plane the Dirichlet condition. In this case $\overline{\Gamma}_S \cap \overline{\Gamma}_D \neq \emptyset$ and therefore we may have singularities where the boundary conditions are changing. So the second statement in Proposition 2 is not valid and we can only prove weak convergence in $H^{\frac{1}{2}}(\Gamma)$ but no error estimates (see [15]). Let Δ_n be a family of uniform grids with meshsize $h = \frac{1}{n}$.
For the numerical integrations in (14) we use techniques from [14] modified with respect to the nonlinearity. To compute the nonlinear integrals in (14) we note that, according to the definition of B, the function Bt_ε^n is also continuous and piecewise linear. However, it has some extra break points inside the pieces Γ_j of the Signorini boundary, where t_ε^n changes its sign. By adding these additional break points to Δ_n and using a corresponding finer subdivision of the integration domain, we can use the same quadrature formulas as in the linear case. However, by the computation of the above break points we have terms of the coefficient vectors γ_j in the limits of integration and, therefore, the resulting system of equations (14)

$$G_n\vec{\gamma} = \vec{F}_n$$

is highly nonlinear. Since B is only Lipschitz continuous but not differentiable, we cannot assume properties needed for a Newton iteration.
Therefore we use Broyden's update [7], which only requires one function evaluation per iteration, for solving the nonlinear system iteratively. This method is a generalization of the secant method. Because of the lack of differentiability we cannot prove local convergence by classical theory. In the case of Lipschitz continuous operators there are weaker conditions which are sufficient for the convergence of Broyden's scheme (see [11]), but here it was not possible to show that such conditions hold. Yet, if the sequence computed by Broyden's method converges, the limit will be the exact solution because the nonlinear operator G_n is Lipschitz continuous and one-to-one.
To speedup the program, we first compute and store the influence matrix for the linear case. Then, for evaluating the function G_n, we check t_ε^n in each Γ_j and use the stored values if there is no change of sign or if Γ_j is a Dirichlet boundary piece.

In our experiments we get convergence by using a continuation method, namely by approximating the solutions of the auxiliary equations

$$(I - sK + s\varepsilon^{-1}VB)t_s = sg$$

where s is increasing from 0 to 1 in small steps. As a first approximation of the matrix corresponding to the Jacobian we chose the matrix by neglecting the nonlinear terms in the system (14).
For the implementation we used a PC-AT and PASCAL; therefore we restricted us to at most 40 degrees of freedom. We also made experiments with different values for n and ε given separately to study the different influence of the parameters.
For testing we set $\frac{\partial u_f}{\partial n} := -\frac{1}{\varepsilon}$. In this case, $t_\varepsilon \equiv -1$ is the exact solution of Problem Q_ε. Having more regularity than required, we get the following L^2-errors and orders o, see Table 1.

n ε	0.1	0.01	0.001	$n^{-\frac{1}{2}}$
10	$9.8152 \cdot 10^{-4}$	$9.8667 \cdot 10^{-4}$	$9.8814 \cdot 10^{-4}$	$9.7978 \cdot 10^{-4}$
o	2.53	2.53	2.53	2.52
20	$1.7037 \cdot 10^{-4}$	$1.7048 \cdot 10^{-4}$	$1.7053 \cdot 10^{-4}$	$1.7036 \cdot 10^{-4}$
o	2.51	2.51	2.51	2.51
40	$2.9997 \cdot 10^{-5}$	$2.9999 \cdot 10^{-5}$	$3.0001 \cdot 10^{-5}$	$2.9997 \cdot 10^{-5}$

Table 1

$t_\varepsilon \equiv 1$ solves Problem Q_ε if we set

$$\frac{\partial u_f}{\partial n} := \begin{cases} 0 & \text{on } \Gamma_S\,, \\ \varepsilon^{-1} & \text{on } \Gamma_D\,. \end{cases}$$

The corresponding L^2-errors are listed in Table 2.

n ε	1	0.1	0.01	$n^{-\frac{1}{2}}$
10	$9.9860 \cdot 10^{-3}$	$3.4637 \cdot 10^{-2}$	$2.0282 \cdot 10^{-1}$	$1.5142 \cdot 10^{-2}$
o	0.61	0.81	0.55	0.43
20	$6.5498 \cdot 10^{-3}$	$1.9771 \cdot 10^{-2}$	$1.3856 \cdot 10^{-1}$	$1.1250 \cdot 10^{-2}$
o	0.93	1.01	0.77	0.67
40	$3.4465 \cdot 10^{-3}$	$9.8467 \cdot 10^{-3}$	$8.1501 \cdot 10^{-2}$	$7.0945 \cdot 10^{-3}$

Table 2

Furthermore we consider the modified problem

$$\begin{aligned} \Delta u &= 0 && \text{in } \Omega\,, \\ u &= 0 && \text{on } \Gamma_D\,, \\ \frac{\partial u}{\partial n}\cdot(u+K) = 0\,,\ \frac{\partial u}{\partial n} &\geq 0\,,\ u \geq K && \text{on } \Gamma_S \end{aligned}$$

with some constants $K \geq 0$, $c \in \mathbb{R}$. Here we compute $u_f(x,y) = c(x^2+y^2-r^2)$ for $f = -4c$ to be the solution of (3). From physical interpretation of the corresponding membrane problem we expect some displacement on the Sigorini boundary for $K = 0$ and $c = -1$, and, hence, this problem corresponds to the linear mixed Dirichlet-Neumann problem (Figure 1). For $K = 0$ and $c = 1$, however, there should be no displacement on the whole boundary. This conjecture gets fully confirmed although singularities appear at the points where the boundary conditions change (see Table 3).

n ε	0.1	0.01	0.001	$n^{-\frac{1}{2}}$
10	$2.5220 \cdot 10^{-2}$	$2.5220 \cdot 10^{-3}$	divergence	$7.9752 \cdot 10^{-3}$
o	0.50	0.50		1.00
20	$1.7875 \cdot 10^{-2}$	$1.7875 \cdot 10^{-3}$	$1.7875 \cdot 10^{-4}$	$3.9970 \cdot 10^{-3}$
o	0.50	0.50	0.50	1.00
40	$1.2647 \cdot 10^{-2}$	$1.2647 \cdot 10^{-3}$	$1.2647 \cdot 10^{-4}$	$1.9996 \cdot 10^{-3}$

Table 3: L^2-error for $K = 0$, $c = 1$

Last we varied the value of $K > 0$ and set $c = 1$ to get a real nonlinear situation (Figures 3–5).

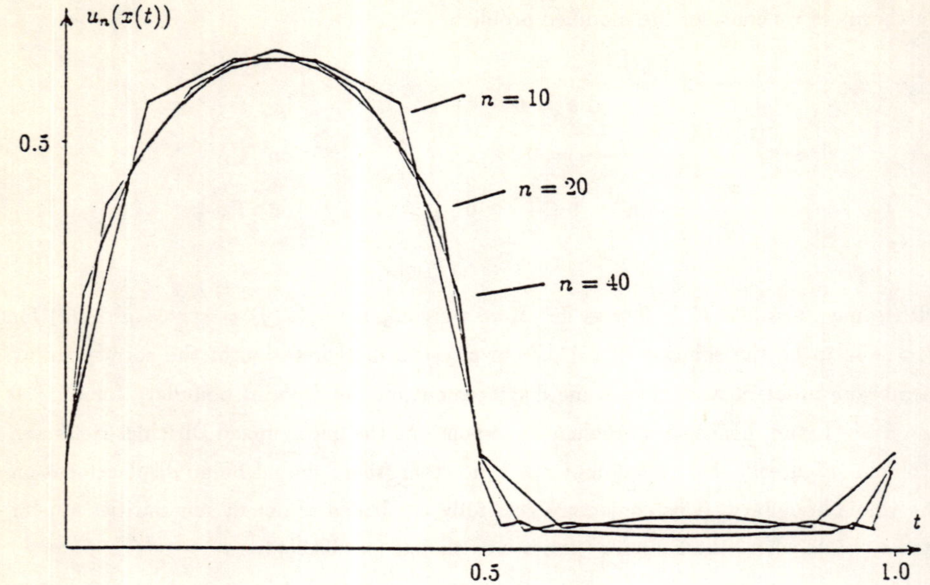

Figure 1 $K = 0$, $c = -1$, $\varepsilon = n^{-\frac{1}{2}}$

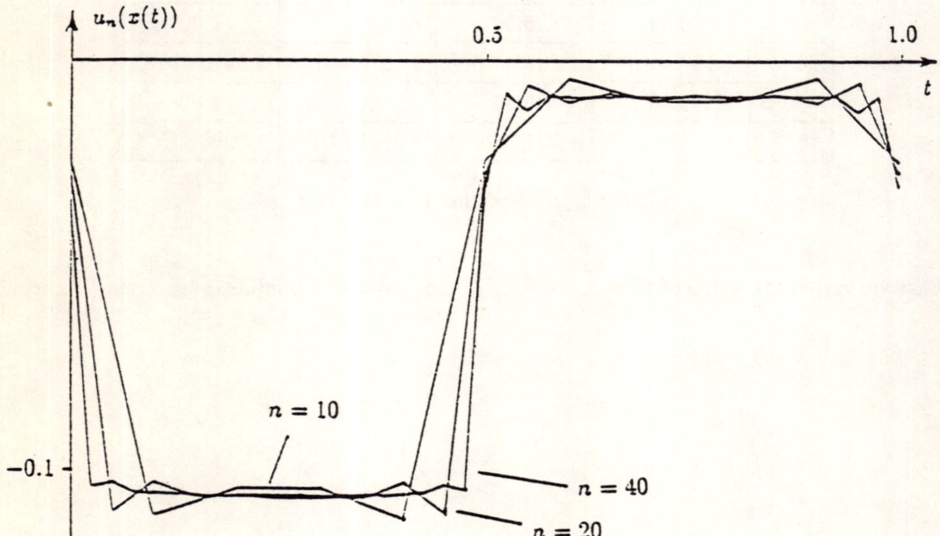

Figure 2 $K = 0.1$, $c = 1$

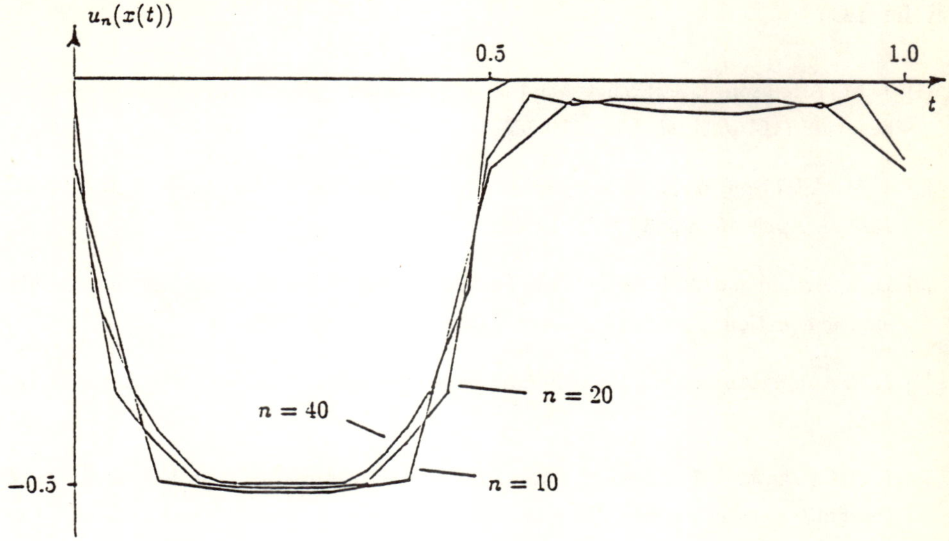

Figure 3 $K = 0.5$, $c = 1$

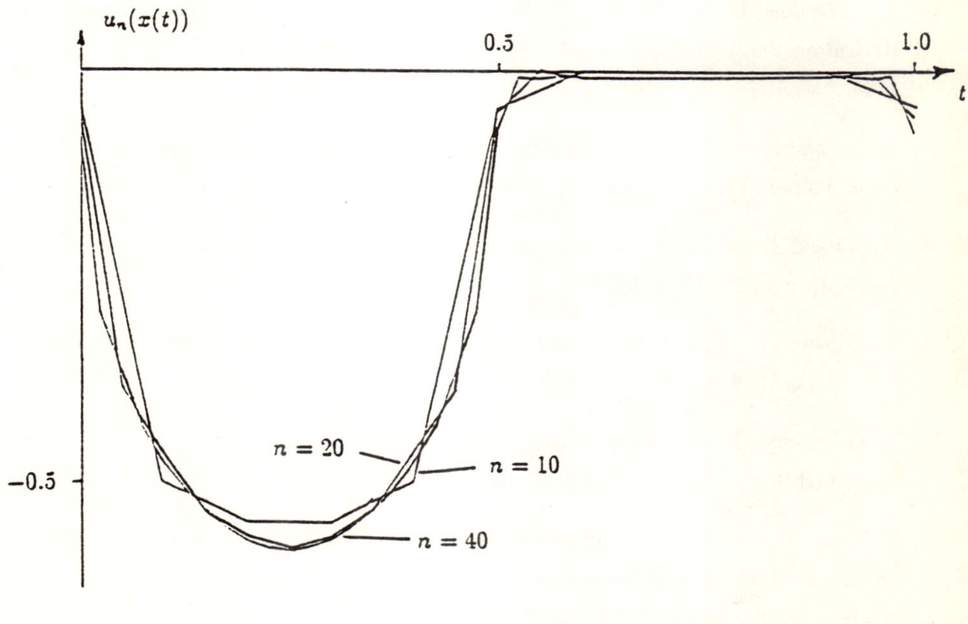

Figure 4 $K = 1.0$, $c = 1$

References

[1] J. M. Aitchison, C. M. Eliott, and J. R. Ockendon. Percolation in gently sloping beaches. *IMA J. Appl. Math.*, 30:269–287, 1983.

[2] J. M. Aitchison, A. A. Lacey, and M. Shillor. A model for an electropaint process. *IMA J. Appl. Math.*, 33:17–31, 1984.

[3] D. N. Arnold and W. L. Wendland. The convergence of spline collocation for strongly elliptic equations on curves. *Numer. Math.*, 47:317–343, 1985.

[4] J. P. Aubin. *Approximation of Elliptic Boundary Value Problems.* Wiley, New York, 1972.

[5] I. Babuška and A. K. Aziz. Survey lectures on the mathematical foundations of the finite element method. In A. K. Aziz, editor, *The Mathematical Foundation of the Finite Element Method with Applications to Partial Differential Equations*, pages 3–359. Academic Press, New York, 1972.

[6] H. Brezis. Problèmes unilateraux. *J. Math. Pures Appl.*, 51:1–168, 1972.

[7] J. E. Dennis Jr. and R. B. Schnabel. *Numerical Methods for Unconstrained Optimization and Nonlinear Equations.* Prentice-Hall, Englewood Cliffs, New Jersey, 1983.

[8] J. Elschner. *Singular Ordinary Differential Operators and Pseudodifferential Equations*, volume 1128 of *Lecture Notes in Math.* Springer, Berlin, 1985.

[9] R. S. Falk. Error estimates for the approximation of a class of variational inequalities. *Math. Comp.*, 28:963–971, 1974.

[10] R. Glowinski. *Numerical Methods for Nonlinear Variational Problems.* Springer, New York, 1980.

[11] A. Griewank. The local convergence of Broyden-like methods on Lipschitzian problems in Hilbert spaces. *SIAM J. Numer. Anal.*, 24:684–705, 1987.

[12] H. Han. The boundary finite element method for Signorini problem. In Y.-I. Zhu and B.-Y. Guo, editors, *Numerical Methods for Partial Differential Equations*, pages 38–49. Springer Lecture Notes 1297, 1987.

[13] H. Han and G. Hsiao. The boundary element method for a contact problem. In *Proceedings of the 2nd China-Japan Symposium on Boundary Element Methods 1988*, 1988.

[14] G. C. Hsiao, P. Kopp, and W. L. Wendland. A Galerkin collocation method for some integral equations of the first kind. *Computing*, 25:89–130, 1980.

[15] B. Kawohl. On nonlinear mixed boundary value problems for second order elliptic differential equations on domains with corners. *Proceedings of the Royal Society of Edinburgh*, 87 A:35–51, 1980.

[16] D. Kinderlehrer and G. Stampacchia. *An Introduction to Variational Inequalities and their Applications*. Academic Press, New York, 1980.

[17] P. D. Panagiotopoulos. A boundary integral inclusion approach to unilateral b.v.ps. in elastostatics. *Mechanics Research Communications*, 10:91–96, 1983.

[18] K. Ruotsalainen and J. Saranen. On the collocation method for a nonlinear boundary integral equation. *J. Comput. Appl. Math.*, 29:339–348, 1989.

[19] K. Ruotsalainen and W. Wendland. On the boundary element method for some nonlinear boundary value problems. *Num. Math.*, 53:299–314, 1988.

[20] G. Schmidt. Galerkin and collocation methods with splines. *Zeitschrift für Analysis und ihre Anwendungen*, 3:371–384, 1984.

[21] H. Schmitz. A collocation method for potential problems with a mixed Dirichlet-Signorini boundary condition. In *Analysis in Domains and on Manifolds with Singularities*. Teubner Leipzig, 1991.

[22] H. Schmitz and G. Schneider. Boundary element solution of the Dirichlet-Signorini problem by a penalty method. Submitted.

[23] W. Spann. *Fehlerabschätzungen zur Randelementmethode beim Signorini-Problem für die Laplace-Gleichung*. PhD thesis, Universität München, 1989.

An Overview of Boundary Element Applications to Nonlinear Heat Transfer Problems

L. C. Wrobel and C. A. Brebbia
Wessex Institute of Technology, Ashurst Lodge, Ashurst,
Southampton SO4 2AA, England

Abstract

This work presents a state-of-the-art review on the application of the Boundary Element Method to nonlinear heat transfer problems. The sources of nonlinearity considered are classified according to the mathematical formulation of the corresponding boundary value problem, and include nonlinear materials, boundary and interface conditions, heat sources and moving boundaries.

Introduction

The Boundary Element Method (BEM) is a well-established numerical technique for the solution of many engineering problems. Its theoretical basis, mathematical formulation and numerical implementation are discussed in a number of textbooks on the subject [1], [2], [3].

The technique basically consists of obtaining an integral equation equivalent to the original partial differential equation describing the problem, and solving it numerically through a discretization procedure. Since the integral equation usually relates boundary values only, the discretization is confined to the boundaries of the region and the dimensionality of the problem is thus reduced by one.

The BEM has been applied to a variety of heat transfer problems in the last twenty years. Some of the early applications already considered certain types of nonlinear features but only in the 80's more consistent formulations for treating nonlinearities have been introduced. These nonlinearities can be classified in many different ways. Based on the mathematical formulation of the corresponding boundary value problems, the classification encompasses four groups as follows:

- Nonlinear materials, as in the case of temperature-dependent thermal conductivity and/or heat capacity. This feature gives rise to a nonlinear partial differential equation;
- Nonlinear boundary conditions, caused by heat radiation or temperature-dependent heat transfer coefficient;
- Nonlinear sources, characteristic of some kind of chemical reaction taking place within the solid medium;
- Moving boundary problems, ocurring in heat transfer problems with phase change.

The present work attempts to review the most important BEM formulations dealing with all the above features. It is shown that the majority of nonlinear phenomena occurring in engineering practice can be efficiently treated by the method.

Material Nonlinearities

This section deals with problems in which the thermal conductivity and heat capacity of the medium is temperature-dependent. The mathematical formulation of the corresponding boundary value problem produces a nonlinear equation which can be written, for the case of steady-state with no internal heat generation, in the form

$$\frac{\partial}{\partial x}\left(K\frac{\partial T}{\partial x}\right)+\frac{\partial}{\partial y}\left(K\frac{\partial T}{\partial y}\right)+\frac{\partial}{\partial z}\left(K\frac{\partial T}{\partial z}\right)=0 \tag{1}$$

in which T is temperature and K the thermal conductivity. Since $K = K(T)$, equation (1) can be expanded as follows:

$$K\nabla^2 T = -\frac{dK}{dT}\left[\left(\frac{\partial T}{\partial x}\right)^2+\left(\frac{\partial T}{\partial y}\right)^2+\left(\frac{\partial T}{\partial z}\right)^2\right] \tag{2}$$

where the nonlinear terms appear explicitly on the right side of the equation.

Onishi and Kuroki [4] have treated this problem in the above form, by considering the right side of equation (2) as a nonlinear source term which was accounted for by domain discretization in an iterative way. One obvious disadvantage of this approach is that domain integrals have to be computed though no real heat sources exist. Furthermore, the resulting system of equations is nonlinear for any kind of boundary condition.

A different approach which removes the nonlinearity associated with the temperature dependence of the conductivity is through the application of Kirchhoff's transform, which replaces the original nonlinear differential equation by a linear one in the transform space [5]. The basic idea is to construct a new variable $U = U(T)$ such that

$$\frac{dU}{dT} = K(T) \tag{3}$$

or, in integral form,

$$U = \int_{T_0}^{T'} K(T)dT \tag{4}$$

where T_0 is an arbitrary reference value. It can be seen by inspection that the direct substitution of expression (3) into equation (1) produces a Laplace equation for the transform variable U, *i.e.*

$$\nabla^2 U = 0 \tag{5}$$

The boundary conditions of the problem also need to be transformed. Boundary conditions of the Dirichlet and Neumann types are linear in the transform space but Robin boundary conditions (*i.e.* convection) become nonlinear. Thus, an iterative scheme is still needed in this formulation but all nonlinearities are now transferred to the boundary conditions.

The Kirchhoff transform formulation has been successfully used by many authors [6]-[9]. An efficient algorithm for the inverse Kirchhoff transformation which approximates the $K(T)$ dependence by a piecewise linear spline has been proposed independently by Azevedo and Wrobel [8] and Bialecki and Nhalik [9].

The important case of multizoned media with nonlinear material behaviour was also examined by the above authors [8], [9]. The added difficulty in this case is that discontinuities arise in the values of the transformed temperature across the interfaces between different regions when continuity of temperature is imposed. Thus, an iterative algorithm is necessary to adjust the transformed temperature along the internal boundaries of the various subregions so that the physical requirement of temperature continuity is satisfied at the final stage. This can be done by writing the continuity condition on the interface in the form [8]

$$U^i = \beta(T^{ij})\,U^j$$

or [9]

$$U^i = U^j + \Delta U(T^{ij})$$

In both equations, the indices i, j refer to the adjoining subregions and T^{ij} is the temperature on the interface between regions i and j. The resulting set of equations can be solved by a Newton-Raphson algorithm [8] or incremental techniques [9].

The Kirchhoff transform technique can also be applied to transient problems governed by the diffusion equation

$$\frac{\partial}{\partial x}\left(K\frac{\partial T}{\partial x}\right) + \frac{\partial}{\partial y}\left(K\frac{\partial T}{\partial y}\right) + \frac{\partial}{\partial z}\left(K\frac{\partial T}{\partial z}\right) = \rho c\frac{\partial T}{\partial t} \tag{6}$$

in which the density ρ and specific heat c may also be temperature-dependent. However, contrary to steady-state problems, this transformation is not sufficient in the sense that the equation for the transform variable still contains a temperature-dependent diffusivity coefficient, in the form

$$\nabla^2 U = \frac{1}{k}\frac{\partial U}{\partial t} \tag{7}$$

where $k = k(T) = K/\rho c$.

Brebbia and Skerget [10],[11], Kikuta *et al.* [12], Pasquetti and Caruso [13] all assumed that the variation of k with T is usually not strong, and employed some kind of mean value of k to linearize equation (7).

A more elegant formulation has been derived by Wrobel and Brebbia [14]. This involves the definition of a modified time variable in the form

$$\tau = \int_0^{t'} k(x, y, z, t)\,dt$$

which leads to

$$\frac{\partial \tau}{\partial t} = k$$

Substituting the above into equation (7), one finally obtains

$$\nabla^2 U = \frac{\partial U}{\partial \tau} \tag{8}$$

Equation (8) can now be readily converted into a boundary integral equation. It is interesting to notice that equation (8) is still nonlinear in that the modified time variable τ is a function of position, and an iterative solution process is again necessary. Wrobel and Brebbia [14] developed an efficient Newton-Raphson scheme and implemented it in conjunction with a dual reciprocity approximation which resulted in a boundary-only formulation.

The efficiency of the above formulation was displayed in several numerical tests. The results for the temperature distribution along the thickness of a wall with nonlinear material properties are reproduced in tables 1 and 2. The wall is 20 cm long, 1 cm high, and is initially at 100°C. The temperature of the left end surface is suddenly raised to 200°C and kept at this value for 10 s, after which it is decreased to 100°C again; the temperature of the right end surface is kept

x	FEM	BEM-Q	BEM-C
0	200.00	200.00	200.00
1	176.16	174.86	175.29
2	153.21	151.03	151.48
3	133.47	131.33	131.74
4	118.60	117.32	117.63
5	108.98	108.74	108.94
6	103.72	104.14	104.27
7	101.29	101.91	102.01
8	100.37	100.87	100.97
9	100.08	100.39	100.50
10	100.01	100.14	100.27

Table 1: Results for $t = 10s$

x	FEM	BEM-Q	BEM-C
0	100.00	100.00	100.00
1	128.53	130.46	130.15
2	139.97	138.70	138.95
3	136.95	132.01	132.47
4	124.72	121.29	121.71
5	114.40	112.37	112.64
6	107.18	106.56	106.71
7	103.24	103.27	103.36
8	101.29	101.56	101.62
9	100.45	100.72	100.77
10	100.13	100.33	100.36

Table 2: Results for $t = 13s$

at 100°C, while the other surfaces are insulated. The thermal conductivity is $K = 2 + 0.01T$ W/(cm °C) and the heat capacity $\rho c = 8$ J/(cm^3 °C).

The tables compare results for boundary element discretizations of 22 equal quadratic and 44 equal constant elements, taking into account symmetry with respect to the x-axis, with a finite element solution obtained with 20 linear elements [15]. The time step value was 1 s in all cases, and the average number of iterations was 4 for the BEM and 3 for the FEM. The agreement of the results can be seen to be very good.

Nonlinear Boundary Conditions

The most common cause of nonlinear boundary conditions is thermal radiation. The complexity introduced by radiation can be quite severe as in the case of a concave portion of the boundary where it produces strong interactions between the temperature at points which can "see" each other.

No serious difficulties arise on the treatment of nonlinear boundary conditions with the BEM, and several formulations have been presented with different iterative algorithms. These include Newton-Raphson solvers [8], [14], Brown's iterative method [10], [11], incremental tech-

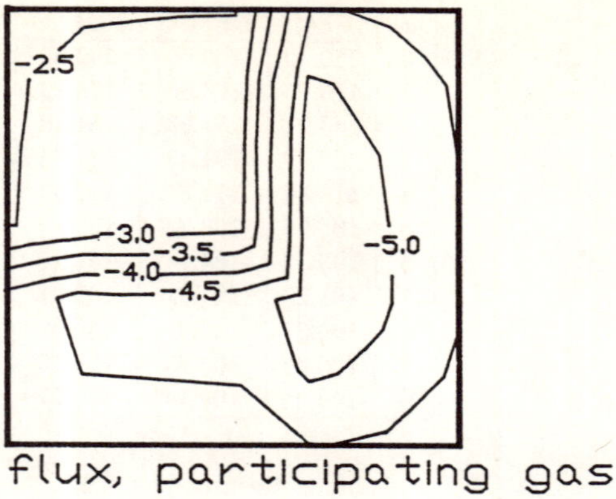

Figure 1: Total incoming radiative heat fluxes (kW/m^2) at base of rectangular cavity

niques [7], [9], "local" linearization [13], and others. The choice of a robust iterative solver is of importance as some simple iterative schemes of Gauss-Seidel type can diverge in the presence of the severe nonlinearities caused by thermal radiation, as pointed out in [7].

The previous papers all dealt with nonlinear boundary conditions considering that the heat flux at a given boundary point depends on the temperature at that point only. In the case when the boundaries are concave the heat flux at a certain position depends on the temperature at all points which can be seen from that position (providing the fluid filling the enclosure is transparent to thermal radiation). The problem of coupling BEM radiative analysis with the conductive part of the mathematical model of a body having concave boundaries has been discussed by Bialecki [16], [17], [18]. Since radiation is one of the few phenomena directly governed by integral equations, the idea of applying the BEM to solve radiative problems arises naturally. Bialecki's formulation has been applied to grey models of both radiating walls and participating isothermal medium [16], monochromatic analysis [17] and non-isothermal, participating non-gray gas radiation [18].

Figure 1, reproduced from [18], shows the radiative heat fluxes at the base of an open rectangular cavity of dimensions $0.7 \times 0.7 \times 0.4m$ filled by a participating gas having known non-uniform temperature. The walls of the enclosure were subdivided into 210 square elements of $0.1 \times 0.1m$ with specified temperature. The base of the cavity is maintained at temperature $700K$ whereas its top is open to an environment at temperature $300K$. The temperature of the other walls varies linearly from $700K$ at the bottom to $300K$ at the top. A 5-band radiating gas is considered with constant emissivity and absorption coefficient within each band. The volume of the enclosure has been divided into 27 cubic isothermal cells consisting of 3 layers of 9 cells each.

Nonlinear Sources

This situation is characteristic of some sort of chemical reaction taking place within the solid medium. Generally, problems with distributed nonlinear sources can be solved by the BEM using domain discretization and iteration. However, approximate techniques have been developed to allow the domain integral to be transformed into equivalent boundary integrals, thus retaining the boundary-only character which is the main feature of the BEM [19], [20].

A recent example of application of one such technique, namely the Dual Reciprocity Method, to a heat conduction problem with a nonlinear source has been presented by Partridge and Wrobel [21] who studied the case of spontaneous ignition of a reactive solid. The partial differential equation describing the problem is the following diffusion-type equation [22]

$$K\nabla^2 T + \rho Q z e^{-E/RT} = \rho c \frac{\partial T}{\partial t} \tag{9}$$

in which Q is the heat of decomposition of the solid, z the collision number, E is the Arrhenius activation energy and R the universal gas constant.

Equation (9) can be simplified by a change of variables. Defining

$$u = E\frac{(T - T_a)}{RT_a^2} \tag{10}$$

where T_a is the ambient temperature, the equation in terms of the new variable becomes

$$\nabla^2 u + \gamma e^u = \frac{1}{k}\frac{\partial u}{\partial t}. \tag{11}$$

with

$$\gamma = \frac{\rho Q E z}{KRT_a^2} e^{-E/RT_a} \tag{12}$$

in which the Frank-Kamenetskii approximation [23] was employed, *i.e.*

$$e^{-E/RT} = e^{-E/RT_a} e^u$$

It is usual to define the non-dimensional Frank-Kamenetskii parameter as $\delta = \ell^2\gamma$, where ℓ is a characteristic problem dimension. The physical significance of δ is discussed next. For this, it is assumed that a given reactive material has an ignition temperature T_m, the ambient temperature is considered to be T_a and the initial temperature is T_0 at time $t = 0$. From equation (10) the corresponding values of u_m, u_a and u_0 can be obtained.

For $\delta < \delta_c$ where δ_c is a critical value, a reactive solid at temperature T_0 can be placed in an environment at temperature T_a ($T_a < T_m$), and the temperature field within the solid will change according to equation (11) until all points are at a temperature $T_a \leq T < T_m$ (not uniform), thus achieving a steady-state situation where ignition does not occur. For $\delta > \delta_c$, the temperature within the solid will continue to increase until some point reaches the ignition temperature T_m.

Thus to model the process of spontaneous ignition a knowledge of δ_c is essential. This value can be obtained for a given geometric shape considering $\partial u/\partial t = 0$ in equation (11) [23], *i.e.*

$$\nabla^2 u = -\gamma e^u \tag{13}$$

Partridge and Wrobel [21] developed a dual reciprocity boundary element formulation and applied it to the calculation of critical values through equation (13) and to simulate the full process of spontaneous ignition using equation (11).

The criticality problem was solved for a series of common geometrical shapes for which solutions are available in the literature, including a square of unit side, a circle of unit radius and an ellipse with major axis = 2 and minor axis = 1. Thus in all cases the characteristic dimension of the problem is $\ell = 1$, such that $\gamma = \delta$.

Numerical results are presented in table 3 for different boundary discretizations with fixed numbers of regularly spaced internal nodes (9 for the square, 17 for the circle and the ellipse).

It can be seen that, in this case, the use of curved quadratic elements produces little difference in the results if the same number of boundary and internal nodes are used. The reason for this behaviour has to do with the boundary conditions of the problem, which impose a constant value of u everywhere. For the circle, this produces a constant boundary flux q. For the ellipse and the square, the flux is variable but its influence on the solution at internal points is of less importance.

A finite element (FEM) solution [23] is also included for comparison. The FEM mesh for each problem consisted respectively of 8, 20 and 34 quadratic isoparametric elements with 37, 75 and 125 node points for half of the region.

shape	32 linear BE	16 quadratic BE	FEM	Exact
square	1.763	1.770	1.703	1.7 [22]
circle	2.032	2.031	2.001	2.0 [24]
ellipse	1.251	1.252	1.234	-

Table 3: DRM results for γ_c for different discretizations

The results given in table 4 are for the same geometric shapes with a fixed boundary discretization and different numbers of regularly spaced internal nodes. It can be seen that, in increasing the number of internal nodes the solution converges towards the exact value; however, solutions within the usual engineering accuracy may be obtained with a reduced number of internal nodes, with little computer time and data preparation effort.

shape	5 nodes	9 nodes	17 nodes	80 nodes
square	-	1.770	-	1.707
circle	2.080	-	2.031	2.004
ellipse	1.276	-	1.252	1.235

Table 4: Results for 16 quadratic boundary elements and different numbers of internal nodes

The dual reciprocity boundary element formulation of Partridge and Wrobel [21] was also applied to the problem of spontaneous ignition of a long cylinder of unit radius, initially at temperature T_0 at all points, exposed at time $t = 0$ to an environment at temperature T_a. The cylinder is of a uniform isotropic reactive material, the ignition temperature of which is T_m. An essential boundary condition $T = T_a$ is imposed at all boundary nodes. The critical value of γ is $\gamma_c = 2.0$ in this case (see table 3).

The cylinder was discretized with 16 linear boundary elements and 19 internal nodes were spaced at intervals of 10cm along a diagonal. The numerical values of the physical parameters for RDX taken from [24] are

- $k = 7.778 \times 10^{-4}$ cm^2/s
- $T_m = 425$ K
- $T_0 = 25°C = 298$ K

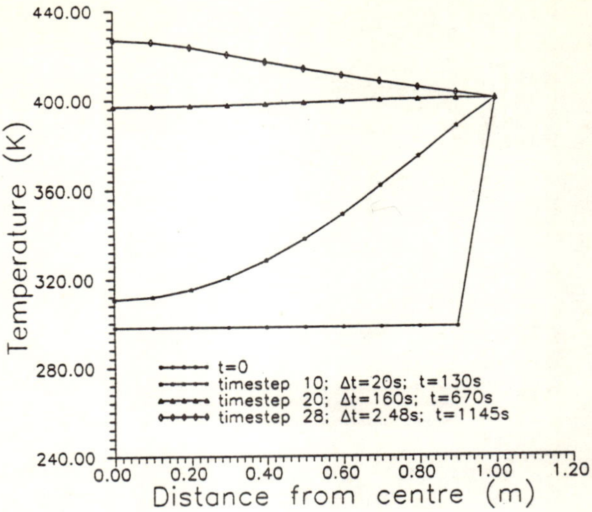

Figure 2: Temperature distribution for $\gamma = 4$

- $T_a = 400$ K
- $E = 47500$ cal/M
- $R = 1.987$ cal/(M K)

such that $T = (u + 59.76)/0.1494$. A variable time step was used because this process has varying stability characteristics. Initially a large Δt may be used, but this must be rapidly reduced as ignition nears. The value of Δt was altered at each time step to maintain the average temperature change at all interior nodes between 5° and 20°.

Results are shown in figure 2 for the case $\gamma = 4$. It can be noticed that ignition is first reached at the centre of the cylinder in this case.

For cases with higher values of γ the ignition point moves out from the centre towards the outer surface and the elapsed time before ignition reduces. This situation is depicted in figure 3 for $\gamma = 50$. The results presented are very similar to those obtained using theoretical considerations in [24].

Moving Boundary Problems

Moving boundaries are an inherent part of phase change phenomena such as solidification, melting, ablation, welding, electrochemical machining, etc. The physics of these phenomena is so complex that no consistent theory exists. The main reason for this is the lack of thermodynamical equilibrium often encountered in phase change phenomena.

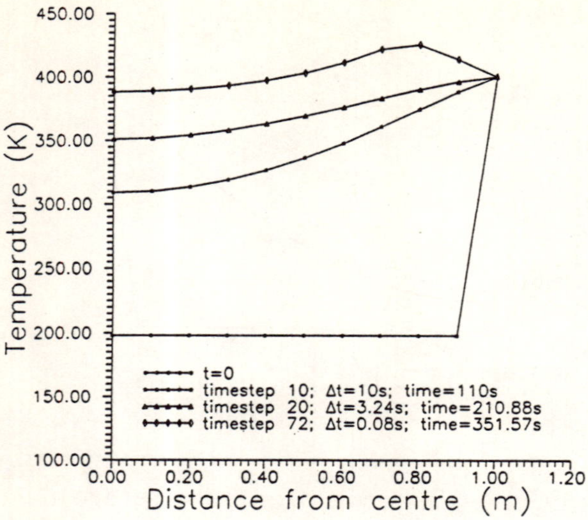

Figure 3: Temperature distribution for $\gamma = 50$

The simplest possible mathematical model is the case of a pure substance with constant temperature of phase change, the so-called Stefan's problem [25]. Even such simple model can become complicated if mushy regions, *i.e.* transition zones filled with a mixture of liquid and crystals, are included in the analysis.

The algorithms employed by the numerous discrete approximation methods for phase change problems can be broadly divided into fixed domain and front tracking schemes. Fixed domain methods have the great advantage of simple numerical implementation and are generally most useful in problems where the temperature distribution is more important than the interface position. Front tracking methods, on the other hand, are more complex due to the need of remeshing during each time step and are mainly used when the primary unknown is the motion of the interface.

The heat transfer equation to be treated here is of the form

$$K\nabla^2 T + \rho L S = \rho c \frac{\partial T}{\partial t} \tag{14}$$

where L is the latent heat of fusion and S the volumetric rate of production of solid due to phase change.

When the solid phase is regarded as immobile, and when the phase constitution is a function of temperature only, one may write

$$S = \frac{\partial h}{\partial T}\frac{\partial T}{\partial t}$$

in which h is the solid volume fraction. The phase change term can then be absorbed into the heat capacity term of equation (14), in the form

$$K\nabla^2 T = \rho\left(c - L\frac{\partial h}{\partial T}\right)\frac{\partial T}{\partial t} \tag{15}$$

The above equation forms the basis of the fixed domain methods and expresses the inherent nonlinearity of the problem through the thermal coefficients.

Many different formulations, mostly based on solving a diffusion equation for the enthalpy rather than temperature, have been proposed in the literature [26]. Regarding the BEM, Hong *et al.* [27] employed the temperature recovery method with application to the solidification of castings in metal and sand moulds. In this method, the temperature variation at all points is first obtained without considering the latent heat. After each time step, the real temperature T at any point can be evaluated using the temperature T' calculated without considering latent heat as follows:

$$\int_{T_1}^{T_2} \rho\, c\, dT - \int_{T_1}^{T_2} \rho\, L \frac{\partial h}{\partial T} dT = \int_{T_1}^{T_2'} \rho\, c\, dT \tag{16}$$

This scheme can be applied not only for congruently melting materials, but also for alloys which solidify over a wide temperature range.

Another possibility which is common in finite element analysis [28] is to treat the term between brackets in equation (15) as an equivalent, temperature-dependent specific heat coefficient. Similar formulations can in principle be developed for the BEM using the algorithms previously discussed for nonlinear materials [14].

Recently, a BEM enthalpy formulation has been derived by Sarler *et al.* [29] which allows for the treatment of each phase separately. The resulting decoupled all-solid and all-liquid geometrically fixed semi-problems were solved by a standard, direct BEM for the diffusion equation. Sarler *et al.* [29] employed the alternating phase truncation method which divides the calculation over a time step into two half-steps: in the all-liquid half-step, enthalpy is locally added to convert the phase change material to the liquid phase; in the all-solid half-step, enthalpy is locally subtracted to convert the phase change material to the solid phase. The enthalpy thermal field calculated in this way can then be converted to the temperature system and, if necessary, the interface location can be approximately determined.

Another enthalpy formulation has been proposed by Tiba [30] whose BEM implementation used the fundamental solution of Laplace's equation and a finite difference approximation for the time derivative term. Domain discretization is necessary in this formulation with the resulting system of algebraic equations solved by a simple iterative procedure.

Generally, phase change is often active only within a temperature and space range which is narrow relative to the overall range considered in a problem. In particular, the phase change zone is often infinitesimally thin, for all practical purposes. Thus the problem features a zone of intense activity, usually more concentrated than the mesh spacing, which moves about in a manner which is unknown "a priori". The problem can then be treated as a moving boundary problem in which the phase change is confined to an infinitesimally thin surface across which the discontinuity of flux of sensible heat and the evolution of latent heat must balance [31]. In this approach, the governing equation is the linear diffusion equation within each mutually exclusive phase domains, with the condition

$$\rho L \frac{\partial X_i}{\partial t} n_{si} = K_s \frac{\partial T_s}{\partial n_s} - K_l \frac{\partial T_l}{\partial n_l} \tag{17}$$

attached to the interface Γ_I between the solid (s) and liquid (l) phases. Thus, all latent heat effects are expressed through (17) which is also used to locate the interface position (X_i are the components of a vector $\mathbf{X}$ defining the location of a point on Γ_I).

The above formulation, which gives rise to front tracking algorithms, was initially developed by Chuang and Szekely [32] for one-dimensional, one-phase problems (only the solid phase was considered in the analysis). This formulation was later extended and improved by Shaw [33] and Wrobel [34]. O'Neill [31] generalized the formulation to multi-dimensional problems, but only

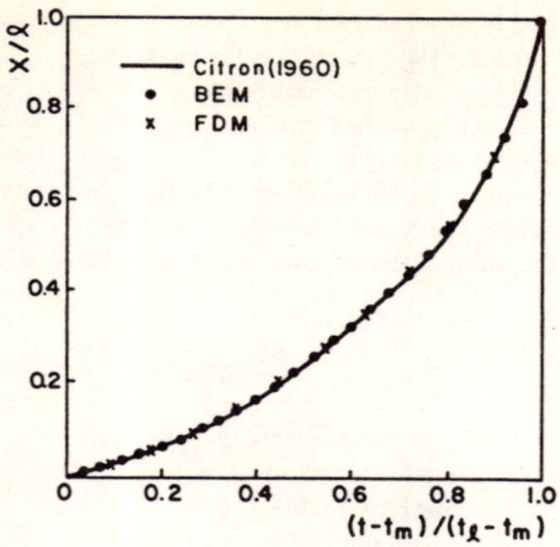

Figure 4: Moving boundary position

solved the simpler quasi-static approximation which is valid when the boundary temperature does not change too rapidly.

Applications of the front tracking BEM formulation to two-dimensional problems have been presented by Coleman [35] and Zabaras and Mukherjee [36], for one-phase and two-phase problems, respectively. Coleman's approach [35] involved a domain integral over the portion of the domain which is melting and was applied to study the melting of a rod of square cross-section, a rod with a L-shaped cross-section and a square rod with a circular hole. Zabaras and Mukherjee [36] applied their formulation to study the solidification of a square region and a rectangular prism, for two distinct situations, considering that the liquid initially filling the domain was at the melting temperature or at a temperature higher than the freezing temperature.

Takahashi *et al.* [37] presented the only BEM formulation to date to include a complete interaction between heat transport and mass diffusion. They solved the two-dimensional problem of a steel slab with a series of circular holes with concentration-dependent melting temperature in conjunction with the diffusion of the alloy components. In this formulation, the latent heat effects behave as a temperature-dependent heat source.

Figures 4 and 5, reproduced from [34], show non-dimensional results for the phase front location and the fixed boundary temperature history for a finite bar of length ℓ, with constant initial temperature T_0, subject to Neumann boundary conditions at both extremities ($q = \overline{q}$ at $x = 0$ and $q = 0$ at $x = \ell$). It was assumed that, after a certain time t_m, the extremity $x = 0$ reached the melting temperature T_m and the bar started to melt. The BEM results of [34] are compared in the figures with an approximate analytical solution [38] and a finite difference solution [39]. The time t_ℓ of the figures represents the time the bar takes to melt completely. The computed values of t_m and t_ℓ were 0.10 and 0.23 for the BEM, compared to analytical values of 0.104 and 0.232, showing the excellent accuracy of the front tracking BEM algorithm.

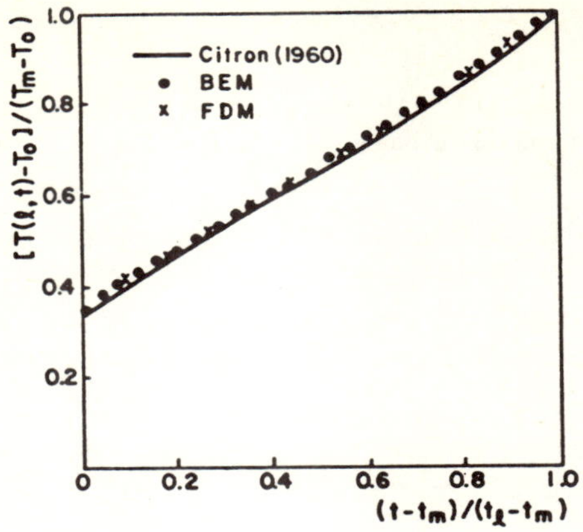

Figure 5: Temperature in fixed boundary

Conclusions

This paper reviewed the most important BEM formulations for solving nonlinear heat transfer problems. It has been shown that the majority of nonlinear phenomena ocurring in situations of practical engineering interest can be treated by the BEM using integral formulations which are equivalent to the original boundary value problem.

Problems including nonlinear sources or the transient analysis of nonlinear materials require some approximations to produce a "boundary-only" integral equation. Nonlinear boundary conditions and moving boundary problems, on the other hand, can be handled in a very efficient way offering many advantages over domain-type numerical techniques. In all cases, BEM formulations have been developed which retain the main attraction of the technique, namely the reduction of the dimensionality of the problem.

References

[1] Brebbia, C.A., *The Boundary Element Method for Engineers*, Pentech Press, London and Halstead Press, New York, 1978.

[2] Brebbia, C.A.; Telles, J.C.F. and Wrobel, L.C., *Boundary Element Techniques*, Springer-Verlag, Berlin, 1984.

[3] Brebbia, C.A. and Dominguez, J., *Boundary Elements: An Introductory Course*, Computational Mechanics Publications, Southampton and McGraw-Hill, New York, 1989.

[4] Onishi, K. and Kuroki, T., Boundary element method in singular and nonlinear heat conduction, in *Boundary Element Methods in Engineering*, Computational Mechanics Publications, Southampton and Springer-Verlag, Berlin, 1982.

[5] Carslaw, H.S. and Jaeger, J.C., *Conduction of Heat in Solids*, 2nd edn, Clarendon Press, Oxford, 1959.

[6] Khader, M.S. and Hanna, M.C., An iterative boundary integral numerical solution for general steady heat conduction problems, *Trans. ASME J. Heat Transfer* **103** (1981) 26-31.

[7] Bialecki, R. and Nowak, A.J., Boundary value problems in heat conduction with nonlinear material and nonlinear boundary conditions, *Appl. Math. Modelling* **5** (1981) 417-421.

[8] Azevedo, J.P.S. and Wrobel, L.C., Non-linear heat conduction in composite bodies: A boundary element formulation, *Int. J. Numer. Meth. Eng.* **26** (1988) 19-38.

[9] Bialecki, R. and Nhalik, R., Solving nonlinear steady state potential problems in inhomogeneous bodies using the boundary element method, *Num. Heat Transfer*, Part B **15** (1989).

[10] Brebbia, C.A. and Skerget, P., Nonlinear time dependent potential problems using BEM, in *Boundary Elements VI*, Computational Mechanics Publications, Southampton and Springer-Verlag, Berlin, 1984.

[11] Skerget, P. and Brebbia, C.A., Time dependent nonlinear potential problems, in *Topics in Boundary Element Research*, Vol. 2, Springer-Verlag, Berlin, 1985.

[12] Kikuta, M.; Togoh, H. and Tanaka, M., Boundary element analysis of nonlinear transient heat conduction problems, *Comp. Methods Appl. Mech. Eng.* **62** (1987) 321-329.

[13] Pasquetti, R. and Caruso, A., Boundary element approach for transient and nonlinear thermal diffusion, *Num. Heat Transfer*, Part B **17** (1990) 83-99.

[14] Wrobel, L.C. and Brebbia, C.A., The dual reciprocity boundary element formulation for nonlinear diffusion problems, *Comp. Methods Appl. Mech. Eng.* **65** (1987) 147-164.

[15] Orivuori, S., Efficient method for solution of nonlinear heat conduction problems, *Int. J. Numer. Meth. Eng.* **14** (1979) 1461-1476.

[16] Bialecki, R., Applying BEM to calculation of temperature fields in bodies containing radiating enclosures, in *BEM VII*, Vol. 1, Computational Mechanics Publications, Southampton and Springer-Verlag, Berlin, 1985.

[17] Bialecki, R., Radiative heat transfer in cavities. BEM solution, in *BEM X*, Vol. 2, Computational Mechanics Publications, Southampton and Springer-Verlag, Berlin, 1988.

[18] Bialecki, R., Solving 3-D heat radiation problems in cavities filled by a participating non-gray medium using BEM, in *Advanced Computational Methods in Heat Transfer*, Vol. 1, Computational Mechanics Publications, Southampton and Springer-Verlag, Berlin, 1990.

[19] Partridge, P.W.; Brebbia, C.A. and Wrobel, L.C., *The Dual Reciprocity Boundary Element Method*, Computational Mechanics Publications, Southampton, 1991.

[20] Nowak, A.J., Solution of transient heat conduction problems using boundary-only formulation, in *BEM IX*, Vol. 3, Computational Mechanics Publications, Southampton and Springer-Verlag, Berlin, 1987.

[21] Partridge, P.W. and Wrobel, L.C., The dual reciprocity boundary element method for spontaneous ignition, *Int. J. Numer. Meth. Eng.* **30** (1990) 953-963.

[22] Boddington, T.; Gray, P. and Harvey, D.I., Thermal theory of spontaneous ignition: Criticality in bodies of arbitrary shape, *Phil. Trans. Royal Soc. London* **270** (1970) 467-506.

[23] Anderson, C.A. and Zienkiewicz, O.C., Spontaneous ignition: Finite element solution for steady and transient conditions, *Trans. ASME J. Heat Transfer* **96** (1974) 398-404.

[24] Zinn, J. and Mader, C.L., Thermal initiation of explosives, *Journal Appl. Physics* **31** (1960) 323-328.

[25] Rubinstein, L.I., *The Stefan Problem*, AMS Translations of Mathematical Monographs, Vol. 27, American Mathematical Society, Providence, R.I., 1971.

[26] Crank, J., *Free and Moving Boundary Problems*, Clarendon Press, Oxford, 1984.

[27] Hong, C.P.; Umeda, T. and Kimura, Y., Application of the boundary element method in two and three dimensional unsteady heat transfer involving phase change: Solidification problems, in *BEM V*, Computational Mechanics Publications, Southampton and Springer-Verlag, Berlin, 1983.

[28] Comini, G.; Del Giudice, R.; Lewis, R.W. and Zienkiewicz, O.C., Finite element solution of non-linear heat conduction problems with special reference to phase change, *Int. J. Numer. Meth. Eng.* **8** (1974) 613-624.

[29] Sarler, B.; Alujevic, A. and Kuhn, G., Treatment of multi-dimensional Stefan problems by the alternating phase truncation boundary element method, in *Boundary Element Methods in Mechanical and Electrical Engineering*, Computational Mechanics Publications, Southampton and Springer-Verlag, Berlin, 1990.

[30] Tiba, M., The boundary element method in two-phase Stefan problems, *Engineering Analysis* **4** (1987) 46-48.

[31] O'Neill, K., Boundary integral equation solution for moving boundary phase change problems, *Int. J. Numer. Meth. Eng.* **19** (1983) 1825-1859.

[32] Chuang, K. and Szekely, J., On the use of Green's functions for solving melting and solidification problems, *Int. J. Heat Mass Transfer* **14** (1971) 1285-1294.

[33] Shaw, R.P., A boundary integral approach to one-dimensional ablation problems, in *Boundary Element Methods in Engineering*, Computational Mechanics Publications, Southampton and Springer-Verlag, Berlin, 1982.

[34] Wrobel, L.C., A boundary element solution to Stefan's problem, in *BEM V*, Computational Mechanics Publications, Southampton and Springer-Verlag, Berlin, 1983.

[35] Coleman, C.J., A boundary integral formulation of the Stefan problem, *Appl. Math. Modelling* **10** (1986) 445-449.

[36] Zabaras, N. and Mukherjee, S., An analysis of solidification problems by the boundary element method, *Int. J. Numer. Meth. Eng.* **24** (1987) 1879-1900.

[37] Takahashi, S.; Onishi, K.; Kuroki, T. and Hayashi, K., Boundary elements to phase change problems, in *BEM V*, Computational Mechanics Publications, Southampton and Springer-Verlag, Berlin, 1983.

[38] Citron, S.J., Heat conduction in a melting slab, *J. Aerospace Sciences* **27** (1960) 219-228.

[39] Mastanaiah, K., On the numerical solution of phase change problems in transient non-linear heat conduction, *Int. J. Numer. Meth. Eng.* **10** (1976) 833-844.

Finite Volumes vs Finite Elements. Is there really a Choice?

O.C. ZIENKIEWICZ †
University College of Swansea, U.K.

and

E. OÑATE
†† Universitat Politecnica de Catalunya
08034 Barcelona, Spain

Summary

The finite volume method appears to be a particular case of finite elements with a non Galerkin weighting. It is of course less accurate for self adjoint problems but has some computationally useful features for first order equations involving only surface integrals. For certain problems this is a substational economy and leads to computationaly useful approximations.

Introduction

The finite volume method evolved in the early seventies via finite difference approximation and today has many proponents in the field of fluid mechanics (Mc Donald [1], Mc. Cormack and Paullay [2], Rizzi and Inouye [3], Patankar [4]). In the field of CSM (Computational Solid Mechanics) it appears to has been introduced by Wilkins [5] as an alternative approximation to derivatives in a *cell*. In this he defines the average gradient of an arbitrary function u in a volume Ω as

$$\left(\frac{\partial u}{\partial x_i}\right)_{av.} \equiv \frac{1}{\Omega}\int_\Omega \frac{\partial u}{\partial x_i} d\Omega = \frac{1}{\Omega}\oint_s u n_i ds \tag{1}$$

using the well known divergence theorem.

Such definition of gradients can be written entirely in terms of function values at the boundaries of a volume and has been used in the early "hydrocodes" of Lawrence Livermore Laboratory.

A comprehensive description of the finite volume techniques is given by Hirsch [6] and lists their apparent merits viz a viz finite element and finite difference approximations. Amongst these he considers as particularly important:

(i) the conservation of fluxes $F = F(u)$ in a *control volume* when the procedure is applied to such conservation equation as

$$\frac{\partial u}{\partial t} + \frac{\partial F_i}{\partial x_i} \equiv \frac{\partial u}{\partial t} + \nabla^T \mathbf{F} = Q \tag{2}$$

(ii) The ease of applying natural (flux) boundary conditions.

In this note we shall show that the above features exist in the general finite element process and that they do not distinguish in any way to the finite volume method. Further, in problems of self-adjoint operators in which

† Currently holding Unesco Chair of Numerical Methods in Engineering at ††

$$F_i = F_i(\frac{\partial u}{\partial x_j}) = -k_{ij}\frac{\partial u}{\partial x_j} \tag{3}$$

it is well known that the finite element method when used with Galerkin weighting provides an optimal approximation and must therefore be, in such cases, more accurate than any alternative.

What then is the distinguishing feature of the finite volume method which can make it attractive?. The answer lies simply in a possible computation advantage of the finite volume procedure in which the major contribution involves only the boundaries of the control volumes. Whether this sole advantage of the finite volume form is sufficient to compensate for a possible accuracy loss, only extensive experiment will show. In the meantime we shall derive below a finite volume format in which a precise interpolation of the variables is used to avoid the fairly arbitrary definitions often used by finite volume proponents.

A General Finite Volume Format

All discrete, generally finite elements, approximations to a differential equation system (such as for instance that of eq (2)) and its boundary conditions can be written for a general equation system.

$$\mathbf{A}\ \mathbf{u} = \mathbf{f} \quad \text{in} \quad \Omega \tag{4a}$$

and the boundary conditions

$$\mathbf{B}\mathbf{u} = \mathbf{t} \quad \text{in} \quad \Gamma \tag{4b}$$

in a unified format:

1) The independent unknowns are approximated as

$$\mathbf{u} \simeq \hat{\mathbf{u}} = \mathbf{N}_i\bar{\mathbf{u}}_i \qquad (i = 1, \cdots, n) \tag{5}$$

where $\bar{\mathbf{u}}_i$ are the unknown parameters and $\mathbf{N}_i$ are the basis functions.

2) The approximating system of equations is written as a set of *algebraic equations*

$$\int_\Omega \mathbf{W}_j^T(\mathbf{A}\hat{\mathbf{u}} - \mathbf{f})d\Omega + \oint_\Gamma \mathbf{W}_j^T(\mathbf{B}\hat{\mathbf{u}} - \mathbf{t})d\Gamma = 0 \tag{6}$$

where $\mathbf{W}_j^T$ $(j = 1, \cdots, n)$ are weighting functions.

Eq.(6) can be written for linear systems, after substituting of the interpolation (5), as

$$\mathbf{K}\bar{\mathbf{u}} = \mathbf{f} \tag{7}$$

where the usual additive property of element or subdomain contributions is preserved whatever the form of the weighting functions.

In self adjoint problems the optimal weighting is the Galerkin one with

$$\mathbf{W}_j = \mathbf{N}_j \tag{8}$$

This leads to minimum energy norm errors and preserves symmetry of matrix $\mathbf{K}$ and is the basis of most frequently used finite element procedures. However other weightings can be used recovering all possible approximation methods (viz Zienkiewicz and Taylor [7]). In what follows we shall use standard finite element interpolations $\mathbf{N}_i$ with $\bar{\mathbf{u}}_i$ standing for nodal values in element subdomains.

The finite volume procedure is in fact a special case of the weighted eq.(6) in which

$$\mathbf{W}_j = \mathbf{I} \;\text{ in }\; \Omega_c \quad (\text{ and } \mathbf{W}_j = \mathbf{0} \text{ elsewhere}) \tag{9}$$

where $\mathbf{I}$ is the unity matrix and Ω_c is a control volume which can be prescribed in various ways.

To fix ideas we shall consider a scalar equation with a variable u as given below

$$\frac{\partial u}{\partial t} + \frac{\partial F_k}{\partial x_k} - \frac{\partial}{\partial x_k} k \frac{\partial u}{\partial x_k} \equiv \frac{\partial u}{\partial t} + \nabla^T \mathbf{F} - \nabla^T k \nabla u = Q \tag{10a}$$

This is to be satisfied in a domain Ω with appropiate boundary conditions

$$\begin{aligned} u = \bar{u} \quad &\text{on} \quad \Gamma_u \\ - k \frac{\partial u}{\partial x_n} = \bar{q} \quad &\text{on} \quad \Gamma_q \end{aligned} \tag{10b}$$

where $\Gamma = \Gamma_u \cap \Gamma_q$ is the boundary of the domain Ω.

The general approximation can now be written, assuming that $F_i = A_i u$, as

$$M_{ji}\dot{\bar{u}}_i + C_{ji}\bar{u}_i + K_{ji}\bar{u}_i + f_j = 0 \tag{11}$$

in which where after using appropriately the Gauss divergence theorem we have†

$$M_{ji} = \int_{\Omega_j} W_j N_i d\Omega \tag{12a}$$

$$C_{ji} = -\int_{\Omega_j} [\nabla W_j]^T A_k N_i d\Omega + \oint_{\Gamma_j} W_j A_k N_i n_k d\Gamma \tag{12b}$$

$$K_{ji} = \int_{\Omega_j} [\nabla W_j]^T k \nabla N_i d\Omega - \oint_{\Gamma_j} W_j k \frac{\partial N_i}{\partial n} d\Gamma \tag{12c}$$

† Note that in (10–11) index k is used for coordinate directions and i, j denotes node numbers. We can write alternatively

$$[\nabla W_j]^T A_k \nabla N_i \equiv W_{j,k} A_k N_{i,k} \quad \text{etc. implying summation in } \; k$$

$$f_j = \int_{\Omega_j} W_j Q d\Omega \tag{12d}$$

In above Ω_j is the *control volume* associated with variable j (to use the finite volume terminology) where $W_j \neq 0$ and which may include the external boundary Γ of the total domain. Providing both W_j and N_j are so chosen that integrals (12) can be evaluated then whatever the external boundary we can specify on it either the normal fluxes $(-k\frac{\partial u}{\partial x_n} = \bar{q})$ or values of $u = \bar{u}$ with equal ease, using either the Galerkin (eq.(8)) or finite volume (eq.(9)) forms.

To illustrate above consider a field of arbitary triangles with linear interpolations with the control volume for node j shown shaded in Figure 1.

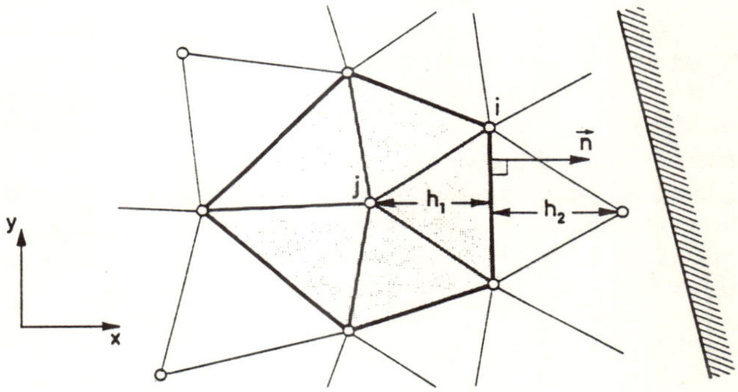

Figure 1. An assembly of finite elements/ finite volumes with shading indicating control volume.

In "standard" finite element assembly the weighting function $W_j = N_j$ is of the form shown in Figure 2a and Ω_j includes all the elements associated with jth node.

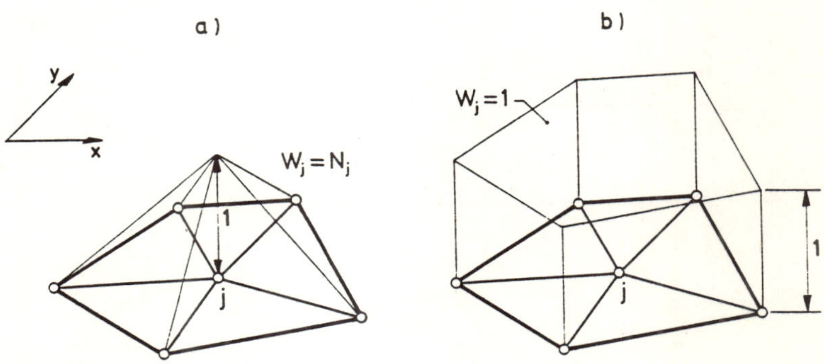

Figure 2. Weighting and basis functions for finite element (a) and finite volume (b)approximations for Figure 1.

As $N_j = 0$ all over the periphery of Ω_j no surface integral exists in expressions for C_{ji} or K_{ji}, thus eliminating the need of evaluating the fluxes on such boundaries. However, on all element interfaces integrals along lines such as $i-j$ (connecting nodes $i-j$) vanish as

$$\begin{aligned} \left(\int_i^j N_j A_k N_i n_k d\Gamma\right)_{left} &= -\left(\int_i^j N_j A_k N_i n_k d\Gamma\right)_{right} \\ \left(\int_i^j N_j k \frac{\partial N_i}{\partial n} d\Gamma\right)_{left} &= -\left(\int_i^j N_j k \frac{\partial N_i}{\partial n} d\Gamma\right)_{right} \end{aligned} \tag{13}$$

for elements left and right of the interface.

This of course means that flux continuity is preserved in a weighted sense. Further direct flux identity on such faces exists in general if we note that on each $N_i + N_j = 1$ with standard interpolations [7]. However, the volume integrals of C_{ji}, K_{ji}, etc. have to be evaluated.

Now if we consider the finite volume situation (Figure 2b) it is evident that all volume (area) integrals dissapear from expressions C_{ji} and K_{ji}. Further the surface integrals in the expression for C_{ji} presents no difficulty in evaluating providing A_k and N_i are continuous in the domain. However inmediately a dificulty arises with the diffusive terms in the K_{ji} integral. Here evaluation of

$$\left(\oint_\Gamma W_j k \frac{\partial N_i}{\partial n} d\Gamma\right) \bar{u}_i$$

presents the difficulty as $\frac{\partial N_i}{\partial n}$ is not continuous.

In the normal direction n as illustrated in Figure 1 between two elements we have a discontinuity shown in Figure 3 with $\frac{\partial N_i}{\partial n}$ jumping from a value of $1/h_1$ to $-1/h_2$ where heights of adjacent triangles are denoted h_1 and h_2.

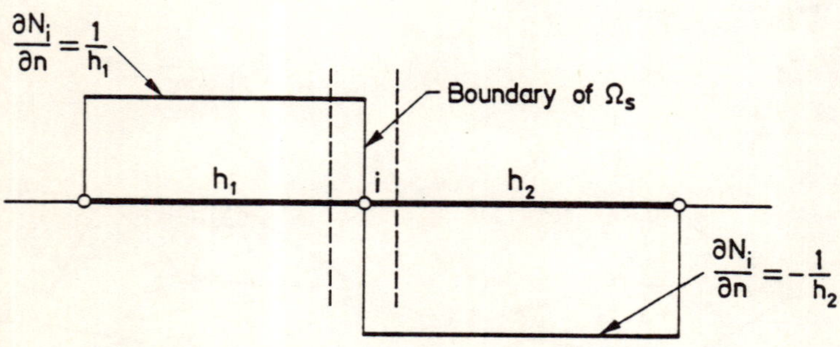

Figure 3. Discontinuity of gradient at boundaries of control volume.

The theory of *distributions* indicates that the jump in $\frac{\partial \hat{u}}{\partial n}$ accros the boundary should be given the average value (though of course in a finite sense position either inside or outside the jump could be justified). It will indeed be found that the mid position is optimal and the value should be then

$$\frac{\partial \hat{u}}{\partial n} = \frac{1}{2}\left(\left.\frac{\partial \hat{u}}{\partial n}\right|_1 + \left.\frac{\partial \hat{u}}{\partial n}\right|_2 \right) \tag{14}$$

where $\left.\frac{\partial \hat{u}}{\partial n}\right|_1$ and $\left.\frac{\partial \hat{u}}{\partial n}\right|_2$ are respectively the values of $\frac{\partial \hat{u}}{\partial n}$ over elements 1 and 2 sharing the interface under consideration.

An illustration of both finite element and finite element approximations can be easily obtained in a one dimensional example of Figure 4 where equal element of size h are used.

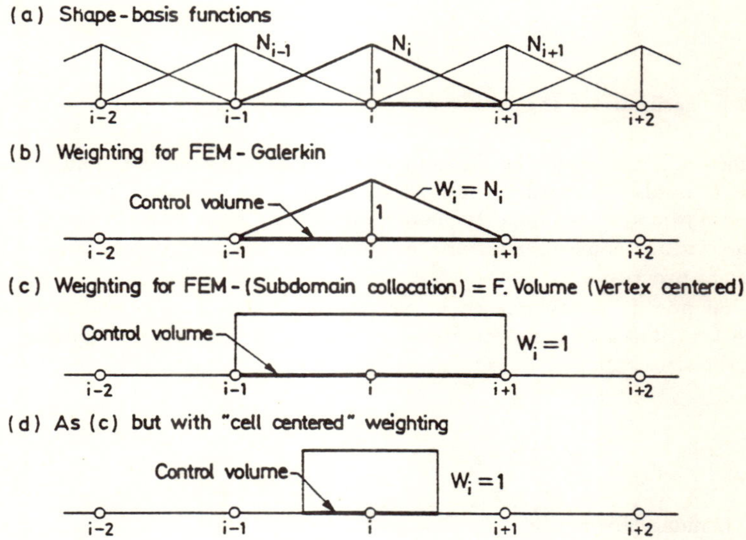

Figure 4. A one dimensional problem.

With linear shape functions used the finite element (Galerkin) approximation to eqs.(10), with $A_i = A = const.$, $Q = Q(x)$, we have a typical assembled equation (which can be easily verified after addition of approximate integrals)

$$\left[\frac{h}{6}\dot{u}_{i+1} + \frac{2h}{3}\dot{u}_i + \frac{h}{6}\dot{u}_{i-1}\right] + \frac{A}{2}[u_{i+1} - u_{i-1}] - \left[\frac{k}{h}u_{i+1} - \frac{2k}{h}u_i + \frac{k}{h}u_{i-1}\right] - f_i = 0 \tag{15}$$

where

$$f_i = \int_{-h}^{h} N_i Q dx \tag{16}$$

The corresponding finite volume equation is obtained using the approximation (14), as

$$\left[\frac{h}{2}\dot{u}_{i+1} + h\dot{u}_i + \frac{h}{2}\dot{u}_{i-1}\right] + A[u_{i+1} - u_{i-1}] - \left[\frac{k}{2h}u_{i+2} - \frac{k}{h}u_i + \frac{k}{2h}u_{i-2}\right] - \bar{f}_i = 0 \tag{17}$$

where

$$\bar{f}_i = \int_{-h}^{h} Q dx \tag{18}$$

It can be easily noted that the two approximations are similar but by no means identical.

The finite volume considered has doubled the mass contained in the finite element. The force f_i is also doubled in case of a constant Q. Further, the mass is not distributed in same proportion at the nodes and neither is the "force" when $Q = Q(x)$.

As expected the convective terms are exactly the double of those obtained by the finite element procedure.

Diffusion now is approximated using external nodes and indeed it is readily recognized that its value is represented by a (curvature) approximation for a mesh of double the size.

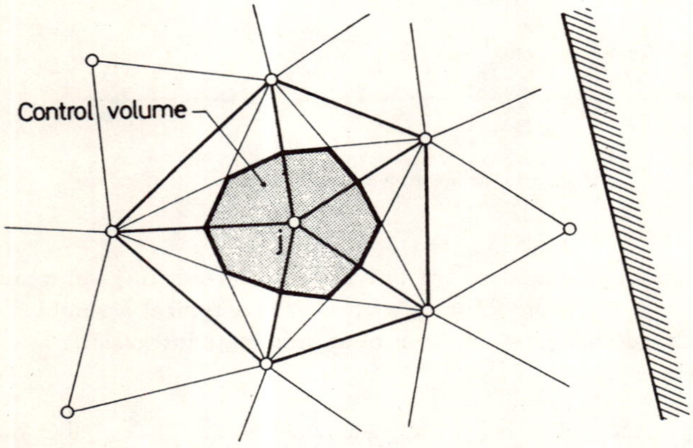

Figure 5. Cell centered control volume.

In the preceeding we have tried to provide "weigthing areas"or control volumes coinciding with elements. Clearly this presents a difficulty and we can now realize why many finite volume methods are applied in so called *cell centered* schemes. Thus if we consider again the discretization of Figure 1 we can, as in Figure 5, assign a control volume to each node without overlapping of the weighted area (the obvious division of each triangle is now indicated). In Figure 4 we show the one dimensional equivalent of above and the reader can readily verify that the finite volume equation now becomes

$$\left[\frac{h}{8}\dot{u}_{i+1} + \frac{6h}{8}\dot{u}_i + \frac{h}{8}\dot{u}_{i-1}\right] + \frac{A}{2}[u_{i+1} - u_{i-1}] - \left[\frac{k}{h}u_{i+1} - \frac{2k}{h}u_i + \frac{k}{h}u_{i-1}\right] - \bar{f}_i = 0 \quad (19)$$

This has the same conectivity as the finite element equations and indeed retrieves here exactly the convective and diffussive terms however showing as before different mass and force distributions (note that now $\bar{f}_i = \int_{-h/2}^{+h/2} Q dx$).

Indeed one must remark here that the use of consistent mass forms which is found to be beneficious in transient Euler fluid dynamics computations using the finite element approximation has never found its way into the finite volume way of thinking. Here the users invariably lump the masses implying for instance in the 1–D example that

$$\dot{u}_{i-1} = \dot{u}_i = \dot{u}_{i+1} \quad (20)$$

If such lumping is done both approximations are identical.

A Quantitative Example

To illustrate the applicability and perfomance of finite volume methods we shall consider a single example of an axially loaded elastic bar of lenght $2l$ (Figure 6). The axial load q will be taken either

as constant

Case I $\qquad q = \text{const.}$

or linear

Case II $\qquad q = cx$

The equilibrium equation and the boundary conditions read now

$$\begin{aligned} &\frac{\partial}{\partial x}\left(EA\frac{\partial u}{\partial x}\right) + q = 0 \qquad 0 \le x \le 2l \\ &u = 0 \quad \text{in} \quad x = 0 \\ &P = EA\frac{\partial u}{\partial x} = 0 \quad \text{in} \quad x = 2l \end{aligned} \quad (21)$$

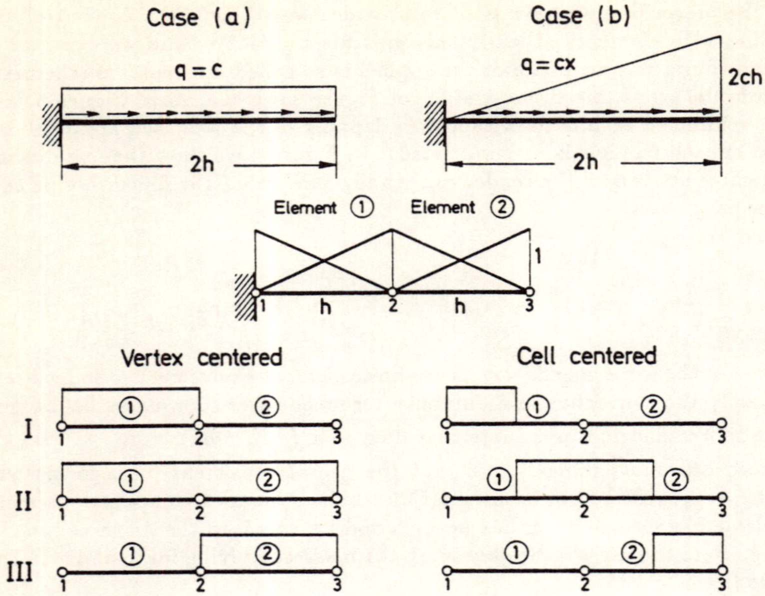

Figure 6. Elastic bar under axial load.

In above u is the axial displacement, P the axial force and E and A are the Young modulus and the area of the bar transverse cross section, respectively.

Following the arguments of proceeding section a typical finite volume equation can be written as

$$\left[EA\frac{\partial u}{\partial x}\right]_R - \left[EA\frac{\partial u}{\partial x}\right]_L + \int_{l_v} q dx = 0 \tag{22}$$

where L and R stand for left and right ends of *volume* of the lenght l_v.

Next we discretize the bar in three two node elements of equal size h. Inside each element the displacements u are linearly interpolated as

$$u = \sum_{i=1}^{2} N_i^{(e)} u_i^{(e)} \tag{23}$$

with $N_i^{(e)} = \frac{1}{2}(1 + \xi\xi_i)$ being the standard linear shape functions of the element.

CASE I, $q = c$

Application of eq.(22) toghether with interpolation (23) to the three volume domains surrounding nodes 1 (element 1), node 2 (element 1 and 2) and node 3 (element 3) gives (taking into account eq.(14))

Volume 1: Element 1

$$\frac{EA}{h}\left[\frac{u_1}{2} + u_2 - \frac{u_3}{2}\right] = ch$$

Volume 2: Elements 1+2

$$\frac{EA}{h}[u_2 - u_1] = 2ch$$

Volume 3: Element 3

$$\frac{EA}{2h}[u_3 - u_1] = ch$$

Note that for volumes 2 and 3 the boundary condition on the traction free end has been imposed exactly.

Solution of above system (with $u_1 = 0$) yields

$$u_2 = u_3 = \frac{2ch^2}{EA}$$

Figure 7 shows the convergence of the vertex centered finite volume solution for meshes of 1, 2, 3 and 4 elements respectively.

It can be easily checked that the finite element and the cell centered solutions for this case are identical and give the exact solution at nodes in all meshes.

CASE II, $q = cx$

Table I shows the percentage error of the vertex centered solution for the nodal axial displacement using three meshes of 1,2 and 3 elements. Note the big error (50%) in the end displacement obtained with the one element mesh. This error reduces to 5.158% if 3 elements are used.

The cell centered solution differs in this case from the nodaly exact finite element values due to the difference in the nodal load (force) vectors as previously explained. Percentage errors in the nodal displacement for the cell centered case are also presented in last column of Table I and show improvement with respect to the vertex centered solution.

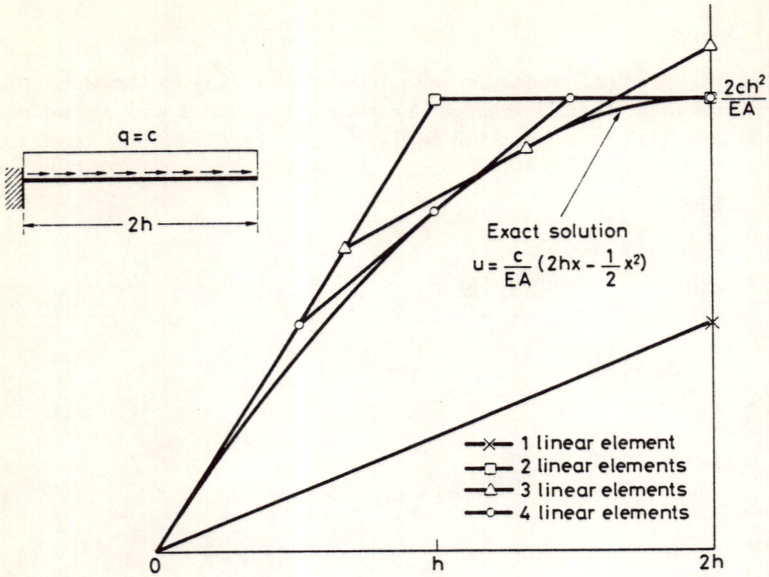

Note: the finite element and cell centered solutions give exact nodal results in all cases

Figure 7. Axially loaded bar. Convervenge of vertex centered finite volume solutions for meshes of 1, 2, 3 and 4 linear elements.

Mesh	Node	Vertex centered		Cell Centered
		Displacement formulation	Mixed formulation	Displacement formulation
1 element	2	50%	25%	12.5%
2 elements	2	9.09%	4.54%	2.27%
	3	12.5%	6.22%	3.5 %
3 elements	2	3.89%	1.85%	1.03%
	3	4.37%	2.14%	1.11%
	4	5.58%	2.77%	1.41%

Note: The finite element solution gives exact nodal results in all cases

Table I. Bar under linearly varying axial load. Percentage errors in nodal displacements for different meshes of linear elements using vertex centered and cell centered finite volume schemes.

Concluding Remarks

From eq.(17) it is easy to observe that for the first order (gradient) terms the approximations given by the vertex centered finite volume process can be readily confined to element assemblies and that evaluations of the approximation coefficients on interfaces can be efficient. The approximation is either similar or identical to that of finite elements and can be extended to higher order interpolations without difficulty. Figure 8 shows some possibilities.

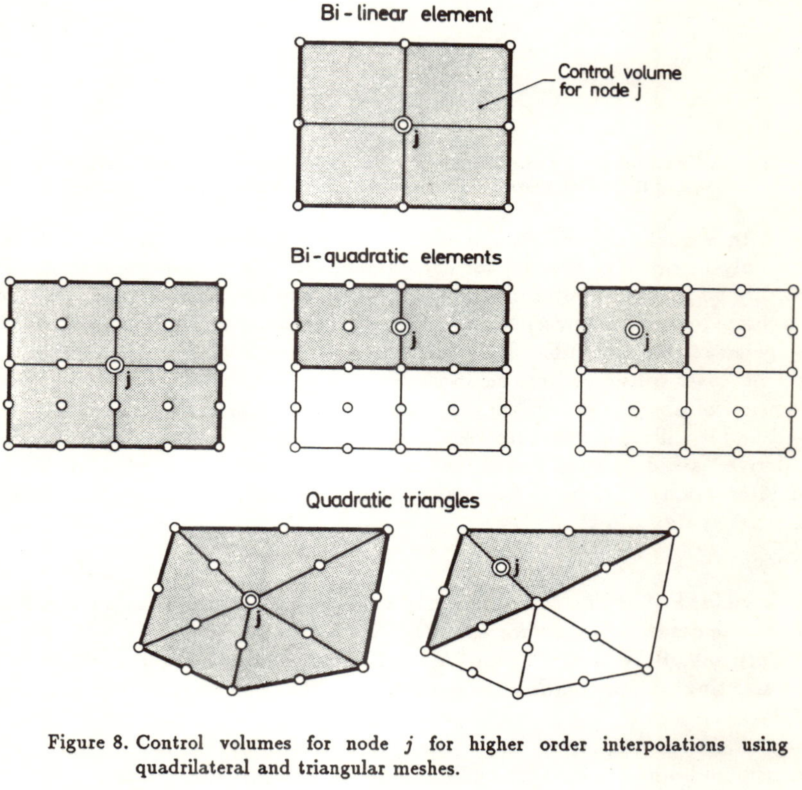

Figure 8. Control volumes for node j for higher order interpolations using quadrilateral and triangular meshes.

This type of approximation does not lend itself simply to second order (diffusion) terms as it involves a wide band of nodes (see Figure 9). This is however possible if explicit, iterative solution methods are used

The problem can be resolved by avoiding second order derivatives by using mixed equation systems. To illustrate the last possibility we shall consider only the diffusive term only in the one dimensional equivalent of eq.(10). writting the relevant equations as

$$\frac{\partial q_i}{\partial x_i} - Q = 0$$

$$q_i + k\frac{\partial u}{\partial x_i} = 0$$

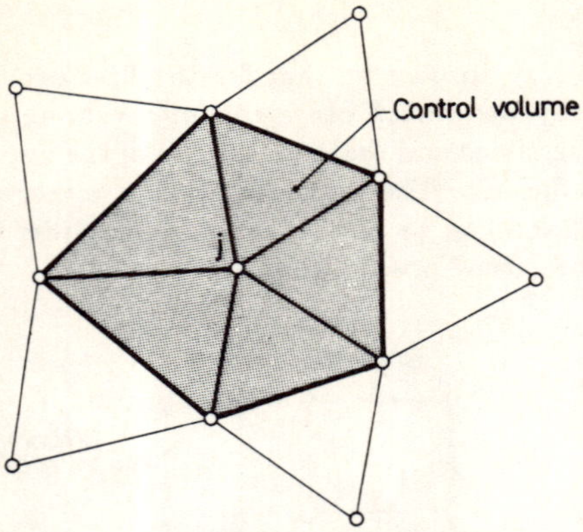

Figure 9. Nodes involved in the discretized equation of an arbitrary control volume using the vertex centered scheme and a "displacement" formulation.

Now with a mixed approximation each of the volumes can be treated as before involving only integration over Ω_j for each assembly. The procedure is simple and could readily be used in many applications in which continuous variation of q_i and u can be assumed with the corresponding gain in accuracy as shown in the elastic bar example given in the Appendix. Here excellent accuracy is obtained and results for the two load cases considered are again shown in Figure 7 and Table I.

Extension of above ideas to various elasticity and plate problems can be easily envisaged and it will be shown that very simple *elements* with simply assembly form can be derived avoiding such difficulties as mechanism formation, etc. inherent in reduced integration procedures frequently used in practice . We shall show such examples elsewhere.

References

[1] Mac Donald, P.W. "The computation of transonic flow through two dimensional gas turbine cascades", *ASME* paper 71-GT-89, (1971).

[2] MacCormack, R.W. and Paullay, A.J. "Computational Efficiency achieved by time splitting of finite difference operators", *AIAA* paper 72–154, San Diego, (1972).

[3] Rizzi, A.W. and Inouye, M. "Time split finite volume method for three dimensional blunt-body flows", *AIAA Journal*, **11**, 1478–85, (1973).

[4] Patankar, S.V., "*Numerical heat transfer and fluid flow*", Series in Computational Methods in Mechanics and Thermal Sciences, W.J. Minkowycz and E. M. Sparrow (Eds.), Hemisphere Publising Corp., (1980).

[5] Wilkins, M.L. "Calculations of elasto-plastic flow". *Methods of Computational Physics* (Eds. B.alder et al.), Vol. **3**, Academic Press, (1964).

[6] Hirsch, C. "*Numerical computation of internal and external flow*", Vol. **I**, J. Wiley, (1989).

[7] Zienkiewicz, O.C. and Taylor, R.L. "*The finite element method*", Vol. **I**. Mc Graw Hill, (1989).

APPENDIX

Finite volume solutions for the bar problem of Figure 6 using a mixed approximation

We consider the bar equilibrium equations written as

$$\begin{aligned} &\frac{dP}{dx} + q = 0 \\ &P - EA\frac{du}{dx} = 0 \qquad \text{in} \quad 0 \leq x \leq 2l \end{aligned} \tag{A.1}$$

with boundary conditions

$$\begin{aligned} u &= 0 \quad \text{in} \quad x = 0 \\ P &= \bar{P} \quad \text{in} \quad x = 2l \end{aligned} \tag{A.2}$$

The weighted residual form of eqs.(A.1) reads

$$\begin{aligned} &\int_0^{2l} W \left[\frac{dP}{dx} + q\right] dx = 0 \\ &\int_0^{2l} \bar{W} \left[P - EA\frac{du}{dx}\right] dx = 0 \end{aligned} \tag{A.3}$$

Choossing $W = \bar{W} = 1$ inside each computation domain of lenght l_v we obtain, after adequate integration by parts,

$$\begin{aligned} &P_R - P_L + \int_{l_v} q dx = 0 \\ &\int_{l_v} P dx - [EAu]_R + [EAu]_L = 0 \end{aligned} \tag{A.4}$$

where indexes R and L refer to right and left ends of the "finite volume" considered.

Next the bar is discretized in linear finite elements where a linear approximations of the axial force P and the displacement u is chosen as

$$P = \sum_{i=1}^{2} N_i^{(e)} P_i^{(e)} \quad , \quad u = \sum_{i=1}^{2} N_i^{(e)} u_i^{(e)} \tag{A.5}$$

Substitution of (A.5) in (A.9) yields the following system of equations for a single two-element domain linking nodes $i-1, i$ and $i+1$.

$$\boxed{\begin{aligned} &P_{i+1} - P_{i-1} + f_i = 0 \\ &\frac{h^{(i+1)}}{2}(P_i + P_{i-1}) + \frac{h^{(i)}}{2}(P_i + P_{i+1}) + EA(u_{i-1} - u_{i-1}) = 0 \end{aligned}} \tag{A.6}$$

Where $h^{(i-1)}$ and $h^{(i)}$ are the lenghts of the two elements considered and

$$f_i = \int_l q dx \qquad \text{with} \qquad l = h^{(i-1)} + h^{(i)} \tag{A.7}$$

Application of eqs.(A.6) to the bar problem of Figure 6 under uniformly distributed axial loading ($q = c$) yields for the two element mesh case:

Volume 1: **Element 1**

$$P_2 - P_1 + ch = 0$$
$$\frac{h}{2}(P_1 + P_2) + EA(u_1 - u_2) = 0$$

Volume 2: **Elements 1+2**

$$P_3 - P_1 + 2ch = 0$$
$$\frac{h}{2}(P_3 + 2P_2 + P_1) + EA(u_1 - u_3) = 0$$

Volume 3: **Element 2**

$$P_3 - P_2 + ch = 0$$
$$\frac{h}{2}(P_2 + P_3) + EA(u_2 - u_3) = 0$$

Solution of above six equations with $P_3 = u_1 = 0$ yields

$$u_2 = \frac{3ch^2}{2EA}, \quad u_3 = \frac{2ch^2}{EA}, \quad P_2 = ch, \quad P_1 = 2ch$$

which coincides precisely with the exact solution. The same preciseness is obtained for the 1 and 3 element meshes as shown in Figure 7.

Results for the linear loading case ($q = cx$) obtained with the mixed approximation are shown in Table I. Note the improvement in accuracy with respect the vertex centered displacement approach, although the cell centered scheme still yields the best approximation for this case.

Use of Additional Incompatible Displacements for Finite Element Formulation

T.H.H. PIAN, Department of Aeronautics & Astronautics, Massachusetts Institute of Technology, Cambridge, MA 02139, USA

C-C. WU, Dept. of Modern Mechanics, University of Science & Technology of China, Hefei, China

Summary

In the formulation of both assumed displacement and assumed stress hybrid finite elements, incompatible displacements may be added for improving their performances. This paper consists of a survey of recent developments in constructing incompatible displacement elements to meet the patch test requirement. The introduction of incompatible displacements in the formulation of hybrid elements results in constraints of initially unconstrained stresses. A review is given of some high performance hybrid elements constructed by using the more desirable stress fields.

Introduction

The performance of finite elements formulated by assumed displacements can be improved by adding displacements with internal parameters that are statically condensed in the element level [1]. Wilson et al [2] formulated 4-node quadrilateral membrane and 8-node hexahedral solid elements by introducing quadratic incompatible displacement modes. In analyzing slender beams in bending by these elements the shear locking difficulty can be avoided. However, for an incompatible element to pass the constant strain patch test, there is a necessary condition that the integration of the strains due to incompatible displacements over the entire element volume must vanish [3]. The Wilson elements can meet this condition only if they are of regular geometries such as rectangular, parallelogram and brick shapes. For elements of general geometry, there exists many alternative measures to meet the convergence requirement. The first part of this paper is to review and compare the various remedies for the incompatible elements.

The assumed stress hybrid finite elements can be formulated conveniently using the Hellinger-Reissner principle which also includes displacements as field variables. A rational procedure for the formulation of such elements is to make balanced assumptions for stresses and displacements. In general, additional incompatible displacements will be needed. As a result, some constraining conditions must be introduced in applying the variational formulation. It turns out that this will lead to certain constraints to initially unconstrained stresses. It is reasonable to think that an element formulated by using such constrained stresses will be of superior quality. The second part of this paper is to present several alternative constraining conditions for stresses and the corresponding hybrid elements.

Incompatible Elements

In the formulation of displacement based finite elements, the displacements $\boldsymbol{u}$ are interpolated in terms of nodal displacements $\boldsymbol{q}$ by compatible shape functions $\boldsymbol{N}_q$ expresses in isoparametric coordinates,

$$u = u_q = N_q(\xi,\eta)q \tag{1}$$

Because u_q are, in general, incomplete polynomials, displacements u_λ are added to make $u = u_q + u_\lambda$ complete polynomials. Here, u_λ are incompatible and are expressed in terms of internal parameters λ, ie.,

$$u_\lambda = N_\lambda(\xi,\eta)\lambda \tag{2}$$

and λ are to be statically condensed in the element level. The strains are then given by,

$$\varepsilon = (Du) = Bq + G\lambda \tag{3}$$

For the resulting incompatible element to pass the patch test under constant strain state, the following condition must be satisfied,

$$\int_{V_n} G dV = \int_{V_n} (DN_\lambda) dV = 0\,; \quad or \quad \int_{V_n} \varepsilon_\lambda dV = 0 \tag{4}$$

Using a 4-node quadrilateral plane stress/strain element as an example, the N_λ functions of the original Wilson element [2], which are usually referred to as Q6, are quadratic functions in ξ and η. In the integral in Eq. 4, the inverse of the Jacobian J between (x,y) and (ξ,η) is involved, with $J^{-1} = J^*(\xi,\eta)/|J|$. The Jacobian is constant only for rectangular and parallelogram shaped elements. Thus, an element of general geometry can not pass the patch test. A simple remedy suggested by Taylor et al [4] involves the replacement of J^{-1} by $J^*(0,0)/|J|$, for the integral which contains the incompatible strains G. This element is often referred to as QM6. It coincides with the Q6 element when the geometry is rectangular. Indeed, in such a case, the incompatible element can lead to a higher rate of convergence because the displacement functions are now complete polynomials in x and y, and the resulting strains and stresses are complete in linear terms.

An alternative approach for constructing incompatible elements that can pass the patch test is to choose u_λ such that the integral of Eq. 4 vanishes. By means of a general formulation of incompatible functions Wu et al [5-7] constructed a series of three quadrilateral elements which are labeled as NQ6, NQ7 and NQ10. They contain, respectively, six, seven and ten terms for each displacement component. This means that, in addition to the four terms for bi-linear interpolation, two, three and six incompatible displacement terms are added respectively. The NQ6 element is complete in quadratic terms only in the isoparametric system. The NQ7 element, on the other hand, is complete in quadratic terms in the Cartesian system. The added terms for the NQ10 element is to maintain cubic completeness in the isoparametric system. The displacement functions for these elements are given in the Appendix. It should be noted that Wachspress [8] also constructed an element QP6 with two incompatible functions which meet the patch test requirement, but they are not complete polynomials.

A comparison of the performance of the various elements is made by an example problem involving a cantilever beam modeled by quadrilateral membrane elements of distorted geometry as shown in Figure 1. The tip deflection v_A and inplane stress σ_x at an upper surface location B due to tip bending moment (case 1) and transverse shear (case 2) are presented in Table 1. Also included are the results obtained by using the compatible element Q4 which is known to have severe locking difficulty for this problem. In most cases, the determination of the inplane stresses are made through differentiating the resulting displacements. Only in the case of the NQ6 element, the stresses are also obtained by an alternative scheme of extrapolating from the Gaussian point solutions. It can be seen that although the Q6 element can not pass the constant strain patch test, it actually yields the most accurate solutions for the tip deflections. The remaining elements all lead to the same results for displacements. The maintaining of quadratic completeness in the Cartesian system and of cubic completeness in the isoparametric system all leads to more accurate stress solutions. It is well-known, that more accurate stress solutions along the boundary of an element can be obtained by extrapolating from the Gaussian points. For the NQ7 element it is not necessary to apply an extrapolating process for stresses, since the assumed displacements are already complete in quadratic terms in the Cartesian coordiantes.

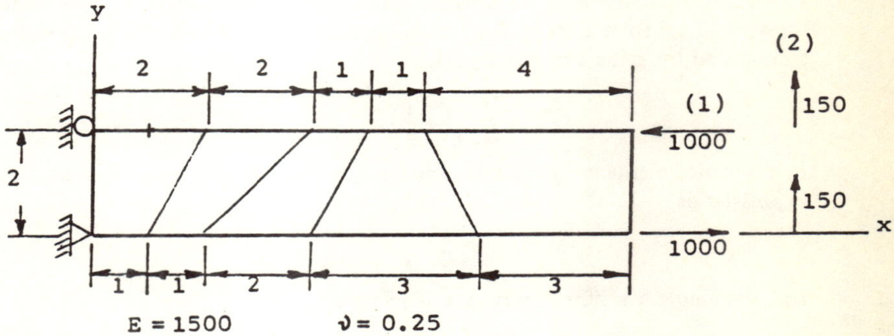

Figure 1. Cantilever beam problems by distorted quadrilateral plane stress elements.

It has been shown [9] that to maintain the stability of the solutions by incompatible elements it is not only necessary to keep the basis functions N_q and N_λ of Eq. 1 and 2 linearly independent, but also to satisfy certain weak connecting conditions along the interelement boundaries. For example, for the plane stress/strain elements, the weak connection pattern along the boundary S_n is shown in Figure 2a. However, to avoid the use of an inplane drilling degree-of-freedom θ_i, one can choose the weak connection pattern shown in Figure 2b. Clearly, all the above incompatible elements which are based on the Q4 element can satisfy this condition.

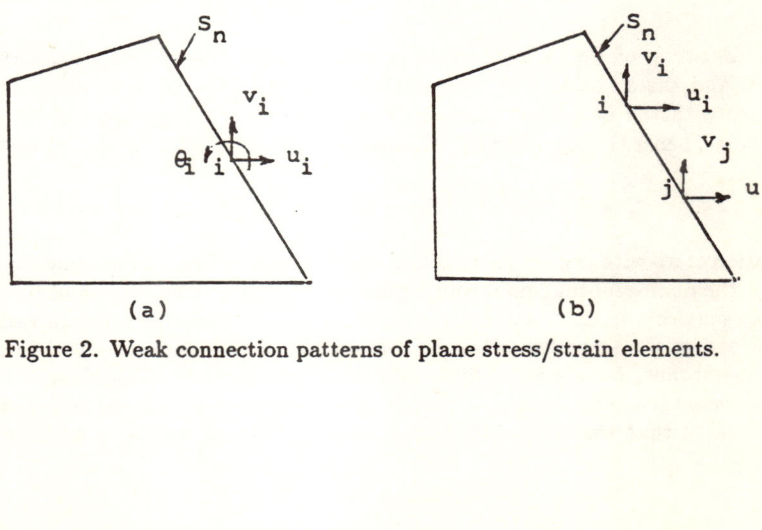

Figure 2. Weak connection patterns of plane stress/strain elements.

Hybrid Stress Elements

When the assumed stress hybrid element is formulated by using the Hellinger-Reissner principle, with assumed incompatible displacements $\boldsymbol{u}_\lambda$, the variational functional of an element with the effect of the external loading excluded can be written as,

$$\pi_R^e = \int_{V_n} \left[\left(-\frac{1}{2}\right) \boldsymbol{\sigma}^T \boldsymbol{S\sigma} + \boldsymbol{\sigma}^T (\boldsymbol{Du}_q + \boldsymbol{Du}_\lambda)\right] dV \tag{5}$$

Because the assumed displacements are incompatible along the boundary of the elements, certain constraining conditions must be introduced so that the stationary condition of the above functional is still valid. The additional incompatible displacements, then, serve as a vehicle to constrain the initially unconstrained stresses to a more desirable assumed stress field. If such a stress field is designated as $\boldsymbol{\sigma}^*$, the element stiffness matrix can be formulated by using the functional,

$$\pi_R^e = \int_{V_n} \left[\left(-\frac{1}{2}\right) \boldsymbol{\sigma}^{*T} \boldsymbol{S\sigma}^* + \boldsymbol{\sigma}^{*T} (\boldsymbol{Du}_q)\right] dV \tag{6}$$

In the finite element formulation, $\boldsymbol{u}_\lambda$ is expressed in the same way as Eq. 1 and $\boldsymbol{\sigma}^*$ is expressed as,

$$\boldsymbol{\sigma}^* = \boldsymbol{P}(\xi, \eta)\boldsymbol{\beta} \tag{7}$$

and the element stiffness matrix $\boldsymbol{k}$ is given by

$$\boldsymbol{k} = \overline{\boldsymbol{G}}^T \boldsymbol{H} \overline{\boldsymbol{G}} \tag{8}$$

where

$$\boldsymbol{H} = \int_{V_n} \boldsymbol{P}^T \boldsymbol{SP} dV \tag{9}$$

and

$$\overline{\boldsymbol{G}} = \int_{V_n} \boldsymbol{P}^T (\boldsymbol{DN}_q) dV \tag{10}$$

Presented in the following are three alternative approaches for obtaining the constraining conditions. The resulting elements are labeled as HS1, HS2 and HS3, respectively.

(1) HS1

The modification of Eq. 1 involves an introduction of independent boundary displacements $\overline{\boldsymbol{u}}$ and the addition of a constraining condition $(\boldsymbol{u} - \overline{\boldsymbol{u}}) = 0$, along the element boundary with the boundary tractions $\boldsymbol{T} = (\boldsymbol{n\sigma})$ as the Lagrange multipliers. Here, $\boldsymbol{u} - \overline{\boldsymbol{u}} = \boldsymbol{u}_\lambda$, hence, by applying the divergence theorem, the functional π_R^e becomes [10],

$$\pi_R^e = \int_{V_n} \left[-\left(\frac{1}{2}\right) \boldsymbol{\sigma}^T \boldsymbol{S\sigma} + \boldsymbol{\sigma}^T (\boldsymbol{Du}_q) - \left(\boldsymbol{D}^T \boldsymbol{\sigma}\right)^T \boldsymbol{u}_\lambda\right] dV \tag{11}$$

A variation of π_R^e with respect to $\boldsymbol{u}_\lambda$ within the element level yields the equation $\boldsymbol{D}^T \boldsymbol{\sigma} = 0$, which is the homogeneous equilibrium equation. Thus, the introduction of the incompatible displacements $\boldsymbol{u}_\lambda$ is to enforce the stress equilibrium condition in the variational or integral sense. In a number of applications which include 4-node and 8-node quadrilateral membrane, 8-node hexahedral solid and some plate bending elements [11-14], it was discovered that the number of the independent constraining equations obtained this way is smaller than that of the λ-parameters used. This, of course, is an indication that the solution for $\boldsymbol{\lambda}$ is singular. Nevertheless, a constrained stress assumption can still be made. The resulting elements, however, are much too rigid. As an example, a 4-node quadrilateral membrane element so constructed will still have the shear locking difficulty. Fortunately, successful remedies have been found. They are either a perturbation of the boundary geometry or a transformation of the assumed stresses to include an arbitrarily small higher order component. The constraining equations obtained this way will lead to a desirable stress field $\boldsymbol{\sigma}^*$ for the construction of elements of superior qualities [11, 14].

(2) HS2

If the energy functional π_R^e are summed over all elements without any adjustment against the interelement incompatibility, then its variation will include a boundary integral term,

$$\sum_n \int_{\partial V_n} \boldsymbol{T}^T \delta \boldsymbol{u}_\lambda dS.$$

Since $\boldsymbol{u}_\lambda$ are arbitrary for individual elements, the boundary tractions $\boldsymbol{T}$ for each element will have to vanish. Alternatively, the above integral should be set to zero. A stronger condition to be satisfied is, for each element,

$$\int_{\partial V_n} \boldsymbol{T}^T \delta \boldsymbol{u}_\lambda dS = 0. \tag{12}$$

This means that the boundary tractions resulting from the assumed stress field must satisfy the condition of vanishing of the virtual work done due to the incompatible displacements $\boldsymbol{u}_\lambda$ along the element boundary. It turns out that the stresses obtained under such condition will have coupling between the constant strain patch test. A suggested remedy is to separate the constant and higher order stresses and express the respective boundary tractions as $\boldsymbol{T}_c$ and $\boldsymbol{T}_h$, and then only maintain

$$\int_{V_n} \boldsymbol{T}_h^T \delta \boldsymbol{u}_\lambda dS = 0. \tag{13}$$

In addition, either of two alternative conditions should be met: (a) by adding

$$-\int_{\partial V_n} \boldsymbol{\sigma}_c^T \boldsymbol{n}^T \boldsymbol{u}_\lambda dS$$

in Eq. 1, and (b) by choosing $\boldsymbol{u}_\lambda$ such that,

$$\int_{\partial V_n} \boldsymbol{n}^T \boldsymbol{u}_\lambda dS = 0. \tag{14}$$

The last condition is equivalent to Eq. 4, the necessary condition for an incompatible element to pass the constant strain patch test. If a stress field is determined by the first alternative, and additional term

$$\int_{V_n} \boldsymbol{\sigma}_h^{*T} (\boldsymbol{D} \boldsymbol{u}_\lambda) dV$$

needs to be included in Eq. 2. This term, however, is of higher order. If, for simplicity, this term is neglected, Eq. 6 can again be used for the finite element formulation. Of course, a stress field determined by the second alternative is immediately adaptable to that equation. These methods will be labeled as HS2a and HS2b respectively. Four-node quadrilateral membrane elements have been constructed by both of the two HS2 methods, while only HS2a elements have been used for constructing an 8-node hexedral solid, a 4-node axisymmetric solid and Reissner-Mindlin plate and shell elements [15-19].

(3) HS3

Another possible method for handling the incompatible displacements in the hybrid element formulation is to choose the assumed stresses such that the integral

$$\int_{V_n} \sigma^T (Du_\lambda) dV$$

vanishes [20]. This means that the assumed stress field must be orthogonal to the strains due to incompatible displacements. However, if this is followed, the resulting element will again fail the constant strain patch test because of the coupling between the constant and higher order stress terms. The remedy, of course, is again the separation of the constant and higher order stresses and to use the vanishing of the integral

$$\int_{V_n} \sigma_h^T (Du_\lambda) dV$$

as the constraining condition of the assumed stress field. Here the displacements u_λ must be so chosen that Eq. 4 is satisfied. This approach has been used in the doctoral thesis study by Sze of the University of Hong Kong [21-23]. Again, 4-node membrane, 8-node solid, 4-node axisymmetric solid and plate and shell elements are included in that study.

Stability of hybrid stress finite elements can be investigated by a study of the zero energy deformation modes. References 24 and 25 have provided detailed dicussions on the analysis and control of zero energy deformation modes in such elements. For an element formulated by the introduction of additional incompatible displacements the necessary condition for preventing zero energy deformation mode is

$$n_\beta \geq (n_q + n_\lambda) - l \tag{15}$$

where n_β = number of β-stress parameters
n_q = number of nodal displacements
n_λ = number of incompatible displacements
and l = number of rigid body degrees of freedom

It should be noted that Chen and Cheung [26-28] have constructed hybrid elements by using a modified three-field Hu-Washizu variational principle with added incompatible displacements to enforce the stress equilibrating condition in the variational sense. In this formulation, the assumed strains are of consistent order as the assumed displacements, but the assumed stress field includes higher order terms such that, unlike the early experience in formulating the HS1 elements, the number of independent constraining equations is the same as that of the displacement parameters $\boldsymbol{\lambda}$. The resulting element, thus, can maintain high performance without the need of a perturbation process. Another advantage of this method is that by appropriate choice of basis functions for stresses and strains, a matrix to be inverted can be made a diagonal one, hence the computing effort can be reduced considerably. Another recent discussion has also been given by Simo and Rifai [29] on the formulation of finite elements using a three-field variational principle with added incompatible displacement modes.

Comparison of Various Hybrid Elements

To compare the hybrid elements formulated by the three approaches based on the above, the same example of a cantilever beam modeled by plane stress elements of distorted geometry shown in Fig. 1 will be used. For these solutions, the assumed displacements are the same as that for the Q6 element, ie., they are complete in quadratic terms in an isoparametric system. For the formulation by the Hellinger-Reissner principle, the initially unconstrained stresses are consistent with the strain field due to assumed displacements, ie., they are complete in linear terms with nine β-parameters as follows:

$$\sigma = \left\{ \begin{array}{c} \sigma_x \\ \sigma_y \\ \tau_{xy} \end{array} \right\} = \begin{bmatrix} 1 & 0 & 0 & \xi & \eta & 0 & 0 & 0 & 0 \\ 0 & 1 & 0 & 0 & 0 & \xi & \eta & 0 & 0 \\ 0 & 0 & 1 & 0 & 0 & 0 & 0 & \xi & \eta \end{bmatrix} \left\{ \begin{array}{c} \beta_1 \\ \vdots \\ \beta_9 \end{array} \right\} \tag{16}$$

It turns out that the constraining equations obtained by methods HS1, HS2a and HS3 are identical. Because all the elements used in this example have one pair of edges in parallel, the resulting HS2b element is also identical to the HS2a element. Thus, the resulting assumed stress field for all the elements is,

$$\sigma^* = \begin{bmatrix} 1 & 0 & 0 & a_1^2\eta & a_3^2\xi \\ 0 & 1 & 0 & b_1^2\eta & b_3^2\xi \\ 0 & 0 & 1 & a_1b_1\eta & a_3b_3\xi \end{bmatrix} \left\{ \begin{array}{c} \beta_1 \\ \vdots \\ \beta_5 \end{array} \right\} \tag{17}$$

where a_i and b_i, $i = 1$ and 3 are given in the Appendix.
For this problem, the resulting displacements by the hybrid elements are the same as that by the incompatible elements QM6 etc., but the resulting stresses are considerably better. It should be pointed out that, in general, method HS2b also reflect the element geometry parameters a_2 and b_2, hence will yield more accurate solutions in both stresses and displacements than that by method HS2a for elements of distorted geometry.

The second example is concerning 4-node quadrilateral axisymmetric elements. For this case, the assumed displacements in isoparametric coordinates are the same as that for the 4-node membrane element. The initially unconstrained stresses are, of course, complete in linear terms. It turns out that when method HS1 is used [30], the constraining condition of equilibrating stresses in the integral sense will cause a coupling between the constant and higher order stress terms. The resulting element, hence, can not pass the constant axial strain patch test. Moreover, it will result in a false shear stress for a problem in which the shear stress τ_{rz} is zero. In fact, incompatible displacement axisymmetric elements also have such shortcomings [31]. Such difficulty, of course, does not exist for the HS2 and HS3 elements. Recently, Tian and Tian [32] suggested a remedy to method HS1 by maintaining the equilibrium of only the higher order stresses, ie.,

$$\int_{V_n} (D^T\sigma_h))^T \delta u_\lambda dV = 0 \tag{18}$$

It turns out that this method is equivalent to the corresponding HS2 method [17]. In [32] an alternative suggestion has also been made for choosing u_λ such that the integral

$$\int_{V_n} (D^T\sigma_h)^T u_\lambda dV$$

vanishes.

An illustrative example given here is an infinitely long thick-wall cylinder under internal pressure. Under the plane strain condition only one row of elements is required in the analysis. Four quadrilateral elements are used as shown in Fig. 3. The degree of geometric distortion for these elements is governed by the parameter e. The purpose is to determine the sensitivities of the element solutions to the degree of geometrical distortion. Comparisons of the various solutions for the radial stress σ_r at the center of the first element are shown in Fig. 4.

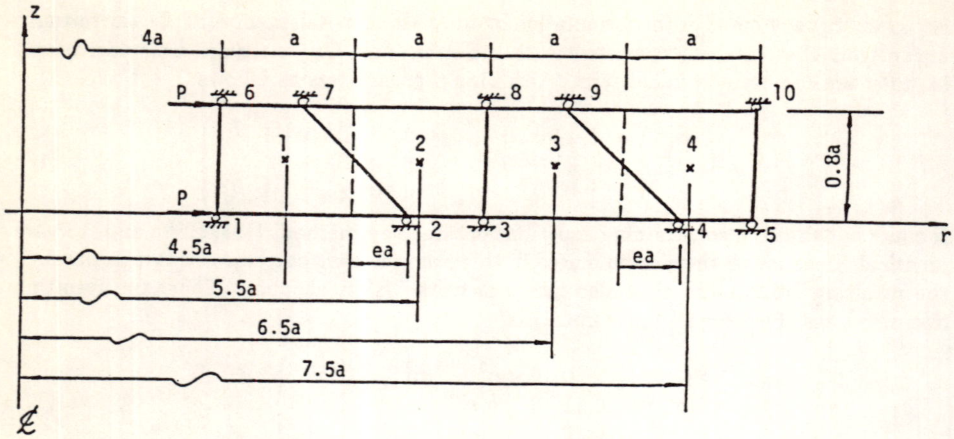

Figure 3. Finite element mesh for thick-walled cylinder under internal pressure.

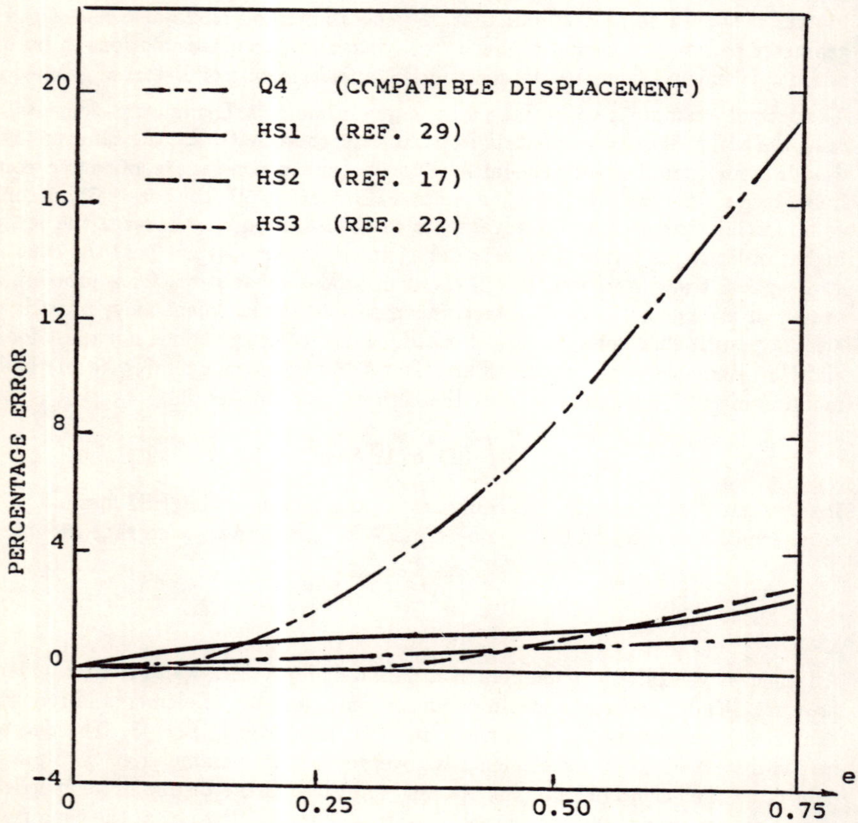

Figure 4. Thick walled cylinder under internal pressure. Variation of stress σ_r vs. distortion parameter e.

Included in these comparisons is the solution by the conventional compatible element. The improvement of the various hybrid elements over the compatible displacement element is clearly demonstrated in this example. It should be emphasized that the HS1 element actually can not pass the patch test, yet it is quite accurate and also insensitive to geometric distortions for this particular problem.

Conclusions

The introduction of additional incompatible displacements not only improves the assumed displacement finite elements, but also serves as a logical method for constraining the assumed stress terms for the formulation of hybrid stress elements. Many different approaches exist which lead to equally superior performance so far as the accuracies of displacements and stresses are concerned. There remains a systematic study to evaluate the various approaches from the point of view of computing efficiency.

Elements	$v_A^{(1)}$	$\sigma_{xB}^{(1)}$	$v_A^{(2)}$	$\sigma_{xB}^{(2)}$
Compatible Element Q4	45.7	-1761	50.96	-2448
Incompatible Elements:				
Q6 (Wilson et al. 1973) [2]	98.4	-2428	100.4	-3355
QM6 (Taylor et al. 1976) [4]	96.1	-2511	97.9	-3388
NQ6 (Wu et al. 1986) [6]	96.	-2439	98.0	-3294
QM6, NQ6, NQ7 } Bilinear Stress Extrapolation		-3014		-4073
NQ7 (jiao & Wu 1988) [7]	96.1	-3007	98.0	-4035
NQ10 (Wu et al. 1987) [5]	96.	-2988	98.0	-4024
Hybrid Stress Elements [9,10]	96.2	-3014	98.2	-4137
Exact	100	-3000	102.6	-4050

Table 1. Comparison of finite element solutions for cantilever beam problems by distorted quadrilateral plane stress elements.

References

1. Pian, T.H.H.: Derivation of Element Stiffness Matrices. J.A.I.A.A. 2, (1964) 576-577.

2. Wilson, E.L.; Taylor, R.L.; Doherty, W.P.; Ghaboussi, J.: Incompatible Displacement Models in Numerical and Computer Methods in Structural Mechanics (Ed. S.J. Fenves, et al), Academic Press, New York, (1973) 43-57.

3. Strang, G.; Fix, G.J.: An Analysis of the Finite Element Method. Prentice Hall, Englewood Cliffs, NJ, (1973).

4. Taylor, R.L.; Beresford, P.J.; Wilson, E.L.: A Non-Conforming Element for Stress Analysis. Int. J. Num. Meth. Engg. 10 (1976) 1211-1219.

5. Wu, C.C.; Huang, M.K.; Pian, T.H.H.: Consistency Condition and Convergence Criteria of Incompatible Elements - General Formulation of Incompatible Functions and its Application. Computers & Structures 27, (1987) 639-644.

6. Pian, T.H.H.; Wu, C.C.: General Formulation of Incompatible Shape Function on an Imcompatible Isoparametric Element. Proc. Invitational China-American Workshop on Finite Element Methods, Chengde, China, June 2-6, (1986) 159-165.

7. Jiao, Z.P.; Wu, C.C.: The Isoparametric Incompatible Element NQ7 - An Improvement on Element NQ9 (in Chinese) J. Hefei University of Technology. 11 (1988) 20-26.

8. Wachspress, E.L.: Incompatible Quadrilateral Basis Functions. Int. J. Num. Meth. Engg. 12 (1978) 589-595.

9. Wu, C.C.; Cheung, Y.K.: Numerical Stability and Constitution Analysis of λ-type Incompatible Elements. Int. J. Num. Meth. Engg. (In Press).

10. Pian, T.H.H.; Chen, D.P.: Alternative Ways for Formulation of Hybrid Stress Elements. Int. J. Num. Meth. Engg. 18, (1982) 1679-1684.

11. Pian, T.H.H.; Sumihara, K.: Rational Approach for Assumed Stress Finite Elements. Int. J. Num. Meth. Engg. 20 (1984) 1685-1695.

12. Pian, T.H.H.: Finite Elements Based on Consistently Assumed Stresses and Displacements. Finite Elements in Analysis and Design 1 (1985) 131-140.

13. Pian, T.H.H.; Kang, D.; Wang, C.: Hybrid Plate Elements Based on Balanced Stresses and Displacements. in Finite Element for Plate and Shell Structures. Vol. 1, Element Technology (Ed. T.J.R. Hughes; E. Hinton) Pineridge Press Ld, Swansea, Wales, U.K. (1986) 244-265.

14. Wang, C.; Pian, T.H.H.: Hybrid Semi-Loof Element for Buckling of Thin-Walled Structures. Computers & Structures 30 (1988) 811-816.

15. Wu, C.C.; Di, S.L.; Huang, M.K.: Model Optimization of Hybrid Elements. (in Chinese) Science Tongbao 31 (1986) 1142-1144.

16. Pian, T.H.H.; Wu, C.C.: A Rational Approach for Choosing Stress Terms for Hybrid Finite Element Formulations. Int. J. Num. Meth. Engg. 26 (1988) 2331-2343.

17. Wu, C.C.; Di, S.L.; Pian, T.H.H.: Optimizing Formulation of Axisymmetric Hyrbid Stress Elements. (in Chinese) Acta Aeronautica et Astronatica Sinica 8 (1987) A439-A448.

18. Di, S.L.; Wu, C.C.; Song, Q.G.: Model Optimization of Hybrid Stress General Shell Element, Acta Mechanica Solida Sinica, 2 (1989) 1-18.

19. Wu, C.C.; Bufler, H.: Multivariable Finite Elements: Consistency and Mode Optimization. (in Chinese) Zongguo kexue (Science in China), series A (1990) 946-956.

20. Pian, T.H.H.; Tong, P.: Relations Between Incompatible Displacement Model and Hybrid Stress Model. Int. J. Num. Meth. Engg. 22 (1986) 173-181.

21. Sze, K.Y.; Chow, C.L.: Efficient Hybrid/Mixed Element Formulations by Free Formulation and Energy Orthogonality, paper presented at Computer Aided Assessment and Control of Localized Damage Conference, Southampton. U.K. June 26-28,1990.

22. Sze, K.Y.: Development of Efficient and Robust Hybrid/Mixed Elements for Solid Structures. Ph.D. Thesis, Dept. of Mechanical Engineering, University of Hong Kong, 1990.

23. Sze, K.Y.; Chow, C.L.; Chen, W.J.: A Rational Formulation of Isoparametric Hybrid Stress Elements for Three-Dimensional Stress Analysis. Finite Elements in Analysis and Design. 7 (1990) 61-72.

24. Pian, T.H.H.; Chen, D.P.: On the Suppression of Zero-Energy Deformation Modes, Int. J. Num. Meth. Engg. 19 (1983) 1741-1752.

25. Wu, C.C.: Dual Zero Energy Modes in Mixed/Hybrid Elements-Definition, Analysis and Control, Comp. Meth. Appl. Mech. Engg. 81 (1990) 39-56.

26. Chen, W.J.; Cheung, Y.K.: A New Approach for the Hybrid Element Method. Int. J. Num. Meth. Engg. 24 (1987) 1697-1709.

27. Chen, W.J.; Cheung, Y.K.: Axisymmetric Solid Elements by the Generalized Hybrid Method. Computers & Structures 27 (1987) 745-792.

28. Cheung, Y.K.; Chen, W.J.: Isoparametric Hybrid Hexahedral Elements for Three Dimensional Stress Analysis. Int. J. Num. Meth. Engg. (1988) 677-693.

29. Simo, J.C.; Rifai, M.S.: A Class of Mixed Assumed Strain Methods and the Method of Incompatible Modes. Int. J. Num. Meth. Engg. 29 (1990) 1595-1638.

30. Tian, Z.S.; Pian, T.H.H.: Axisymmetric Solid Elements by a Rational Hybrid Stress Method. Computers & Structures 20 (1985) 141-149.

31. Cook, R.D.: A Note on Certain Incompatible Elements. Int. J. Num. Meth. Engg. 6 (1973) 146-147.

32. Tian, Z.S.; Tian, Z.R.: Further Study of Construction of Axisymmetric Finite Element by Hybrid Stress Method. Proc. Third Int. Conf. on Education, Practice and Promotion of Computational Methods in Engineering using Small Computers (EPMESC III), Macau, (1990) 549-558.

Appendix

Displacement Functions for Incompatible Elements

$$\boldsymbol{u} = \boldsymbol{u}_q + \boldsymbol{u}_\lambda$$

$$\boldsymbol{u}_q = \left\{ \begin{array}{c} u_q \\ v_q \end{array} \right\} = \sum_{i=1}^{4}(1+\xi_i\xi)(1+\eta_i\eta)\left\{ \begin{array}{c} u_i \\ v_i \end{array} \right\}$$

$$\boldsymbol{u}_\lambda = \left\{ \begin{array}{c} u_\lambda \\ v_\lambda \end{array} \right\} = \left[\begin{array}{cc} \boldsymbol{f}_n & 0 \\ 0 & \boldsymbol{f}_n \end{array} \right] \left\{ \begin{array}{c} \lambda_1 \\ \vdots \\ \lambda_{2n} \end{array} \right\}$$

a_i and b_i, where $i = 1$ to 4, are determined by

$$x = a_4 + a_1\xi + a_3\eta + a_2\xi$$
$$y = b_4 + b_1\xi + b_3\eta + b_2\xi$$

Q6, MQ6:

$$\boldsymbol{f}_2 = [\xi^2 \quad \eta^2]$$

NQ6:

$$\begin{aligned} \boldsymbol{f}_2 &= [\xi^2 + \frac{2}{3}(r\xi - s\eta) \quad \eta^2 - \frac{2}{3}(r\xi - s\eta)] \\ r &= (a_1b_2 - b_1a_2)/(a_3b_1 - b_3a_1) \\ s &= (a_2b_3 - b_2a_3)/(a_3b_1 - b_3a_1) \end{aligned}$$

NQ10:

$$\boldsymbol{f}_6 = \begin{array}{ccc} [\xi^2 + \frac{2}{3}(r\xi - s\eta) & \eta^2 - \frac{2}{3}(r\xi - s\eta) & \xi^3 - \xi \\ \xi^2\eta - \frac{1}{3}\eta & \xi\eta^2 - \frac{1}{3}\xi & \eta^3 - \eta] \end{array}$$

NQ7:

$$\begin{aligned} \boldsymbol{f}_3 = [a_1^2\xi^2 + a_3^2\eta^2 + 2a_1a_3\xi^2\eta + 2a_2a_3\xi\eta^2 + a_2^2\xi^2\eta^2 + 2(a_1\xi + a_3\eta)\frac{r_2}{s_0} \\ b_1^2\xi^2 + b_3^2\eta^2 + 2b_1b_3\xi^2\eta + 2b_2b_3\xi\eta^2 + b_2^2\xi^2\eta^2 + 2(b_1\xi + b_3\eta)\frac{r_2}{s_0} \\ a_1b_1\xi^2 + a_3b_3\eta^2 + (a_1b_2 + a_2b_1)\xi^2\eta + (a_3b_2 + a_2b_3)\xi\eta^2 + a_2b_2\xi^2\eta + c_1\xi + c_2\eta] \end{aligned}$$

$$\begin{aligned} r_1 &= b_2(a_1^2 - a_3^2) + a_2(a_3b_3 - a_1b_1) \\ r_2 &= a_2(b_3^2 - b_1^2) + b_2(a_1b_1 - a_3b_3) \\ s_0 &= -3(a_1b_3 - a_3b_1) \\ c_1 &= (a_1r_2 + b_1r_1)/s_0 \\ c_2 &= (a_3r_2 + b_3r_1)/s_0 \end{aligned}$$

Hybrid Analytical Technique for the Nonlinear Analysis of Curved Beams

AHMED K. NOOR
Center for Computational Structures Technology
University of Virginia
NASA Langley Research Center
Hampton, VA

and
CARL M. ANDERSEN
College of William and Mary in Virginia
Williamsburg, VA

Summary

The application of a two-step hybrid technique to the geometrically nonlinear analysis of curved beams is used to demonstrate the potential of hybrid analytical techniques in nonlinear structural mechanics. The hybrid technique is based on successive use of the perturbation method and a classical direct variational procedure. The functions associated with the various-order terms in the perturbation expansion of the fundamental unknowns, and their sensitivity derivatives with respect to material and geometric parameters of the beam, are first obtained by using the perturbation method. These functions are selected as coordinate functions (or modes) and the classical direct variational technique is then used to compute their amplitudes. The potential of the proposed hybrid technique for nonlinear analysis of structures is discussed. The effectiveness of the hybrid technique is demonstrated by means of numerical examples. The symbolic computation system Mathematica is used in the present study. The tasks performed on Mathematica include: 1) generation of algebraic expressions for the perturbation functions of the different response quantities and their sensitivity derivatives; and 2) determination of the radius of convergence of the perturbation series.

Introduction

Prior to the advent of powerful digital computers, approximate analytical techniques such as the perturbation method, the Bubnov-Galerkin, and the Rayleigh-Ritz techniques were standard tools for the nonlinear analysis of structures. The widespread availability of powerful computers and the fascination with the finite element method due to its versatility in handling complex geometries, and simplicity of computer implementation, has resulted in a relative stagnation in the development of effective analytical techniques.

Analytical techniques have the major advantage over numerical discretization techniques (such as finite element method) of providing physical insight into the nature of the response.

In the perturbation method the fundamental unknowns are expanded in perturbation series in terms of unknown functions with preassigned coefficients. The unknown functions are obtained by solving a recursive set of differential equations which are, in general, simpler than the original governing differential equation of the problem (see, for example, Refs. 5 and 13). By contrast, in the Bubnov-Galerkin and Rayleigh-Ritz techniques, the fundamental unknowns are sought in the form of a series of *a priori* chosen coordinate functions (or modes) with known coefficients. Reviews of the many applications of these techniques to nonlinear structural mechanics problems are given in Ref. 11.

Despite their usefulness in solving nonlinear structural problems, the aforementioned analytical techniques have a number of drawbacks. The perturbation method has two major drawbacks. The first stems from the fact that as the number of terms in the perturbation series increases, the mathematical complexity of the differential equations builds up rapidly. Therefore, for practical applications, the perturbation series has to be restricted to a few terms. The second drawback is the requirement of restricting the perturbation parameter to small values in order to obtain solutions of acceptable accuracy. The main difficulty of both the Bubnov-Galerkin and Rayleigh-Ritz techniques, from a practical point of view, is the difficulty of selecting good coordinate functions (or modes) for structures with complicated geometry and/or complex response.

The aforementioned drawbacks of the regular perturbation technique have been recognized and a number of remedial actions were proposed. These include: 1) using a small number of terms (e.g., two or three) in the perturbation expansion and generating "mimic functions" which give accurate numerical estimates of the solution over the entire physical domain (see Ref. 4), or 2) applying a nonlinear transformation (e.g., Shanks transformation) to estimate the solution as the number of terms goes to infinity. However, as shown in Ref. 12, the success of these methods cannot be guaranteed in general, and the remedial actions may fail to produce satisfactory results.

In an attempt to overcome the major drawbacks of the perturbation method and the classical Bubnov-Galerkin/Rayleigh-Ritz technique, a hybrid technique that combines both the two techniques was proposed by the first author and his coworkers in 1984 (see Refs. 6 and 7). The technique was shown to provide a more effective analysis technique for nonlinear thermal and structural problems than either of the two parent techniques. The technique was later applied to other nonlinear two-point boundary-value problems in Refs. 2 and 3. However, the full potential of the hybrid technique has not yet been realized. The widespread use and increasing capabilities of symbolic computation software systems (such as MACSYMA,

Mathematica and Scratchpad II) should help in realizing this potential. The present study is a step in that direction and focuses on the effective use of symbolic computations in conjunction with the hybrid technique for nonlinear analysis of curved beams. The symbolic computation system Mathematica is used in the present study (see Ref. 15).

Mathematical Formulation

The analytical formulation is based on the moderate-rotation, geometrically nonlinear, shallow curved beam theory with the effects of transverse shear deformation included. A total Lagrangian description of the beam deformation is used, and the beam deformations at different load levels are referred to the initial coordinate system of the undeformed beam. A mixed formulation is used in which the fundamental unknowns consist of the six beam quantities the generalized displacements u, w and ϕ, and the internal forces N, Q and M (see Fig. 1 for sign convention). The fundamental equations of the theory used herein are summarized in Appendix A.

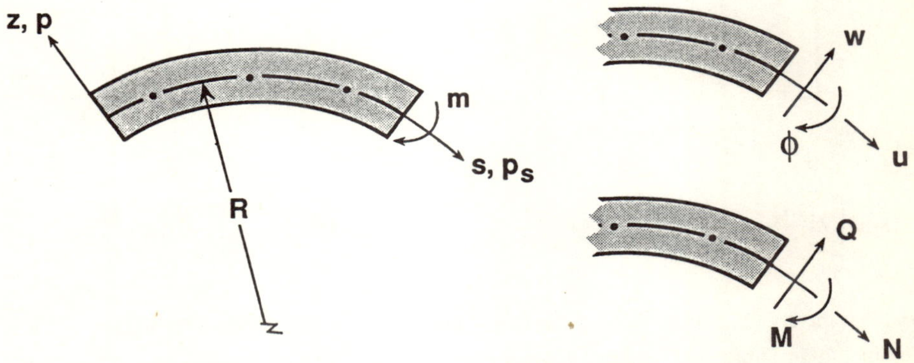

Figure 1 - Sign convention for generalized displacements, internal and external forces.

Governing Differential Equations. The governing equations are first-order nonlinear ordinary differential equations which can be cast in the following compact form:

$$\partial_s\{\psi\} - [I_1]\{\psi\} - \{G(P,\psi)\} + \lambda\{P\} = 0 \tag{1}$$

where $\{\psi\}$ is the vector of fundamental unknowns; $\{P\}$ is the vector of external forces and moments; λ is a load parameter; $[I_1]$ is a lower triangular matrix whose coefficients are the geometric and material parameters of the beam; $\{G(P,\psi)\}$ is the vector of nonlinear terms; s is the axial coordinate of the beam; and $\partial_s \equiv d/ds$. The explicit forms of $\{\psi\}$, $\{P\}$, $[I_1]$ and $\{G(P,\psi)\}$ are given by (see Fig. 1 and Appendix A):

$$\{\psi\} = \begin{Bmatrix} N \\ Q \\ M \\ \phi \\ w \\ u \end{Bmatrix}, \quad \{P\} = \begin{Bmatrix} p_s \\ p \\ m \\ \cdot \\ \cdot \\ \cdot \end{Bmatrix} \tag{2}$$

$$[I_1] = \begin{bmatrix} \cdot & \cdot & \cdot & \cdot & \cdot & \cdot \\ \frac{1}{R} & \cdot & \cdot & \cdot & \cdot & \cdot \\ \cdot & 1 & \cdot & \cdot & \cdot & \cdot \\ \cdot & \cdot & \frac{1}{EI} & \cdot & \cdot & \cdot \\ \cdot & \frac{1}{GA} & \cdot & -1 & \cdot & \cdot \\ \frac{1}{EA} & \cdot & \cdot & \cdot & \frac{-1}{R} & \cdot \end{bmatrix}, \quad \{G(P,\psi)\} = \begin{Bmatrix} \cdot \\ G_2 \\ \cdot \\ \cdot \\ \cdot \\ -\frac{1}{2}\left(\frac{Q}{GA} - \phi\right)^2 \end{Bmatrix} \tag{3}$$

and

$$G_2 = \left[\lambda\left(\frac{Q}{GA} - \phi\right) p_s + \lambda \frac{N}{GA} p - \frac{1}{R}\frac{N^2}{GA} + N\frac{M}{EI} - NQ\,\partial_s\left(\frac{1}{GA}\right)\right] \Big/ \left(1 + \frac{N}{GA}\right). \tag{4}$$

Note that the effects of concentrated forces and moments can be accounted for in Eqs. 2, 3 and 4 by including Dirac delta function terms in the expressions of p_s, p and m. Equations 1 are in the same format used by a variety of state-space solution procedures including transfer matrices (or matrix progression); the method of initial functions; as well as various numerical integration procedures.

Sensitivity Derivatives. Sensitivity derivatives can be used to study the sensitivity of the nonlinear response to variations in the different geometric and material parameters of the beam. The equations used in evaluating the sensitivity derivatives are obtained by differentiating the governing functional used in evaluating the response of the beam $\pi(\psi)$ (see Eq. A.4, Appendix A) with respect to a typical geometric or stiffness parameter, β_ι, viz.

$$\dot{\pi}\{\psi,\dot{\psi}\} = \frac{\partial \pi}{\partial \beta_\iota} + \frac{\partial \pi}{\partial\{\psi\}}\{\dot{\psi}\} + \frac{\partial \pi}{\partial\{P\}}\{\dot{P}\} \tag{5}$$

where $(\dot{\ }) = \frac{d}{d\beta_\iota}(\ )$. The quantities $\{\psi\}$ and $\{\dot{\psi}\}$ in $\dot{\pi}$ are then given independent and simultaneous variations. The stationary conditions for $\dot{\pi}$ include Eqs. 1 and the following equations for the sensitivity derivatives of the response quantities

$$\partial_s\{\dot{\psi}\} - \left[[I_1] + \left[\frac{\partial G_i}{\partial \psi_j}\right]\right]\{\dot{\psi}\} - \left[\frac{\partial I_1}{\partial \beta_\iota}\right]\{\psi\} - \left\{\frac{\partial G}{\partial \beta_\iota}\right\} - \left[\frac{\partial G_i}{\partial P_j}\right]\{\dot{P}\} + \lambda\{\dot{P}\} = 0 \tag{6}$$

Equations 6 can also be obtained by differentiating Eqs. 1 with respect to β_ι.

Solution of Governing Equations, Eqs. 1 and 6

Basic Idea of the Technique. The hybrid analytical technique used in the present study is basically a two-step technique. The first step involves the generation of coordinate functions (or modes) for the fundamental unknowns using the standard regular perturbation method. The second step consists of computing the amplitudes of the coordinate functions (or modes) via a direct variational procedure.

The hybrid technique was shown in Refs. 2, 3, 6 and 7 to exploit the best elements of both the regular perturbation method and the direct variational procedure as follows:

a) The perturbation method is used as a systematic and general approach for generating coordinate functions.

b) The direct variational procedure is used as an efficient approach for minimizing and distributing the error, in the field variables, throughout the domain.

In addition, the hybrid technique was shown to extend the range of applicability of the perturbation method and to enhance the effectiveness of the direct variational procedure. It also alleviates the following major drawbacks of the two classical techniques by:

a) Relaxing the restriction on the size of the parameter in the perturbation expansion.

b) Providing a systematic selection of the coordinate functions (or modes) needed in the direct variational procedure.

The details of the hybrid technique are given in Refs. 6 and 7. Its application to the geometrically nonlinear analysis of curved beams is described subsequently. The technique is *applied to the generation of both the nonlinear response of the beam, and the sensitivity derivatives of the different response quantities.*

Table 1 - Explicit form of the right-hand-side vectors in Eqs. 8

Equation	$\{R\}_1$	$\{R\}_2$	$\{R\}_3$
1	$-p_s$	0	0
2	$-p+\frac{N_1}{R}$	$\frac{N_2}{R}+R_{Q_2}$	$\frac{N_3}{R}+R_{Q_3}$
3	$-m+Q_1$	Q_2	Q_3
4	$\frac{M_1}{EI}$	$\frac{M_2}{EI}$	$\frac{M_3}{EI}$
5	$\frac{Q_1}{GA}-\phi_1$	$\frac{Q_2}{GA}-\phi_2$	$\frac{Q_3}{GA}-\phi_3$
6	$\frac{N_1}{EA}-\frac{w_1}{R}$	$\frac{N_2}{EA}-\frac{w_2}{R}$ $-\frac{1}{2}\left(\frac{Q_1}{GA}-\phi_1\right)^2$	$\frac{N_3}{EA}-\frac{w_3}{R}$ $-\left(\frac{Q_1}{GA}-\phi_1\right)\left(\frac{Q_2}{GA}-\phi_2\right)$

where

$$R_{Q_2}=N_1\left[\frac{M_1}{EI}+\frac{1}{GA}\left(-\frac{N_1}{R}+p\right)-Q_1\,\partial_s\left(\frac{1}{GA}\right)\right]+\left(\frac{Q_1}{GA}-\phi_1\right)p_s$$

$$R_{Q_3}=\frac{N_1}{GA}\left[-2\,\frac{N_2}{R}+N_1\left(\frac{1}{R}\,\frac{N_1}{GA}-\frac{M_1}{EI}\right)\right]+\frac{1}{EI}\,(N_1M_2+N_2M_1)$$

$$-\left[N_1\left(Q_2-N_1\frac{Q_1}{GA}\right)+N_2Q_1\right]\partial_s\left(\frac{1}{GA}\right)+\frac{1}{GA}\left(N_2-\frac{N_1^2}{GA}\right)p+\left[\frac{Q_2}{GA}-\phi_2-\frac{N_1}{GA}\left(\frac{Q_1}{GA}-\phi_1\right)\right]p_s$$

and the subscript of each of the response quantities denote the order of the perturbation function.

Generation of Coordinate Functions. The vectors of fundamental unknowns $\{\psi\}$ and $\{\dot{\psi}\}$ are represented by the regular perturbation expansion

$$\begin{Bmatrix} \psi \\ \dot{\psi} \end{Bmatrix} = \sum_{j=1}^{n} \lambda^j \begin{Bmatrix} \psi \\ \dot{\psi} \end{Bmatrix}_j \tag{7}$$

where $\{\psi\}_j$ and $\{\dot{\psi}\}_j$ are vectors of perturbation functions which represent modes and n is the total number of terms in the expansion.

The differential equations used in generating $\{\psi\}_j$ and $\{\dot{\psi}\}_j$ are obtained by substituting the expansions for $\{\psi\}$ and $\{\dot{\psi}\}$, Eqs. 7, into the governing differential equations, Eqs. 1 and 6, and setting the coefficients of λ^j successively equal to zero. This leads to a recursive set of equations for $\{\psi\}_j$ and $\{\dot{\psi}\}_j$. The equations for $\{\psi\}_j$ have the following form:

$$\partial_s \{\psi\}_j = \{R\}_j \qquad j = 1 \text{ to } n \tag{8}$$

The explicit forms of the first three right-hand side vectors $\{R\}_j$ are given in Table 1. Note that whereas the original differential equations, Eqs. 1, are nonlinear, the recursive set of equations, Eqs. 8, are all linear.

For each order j of the perturbation analysis, the form of the vector $\{R\}_j$ is such that the six components of $\{\psi\}_j$ can be determined one at a time in the following order: $N_j, Q_j, M_j, \phi_j, w_j, u_j$. In the solution process, six constants of integration are temporarily introduced and subsequently evaluated when the boundary conditions are applied. When *all* the loads are concentrated at a finite set of load of points, then Eqs. 8 have a particularly simple structure and the components of the solution vector, $\{\Psi\}$, are piecewise polynomials. Note, however, that the presence of concentrated axial and normal forces at the same points results in terms containing products of a Dirac delta function times heaviside functions in $R_{Q_2}, R_{Q_3}, \ldots$ (see Table 1). These terms affect the discontinuities in $Q_2, Q_3, \ldots$, respectively.

Computation of the Amplitudes of the Coordinate Functions. In the second step of the hybrid technique the perturbation functions $\{\psi\}_j$ and $\{\dot{\psi}\}_j$ are chosen as coordinate functions. The vectors of fundamental unknowns $\{\psi\}$ and $\{\dot{\psi}\}$ are then expressed as linear combinations of these functions, as follows:

$$\{\psi\} = \sum_{j=1}^{n} c_j \{\psi\}_j \tag{9}$$

$$\{\dot{\psi}\} = \sum_{j=1}^{n} \bar{c}_j \, \{\dot{\psi}\}_j \tag{10}$$

where c_j and $\bar{c}_j$ are unknown parameters which represent amplitudes of the coordinate functions, $\{\psi\}_j$ and $\{\dot{\psi}\}_j$, respectively.

The parameters c_j and $\bar{c}_j$ are obtained by applying the direct variational technique to the governing functionals π and $\dot{\pi}$ (Eqs. A.4, Appendix A, and Eq. 5).

Determination of the Radius of Convergence of the Perturbation Expansion. The radius of convergence of the perturbation expansion of a response function can be determined by finding the nearest singularity to the origin of the analytic continuation, in the complex λ-plane, for that function. In the curved beam problem considered herein, the radius of convergence appears to be independent of both the axial coordinate s, as well as the particular response function used. In the present study two different techniques were applied to the determination of the radius of convergence; namely, Pade´ approximants (Refs. 1, 12 and 14) and the ratio test.

Application of Symbolic Computations. The symbolic computation system Mathematica was used in the implementation of the foregoing hybrid analytical technique. A flowchart of the Mathematica Program is shown in Fig. 2 and a listing of the Program is given in Ref. 10. The tasks which were performed using Mathematica are:

1. Generation of governing differential equations - Euler-Lagrange equations of the functional.
2. Generation of various-order perturbation equations and solution of these equations to generate perturbation functions.
3. Application of direct variational procedure for the determination of amplitudes of coordinate functions and the generation of the nonlinear response.
4. Evaluation of derivatives of response quantities with respect to geometric and stiffness parameters of the beam (this includes the generation of the perturbation functions and the determination of their amplitudes.
5. Determination of the radius of convergence of the perturbation series.

Potential of the Hybrid Analytical Technique

The hybrid analytical technique has high potential for solution of nonlinear structural problems, especially for structures with simple geometry but complex construction. Examples of these structures are provided by ring- and stringer-stiffened closed cylindrical shells, and

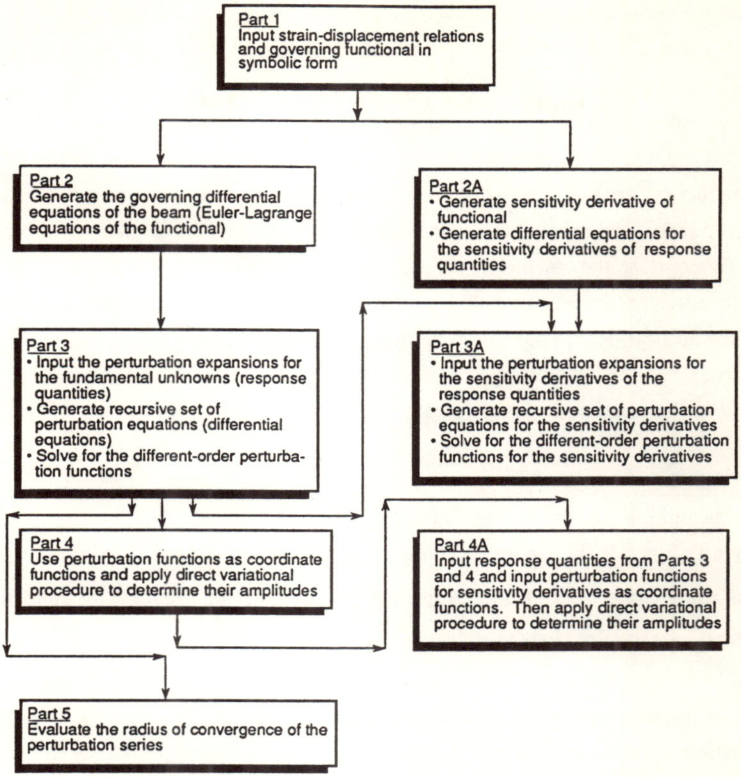

Figure 2 - Flow chart for the Mathematica Program used in the present study.

shell panels with discrete stiffeners and rectangular or circular planform. The numerical studies conducted clearly demonstrated the accuracy and effectiveness of the technique. In particular, the following two points are worth mentioning:

1. The hybrid technique can be thought of as either of the following:

a) A generalized perturbation method in which i) the perturbation expansions of the field variables contain free parameters rather than fixed coefficients, and ii) the perturbation parameters need not be small. Since the accuracy of the solutions obtained with the hybrid technique appears to be insensitive to the choice of perturbation parameters, they may be introduced artificially to simplify the form of the recursive set of differential equations used in evaluating the various-order perturbation solutions.

b) An extended direct variational technique with the coordinate functions generated by using a perturbation technique rather than chosen *a priori*.

2. The hybrid technique presented herein is the analytic counterpart of the reduction method presented in Refs. 8 and 9. In the reduction method the initial discretization is done via finite elements, then the vector of fundamental unknowns is expressed as a linear combination of a small number of basis vectors or global modes, and the Bubnov-Galerkin technique is used to compute the coefficients in the linear combination. The primary objective of using the reduction method is to reduce considerably the number of degrees of freedom in the initial discretization and, hence, reduce the computational effort involved in the solution of the problem. By contrast, the objectives of the hybrid analytical technique are:

a) to enhance the effectiveness of the direct variational technique by removing (or reducing) the arbitrariness in the selection of the coordinate functions.

b) to extend the range of applicability of the regular perturbation method by removing the restriction of a small-perturbation parameter.

Numerical Studies

The hybrid technique was used to generate the geometrically nonlinear response of the curved beam shown in Fig. 3. The beam has clamped edges and is subjected to a transverse concentrated load at the center, $P = \tilde{Q}_c$ (see Appendix A). Numerical results were generated, using the Mathematica program, for the response and the sensitivity derivatives, obtained by both the hybrid technique and the perturbation method. The radius of convergence, $\hat{P}$, was also calculated using Mathematica. Both Pade´ approximants and the ratio test were used to determine the radius of convergence in the range $0 < | R_o/R | < 3.5$. The response function selected for computing the radius of convergence is the normal displacement at the center, w_c. A fifteen-term perturbation series for w_c was used. For $| R_o/R | \lesssim 1$, the nearest singularities are off the real axis (Ref. 14) and Pade´ approximants are used in calculating the loadings, $\hat{P}$,

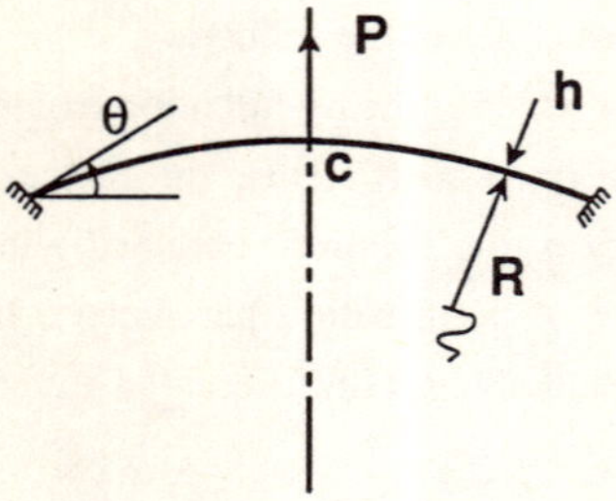

$E = 6.895 \times 10^{10}$ Pa

$G = 2.758 \times 10^{10}$ Pa

$A = 5.08 \times 10^{-2}\ m^2$

$h = 5.08 \times 10^{-2}$ m

$I = h^3/12$

$R = 2.54$ m

$\theta = 0.245$ rad.

Figure 3 - Shallow curved beam used in the present study.

near these singularities. On the other hand, for $|R_o/R| \gtrsim 1$, the nearest singularities are on the real axis, and $\hat{P}$ is calculated using the ratio test for the perturbation series.

An indication of the accuracy and convergence of the solutions obtained by the hybrid technique and the regular perturbation method is shown in Fig. 4. The standard of comparison is taken to be the converged finite element solution. As can be seen from Fig. 4, the perturbation solutions are grossly in error for λ outside the radius of convergence of the perturbation series, i.e., for $|\lambda| = \left|\dfrac{PR^2\theta}{\pi^2 EI}\right| > 1.0$. By contrast, the solutions obtained by the hybrid technique, using five or more approximation functions, are indistinguishable from the converged finite element solution everywhere within the range of loading considered.

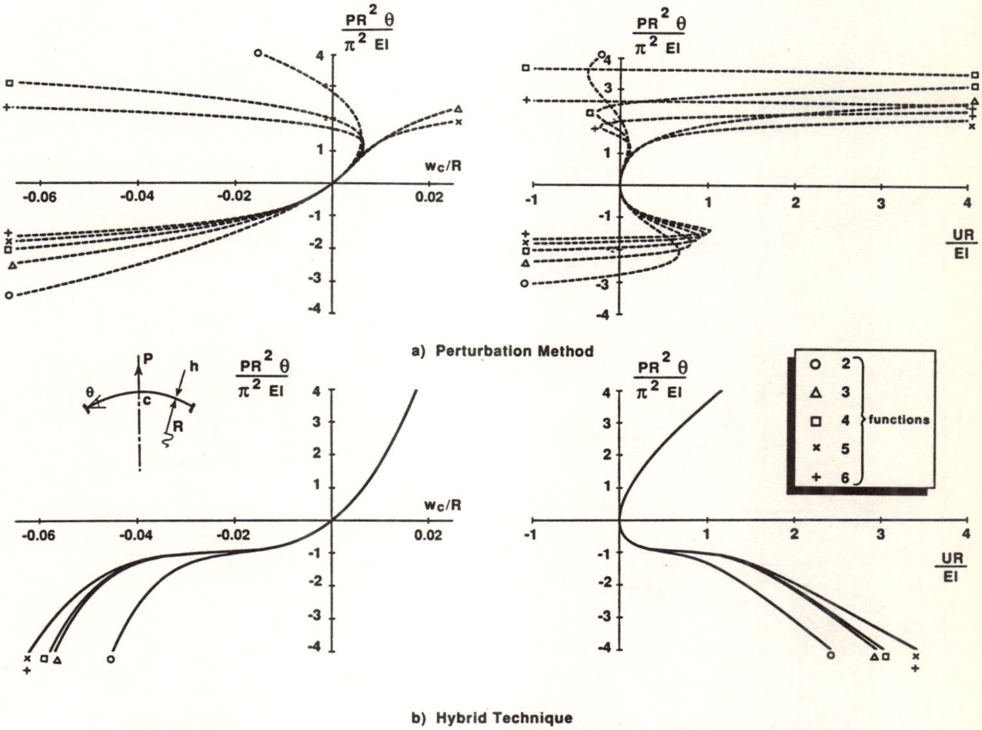

Figure 4 - Accuracy and convergence of solutions obtained by perturbation method and hybrid technique. Clamped shallow curved beam subjected to a central concentrated load (see Fig. 3).

To illustrate the use of sensitivity analysis, β was selected to be the bending stiffness EI. In Fig. 5, plots are shown for the sensitivity of the center normal displacement, w_c, and total complementary strain energy, U, with respect to variations in EI. The normalization parameter used in Fig. 5 includes $(EI)_o$, which is the value of EI given in Fig. 3. Again, the sensitivity derivatives obtained by the hybrid technique, using five or more approximation functions, are indistinguishable from the converged finite element results, and are considerably more accurate than the corresponding ones obtained by the perturbation method.

The variation of the normalized radius of convergence of the perturbation series, $\frac{\hat{P}R_o^2\theta}{\pi^2 EI}$, with

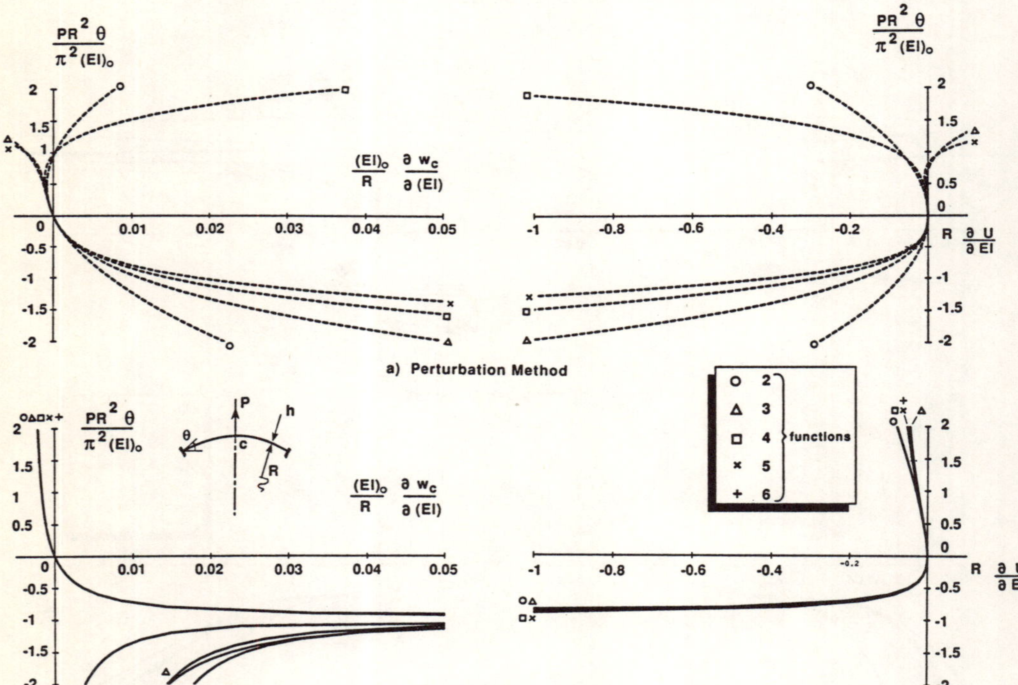

Figure 5 - Accuracy and convergence of the sensitivity derivatives of the center deflection, w_c, and total complementary strain energy, U, obtained by the perturbation method and the hybrid technique. Clamped shallow curved beam subjected to a central concentrated load (see Fig. 3). $(EI)_o$ refers to the numerical value of EI given in Fig. 3.

the curvature of the beam is depicted in Fig. 6. The quantity R_o refers to the numerical value of the radius of curvature given in Fig. 3. As can be seen from Fig. 6, the radius of convergence increases with the increase in the beam curvature.

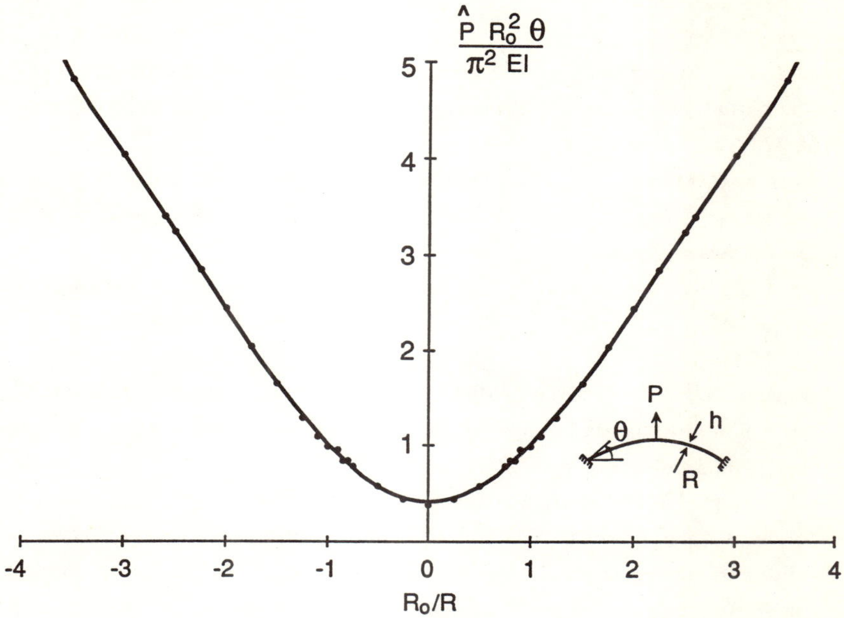

Figure 6 - Variation of the radius of convergence of the perturbation series with the curvature of the beam. Clamped shallow curved beam subjected to a central concentrated load (see Fig. 3). R_o refers to the numerical value of R given in Fig. 3.

Concluding Remarks

The application of a two-step hybrid analytical technique to the geometrically nonlinear analysis of curved beams is used to demonstrate the potential of symbolic computations in structural mechanics. The hybrid technique is based on successive use of the regular perturbation method and a classical direct variational procedure. The tasks which can be efficiently performed by symbolic computations are: 1) generation of algebraic expressions for the perturbation functions for the different response quantities and their sensitivity derivatives (with respect to both geometric and material parameters); and 2) determination of the radius of convergence of the perturbation series. The entire study was carried out using the symbolic computation system Mathematica.

Acknowledgment

The work of the first author was partially supported by a NASA Cooperative Agreement NCCW-0011, and by an Air Force Office of Scientific Research Grant AFOSR-90-0369.

References

1. Andersen, C. M. and Geer, J. F., "Power Series Expansions for the Frequency and Period of the Limit Cycle of the van der Pol Equation," SIAM Journal of Applied Mathematics, Vol. 42, 1982, pp. 678-693.
2. Andersen, C. M. and Geer, J. F., "Investigating a Hybrid Perturbation-Galerkin Technique Using Computer Algebra," ICASE Report 88-65, 1988; Journal of Symbolic Computation (to appear).
3. Geer, J. F. and Andersen, C. M., "A Hybrid Perturbation-Galerkin Technique Which Combines Multiple Expansions," SIAM Journal of Applied Mathematics, Vol. 50, 1990, pp. 1474-1495.
4. Guttmann, A. J., "Derivative of 'Mimic Functions' From Regular Perturbation Expansions in Fluid Mechanics," Journal of the Institute of Mathematics and Its Applications, Vol. 15, No. 3, June 1975, pp. 307-317.
5. Nayfeh, A. H., Introduction to Perturbation Techniques, John Wiley, New York, 1981.
6. Noor, A. K. and Balch, C. D., "Hybrid Perturbation/Bubnov-Galerkin Technique for Nonlinear Thermal Analysis," AIAA Journal, Vol. 22, No. 2, Feb. 1984, pp. 287-294.
7. Noor, A. K., "Hybrid Analytical Technique for Nonlinear Analysis of Structures," AIAA Journal, Vol. 23, No. 6, 1984, pp. 938-946.
8. Noor, A. K., Balch, C. D. and Shibut, M. A., "Reduction Methods for Nonlinear Steady-State Thermal Analysis," International Journal for Numerical Methods in Engineering, Vol. 20, No. 7, 1984, pp. 1323-1348.
9. Noor, A. K., "On Making Large Nonlinear Problems Small," Computer Methods in Applied Mechanics and Engineering, Vol. 34, 1982, pp. 955-985.
10. Noor, A. K. and Andersen, C. M., "Hybrid Analytical Technique for the Nonlinear Analysis of Curved Beams," in Nonlinear Computational Mechanics - State-of-the-Art, ed. by P. Wriggers and W. Wagner, Springer-Verlag, NY, 1991.
11. Svirski, I. V., Methods of the Type of Bubnov-Galerkin and Successive Approximations, Nauka, Moscow, 1968 (in Russian).
12. Van Dyke, M., "Analysis and Improvement of Perturbation Series," Quarterly Journal of Mechanics and Applied Mathematics, Vol. 27, Pt. 4, Nov. 1974, pp. 423-450.
13. Van Dyke, M., Perturbation Methods in Fluid Mechanics, Parabolic Press, Stanford, CA, 1975.

14. Van Dyke, M., "Fluid Mechanics Off Real Axis," in Engineering Science, Fluid Dynamics, Yates, G.T. (ed.), World Scientific Publishing Co., Teaneck, NJ, 1990, pp. 355-366.

15. Wolfram, S., Mathematica: A System for Doing Mathematics by Computer, Addison-Wesley, Redwood City, CA, 1988.

Appendix A - Fundamental Equations for the Geometrically Nonlinear Bean Theory Used in the Present Study

A total Lagrangian description of the beam deformation is used, and the beam configurations at different load levels are referred to the initial coordinate system of the undeformed beam. A mixed formulation is used with the fundamental unknowns consisting of the six beam quantities characterizing the response: generalized displacements u, w and ϕ, and internal forces N, Q and M.

Strain Displacement Relations

$$\varepsilon = \partial_s u + \frac{w}{R} + \underline{\frac{1}{2}(\partial_s w)^2} \tag{A.1}$$

$$\kappa = \partial_s \phi \tag{A.2}$$

$$\gamma = \partial_s w + \phi \tag{A.3}$$

where ε, κ and γ are the extensional, bending and transverse shear strains, respectively; R is the local radius of curvature of the centerline of the beam; and $\partial_s \equiv d/ds$.

Governing Functional. The functional for the Hellinger-Reissner mixed variational principle used in the present study is given by:

$$\pi(N, Q, M, \phi, w, u) = V - U - W \tag{A.4}$$

$$V = \int \left[\left\{ \partial_s u + \frac{w}{R} + \underline{\frac{1}{2}(\partial_s w)^2} \right\} N + (\partial_s \phi) M + (\partial_s w + \phi) Q \right] ds \tag{A.5}$$

$$U = \frac{1}{2} \int \left(\frac{N^2}{EA} + \frac{M^2}{EI} + \frac{Q^2}{GA} \right) ds \tag{A.6}$$

$$W = \int (p_s u + pw + m\phi)\, ds + \sum (\tilde{N}u + \tilde{Q}w + \tilde{M}\phi)$$

$$+ \sum [N(-\tilde{u}+u) + Q(-\tilde{w}+w) + M(-\tilde{\phi}+\phi)] \tag{A.7}$$

In Eqs. A.6 and A.7, EA, EI and GA are the extensional, bending and shear stiffnesses of the beam; p_s and p are external load components in the s and z coordinate directions, respectively; and m is the intensity of the applied moment. The underlined terms in Eqs. A.1 and A.5 represent the nonlinear contributions, and quantities with a tilde (~) denote prescribed displacements and forces. The summation sign in Eqs. A.7 extends over all applied external forces and prescribed displacements in the beam. Note that $\tilde{N}$, $\tilde{Q}$ and $\tilde{M}$ can be represented by the Dirac delta function terms in the expressions for p_s, p and m.

A Congruential Formulation of Nonlinear Finite Elements

C. A. FELIPPA and L. A. CRIVELLI

Department of Aerospace Engineering Sciences
and Center for Space Structures and Controls
University of Colorado at Boulder
Boulder, CO 80309-0429, USA

Summary

The core-congruential formulation (CCF) of geometrically nonlinear finite elements based on the Total Lagrangian (TL) description is presented. Although the key ideas behind the CCF can be traced back to Rajasekaran and Murray in 1973, it has not subsequently received serious attention. The CCF is first described in general form and then applied to the derivation of geometrically nonlinear bar, plane stress, and Timoshenko beam elements using the Green-Lagrange strain measure. Several advantages of the CCF, notably the physically clean separation of material and geometric stiffnesses, and its independence with respect to the ultimate choice of shape functions and element degrees of freedom, are noted. Two problems involving very large motions solved with the beam element display the range of applicability of this formulation, which trascends the kinematic limitations commonly attributed to the TL description.

1 Introduction

There is an elegant Total Lagrangian (TL) formulation of geometrically nonlinear finite elements that has received little attention in the literature. This will be referred to as the *core-congruential formulation*, or CCF, in the sequel. The key concepts, presented by Rajasekaran and Murray in 1973 [1], evolved from the analysis and reinterpretation of the pioneer work of Mallet and Marcal [2], as well as Murray's previous work in geometrically nonlinear finite element analysis [3]. The discussion of [1] by Felippa [4] provided parametric expressions for the stiffness matrices that appear at various levels of the discrete governing equations.

In 1987 a course in nonlinear finite elements offered by the first author presented the derivation of several elements using the CCF. Preparation of homework assignments and feedback from students helped to streamline the material. Subsequently the doctoral thesis by Crivelli [5] made extensive use of the CCF in the systematic development of a three-dimensional nonlinear Timoshenko beam element capable of undergoing arbitrarily large rotations. This particular application pushed this formulation beyond frontiers hitherto deemed unpassable by a TL element with rotational degrees of freedom.[1] Experience has shown that the beam formulation exhibits computational robustness lacking in other TL formulations that incorporate arbitrary kinematic approximations *ab initio.*

The central idea of the CCF is the construction of TL stiffness matrices for geometrically nonlinear analysis through the scheme

$$\mathbf{K}^{level} = \int_{V_0} \mathbf{G}^T \mathbf{S}^{level} \mathbf{G}\, dV_0, \tag{1}$$

where $\mathbf{S}$ is the core stiffness matrix, $\mathbf{K}$ the physical stiffness in terms of nodal degrees of freedom $\mathbf{v}$, $\mathbf{G}$ a core-to-physical-freedom transformation matrix independent of $\mathbf{v}$, V_0 the appropriate reference integration volume, and in which "level" identifies the *governing equation level at which the stiffness matrix is used.* The three levels of main interest in practice are: energy, force equilibrium, and first-order incremental equilibrium. The core stiffness matrix is expressed in terms of the *displacement gradients* at each material point. Displacement gradients $\mathbf{g}$ make a better choice of core variables than finite strains because for many element types they can be expressed linearly in terms of node displacements $\mathbf{v}$ as $\mathbf{g} = \mathbf{G}\mathbf{v}$, a property that validates (1) for all levels.[2]

The qualifier "core" emphasizes the *independence of* $\mathbf{S}^{level}$ *with respect to discretization decisions* such as element geometry, shape functions, and choice of nodal degrees of freedom. Such a dependence is introduced by the congruential transformation indicated in (1) and the integration over element volumes.

The CCF has several advantages noted later. But perhaps the most important one is the clean separation of physical effects. The importance of this factor should not be underestimated, because physical transparency is the key to success in nonlinear analysis.

As a final general remark, the specialization of (1) to linear analysis reduces to the well known $\mathbf{B}^T\mathbf{E}\mathbf{B}$ integrand of the standard formulation of linear stiffness matrices only after the displacement gradients are eliminated as intermediate variables. This shows that such an alternative congruential form does not extend naturally to geometrically nonlinear analysis.[3]

---

[1] The conventional wisdom is that TL-based elements would utterly fail beyond moderate rotations, and an updated Lagrangian or corotational description is necessary for handling truly large motions. For example, recently Mathiasson *et.al.* [6] concluded that "The TL formulation can only be used in problems with small or moderate displacements."

[2] If $\mathbf{G}$ depends on $\mathbf{v}$, the transformation in (1) is more complex and varies with level, as discussed in Section 6.

[3] Indeed, naive generalization of the $\mathbf{B}^T\mathbf{E}\mathbf{B}$ form to nonlinear analysis led several investigators to incorrect results during the 60's and 70's.

2 Historical Background

In 1968 Mallet and Marcal [2] presented a standard nomenclature for geometrically nonlinear finite element structural analysis based on the Total Lagrangian (TL) description. Consider a discrete, finite element model of a static structural system under dead loading with nodal displacement degrees of freedom collected in array $\mathbf{v}$. Displacements are measured from a fixed reference configuration $\mathcal{C}_0$ to a current configuration $\mathcal{C}$. The virtual-work conjugate forces, independent of $\mathbf{v}$, are collected in array $\mathbf{p}$. The system has a total potential energy function $J = U - W$ that is the difference between the strain energy U and the loads potential $W = \mathbf{p}^T\mathbf{v}$. The residual node forces are $\mathbf{r} = \partial J/\partial \mathbf{v}$, and the symbol Δ denotes increment associated with the variation of the current configuration.

Mallet and Marcal expressed the total potential energy, the residual (force-balance) equilibrium equations, and the incremental equilibrium equations as follows:

$$J = U - W = \tfrac{1}{2}\mathbf{v}^T\left[\mathbf{K}_0 + \tfrac{1}{3}\mathbf{N}_1 + \tfrac{1}{6}\mathbf{N}_2\right]\mathbf{v} - \mathbf{p}^T\mathbf{v}, \tag{2}$$

$$\mathbf{r} = \frac{\partial J}{\partial \mathbf{v}} = \left[\mathbf{K}_0 + \tfrac{1}{2}\mathbf{N}_1 + \tfrac{1}{3}\mathbf{N}_2\right]\mathbf{v} - \mathbf{p} = \mathbf{0}, \tag{3}$$

$$\Delta\mathbf{r} = \left[\mathbf{K}_0 + \mathbf{N}_1 + \mathbf{N}_2\right]\Delta\mathbf{v} - \Delta\mathbf{p} = \mathbf{0}. \tag{4}$$

Here $\mathbf{K}_0$ is the *linear stiffness matrix* evaluated at the reference configuration, whereas $\mathbf{N}_1$ and $\mathbf{N}_2$ are *nonlinear stiffness matrices*, also evaluated at the reference configuration, that depend linearly and quadratically, respectively, on the node displacements $\mathbf{v}$. The N matrices were said "to repeat" in the foregoing expressions.

Five years later Rajasekaran and Murray [1] examined more critically the structure of the matrices that appear in the above equations. In that investigation they chose to start from the "core" stiffness matrices corresponding to $\mathbf{K}$, $\mathbf{N}_1$ and $\mathbf{N}_2$ expressed in terms of displacement gradients, and in doing so laid down the main idea of the CCF. Working with specific elements they showed that the nonlinear stiffness matrices $\mathbf{N}_1$ and $\mathbf{N}_2$ are *not uniquely determined.* Indeed (2)–(4) as written are unique only for a single degree of freedom. They did not present, however, a general expression valid for arbitrary elements. This was partly done by Felippa [4], who in the discussion of [1] considered again those equations, rewritten here in a more general and compact form:

$$J = \tfrac{1}{2}\mathbf{v}^T\mathbf{K}^U\mathbf{v} + \left(\mathbf{p}^0 - \mathbf{p}\right)^T\mathbf{v}, \tag{5}$$

$$\mathbf{r} = \mathbf{K}^r\,\mathbf{v} + \mathbf{p}^0 - \mathbf{p} = \mathbf{f} - \mathbf{p} = \mathbf{0}, \tag{6}$$

$$\Delta\mathbf{r} = \mathbf{K}\,\Delta\mathbf{v} - \Delta\mathbf{p} = \mathbf{0}, \tag{7}$$

in which the notation of this paper — rather than that of [4] — is used. Here $\mathbf{K}^U$,

$\mathbf{K}^r$ and $\mathbf{K}$ denote the *energy*, *secant* and *tangent* stiffness matrices, respectively[4]. In addition, $\mathbf{p}^0$ is the *prestress force vector*, which vanishes if the reference configuration is stress free and was omitted in [4], and $\mathbf{f} = \mathbf{K}^r\mathbf{v} + \mathbf{p}^0$ is the internal force vector. The tangent stiffness is of course fundamental in incremental-iterative solution methods and stability analysis, while the secant stiffness (by itself or in the internal-force form $\mathbf{K}^r\mathbf{v} + \mathbf{p}^0$) is important in pseudo-force methods. The energy stiffness enjoys limited application *per se* but has theoretical importance as source for the other two.

In linear problems $\mathbf{K}^U = \mathbf{K}^r = \mathbf{K} = \mathbf{K}_0$ and the three stiffness matrices coalesce. But in nonlinear problems not only do the matrices differ but, as shown in the next section, $\mathbf{K}^U$ and $\mathbf{K}^r$ may involve arbitrary scalar coefficients. Such parametrized expressions were given in [4] under the following restrictions:

(R1) $\mathbf{K}^r$ is symmetric.
(R2) The reference configuration is stress free.
(R3) The finite strain measure is quadratic in the displacement gradients.
(R4) The transformation between core and physical freedoms is linear.

The following treatment eliminates restrictions (R1) and (R2) altogether, and the other two selectively.

3 Core Stiffness Equations

A conservative, geometrically nonlinear structure under dead loading is viewed as a continuum undergoing finite displacements $\mathbf{u}$. These displacements are measured from a fixed *reference* configuration $\mathcal{C}_0$ to a variable *current* configuration $\mathcal{C}$. *No discretization into finite elements is implied at this stage.* We confine our attention to the case in which the structure behavior stays within the linear elastic range, thus implying small deformational strains but arbitrarily large rotations. Corresponding points or *particles* in the reference and current configuration are referred to a fixed Cartesian coordinate system and have the coordinates X_i and x_i $(i = 1, \dots n_d)$, respectively, where n_d is the number of space dimensions. The displacement field components are $u_i = x_i - X_i$.

Let the state of strain at a particle in the current configuration be characterized by n_s strains e_i $(i = 1, 2, \dots n_s)$ collected in an array $\mathbf{e}$, and let the corresponding conjugate stresses be s_i $(i = 1, 2, \dots n_s)$, collected in an array $\mathbf{s}$. Using the summation convention the elastic stress-strain relations are written

$$s_i = s_i^0 + E_{ij}e_j, \quad \text{with} \quad E_{ij} = E_{ji}, \qquad \text{or} \quad \mathbf{s} = \mathbf{s}^0 + \mathbf{E}\mathbf{e}, \tag{8}$$

[4] Energy and secant stiffnesses are not denoted by $\mathbf{K}^e$ and $\mathbf{K}^s$ because such symbols are used for other purposes in the finite element course noted in the Introduction.

where s_i^0 are stresses in the reference configuration (stresses that remain if $e_i = 0$, also called prestresses) and E_{ij} are elastic moduli arranged as a $n_s \times n_s$ square array in the usual manner.

Let $\mathcal{J}$, $\mathcal{U}$, $\mathcal{W}$, $\mathbf{\Psi}$, $\mathbf{\Phi}$ and $\mathbf{\Upsilon}$ denote the analogues of J, U, W, $\mathbf{p}$, $\mathbf{f}$ and $\mathbf{r}$, respectively, at the particle level.[5] The strain energy density can be expressed as

$$\mathcal{U} = e_i s_i^0 + \tfrac{1}{2} e_i E_{ij} e_j = \mathbf{e}^T \mathbf{s}^0 + \tfrac{1}{2} \mathbf{e}^T \mathbf{E} \mathbf{e}. \tag{9}$$

The total strain energy U is obtained by integrating (9) over the structure volume: $U = \int_{V_0} \mathcal{U}\, dV_0$; the integration taking place — as can be expected in a TL description — over the reference configuration geometry.

Next, introduce the n_g displacement gradients $g_{mn} = \partial u_m / \partial X_n$. These are subsequently identified as g_i ($i = 1, 2, \ldots n_g$) so they can be conveniently arranged in a one-dimensional array $\mathbf{g}$. Following [2], assume that the strains e_i are linked to the displacement gradients through matrix relations of the form

$$\mathbf{e}_i = \mathbf{h}_i^T \mathbf{g} + \tfrac{1}{2} \mathbf{g}^T \mathbf{H}_i \mathbf{g}, \qquad i = 1, 2, \ldots n_s \tag{10}$$

where $\mathbf{h}_i$ and $\mathbf{H}_i$ are arrays of dimension $n_g \times 1$ and $n_g \times n_g$, respectively, with $\mathbf{H}_i$ symmetric. In [2] and [4] it was assumed that $\mathbf{H}_i$ is independent of $\mathbf{g}$, which is the case for the Green-Lagrange strain measure. This restriction, labeled (R3) in Section 2, will be enforced below except in Section 5. As noted previously, for deriving core equations we regard the displacement gradients $\mathbf{g}$ as degrees of freedom. On substituting (8) and (10) into (9) we obtain the "core counterparts" of (5)–(7), in which $\mathbf{v}$ has become $\mathbf{g}$:

$$\mathcal{J} = \mathcal{U} - \mathcal{W} = \tfrac{1}{2} \mathbf{g}^T \mathbf{S}^U \mathbf{g} + (\mathbf{\Psi}^0 - \mathbf{\Psi})^T \mathbf{g}, \tag{11}$$

$$\mathbf{\Upsilon} = \frac{\partial \mathcal{J}}{\partial \mathbf{g}} = \mathbf{S}^r \mathbf{g} + \mathbf{\Psi}^0 - \mathbf{\Psi} = \mathbf{\Phi} - \mathbf{\Psi} = 0, \tag{12}$$

$$\Delta \mathbf{\Upsilon} = \mathbf{S} \Delta \mathbf{g} - \Delta \mathbf{\Psi} = \mathbf{0}. \tag{13}$$

Here $\mathbf{S}^U$, $\mathbf{S}^r$ and $\mathbf{S}$ denote the energy, secant and tangent core stiffness matrices, and $\mathbf{\Psi}^0$, which is independent of $\mathbf{g}$, is the core counterpart of $\mathbf{p}^0$. With this notation the first and second variations of the strain energy density can be expressed as

$$\delta \mathcal{U} = \delta \mathbf{g}^T (\mathbf{S}^U \mathbf{g} + \mathbf{\Psi}^0) + \tfrac{1}{2} \mathbf{g}^T \delta \mathbf{S}^U \mathbf{g} = \delta \mathbf{g}^T \left(\mathbf{S}^r \mathbf{g} + \mathbf{\Psi}^0 \right) = \delta \mathbf{g}^T \mathbf{\Phi}, \tag{14}$$

$$\delta^2 \mathcal{U} = \delta \mathbf{g}^T \mathbf{S}^r \, \delta \mathbf{g} + \delta \mathbf{g}^T \delta \mathbf{S}^r \mathbf{g} + (\delta^2 \mathbf{g})^T \mathbf{\Phi} = \delta \mathbf{g}^T \mathbf{S} \, \delta \mathbf{g}. \tag{15}$$

---

[5] The first three acquire the meaning of energy densities, whereas $\mathbf{\Psi}$ is a dead-loading applied force density independent of $\mathbf{u}$.

These variational equations implicitly determine $\mathbf{S}^r$, $\mathbf{\Phi}$ and $\mathbf{S}$ from $\mathbf{S}^U$ and $\mathbf{\Psi}^0$. If restriction (R4) holds, the term $(\delta^2\mathbf{g})^T\mathbf{\Phi}$ drops out.[6] For convenience introduce the following $n_g \times n_g$ matrices (with summation convention implied):

$$\begin{aligned} \mathbf{S}_0 &= E_{ij}\,\mathbf{h}_i\mathbf{h}_j, \quad \mathbf{S}_1 = E_{ij}\,\mathbf{h}_i\mathbf{g}^T\mathbf{H}_j, \quad \mathbf{S}_1^* = E_{ij}\,(\mathbf{h}_i^T\mathbf{g})\,\mathbf{H}_j, \\ \mathbf{S}_2 &= E_{ij}\,\mathbf{H}_i\,\mathbf{g}\mathbf{g}^T\mathbf{H}_j, \qquad \mathbf{S}_2^* = E_{ij}\,(\mathbf{g}^T\mathbf{H}_i\mathbf{g})\,\mathbf{H}_j, \end{aligned} \tag{16}$$

in which parentheses are used to emphasize the grouping of *scalar* quantities such as $\mathbf{g}^T\mathbf{H}_i\mathbf{g}$. It may be then verified[7] that, if assumptions (R3)-(R4) of Section 2 hold, the core stiffnesses and prestress vector in (12)–(14) possess the general form:

$$\begin{aligned} \mathbf{S}^U(\alpha,\beta) &= \mathbf{S}_0 + \tfrac{1}{2}\alpha(\mathbf{S}_1 + \mathbf{S}_1^T) + (1-\alpha)\mathbf{S}_1^* + \tfrac{1}{4}\beta\mathbf{S}_2 + \tfrac{1}{4}(1-\beta)\mathbf{S}_2^* + s_i^0\mathbf{H}_i, \\ \mathbf{S}^r(\phi,\psi) &= \mathbf{S}_0 + \tfrac{1}{2}\mathbf{S}_1 + \phi\mathbf{S}_1^T + (1-\phi)\mathbf{S}_1^* + \tfrac{1}{4}(2-\psi)\mathbf{S}_2 + \tfrac{1}{4}\psi\mathbf{S}_2^* + s_i^0\mathbf{H}_i, \\ \mathbf{S} &= \mathbf{S}_0 + \mathbf{S}_1 + \mathbf{S}_1^T + \mathbf{S}_1^* + \mathbf{S}_2 + \tfrac{1}{2}\mathbf{S}_2^* + s_i^0\mathbf{H}_i, \qquad \mathbf{\Psi}^0 = s_i^0\mathbf{h}_i. \end{aligned} \tag{17}$$

Here α, β, ϕ and ψ are arbitrary scalar coefficients in the sense that $\mathbf{g}^T\mathbf{S}^U\mathbf{g}$ and $\mathbf{S}^r\mathbf{g}$ are independent of them. In fact,

$$\mathbf{\Phi} = \mathbf{S}^r\mathbf{g} + \mathbf{\Psi}^0 = s_i\mathbf{b}_i, \tag{18}$$

where $\mathbf{b}_i$ is defined in (20). The expressions (17) are more general than those given in [4] because restrictions (R1)-(R2) of Section 2 are no longer enforced. Note that $\mathbf{S}^r$ becomes symmetric if $\phi = 1/2$.

The "repeatable forms" of equations (2)–(4) are obtained if $\alpha = \beta = \psi = 2/3$ and $\phi = 1/2$, in which case the combinations $\mathbf{S}_1 + \mathbf{S}_1^T + \mathbf{S}_1^*$ and $\mathbf{S}_2 + \frac{1}{2}\mathbf{S}_2^*$ become the core counterparts of $\mathbf{N}_1$ and $\mathbf{N}_2$, respectively. But this observation has largely historical interest. More physically relevant are the following combinations:

$$\begin{aligned} \mathbf{S}_D &= \mathbf{S}_1 + \mathbf{S}_1^T + \mathbf{S}_2, \qquad \mathbf{S}_M = \mathbf{S}_0 + \mathbf{S}_D, \\ \mathbf{S}_G &= \mathbf{S}_1^* + \tfrac{1}{2}\mathbf{S}_2^* + s_i^0\mathbf{H}_i = s_i\mathbf{H}_i. \end{aligned} \tag{19}$$

These are the core versions of the *initial-displacement*, *material* and *geometric* stiffness, respectively. The core tangent stiffness is $\mathbf{S} = \mathbf{S}_0 + \mathbf{S}_D + \mathbf{S}_G = \mathbf{S}_M + \mathbf{S}_G$.

---

[6] If $\mathbf{g} = \mathbf{G}\mathbf{v}$ with $\mathbf{G}$ independent of $\mathbf{v}$, $\delta^2\mathbf{g} = \mathbf{G}\,\delta^2\mathbf{v} = \mathbf{0}$ because $\mathbf{v}$ are independent variables. On the other hand, if displacement gradients are nonlinear functions of node displacements expressable as $g_i = g_i(v_j)$, then

$$\delta g_i = \frac{\partial g_i}{\partial v_j}\delta v_j = G_{ij}\,\delta v_j, \qquad \delta^2 g_i = \frac{\partial^2 g_i}{\partial v_j\partial v_k}\delta v_j\delta v_k + \frac{\partial g_i}{\partial v_j}\cancelto{0}{\delta^2 v_j} = F_{ijk}\,\delta v_j\,\delta v_k.$$

Thus $\delta\mathbf{g}$ is still $\mathbf{G}\,\delta\mathbf{v}$ but $\delta^2\mathbf{g} = (\mathbf{F}\,\delta\mathbf{v})\,\delta\mathbf{v}$, where $\mathbf{F}$ is a cubic array.

[7] The verification was done by hand for [4], and rechecked with the MACSYMA symbolic algebra computer system 14 years later.

4 Spectral Expression of Core Stiffnesses

There is a more compact alternative expression of the core stiffnesses, not presented in [4], that offers theoretical as well as implementational advantages at the cost of some generality. Define vectors $\mathbf{b}_i$ and $\mathbf{c}_i$ as

$$e_i = \mathbf{c}_i^T\mathbf{g}, \quad \mathbf{c}_i = \mathbf{h}_i + \tfrac{1}{2}\mathbf{H}_i\mathbf{g}, \qquad \mathbf{b}_i = \frac{\partial e_i}{\partial \mathbf{g}} = \mathbf{h}_i + \mathbf{H}_i\mathbf{g}. \tag{20}$$

Then the spectral forms[8] are

$$\mathbf{S}^U(1,1) = \mathbf{S}^U\Big|_{\alpha=\beta=1} = E_{ij}\,\mathbf{c}_i\mathbf{c}_j^T + s_i^0\mathbf{H}_i, \tag{21}$$

$$\mathbf{S}^r(0,0) = \mathbf{S}^r|_{\phi=\psi=0} = E_{ij}\,\mathbf{b}_i\mathbf{c}_j^T + s_i^0\mathbf{H}_i, \tag{22}$$

$$\mathbf{S}^r(\tfrac{1}{2},1) = \mathbf{S}^r\big|_{\phi=\frac{1}{2},\psi=1} = E_{ij}\,\mathbf{c}_i\mathbf{c}_j^T + \tfrac{1}{2}(s_i + s_i^0)\mathbf{H}_i, \tag{23}$$

$$\mathbf{S} = E_{ij}\,\mathbf{b}_i\mathbf{b}_j^T + s_i\mathbf{H}_i = \mathbf{S}_M + \mathbf{S}_G. \tag{24}$$

Note that $\mathbf{S}^r(\frac{1}{2},1)$ is symmetric but $\mathbf{S}^r(0,0)$ is not. It is seen that for energy and secant stiffnesses, compactness is paid in terms of settling for specific coefficient values. The foregoing relations may be obtained by repeated differentiation of $\mathcal{U}$ with respect to $\mathbf{g}$. A lesson from this observation is that direct differentiation of nonlinear matrix relations may conceal the richness of possible expressions as revealed in equations (17).

5 Generalization to H(g)

An advantage of the spectral forms is that the complication introduced by $\mathbf{H}_i$ being functions of $\mathbf{g}$ (as it happens for all nonlinear strain measures except Green-Lagrange's) can be accommodated by redefining some of the previous relations as follows:

$$\begin{gathered} \mathbf{b}_i^* = \frac{\partial e_i}{\partial \mathbf{g}} = \mathbf{h}_i + \mathbf{H}_i^*\mathbf{g}, \qquad \mathbf{H}_i^* = \mathbf{H}_i + \tfrac{1}{2}\mathbf{g}^T\frac{\partial \mathbf{H}_i}{\partial \mathbf{g}}, \\ \mathbf{H}_i^{**} = \frac{\partial^2 e_i}{\partial \mathbf{g}\partial \mathbf{g}} = \frac{\partial \mathbf{b}_i}{\partial \mathbf{g}} = \mathbf{H}_i + \frac{\partial\left(\frac{1}{2}\mathbf{g}^T\frac{\partial \mathbf{H}_i}{\partial \mathbf{g}}\mathbf{g}\right)}{\partial \mathbf{g}}, \\ \mathbf{S}^r(0,0) = E_{ij}\,\mathbf{b}_i^*\mathbf{c}_j^T + s_i^0\mathbf{H}_i^*, \qquad \mathbf{S} = E_{ij}\,\mathbf{b}_i^*(\mathbf{b}_j^*)^T + s_i\mathbf{H}_i^{**}. \end{gathered} \tag{25}$$

There is no simple form of $\mathbf{S}^r(\frac{1}{2},1)$ equivalent to (23), and indeed symmetric forms of $\mathbf{S}^r$ do not generally exist if strain measures other than Green-Lagrange's are used.

---

[8] So called because of the formal similarity of equations (21)–(24) with the spectral decomposition of a matrix as the sum of rank-one matrices.

6 Transformation to Physical Coordinates

Subdivide the continuum into finite elements. Over each individual element[9] the displacement field $\mathbf{u}^T = (u_1, u_2, u_3)$ is given by $\mathbf{u} = \mathbf{P}\mathbf{v}$, where $\mathbf{v}$ now collects the element node-displacement degrees of freedom and $\mathbf{P}$ is a matrix of shape functions. If the displacement gradients are linear in $\mathbf{v}$, differentiation yields $\mathbf{g} = \mathbf{G}\mathbf{v}$, $\delta\mathbf{g} = \mathbf{G}\,\delta\mathbf{v}$ and $\delta^2\mathbf{g} = \mathbf{0}$ (cf. footnote 6). Invariance of the strain energy variations δU and $\delta^2 U$ obtained by integrating (14) and (15) over V yields

$$\mathbf{K}^U = \int_{V_0} \mathbf{G}^T\mathbf{S}^U\mathbf{G}\,dV_0,\ \mathbf{K}^r = \int_{V_0} \mathbf{G}^T\mathbf{S}^r\mathbf{G}\,dV_0,\ \mathbf{K} = \int_{V_0} \mathbf{G}^T\mathbf{S}\mathbf{G}\,dV_0,\ \mathbf{p} = \int_{V_0} \mathbf{G}^T\boldsymbol{\Psi}\,dV_0, \tag{26}$$

where V_0 stands now for the *element* volume, and the last transformation also applies to $\boldsymbol{\Psi}^0$. Although the dependency of $\mathbf{S}^{level}$ and $\boldsymbol{\Psi}$ on $\mathbf{g}$ is not made implicit in these equations, it must be remembered that the transformation $\mathbf{g} = \mathbf{G}\mathbf{v}$ also appears there. Because of the ensuing algebraic complexity, numerical integration is generally required unless the gradients are constant over the element.

Often $\mathbf{G}$ is expressed as a chain of transformations, some of which are position dependent and remain inside the integral whereas others are not and may be taken outside. For example, in the bar element studied below, $\mathbf{G} = \mathbf{T}\overline{\mathbf{G}}$, where $\overline{\mathbf{G}}$ transforms $\mathbf{g}$ to local node displacements and $\mathbf{T}$ transforms local to global node displacements.

If the relation between $\mathbf{g}$ and $\mathbf{v}$ is nonlinear so that $\mathbf{g} = \mathbf{g}(\mathbf{v})$, then $\mathbf{G} = \partial\mathbf{g}/\partial\mathbf{v}$ depends on $\mathbf{v}$, and transformation rules change with level. This complication occurs in elements with rotational freedoms, such as beams, plates and shells, if finite rotations are exactly treated. Both residual and incremental levels are affected. At the residual level, $\mathbf{K}^r$ cannot be extracted, except artificially, and it is better to work with the internal force vector $\mathbf{f} = \int_{V_0} \mathbf{G}^T\boldsymbol{\Phi}\,dV_0 = \int_{V_0} \mathbf{G}^T s_i \mathbf{b}_i\,dV_0$ directly. At the incremental level, the term $(\delta^2\mathbf{g})^T\boldsymbol{\Phi}$ in (15) must be retained because $\delta^2\mathbf{g}$ does not vanish. Invariance of $\delta^2 U$ yields

$$\mathbf{K} = \int_{V_0} (\mathbf{G}^T\mathbf{S}\mathbf{G} + \mathbf{Q})\,dV_0, \tag{27}$$

where $\mathbf{S}$ is that given in (17), and the entries of the symmetric matrix $\mathbf{Q}$ are, from footnote 6 and equation (18):

$$Q_{jk} = Q_{kj} = F_{mjk}\Phi_m = \frac{\partial^2 g_m}{\partial v_j \partial v_k}\Phi_m = \frac{\partial^2 g_m}{\partial v_j \partial v_k} s_i b_{im}, \tag{28}$$

in which b_{im} denotes the mth entry of $\mathbf{b}_i$. It is seen that $\mathbf{Q}$ vanishes in a stress-free state and can therefore be viewed as a correction to the geometric stiffness matrix.

---

[9] For simplicity elements will not be identified by an element index in the sequel.

7 Example 1: 3D Bar Element ($n_d = 3$, $n_s = 1$, $n_g = 3$)

The first CCF application example is to a two-node prismatic, straight bar element of reference length L_0 and cross section area A_0 in 3D space. To simplify node subscripting notations, Cartesian reference systems and displacement components will be denoted by (X, Y, Z), (x, y, z) and (u_X, u_Y, u_Z) rather than (X_1, X_2, X_3), (x_1, x_2, x_3) and (u_1, u_2, u_3), respectively. The bar node displacements are (v_{Xn}, v_{Yn}, v_{Zn}), $n = 1, 2$.

In the reference configuration C_0 the bar element is referred to a *local* Cartesian system $(\overline{X}, \overline{Y}, \overline{Z})$, with $\overline{X}$ located along the bar axis, so that the element extends from node 1 at (0,0,0) to node 2 at (L_0,0,0). The current configuration is also straight and extends from $(\bar{x}_1 = \bar{v}_{X1}, \bar{y}_1 = \bar{v}_{Y1}, \bar{z}_1 = \bar{v}_{Z1})$ to $(\bar{x}_2 = L_0 + \bar{v}_{X2}, \bar{y}_2 = \bar{v}_{Y2}, \bar{z}_2 = \bar{v}_{Z2})$. The element displacement field in local coordinates may be expressed as

$$\bar{\mathbf{u}} = \begin{Bmatrix} \bar{u}_X \\ \bar{u}_Y \\ \bar{u}_Z \end{Bmatrix} = \begin{bmatrix} 1-\zeta & 0 & 0 & \zeta & 0 & 0 \\ 0 & 1-\zeta & 0 & 0 & \zeta & 0 \\ 0 & 0 & 1-\zeta & 0 & 0 & \zeta \end{bmatrix} \begin{Bmatrix} \bar{v}_{X1} \\ \bar{v}_{Y1} \\ \bar{v}_{Z1} \\ \bar{v}_{X2} \\ \bar{v}_{Y2} \\ \bar{v}_{Z2} \end{Bmatrix} = \mathbf{P}\bar{\mathbf{v}}, \tag{29}$$

where $\zeta = \overline{X}/L_0$ is a natural axial coordinate varying from 0 to 1. The three displacement gradients that intervene in the definition of nonlinear strains are

$$\mathbf{g} = \begin{Bmatrix} g_1 \\ g_2 \\ g_3 \end{Bmatrix} = \begin{Bmatrix} \partial \bar{u}_X/\partial \bar{X} \\ \partial \bar{u}_Y/\partial \bar{X} \\ \partial \bar{u}_Z/\partial \bar{X} \end{Bmatrix} = \frac{1}{L_0} \begin{bmatrix} -1 & 0 & 0 & 1 & 0 & 0 \\ 0 & -1 & 0 & 0 & 1 & 0 \\ 0 & 0 & -1 & 0 & 0 & 1 \end{bmatrix} \bar{\mathbf{v}} = \frac{1}{L_0} \overline{\mathbf{G}}\bar{\mathbf{v}} \tag{30}$$

As strain measure we adopt the Green-Lagrange axial strain, defined as

$$\begin{aligned} e \equiv e_1 &= \frac{\partial \bar{u}_X}{\partial \overline{X}} + \tfrac{1}{2}\left[\left(\frac{\partial \bar{u}_X}{\partial \overline{X}}\right)^2 + \left(\frac{\partial \bar{u}_Y}{\partial \overline{X}}\right)^2 + \left(\frac{\partial \bar{u}_Z}{\partial \overline{X}}\right)^2\right] = g_1 + \tfrac{1}{2}(g_1^2 + g_2^2 + g_3^2) \\ &= \begin{bmatrix} 1 \\ 0 \\ 0 \end{bmatrix}^T \begin{Bmatrix} g_1 \\ g_2 \\ g_3 \end{Bmatrix} + \tfrac{1}{2} \begin{Bmatrix} g_1 \\ g_2 \\ g_3 \end{Bmatrix}^T \begin{bmatrix} 1 & 0 & 0 \\ 0 & 1 & 0 \\ 0 & 0 & 1 \end{bmatrix} \begin{Bmatrix} g_1 \\ g_2 \\ g_3 \end{Bmatrix} = \mathbf{h}^T\mathbf{g} + \tfrac{1}{2}\mathbf{g}^T\mathbf{H}\mathbf{g}. \end{aligned} \tag{31}$$

Thus for this choice of strain, $\mathbf{h}^T = [1 \;\; 0 \;\; 0]$ and $\mathbf{H}$ is the 3×3 identity matrix. The conjugate stress measure $s_1 \equiv s$ is the second Piola-Kirchhoff (PK2) axial stress. The stress-strain relation is $s = s^0 + Ee$, where s^0 and s are PK2 axial stresses in the reference and current configurations, respectively, and E is Young's modulus.

Since $\mathbf{H}$ is independent of $\mathbf{g}$, to form the core stiffnesses in local coordinates we can use the spectral expressions of Section 4. First construct the vectors

$$\mathbf{c} \equiv \mathbf{c}_1 = \begin{Bmatrix} 1 + \frac{1}{2}g_1 \\ \frac{1}{2}g_2 \\ \frac{1}{2}g_3 \end{Bmatrix}, \qquad \mathbf{b} \equiv \mathbf{b}_1 = \begin{Bmatrix} 1 + g_1 \\ g_2 \\ g_3 \end{Bmatrix}, \tag{32}$$

which inserted into (21), (23) and (24) yield

$$\mathbf{S}^U(1,1) = E\mathbf{c}\mathbf{c}^T + s^0\mathbf{H} = E\begin{bmatrix} (1+\frac{1}{2}g_1)^2 & \frac{1}{2}g_2(1+\frac{1}{2}g_1) & \frac{1}{2}g_3(1+\frac{1}{2}g_1) \\ & \frac{1}{4}g_2^2 & \frac{1}{4}g_2g_3 \\ symm & & \frac{1}{4}g_3^2 \end{bmatrix} + s^0\begin{bmatrix} 1 & 0 & 0 \\ 0 & 1 & 0 \\ 0 & 0 & 1 \end{bmatrix}, \tag{33}$$

$$\mathbf{S}^r(\tfrac{1}{2},1) = E\mathbf{c}\mathbf{c}^T + s^m\mathbf{H} = E\begin{bmatrix} (1+\frac{1}{2}g_1)^2 & \frac{1}{2}g_2(1+\frac{1}{2}g_1) & \frac{1}{2}g_3(1+\frac{1}{2}g_1) \\ & \frac{1}{4}g_2^2 & \frac{1}{4}g_2g_3 \\ symm & & \frac{1}{4}g_3^2 \end{bmatrix} + s^m\begin{bmatrix} 1 & 0 & 0 \\ 0 & 1 & 0 \\ 0 & 0 & 1 \end{bmatrix}, \tag{34}$$

$$\mathbf{S} = E\mathbf{b}\mathbf{b}^T + s\mathbf{H} = E\begin{bmatrix} (1+g_1)^2 & g_2(1+g_1) & g_3(1+g_1) \\ & g_2^2 & g_2g_3 \\ symm & & g_3^2 \end{bmatrix} + s\begin{bmatrix} 1 & 0 & 0 \\ 0 & 1 & 0 \\ 0 & 0 & 1 \end{bmatrix}. \tag{35}$$

where $s^m = \frac{1}{2}(s^0 + s)$. The clean separation into material and geometric (initial-stress) stiffnesses should be noted. As (30) is linear in $\mathbf{v}$, the physical stiffnesses in local coordinates are obtained through (26) in which $V_0 = A_0 L_0$. For the tangent stiffness this transformation yields

$$\overline{\mathbf{K}} = \frac{1}{L_0^2}\int_{V_0} \overline{\mathbf{G}}^T\mathbf{S}\overline{\mathbf{G}}\,dV_0 = \frac{A_0}{L_0}\overline{\mathbf{G}}^T(E\mathbf{b}\mathbf{b}^T + s\mathbf{H})\,\overline{\mathbf{G}}, \tag{36}$$

with similar expressions for the other local stiffnesses. Given the simplicity of $\overline{\mathbf{G}}$, $\overline{\mathbf{K}}$ could easily be expressed in closed form, and similarly for $\overline{\mathbf{K}}^U$ and $\overline{\mathbf{K}}^r$. Finally, transformation to displacements u_X, u_Y, u_Z in the global Cartesian system is handled in the usual manner. If $\overline{\mathbf{u}} = \mathbf{T}\mathbf{u}$, the global stiffness matrix is given by

$$\mathbf{K} = \frac{A_0}{L_0}\begin{bmatrix} \mathbf{T}^T & \mathbf{0} \\ \mathbf{0} & \mathbf{T}^T \end{bmatrix}\overline{\mathbf{G}}^T(E\mathbf{b}\mathbf{b}^T + s\mathbf{H})\,\overline{\mathbf{G}}\begin{bmatrix} \mathbf{T} & \mathbf{0} \\ \mathbf{0} & \mathbf{T} \end{bmatrix}. \tag{37}$$

For efficient computer implementation, a closed form for $\mathbf{K}$ is easily obtained from (37) by symbolic manipulation and automatically converted to program statements.

8 Example 2: N-Node Plane Stress Element ($n_d = 2$, $n_s = 3$, $n_g = 4$)

As second example we consider an arbitrary N-node isoparametric element that models a plane stress problem. The Cartesian reference system and displacement components will be denoted by (X, Y), (x, y) and (u_X, u_Y) rather than (X_1, X_2), (x_1, x_2) and (u_1, u_2), respectively. The element nodes are located at X_n, Y_n, $(n = 1, \ldots N)$ in the reference configuration and move to $(x_n = X_n + u_{Xn},\ y_n = Y_n + u_{Yn})$, $(n = 1, \ldots N)$ in the current configuration.

The element displacement field is expressed as

$$\begin{Bmatrix} u_X \\ u_Y \end{Bmatrix} = \begin{bmatrix} P_1 & 0 & P_2 & \dots & 0 \\ 0 & P_1 & 0 & \dots & P_N \end{bmatrix} \begin{Bmatrix} v_{X1} \\ v_{Y1} \\ v_{X2} \\ \vdots \\ v_{YN} \end{Bmatrix} = \mathbf{Pv}, \tag{38}$$

where $P_n(\xi,\eta)$ are the appropriate isoparametric shape functions. In this case four displacement gradients appear, and these are arranged as

$$\mathbf{g} = \begin{Bmatrix} g_1 \\ g_2 \\ g_3 \\ g_4 \end{Bmatrix} = \begin{Bmatrix} \partial u_X/\partial X \\ \partial u_Y/\partial X \\ \partial u_X/\partial Y \\ \partial u_Y/\partial Y \end{Bmatrix} = \mathbf{Gv}, \tag{39}$$

where $\mathbf{G}$ is a sparse $4\times 2N$ matrix that contains the shape function derivatives $\partial P_n/\partial X$ and $\partial P_n/\partial Y$. The strain measures chosen are the three components e_i $(i = 1,2,3)$ of the Green-Lagrange strains defined in the usual manner:

$$e_1 = e_{XX} = g_1 + \tfrac{1}{2}(g_1^2 + g_2^2) = \begin{Bmatrix} 1 \\ 0 \\ 0 \\ 0 \end{Bmatrix}^T \mathbf{g} + \tfrac{1}{2}\mathbf{g}^T \begin{bmatrix} 1 & 0 & 0 & 0 \\ 0 & 1 & 0 & 0 \\ 0 & 0 & 0 & 0 \\ 0 & 0 & 0 & 0 \end{bmatrix} \mathbf{g}, \tag{40}$$

$$e_2 = e_{YY} = g_4 + \tfrac{1}{2}(g_3^2 + g_4^2) = \begin{Bmatrix} 1 \\ 0 \\ 0 \\ 0 \end{Bmatrix}^T \mathbf{g} + \tfrac{1}{2}\mathbf{g}^T \begin{bmatrix} 0 & 0 & 0 & 0 \\ 0 & 0 & 0 & 0 \\ 0 & 0 & 1 & 0 \\ 0 & 0 & 0 & 1 \end{bmatrix} \mathbf{g}, \tag{41}$$

$$e_3 = e_{XY} + e_{YX} = g_2 + g_3 + g_1 g_3 + g_2 g_4 = \begin{Bmatrix} 0 \\ 1 \\ 1 \\ 0 \end{Bmatrix}^T \mathbf{g} + \tfrac{1}{2}\mathbf{g}^T \begin{bmatrix} 0 & 0 & 1 & 0 \\ 0 & 0 & 0 & 1 \\ 1 & 0 & 0 & 0 \\ 0 & 1 & 0 & 0 \end{bmatrix} \mathbf{g}, \tag{42}$$

from which expressions for $\mathbf{h}_i$ and $\mathbf{H}_i$ $(i = 1,2,3)$ follow. For brevity, only the derivation of the tangent stiffness matrix will be described. Begin by forming the vectors

$$\mathbf{b}_1 = \begin{Bmatrix} 1+g_1 \\ g_2 \\ 0 \\ 0 \end{Bmatrix}, \quad \mathbf{b}_2 = \begin{Bmatrix} 0 \\ 0 \\ g_3 \\ 1+g_4 \end{Bmatrix}, \quad \mathbf{b}_3 = \begin{Bmatrix} g_3 \\ 1+g_4 \\ 1+g_1 \\ g_2 \end{Bmatrix}. \tag{43}$$

Then from (24) we get the core stiffness

$$\mathbf{S} = E_{ij}\mathbf{b}_i\mathbf{b}_i^T + s_i\mathbf{H}_i = \mathbf{S}_M + \mathbf{S}_G, \tag{44}$$

where $s_i = s_i^0 + E_{ij}e_j$, $(i,j = 1,2,3)$, are the PK2 stresses in the current configuration.

In full and denoting $a_1 = 1 + g_1$, $a_4 = 1 + g_4$:

$$\mathbf{S}_M = \begin{bmatrix} E_{11}a_1^2 + E_{13}a_1g_3 + E_{33}g_3^2 & E_{11}a_1g_2 + E_{13}(a_1a_4 + g_2g_3) + E_{33}a_4g_3 & E_{12}a_1g_3 + E_{13}a_1^2 + E_{23}g_3^2 + E_{33}a_1g_3 & E_{12}a_1a_4 + E_{13}a_1g_2 + E_{23}a_4g_3 + E_{33}g_2g_3 \\ & E_{11}g_2^2 + E_{13}a_4g_2 + E_{33}a_4^2 & E_{12}g_2g_3 + E_{13}a_1g_2 + E_{23}a_4g_3 + E_{33}a_1a_4 & E_{12}a_4g_2 + E_{13}g_2^2 + E_{23}a_4^2 + E_{33}a_4g_2 \\ & & E_{22}g_3^2 + E_{23}a_1g_3 + E_{33}a_1^2 & E_{22}a_4g_3 + E_{23}(a_1a_4 + g_2g_3) + E_{33}a_1g_2 \\ symm & & & E_{22}a_4^2 + E_{23}a_4g_2 + E_{33}g_2^2 \end{bmatrix} \tag{45}$$

$$\mathbf{S}_G = \begin{bmatrix} s_1 & 0 & s_3 & 0 \\ & s_1 & 0 & s_3 \\ & & s_2 & 0 \\ symm & & & s_2 \end{bmatrix}. \tag{46}$$

The physical tangent stiffness is

$$\mathbf{K} = \int_{V_0} \mathbf{G}^T(\mathbf{S}_M + \mathbf{S}_G)\mathbf{G}\, dV_0, \tag{47}$$

which is conveniently evaluated by numerical integration. Since several of the integrand matrices are quite sparse, for efficiency in the computer implementation the integrand should be evaluated symbolically and converted to program statements before being encapsulated in the Gauss quadrature loop.

9 Example 3: 3D Timoshenko Beam ($n_d = 3$, $n_s = 3$, $n_g = 9$)

As final CCF application we describe the derivation of a new TL Timoshenko beam element capable of undergoing arbitrarily large rotations in three dimensions. Only a brief outline of the derivation is given; full details may be found in Crivelli's thesis [5].

The beam kinematics is based on two assumptions: (1) the TL description, and (2) the Timoshenko beam hypothesis: planes sections remain plane although not necessarily normal to the deformed longitudinal axis. The beam is assumed isotropically elastic, with Young's modulus E and shear modulus $G = \frac{1}{2}E/(1+\nu)$. The reference configuration of the beam is straight, prismatic and stress free, with cross section A and length L. A local reference frame $\mathbf{n}_i$ is defined on it, with $\mathbf{n}_1$ directed along the longitudinal axis (locus of cross section centroids). Vectors $\mathbf{n}_2$ and $\mathbf{n}_3$ are in the plane of the left-end cross section; these are eventually aligned with principal inertia axes to simplify some expressions. Along these vectors we attach the reference coordinate system (X_1, X_2, X_3). We further define a set of *moving* frames, denoted by $\mathbf{a}_1$, $\mathbf{a}_2$ and $\mathbf{a}_3$, parametrized by the longitudinal coordinate X_1. These frames displace rigidly attached to the beam cross sections as shown in Figure 1.

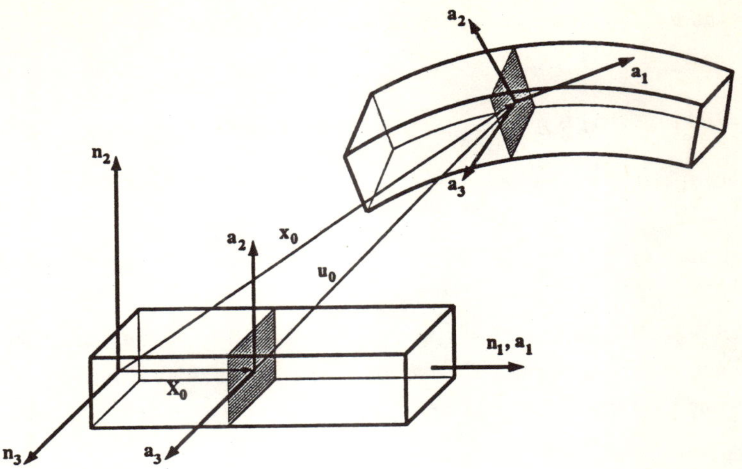

Fig. 1. Reference frames describing beam kinematics.

A beam particle originally at (X_1, X_2, X_3) displaces to

$$\mathbf{x}(\mathbf{X}) = \mathbf{x}_0(X_1) + \mathbf{R}^T(X_1)\,\boldsymbol{\zeta}(X_2, X_3) \tag{48}$$

where $\mathbf{x}_0$ is the position of the cross section centroid, $\mathbf{R}$ is a 3-by-3 orthogonal matrix function that orients the displaced cross section, and $\boldsymbol{\zeta}$ is the cross-section coordinate system given by $\boldsymbol{\zeta}^T = [\,0 \quad X_2 \quad X_3\,]$. The displacement field is

$$\mathbf{u} = \mathbf{x} - \mathbf{X} = \mathbf{u}_0 + (\mathbf{R}^T - \mathbf{I})\boldsymbol{\zeta}, \tag{49}$$

where $\mathbf{u}_0(X_1) = \mathbf{x}_0(X_1) - \mathbf{X}_0(X_1)$. The skew-symmetric curvature matrix is defined as $\tilde{\boldsymbol{\kappa}} = \mathbf{R}\left(d\mathbf{R}^T/dX_1\right)$. The curvature vector $\boldsymbol{\kappa}$ is the 3-vector extracted from $\tilde{\boldsymbol{\kappa}}$ as $\boldsymbol{\kappa} = \mathbf{axial}(\tilde{\boldsymbol{\kappa}})$. The finite strains that contribute to the internal energy are the axial strain e_{11} and the two shear strains γ_{12} and γ_{13}:

$$\begin{aligned} e_{11} &= \mathbf{h}_1^T\mathbf{g}_1 + \tfrac{1}{2}\mathbf{g}_1^T\mathbf{H}\mathbf{g}_1, \\ \gamma_{12} &= 2e_{12} = \mathbf{h}_2^T\mathbf{g}_1 + \tfrac{1}{2}\mathbf{g}_2^T\mathbf{H}\mathbf{g}_1 + \mathbf{h}_1^T\mathbf{g}_2 + \tfrac{1}{2}\mathbf{g}_1^T\mathbf{H}\mathbf{g}_2, \\ \gamma_{13} &= 2e_{13} = \mathbf{h}_3^T\mathbf{g}_1 + \tfrac{1}{2}\mathbf{g}_3^T\mathbf{H}\mathbf{g}_1 + \mathbf{h}_1^T\mathbf{g}_3 + \tfrac{1}{2}\mathbf{g}_1^T\mathbf{H}\mathbf{g}_3. \end{aligned} \tag{50}$$

Here the 3×3 matrix $\mathbf{H}$ depends on the strain measured selected (for Green-Lagrange strains, $\mathbf{H} = \mathbf{I}$) and $\mathbf{g}_i$ are the displacement gradient 3-vectors

$$\mathbf{g}_1 = \frac{\partial \mathbf{u}_0}{\partial X_1} + \mathbf{R}^T\tilde{\boldsymbol{\zeta}}^T\boldsymbol{\kappa}, \qquad \mathbf{g}_2 = (\mathbf{R}^T - \mathbf{I})\,\mathbf{h}_2, \qquad \mathbf{g}_3 = (\mathbf{R}^T - \mathbf{I})\,\mathbf{h}_3, \tag{51}$$

in which $\mathbf{h}_j$ denotes the 3-vector whose j^{th} component is 1 and the other two are zero.

Introducing the *angular distortion vector* $\boldsymbol{\phi}$ and its associated skew-symmetric matrix $\tilde{\boldsymbol{\phi}}$ as

$$\boldsymbol{\phi} = \mathbf{R}\mathbf{b}_0, \qquad \mathbf{b}_0 = \mathbf{h}_1 + \frac{\partial \mathbf{u}_0}{\partial X_1}, \qquad \boldsymbol{\phi} = \mathbf{axial}(\tilde{\boldsymbol{\phi}}), \qquad \tilde{\boldsymbol{\phi}} = \mathbf{spin}(\boldsymbol{\phi}), \tag{52}$$

we can further split the strains as

$$e_b = \left(\frac{\partial \mathbf{u}_0}{\partial X_1}\right)^T \mathbf{c}_0, \quad e_f = \boldsymbol{\zeta}^T \tilde{\boldsymbol{\phi}} \boldsymbol{\kappa} + \tfrac{1}{2} \boldsymbol{\kappa}^T \tilde{\boldsymbol{\zeta}} \tilde{\boldsymbol{\zeta}}^T \boldsymbol{\kappa}, \quad \gamma_{12} = \mathbf{h}_2^T \boldsymbol{\phi} + \mathbf{h}_2^T \tilde{\boldsymbol{\zeta}}^T \boldsymbol{\kappa}, \quad \gamma_{13} = \mathbf{h}_3^T \boldsymbol{\phi} + \mathbf{h}_3^T \tilde{\boldsymbol{\zeta}}^T \boldsymbol{\kappa}, \tag{53}$$

where $\mathbf{c}_0 = \mathbf{h}_1 + \frac{1}{2}(\partial \mathbf{u}_0/\partial X_1)$ and $\tilde{\boldsymbol{\zeta}} = \mathbf{spin}(\boldsymbol{\zeta})$. Here e_b and e_f measure mean (bar-like) and flexural normal deformations. The two transverse shear strains are $\gamma_2 = \mathbf{h}_2^T \boldsymbol{\phi}$ and $\gamma_3 = \mathbf{h}_3^T \boldsymbol{\phi}$, while the torsional shear strain is composed of $\mathbf{h}_2 \tilde{\boldsymbol{\zeta}}^T \boldsymbol{\kappa}$ and $\mathbf{h}_3^T \tilde{\boldsymbol{\zeta}}^T \boldsymbol{\kappa}$.

The strain energy stored in the current configuration is

$$U = \tfrac{1}{2} \int_{L_0} \int_{A_0} \mathcal{U} \, dA_0 \, dX_1, \qquad \text{with } \mathcal{U} = E e_{11}^2 + \mu G \left(\gamma_{12}^2 + \gamma_{13}^2\right), \tag{54}$$

where μ is a shear correction factor.[10] The 9×9 core stiffnesses are expressed in terms of the the 9-component gradient vector $\mathbf{g}^T = (\mathbf{g}_1^T \quad \mathbf{g}_2^T \quad \mathbf{g}_3^T)$. Defining $\mathbf{c}_i = \mathbf{h}_i + \frac{1}{2}\mathbf{H}\mathbf{g}_i$ and $\mathbf{b}_i = \mathbf{h}_i + \mathbf{H}\mathbf{g}_i$ for $i = 1, 2, 3$, and applying the formulas of Section 4 we obtain for the spectral form at the energy level

$$\mathbf{S}^U = \begin{bmatrix} E\mathbf{S}_1^U + \mu G(\mathbf{S}_2^U + \mathbf{S}_3^U) & \mu G\mathbf{S}_4^U & \mu G\mathbf{S}_5^U \\ \mu G\mathbf{S}_4^{U^T} & \mu G\mathbf{S}_1^U & 0 \\ \mu G\mathbf{S}_5^{U^T} & 0 & \mu G\mathbf{S}_1^U \end{bmatrix}, \tag{55}$$

where $\mathbf{S}_1^U = \mathbf{c}_1\mathbf{c}_1^T$, $\mathbf{S}_2^U = \mathbf{c}_2\mathbf{c}_2^T$, $\mathbf{S}_3^U = \mathbf{c}_3\mathbf{c}_3^T$, $\mathbf{S}_4^U = \mathbf{c}_2\mathbf{c}_1^T$ and $\mathbf{S}_5^U = \mathbf{c}_3\mathbf{c}_1^T$. At the residual level we obtain for the spectral form $\mathbf{S}^r = \mathbf{S}^r(0, 0)$

$$\mathbf{S}^r = \begin{bmatrix} E\mathbf{S}_1^r + \mu G(\mathbf{S}_2^r + \mathbf{S}_3^r) & \mu G\mathbf{S}_4^r & \mu G\mathbf{S}_5^r \\ \mu G\mathbf{S}_6^r & \mu G\mathbf{S}_1^r & 0 \\ \mu G\mathbf{S}_7^r & 0 & \mu G\mathbf{S}_1^r \end{bmatrix}, \tag{56}$$

where $\mathbf{S}_1^r = \mathbf{b}_1\mathbf{c}_1^T$, $\mathbf{S}_2^r = \mathbf{b}_2\mathbf{c}_2^T$, $\mathbf{S}_3^r = \mathbf{b}_3\mathbf{c}_3^T$, $\mathbf{S}_4^r = \mathbf{b}_2\mathbf{c}_1^T$, $\mathbf{S}_5^r = \mathbf{b}_3\mathbf{c}_1^T$, $\mathbf{S}_6^r = \mathbf{b}_1\mathbf{c}_2^T$ and $\mathbf{S}_7^r = \mathbf{b}_1\mathbf{c}_3^T$. The internal force vector conjugate to $\delta\mathbf{g}$ is $\boldsymbol{\Phi} = \mathbf{S}^r\mathbf{g} = \boldsymbol{\Phi}_\sigma + \boldsymbol{\Phi}_\tau$, in which

$$\boldsymbol{\Phi}_\sigma = E \begin{Bmatrix} \mathbf{b}_1 e_{11} \\ 0 \\ 0 \end{Bmatrix}, \qquad \boldsymbol{\Phi}_\tau = \mu G \begin{Bmatrix} \mathbf{b}_2\gamma_{12} + \mathbf{b}_3\gamma_{13} \\ \mathbf{b}_1\gamma_{12} \\ \mathbf{b}_1\gamma_{13} \end{Bmatrix}, \tag{57}$$

represent the contribution of the normal and shear stresses, respectively.

---

[10] This factor is initially assumed to be the same for bending-shear and torsion for simplicity in the foregoing expressions. Distinction can be made if necessary after passing to stress resultants in the over-the-element integration process.

At the incremental level we operate with the second variation (15). Terms $\delta\mathbf{g}^T\mathbf{S}^r\,\delta\mathbf{g} + \delta\mathbf{g}^T\delta\mathbf{S}^r\mathbf{g}$ provide two components of the core tangent stiffness: $\mathbf{S}_M$ and $\mathbf{S}_{GI}$, which are independent of the choice of element degrees of freedom. On the other hand, the term $(\delta^2\mathbf{g})^T\boldsymbol{\Phi}$, which has to be retained because the relation between $\mathbf{g}$ and rotational degrees of freedom is nonlinear, leads to a geometric stiffness component $\mathbf{S}_{GD}$, which depends on the parametrization chosen to describe rotations (for example, Euler angles or Euler parameters).

The core material stiffness matrix is

$$\mathbf{S}_M = \begin{bmatrix} E\mathbf{S}_1 + \mu G(\mathbf{S}_2 + \mathbf{S}_3) & \mu G\mathbf{S}_4 & \mu G\mathbf{S}_5 \\ \mu G\mathbf{S}_4^T & \mu G\mathbf{S}_1 & 0 \\ \mu G\mathbf{S}_5^T & 0 & \mu G\mathbf{S}_1 \end{bmatrix}, \tag{58}$$

where $\mathbf{S}_1 = \mathbf{b}_1\mathbf{b}_1^T$, $\mathbf{S}_2 = \mathbf{b}_2\mathbf{b}_2^T$, $\mathbf{S}_3 = \mathbf{b}_3\mathbf{b}_3^T$, $\mathbf{S}_4 = \mathbf{b}_2\mathbf{b}_1^T$ and $\mathbf{S}_5 = \mathbf{b}_3\mathbf{b}_1^T$. The parametrization-independent geometric stiffness is

$$\mathbf{S}_{GI} = \begin{bmatrix} Ee_{11}\mathbf{H} & \mu G\gamma_{12}\mathbf{H} & \mu G\gamma_{13}\mathbf{H} \\ \mu G\gamma_{12}\mathbf{H} & 0 & 0 \\ \mu G\gamma_{13}\mathbf{H} & 0 & 0 \end{bmatrix}. \tag{59}$$

The computation of $\mathbf{S}_{GD}$ depends on the rotation parametrization; we refer to [5] for details. The core tangent stiffness matrix is the combination $\mathbf{S} = \mathbf{S}_M + \mathbf{S}_{GI} + \mathbf{S}_{GD}$. Completion of the element derivation require the following transformation steps, which are only outlined for brevity.

1. The variations $\delta\mathbf{g}$ are related to variations of a vector $\mathbf{q}$ of generalized coordinates:

$$\delta\mathbf{g} = \begin{bmatrix} \mathbf{I} & \mathbf{R}^T\tilde{\zeta}^T & \mathbf{R}^T\tilde{\kappa}\tilde{\zeta}^T \\ 0 & 0 & \mathbf{R}^T\tilde{\mathbf{h}}_2^T \\ 0 & 0 & \mathbf{R}^T\tilde{\mathbf{h}}_3^T \end{bmatrix}\delta\mathbf{q}, \qquad \text{in which} \quad \delta\mathbf{q} = \begin{Bmatrix} \dfrac{d\delta\mathbf{u}_0}{dX_1} \\ \dfrac{d\delta\Theta}{dX_1} \\ \delta\Theta \end{Bmatrix}, \tag{60}$$

   where $\mathbf{I}$ is the 3×3 identity matrix and $\delta\Theta$ is the axial vector associated with the skew-symmetric matrix $\widetilde{\delta\Theta} = \mathbf{R}\,\delta\mathbf{R}^T$.

2. The variations of $\mathbf{q}$ are related to variations of nodal degrees of freedom $\mathbf{v}$ of the beam element. Euler parameters are chosen as rotational parameters and related to angular coordinates. As usual for Timoshenko beam elements, translational displacements are linearly interpolated. For the rotations, however, the shape functions contain trigonometric terms so that constant curvature states can be exactly represented for arbitrary node rotations. The physical stiffnesses are obtained by the method explained in Section 6. Finally, integrated stress-resultant expressions are simplified by introducing principal and polar moments of inertia.

Procedural details pertaining to these two steps are given in [5]. Two highly nonlinear problems solved with this formulation are reported below.

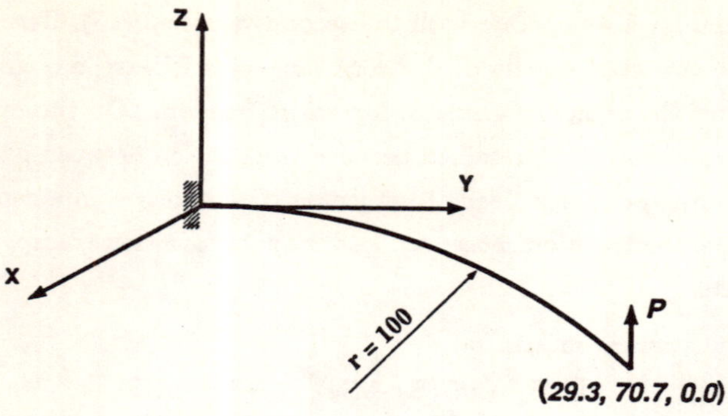

Fig. 2. Curved cantilever bend under tip load: Problem definition.

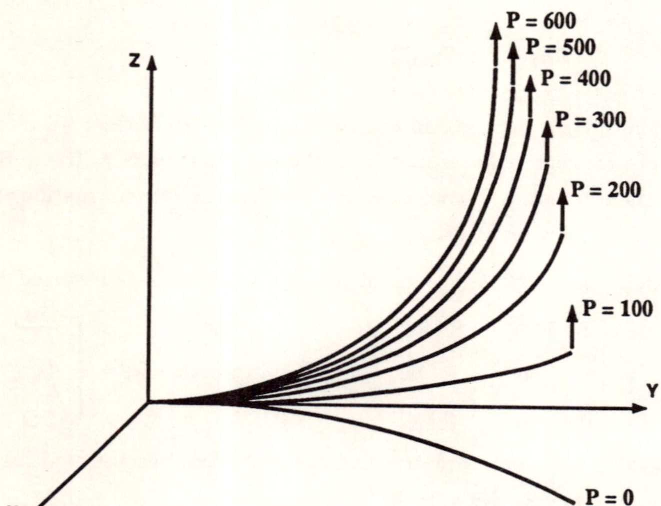

Fig. 3. Curved cantilever bend under tip load: Deflected shapes.

10 Application Problem 1: Large Displacement of a 45° Cantilever Bend

The first problem concerns the large displacement analysis of a 45° cantilever bend initially lying on the horizontal (X,Y) plane. The bend is an arc of a circle of radius $r = 100$ and the beam cross-section is a square with sides of unit length. The beam has modulus $E = 10^7$ and Poisson's ratio $\nu = 0$. It is subjected to an end load $P = 600$ normal to the (X, Y) plane as shown in Figure 2. The model consists of 8 straight beam elements, and the full load was applied in 6 equal increments. The solution method is incremental-iterative with full Newton iteration used in the corrective phase.

This problem was treated by Bathe and Bolourchi [7] with 3D brick elements, and subsequently by Simo and Vu-Quoc [8] and Cardona [9] with beam elements based on other formulations. Results for the 3 tip displacement components are compared below.

	Load $P = 300$	Load $P = 450$	Load $P = 600$
Bathe *et.al.* [7]	22.33, 58.84, 40.08	18.62, 53.32, 48.39	15.79, 47.23, 53.37
Simo *et.al.* [8]	22.50, 59.20, 39.50		15.90, 47.20, 53.40
Cardona [9]	22.14, 58.64, 40.35	18.38, 52.11, 48.59	15.55, 47.04, 53.50
Present	22.31, 58.85, 40.08	18.59, 53.34, 48.39	15.75, 47.25, 53.37

As can be observed the present results compare well with those obtained with 3D elements in [7]. Deflected shapes for selected load levels are shown in Figure 3.

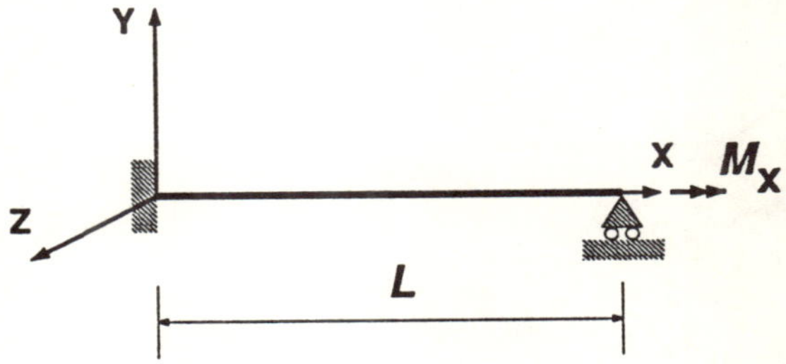

Length	$L = 240\ mm$
Polar Moment	$I_x = 2.16\ mm^4$
Second Moments	$I_y = I_z = 0.0833\ mm^4$
Young's Modulus	$E = 71240\ N/mm^2$
Shear Modulus	$G = 27190\ N/mm^2$

Fig. 4. Cable hockling: Problem definition.

11 Application Problem 2: Cable Hockling

The second problem, *cable hockling*, is more computationally demanding. An initially straight cable, modelled as a beam, is subjected to a tip torsional moment. The geometry and physical properties are given in Figure 4. The cable is clamped at one end and supported at the other so that the only motions allowed at that point are axial displacement and torsional rotation about the longitudinal X axis. No rotation is allowed about Y or Z. The purpose of these restrictions is to keep the problem conservative, because

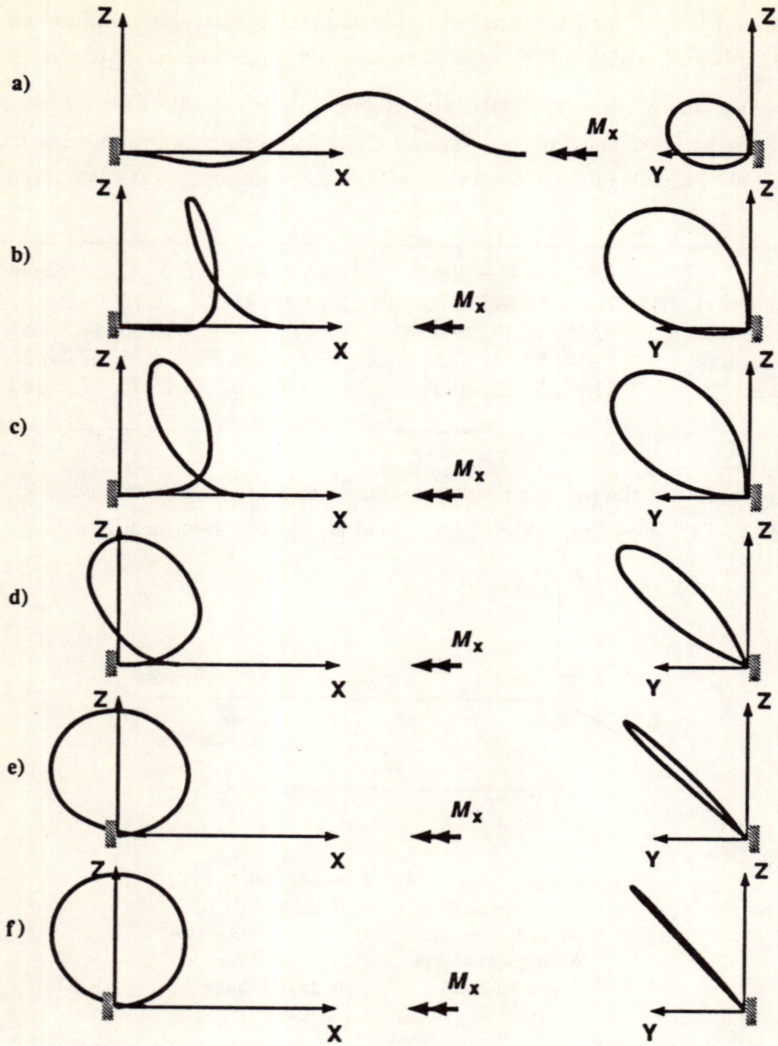

Fig. 5. Cable hockling: deformed shapes at different load levels:
a) After bifurcation, b) $M_x = 50$, c) $M_x = 0$, d) $M_x = -50$,
e) At the limit point, f) $M_x = 0$.

if the torque is allowed to rotate about Y or Z, the problem becomes nonconservative and dynamical methods are required to assess its stability [10,11].

This problem has received a great deal of attention from the engineering community due to its practical importance. The main objective is to estimate the critical applied torque at which the cable departs (bifurcates) from its straight configuration, resulting in the formation of a loop or hockle. This has direct application to marine cables used in tasks such as lifting objects from the ocean floor, for which structural failure could be disastrous. Under the assumptions of infinitesimal bending deformations, Greenhill obtained an analytical formula to predict the critical torque; see *e.g.* pp. 417–418 of Love [12]. The post-bifurcation response analysis of this problem, however, has not been pursued until recently, as discussed in the research conducted by Nour-Omid and Rankin [13]. This post-critical response has also been analyzed by using the present TL formulation. The structure is discretized by twenty equally-spaced beam elements.

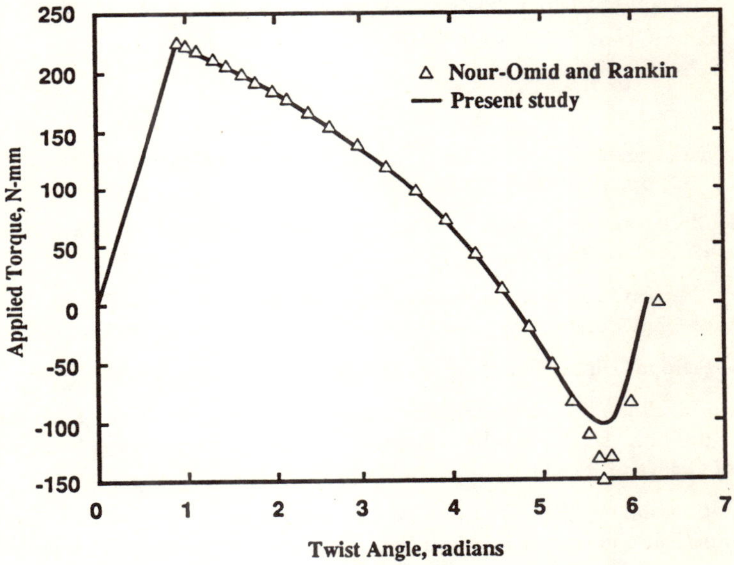

Fig. 6. Cable hockling: Tip rotation vs. applied moment.

The deformed shapes at different load levels are shown in Figure 5, which displays the loop-formation process previously mentioned. The curves on the left and right side show deformed shapes looking along the Y and Z axes, respectively.

If the torque is held under the critical value, the beam twists without lateral deflection. Along this fundamental path the response is linear. At the critical torque a bifurcation point is reached, at which the fundamental path becomes unstable. The beam acquires a helical shape with the free end moving towards the clamped end. This new equilibrium branch is unstable and the cable undergoes large displacements and rotations as the moment decreases. The unloading process continues until reaching a sharp limit point which corresponds to a *negative* value of the applied torque (it should be mentioned that

Nour-Omid and Rankin [13], who use a Hermitian beam element based on corotational description, characterize this critical point as a secondary bifurcation; evidence for categorization is presently inconclusive). After traversing this point, the torque reverses again until the cable reaches a circular-shaped unloaded deformed configuration.

The computed variation of the twist angle at the moving end versus the applied torque is given in Figure 6. Results are compared to those given by Nour-Omid and Rankin.

Acknowledgement

The research reported herein has been supported by the Air Force Office of Scientific Research under Grant F49620-87-C-0074.

References

1. S. Rajasekaran and D. W. Murray, Incremental finite element matrices, *J. Str. Div. ASCE*, **99**, pp. 2423–2438, 1973.
2. R. H. Mallet and P. V. Marcal, Finite element analysis of nonlinear structures, *J. Str. Div. ASCE*, **94**, pp. 2081–2105, 1968.
3. D. W. Murray, Finite element nonlinear analysis of plates, Ph. D. Dissertation, Dept. of Civil Engineering, University of California, Berkeley, California, 1967.
4. C. A. Felippa, Discussion of [1], *J. Str. Div. ASCE*, **100**, pp. 2519–2521, 1974.
5. L. A. Crivelli, A Total-Lagrangian beam element for analysis of nonlinear space structures, Ph. D. Dissertation, Dept. of Aerospace Engineering Sciences, University of Colorado, Boulder, 1990.
6. K. Mathiasson, A. Bengston and A. Samuelsson, On the accuracy and efficiency of numerical algorithms for geometrically nonlinear structural analysis, in P. G. Bergan, K.-J. Bathe and W. Wunderlich (eds.) *Finite Element Methods for Nonlinear Problems*, Springer-Verlag, Berlin, pp. 3–24, 1986.
7. K. J. Bathe and S. Bolourchi, Large displacement analysis of three-dimensional beam structures, *Int. J. Numer. Meth. Engrg.*, **14**, pp. 961–986, 1979.
8. J. C. Simo and L. Vu-Quoc, A three-dimensional finite strain rod model. Part II: Computational aspects, *Comp. Meths. Appl. Mech. Engrg.*, **58**, pp. 79–116, 1986.
9. A. Cardona, An integrated approach to mechanism analysis, Ph. D. Dissertation, LTAS, Université de Liège, Belgium, 1989.
10. V. V. Bolotin, *Nonconservative Problems of the Theory of Elastic Stability*, Pergamon Press, New York, 1963.
11. H. Ziegler, *Principles of Structural Stability*, Blaisdell, Massachusetts, 1968.
12. A. E. H. Love, *The Mathematical Theory of Elasticity*, Dover, New York, 1944.
13. B. Nour-Omid and C. C. Rankin, Finite rotation analysis and consistent linearization using projectors, submitted to *Comp. Meths. Appl. Mech. Engrg.*, 1990.

Large Displacement Analysis of Highly Flexible Off-Shore Structures

Pål G. BERGAN[1]
A.S. Veritas Research, Det norske Veritas, N-1322 Høvik, Norway

Kjell Magne MATHISEN[2]
Division of Structural Engineering, The Norwegian Institute of Technology, N-7034 Trondheim-NTH, Norway

Summary

A new three-dimensional beam element based on a mixed variational formulation has been developed. This element utilizes a co-rotational "ghost" reference frame to which all element quantities are referred except for the internal axial forces. The beam element is formulated such that different order of interpolation of axial displacements and axial forces may be prescribed. Successful use of the Hilber, Hughes and Taylor's time integrator is demonstrated for this special element which includes internal displacement constraints. A general technique for obtaining the static equilibrium configuration of extremely nonlinear, flexible structures is outlined.

Two problems involving very large displacements are being solved with the new element. One of these examples includes static and dynamic analyses performed for a realistic highly flexible riser exposed to regular waves. The examples demonstrate the range of applicability for this formulation, which exceeds the common limitations of standard displacement formulations.

1 Introduction

An important trend in the present off-shore oil and gas technology is the move towards production of hydrocarbons in increasingly deeper waters. The development of new deep water production systems based on compliant and flexible types of structures has thereby been stimulated, particularly in connection with articulated anchoring systems, tension leg platforms and flexible riser systems. The overall behavior of these types of structures basically involves large displacements, nonlinear constraints and nonlinear interaction between the environmental loads and the structural system.

The application of flexible risers and slender pipelines in the development of innovative, floating deep water production systems exposed to very severe weather conditions has created a need for new, comprehensive three-dimensional analysis capabilities. The problems of flexible risers are inherently nonlinear and three-dimensional in nature owing to the large amplitude response. There is a need for comprehensive and efficient

[1]Also Professor, Division of Structural Engineering, The Norwegian Institute of Technology, Norway.

[2]Currently Visiting Scholar, Center for Space Structures and Controls, University of Colorado at Boulder, U.S.A.

three-dimensional analysis methods for the computations of motion and forces in flexible pipelines. It is noted that the analysis of flexible risers is much more difficult than that of traditional rigid steel risers.

Most of the present flexible riser design practice is based on approximating the system by a cable structure. This simplifies the problem because rotational freedoms may be omitted and because the bending effects, which give rise to ill-conditioning due to the inherently large difference in axial and bending stiffnesses, are removed. In order to get around this difficulty the beam element formulation must be capable of controlling the ill-conditioning and it must lead to a well-conditioned system of linear equations that may be readily solved.

In order to improve the stability of potentially numerically sensitive systems, an alternative mixed approach has been adopted. The mixed formulation avoids such instabilities since axial forces are an independent part of the solution process and they are decoupled from the axial displacements. The mixed approach for beams has earlier been presented by Hibbitt et al. [1] and McNamara et al. [2, 3]. However, the present approach is more general, since the formulation is taking into account the relative axial and bending stiffness and it also accounts for torsional deformations and it allows for any type of cross-section.

Most large-scale structures today are being analyzed by means of the finite element method [4]. A fully integrated fluid-structure analysis requires that the computation of hydrostatic forces is formulated in such a way that it fits directly into the nonlinear structural model. This implies that nodal reaction forces and incremental force-displacement matrices must be established for any response state during a loading history. Typically hydrodynamic forces are the most important type of loading for floating or submerged structures and it is required that the computation of hydrodynamic loads be integrated with the structural analysis.

Other crucial problems to be addressed are finding the initial equilibrium configuration for highly flexible systems and for systems that are floating. These problems as well as the following simulation of dynamic response pose great demands on the numerical solution algorithms for solving the nonlinear equations.

2 Finite Element Formulation

2.1 Large Rotation Kinematics

The present formulation uses a fixed global reference system for describing the motion of a point

$$\mathbf{x} = \mathbf{X} + \mathbf{v} \tag{1}$$

where $\mathbf{X}$ is the initial position vector, $\mathbf{x}$ the current position vector and $\mathbf{v}$ the displacement vector. The finite element nodal points have six degrees of freedom, i.e. three translations and three rotations. Since only the translational displacement components are true vectors in large-dispacement analysis, the rotational motion is defined by use of an orthogonal base vector triad $\mathbf{i}_i$, whose orientation is defined by way of the rotation matrix T_{ij}

$$\mathbf{i}_i^n = T_{ij}^n \mathbf{I}_j \tag{2}$$

where the superscript n signifies configuration C_n and $\mathbf{I}_j$ represents the orthogonal base vectors of the fixed global reference system. Each node has a local coordinate system $\mathbf{i}_i$ attached to it. This system is initially parallel to the global base vector system $\mathbf{I}_j$, and it moves along with the node during deformation.

The rotation of a node is thus not given by *Euler* angles or by angular parameters (e.g. *Euler* or *Rodrigues* parameters) [5, 6], which are often used, but rather by the orthonormal components T_{ij} of the rotation matrix.

The incremental-iterative solution procedure involves incremental update of translations as well as rotations at every node. The incremental translations $\Delta\mathbf{v}$ may easily be updated by simple addition while the incremental rotations $\Delta\Theta_i$, about the global base vectors, cannot be accumulated in the same simple manner. This is because they cannot be treated vectorially for arbitrary large rotations in space. Thus a main problem is to establish (2) in configuration C_{n+1} on basis of the known incremental rotations $\Delta\Theta_i$.

The updated rotation matrix in C_{n+1} may be formulated as a transformation of T_{ij}^n, hence

$$\mathbf{i}_i^{n+1} = T_{ik}^n \Delta T_{kj}^{n+1} \mathbf{I}_j = T_{ij}^{n+1} \mathbf{I}_j \tag{3}$$

The incremental transformation matrix ΔT_{kj} may be obtained in different ways. Argyris [7] and Rankin and Brogan [8] established the incremental transformation matrix on basis of the rotation matrix for small rotations. However, a more direct and simpler form of ΔT_{ij} has been derived by Bergan and Nygård [9, 10].

$$\Delta T_{ij}^{n+1} = \delta_{ij}\cos(\Delta\Theta) + \frac{\Delta\Theta_i}{\Delta\Theta}\frac{\Delta\Theta_j}{\Delta\Theta}\left[1-\cos(\Delta\Theta)\right] + e_{ijk}\frac{\sin(\Delta\Theta)}{\Delta\Theta}\Delta\Theta_k \tag{4}$$

where

$$\Delta\Theta = \sqrt{\sum_{i=1}^{3}\Delta\Theta_i^2} \tag{5}$$

e_{ijk} is the permutation symbol [11].

Note, that the simple procedure by Bergan and Nygård [9, 10] is invariant with respect to orientation of global base vectors and it retains an orthonormal form of T_{ij}^{n+1}.

2.2 3D-Beam Element for Highly Slender Structures

2.2.1 Beam Kinematics

It is commonly known that in a step by step Total Lagrangian (TL) formulation with C_0 as a fixed reference, see Fig. 1, artificial strains easily arise in beam elements due to non-homogenuities in the displacement expansions in transverse and longitudinal directions. For this reason the TL-formulation is primarily suited for problems involving beams, plates and shells with only small or moderate rotations, see e.g. [12]. As an alternative to the TL-formulation, Bergan [13] and others [14, 15, 16] introduced the concept of co-rotated 'ghost' reference. The TL-formulation may be referred to as a continuum approach, since the finite element equations are obtained by discretization of the equations of motion for the whole continuum. In contrast, the attention is now focused on a single finite element of the body when deriving the equations for the co-rotational Lagrangian (CL) formulation.

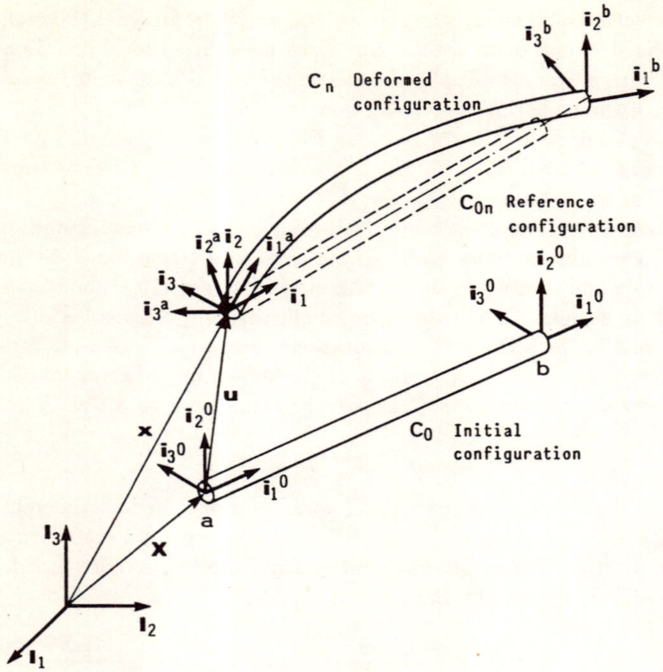

Figure 1: Reference configuration for the mixed beam element.

The motion of a beam element is described in Fig. 1. Each beam element has a local coordinate system $\mathbf{i}_i^0$ attached to the beam ends in the initial configuration C_0. $\mathbf{i}_1^0$ follows the longitudinal axis and $\mathbf{i}_2^0$ and $\mathbf{i}_3^0$ are oriented along the principal axes of the cross-section. The initial configuration C_0 in Fig. 1 is moved along with the element as a rigid body C_{0n} and forms the reference configuration for strains and stresses. The figure also shows the element in a deformed configuration C_n. At the same time the $\mathbf{i}_i^0$ systems attached to the two ends have rotated along with the end nodal points a and b and are denoted $\mathbf{i}_i^a$ and $\mathbf{i}_i^b$. This deformed configuration is measured in relation to a co-rotated 'ghost' reference C_{0n}. This reference element has one end at node a with the longitudinal axis $\mathbf{i}_1$ pointing in the direction of node b. The cross-sectional axes $\mathbf{i}_2$ and $\mathbf{i}_3$ of the ghost reference are established from averages of $\mathbf{i}_2^a$ and $\mathbf{i}_2^b$ and of $\mathbf{i}_3^a$ and $\mathbf{i}_3^b$, respectively [17].

The CL-formulation with strains and stresses referred to the co-rotated reference body C_{0n} is in principle identical to the TL-formulation with C_0 as reference, since only rigid-body motions have taken place between C_0 and C_{0n} (see e.g. [10, 18]). However, there are differences when it comes to the local computation of strains due to simplifications that may be made.

The *basic* mixed 3D-beam element has three translational and three rotational nodal displacement degrees of freedom at each end node. The displacements of the reference axis of C_{0n} are defined by the interpolations

$$\begin{aligned} u_o &= \mathbf{N}_u \mathbf{v}_u \quad \text{where} \quad \mathbf{v}_u^T = [v_{xa}, v_{xb}] \\ v_o &= \mathbf{N}_v \mathbf{v}_v \quad \text{where} \quad \mathbf{v}_v^T = [v_{ya}, \theta_{za}, v_{yb}, \theta_{zb}] \\ w_o &= \mathbf{N}_w \mathbf{v}_w \quad \text{where} \quad \mathbf{v}_w^T = [v_{za}, \theta_{ya}, v_{zb}, \theta_{yb}] \\ \theta &= \mathbf{N}_\theta \mathbf{v}_\theta \quad \text{where} \quad \mathbf{v}_\theta^T = [\theta_{xa}, \theta_{xb}] \end{aligned} \tag{6}$$

Cubic interpolation polynomials $\mathbf{N}_v$ and $\mathbf{N}_w$ are used for bending in the local x-y and x-z planes whereas linear interpolation polynomials $\mathbf{N}_u$ and $\mathbf{N}_\theta$ are used for the local axial displacement and torsion. The local rotational degrees of freedom express deformational rotations in relation to the co-rotated C_{0n} configuration, defined by the base vectors $\mathbf{i}_i$. The local rotation at node a, may be obtained through the rotation matrix of the node, T_{ij}^a, referred to the C_{0n} system, T_{ij}, as follows:

$$\mathbf{i}_i^a = T_{ij}^a \mathbf{I}_j = T_{ij}^a T_{jk} \mathbf{i}_k = \tilde{T}_{ij}^a \mathbf{i}_j \tag{7}$$

The local rotations θ_{ia}, at node a, may now be retrieved from the rotation matrix $\tilde{T}_{ij}^a$ [9, 10]

$$\begin{aligned} \theta_{xa} &= \frac{1}{2} \frac{\theta}{\sin(\theta)} \left(\tilde{T}_{23}^a - \tilde{T}_{32}^a \right) \\ \theta_{ya} &= \frac{1}{2} \frac{\theta}{\sin(\theta)} \left(-\tilde{T}_{13}^a + \tilde{T}_{31}^a \right) \\ \theta_{za} &= \frac{1}{2} \frac{\theta}{\sin(\theta)} \left(\tilde{T}_{12}^a - \tilde{T}_{21}^a \right) \end{aligned} \tag{8}$$

where

$$\theta = \arccos\left[\frac{1}{2} \left(\tilde{T}_{kk}^a - 1 \right) \right] \tag{9}$$

These relative, deformational rotations are assumed to be small or moderately small, since all large rotational effects are extracted from the local element degrees of freedom through the 'ghost'-reference concept.

A higher order axial displacement interpolation may easily be constructed by adopting the hierarchical approach. To this end, the *basic* linear axial displacement functions of (6) are simply augmented by adding 'relative displacement modes' to produce

$$u_o = \mathbf{N}_u \mathbf{v}_u + \hat{\mathbf{N}}_u \hat{\mathbf{v}}_u \tag{10}$$

where $\hat{\mathbf{N}}_u$ contains the higher-order modes while $\hat{\mathbf{v}}_u$ contains the 'relative nodal point displacements' at the internal nodal points.

2.2.2 Basic Assumptions

The relative rotations are assumed to be small such that the Bernoulli-Euler and Navier assumptions apply. Coupling effects between torsion and bending about the principal axes may be neglected if provided that the longitudinal axis is placed at the shear center [19]. Simple St.Venant torsion is assumed, and coupling effects between torsion and axial displacements are neglected. The torsional problem thus becomes completely uncoupled from the other deformations. Warping resistance and torsional stability problems are not considered.

The beam is supposed to be so slender that transverse shear deformations can be neglected in comparison with bending deformations. This assumption is totally justified for the types of flexible structural systems considered herein. Lateral contraction caused by elongation is also neglected. The Bernoulli-Euler and Navier assumptions imply that the cross-sectional planes remain plane and perpendicular to the x-axis during deformations. The displacements of an arbitrary point P, defined by coordinates x, y, z in the C_{0n} configuration may then be expressed as

$$\begin{aligned} u(x,y,z) &= u_o - y v_{o,x} - z w_{o,x} \\ v(x,y,z) &= v_o - z\theta \\ w(x,y,z) &= w_o + y\theta \end{aligned} \tag{11}$$

The Green strain tensor yields the following expression for the axial fiber strain [11]

$$E_{xx} = u_{,x} + \frac{1}{2}\left(u_{,x}^2 + v_{,x}^2 + w_{,x}^2\right) \tag{12}$$

According to studies made by Remseth [12] and Søreide [20], and later also by Mattiasson et al. [21, 22], the squares of the in-plane displacement gradients should be neglected in the CL-formulation to avoid self-straining effects. This is also consistent with von Kármán theory for thin elastic plates [23].

By substituting the expressions for the displacements (11) into (12), E_{xx} takes the following form:

$$\begin{aligned} E_{xx} &= u_{o,x} - y v_{o,xx} - z w_{o,xx} + \frac{1}{2}\left(v_{o,x}^2 + w_{o,x}^2\right) + \\ &\quad \theta_{,x}\left(y w_{o,x} - z v_{o,x}\right) + \frac{1}{2}\theta_{,x}^2\left(y^2 + z^2\right) \end{aligned} \tag{13}$$

Engseth [24] experienced that the coupling between the gradients of the transverse displacements $v_{o,x}$, $w_{o,x}$ and $\theta_{,x}$ were of minor importance, and they may be neglected. The rotation gradient $\theta_{,x}$ is generally much smaller then $v_{o,x}$ and $w_{o,x}$. The gradient terms of $\theta_{,x}^2$ are therefore also neglected.

The following approximation for the normal Green strain remains

$$E_{xx} = u_{o,x} - y v_{o,xx} - z w_{o,xx} + \frac{1}{2}\left(v_{o,x}^2 + w_{o,x}^2\right) \tag{14}$$

Introducing curvatures κ_y and κ_z, one may write

$$E_{xx} = E_{xxo} - y\kappa_y - z\kappa_z \tag{15}$$

where

$$E_{xxo} = u_{o,x} + \frac{1}{2}\left(v_{o,x}^2 + w_{o,x}^2\right) \tag{16}$$

and

$$\kappa_y = v_{o,xx} \tag{17}$$

and

$$\kappa_z = w_{o,xx} \tag{18}$$

The strain at the centroidal axis is denoted by subscript o.

2.2.3 Mixed Finite-Element Matrix Representation of the Variational Equations of Motion

The internal virtual work for the element may be written

$$\delta W_{\text{int}} = \int_{L_0} (\delta E_{xxo} N_{xx} + \delta\kappa_y M_y + \delta\kappa_z M_z + \delta\theta_{,x} M_\theta)\, dx \tag{19}$$

where δ denotes 'variation of', and the following expressions for the cross-sectional force resultants have been introduced

$$\begin{aligned} N_{xx} &= \textstyle\int_{A_0} S_{xx} dA_0 \\ M_z &= -\textstyle\int_{A_0} y S_{xx} dA_0 \\ M_y &= -\textstyle\int_{A_0} z S_{xx} dA_0 \\ M_\theta &= GI_T\theta_{,x} \end{aligned} \tag{20}$$

L_0 is the initial length and A_0 is the initial cross-sectional area. M_θ is the torsional moment and GI_T is the torsional stiffness of the cross-section, which is assumed to be constant. S_{xx} is the 2. Piola-Kirchhoff stress component which is energy conjugate to the Green fiber strain component E_{xx}. S_{xx} is computed from the current strain or total strain history, using the appropriate material law. It is seen from expression (20) that the internal virtual work rate is associated with axial strain and torsional shear only.

In order to analyze very slender members for which the axial to bending stiffness ratio is very high, the axial force, N_{xx}, may be regarded as an independent variable. A *mixed* finite element formulation [25, 26] is adopted where the axial force is independently interpolated and combined with the corresponding axial displacements via a Lagrangian constraint

$$E_{xxo} = \frac{\tilde{N}}{N_T} \tag{21}$$

where N_T is the axial cross-sectional modul depending on the tangential material law C_T

$$N_T = \int_{A_0} C_T\, dA_0 \tag{22}$$

Thus, within each discrete element the axial force $\tilde{N}$ is expressed in terms of a set of undetermined parameters $\boldsymbol{\lambda}_N$ and simple polynomial functions $\mathbf{P}_N$

$$\tilde{N} = \mathbf{P}_N \boldsymbol{\lambda}_N \tag{23}$$

It is easily shown by variational calculus that the Lagrangian multiplier $\tilde{N}$ is in fact the true axial force [27]. McNamara et al. [2], characterized the constraint (21) as an inextensibility condition, however such an interpretation is not correct.

The augmented internal virtual work for the element may then take the form, see e.g. reference [4]

$$\delta\tilde{W}_{\text{int}} = \int_{L_0} \left[\delta E_{xxo}\tilde{N} + \delta\kappa_y M_y + \delta\kappa_z M_z + \delta\theta_{,x} M_\theta + \delta\tilde{N}(E_{xxo} - \frac{\tilde{N}}{N_T})\right] dx \tag{24}$$

Prathap [28] studied the errors due to constraints in mixed formulations, and denoted them as 'errors of second kind'. Following [28, 29], the axial strain at the centroidal axis,

E_{xxo}, and the axial force, $\tilde{N}$, must be of the same order of interpolation, if not, so-called "*membrane-locking*" will occur and destroy the convergence even for small element sizes.

The interpolation functions for the transverse displacements (6) are based on Hermitian cubic interpolation whereas the interpolation of the axial displacements depends upon the number of hierarchical modes included at internal points. Starting with the *basic* element of (6), we have linear interpolation, while one, two and four internal points yields quadratic, cubic and quintic interpolation of axial displacements, respectively. In order to match the quadratic terms $v_{o,x}^2$ and $w_{o,x}^2$ (fourth order polynomials) of (16), the interpolation functions for the axial displacements should be of fifth order. This may only be obtained using four internal nodal points for the axial displacements, and fourth degree interpolation functions, $\mathbf{P}_N$, for the axial-force, $\tilde{N}$. Several examples are included in reference [17] to compare constant, linear, cubic and even fourth order interpolation for $\tilde{N}$ for the whole family of hierarchical elements including the *basic* element with only linear axial displacement functions.

In order to avoid pivoting in the Gaussian elimination scheme, a linear combination of the augmented internal virtual work (24) and the internal virtual work for the standard displacement formulation (19) is adopted, hence

$$\delta \bar{W}_{\text{int}} = (1 - \alpha_N)\delta \tilde{W}_{\text{int}} + \alpha_N \, \delta W_{\text{int}} \tag{25}$$

where α_N is a parameter describing the ratio between the axial and bending stiffness for the element. In this work, α_N is given by the following expression

$$\alpha_N = \begin{cases} \frac{I_0}{A_0 L_0^2} & \text{if } \frac{I_0}{A_0 L_0^2} < 1 \\ 1 & \text{otherwise} \end{cases} \tag{26}$$

where I_0 is the smallest of the quadratic area moments about the principal axes. Choosing $\alpha_N = 1$, we have a standard displacement formulation, otherwise a mixed formulation.

The following expressions for the variation of the strain at the centroidal axis and the curvatures apply

$$\begin{aligned} \delta E_{xxo} &= \delta u_{o,x} + v_{o,x}\delta v_{o,x} + w_{o,x}\delta w_{o,x} \\ \delta \kappa_y &= \delta v_{o,xx} \\ \delta \kappa_z &= \delta w_{o,xx} \end{aligned} \tag{27}$$

By substituting this into (24) together with the interpolation assumptions (6), (10) and (23), it is seen that the augmented internal reaction forces are given by

$$\delta \bar{W}_{\text{int}} = \delta \mathbf{v}_u^T \mathbf{S}_u + \delta \mathbf{v}_v^T \mathbf{S}_v + \delta \mathbf{v}_w^T \mathbf{S}_w + \delta \mathbf{v}_\theta^T \mathbf{S}_\theta + \delta \hat{\mathbf{v}}_u^T \hat{\mathbf{S}}_u + \delta \boldsymbol{\lambda}_N \mathbf{S}_\lambda = \delta \bar{\mathbf{v}}^T \bar{\mathbf{S}}^S \tag{28}$$

where

$$\bar{\mathbf{S}}^S = \begin{bmatrix} \mathbf{S}^S \\ \hat{\mathbf{S}}_u \\ \mathbf{S}_\lambda \end{bmatrix} \quad \text{and} \quad \bar{\mathbf{v}} = \begin{bmatrix} \mathbf{v} \\ \hat{\mathbf{v}}_u \\ \boldsymbol{\lambda}_N \end{bmatrix} \tag{29}$$

Combining (24) and (28), the reaction forces become

$$\mathbf{S}^S = \begin{bmatrix} \mathbf{S}_u \\ \mathbf{S}_v \\ \mathbf{S}_w \\ \mathbf{S}_\theta \end{bmatrix} = \begin{bmatrix} \int_{L_0} \mathbf{N}_{u,x}^T \bar{N} \, dx \\ \int_{L_0} (\mathbf{N}_{v,xx}^T M_y + \mathbf{N}_{v,x}^T v_{o,x} \bar{N}) \, dx \\ \int_{L_0} (\mathbf{N}_{w,xx}^T M_z + \mathbf{N}_{w,x}^T w_{o,x} \bar{N}) \, dx \\ \int_{L_0} \mathbf{N}_{\theta,x}^T M_\theta \, dx \end{bmatrix} \tag{30}$$

and

$$\hat{\mathbf{S}}_u = \int_{L_0} \hat{\mathbf{N}}_{u,x}^T \bar{N}\, dx \tag{31}$$

and

$$\mathbf{S}_\lambda = (1-\alpha_N) \int_{L_0} \mathbf{P}_N^T (E_{xxo} - \frac{\tilde{N}}{N_T})\, dx \tag{32}$$

The equivalent axial force in the expression for the vectors of internal forces $\mathbf{S}^S$ and $\hat{\mathbf{S}}_u$ in (30) and (31) is

$$\bar{N} = (1-\alpha_N)\tilde{N} + \alpha_N N_{xx} \tag{33}$$

where $\tilde{N}$ is the independent axial force variable, while N_{xx} is computed on basis of the current axial stress S_{xx}.

In addition to the internal forces $\bar{\mathbf{S}}^S$, inertia and external forces may be established from the expression for the internal virtual work for elastodynamics (see e.g. [30])

$$\int_{V_0} (\mathbf{f}_0 - \rho_0 \ddot{\mathbf{u}}) \cdot \delta\mathbf{u}\, dV_0 + \int_{S_0} \mathbf{t}_0 \cdot \delta\mathbf{u}\, dS_0 = \delta\mathbf{v}^T(\mathbf{S}^I + \mathbf{S}^E) \tag{34}$$

The consistent inertia force vector and external force vector may formally be written

$$\mathbf{S}^I = \int_{V_0} \rho_0 \mathbf{N}^T \mathbf{N}\, dV_0\, \ddot{\mathbf{v}} = \mathbf{m}\ddot{\mathbf{v}} \tag{35}$$

$$\mathbf{S}^E = \int_{V_0} \mathbf{N}^T \mathbf{f}_0\, dV_0 + \int_{S_0} \mathbf{N}^T \mathbf{t}_0\, dS_0 \tag{36}$$

where

$$\mathbf{N}^T = [\mathbf{N}_u\ \mathbf{N}_v\ \mathbf{N}_w\ \mathbf{N}_\theta]^T \tag{37}$$

$\mathbf{m}$ is the consistent mass matrix and $\ddot{\mathbf{v}}$ is the second time derivative of the nodal displacement vector $\mathbf{v}$.

The augmented dynamic equilibrium equations for an element at time t may then be stated as

$$\mathbf{S}^I + \mathbf{S}^S = \mathbf{S}^E \tag{38}$$

subject to the nonlinear constraints

$$\mathbf{S}_\lambda = 0 \tag{39}$$

where $\mathbf{S}^E$ includes conservative as well as nonconservative external force contributions.

In a strict sense, Eqs. (34) to (37) should only be used in conjunction with the *basic* element of (6) since the consistent inertia and external force contribution for the internal nodal points have not been included in the expressions above. However, these terms are normally of minor importance and may therefore be neglected. Also note that for higher order axial displacement interpolation elements, $\mathbf{S}^S$ is replaced by a reduced internal force vector in (38). This is due to static condensation of hierarchical nodal degrees of freedom at the internal nodal points.

The solution of the nonlinear dynamic equilibrium equations calls for an incremental form of the variational equations of motion. Using a co-rotational formulation, the increment in the augmented internal virtual work may be derived directly from the equations for the TL-formulation. For a beam element this implies

$$\begin{aligned}\delta\Delta\bar{W}_{\text{int}} &= \int_{L_0}\big[\delta\Delta E_{xxo}\bar{N} + \delta E_{xxo}\Delta\bar{N} + \delta\kappa_y\Delta M_y + \delta\kappa_z\Delta M_z \\ &\quad + \delta\theta_{,x}\Delta M_\theta + (1-\alpha_N)\delta\tilde{N}(\Delta E_{xxo} - \frac{\Delta\tilde{N}}{N_T})\,\big]\,dx \end{aligned} \tag{40}$$

By substituting the interpolation assumptions (6), (10) and (23) into (40) we may write it on matrix form

$$\delta\Delta\bar{W}_{\text{int}} = \begin{bmatrix}\delta\mathbf{v}_u \\ \delta\mathbf{v}_v \\ \delta\mathbf{v}_w \\ \delta\mathbf{v}_\theta \\ \delta\hat{\mathbf{v}}_u \\ \delta\boldsymbol{\lambda}_N\end{bmatrix}^T (\bar{\mathbf{k}}^G + \bar{\mathbf{k}}^M) \begin{bmatrix}\Delta\mathbf{v}_u \\ \Delta\mathbf{v}_v \\ \Delta\mathbf{v}_w \\ \Delta\mathbf{v}_\theta \\ \Delta\hat{\mathbf{v}}_u \\ \Delta\boldsymbol{\lambda}_N\end{bmatrix} = \delta\bar{\mathbf{v}}^T\bar{\mathbf{k}}_T\Delta\bar{\mathbf{v}} \tag{41}$$

This may also be arranged in the form

$$\delta\Delta\bar{W}_{\text{int}} = \begin{bmatrix}\delta\mathbf{v} \\ \delta\hat{\mathbf{v}}_u \\ \delta\boldsymbol{\lambda}_N\end{bmatrix}^T \begin{bmatrix}\bar{\mathbf{k}}_{bb} & \bar{\mathbf{k}}_{bh} & \bar{\mathbf{k}}_{b\lambda} \\ \bar{\mathbf{k}}_{hb} & \bar{\mathbf{k}}_{hh} & \bar{\mathbf{k}}_{h\lambda} \\ \bar{\mathbf{k}}_{\lambda b} & \bar{\mathbf{k}}_{\lambda h} & \bar{\mathbf{k}}_{\lambda\lambda}\end{bmatrix}_T \begin{bmatrix}\Delta\mathbf{v} \\ \Delta\hat{\mathbf{v}}_u \\ \Delta\boldsymbol{\lambda}_N\end{bmatrix} \tag{42}$$

where the subscripts b and h refer to the *basic* and *hierarchical* nodal degrees of freedom, $\mathbf{v}$ and $\hat{\mathbf{v}}_u$, respectively, and λ refers to the Lagrangian multipliers, $\boldsymbol{\lambda}_N$. Each submatrix consists of a geometric and material stiffness part, where the geometric stiffness is derived from the first term in (40)

$$\bar{\mathbf{k}}_{bb}^G = \int_{L_0}\bar{N}_{\text{eff}}\begin{bmatrix}\mathbf{0} & \mathbf{0} & \mathbf{0} & \mathbf{0} \\ \mathbf{0} & \mathbf{N}_{v,x}^T\mathbf{N}_{v,x} & \mathbf{0} & \mathbf{0} \\ \mathbf{0} & \mathbf{0} & \mathbf{N}_{w,x}^T\mathbf{N}_{w,x} & \mathbf{0} \\ \mathbf{0} & \mathbf{0} & \mathbf{0} & \mathbf{0}\end{bmatrix}dx \tag{43}$$

and

$$\bar{\mathbf{k}}_{bh}^G = \bar{\mathbf{k}}_{b\lambda}^G = \bar{\mathbf{k}}_{hb}^G = \bar{\mathbf{k}}_{hh}^G = \bar{\mathbf{k}}_{h\lambda}^G = \bar{\mathbf{k}}_{\lambda b}^G = \bar{\mathbf{k}}_{\lambda h}^G = \bar{\mathbf{k}}_{\lambda\lambda}^G = \mathbf{0} \tag{44}$$

while the remaining terms in (40) form the basis for the augmented material stiffness

$$\bar{\mathbf{k}}_{bb}^M = \int_{L_0}\begin{bmatrix}
\alpha_N N_T\mathbf{N}_{u,x}^T\mathbf{N}_{u,x} & \begin{matrix}\alpha_N(M_{yT}\mathbf{N}_{u,x}^T\mathbf{N}_{v,xx} \\ +v_{o,x}N_T\mathbf{N}_{u,x}^T\mathbf{N}_{v,x})\end{matrix} & \begin{matrix}\alpha_N(M_{zT}\mathbf{N}_{u,x}^T\mathbf{N}_{w,xx} \\ +w_{o,x}N_T\mathbf{N}_{u,x}^T\mathbf{N}_{w,x})\end{matrix} & \mathbf{0} \\
\begin{matrix}\alpha_N(M_{yT}\mathbf{N}_{v,xx}^T\mathbf{N}_{u,x} \\ +v_{o,x}N_T\mathbf{N}_{v,x}^T\mathbf{N}_{u,x})\end{matrix} & \begin{matrix}B_{yT}\mathbf{N}_{v,xx}^T\mathbf{N}_{v,xx} \\ +v_{o,x}M_{yT}(\mathbf{N}_{v,xx}^T\mathbf{N}_{v,x} \\ +\mathbf{N}_{v,x}^T\mathbf{N}_{v,xx}) \\ +v_{o,x}^2N_T\mathbf{N}_{v,x}^T\mathbf{N}_{v,x}\end{matrix} & \begin{matrix}B_{yzT}\mathbf{N}_{v,xx}^T\mathbf{N}_{w,xx} \\ +w_{o,x}M_{yT}\mathbf{N}_{v,xx}^T\mathbf{N}_{w,x} \\ +v_{o,x}M_{zT}\mathbf{N}_{v,x}^T\mathbf{N}_{w,xx} \\ +v_{o,x}w_{o,x}N_T\mathbf{N}_{v,x}^T\mathbf{N}_{w,x}\end{matrix} & \mathbf{0} \\
\begin{matrix}\alpha_N(M_{zT}\mathbf{N}_{w,xx}^T\mathbf{N}_{u,x} \\ +w_{o,x}N_T\mathbf{N}_{w,x}^T\mathbf{N}_{u,x})\end{matrix} & \begin{matrix}B_{yzT}\mathbf{N}_{w,xx}^T\mathbf{N}_{v,xx} \\ +v_{o,x}M_{zT}\mathbf{N}_{w,xx}^T\mathbf{N}_{v,x} \\ +w_{o,x}M_{yT}\mathbf{N}_{w,x}^T\mathbf{N}_{v,xx} \\ +v_{o,x}w_{o,x}N_T\mathbf{N}_{w,x}^T\mathbf{N}_{v,x}\end{matrix} & \begin{matrix}B_{zT}\mathbf{N}_{w,xx}^T\mathbf{N}_{w,xx} \\ +w_{o,x}M_{zT}(\mathbf{N}_{w,xx}^T\mathbf{N}_{w,x} \\ +\mathbf{N}_{w,x}^T\mathbf{N}_{w,xx}) \\ +w_{o,x}^2N_T\mathbf{N}_{w,x}^T\mathbf{N}_{w,x}\end{matrix} & \mathbf{0} \\
\mathbf{0} & \mathbf{0} & \mathbf{0} & GI_T\mathbf{N}_{\theta,x}^T\mathbf{N}_{\theta,x}
\end{bmatrix}dx \tag{45}$$

$$\bar{\mathbf{k}}_{bh}^{M} = \left(\bar{\mathbf{k}}_{hb}^{M}\right)^{T} = \int_{L_0} \begin{bmatrix} \alpha_N N_T \mathbf{N}_{u,x}^T \hat{\mathbf{N}}_{u,x} \\ \alpha_N M_{yT} \mathbf{N}_{v,xx}^T \hat{\mathbf{N}}_{u,x} \\ \alpha_N M_{zT} \mathbf{N}_{w,xx}^T \hat{\mathbf{N}}_{u,x} \\ \mathbf{0} \end{bmatrix} dx \tag{46}$$

$$\bar{\mathbf{k}}_{b\lambda}^{M} = \left(\bar{\mathbf{k}}_{\lambda b}^{M}\right)^{T} = \int_{L_0} \begin{bmatrix} \alpha_N \mathbf{N}_{u,x}^T \mathbf{P}_N \\ \mathbf{0} \\ \mathbf{0} \\ \mathbf{0} \end{bmatrix} dx \tag{47}$$

$$\bar{\mathbf{k}}_{hh}^{M} = \int_{L_0} \alpha_N N_T \hat{\mathbf{N}}_{u,x}^T \hat{\mathbf{N}}_{u,x}\, dx \tag{48}$$

$$\bar{\mathbf{k}}_{h\lambda}^{M} = \left(\bar{\mathbf{k}}_{\lambda h}^{M}\right)^{T} = \int_{L_0} \alpha_N \hat{\mathbf{N}}_{u,x}^T \mathbf{P}_N\, dx \tag{49}$$

$$\bar{\mathbf{k}}_{\lambda\lambda}^{M} = \int_{L_0} \frac{1}{N_T}(\alpha_N - 1)\mathbf{P}_N^T \mathbf{P}_N\, dx \tag{50}$$

It should be noted that the geometric stiffness $\bar{\mathbf{k}}_{bb}^{G}$ is calculated by use of an effective axial force resultant (see [17])

$$\bar{N}_{\text{eff}} = \bar{N} + Q_e - Q_i \tag{51}$$

It is assumed that the element may be subjected an external hydrostatic pressure; it may also be a pipe with internal fluid pressure as well. Q_e and Q_i are then the axial force resultants due to external and internal hydrostatic pressure, while $\bar{N}$ is defined in (33). The tangential cross-sectional moduli depend directly on the material law C_T, and they are given by Eq. (22) and

$$\begin{aligned} M_{yT} &= -\textstyle\int_{A_0} y C_T\, dA_0 \\ M_{zT} &= -\textstyle\int_{A_0} z C_T\, dA_0 \end{aligned} \qquad \begin{aligned} B_{yT} &= \textstyle\int_{A_0} y^2 C_T\, dA_0 \\ B_{zT} &= -\textstyle\int_{A_0} z^2 C_T\, dA_0 \\ B_{yzT} &= -\textstyle\int_{A_0} yz C_T\, dA_0 \end{aligned} \tag{52}$$

The off-diagonal terms M_{yT}, M_{zT} and B_{yzT} vanish provided C_T is constant over the cross-section, the x-axis goes through the centroid and y and z follow the principal axes. Hence, it is seen from (45) that there is no coupling between axial, bending and torsional material stiffness terms in such cases.

The reason for introducing the parameter α_N, see (25) and (26), should now be evident, since it ensures that the material stiffness terms coupled to axial and bending deformations always will be of the same order of magnitude, and, thus, numerical illconditioning is avoided. However, a stable solution scheme requires that the undetermined parameters $\boldsymbol{\lambda}_N$ must be eliminated after the elimination of the displacement variables v within each discrete element. A more detailed discussion of the numerical aspects and implementation is given in [17].

An incremental linearized form of the equation of motion subject to the nonlinear constraints (39) is readily obtained by equilibrating the increments of inertia and external forces with the incremental internal forces (41)

$$\bar{\mathbf{m}}\Delta\ddot{\bar{\mathbf{v}}} + \bar{\mathbf{k}}_T \Delta\bar{\mathbf{v}} = \Delta\bar{\mathbf{S}} \tag{53}$$

where

$$\Delta\bar{\mathbf{S}} = \bar{\mathbf{S}}^E - \bar{\mathbf{S}}^S - \bar{\mathbf{m}}\ddot{\bar{\mathbf{v}}} \tag{54}$$

The unknowns are sorted as shown in (29), hence

$$\bar{\mathbf{m}} = \begin{bmatrix} \mathbf{m} & \mathbf{0} & \mathbf{0} \\ \mathbf{0} & \mathbf{0} & \mathbf{0} \\ \mathbf{0} & \mathbf{0} & \mathbf{0} \end{bmatrix}, \quad \bar{\mathbf{S}}^E = \begin{bmatrix} \mathbf{S}^E \\ \mathbf{0} \\ \mathbf{0} \end{bmatrix} \quad \text{and} \quad \ddot{\bar{\mathbf{v}}} = \begin{bmatrix} \ddot{\mathbf{v}} \\ \mathbf{0} \\ \mathbf{0} \end{bmatrix} \tag{55}$$

It should be noted that all quantities (element forces, masses and stiffnesses) are referred to the co-rotated 'ghost' reference C_{0n}.

3 Solution of the Dynamic Equations

The equations of motion may be obtained through the principle of virtual work. This is combined with standard finite element assumptions of interpolation of displacements between nodal points, leading to the semidiscrete global equations of motion

$$\mathbf{R}^I + \mathbf{R}^D + \mathbf{R}^S = \mathbf{R}^E \tag{56}$$

subject to a set of nonlinear holonomic constraints

$$\mathbf{C}(\mathbf{r}, t) = \mathbf{0} \tag{57}$$

$\mathbf{R}^I$ is the inertia force vector, $\mathbf{R}^D$ the damping force vector, $\mathbf{R}^S$ the internal force vector and $\mathbf{R}^E$ the external force vector that consists of conservative $\mathbf{R}^C$ as well as nonconservative forces $\mathbf{R}^{NC}$. $\mathbf{C}$ is the vector of linearly independent constraints. The corresponding incremental form is obtained as

$$\begin{bmatrix} \mathbf{M} \\ \mathbf{0} \end{bmatrix}_n \Delta\ddot{\mathbf{r}}_{n+1} + \begin{bmatrix} \mathbf{C}_T \\ \mathbf{0} \end{bmatrix}_n \Delta\dot{\mathbf{r}}_{n+1} + \begin{bmatrix} \mathbf{K}_T & \mathbf{B}^T \\ \mathbf{B} & \mathbf{0} \end{bmatrix}_n \begin{bmatrix} \Delta\mathbf{r} \\ \Delta\boldsymbol{\lambda} \end{bmatrix}_{n+1} = \begin{bmatrix} \mathbf{R}^R_{n+1} - \mathbf{B}^T_n \boldsymbol{\lambda}_n \\ \boldsymbol{\gamma}^R_{n+1} \end{bmatrix} \tag{58}$$

where $\mathbf{R}^R$ and $\boldsymbol{\gamma}^R$ represents residual or out-of-balance forces of the equilibrium equations and the nonlinear constraints, respectively. $\Delta\mathbf{r}$ is the incremental nodal displacements and superposed dots signifies differentiation with respect to time. $\Delta\boldsymbol{\lambda}$ corresponds to an increment of the Lagrangian multipliers and $\mathbf{B}$ are the gradients of the constraints (57). $\mathbf{M}$ and $\mathbf{C}_T$ is the assembled mass and damping matrices which in general are time dependent due to the hydrodynamic mass and damping contributions.

The incremental equations (58) may be written on the form [17]

$$\begin{bmatrix} \hat{\mathbf{K}}_n & \mathbf{B}^T_n \\ \mathbf{B}_n & \mathbf{0} \end{bmatrix} \begin{bmatrix} \Delta\bar{\mathbf{r}}_{n+1} \\ \Delta\bar{\boldsymbol{\lambda}}_{n+1} \end{bmatrix} = \begin{bmatrix} \Delta\hat{\mathbf{R}}_{n+1} - \mathbf{B}^T_n \bar{\boldsymbol{\lambda}}_n \\ \Delta\boldsymbol{\gamma}_{n+1} \end{bmatrix} \tag{59}$$

$\hat{\mathbf{K}}_n$ is the effective stiffness matrix, which, depending on the specific time integrator used, consists of some form of linear combination of $\mathbf{M}$, $\mathbf{C}_T$, $\mathbf{K}_T$. Note that the effective stiffness $\hat{\mathbf{K}}_n$ in this way depends upon the time step, and that it may lack symmetry due to a nonsymmetric load correction stiffness contribution

$$\mathbf{K}^{NC} = -\frac{\partial \mathbf{R}^{NC}}{\partial \mathbf{r}} \tag{60}$$

$\Delta\hat{\mathbf{R}}_{n+1}$ is the associated effective incremental load vector that accounts for the unbalance at time step n in addition to the increment of the external loads for time step $(n+1)$. The nonlinear dynamic equations (56) may be solved in an incremental-iterative fashion utilizing some form of Newton-Raphson iteration procedure.

A thorough description of the implementation of the Hilber, Hughes and Taylor's α-method [31] is given in references [17].

4 Numerical Studies

The methods described herein have been implemented in the general purpose finite element program FENRIS (Finite Element NonlineaR Integrated System). The capabilities of this program are outlines in references [32, 33, 34] and the post-processing has been performed employing the program SELPLOT [35].

4.1 Large Deflection of a Cantilever Beam Subjected to an End Moment

The present example studies the significance of higher order axial interpolation polynomials with respect to alternative simplifications made for the strain measure, E_{xx}. This example has been studied by many investigators, e.g. Remseth [12] and Søreide [20]. Both investigators reported a considerable gain in accuracy by including higher order axial displacement interpolation polynomials in the beam element formulation. However, their studies were based on the Total and Updated Lagrangian formulation, while the present investigation is based on the Co-rotational formulation. have been used.

The exact solution for the end displacement of an elastic cantilever with a concentrated end moment, as shown in Fig. 2, may be written on the form

$$\frac{u}{L} = 1 - \frac{\sin\theta}{\theta} \tag{61}$$

$$\frac{w}{L} = \frac{1-\cos\theta}{\theta} \tag{62}$$

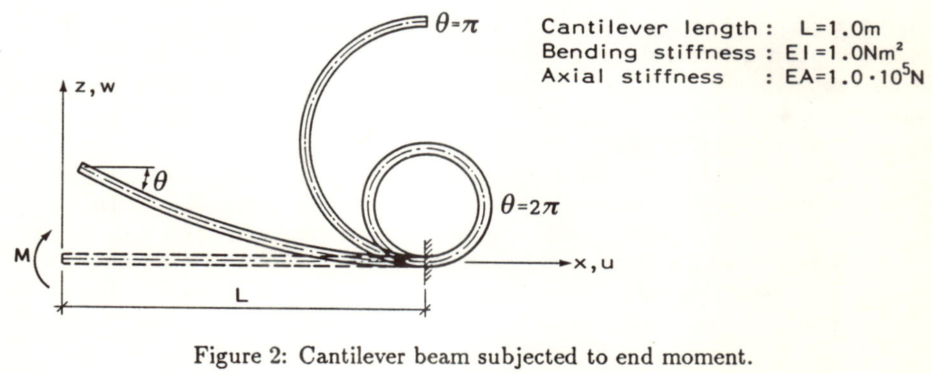

Figure 2: Cantilever beam subjected to end moment.

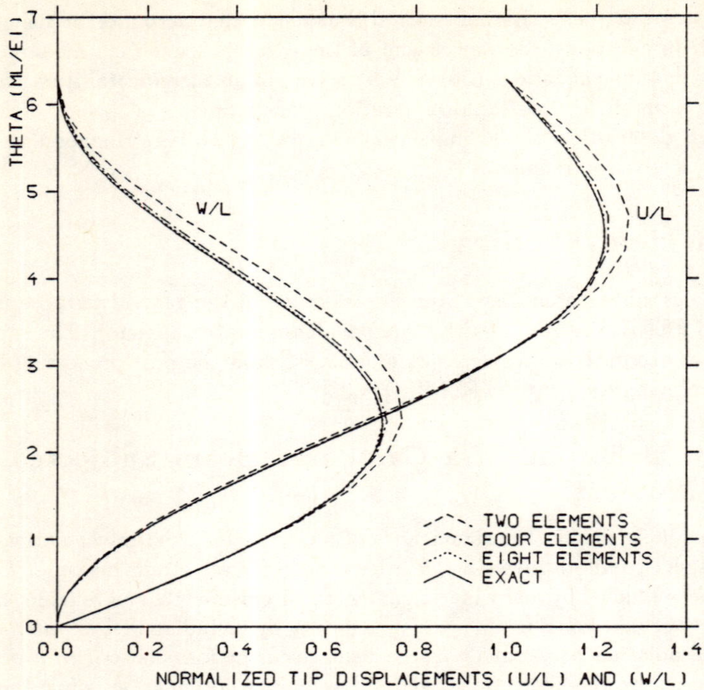

Figure 3: Load deflection curves for cantilever beam subjected to end moment. Linear axial displacements and linear strains.

u and w are the horizontal and vertical tip deflections, respectively. The end slope θ of the cantilever depends on the end moment M as follows

$$\theta = \frac{ML}{EI} \tag{63}$$

L is the length of the undeformed cantilever and EI is the cross-sectional bending stiffness.

The solution is obtained using prescribed load increments, $\Delta\theta = \frac{\Delta ML}{EI} = \frac{\pi}{8}$, i.e. 16 increments are required to bend the cantilever 360 degrees. The problem is analyzed using true Newton-Raphson (NR) equilibrium iterations and with an energy norm $\epsilon_E = 1.0 \times 10^{-4}$ as convergence criterion.

Results in terms of normalized tip deflections $\frac{u}{L}$ and $\frac{w}{L}$ against the tip slope θ for the cantilever modeled by 2, 4 and 8 elements are shown in Fig. 3 and Fig. 4. The curves of Fig. 3 show results for linear axial displacement modeling and linear strains, while Fig. 4 displays results for cubic axial displacement interpolation and nonlinear strain expressions. Linear, quadratic, cubic as well as quintic axial displacement interpolation have been compared, and both linear

$$E_{xx} = u_{,x} \tag{64}$$

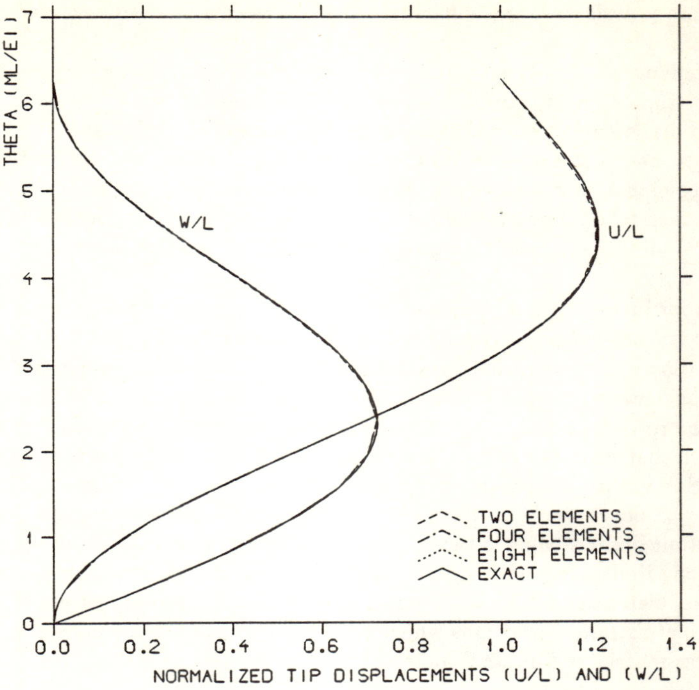

Figure 4: Load deflection curves for cantilever beam subjected to end moment. Cubic axial displacements and nonlinear strains.

and nonlinear strain expressions

$$E_{xx} = u_{,x} + \frac{1}{2}(v_{,x}^2 + w_{,x}^2) \tag{65}$$

It should be noted that the square of the (in-plane) axial displacement gradients, $u_{,x}$, has been neglected in the expression for the nonlinear strain of (65) in the CL-formulation, due to self-straining effects.

For the case of linear strains (64), the results obtained and the corresponding number of NR-iterations per step are exactly the same for all cases analyzed. The use of two, three, four and five Gauss points for integration along the length of the beam are compared for the different element interpolations and meshes. For the case of linear strains, two Gauss points is sufficient for the linear and quadratic interpolation while three and five points have to be used for the cubic and quintic interpolation. This is necessary in order to avoid singularity in the condensation of the axial displacement degrees of freedom at the internal nodal points.

For the case with nonlinear strain expression (65), both linear and quadratic interpolation elements diverge, while the elements with cubic and quintic interpolation give results which are not significantly different. The divergence of the linear and cubic elements is primarily due to self-straining. For the element with cubic interpolation, the highest order polynomial terms in the numerical integration in the length direction are of the order eight. Hence, the integrations in longitudinal direction should be carried out by means of five Gauss points to be exact. However, the numerical studies show that five and even four points lead to divergence for the coarsest mesh which contains two elements. The use of three Gauss points improves the performance of the element considerably and convergence is thereby obtained for all meshes. For the element with four internal nodes, the best performance is obtained using five Gauss points, i.e. the same as for the case with linear strains.

The reason for including axial displacements at four internal nodal points is to reproduce the quadratic strain terms contributed by the slopes $v_{,x}^2$ and $w_{,x}^2$ in accordance with the expression (65), and thereby artificial straining in the longitudinal direction is avoided. However, for the present example it does not pay to include more than two internal nodal points, since the results for the cubic and quintic elements are nearly indistinguishable.

Fig. 4 shows that even two elements with cubic interpolation polynomials are highly accurate and give results comparable with those obtained by eight linear axial interpolation elements with linear strain measure. On the other hand, the convergence rate for the elements based on linear strains is twice as good as for the elements based on nonlinear strain expression. Hence, the difference in computing time may be marginal, since the cubic interpolation elements include static condensation of the internal degrees of freedoms as well as complete strain expressions which yield more computations in establishing the element stiffness matrix and internal force vector.

It is noted that constant, linear, quadratic, cubic and quartic interpolation of the axial force, $\tilde{N}$ has also been compared. For this example, it is found that the choice of interpolation for $\tilde{N}$ is of little importance since it does not influence the convergence rate nor the results. However, this example involves bending moments only, and, hence, the axial force should in effect be zero when integrating over the element length.

4.2 Analysis of a Steep Wave Flexible Riser

4.2.1 Problem Definition

A flexible, steep wave, single riser system is studied in order to investigate the capabilities of the present methods in simulating the behavior of highly flexible systems subjected to environmental loads. This test problem includes realistic parameters for examining the various solution techniques and the different formulations for parameters relevant to riser analyses.

The upper point of the flexible riser is connected to a floating barge and it follows the motion of the platform when exposed to current and waves. The other end point is fixed 2 meters above the sea floor. Both end points are considered to be clamped such that the longitudinal direction of the riser is perpendicular to the horizontal plane (the water plane). The riser system is placed at 45 meters water depth (MWL) and the environmental conditions are typical of those of the southern part of the North Sea.

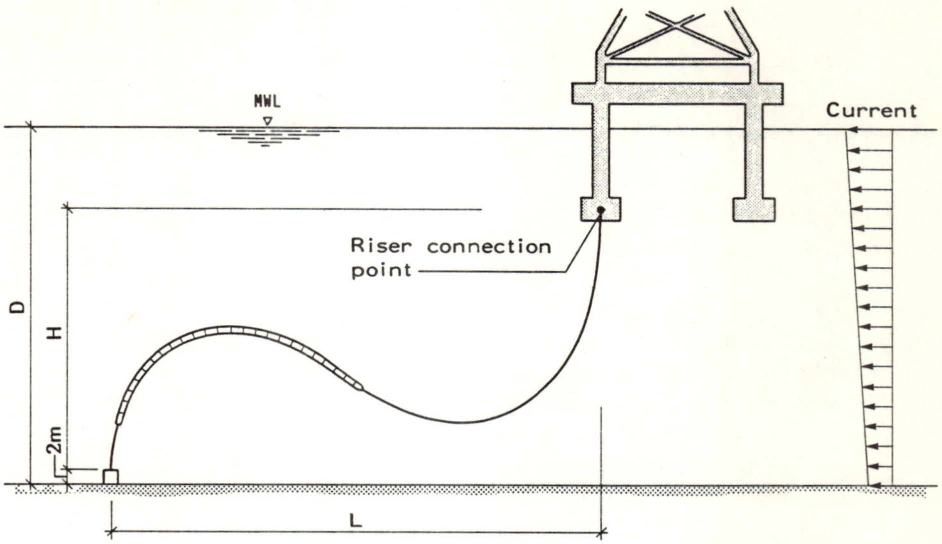

Figure 5: Structural system for a flexible steep wave single riser system.

Additional buoyancy is considered for the lower part of the riser in order to keep the riser clear of the sea bed. These buoyancy modules are clamped to the outside of the riser; in the computational model they are represented by equivalent, discrete buoyancy modules attached to the corresponding nodes. Fig. 6 gives geometric and material data of the riser and the finite element model is shown for a straight resting position. The additional buoyancy volumes are attached to nodes 5 through 30, resulting in a total of 26 buoyancy volumes. Note that the added mass coefficients considered for the buoyancy section account for the difference in outer diameter.

The riser considered has conically shaped stiffeners at each end. The stiffeners extend 3 meters from each end point of the riser. In order to account for the additional stiffness the two first and the two last beam elements are prescribed with bending, torsional and axial stiffness corresponding to approximately 5 (for element 2 and 58) and 10 (for element 1 and 59) times the stiffness of the riser itself.

It is noted that the finite element model shown in Fig. 6 represents a very fine spatial discretization. The reason for using such a fine model is that the comparisons which are made regarding time steps and solution strategy, should not be influenced by inaccuracies in the spatial discretization.

A feasibility study of a riser system requires that both environmental and functional loads are considered; hence, various combinations of water levels and barge drafts must be studied for different wave and current directions. Functional loads on both intact and damaged mooring systems is investigated as well. The results are subject to be checking against requirements set by approving authorities and the manufacturer.

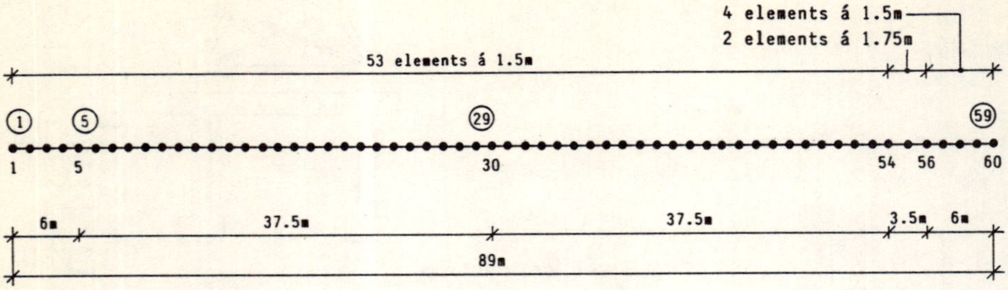

Environmental data :

Mass density of water	: $\rho_e = 1025\,kg/m^3$
Mass density of internal fluid	: $\rho_i = 870\,kg/m^3$
Overpressure at internal fluid surface	: $p_f = 29.6 \cdot 10^6\,N/m^2$
Distance to internal fluid surface	: $h_f = 0$
Water depth (mean water level)	: $D = 45\,m$
Acceleration of gravity	: $g = 9.81\,m/s^2$

Flexible riser:

Total length	: $L_r = 89\,m$
Bending stiffness	: $EI = 2.935 \cdot 10^4\,Nm^2$
Torsional stiffness	: $EI_T = 4.815 \cdot 10^6\,Nm^2$
Axial stiffness	: $EA = 5.260 \cdot 10^8\,N$
Outer cross-sectional area	: $A_e = 0.06592\,m^2$
Internal cross-sectional area	: $A_i = 0.03243\,m^2$
Mass per unit length (in air)	: $m = 154.6\,kg/m$
Drag products :	
axial	: $C_{Dx}D = 0.029\,m$
lateral	: $C_{Dy}D = C_{Dz}D = 0.23\,m$
Added mass coefficients :	
axial	: $C_{mx} = 0.01$
lateral	: $C_{my} = C_{mz} = 1.0$

Buoyancy section:

Additional mass per node	: $m_B = 125.0\,kg$
Additional buoyancy per node	: $f_B = 3.968\,kN$
Drag products:	
axial	: $C_{Dx}D = 0.092\,m$
lateral	: $C_{Dy}D = C_{Dz}D = 0.43\,m$
Added mass coefficients:	
axial	: $C_{mx} = 1.18$
lateral	: $C_{my} = C_{mz} = 5.15$

Figure 6: Computational model for a steep wave, flexible riser.

			Wave			*Barge Motion*					
						Longitudinal			*Vertical*		
Case	H^1	L^2	a^3	θ^4	T^5	a^3	θ^4	T^5	a^3	θ^4	T^5
1	32.5	54.9	7.8	0	12,6	4.0	90	12.6	8.2	71	12.6
2	32.5	65.5	10.4	0	14.5	7.4	90	14.5	11.4	63	14.5
3	32.5	60.2	9.2	0	13.6	5.8	90	13.6	9.9	67	13.6

Table 1: Wave and floater characteristics.

The following results are of vital importance and need to be checked for a flexible riser system:

- bending radius must not exceed a minimum bending radius
- tension must not exceed a maximum tension
- torsional moment must not exceed a maximum torsional moment
- compressive longitudinal forces are not allowed
- there must be no contact between riser and sea bed

Fatigue of the riser should be considered as well.

In the present study the following three design conditions are considered:

Case 1: The floating barge is in a "near" position with the wave direction parallel with the riser plane and current flow parallel with wave direction.

Case 2: The floating barge is in a "far" position with the wave direction parallel with the riser plane and current flow parallel with wave direction.

Case 3: The floating barge in "middle" position with the wave direction perpendicular to the riser plane and current flow parallel with wave direction.

The three positions considered for the floating barge are studied along with second order effects due to wave drift forces (including mean and slowly varying drift forces), current forces and wind forces on the barge. The three cases are analyzed with wave and barge characteristics according to Table 1. The values considered for H and L, see Table 1, thus represent extreme quasi-static excursions from the mean position without wind, current and wave loading. The velocity profile of the current is given in Table 2. The design conditions are assumed to be a combination of a 100-year wave and a 1-year current.

In this example the concept of effective tension is used to model buoyancy and internal fluid forces for both the static and the dynamic analyses [17]. Hydrodynamic forces due to waves and current are modeled by the modified Morison formula, see [36].

[1] H : Vertical distance between upper and lower riser connection points, see Fig. 5.

[2] L : Horizontal distance between upper and lower riser connection points, see Fig. 5.

[3] a : Amplitude in meters.

[4] θ : Phase angle in degrees.

[5] T : Period in seconds.

Water depth (m)	0	45
Current (m/s)	1.18	0.72

Table 2: Velocity profile of current.

4.2.2 Static Equilibrium Configuration

The static equilibrium configuration of the riser is established by starting a simulation from an initially straight horizontal position. It is noted that other initial configurations may also be considered, see reference [37]. Based on the experience from the previous study [37], the static analysis is carried out in four stages.

The first stage of analysis consists of one increment with application of buoyancy, internal fluid and gravitational forces. The bending stiffness is scaled by a factor 10^6 for the first increment in which the end points of the riser is fixed in their initial position.

In the second stage of analysis a displacement history is prescribed for the right end of the riser, causing it to move to its resting position at the same time as the two end points are gradually rotated to a vertical position. A total of 5 equal steps are employed during this stage at the same time as the scaling factor for the bending stiffness is reduced to a value of 10^5.

During the third stage the scaling factor for the bending stiffness is step-wise reduced from 10^5 to 10^4, 10^3, 10^2, 10 and, finally, the correct value. The last value corresponds to a scaling factor equal to one. The end points are kept fixed during this stage.

During the final stage current forces are applied during one increment, see Table 2.

A total of 42 Newton-Raphson equilibrium iterations are required for the 12 incremental steps to get the final resting position. This yields an average of 3.5 iterations per increment. It is noted that none of the increments require more than 5 equilibrium iterations.

It is noted that a relatively strict convergence criterion is needed in that the bending stiffness vary with a factor of 10^6. When applying an energy criterion, see [17], this corresponds to a reduction of the energy norm, ϵ_E, with a factor of approximately 10^{-3} in order to obtain a sufficient accuracy throughout all stages of the analysis. $\epsilon_E = 1.0 \times 10^{-5}$ gives good accuracy. However, setting $\epsilon_E = 1.0 \times 10^{-6}$ yields only one or two more iterations per increment, and this in fact represents a safer choice for most practical purposes. It is also noted that this rather strict convergence criterion may be relaxed to $\epsilon_E = 1.0 \times 10^{-3}$ for subsequent dynamic analyses. Also note that an displacement criterion (see e.g. reference [38]) should not be used because it needs to be tightened to approximately 1.0×10^{-9} in order to give the same accuracy as obtained above with the energy criterion. However, this causes twice as many iterations at some intermediate positions.

The static analysis is carried out with the mixed beam element with linear axial displacement interpolation and constant axial force (one Lagrangian multiplier per element). The condition number for the system in its final resting position is of order 10^8. It is noted that the standard displacement beam element may also be used for this case. However, severe oscillations occur for the axial forces, and, thus, more iterations are required in order to converge to the same amount of accuracy as for the mixed beam element solution.

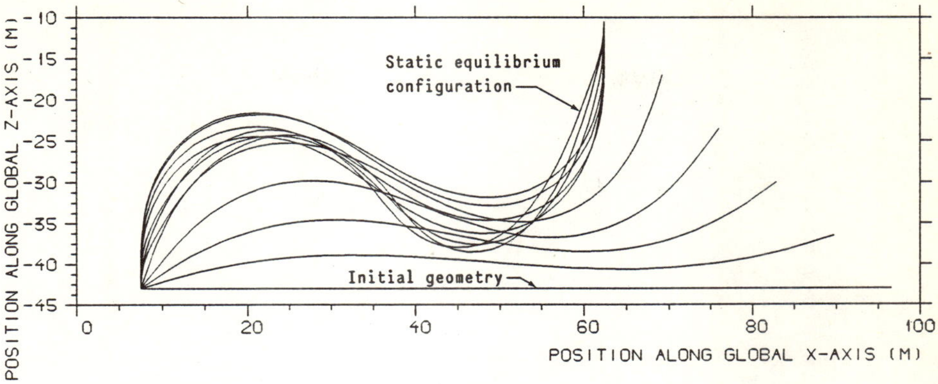

Figure 7: Sequence of deformed configurations obtained with the bending stiffening approach starting from a horizontal position.

Fig. 7 shows the initial and final positions as well as the 10 intermediate positions obtained when the bending stiffening approach is applied to case 1, with the floating barge in the "near" position.

4.2.3 Nonlinear Dynamic Analyses in Regular Waves

Dynamic analysis of the three cases shown in Table 1 is performed starting from the static equilibrium configuration with applied loads from current.

In order to find a suitable time step a comparison of four different time step sizes is first carried out for case 1 using lumped mass matrices.

The dynamic loads are defined in terms of the data for the regular wave and the floater motion shown in Table 1. In case 1 the wave load is acting in the XZ-plane (the riser plane). The wave loads are applied smoothly during two wave periods by the use of a ramp function. The HHT α-method with constant average acceleration and $\alpha_H = -0.05$ is used for this simulation. Rayleigh proportional damping with $\alpha_1 = 0.01$ and $\alpha_2 = 0.05$, is also introduced.

Fig. 8a shows time history plots of the horizontal displacement of node 30 for the total simulation time of 63 seconds, corresponding to a total of five wave cycles. It is seen that a near steady state response is obtained within the fourth wave cycle. Examining the displacements only, no numerical instability is observed. This applies even for the case of the largest time step, which diverges after 57 seconds. It is therefore noted that in order to study the instability characteristics of an integrator, both displacements and forces should be scrutinized. Fig. 8b shows the effective axial tension in element 30 during the fourth and the fifth wave cycle. This clearly demonstrates that the two largest time step solutions of $\Delta t = 0.5$ second and $\Delta t = 1.0$ second lead to instable response for the internal forces.

Fig. 9 shows the required number of Newton-Raphson equilibrium iterations for the various time steps corresponding to a convergence criterion $\epsilon_E = 1.0 \times 10^{-4}$. It is seen that the case of the largest time step does not converge within the prescribed maximum of 20 iterations per incremental step. In contrast, the two smallest time step solutions converge

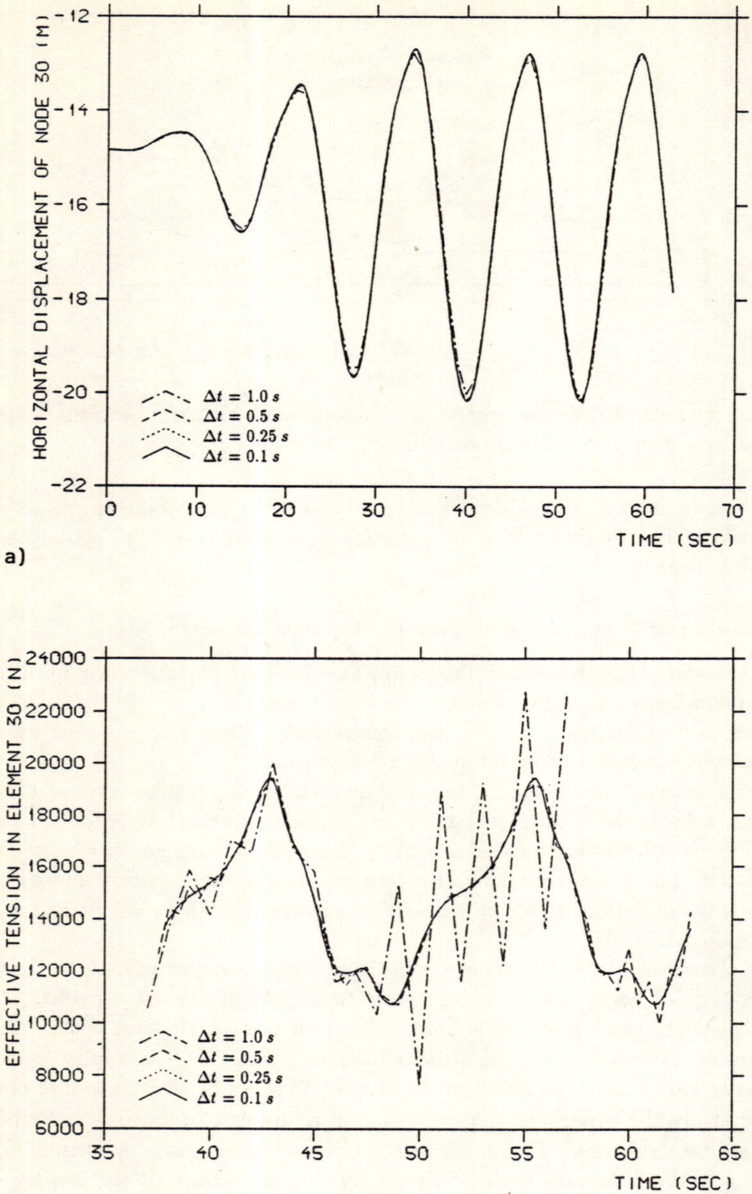

Figure 8: Time history plots from a dynamic analysis with a regular wave corresponding to case 1: a) Horizontal displacement of node 30, and, b) Effective axial tension in element 30.

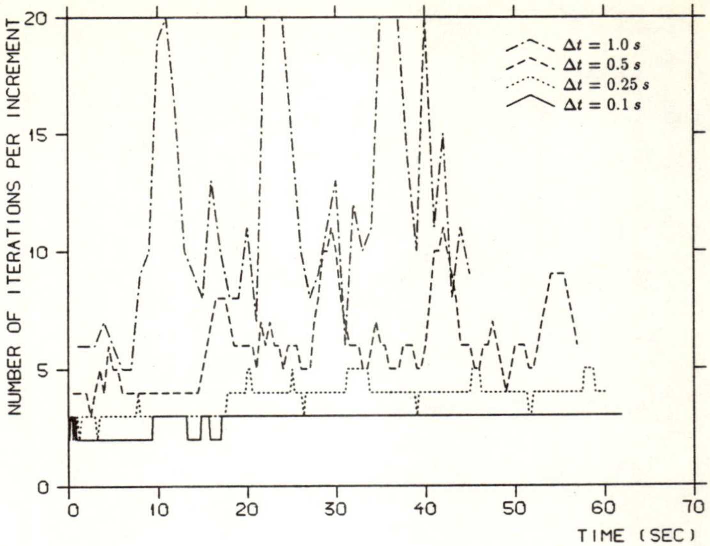

Figure 9: Number of required equilibrium iterations in dependence of the time step.

within a maximum of 5 and 3 iterations, respectively. It is seen that there is virtually no difference between the solutions corresponding to $\Delta t = 0.1$ second and $\Delta t = 0.25$ second. The CPU-time[1] required for the two solutions are 3100 and 1800 seconds, respectively; thus, in terms of computational efficiency, a time step size of $\Delta t = 0.25$ second is preferable for the present case.

In reference [17], the influence of mass representation for a fixed time step size of $\Delta t = 0.25$ second is investigated. The study shows that the mass representation has a relatively small influence on the displacements, however, it is found that the lumped mass representation does not capture the peak values for the effective axial tension. For this reason the consistent mass is adopted for the rest of this particular example.

Simulations of the regular wave cases according to Table 1 are carried out with a time step of $\Delta t = 0.25$ second and with consistent mass representation for a total simulation time of five wave cycles. For case 1 and 2 the loads are acting in the riser plane, thus yielding only in-plane curvatures and forces and no torsional moment. The riser model is in fact fixed against motion in global Y-direction and against rotation about global X- and Z-axis.

Fig. 10a shows a sequence of deformed shapes throughout the fifth wave cycle of case 1. The corresponding results obtained for case 2 are shown in Fig. 10b.

In contrast to case 1 and 2, case 3 considers waves and current coming in at 90° to the riser plane. However, the waves are still parallel to the prescribed motion of the floating barge. Figs. 11a and 11b display a sequence of deformed shapes projected onto the XZ-

[1]The computer time corresponds to a DECstation 3100 configuration.

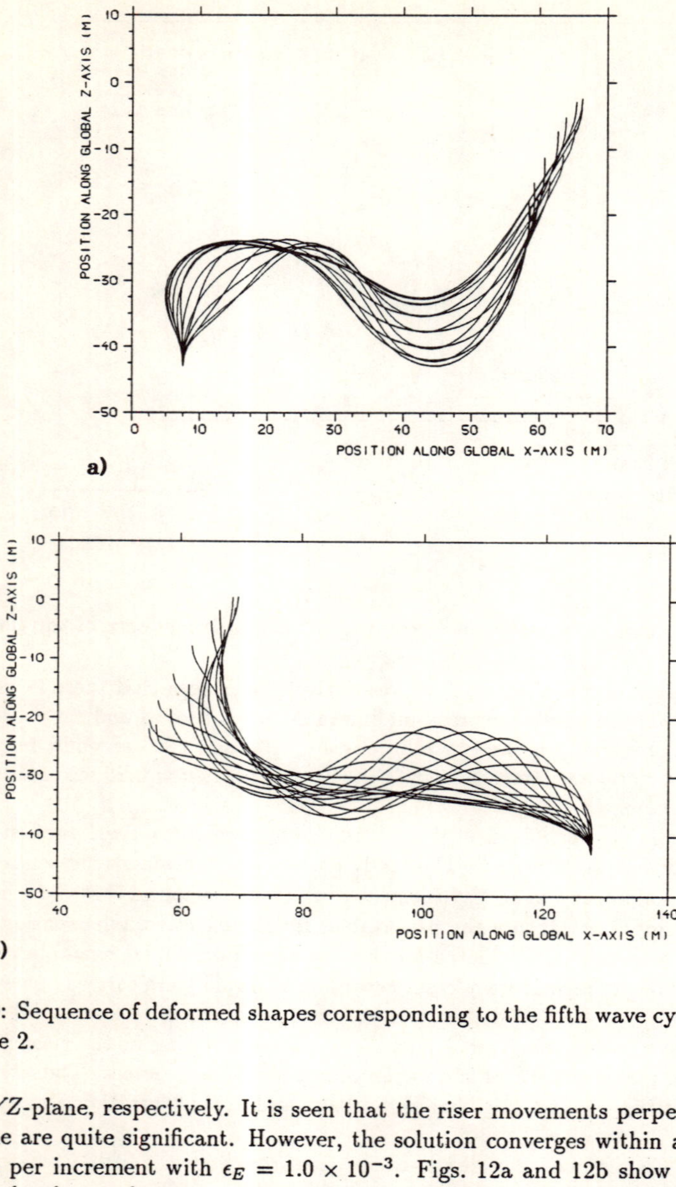

Figure 10: Sequence of deformed shapes corresponding to the fifth wave cycle for: a) Case 1, b) Case 2.

and the YZ-plane, respectively. It is seen that the riser movements perpendicular to the riser plane are quite significant. However, the solution converges within an average of 3 iterations per increment with $\epsilon_E = 1.0 \times 10^{-3}$. Figs. 12a and 12b show the curvatures about the local y- and z-axes, respectively. The envelope of the principal curvatures are shown in Fig. 12c. Finally, Figs. 12d, 12e and 12f show envelopes of the internal axial force, the effective axial tension and the torsional moment, respectively.

The results presented correspond to the fifth wave cycle and only envelopes for elements 3 through 57 are shown.

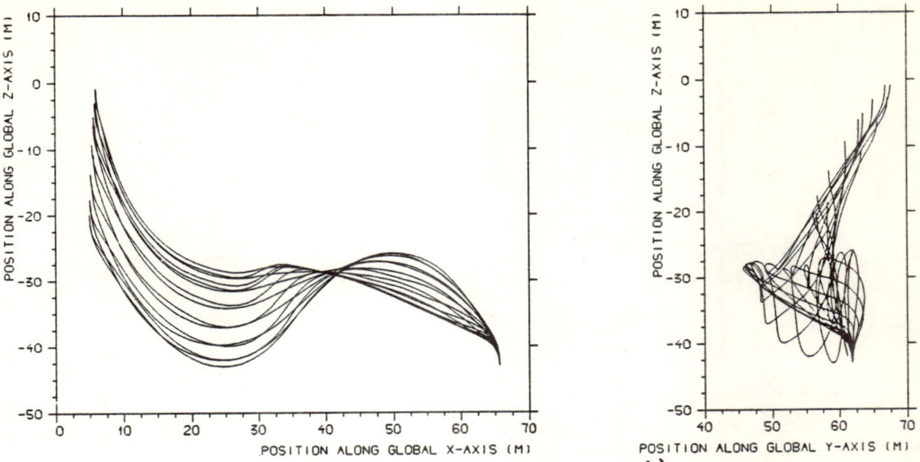

a) b)

Figure 11: Sequence of deformed shapes corresponding to the fifth wave cycle for case 3: a) XZ-projection, and, b) YZ-projection.

No compression is observed anywhere in the riser for any of the cases. A minimum clearance between the riser and the sea bed occurs for case 1, this distance is approximately 2 meters. The maximum tension, the maximum torsional moment and the maximum radius of curvature may be taken from the envelope curves and compared with the actual design criterias.

5 Conclusions

The main conclusions from the present study may be summarized as follows:

- The current mixed beam element is significantly more reliable and robust than the standard displacement formulation for slender and very slender structures. Even though the numerical formulation of the standard beam element is about 25% cheaper than the mixed beam element the mixed element is preferable for slender structures because of faster convergence and reliability in obtaining convergence.

- Inclusion of higher order axial interpolation polynomials seems to be important when a crude finite element model is used in connection with problems that display relatively large local deformations, see Section 4.1. However, the degree of the polynomial should not exceed third degree when considering accuracy against computer costs.

- A three point Gaussian integration scheme is recommended for mixed beam elements in which the axial displacement is expressed with a polynomial of degree three or less.

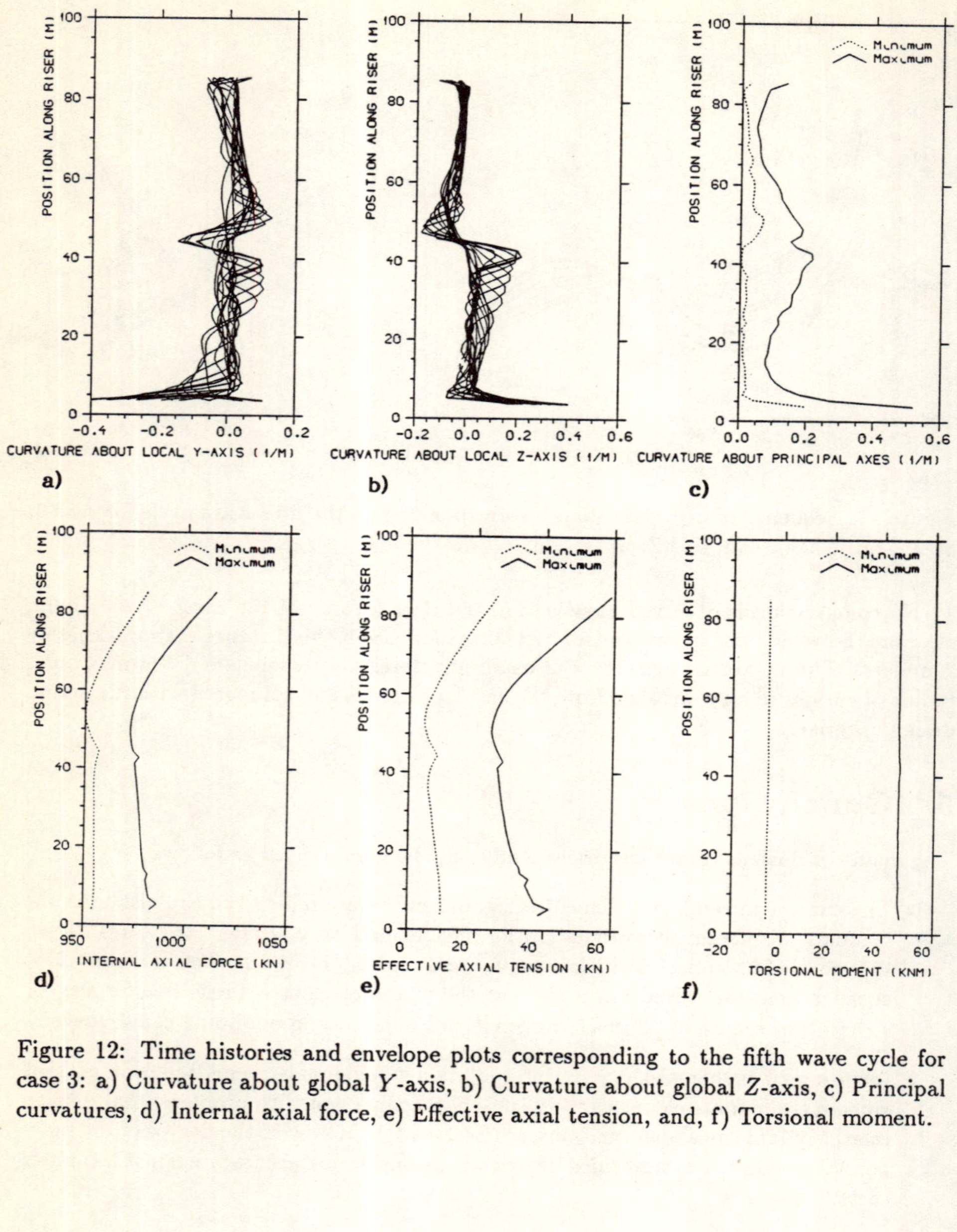

Figure 12: Time histories and envelope plots corresponding to the fifth wave cycle for case 3: a) Curvature about global Y-axis, b) Curvature about global Z-axis, c) Principal curvatures, d) Internal axial force, e) Effective axial tension, and, f) Torsional moment.

- The HHT α-method with controllable numerical damping has been used successfully for time integration of the equations of motion and it is strongly recommended. Even for problems that do not involve constraints this method is recommended. It is also well suited for use in connection with automatic time stepping algorithms [39]. Furthermore, a value of α_H between -0.05 and -0.1 is recommended for the type of problems studied herein.
- Both energy and force norms are well suited as termination criteria for nonlinear general purpose beam analysis. However, due to the natural scaling of translations and rotations the energy norm is in general preferred. The values $\epsilon_E = 1.0 \times 10^{-3}$ and $\epsilon_F = 5.0 \times 10^{-2}$ are recommended for most purposes. However, certain restrictions when using the bending-stiffening approach must be observed.
- The bending-stiffening approach has been found to be effective in finding the static equilibrium configuration of highly flexible beam type of structures, such as flexible risers. Moreover, the bending stiffness should be scaled such that the starting value of the bending stiffness corresponds to a slenderness ratio of about 50.
- The consistent mass representation has been found to be preferable, in particular for analysis of flexible risers (see Section 4.2). It is noted that when using the finite element generation scheme described in reference [17], the increase in computer costs associated with use of consistent mass is marginal.

Acknowledgement

The authors would like to thank their colleagues in the FENRIS project for developing parts of the computer program that was used. We also acknowledge the support, financial or otherwise from NTH, SINTEF, Det norske Veritas, NTNF, Conoco and Statoil.

References

[1] Hibbitt, H.D., Becker, E.B. and Taylor, L.M., 'Nonlinear Analysis of some Slender Pipelines', *Computer Methods in Applied Mechanics and Engineering*, Vol. 17/18, pp. 203–225, 1979.

[2] McNamara, J.F., O'Brien, P.J. and Gilroy, S.G., 'Nonlinear Analysis of Flexible Risers using Hybrid Finite Elements', *Proc. of the 5th Int. OMAE Symp.*, Vol. III, pp. 371–377, 1986.

[3] O'Brien, P.J. and McNamara, J.F., 'Analysis of Flexible Riser Systems Subjected to Three-Dimensional Seastate Loading', *Proc. of the Int. Conf. on Behaviour of Offshore Structures (BOSS), Trondheim, Norway June 1988*, Vol. 3, pp. 1373–1388, Tapir, Trondheim, 1988.

[4] Zienkiewicz, O.C. and Taylor, R.L., *The Finite Element Method*, McGraw–Hill, 4th ed., 1989.

[5] Shabana, A.A., *Dynamics of Multibody Systems*, Wiley, 1989.

[6] Roberson, R.E. and Schwertassek, R., *Dynamics of Multibody Systems*, Springer–Verlag, 1988.

[7] Argyris, J.H., 'An Excursion into Large Rotations', *Computer Methods in Applied Mechanics and Engineering*, Vol. 32, pp. 85–155, 1982.

[8] Rankin, C.C. and Brogan, F.A., 'An Element-independent Corotational Procedure for the Treatment of Large Rotations', in L.H. Sobel and K. Thomas (Eds.), *Collapse Analysis of Structures*, pp. 85–100, New York, ASME, 1984.

[9] Bergan, P.G. and Nygård, M.K., 'Nonlinear Shell Analysis Using Free Formulation Finite Elements', in P.G. Bergan, K.J. Bathe and W. Wunderlich (Eds.), *Finite Element Methods for Nonlinear Problems*, Springer–Verlag, 1985.

[10] Nygård, M.K., 'The Free Formulation for Nonlinear Finite Elements with Applications to Shells', *Report No. 86–2*, Division of Structural Mechanics, The Norwegian Institute of Technology, Trondheim, 1986.

[11] Malvern, L.E., *Introduction to the Mechanics of a Continuous Medium* , Prentice–Hall, Englewood Cliffs, N.J., 1969.

[12] Remseth, S.N., 'Nonlinear Static and Dynamic Analysis of Space Structures', *Report No. 78–2*, Division of Structural Mechanics, The Norwegian Institute of Technology, Trondheim, 1978.

[13] Bergan, P.G., *FENRIS notat 4: Tredimensjonalt Bjelkeelement*, Division of Structural Mechanics, The Norwegian Institute of Technology, Trondheim, 1980 (in Norwegian).

[14] Mattiasson, K., 'On the Co–Rotational Finite Element Formulation for Large Deformation Problems', *Publication 83:1*, Department of Structural Mechanics, Chalmers University of Technology, 1983.

[15] Belytschko, T. and Hsieh, B.J., 'Nonlinear Transient Finite Element Analysis with Convected Coordinates', *Int. J. for Numerical Methods in Engineering*, Vol. 7, pp. 255–271, 1973.

[16] Belytschko, T. and Glaum, L.W., 'Application of Higher Order Corotational Stretch Theories to Nonlinear Finite Element Analysis', *Computers and Structures*, Vol. 10, pp. 175–182, 1979.

[17] Mathisen, K.M., 'Large Displacement Analysis of Flexible and Rigid Systems Considering Displacement–Depending Loading and Nonlinear Constraints', *Doktor Ingeniøravhandling 1990–33*, Division of Structural Engineering, The Norwegian Institute of Technology, Trondheim, 1990.

[18] Mollestad, E., 'Techniques for Static and Dynamic Solution of Nonlinear Finite Element Problems', *Report No. 83–1*, Division of Structural Mechanics, The Norwegian Institute of Technology, Trondheim, 1983.

[19] Vlasov, V.Z., *Thin-Walled Elastic Beams*, Israel Program for Scientific Translations, Jerusalem, 1961.

[20] Søreide, T.H., 'Collapse Behaviour of Stiffened Plates Using Alternative Finite Element Formulations', *Report No. 77–3*, Division of Structural Mechanics, The Norwegian Institute of Technology, Trondheim, 1977.

[21] Mattiasson, K. and Samuelsson, A., 'Total and Updated Lagrangian Forms of the Co-rotational Finite Element Formulation in Geometrically and Materially Nonlinear Analysis', *Proc. of the Int. Conf. on Numerical Methods for Non-Linear Problems*, Vol. 2, Taylor et. al. (Eds.), Pineridge Press, Swansea, pp. 134–151, 1984.

[22] Mattiasson, K., Bengtsson, A. and Samuelsson, A., 'On the Accuracy and Efficiency of Numerical Algorithms for Geometrically Nonlinear Structural Analysis', in P.G. Bergan, K.J. Bathe and W. Wunderlich (Eds.), *Finite Element Methods for Nonlinear Problems*, Springer–Verlag, 1985.

[23] Von Karman, T., 'Festighetsprobleme in Maschinenbau', *Encyklopädie der Matematischen Wissenschaften*, Vol. 4, 1910.

[24] Engseth, A.G., 'Finite Element Collapse Analysis of Tubular Steel Offshore Structures', *Report UR-85–46*, Division of Marine Structures, The Norwegian Institute of Technology, Trondheim, 1985.

[25] Pian, T.H.H. and Tong, P., 'Basis of Finite Element Methods for Solid Continua', *Int. J. for Numerical Methods in Engineering*, Vol. 1, pp. 3–28, 1969.

[26] Pian, T.H.H. and Tong, P., 'Finite Element Methods in Continuum Mechanics', *in Advances in Applied Mechanics*, by C.S. Yih (Editor), Vol. 12, Academic Press, pp. 1–58, 1972.

[27] Bourgat, J.F., Dumay, J.M. and Glowinsky, R., 'Large Displacement Calculations of Flexible Pipe-Lines by Finite Element Method and Non-Linear Programming Methods', Computational Methods in Non-Linear Mechanics, *Proc. of the TICOM 2nd Int. Conf.*, Austin, Texas, March 1979, J.T. Oden (Ed.), North–Holland Publishing Company, 1980.

[28] Prathap, G., 'An Additional Stiffness Parameter Measure of Error of the Second Kind in the Finite Element Method', *Int. J. Numerical Methods in Engineering*, Vol. 21, pp. 1001–1012, 1985.

[29] Zienkiewicz, O.C., Qu, S., Taylor, R.L. and Nakazawa, S., 'The Patch Test for Mixed Formulations', *Int. J. for Numerical Methods in Engineering*, Vol. 23, pp. 1873–1883, 1986.

[30] Washizu, K, *Variational Methods in Elasticity and Plasticity*, Pergamon Press, 3rd ed., 1982.

[31] Hilber, H.M., Hughes, T.J.R. and Taylor, R.L., 'Improved Numerical Dissipation for Time Integration Algorithms in Structural Dynamics', *Earthquake Engineering and Structural Dynamics*, Vol. 5, pp. 283–292, 1977.

[32] FENRIS - Finite Element Nonlinear Integrated System, System Manual, Theory - Program Outline - Data Input, NTH–SINTEF–VERITAS, Veritas Sesam Systems A.S., Høvik, Norway, 1986.

[33] Bergan, P.G. and Arnesen, A., 'FENRIS – A General Purpose Nonlinear Finite Element Program', in *Proc. 4th Int. Conf. on Finite Element Systems*, Southampton, 1983.

[34] Bergan, P.G., 'Nonlinear Finite Element Analysis by use of FENRIS', in *Proc. 7th Int. Seminar on Computational Aspects of Finite Element Methods (CAFEM-7)*, Chicago, 1983.

[35] SELPLOT - Tailor Made Interactive Postprocessor for Results from SESAM Non-Linear Riser Package, User's Manual, A.S. Veritas Research, VR Report No. 87–2034, Høvik, Norway, 1987.

[36] Morison, J.R., O'Brian, M.P., Johnson, J.W. and Schaaf, S.A., 'The Forces exerted by Surface Waves on Piles', *Petroleum Trans.*, AIME, Vol. 189, pp. 149-154, 1950.

[37] Mathisen, K.M. and Bergan P.G., 'Nonlinear Static Analysis of Flexible Risers', *Proc. of the 5th Int. OMAE Symp.*, ASME, Tokyo, Vol. III, pp. 337–345, 1986.

[38] Bergan, P.G. and Clough, R.W., 'Convergence Criteria for Iterative Processes', *AIAA Journal*, Vol. 10, no. 8, pp. 1107–1108, 1972.

[39] Bergan, P.G. and Mollestad, E., 'An Automatic Time Stepping Algorithm for Dynamic Problems', *Computer Methods in Applied Mechanics and Engineering*, Vol. 49, pp. 299–318, 1985.

Nonlinear Finite Element Analysis of Composite and Sandwich Shell Structures

F.G.Rammerstorfer[1], K.Dorninger[2] and A.Starlinger[3]

[1]Institute of Lightweight Structures and Aerospace Engineering,
Vienna Technical University, Vienna, Austria

[2]Saturn Corp., Troy, MI, USA

[3]NASA Lewis Research Center, Structural Integrity Branch, Cleveland, OH, USA

ABSTRACT

Finite shell elements for geometrically and materially nonlinear stress and stability analyses of layered fiber reinforced composite shells as well as sandwich shells with orthotropic core material are presented. The element formulations are based on the degeneration principle including large displacements with an efficient analytical thickness integration. Examining suitable failure criteria and taking into account post-cracking, post-wrinkling and post-buckling stiffnesses allow for the investigation of progressive damage. Some examples illustrate that an extended range of composite specific problems can be covered by these laminated and sandwich elements.

1. INTRODUCTION

In designing composite shell structures the prediction of the limit load is crucial for fully utilizing the material capabilities. Either buckling - local or global - of the structure or the strength of the material (or both) limit the sustainable load of the shell. In this paper special emphasis will be put on investigating these two limiting factors, especially for layered composite and sandwich shells.

By using special shell finite elements developed for laminated fiber-reinforced composites (LFC-element [1]) and for sandwich shells (LM-element [2]) these limit load analyses including the post failure and post buckling regime can be conducted for arbitrary shell structures.

The onset of failure, particularly matrix or fiber cracking in fiber reinforced plastic (FRP) shells as well as wrinkling or intracell buckling of the face layers in sandwich shells, and the effective safety margin to overall collapse can be computed, which is of considerable practical relevance.

The complexity of the internal stress state typically found in layered composite materials and, in addition, the number of failure modes possible necessitate the introduction of some simplifications and assumptions in order to make post-failure analyses feasible.

2. LFC-ELEMENT FOR FRP

A detailed description of the theoretical background of the LFC-element along with a number of illustrative examples can be found in [1,3]. In this chapter only the outline of the element formulation is presented with special emphasis on the failure analysis modelling for FRP shells.

2.1 Basics

The element formulation is based on the well known degeneration principle, see e.g. [4,5]. By using objective strain and stress measures, geometrically nonlinear behavior in terms of large deformations is included.

The incremental finite element equation has the well known form:

$$[{}^{m}\underset{\approx}{\mathbf{K}}_{e} + {}^{m}\underset{\approx}{\mathbf{K}}_{g}](\Delta \mathbf{u})^{1} = {}^{m+1}\underset{\sim}{\mathbf{r}} - ({}^{m}\underset{\sim}{\mathbf{f}} - \Delta \underset{\sim}{\mathbf{f}}_{th}) \quad (1).$$

The usual iterative application of eqn. (1) in each increment improves the result up to a given accuracy.

For the updated Lagrange formulation the stiffness matrices and the nodal force vectors for one element (e) are given by the following integrals over the element volume (${}^{m}\mathrm{V}$):

$${}^{m}\underset{\approx}{\mathbf{K}}_{e}^{(e)} = \int_{{}^{m}V} {}^{m}\underset{\approx}{\mathbf{B}}_{l}^{T}\, {}^{m}\underset{\approx}{\mathbf{C}}\, {}^{m}\underset{\approx}{\mathbf{B}}_{l}\, d^{m}\mathrm{V} \quad (2),$$

$${}^{m}\underset{\approx}{\mathbf{K}}_{g}^{(e)} = \int_{{}^{m}V} {}^{m}\underset{\approx}{\mathbf{B}}_{nl}^{T}\, {}^{m}\underset{\approx}{\mathcal{T}}\, {}^{m}\underset{\approx}{\mathbf{B}}_{nl}\, d^{m}\mathrm{V} \quad (3),$$

$${}^{m}\underset{\sim}{\mathbf{f}}^{(e)} = \int_{{}^{m}V} {}^{m}\underset{\approx}{\mathbf{B}}_{l}^{T}\, {}^{m}\underset{\sim}{\tau}\, d^{m}\mathrm{V} \quad (4),$$

$$\Delta \underset{\sim}{\mathbf{f}}_{th}^{(e)} = \int_{{}^{m}V} {}^{m}\underset{\approx}{\mathbf{B}}_{l}^{T}\, {}^{m}\underset{\approx}{\mathbf{C}}\, \underset{\sim}{\alpha} \Delta\vartheta\, d^{m}\mathrm{V} \quad (5).$$

In eqn. (1) the sum in square brackets is often referred to as the current global tangent stiffness matrix ${}^{m}\underset{\approx}{\mathbf{K}}$ at state m, comprised of ${}^{m}\underset{\approx}{\mathbf{K}}_{e}$ (material stiffness matrix, depending on the current material matrix ${}^{m}\underset{\approx}{\mathbf{C}}$ which, in turn, depends on the local state of damage) and ${}^{m}\underset{\approx}{\mathbf{K}}_{g}$ (initial stress or geometrical stiffness matrix, depending explicitly on the Cauchy stress tensor, ${}^{m}\underset{\approx}{T}$). The external load vector at the current state $m+1$, ${}^{m+1}\underset{\sim}{\mathbf{r}}$, is given by the surface and body forces, concentrated or distributed. ${}^{m}\underset{\sim}{\mathbf{f}}$ is the vector of internal forces corresponding to the stresses at state m (vector ${}^{m}\underset{\sim}{\tau}$) and $\Delta\underset{\sim}{\mathbf{f}}_{th}$ is the vector of internal nodal forces equivalent to stress increments resulting from a temperature increment $\Delta\vartheta$ and computed by using the direction dependent coefficients of linear thermal expansion (vector $\underset{\sim}{\alpha}$).

The general theory used to describe the degenerated shell element's geometry and its deformations as well as the assumptions with respect to modified plane stress conditions are in close analogy to [5] and are described in detail in [1]. For example, the interpolation of the geometry is performed by

$$ {}^{m}x_i(r,s,t) = \sum_{k=1}^{M} \phi^{(k)}(r,s)[{}^{m}x_i^{(k)} + t\frac{\mathrm{h}^{(k)}}{2}\cos{}^{m}\Psi_i^{(k)}] \tag{6} $$

where $\phi^{(k)}(r,s)$ represent the standard 2/D shape functions (i.e. Lagrangian polynomials); M is the number of nodes forming the element, $\mathrm{h}^{(k)}$ is the thickness of the shell at node k and the angles $\Psi_i^{(k)}$ are used to determine the position of the shell's normal at node k.

2.2 Material Description of the Multilayer FRP-Composite

Due to the anisotropic and layered setup of composite shells the overall material matrix ${}^{m}\underset{\approx}{\mathbf{C}}$ in eqns. (2,5) and the vector of coefficients of thermal expansion $\underset{\sim}{\alpha}$ in eqn. (5) become position and orientation dependent. Each layer is assumed to have orthotropic material behavior with respect to its individual fiber-fixed local coordinate system l, q. The definition of the local system and the nomenclature for the material setup of the LFC-element are shown in Fig. 1.

Assuming a linear elastic material (in terms of Cauchy stresses and Almansi strains) the elasticity matrix and the vectors of stresses and strains can be defined corresponding to the modified plane stress conditions ($\tau'_{33} = 0$) resulting from the degeneration principle:

$$ {}^{m}\underset{\sim}{\tau}' = {}^{m}\underset{\approx}{\mathbf{C}}'({}^{m}\underset{\sim}{\varepsilon}' - \underset{\sim}{\alpha}'\,{}^{m}\vartheta) \tag{7} $$

with ${}^{m}\vartheta$ being the temperature difference with respect to a stress-free reference temperature, and

$$ \underset{\sim}{\tau}'^{T} = (\tau'_{11}, \tau'_{22}, 0, \tau'_{12}, \tau'_{13}, \tau'_{23}), \qquad \underset{\sim}{\varepsilon}'^{T} = (\varepsilon'_{11}, \varepsilon'_{22}, 0, \gamma'_{12}, \gamma'_{13}, \gamma'_{23}) \qquad (8). $$

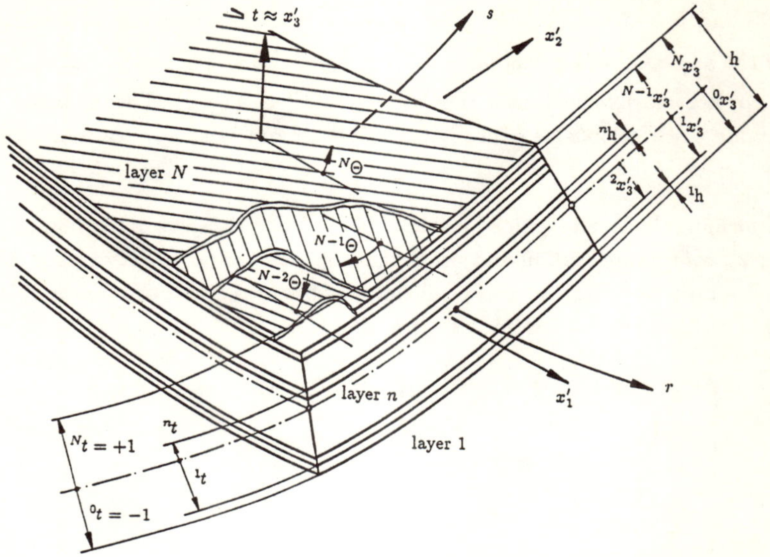

Figure 1. Geometry and Material Setup of the LFC-Element

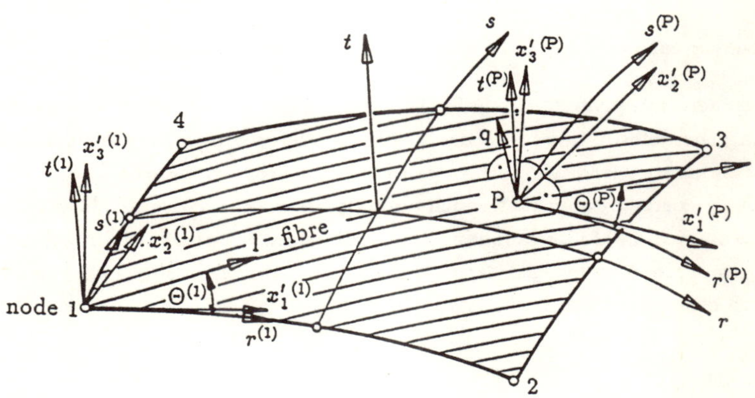

Figure 2. Coordinate Systems of the LFC-Element

Equations (7,8) are referred to the shell's local coordinate system $\underset{\sim}{x}'$ (x_1' being tangential to the natural coordinate r and x_3' being normal to the shell's midsurface), see Fig. 2.

The material matrix and the vector of coefficients of linear thermal expansion of the unidirectionally reinforced layer (UD-layer) are defined in the local layer coordinate system denoted by subscript L. A rotation transformation serves for computing $\underset{\approx}{C}'$ and

$\underset{\sim}{\alpha}'$ from $\underset{\approx}{C}_L$ and $\underset{\approx}{\alpha}_L$. Since element geometry and fiber-direction are independent of each other the angle Θ, which denotes the fiber angle with respect to the $\underset{\sim}{x}'$ coordinate system, is not necessarily constant within the element (see Fig. 2). The variation of the fiber-angle due to deformations is negligibly small for practical applications as shown in [1,3].

After transforming the locally defined material laws of all UD-layers of the laminate into the x_1', x_2', x_3' coordinate system and using the isoparametric concept one can express the material matrix and the vector of coefficients of linear thermal expansion of layer n as a function of the natural coordinates r, s and obtain

$$\underset{\approx}{C}'(r,s,t) = \begin{cases} {}^1\underset{\approx}{C}'(r,s) \\ {}^2\underset{\approx}{C}'(r,s) \\ \vdots \\ {}^N\underset{\approx}{C}'(r,s) \end{cases} \qquad \underset{\sim}{\alpha}'(r,s,t) = \begin{cases} {}^1\underset{\sim}{\alpha}'(r,s) & -1 \leq \ t \leq {}^1t \\ {}^2\underset{\sim}{\alpha}'(r,s) & {}^1t < \ t \leq {}^2t \\ \vdots & \vdots \\ {}^N\underset{\sim}{\alpha}'(r,s) & {}^{N-1}t < \ t \leq +1 \end{cases} \tag{9}$$

with

$$^nt = -1 + \frac{2}{\mathrm{h}} \sum_{j=1}^{n} {}^j\mathrm{h} \qquad \text{(definition of } {}^j\mathrm{h} \text{ see Fig. 1)} \tag{10}$$

for the multi-layer compound with layer number $n = 1, \ldots, N$.

The usual degeneration principle includes the assumption that normal vectors remain straight during the deformation. In order to reduce the error resulting from this kinematic restriction shear-correction factors can be introduced. *Noor and Peters* [7] proposed a predictor-corrector approach which not only improves the overall shear response but also the computation of interlaminar stress quantities. *Dorninger* [6] introduced additional degrees of freedom for each layer. As long as rather thin shells are considered the use of such corrections can be omitted.

2.3 Stiffness Expressions

Computation of the element stiffness matrix requires a three-dimensional integration (see eqns. (2–5)) which usually is performed by some sort of numerical integration technique. Since we are dealing with multi-layer shells (with a very large number of layers allowed), where each layer requires at least one (even better two or more) integration points over its thickness, the numerical effort increases rapidly with the number of layers. An effective way to overcome this difficulty is the use of a quasi-analytical thickness integration as described for homogeneous shells e.g. in [8,9]. This, however, requires an assumption on the $t-$dependence of the Jacobian matrix (see e.g. [10]). In the present work the $t-$dependence is neglected. This approximation is acceptable as long as the shells are thin, their curvature is moderate and their thickness does not vary too much.

The displacements follow from the description of the geometry (eqn. (6)) and, therefore, the t–dependence of the displacement derivatives can be formulated explicitely:

$$ {}^m\mathrm{U}_{i,j}(r,s,t) = {}^m x_{i,j} - {}^0 x_{i,j} = {}^m\overline{\mathrm{U}}_{i,j}(r,s) + t\,{}^m\widehat{\mathrm{U}}_{i,j}(r,s) \qquad i,j = 1,2,3 \tag{11} $$

Using this relation the Almansi strains ε_{ij} can be decomposed into a t–independent part ($\rightarrow \overline{\varepsilon}$), a part linear in t ($\rightarrow \widehat{\varepsilon}$) and a t^2–dependent part ($\rightarrow \widetilde{\varepsilon}$). If the temperature field is assumed to be linearly distributed over the thickness of the shell, with $\overline{\vartheta}$ being the temperature load of the midsurface and $\widehat{\vartheta}$ the temperature difference between opposite points on the two surfaces of the shell the vector of Cauchy stress components (eqn. (7)) becomes:

$$ \begin{aligned} {}^m\underset{\sim}{\tau}(r,s,t) = {}^m\underset{\approx}{C}(r,s,t)\Big(&\big[{}^m\overline{\underset{\sim}{\varepsilon}}(r,s) - {}^m\underset{\sim}{\alpha}(r,s,t)\,{}^m\overline{\vartheta}(r,s)\big] + \\ &+ t\big[{}^m\widehat{\underset{\sim}{\varepsilon}}(r,s) - {}^m\underset{\sim}{\alpha}(r,s,t)\,\frac{{}^m\widehat{\vartheta}(r,s)}{2}\big] + t^2\,{}^m\widetilde{\underset{\sim}{\varepsilon}}(r,s)\Big) \end{aligned} \tag{12} $$

where

$$ {}^m\underset{\approx}{C}(r,s,t) = {}^m\underset{\approx}{G}^T(r,s)\,\underset{\approx}{C}'(r,s,t)\,{}^m\underset{\approx}{G}(r,s) \tag{13} $$

$$ {}^m\underset{\sim}{\alpha}(r,s,t) = {}^m\underset{\approx}{G}^{-1}(r,s)\,\underset{\sim}{\alpha}'(r,s,t) \tag{14} $$

The matrix ${}^m\underset{\approx}{G}$ represents the transformation from the global $\underset{\sim}{x}$ system to the local $\underset{\sim}{x}'$ system. $\underset{\approx}{G}$ is composed of the elements of the Jacobian matrix and, therefore, if $\underset{\approx}{J}$ is independent of t so is $\underset{\approx}{G}$. The $\underset{\approx}{B}_l$ and $\underset{\approx}{B}_{nl}$ matrices are constructed from derivatives of the shape functions and, therefore, they can be decomposed into a t–independent part ($\rightarrow \overline{\underset{\approx}{B}}_l, \overline{\underset{\approx}{B}}_{nl}$) and a part linear in t ($\rightarrow \widehat{\underset{\approx}{B}}_l, \widehat{\underset{\approx}{B}}_{nl}$). Now the stiffness matrices and nodal force vectors of eqns. (2–5) can be rewritten and the quasi-analytical thickness integration can be performed.

By introducing some abbreviations $\underset{\approx}{K}_e^{(e)}, \underset{\approx}{K}_g^{(e)}, \underset{\sim}{f}^{(e)}, \Delta\underset{\sim}{f}_{th}^{(e)}$ follow as:

$$ {}^m\underset{\approx}{K}_e^{(e)} = \int_{-1}^{+1}\!\!\int_{-1}^{+1} \big({}^m\overline{\underset{\approx}{B}}_l^T({}^m\underset{\approx}{C}1\,{}^m\overline{\underset{\approx}{B}}_l + {}^m\underset{\approx}{C}2\,{}^m\widehat{\underset{\approx}{B}}_l) + {}^m\widehat{\underset{\approx}{B}}_l^T({}^m\underset{\approx}{C}2\,{}^m\overline{\underset{\approx}{B}}_l + {}^m\underset{\approx}{C}3\,{}^m\widehat{\underset{\approx}{B}}_l)\big)\det|{}^m\underset{\approx}{J}|\,dr\,ds \tag{15} $$

$$ {}^m\underset{\approx}{K}_g^{(e)} = \int_{-1}^{+1}\!\!\int_{-1}^{+1} \big({}^m\overline{\underset{\approx}{B}}_{nl}^T({}^m\underset{\approx}{S}1\,{}^m\overline{\underset{\approx}{B}}_{nl} + {}^m\underset{\approx}{S}2\,{}^m\widehat{\underset{\approx}{B}}_{nl}) + {}^m\widehat{\underset{\approx}{B}}_{nl}^T({}^m\underset{\approx}{S}2\,{}^m\overline{\underset{\approx}{B}}_{nl} + {}^m\underset{\approx}{S}3\,{}^m\widehat{\underset{\approx}{B}}_{nl})\big)\det|{}^m\underset{\approx}{J}|\,dr\,ds \tag{16} $$

$$ {}^{m}\underset{\sim}{f}^{(e)} = \int_{-1}^{+1}\int_{-1}^{+1} ({}^{m}\overline{\underset{\approx}{B}}_{l}^{T}\, {}^{m}\underset{\sim}{S}1 + {}^{m}\widehat{\underset{\approx}{B}}_{l}^{T}\, {}^{m}\underset{\sim}{S}2) \det|{}^{m}\underset{\approx}{J}|\, dr\, ds \tag{17} $$

$$ {}^{n}\Delta \underset{\sim}{f}_{th}^{(e)} = \int_{-1}^{+1}\int_{-1}^{+1} ({}^{m}\overline{\underset{\approx}{B}}_{l}^{T}({}^{m}\underset{\sim}{C}1_{th}\, \Delta\overline{\vartheta} + {}^{m}\underset{\sim}{C}2_{th} \frac{\Delta\widehat{\vartheta}}{2}) + {}^{m}\widehat{\underset{\approx}{B}}_{l}^{T}({}^{m}\underset{\sim}{C}2_{th}\, \Delta\overline{\vartheta} + {}^{m}\underset{\sim}{C}3_{th} \frac{\Delta\widehat{\vartheta}}{2})) \det|{}^{m}\underset{\approx}{J}|\, dr\, ds \tag{18} $$

where the following abbreviations are used:

$$ \begin{aligned} {}^{m}\underset{\sim}{S}1 &= {}^{m}\underset{\approx}{C}1\, {}^{m}\overline{\underset{\sim}{\varepsilon}} - {}^{m}\underset{\sim}{C}1_{th}\, {}^{m}\overline{\vartheta} + {}^{m}\underset{\approx}{C}2\, {}^{m}\widehat{\underset{\sim}{\varepsilon}} - {}^{m}\underset{\sim}{C}2_{th} \frac{{}^{m}\widehat{\vartheta}}{2} + {}^{m}\underset{\approx}{C}3\, {}^{m}\widetilde{\underset{\sim}{\varepsilon}} \\ {}^{m}\underset{\sim}{S}2 &= {}^{m}\underset{\approx}{C}2\, {}^{m}\overline{\underset{\sim}{\varepsilon}} - {}^{m}\underset{\sim}{C}2_{th}\, {}^{m}\overline{\vartheta} + {}^{m}\underset{\approx}{C}3\, {}^{m}\widehat{\underset{\sim}{\varepsilon}} - {}^{m}\underset{\sim}{C}3_{th} \frac{{}^{m}\widehat{\vartheta}}{2} + {}^{m}\underset{\approx}{C}4\, {}^{m}\widetilde{\underset{\sim}{\varepsilon}} \\ {}^{m}\underset{\sim}{S}3 &= {}^{m}\underset{\approx}{C}3\, {}^{m}\overline{\underset{\sim}{\varepsilon}} - {}^{m}\underset{\sim}{C}3_{th}\, {}^{m}\overline{\vartheta} + {}^{m}\underset{\approx}{C}4\, {}^{m}\widehat{\underset{\sim}{\varepsilon}} - {}^{m}\underset{\sim}{C}4_{th} \frac{{}^{m}\widehat{\vartheta}}{2} + {}^{m}\underset{\approx}{C}5\, {}^{m}\widetilde{\underset{\sim}{\varepsilon}} \end{aligned} \tag{19} $$

are "stress integrals" and

$$ {}^{m}\underset{\approx}{C}i(r,s) = {}^{m}\underset{\approx}{G}^{T}(r,s) \sum_{n=1}^{N} {}^{n}\underset{\approx}{C}'(r,s) \frac{{}^{n}t^{i} - {}^{n-1}t^{i}}{i}\, {}^{m}\underset{\approx}{G}(r,s) \qquad i = 1,\ldots,5 \tag{20} $$

$$ {}^{m}\underset{\sim}{C}i_{th}(r,s) = {}^{m}\underset{\approx}{G}^{T}(r,s) \sum_{n=1}^{N} {}^{n}\underset{\approx}{C}'(r,s)\, {}^{n}\underset{\sim}{\alpha}'(r,s) \frac{{}^{n}t^{i} - {}^{n-1}t^{i}}{i} \qquad i = 1,\ldots,4 \tag{21} $$

are thickness integrals of the material matrix and the vector of linear thermal expansion. ${}^{m}\underset{\approx}{S}1$, ${}^{m}\underset{\approx}{S}2$, ${}^{m}\underset{\approx}{S}3$ are equivalent to ${}^{m}\underset{\sim}{S}1$, ${}^{m}\underset{\sim}{S}2$, ${}^{m}\underset{\sim}{S}3$, but the elements are rearranged in special matrix form, see [1].

The summations in eqns. (20,21) are independent of the state m (as long as local failure is not taken into account), thus they only have to be computed once (prior to the incremental analysis) for the complete prefailure analysis. This results in a decrease in the numerical effort in each increment due to the reduction of the 3/D numerical integration to a 2/D one. The more layers there are within the laminate the more efficient the analytical thickness integration becomes compared to the fully 3/D numerical integration.

2.4 Effects of Local Failure in FRP shells

Due to the nature of FRP composite material the failure behavior is very complex with numerous different failure modes possible. Therefore, an accurate prediction of failure with reasonable effort is almost impossible so that some restrictions have to be made. In this paper a simple way of dealing with this kind of behavior was chosen:

Failure is considered to occur within the layers only and delamination is not taken into account. With respect to this and by assuming a linear elastic stress-strain relationship up to failure (which is valid for many FRPs) one can use proper strength criteria to determine onset of failure. Although the following procedures are based on "homogenized" material, the local damage of the composite, i.e. matrix or fiber cracking, can be estimated. In this paper FRP failure is indicated by a combination of two failure criteria, and two distinct failure modes are assumed:

A quadratic strength criterion, the Tsai-Wu-criterion [11] serves for predicting failure for stress states with relatively large transverse stress components which affect the matrix material rather than the fibers. Therefore, violation of this criterion can be related to matrix failure. Endurable stress states lie within a failure surface in the stress space

$$F_{01}\sigma_{ll} + F_{11}\sigma_{ll}^2 + F_{12}\sigma_{ll}\sigma_{qq} + F_{02}\sigma_{qq} + F_{22}\sigma_{qq}^2 + F_{44}\tau_{lq}^2 < 1 \qquad (22),$$

with

$$F_{01} = \frac{1}{\sigma_{lTu}} + \frac{1}{\sigma_{lCu}} \qquad F_{11} = \frac{-1}{\sigma_{lTu}\sigma_{lCu}} \qquad F_{44} = \frac{1}{\tau_{lqu}^2}$$
$$F_{02} = \frac{1}{\sigma_{qTu}} + \frac{1}{\sigma_{qCu}} \qquad F_{22} = \frac{-1}{\sigma_{qTu}\sigma_{qCu}} \qquad F_{12} = -\sqrt{F_{11}F_{22}} \qquad (23).$$

$\sigma_{ll}, \sigma_{qq}, \tau_{lq}$ are the inplane engineering stresses. With respect to material damage we use the same laws for Cauchy stresses, too. The following notations are used:

σ_{lCu} maximum endurable, i.e. ultimate uniaxial compression stress in fiber direction,
σ_{lTu} ultimate tensile stress in fiber direction,
σ_{qCu} ultimate compression stress normal to fiber direction,
σ_{qTu} ultimate tensile stress normal to fiber direction,
τ_{lqu} ultimate shear stress.

Since in many cases the Tsai-Wu-criterion overestimates the strength of the UD-layer in the case of stresses acting predominantly in fiber direction, a maximum stress limit in fiber direction is imposed:

$$\sigma_{lCu} < \sigma_{ll} < \sigma_{lTu} \qquad (24).$$

Post-cracking stiffnesses are introduced according to the two assumed failure modes. In the case of matrix cracking they take the form:

$$
\underset{\approx}{C}_L = \begin{pmatrix} E_{11} & E_{12} & 0 & 0 & 0 & 0 \\ & E_{22} & 0 & 0 & 0 & 0 \\ & & 0 & 0 & 0 & 0 \\ & sym. & & E_{44} & 0 & 0 \\ & & & & E_{55} & 0 \\ & & & & & E_{66} \end{pmatrix} \xrightarrow{\text{matrix-cracking}} \begin{pmatrix} \beta_E E_{11} & 0 & 0 & 0 & 0 & 0 \\ & 0 & 0 & 0 & 0 & 0 \\ & & 0 & 0 & 0 & 0 \\ & sym. & & 0 & 0 & 0 \\ & & & & E_{55} & 0 \\ & & & & & E_{66} \end{pmatrix} \quad (25).
$$

E_{12}, E_{22} and E_{44} are set to zero, which is a simple representation of the matrix stiffness being removed. The reason for introducing a correction factor β_E is twofold: First, a damaged matrix also reduces the stiffness in fiber direction and, second, for nonstraight (e.g. wavy) fibers the reduced support by the damaged matrix can lead to a further loss of stiffness. β_E depends on the composition of the layer and must be given as an additional material parameter.

In the case of fiber failure the reduced stiffnesses are modeled by:

$$
\underset{\approx}{C}_L = \begin{pmatrix} E_{11} & E_{12} & 0 & 0 & 0 & 0 \\ & E_{22} & 0 & 0 & 0 & 0 \\ & & 0 & 0 & 0 & 0 \\ & sym. & & E_{44} & 0 & 0 \\ & & & & E_{55} & 0 \\ & & & & & E_{66} \end{pmatrix} \xrightarrow{\text{fiber-failure}} \begin{pmatrix} E_{22} & E_{12} & 0 & 0 & 0 & 0 \\ & E_{22} & 0 & 0 & 0 & 0 \\ & & 0 & 0 & 0 & 0 \\ & sym. & & E_{44} & 0 & 0 \\ & & & & E_{55} & 0 \\ & & & & & E_{66} \end{pmatrix} \quad (26).
$$

As can be seen from eqn. (26), E_{11} is set equal to E_{22}. This is a simple representation of the assumption that locally after fiber failure only the matrix stiffness remains.

By assuming thin layers it is sufficient to investigate the stress state in the layers' midsurface only. In addition, if cracking occurs the corresponding failure mode is taken to be valid through the entire thickness of the layer. Thick layers with significant stress gradients in thickness direction can be treated by subdividing them into several "thin" sublayers.

Equations (25,26) represent "secant"-stiffnesses of the material and, therefore, stresses can be computed directly from the total strains. The following procedure, applied at each load step m at each stiffness sampling point r_i, s_j (= 2/D integration points) for all N layers of the laminate, accounts for stiffness changes due to material cracking:

The local inplane strains ${}^n\underset{\sim}{\varepsilon}_L$ at the midsurface of layer n with respect to the layer's local axis are derived from the global strain vector ${}^m\underset{\sim}{\varepsilon}$.

With ${}^n\underset{\sim}{\varepsilon}_L$ and the corresponding local elasticity matrix ${}^n\underset{\approx}{C}_L$ of the previous step an estimate of the stresses is computed.

In the next step the combined failure criterion is examined. If failure is indicated by a violation of eqn. (22) or eqn. (24) the local stiffness matrix is changed according to eqn. (25) or eqn. (26), respectively. The elements of ${}^{n}\underset{\approx}{C}_L$ to be reduced are properly updated in the $\underset{\approx}{C}i$ matrices (eqn. (20)) and the $\underset{\sim}{C}i_{th}$ vectors (eqn. (21)).

2.5 Numerical Examples

a) Buckling and Postbuckling of a Multilayer Square Plate

The practical purpose of this example can be seen in optimizing an inplane loaded plate (compression in $\Theta = 0$ direction) with regard to buckling. The optimization parameter is the fiber angle Θ.

Lay-up: $[(+\Theta/-\Theta)_{6s}]$... 24-layer angle-ply, layer thickness = 0.0529 mm

Material: Graphite/Epoxy

$E_l = 127.5\,\mathrm{kN/mm^2}$	$E_q = 11.0\,\mathrm{kN/mm^2}$	$\nu_{ql} = 0.35$
$G_{lq} = 5.5\,\mathrm{kN/mm^2}$	$G_{lt} = 5.5\,\mathrm{kN/mm^2}$	$G_{qt} = 4.6\,\mathrm{kN/mm^2}$

The whole plate has been modelled by sixteen 16-node LFC-elements, the width of the square being 900 mm; boundary conditions: (a) simply supported, (b) clamped.

For several fiber angles Θ the buckling load has been computed by using buckling procedures as outlined e.g. in [12]. Figure 3 indicates a strong dependence of the buckling load and the corresponding buckling mode on the fiber angle. To verify this result a similar FE-analysis by *Nemeth* [13] has been used for comparison. Figure 3 shows the very close agreement between these two analyses.

Figure 4 shows the computed postbuckling behavior for fiber angle $\Theta = 60°$. The corresponding deformation figures indicate that, after reaching the buckling load, out-of-plane deformations (corresponding to the shape of the buckling mode) develop, which, in turn, drastically lower the overall stiffness of the plate.

b) Bending of a Flexural Specimen

In this example the procedure for taking into account the local stiffness degradation due to progressive failure has been applied. For testing the ultimate flexural strength and modulus of an FRP laminate of the type typically used in aircraft structures a simple flexural specimen (i.e. a three point bending bar) is used. This bar is loaded up to complete failure and the load-displacement path is recorded. A comparison of the measured results with the finite element investigations shows the applicability of the strength calculations included in the LFC formulation.

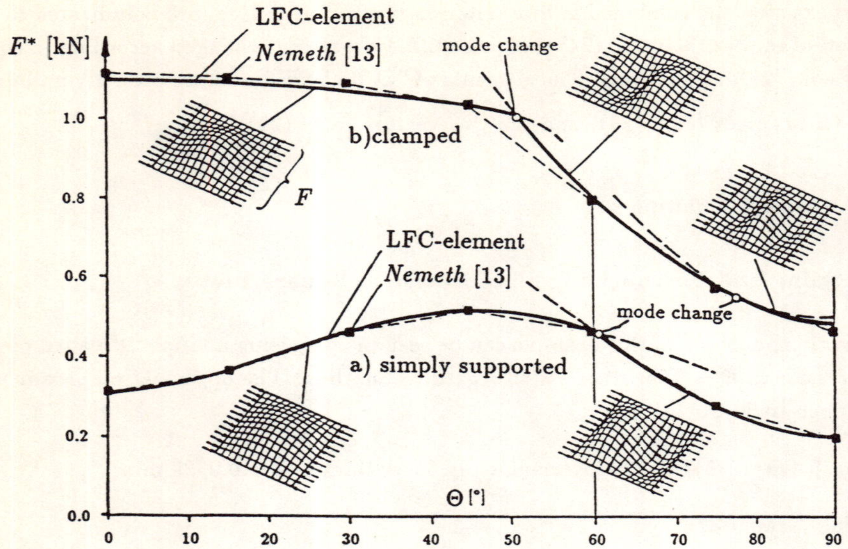

Figure 3. Influence of Fiber Angle on Buckling Loads and Buckling Modes

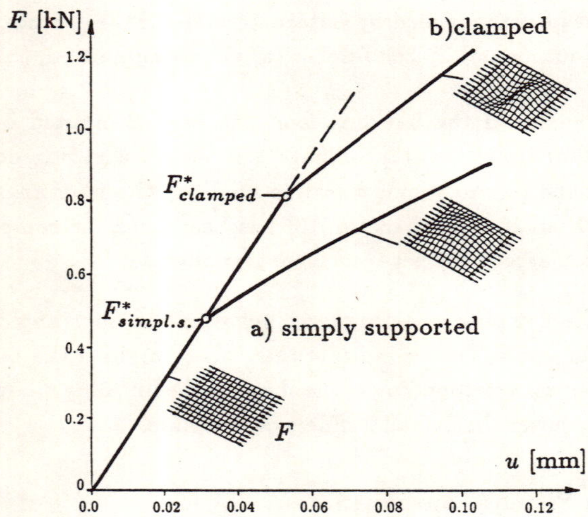

Figure 4. Pre and Post-Buckling Behavior of the Plate with $\Theta = 60°$

Lay-up: 30 layers of Kevlar29 fabric, effective layer thickness = 0.09 mm
For the FE model each fabric layer is subdivided into two sublayers in order to approximate the woven reinforcement by UD-layers.

→ assumed lay-up: $[(0/90)_{15s}]$... cross-ply, layer thickness = 0.045 mm

Model: one quarter of the bar has been modeled by eight 16-node elements

Material: Kevlar29 (UD)

$E_l = 57.2\ kN/mm^2$	$E_q = 3.9\ kN/mm^2$	$\nu_{ql} = 0.35$
$G_{lq} = 2.3\ kN/mm^2$	$G_{lt} = 2.3\ kN/mm^2$	$G_{qt} = 2.3\ kN/mm^2$
$\sigma_{lTu} = 1300.\ N/mm^2$	$\sigma_{lCu} = 227.\ N/mm^2$	$\tau_{lqu} = 34.\ N/mm^2$
$\sigma_{qTu} = 12.\ N/mm^2$	$\sigma_{qCu} = 53.\ N/mm^2$	$\beta_E = 0.2$

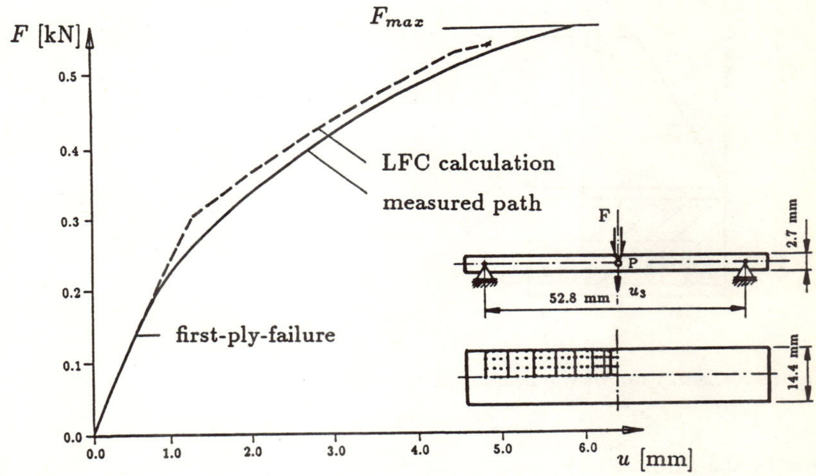

Figure 5. Load-Displacement Path of a Flexural FRP Specimen

Figure 5 shows the calculated and the measured load-displacement path of the flexural specimen. Reasonably good agreement is obtained. First-ply-failure occurs at a relatively low load level and might be interpreted as the beginning of the nonlinear deformations of the experimentally obtained curve. However, in the FE calculations a noticable deviation from the linear path is observed only at a much higher load level. The calculated deformations and the calculated ultimate load are slightly lower than the measured values but for estimating the nonlinear behavior of the material the procedure appears to be satisfactory.

3. LM-ELEMENTS FOR SANDWICH SHELLS

Sandwich plates and shells becomes more and more important in lightweight constractions. Frequently, combinations of layered composite shells with sandwich shells are used to achieve optimized configurations.

3.1 Modifications to the LFC-Element

Under the assumption of *antiplane core* conditions a finite element formulation is derived for a layer model of the sandwich shell. Based on the LFC-element described in section 2 the features typical for sandwich structures are introduced into the finite element formulation: The local material matrix $\underset{\approx}{C}$ is defined independently for each layer.

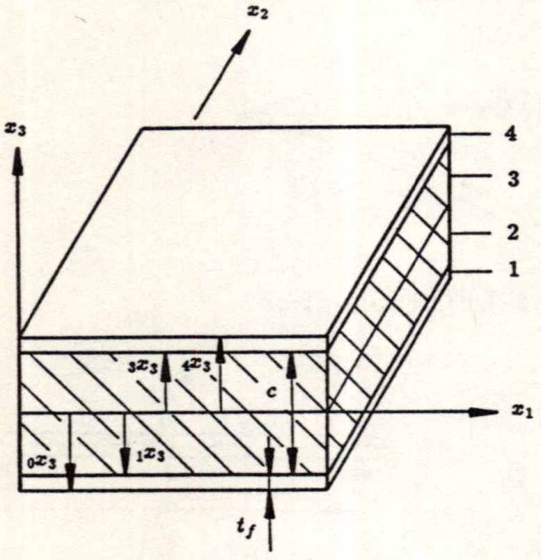

$$
\begin{aligned}
{}^0t &= -1.0\\
{}^1t &= -\frac{c}{2\,(c+2\,t_f)}\\
{}^2t &= 0.0\\
{}^3t &= \frac{c}{2\,(c+2\,t_f)}\\
{}^4t &= 1.0
\end{aligned}
$$

Figure 6. Definition of the Sandwich Layer Model

Due to the assumptions of antiplane core conditions the material matrix $\underset{\approx}{C}_L$ for the isotropic face layer material

$$
{}^1\underset{\approx}{C}_L = {}^4\underset{\approx}{C}_L = \underset{\approx}{C}_f = \begin{pmatrix} \frac{E_f}{1-\nu_f^2} & \frac{\nu_f\,E_f}{1-\nu_f^2} & 0 & 0 & 0 & 0 \\ \frac{\nu_f\,E_f}{1-\nu_f^2} & \frac{E_f}{1-\nu_f^2} & 0 & 0 & 0 & 0 \\ 0 & 0 & 0 & 0 & 0 & 0 \\ 0 & 0 & 0 & G_{f,LW} & 0 & 0 \\ 0 & 0 & 0 & 0 & 0 & 0 \\ 0 & 0 & 0 & 0 & 0 & 0 \end{pmatrix} \qquad (27),
$$

and the material matrix $\underset{\approx}{C}_c$ for the orthotropic core material are defined with respect to the orthotropy axes:

$$
{}^2\underset{\approx}{C}_L = {}^3\underset{\approx}{C}_L = \underset{\approx}{C}_c = \begin{pmatrix} 0 & 0 & 0 & 0 & 0 & 0 \\ 0 & 0 & 0 & 0 & 0 & 0 \\ 0 & 0 & 0 & 0 & 0 & 0 \\ 0 & 0 & 0 & 0 & 0 & 0 \\ 0 & 0 & 0 & 0 & G_{c,LT} & 0 \\ 0 & 0 & 0 & 0 & 0 & G_{c,WT} \end{pmatrix} \qquad (28).
$$

As the out-of-plane shear stresses are assumed to remain constant over the core thickness, no correction factors are required for the shear rigidities. If the local coordinate system defined for each layer does not coincide with the x'_1,x'_2,x'_3-coordinate system a transformation of the material matrices is needed leading to $\underset{\approx}{C}'$ matrices, see section 2.

3.2 Accounting for Local Instability Phenomena

As the stiffness of the sandwich structure may considerably be changed by local buckling phenomena, the influence of regions with buckled face layers on the global deformation and stability behavior has to be considered for further postcritical loading. The steps in accounting for local instability phenomena:

- Calculation of membrane forces in the face layers,
- Determination of critical loads inducing local buckling, i.e. buckling of the face layers at short wavelengths including shear buckling and intracell buckling in the case of honeycomb cores,
- Check for the occurence of local instability: The smaller of the two critical loads is taken for checking the safety against local buckling:

$$P_{crit}^{buckl} = Min\left(P_{crit}^{wrinkl}, P_{crit}^{icb}\right) \qquad (29),$$

- Update of the vector of out-of-balance loads and of the tangential stiffness matrices in case of appearance of local buckling.

Since the determination of local stability limits is influenced by the actual loading state (biaxial loading, bending moments, ...), the critical loads have to be calculated for each state m.

a) Wrinkling of the Face Layers

Under the assumption of linear elastic isotropic face layers and linear elastic orthotropic core material the critical face layer force per unit width P_{crit}^{buckl} inducing short wavelength buckling of the face layers is derived in [2,14].

In the derivations of the wrinkling load the follwing conventions have been used: The $\tilde{x}$-direction of the local coordinate system ($\tilde{x}$, $\tilde{y}$, $\tilde{z}$) follows the direction of the larger of the two compressive principal force components, $\tilde{z} = 0$ denotes the face layer in which the maximum compressive principal force component of both face layers is found. As a consequence of the small wavelengths of the wrinkles as compared to the dimensions of a typical sandwich structure the analysis of critical wrinkling forces can be reduced to the analysis of the infinite plate. In the case of shells the influence of curvature on the local instability limit can be neglected, if the radii of shell curvature are essentially larger than the wavelengths of the local buckles. With regard to *St. Venant's* principle the influence of global boundary conditions of the structure can be neglected on analyzing wrinkling. For the same reasons axial forces in the face layers can be treated as constant in the local region of interest.

If sinusoidal wrinkling deformations of the face layers are assumed the nontrivial core deformations may be described by:

$$u(\bar{x},\bar{y},\bar{z}) = \bar{u}(\bar{z}) \cos \frac{\pi \bar{x}}{a_{\bar{x}}} \sin \frac{\pi \bar{y}}{a_{\bar{y}}} \quad (30),$$

$$v(\bar{x},\bar{y},\bar{z}) = \bar{v}(\bar{z}) \sin \frac{\pi \bar{x}}{a_{\bar{x}}} \cos \frac{\pi \bar{y}}{a_{\bar{y}}} \quad (31),$$

$$w(\bar{x},\bar{y},\bar{z}) = \bar{w}(\bar{z}) \sin \frac{\pi \bar{x}}{a_{\bar{x}}} \sin \frac{\pi \bar{y}}{a_{\bar{y}}} \quad (32).$$

In order to characterize the wrinkling deformation pattern the following parameters are introduced:

$$\alpha = \frac{\pi}{a_{\bar{x}}}, \qquad \beta = \frac{\pi}{a_{\bar{y}}}.$$

The elasticity matrix $\underset{\approx}{C}_{orth}$ for orthotropic core materials (axes of orthotropy L, W, T) is generally described by:

$$\underset{\approx}{C}_{orth} = \begin{pmatrix} C_{LL} & C_{LW} & C_{LT} & 0 & 0 & 0 \\ C_{LW} & C_{WW} & C_{WT} & 0 & 0 & 0 \\ C_{LT} & C_{WT} & C_{TT} & 0 & 0 & 0 \\ 0 & 0 & 0 & 2G_{WT} & 0 & 0 \\ 0 & 0 & 0 & 0 & 2G_{LT} & 0 \\ 0 & 0 & 0 & 0 & 0 & 2G_{LW} \end{pmatrix} \quad (33),$$

$C_{LL}, C_{LW}, \ldots, C_{TT}$ being defined by the direction dependent moduli of elasticity and the direction dependent Poisson ratios; see [2].

In the present analysis the local coordinate system ($\bar{x}$,$\bar{y}$,$\bar{z}$) is assumed to coincide with the axes of orthotropy L, W, T. By using *Hooke's Law* the assumed wrinkling pattern in combination with the local equilibrium condition leads to the following set of coupled linear differential equations dependent on $\bar{z}$ ($'$ defining derivatives with respect to $\bar{z}$):

$$\underset{\approx}{\mathbf{U}}_1 \underset{\sim}{\mathbf{y}}'' - \underset{\approx}{\mathbf{U}}_2 \underset{\sim}{\mathbf{y}}' - \underset{\approx}{\mathbf{U}}_3 \underset{\sim}{\mathbf{y}} = \underset{\sim}{\mathbf{0}} \qquad (34),$$

where $\underset{\sim}{\mathbf{y}}$ represents the amplitudes $\underset{\sim}{\mathbf{y}}^T = (\tilde{u}, \tilde{v}, \tilde{w})$ depending on $\bar{z}$. The coefficient matrices $\underset{\approx}{\mathbf{U}}_i$ are:

$$\underset{\approx}{\mathbf{U}}_1 = \begin{pmatrix} G_{\bar{x}\bar{z}} & 0 & 0 \\ 0 & G_{\bar{y}\bar{z}} & 0 \\ 0 & 0 & C_{\bar{z}\bar{z}} \end{pmatrix} \qquad (35),$$

$$\underset{\approx}{\mathbf{U}}_2 = \begin{pmatrix} 0 & 0 & -\alpha(C_{\bar{x}\bar{z}} + G_{\bar{x}\bar{z}}) \\ 0 & 0 & -\beta(C_{\bar{y}\bar{z}} + G_{\bar{y}\bar{z}}) \\ \alpha(C_{\bar{x}\bar{z}} + G_{\bar{x}\bar{z}}) & \beta(C_{\bar{y}\bar{z}} + G_{\bar{y}\bar{z}}) & 0 \end{pmatrix} \qquad (36),$$

$$\underset{\approx}{\mathbf{U}}_3 = \begin{pmatrix} \alpha^2 C_{\bar{x}\bar{x}} + \beta^2 G_{\bar{x}\bar{y}} & \alpha\beta(C_{\bar{x}\bar{y}} + G_{\bar{x}\bar{y}}) & 0 \\ \alpha\beta(C_{\bar{x}\bar{y}} + G_{\bar{x}\bar{y}}) & \alpha^2 G_{\bar{x}\bar{y}} + \beta^2 C_{\bar{y}\bar{y}} & 0 \\ 0 & 0 & \alpha^2 G_{\bar{x}\bar{z}} + \beta^2 G_{\bar{y}\bar{z}} \end{pmatrix} \qquad (37).$$

The following substitution can be used for solving this set of differential equations:

$$\underset{\sim}{\mathbf{y}} = \underset{\sim}{\mu}\, e^{\varrho \bar{z}} \qquad (38).$$

After inserting eqn. (38) into eqns. (35-37) one obtains

$$\underset{\approx}{\mathbf{U}}\, \underset{\sim}{\mu}\, e^{\varrho \bar{z}} = \underset{\sim}{\mathbf{0}} \qquad (39),$$

where $\underset{\approx}{\mathbf{U}}$ is defined by the eigenvalue problem:

$$\underset{\approx}{\mathbf{U}}\, \underset{\sim}{\mu} = \left(\underset{\approx}{\mathbf{U}}_1\, \varrho^2 - \underset{\approx}{\mathbf{U}}_2\, \varrho - \underset{\approx}{\mathbf{U}}_3\right) \underset{\sim}{\mu} = \underset{\sim}{\mathbf{0}} \qquad (40),$$

and $\underset{\sim}{\mu}$, ϱ are an eigenvector and the corresponding eigenvalue, respectively.

With

$$\Lambda = \varrho^2 \qquad (41)$$

the characteristic equation of the eigenvalue problem can be found to be

$$det\ \underset{\approx}{U} = \Theta_6\Lambda^3 + \Theta_4\Lambda^2 + \Theta_2\Lambda + \Theta_0 = 0 \qquad (42),$$

with parameters Θ_i depending on the material properties and the buckling wavelength parameters α and β. The roots of this cubic polynomial can be found analytically. The six free coefficients in the individual solutions for $\underset{\sim}{y}$ can be determined by the boundary conditions which depend on the local loading situation, i.e. bending or pure compression of the shell, and on the wrinkling mode, i.e. symmetrical or antisymmetrical wrinkling; see Fig. 7.

In any case fife of the six boundary conditions are:

$$u(\bar{x},\bar{y},\bar{z}=0) = v(\bar{x},\bar{y},\bar{z}=0) = u(\bar{x},\bar{y},\bar{z}=c) = v(\bar{x},\bar{y},\bar{z}=c) = 0 \qquad (43),$$

$$\sigma_z(\bar{x},\bar{y},\bar{z}=0) = -\sigma_0 \sin\alpha\bar{x}\sin\beta\bar{y} \qquad (44).$$

The sixth boundary condition depends on the wrinkling mode; see Fig. 7.

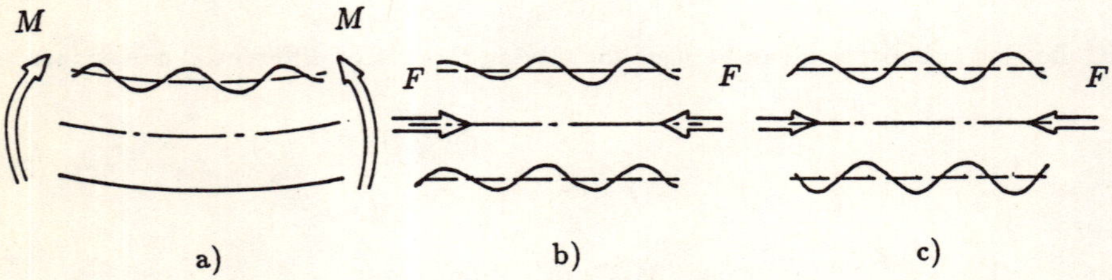

Figure 7. a) One-Face Layer Wrinkling, b) Antisymmetrical, c) Symmetrical Wrinkling

In the case of pure bending (see Fig. 7(a)) the $\bar{z}$-displacements are supposed to vanish completely at the opposite side of the core, because the resulting tensile forces do not induce buckling at the opposite face layer:

$$w(\bar{x},\bar{y},\bar{z}=c) = 0 \qquad (45).$$

In the case of pure compressive forces the sixth boundary condition for antisymmetrical wrinkling is

$$\sigma_z(\bar{x},\bar{y},\bar{z}=c) = \sigma_0 \sin\alpha\bar{x}\sin\beta\bar{y} \qquad (46).$$

If a symmetrical wrinkling pattern is assumed, the remaining boundary condition is determined by the condition of symmetry:

$$w(\bar{x}, \bar{y}, \bar{z} = c/2) = 0 \tag{47}$$

The verification of eqns. (43 - 47) yields to a system of 6 linear equations for the determination of the six free coefficients in the nontrivial solution $\underset{\sim}{y}$.

The vector $\underset{\sim}{y}$ is now determined and the deformation behavior of the core is known in dependence on the wavelengths $a_{\bar{x}}$ and $a_{\bar{y}}$. If the deformations of the face layers and of the core are supposed to be identical at the interface $\bar{z} = 0$, i.e. $w_f(\bar{x}, \bar{y}, \bar{z} = 0) = w(\bar{x}, \bar{y}, \bar{z} = 0)$, corresponding to a perfect interface between the face layers and the core, the equivalent stiffness of the core C_{core} can be calculated as:

$$C_{core} = \frac{\sigma_0 \sin \dfrac{\pi \bar{x}}{a_{\bar{x}}} \sin \dfrac{\pi \bar{y}}{a_{\bar{y}}}}{\bar{w}(\bar{z} = 0) \sin \dfrac{\pi \bar{x}}{a_{\bar{x}}} \sin \dfrac{\pi \bar{y}}{a_{\bar{y}}}} = \frac{\sigma_0}{\bar{w}(\bar{z} = 0)} \tag{48}$$

If the loading conditions differ from pure bending and pure compression the following approximation for C_{core} is proposed in analogy to [15]:

$$C_{core} = \frac{(1 + \eta)\, C_{core}^{compr} + (1 - \eta)\, C_{core}^{bending}}{2} \tag{49}$$

where

$$\eta = \frac{P_{\bar{x}}^{lf}}{P_{\bar{x}}^{uf}} \tag{50}$$

with $P_{\bar{x}}^{uf}$ being the membrane force in the upper (i.e. $\bar{z} = 0$) face layer relevant for wrinkling, and $P_{\bar{x}}^{lf}$ is the membrane force in the lower face layer.

Under general loading conditions the equivalent stiffness of the core depends on the parameter η and on the wavelengths $a_{\bar{x}}$ and $a_{\bar{y}}$.

Now the equation for the considered face layer modelled as a plate on an elastic foundation can be formulated:

$$K_f^{local} \nabla\nabla w_f + P_{\bar{x}} \frac{\partial^2 w_f}{\partial \bar{x}^2} + P_{\bar{y}} \frac{\partial^2 w_f}{\partial \bar{y}^2} + C_{core}(a_{\bar{x}}, a_{\bar{y}})\, w_f = 0 \tag{51}$$

Since the axes $\tilde{x}$ and $\tilde{y}$ are assumed to coincide with the principal stress axes at the considered point on the face layer, the shear force term is not present in eqn. (51). Introduction of the definition

$$\xi = \frac{P_{\tilde{y}}}{P_{\tilde{x}}} \tag{52}$$

allows the reduction of the description of biaxial loading by one parameter, and one obtains:

$$P_{\tilde{x}}^{c} = \frac{\left(\alpha^2+\beta^2\right)^2 K_f^{local} + C_{core}(\alpha,\beta,\eta)}{\alpha^2+\xi\beta^2} \tag{53}.$$

This equation for $P_{\tilde{x}}^{c}$ still contains the wavelength parameters α and β. The smallest value of $P_{\tilde{x}}^{c}$ defined as P_{crit}^{wrinkl} has to be determined by minimization in the parameter space $(\alpha \times \beta)$. Since the equivalent stiffness of the core C_{core} is not an explicit function of these parameters, the extremum cannot be determined by a closed formula but by a numerical procedure, see [2].

As a typical example Fig. 8 shows the results of the wrinkling analysis for a uniaxially loaded sandwich with $E_f = 70000N/mm^2, E_c = 10N/mm^2, \nu_f = \nu_c = 0.3, t_f = 0.1mm$, the core thickness c is varied. For comparison reasons in this figure also the results obtained by approximations according to [16] are shown.

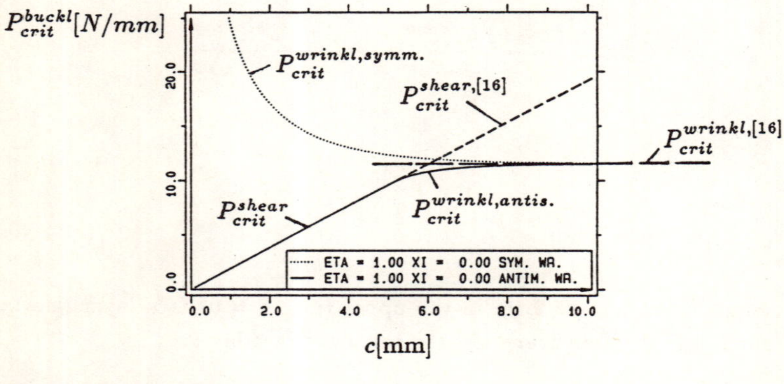

Figure 8. P_{crit}^{wrinkl} over Core Height c

b) Intracell Buckling of the Face Layers

In honeycomb cores a further local instability phenomenon can occur: intracell buckling of the face layers. Supported by the cell walls the face layers can show a pattern of buckles (dimples) in the size of the honeycomb cells. Since the out-of-plane buckling deformations have to vanish at the cell walls, the analysis can be confined to the hexagonal area described by the supporting cell walls.

Stamm and Witte [15] investigated the phenomenon of intracell buckling for honeycomb cores with a nonregular hexagonal shape of the cells under biaxial loading conditions: In order to include the various forms of hexagons used in practical design, the following cell size parameters are introduced (see Fig. 9) and characterized by γ_{icb}:

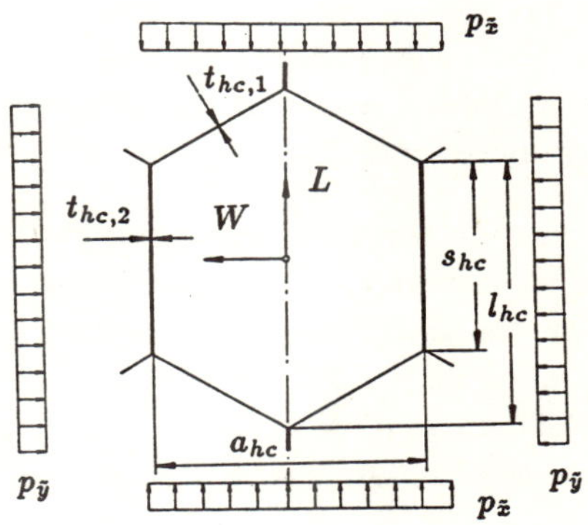

Figure 9. Cell Size Parameters for Nonregular Hexagonal Honeycomb Cells

$$\gamma_{icb} = \frac{a_{hc}}{l_{hc}} \qquad (54).$$

Assuming a local buckling pattern

$$w_f = \sin\frac{\pi\, x\, m}{l_{hc}}\ \sin\frac{\pi\, y\, n}{a_{hc}} \qquad (55)$$

the following equation can be obtained for the critical force:

$$P_{crit}^{icb} = \Gamma_{icb}\, \frac{K_f^{local}}{a_{hc}^2} \qquad (56),$$

where

$$\Gamma_{icb} = \pi^2\, \frac{\left(\gamma_{icb}^2\, m^2 + n^2\right)^2}{\gamma_{icb}^2\, m^2 + \xi^p\, n^2} \qquad m,n = 1,2,\ldots$$

$$\xi^p = \frac{p_{\bar{y}}}{p_{\bar{x}}} \qquad (57).$$

Since eqn. (57) still contains the buckling wavenumbers (m, n), the critical value has to be determined by minimization with respect to these parameters.

c) Postcritical Analysis

The values of the principal forces and their directions can be calculated for each face layer. After the evaluation of the critical loads, P_{crit}^{buckl}, which are dependent on the deformation state m, the safety against local buckling is checked. If local buckling occurs, the stiffness matrix of the locally buckled shell element has to be updated. The postbuckling behavior of the wrinkled face layer is approximated by the assumption of a constant post-critical membrane compression force perpendicular to the wrinkles.

As in the case of local buckling the tangential membrane rigidity of the face layer is reduced, the elasticity parameters have to be adapted. Thus, in effect, a nonlinear material law is introduced for the shell element. By setting the tangential Young's modulus of elasticity of the facing material E_f^ℓ to zero in the direction ℓ, where local buckling occurred, the material matrix can be updated. The shear terms in the material matrix are assumed not to be influenced by local buckling.

In the case of local buckling in direction ℓ the updated tangential material matrix ${}^m\check{\underset{\approx}{C}}_f$ for the buckled face layer takes the form

$$\check{\underset{\approx}{C}}_f^\ell = \begin{pmatrix} 0 & 0 & 0 & 0 & 0 & 0 \\ 0 & \frac{E_f}{1-\nu_f^2} & 0 & 0 & 0 & 0 \\ 0 & 0 & 0 & 0 & 0 & 0 \\ 0 & 0 & 0 & G_{f,LW} & 0 & 0 \\ 0 & 0 & 0 & 0 & 0 & 0 \\ 0 & 0 & 0 & 0 & 0 & 0 \end{pmatrix} \qquad (58).$$

Since this matrix is defined in a local coordinate system of the face layer with the main axis ℓ pointing into the direction of the local buckling load (i.e. the extension of wrinkles is perpendicular to this direction ℓ), a transformation into the shell's local coordinate system x_1', x_2', x_3' is necessary.

In the case of biaxial local buckling in one face layer (i.e. buckling in both principal directions, which is only possible for pure biaxial compression: $\xi = 1.0$) all material parameters are set to zero, except the shear modulus $G_{f,LW}$.

The updated tangential material matrix $\underset{\approx}{C}'$ for an integration point, at which local buckling occurred, then takes the form

$$\breve{\underset{\approx}{C}}'(r,s,t) = \begin{cases} {}^1\breve{\underset{\approx}{C}}'(r,s) & {}^0t \leq \ t \leq {}^1t \\ {}^2\underset{\approx}{C}'(r,s) & {}^1t < \ t \leq {}^2t \\ {}^3\underset{\approx}{C}'(r,s) & {}^2t < \ t \leq {}^3t \\ {}^4\breve{\underset{\approx}{C}}'(r,s) & {}^3t < \ t \leq 1 \end{cases} \tag{59}$$

3.3 Accounting for Thickness Reduction Induced by Loading

Due to the limited resistance of the sandwich core to deformations perpendicular to the face layers the effect of local thickness reduction may result in a loss in stiffness, too. This local phenomenon is similar to the *Brazier effect*. In order to investigate the influence of this effect in sandwich shells, the strains in the direction of the shell's normal have to be evaluated:

With M being the global bending moment of the shell and R the radius of shell curvature in the deformed state one obtaines the normal stress in the core in the direction of the current shell's normal

$$\sigma_n = -\frac{M}{R\ (c + t_f)} \tag{60}$$

Since there exist two principal curvatures in each shell point, the resulting stress in the shell's normal direction can be approximated by their sum

$$\sigma_n^{bend} = \sigma_{n,1} + \sigma_{n,2} \tag{61}$$

Thus, these normal stresses in the core can be evaluated at any point in the shell. As the determination of the stresses σ_n^{bend} requires the knowledge of the current bending moments acting on the shell, the thickness of the core is reduced by an *a-posteriori*-procedure after each iteration. The evaluation of the radii of curvature has to take into account the effect of initial curvature of a structure as well as the influence of deformations induced by loading.

In the case of a sandwich structure loaded by external pressure additional normal stresses appear in the core. Due to the constant shear stresses in the core these normal stresses decrease linearly from the loaded face layer ($\hat{z} = 0$)to the unloaded one ($\hat{z} = c$):

$${}^m\sigma_z^{pr} = {}^mp\left(\frac{\hat{z}}{c} - 1\right) \tag{62}$$

The compressive core stress in the direction of the shell's normal is determined by superposition of the contributions due to bending and pressure loading. The updated core thickness results from

$$^{m+1}c = (1 + {}^{m}\epsilon_n)\ {}^{0}c \tag{63),}$$

with

$$^{m}\epsilon_n = \frac{{}^{m}\sigma_n^{bend} + {}^{m}\sigma_z^{pr,ave}}{C_{TT}} \tag{64),}$$

where ${}^{m}\sigma_z^{pr,ave} = -{}^{m}p/2$.

The thickness reduction induced by loading has then to be taken into account in the next state $m+1$ to be considered. Thus, the update of the core thickness c lags behind the other kinematical quantities.

3.4 Numerical Examples

a) Stability of a Clamped Beam

The analysis of the stability behavior of a clamped sandwich beam under compression shows the applicability of the algorithms derived in the present work to the case of global as well as local instabilities. Since there exist formulas for the analytical approximative determination of the global buckling load accounting for the shear effects, the results of the numerical analysis can be checked, provided the sandwich parameters fulfill the conditions assumed for the derivation of the analytical method.

The sandwich beam modelled for the FE analysis is shown schematically in Fig. 10.

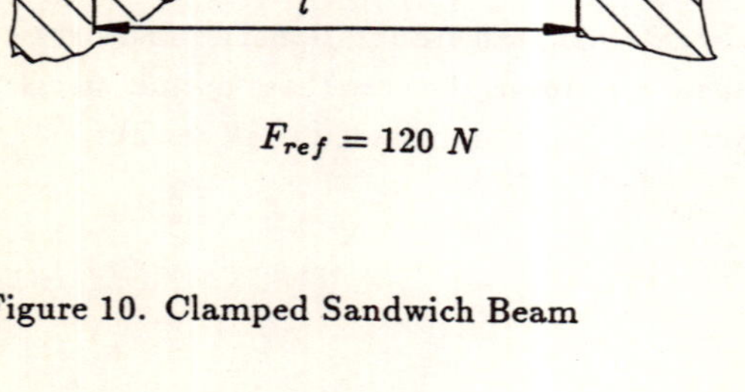

$F_{ref} = 120\ N$

$l = 500\ mm$

$b = 60\ mm$

$c = 30\ mm$

$t_f = 0.5\ mm$

$E_f = 70000\ N/mm^2$

E_c : to be varied

$\nu_f = 0.3$

$\nu_c = 0.3$

Figure 10. Clamped Sandwich Beam

In order to analyze the influence of an isotropic low-density core on the stability behavior, the Young's modulus of the core is varied in a series of computations. In Fig. 11 the critical loads corresponding to the individual kinds of stability loss are plottet in dependence on the ratio of the Young's moduli E_c/E_f.

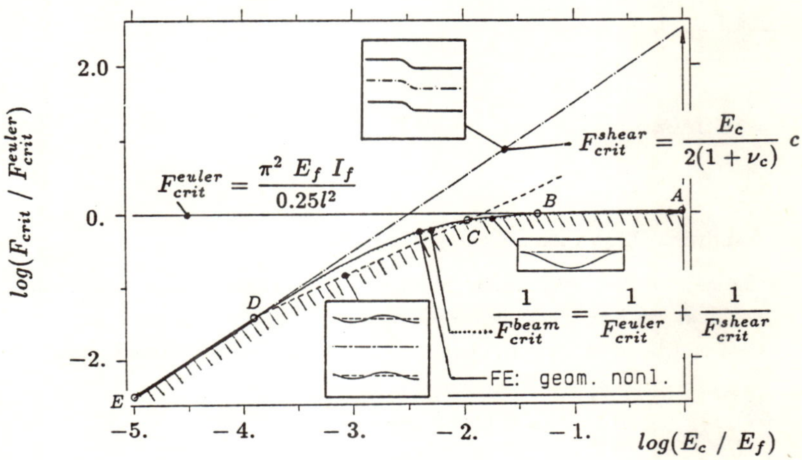

Figure 11. Critical Load Ratio $F_{crit}/F_{crit}^{euler}$ in Dependence on E_c/E_f

Figure 11 shows how increased core stiffness, i.e. increased E_c, leads to transitions from shear buckling of the core (point E to D) to antisymmetrical wrinkling of the face layers (D to C), global buckling influenced by the small shear stiffness of the core (C to B) and, finally, to pure Euler buckling (B to A). The good correspondence with some approximative analytical formulas taken from [16] becomes obvious.

b) Wrinkling in a Square Sandwich Plate with a Central Circular Hole

In this example the influence of local instabilities on the deformation behavior of a square sandwich plate with a central circular hole is investigated. The plate is uniaxially loaded by compressive loads.

Due to the central cylindrical hole a stress concentration is induced near the hole, where the local instability will occur first. The plate configuration is shown schematically in Fig. 12.

Due to the double symmetry conditions only one quarter of the sandwich shell is modelled; 36 16-noded LM sandwich shell elements are used.The load-displacement developments for the edge at which the compressive loads are applied are plotted in Fig.

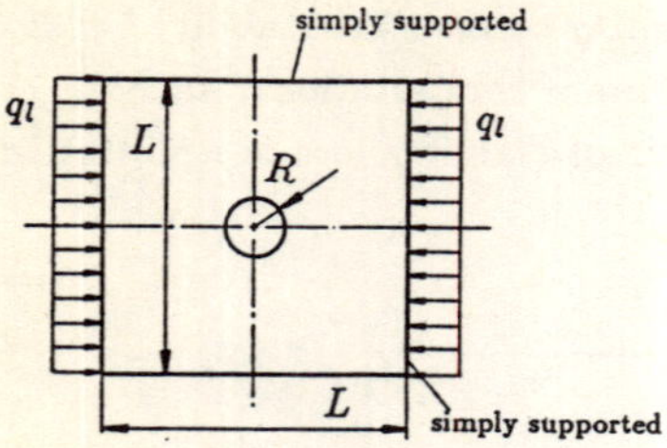

$$L = 1000\ mm$$
$$R = 100\ mm$$
$$c = 20\ mm$$
$$t_f = 0.1\ mm$$

$$q_{l,ref} = 1.6\ N/mm \qquad E_f = 70000\ N/mm^2 \qquad E_c = 10\ N/mm^2$$
$$\nu_f = 0.3 \qquad \nu_c = 0.3$$

Figure 12. Sandwich Plate with a Central Circular Hole

13. In this figure also the development of wrinkling is shown by hatching the wrinkled areas for two post critical load levels. Of course, wrinkling leads to a considerable stiffness loss as can be seen from the behavior of the load displacement path.

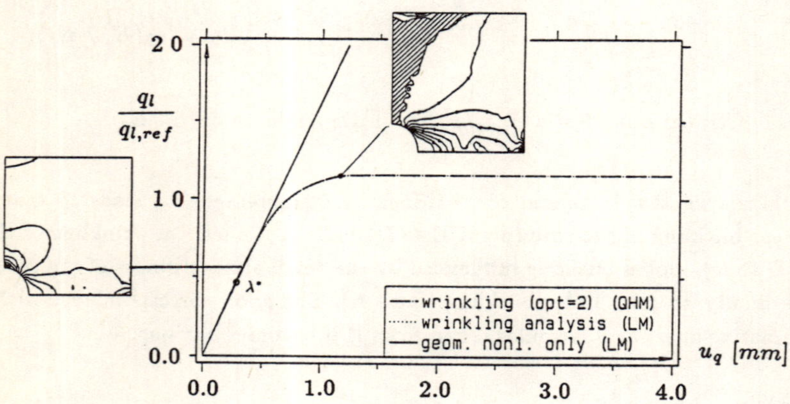

Figure 13. Load-Displacement Developments for a Square Sandwich Plate with a Central Cylindrical Hole

c) Stability Analysis of a Cylindrical Sandwich Shell

In this example the stability behavior of a cylindrical sandwich shell under external pressure is analyzed. As the core consists of honeycomb material, local instability phenomena including intracell buckling are taken into account.

The geometrical and the material parameters of the shell are shown in Fig. 14.

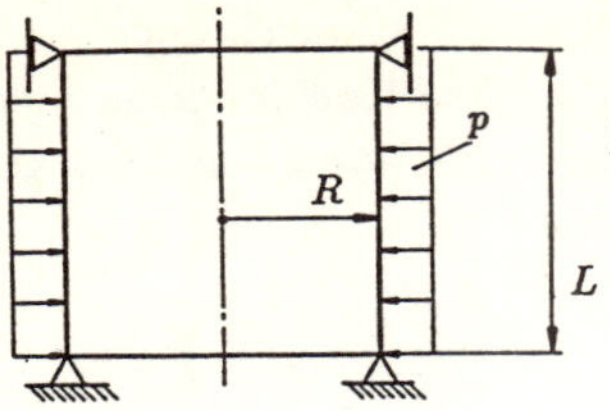

$p_{ref} = 0.002\ N/mm^2$

$R = 2000\ mm$ $\quad t_f = 0.2\ mm$

$L = 4000\ mm$ $\quad t_{hc,min} = 0.2\ mm$

$c = 35.6\ mm$ $\quad a_{hc} = s_{hc} = 9.525\ mm$

$l_{hc} = 14.29\ mm$

$E_f = 70000\ N/mm^2$

$\nu_f = 0.3$

$E_{c,L} = E_{c,W} = 1\ N/mm^2$

$E_{c,T} = 69\ N/mm^2$

$\nu_{c,LW} = \nu_{c,LT} = \nu_{c,WT} = 0.3$

$G_{c,LT} = 82.7\ N/mm^2$

$G_{c,WT} = 69.0\ N/mm^2$

$G_{c,LW} = 1.0\ N/mm^2$

$E_c^{orig} = 70000\ N/mm^2$

$\nu_c^{orig} = 0.3$

Figure 14. Cylindrical Sandwich Shell

The major orthotropy axis L of the honeycomb material is assumed to be parallel to the axis of revolution.

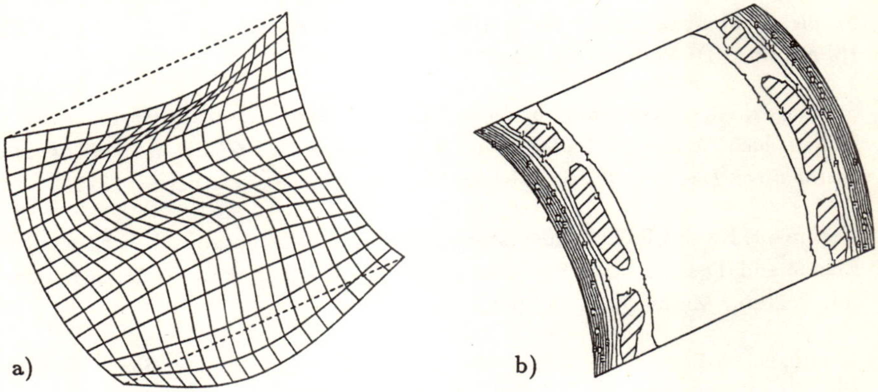

Figure 15. a) Global Buckling Mode, b) Intracell Buckled Areas

In the nonlinear finite element analyses only one quarter of the cylindrical sandwich shell is modelled; 30 LM-elements are used. First, no local phenomena are taken into consideration in the nonlinear analyses, while the pressure is continuously increased. According to the method of accompanying eigenvalue analyses the critical buckling factors and modes are determined for several load steps. The corresponding critical buckling pressures, however, remain constant and are: $p_{crit}^{FE} = 0.05556 N/mm^2$. Figure 15(a) shows the buckling mode of the quarter cylinder.

If local buckling phenomena are taken into account, intracell buckling occurs before the buckling load of the previous analysis is reached. In Fig. 15(b) the locally buckled regions for a local postbuckling state are marked by hatching. The local intracell buckling is initiated in the region of the maximum bending moments. The load inducing local buckling cannot be surpassed, since the stiffness reduction is performed all over the circumference of the model.

Acknowledgement

The financial support of this project by the Federal Republic of Austria (Ministry of Science and Research) and the Fonds zur Förderung der wissenschaftlichen Forschung (by Schrödinger Stipendium) is gratefully acknowledged.

References

[1] Dorninger, K., *Entwicklung von nichtlinearen FE-Algorithmen zur Berechnung von Schalenkonstruktionen aus Faserverbundstoffen*, Fortschritt-Berichte, VDI Reihe 18, Nr. 65, VDI Verlag, Düsseldorf, FRG, 1989.

[2] Starlinger, A., *Development of Efficient Finite Shell Elements for the Analysis of Sandwich Structures Under Large Deformations and Global as Well as Local Instabilities*, Doctoral Thesis, Vienna Technical University, Austria, 1990.

[3] Dorninger, K. and F.G. Rammerstorfer, 'A Layered Composite Shell Element for Elastic and Thermoelastic Stress and Stability Analysis at Large Deformations', Int. J. Num. Meths. Eng. - in press.

[4] Ramm, E., 'A Plate/Shell Element for Large Deflections and Rotations', *Formulations and Computational Algorithms in Finite Element Analysis*, Ed. K.J. Bathe, J.T. Owen, W. Wunderlich, Proc. U.S.-German Symp., MIT, Cambridge, MA, 1977.

[5] Bathe, K.J., *Finite Element Procedures in Engineering Analysis*, Prentice-Hall, Englewood Cliffs, NJ, 1982.

[6] Dorninger, K., 'A Nonlinear Layered Shell Finite Element With Improved Transverse Shear Behavior', Rep., Dept.Civil Engng.,RPI, Troy, NY, 1990.

[7] Noor, A.K. and J.M. Peters, 'A Posteriori Estimates for Shear Correction Factors in Multilayered Composite Cylinders', J. Eng. Mechs. **115**, No. 6, 1225–1244 (1989).

[8] Stanley, G.M., *Continuum-Based Shell Elements*, Dissertation, Stanford University, Stanford, CA, 1985.

[9] Ramm, E. and A. Matzenmiller, 'Large Deformation Shell Analysis Based on the Degeneration Concept', *State-of-the-Art Texts on FEM for Plate and Shell Structures*, Eds. T.J.R. Hughes, E. Hinton, Pineridge Press, Swansea, UK, 1986.

[10] Laschet, G., J.P. Jeusette and P. Beckers, 'Homogenization and Pre-Integration Techniques for Multilayer Composites and Sandwich Finite Element Models', Int. J. Num. Meths. Eng. **27**, 257–269 (1989).

[11] Chawla, K.K., *Composite Materials*, Springer Verlag, New York, NY, 1987.

[12] Rammerstorfer, F.G., 'Jump Phenomena Associated with the Stability of Geometrically Nonlinear Structures', *Recent Advances in Non-Linear Computational Mechanics*, Eds. E. Hinton, D.R.J. Owen, C. Taylor, Pineridge Press, Swansea, UK, 1982.

[13] Nemeth, M.P., 'Importance of Anisotropy on Buckling of Compression-Loaded Symmetric Composite Plates', AIAA Journal **24**, No. 11, 1831–1835 (1986).

[14] Starlinger, A. and F.G. Rammerstorfer, 'Berücksichtigung von lokalem kurzwelligem Deckschichtbeulen in speziellen finiten Sandwich-Schlenelementen', ZAMM **71** - in press.

[15] Stamm K. and H. Witte, *Sandwichkonstruktionen - Berechnung, Fertigung, Ausführung*. Springer-Verlag, Wien, 1974.

[16] Wiedemann J., *Leichtbau: Band 1 - Elemente*. Springer-Verlag, Berlin, 1986.

A Family of Mixed and Hybrid Finite Elements for the Numerical Analysis of Viscoplastic Shells of Revolution

F.G. KOLLMANN

D. CORDTS[1]

H.-P. HACKENBERG

Technische Hochschule Darmstadt
Fachgebiet Maschinenelemente und Maschinenakustik
Darmstadt, Germany

Summary

In this paper we first very briefly describe a theory for inelastically deformed thin shells by Kollmann and Mukherjee [1]. Main feature of this shell theory is a variational principle which contains displacements (or velocities) and strains (or strain rates) as primary unknowns to be varied independently. The theory is formulated under the assumption of vanishing transverse normal stress and normal strain, respectively. Next, we indicate the implementation of our mixed finite elements. Further, we perform intensive numerical tests for the PSS-family (plane stress and plane strain). We show that these elements are free of locking im the thin shell limit. Next, we investigate the convergence of the elements. For this purpose we compute the dependence of the error between our numerical and a closed form solution in the L_2- and H_1-norm on mesh refinements for cylindrical shells under two different loading conditions. As essential result we find that stress like quantities exhibit the same rate of convergence as the radial deflection. This is an important improvement upon displacement based finite elements, where the convergence rate of the bending moment is one order lower than that of the deflection. Finally, we present some numerical results for a viscoplastic cylindrical shell loaded by internal pressure where the inelastic material behavior is governed by Bodner and Partom's inelastic constitutive model.

1 Introduction

In this paper a family of mixed finite elements (FE's) is presented which are designed for elastic and inelastic computation of axisymmetric shells. This family of mixed

[1]D. Cordts is now affiliated to Carl Freudenberg, D 6940 Weinheim. He contributed to this research when he was associated with Technische Hochschule Darmstadt.

elements is especially designed for the numerical computation of inelastic shells. It is presupposed that the inelastic material behavior is governed by a rate type unified constitutive model with internal state variables. Such models have been proposed by Anand [2], Bodner and Partom [3], Hart [4], Miller [5,6] and Robinson [7], to mention only a few contributions. All these models are not only highly nonlinear but more or less mathematically stiff. Beyond known stability problems these models cause additional difficulties in the FE-analysis of plates and shells. The inelastic strain rates are functions of the current values of the stresses and a suitably selected set of internal state variables. Most of the inelastic constitutive models predict large changes of the inelastic strain rates for only small changes of the stresses in some critical regions, which are essential for practical applications. Therefore, it is extremely important to compute the stresses with utmost accuracy. However, it is well known that in the standard finite element method (FEM) the stresses are obtained with less accuracy than the displacements. To our kowledge this seems to be an important reason why until very recent no FE's for inelastic analysis of plates and shells using the constitutive models mentioned above have been proposed. A detailed discussion of prior research of computation of inelastic shells by the FEM is given in [8].

The starting point of the present work is a general, geometrically linear theory of inelastic shells by Kollmann and Mukherjee [1]. As a basic kinematic assumption they suppose that both displacements and strains are linearly distributed over the shell thickness. They further start from a three-dimensional two-field variational principle which originally has been published by Oden and Reddy [9] and also by Lee and Pian [10] for elastic three-dimesional bodies. Lee and Pian mention the applicability of this principle for inelastic analysis of shells. However, they do not pursue this topic further. The elastic principle has been rediscovered independently by Mukherjee and Kollmann [11] and extended to inelastic analysis. The functional of this principle contains only kinematic quantities, velocities and total strain rates, as primary variables to be varied independently. Kollmann and Bergmann [8] have developed a hybrid strain FE for the computation of axisymmetric inelastic shells and plates based on [1]. The outline of the present paper is as follows.

First, we present a brief summary of the underlying inelastic shell theory. Next, we indicate the implementation of a family of mixed FE's with conical geometry as originally proposed in the context of a pure displacement formulation by Zienkiewicz [12] and coworkers. For these elements we first investigate the computational behavior of *elastic* cylindrical shells. We study the performance of the elements in the thin shell limit. Next, we investigate the dependence of the error between known closed form and our numerical solutions in the L_2- and H_1-norm on mesh refinements. Finally, we present some numerical results for *inelastically* deformed shells.

2 Inelastic shell theory

In this section we give a very brief description of the underlying inelastic shell theory [1]. We start with the frame of the inelastic constitutive equations considered. Many of the so-called unified viscoplastic models fit into the following structure. Only such materials are considered which are elastically isotropic and homogeneous. A fundamental constitutive assumption presupposes that the total strain rate tensor $\dot{\boldsymbol{\epsilon}}$ can be decomposed additively into an elastic and inelastic part

$$\dot{\boldsymbol{\epsilon}} = \dot{\boldsymbol{\epsilon}}^{\mathrm{E}} + \dot{\boldsymbol{\epsilon}}^{\mathrm{N}} . \tag{1}$$

The tensor $\dot{\boldsymbol{\epsilon}}^{\mathrm{E}}$of the elastic strains is related to the stress rate tensor $\dot{\boldsymbol{\sigma}}$ by Hooke's generalized law

$$\dot{\boldsymbol{\sigma}} = 2G\dot{\boldsymbol{\epsilon}}^{\mathrm{E}} + \lambda(\mathrm{tr}\dot{\boldsymbol{\epsilon}}^{\mathrm{E}})\boldsymbol{I} . \tag{2}$$

Here G and λ are the Lamé constants, tr denotes the trace operator and $\boldsymbol{I}$ is the identity tensor. The inelastic strain rate tensor $\dot{\boldsymbol{\epsilon}}^{\mathrm{N}}$obeys in the isothermal case an evolution equation of the following form

$$\dot{\boldsymbol{\epsilon}}^{\mathrm{N}} = \boldsymbol{f}(\boldsymbol{\sigma}, \boldsymbol{q}^{(\mathrm{k})}) , \qquad \mathrm{k} = 1, 2, \ldots \mathrm{n} . \tag{3}$$

Here $\boldsymbol{f}$ is a tensorvalued function of the current values of the stress tensor $\boldsymbol{\sigma}$ and a set $\boldsymbol{q}^{(\mathrm{k})}$, $\mathrm{k} = 1, 2, \ldots n$ of suitably selected internal state variables. The internal state variables are either scalars or second order tensors. For these otherwise unspecified state variables evolution equations also exist

$$\dot{\boldsymbol{q}}^{(\mathrm{k})} = \boldsymbol{g}^{(\mathrm{k})}(\boldsymbol{\sigma}, \boldsymbol{q}^{(\mathrm{r})}) . \qquad \mathrm{k}, \mathrm{r} = 1, 2, \ldots n . \tag{4}$$

Here again $\boldsymbol{g}^{(\mathrm{k})}$, $\mathrm{k} = 1, 2, \ldots$ denote functions wich are depending on the corresponding internal variable $\boldsymbol{q}^{(\mathrm{k})}$ either scalarvalued or tensorvalued.

The underlying inelastic shell theory has been extensively presented in references [1,8]. For the sake of brevity we only give a summary of this theory as far as it is used in the FE discretizsation.

$$\boldsymbol{u}^* = \boldsymbol{u} + \xi\, \boldsymbol{w} . \tag{5}$$

$\boldsymbol{u}$ is the displacement vector of $\mathcal{S}$ and $\boldsymbol{w}$ the difference vector. The following component representations for the strain tensor ϵ in the shell space $\mathcal{B}$ are available [1]

$$\mu\, \epsilon_{\alpha\beta} = e_{\alpha\beta} + \xi\, \kappa_{\alpha\beta} , \tag{6}$$

$$\mu\, \epsilon_{\alpha 3} = \psi_\alpha + \xi\, \rho_\alpha , \tag{7}$$

$$\epsilon_{33} = e_{33} . \tag{8}$$

The quantities $e_{\alpha\beta}$, $\kappa_{\alpha\beta}$, ψ_α and ρ_α are components of tensor and vector fields, respectively, which are defined on $\mathcal{S}$.

We introduce vectors of generalized strain rates $\dot{\gamma}$ and of generalized velocities $\dot{\boldsymbol{v}}$. The relation between the strain rates and the velocities is

$$\dot{\gamma} = \boldsymbol{L}_{\gamma v}\ \dot{\boldsymbol{v}} , \tag{9}$$

where $L_{\gamma v}$ is a generalized strain rate - velocity operator. A concise version of the variational principle reads [8]

$$\delta\left\{\int_S \left[\frac{1}{2}\dot{\gamma}^T\ D_{\gamma\gamma}\ \dot{\gamma} - \dot{\gamma}^T\ D_{\gamma\gamma} L_{\gamma v}\ \dot{v} + \dot{F}_L{}^T\ \dot{v} + \dot{f}^{NT}\ L^N_{\gamma v}\ \dot{v}\right] \mathrm{dS}\right\} = 0\,. \tag{10}$$

Here S is the area of the shell midsurface and $\mathrm{d}S$ the area element. Further, $D_{\gamma\gamma}$ is a generalized elasticity matrix for the shell, $\dot{F}_L{}^T$ a generalized load rate vector, $\dot{f}^{NT}$ a generalized vector of inelastic pseudo-force rates and $L^N_{\gamma v}$ a linear operator. Details of these objects can be found in [1,8,13].

We mention that $\dot{\gamma}$ is the assumed strain rate field while $\dot{v}$ is the velocity field. The assumed strain rate filed can be discontinuous between interelement boundaries.

3 Implementation of mixed finite elements

We start with the discretization of the variational principle (10). Kollmann and Bergmann [14] have introduced different shape functions for the generalized velocity vector $\dot{v}$ and the generalized strain rate vector $\dot{\gamma}$

$$\dot{v} = N\hat{\dot{v}}\,, \tag{11}$$

$$\dot{\gamma} = \overline{N}\hat{\dot{\gamma}}\,, \tag{12}$$

where a hat ($\hat{}$) indicates nodal values.

Then, the mixed FE model can be derived from the variational principle equ. (10) by standard procedures

$$K_{\gamma\gamma}\hat{\dot{\gamma}} + K_{\gamma v}\hat{\dot{v}} = \mathbf{0}\,, \tag{13}$$

$$K^T_{\gamma v}\hat{\dot{\gamma}} = -\dot{F}_{\mathrm{L}} - \dot{F}_{\mathrm{N}}\,, \tag{14}$$

where the following quantities have been defined

$$K_{\gamma\gamma} := \int_S \overline{N}^T D_{\gamma\gamma}\overline{N}\mathrm{d}S\,, \tag{15}$$

$$K_{\gamma v} := \int_S \overline{N}^T D_{\gamma\gamma}L_{\gamma v}N\mathrm{d}S\,, \tag{16}$$

$$\dot{F}_{\mathrm{L}} := \int_S N^T\,\dot{f}_{\mathrm{L}}\,\mathrm{d}S\,, \tag{17}$$

$$\dot{F}_{\mathrm{N}} := \int_S \left(L^N_{\gamma v}N\right)^T\,\dot{f}_{\mathrm{N}}\,\mathrm{d}S\,. \tag{18}$$

It is possible to find an implementational scheme of the mixed model which fits into any FE-code based on the standard displacement formulation. Details of this implementation can be found in [13].

Element	Cone		Cylinder		Plate	
	1	2	1	2	1	2
1PSS1	1	1	1	1	1	1
2PSS1	1	2	1	3	1	2
3PSS1	2	6	4	9	2	6
1PSS2	1	1	1	1	1	1
2PSS2	1	1	1	1	1	1
3PSS2	1	2	1	2	1	2
1PSS3	1	1	1	1	1	1
2PSS3	1	1	1	1	1	1
3PSS3	1	1	1	1	1	1

Table 1: Number of zero eigenvalues for PSS-elements

In the following we desribe results obtained for a conical element as originally proposed in the context of the displacement method by Zienkiewicz and coworkers [12] and extended to a mixed hybrid strain formulation by Kollmann and Bergmann [8]. We introduce the following notation:

PSS: Element derived under assumpion of plane strain and plane stress (classical shell theory)

We attribute to each abbreviation two numbers. The first one indicates the degree V of the polynomial used for the interpolation of the velocitites and the second one gives this information S for the strain rates. E.g. 1PSS3 indicates a plane-strain-plain-stress shell element with linear approximation of the velocities and cubic one of the strain rates.

We first tested our mixed elements for correct rank of the element matrices. For this purpose an eigenvalue analysis of one and two assembled elements has been performed. Since for the axisymmetric case there is only one rigid body mode (along the axis of rotation), any correctly defined element or patch of such elements must yield one and only one zero eigenvalue. Table 1 shows our results for the PSS-family. Results for general conical, cylindrical and plate elements are presented, where the latter two are degenerate cases. For each geometry the first column gives the number of zero eigenvalues for one and the second column for two elements.

In all cases $V > S$ more than one zero eigenvalue for the patches of two elements are present. This is a clear indication that either the discrete BB-condition or the discrete K-ellipticity-condition is violated [15]. In the following we, therefore, confine to elements with $V \leq S$, which give correct rank.

4 Numerical tests of PSS-elements for elastic cylindrical shells

In this section we present intensive numerical tests for elastic cylindrical shells using the PSS-family. All shells are simply supported at their edges. We used for comparison with our numerical results two known solutions [16] in closed form. One model problem is a cylinder under uniform internal pressure and the other one a long cylinder loaded in its middle section by a ring force. Both solutions are derived under the classical Love-Kirchhoff hypothesis. Therefore, these solutions can be used as benchmark problems for our finite elements in the thin shell limit. A parameter used in the literature to measure the thiness of a cylindrical shell is the ratio h/R, where R is the radius of the midsurface of the cylindrical shell. If not specified otherwise, we compute our numerical results for a value of $h/R = 4.0 \cdot 10^{-5}$.

From the work of Loula and associates [17] it is known that no prove of stability of a mixed finite element method for cylindrical shells can be given in the limit case $h \to 0$. Therefore, we can only perform numerical tests for stability and convergence.

We first investigate our elements for locking. For this purpose we compute results for a shell loaded by internal pressure until a value of $h/R = 4.0 \cdot 10^{-10}$. Further, we study the behavior of the elements in the incompressible limit ($\nu = 0.5$). Finally, we investigate the convergence behavior of the mixed elements with refinements of the uniform finite element grid. We compute the dependence of the error between the closed form and the numerical solution measured in the L_2- and H_1-norm, respectively, on the number of elements. This study of convergence has been performed for our two model problems.

We find it useful to nondimensionalize all quantities. For this purpose we introduce a nondimensional shell parameter[2] $\overline{\alpha}$

$$\overline{\alpha} := \sqrt[4]{\frac{3(1-\nu^2)}{R^2 t^2}}\, L \tag{19}$$

and a normalized axial coordinate $\overline{z} = z/L$, where L is the axial length of the shell. Definitions of further nondimensional quantities for both model problems are given in [13].

4.1 Test for locking in the thin shell limit

We investigate the performance with decreasing values of h/R where the shell is loaded by internal pressure and the edges are simply supported. For these tests we keep the values of the parameter $\overline{\alpha} = 25.7091$ and of the radius $R = 250$ mm constant. Our computations have been performed with 50 elements spaced uniformly over the positive half of the shell. The results of our numerical studies for decreasing values of the shell thickness are presented in table 1. The displacement $\overline{w}$ is given at the points $\overline{z} = 0$ and $\overline{z} = 0.4100$. At the the point $\overline{z} = 0$ the shell shows pure membrane

[2] All nondimensional quantities are indicated by a superposed bar.

	$\overline{w}$		$\overline{N}_{\varphi\varphi}$		$\overline{M}_{zz}$
	$\overline{z}=0$	$\overline{z}=0.4100$	$\overline{z}=0.0011$	$\overline{z}=0.4094$	$\overline{z}=0.4694$
C-F	1.0000	1.069	1.0000	1.0570	0.5487
FE: h/R					
$4\cdot10^{-2}$	0.9990	1.0665	0.9999	1.0556	0.5465
$4\cdot10^{-4}$	1.0000	1.0680	1.0000	1.0576	0.5525
$4\cdot10^{-6}$	1.0000	1.0680	1.0000	1.0576	0.5525
$4\cdot10^{-8}$	1.0000	1.0680	1.0000	1.0576	0.5525
$4\cdot10^{-10}$	1.0000	1.0680	1.0000	1.0576	0.5525

Table 2: Behavior of 1PSS1-element in the thin shell limit ($h/R \to 0$) C-F = closed form solution

deformation. At $\overline{z}=0.4100$ the maximal radial deflection due to bending effects is achieved. Both the points are exact coordinates in the finite element grid. For the bending moment the maximal value is presented in table 2. It can be seen that down to the very small value $h/R = 4\cdot10^{-10}$ the correspondence between the closed form and the FE solution is excellent. From these tests it can be concluded that the element is free of locking. We mention that exactly the same numerical values are obtained for the 2pss2-element. Finally, it can be shown from numerical simulations that our finite element solution matches the closed form one in the incompressible limit [13].

4.2 Convergence tests

In this subsection we study the convergence of the 1PSS1- and 2PSS2-elements for case that the long shell[3] is loaded by a cylindrical ring force. For comparison we also compute the dependence of the respective error-norms for the displacement based element by Zienkiewicz and coworkers [12], which we call for brevity the Zienkiewicz-element.

Fig. 1 presents the dependence of the L_2-error-norms of the deflection, the membrane force and the bending moment for the 1PSS1-elements on the number of elements which is proportional to $1/h$. Fig. 2 gives the same results for the 2PSS2-element. It can be clearly seen, that for all quantities quadratic convergence is obtained. This is a significant advantage against conventional displacement based FEs. In Fig. 7 we give the dependence of the respective L_2-error-norms for the Zienkiewicz-element [12] on mesh refinements. While the deflection and the membrane force show quadratic convergence, for the bending moment only linear convergence is obtained.

[3]The notion of a "long shell" is made precise in [13].

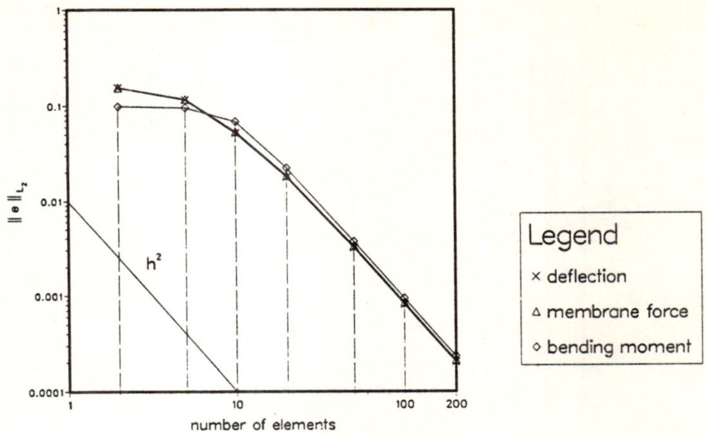

Figure 1: Dependence of L_2-error-norms of deflection, membrane force and bending moment for 1PSS1-element on mesh refinment

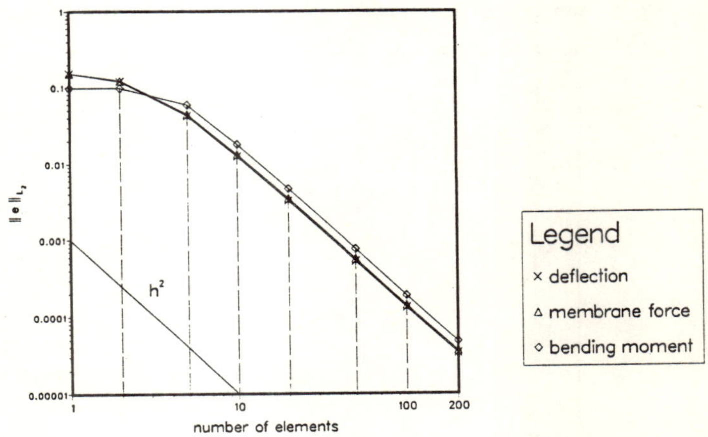

Figure 2: Dependence of L_2-error-norms of deflection, membrane force and bending moment for 2PSS2-element on mesh refinment

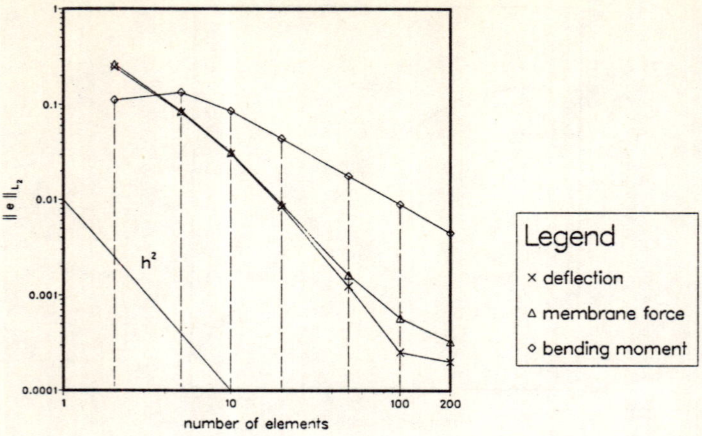

Figure 3: Dependence of L_2-error-norms of deflection, membrane force and bending moment for Zienkiewicz-element on mesh refinment

Since in the analytical solution (compare [16]) the membrane force is equal to the radial deflection (minus a constant) it has to be expected that both quantitites show the same convergence behavior. Finally, Figs. 4 to 6 give a comparison of the H_1-errors for the 1PSS1- and 2PSS2-elements. As expected, the rate of convergence de-

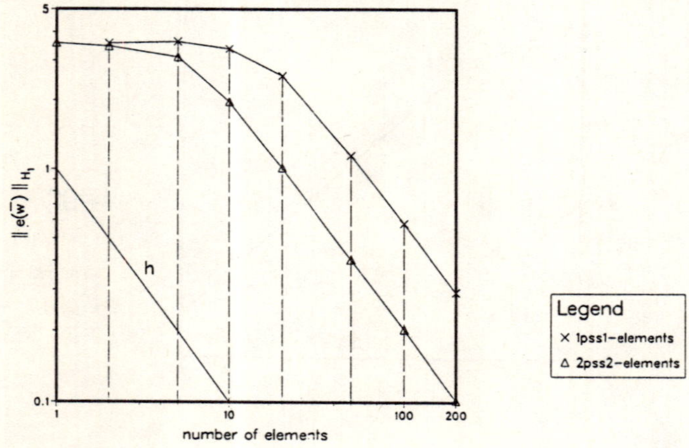

Figure 4: Comparison of dependence of H_1-error-norm for deflection on mesh refinement for 1PSS1 and 2PSS2 elements

creases from second (in the L_2-norm) to first order in the H_1-norm. Unexpected is the fact that the rate of convergence is equal for 1PSS1- and 2PSS2-elements. Intuitively, one would expect that the rate of convergence should be higher for the higher order

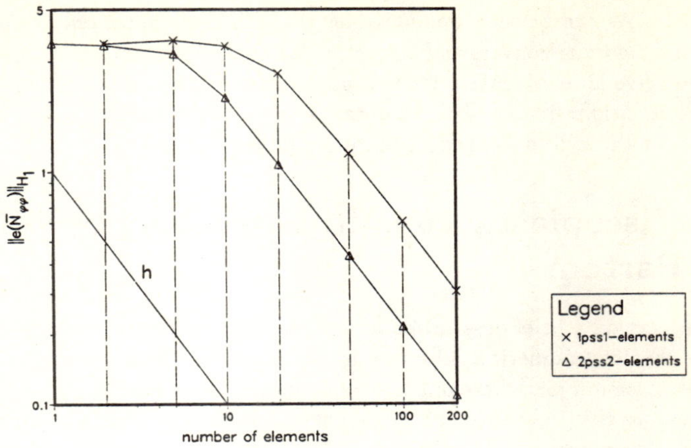

Figure 5: Comparison of dependence of H_1-error-norm for membrane force on mesh refinement for 1PSS1 and 2PSS2 elements

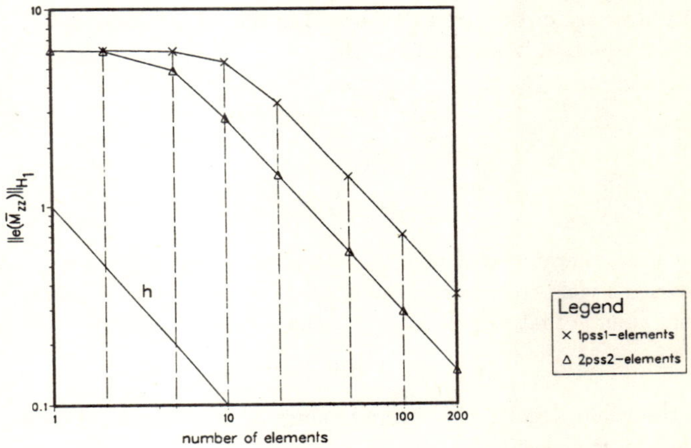

Figure 6: Comparison of dependence of H_1-error-norm for bending moment on mesh refinement for 1PSS1- and 2PSS2-elements

elements. As can be seen we obtain for the 2PSS2-elements only a smaller constant. Since no rigorous convergence and stability analysis for our shell elements exists we can not give an explanation for this unexpected behavior.

We mention, finally, that we obtain a completely corresponding convergence behavior for the cylinder loaded by internal pressure.

5 Viscoplastic constitutive model by Bodner and Partom

In this section a brief description of the viscoplastic model by Bodner and Partom is given. For our numerical inbestigations of inelastic shells we used this model in its simplest version [3], where only isotropic hardening is encluded.

Here we give the version of the Bodner-partom-model for infinitesimal strains. We use the decomposition (1). The flow rule for the inelastic strain tensor takes the form

$$\dot{\boldsymbol{\epsilon}}^{\mathrm{N}} = \lambda\, \boldsymbol{s}\,, \tag{20}$$

where λ is a scalar and $\boldsymbol{s}$ the stress deviator. Squaring eq. (20) gives

$$\lambda^2 = \frac{I_2(\dot{\boldsymbol{\epsilon}}^{\mathrm{N}})}{J_2}\,, \tag{21}$$

where $I_2(\dot{\boldsymbol{\epsilon}}^{\mathrm{N}})$ is the second invariant of the inelastic strain rate tensor $\dot{\boldsymbol{\epsilon}}^{\mathrm{N}}$ and J_2 is the second invariant of the stress deviator. For the second invariant $I_2(\dot{\boldsymbol{\epsilon}}^{\mathrm{N}})$ the following evolution equation is assumed to hold

$$I_2(\dot{\boldsymbol{\epsilon}}^{\mathrm{N}}) = D_0^2 \exp\left[-\left(\frac{A^2}{J_2}\right)^n\right] \tag{22}$$

with

$$A^2 = \frac{1}{3}\, Z^2 \left(\frac{n+1}{n}\right)^{1/n} \tag{23}$$

The material parameter Z in eq. (23) corresponds in a general way to a yield stress. However, no yield condition in the classical sense is given in this theory. Bodner and Partom assume that the parameter Z depends on the plastic work W_{p}, where

$$\mathrm{d}W_{\mathrm{p}} = \boldsymbol{\sigma} : \mathrm{d}\boldsymbol{\epsilon}^{n} \tag{24}$$

Here the colon denotes the scalar product of two tensors. Finally, the quantity Z depends on the plastic work W_{p} as

$$Z = Z_1 + (Z_0 - Z_1)\exp\left(\frac{-m\,W_{\mathrm{p}}}{Z_0}\right)\,. \tag{25}$$

The quantities n, m, Z_0, Z_1 and D_0^2 are material parameters, which have to be identified from experimental tests. Bodner and Partom have identified this material parameters at room temperature for a titanium alloy[4] We take from [3] the following

[4]The chemical composition can be found in [3].

Parameter	Value	Dimension
E	118 000	MPa
ν	0.34	-
Z_0	1150	MPa
Z_1	1400	MPa
D_0^2	$1 \cdot 10^8$	s^{-2}
n	1	-
m	100	-

Table 3: Material parameters for titanium alloy

values of the parameters which have been used for our numerical work in the following section. For time integration of the inelastic initial value problem an implicit algorithm by Cordts and Kollmann [18] has been used.

6 Numerical results of inelastic computation

As a test example we have computed a cylindrical shell made of the titanium alloy referenced in section 5. The shell has an axial length of 1 000 mm, a radius of 250 mm and a thickness of 10 mm. It is closed at its ends. The shell has been discretized using 70 equally spaced 1PSS1 elemts. The loading history of the shell is depicted in Fig. 7. The internal pressure is increased within 1 s linearly from zero to the maximal

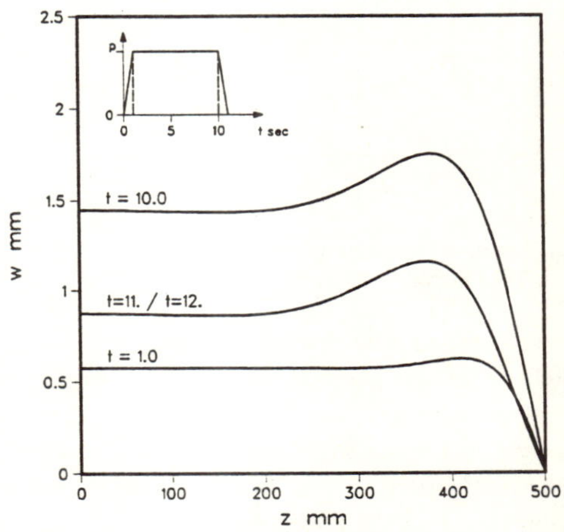

Figure 7: Axial deflection of cylindrical shell

pressure of 13.0 MPa. Then this pressure is held constant over a hold time of 9 s. Finally, the pressure drops linearly within 1 s to zero. In Fig. 7 the deflection of the shell is depicted for the following times: $t = 1.00$ s, $t = 10$ s , $t = 11$ s and $t = 12$ s. At $t = 1$ s the loading of the shell is completed. In the middle section a pure membrane deformation prevails. Only at the axial edge in the region 350mm $\leq z \leq$ 400mm bending effects can be noticed. During the hold time 1s $\leq t \leq$ 100s considerable additional deformation due to viscoplastic effects takes place. It is remarkable that the region of noticeable bending effects spreads into the interior of the shell. During unloading (10s $\leq t \leq$ 11s) the elastic part of the deformation is recovered. After unloading no additional inelastic deformation can be observed.

Fig. 8 sows the recorded history of the development of the the second invariant of the inelastic strain rate tensor (compare eq. (22)) at a point of the shell ($z = 408.943$ mm, $r = 254.840$ mm), where bending effects dominate. During the first second of the loading history the equivalent inelastic strain rate is very small. Then it increases sharply with time. At the beginning of the hold time the second invariant of the inelastic strain rate drops continuously. With unloading it sharply drops to zero.

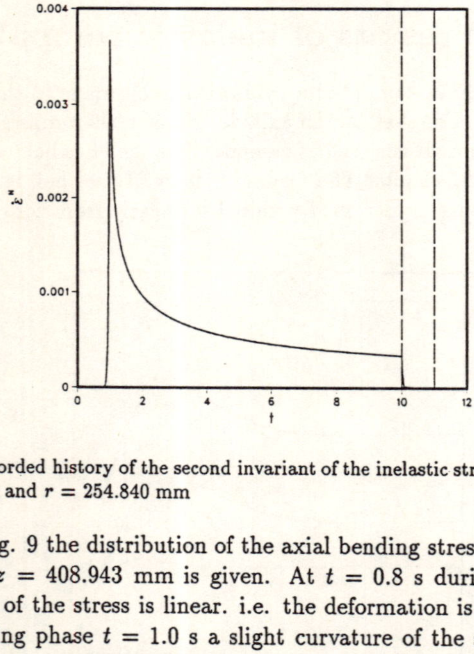

Figure 8: Recorded history of the second invariant of the inelastic strain rate at $z = 408.943$ mm and $r = 254.840$ mm

Finally in Fig. 9 the distribution of the axial bending stress σ_{zz} over the thicknes of the shell at $z = 408.943$ mm is given. At $t = 0.8$ s during the loading period the distribution of the stress is linear. i.e. the deformation is purely elastic. At the end of the loading phase $t = 1.0$ s a slight curvature of the stress distribution can be observed. During the hold time until $t = 10$ s a redistribution of the bending stress occurs and the curvature of the stress distribution is increased. Finally after unloading at $t = 11$ s the residual stress has developed. It has to be noticed that in the theory of Bodner and Partom no loading or unloading conditions nor a yield surface are present.

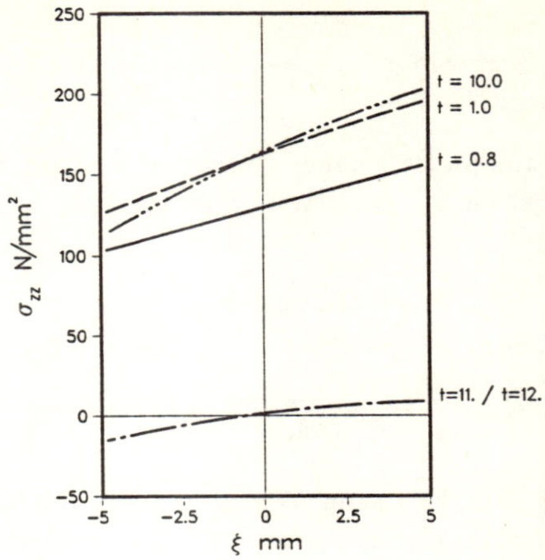

Figure 9: History distribution of axial bending stress at z = 408.943 mm

7 Acknowledgements

This research has been supported by Deutsche Forschungsgemeinschaft, Bad Godesberg. Mrs. Gallowsky has written the FORTRAN code.

References

[1] F. G. Kollmann and S. Mukherjee. A general, geometrically linear theory of inelastic thin shells. *Acta Mechan.*, 57:41–67, 1985.

[2] L. Anand. Constitutive equations for hot-working of metals. *Int.J.Plasticity*, 1:213–232, 1985.

[3] S. R. Bodner and Y. Partom. Constitutive equations for elastic-viscoplastic strain-hardening materials. *Trans. ASME, J.Appl.Mech.*, 42:385 – 389, 1975.

[4] E. W. Hart. Constitutive relations for the non-elastic deformation of metals. *Trans. ASME J.Engng.Mat.Technol.*, 98:97–105, 1976.

[5] A. K. Miller. An inelastic constitutive model for monotonic, cyclic, and creep deformation. part 1: equations develpoment and analytical procedures. *Trans. ASME J.Engng.Technol.*, 98:97–105, 1976.

[6] A. K. Miller. An inelastic constitutive model for monotonic, cyclic, and creep deformation. part 2: applications to type 304 stainless steel. *Trans. ASME J.Engng.Technol.*, 98:106–113, 1976.

[7] D. N. Robinson. *A candidate recovery model for 2-1/4 Cr-1 Mo steel and its experimental implementation.* Technical Report ORNL-TM-5510, Oak Ridge National Laboratory, 1975.

[8] F. G. Kollmann and V. Bergmann. Numerical analysis of viscoplastic axisymmetric shells based on a hybrid strain finite element. *Comp.Mech.*, 1990. To appear.

[9] J. T. Oden and J. N. Reddy. On dual complementary variational principles in mathematical physics. *Int.J.Engng.Sci.*, 12:1–29, 1974.

[10] S. W. Lee and T. H. H. Pian. Improvement of plate and shell finite elements by mixed formulations. *AIAA J.*, 16:29–34, 1978.

[11] S Mukherjee and F. G. Kollmann. A new rate principle for suitable for analysis of inelastic deformation of plates and shells. *J.Appl.Mech.*, 52:533–535, 1985.

[12] O. C. Zienkiewicz, J. Bauer, K. Morgan, and E. Onate. A simple and efficient element for axisymmetric shells. *Int. J. Num. Meth. Eng.*, 11:1545–1558, 1977.

[13] F. G. Kollmann, D. Cordts, and H.-P. Hackenberg. *Implementation and numerical tests of a family of mixed and hybrid strain finite elements for the computation of axisymmetric viscoplastic shells.* Unpublished report MuM-Report 90/1, Technische Hochschule Darmstadt, Darmstadt, Federal Republic of Germany, 1990.

[14] F. G. Kollmann and V. Bergmann. *A new finite element for geometrically linear, inelastic analysis of axisymmetric plates and shells.* Unpublished report, Cornell University, Theoretical and Applied Mechanics, Ithaca, N.Y., 1986.

[15] W. Xue and S. Atluri. Existence and stability, and discrete bb and rank conditions for general mixed-hybrid finite elements in elasticity. In K. Reed and R. Spilker, editors, *Mixed Finite Element Methods*, pages 91 – 112, ASME American Society of Mechanical Engineers, New York, 1985. Vol. 73 Series AMD.

[16] S. P. Timoshenko and P. Woinowsky-Krieger. *Theory of plates and shells.* McGraw-Hill International Book Company, Tokyo, 2 edition, 1970. International student edition.

[17] A. F. D. Loula, I. Miranda, T. J. R. Hughes, and L. P. Franca. On finite element methods for axisymmetric shell analysis. *Comp.Meth.Appl.Mech.Engng.*, 72:201–231, 1989.

[18] D. Cordts and F. G. Kollmann. An implicit time integration scheme for inelastic constitutive equations with internal state variables. *Int. J. Num. Meth. Eng.*, 23:533–554, 1986.

Reliability of Computational Analysis of Plasticity Problems

by

I. BABUŠKA[1], Y. LI[2]
Institute for Physical Science and Technology
University of Maryland
College Park, Maryland 20742

and

K. L. JERINA[3]
Mechanical Engineering
Campus Box 1185
Washington University
St. Louis, MO 63130

[1]The work was partially supported by the Office of Naval Research Grant #N00014-90-J-1030

[2]The work was partially supported by the National Science Foundation Grant #CCR-88-20279

[3]The work was partially supported by the Office of Naval Research Grant #N00014-87-K-0429

Abstract. The paper addresses the reliability of computational analysis of plasticity problems in light of the experimental work and its mathematical formulation. The paper also addresses some basic constitutive laws used in engineering and proposes one which leads to a well posed mathematical problem and agrees well with the experimental data. The data of the experimental project for 5454 aluminum alloy in the H32 condition for periodic and random strains are presented. Various constitutive laws are assessed in light of the experimental data. Outstanding problems related to the reliability of the analysis are mentioned.

1) Introduction

The problem of the reliability of the computational analysis of solid mechanics relies on the assumptions made. In this paper we will discuss the question of nonlinear constitutive laws especially related to the plasticity problem.

The study of plasticity and cyclic loading has generated a large amount of literature. We refer here for example to [1] and an extensive list of references there. We will be especially concerned about proposed models describing plastic strain hardening during general cyclic loading (e.g., of random character). These models are either empirical or are related to some mechanical ideas. As examples we refer to [1-7]. The models used in practice are usually one dimensional, are tested in very limited sets of experimental data and are extended to higher dimensional problems by some mechanical hypotheses (e.g., v. Mises, Tresca, Hencky, etc.).

In this paper we will address the problem of the selection of the constitutive law especially in one dimension. Also, the generalization to higher dimension will be mentioned briefly.

2) The one-dimensional model problem

The typical quasistatic model problem (in one-dimension) in solid mechanics is the following: Let $I = [0,1]$, $W = [0,\infty)$ and $f(x,t)$, $f(x,0) = 0$ piecewise analytic in t and $L_2(I)$ for every t, be given. The problem is to find $u(x,t)$ and $\sigma(x,t)$, $(x,t) \in I \times W$ such that

$$\frac{\partial \sigma}{\partial x}(x,t) = f(x,t) \tag{2.1}$$

$$\frac{\partial u}{\partial x}(x,t) = \varepsilon(x,t) \tag{2.2}$$

$$\dot{\sigma}(x,t) = A(\dot{\varepsilon}(x,t)) \tag{2.3}$$

$$u(0,t) = u(1,t) = 0 \tag{2.4}$$

$$u(x,0) = 0 \tag{2.5}$$

where

$$\dot{\sigma}(x,t) = \frac{\partial \sigma}{\partial t}(x,t)$$

$$\dot{\varepsilon}(x,t) = \frac{\partial \varepsilon}{\partial t}(x,t)$$

and A is an operator which maps any function $e(t)$ (strain) continuous on $[0,\infty)$, $e(0) = 0$ into the function $s(t)$ (stress) which is continuous on $[0,\infty)$ with $s(0) = 0$ and $s(t_0)$, $t_0 > 0$ depends only on $e(t)$ for $0 \le t \le t_0$. (Hence we assume that A in (2.3) is x-independent). Operator A is the *constitutive law operator*. One of the main problems is to have an operator which *guarantees* reasonable properties of the solution of (2.1)-(2.5) such as the existence and uniqueness. In addition the operator has to approximate the material behavior well.

Let us now formulate a specific class of constitutive operators A which describe the plastic behavior and guarantee that the problem (2.1)-(2.5) is a well posed problem. We will introduce the family of the constitutive laws which are based on two following basic assumptions:

1) Existence of a convex yield surface.

2) The normality condition: the plastic increment is proportional to the outward normal to the yield surface during plastic flow.

Let $\mathfrak{U}$ be a convex set in $\mathbb{R}^n$, and let $F(s,\alpha)$ be defined on $\mathbb{R}\times\mathfrak{U}$ so that

(2.6) $F(s,\alpha)$ is convex and C^1 functional on $\mathbb{R}\times\mathfrak{U}$.

(2.7) $$F(0,0) = 0.$$

(2.8) There exist constants γ, Γ such that $0 < \gamma \le \left|\frac{\partial F}{\partial s}\right|, \left|\frac{\partial F}{\partial \alpha}\right| \le \Gamma$ uniformly on the set $\{(x,\alpha)|F(s,\alpha) = Z_0\}$ for some $Z_0 > 0$.

The function $F(s,\alpha) = Z_0$ is the yield surface. Let us now derive the constitutive law using the above two postulates.

Let (s,α) lie inside the yield surface, i.e., $F(s,\alpha) < Z_0$. Then the material will be assumed elastic

(2.9a) $$\dot{s} = E\dot{e}$$

(2.9b) $$\dot{\alpha} = 0.$$

In (2.9a) E denotes the model of elasticity.

If (s,α) lies on the yield surface, i.e., $F(s,\alpha) = Z_0$ then we will split the strain increment $\dot{e}$ into elastic and plastic parts

(2.10) $$\dot{e} = \dot{e}^E + \dot{e}^P$$

and

(2.11) $$\dot{s} = E\dot{\varepsilon}^E = E(\dot{\varepsilon}-\dot{\varepsilon}^P)$$

Using the normality condition we get the following equation

(2.11a) $$\dot{s} = \left\{E - \frac{E^2\left(\frac{\partial F}{\partial s}\right)^2}{\left(\frac{\partial F}{\partial \alpha}\right)^T\left(\frac{\partial F}{\partial \alpha}\right) + E\left(\frac{\partial F}{\partial s}\right)^2}\right\}\dot{\varepsilon}$$

(2.11b) $$\dot{\alpha} = -\frac{\frac{\partial F}{\partial \alpha}}{\left(\frac{\partial F}{\partial \alpha}\right)^T \left(\frac{\partial F}{\partial \alpha}\right)} \left(\frac{\partial F}{\partial s} \dot{s}\right)$$

where $\left(\frac{\partial F}{\partial \alpha}\right)^T = \left(\frac{\partial F}{\partial \alpha_1}, \cdots, \frac{\partial F}{\partial \alpha_m}\right)$.

Hence, the constitutive law under consideration is

$$\left.\begin{aligned} &\dot{s} = E\dot{\varepsilon} &\quad (2.12a)\\ &\dot{\alpha} = 0 &\quad (2.12b)\end{aligned}\right\} \text{ for } (s,\alpha) \in \mathcal{E}$$

$$\left.\begin{aligned} &\dot{s} = \left[E - \frac{E^2\left(\frac{\partial F}{\partial s}\right)^2}{\left(\frac{\partial F}{\partial \alpha}\right)^T \frac{\partial F}{\partial \alpha} + E\left(\frac{\partial F}{\partial s}\right)^2}\right]\dot{e} &\quad (2.13a)\\ &\dot{\alpha} = -\frac{\left(\frac{\partial F}{\partial \alpha}\right)}{\left(\frac{\partial F}{\partial \alpha}\right)^T \left(\frac{\partial F}{\partial \alpha}\right)} \left(\frac{\partial F}{\partial s} \dot{s}\right) &\quad (2.13b)\end{aligned}\right\} \text{ for } (s,\alpha) \in \mathcal{P}$$

where

(2.14a) $$\mathcal{E} = \left\{(s,\alpha) \,|\, F(s,\alpha) < Z_0 \text{ or } F(s,\alpha) = Z_0 \text{ and } \frac{\partial F}{\partial s}\dot{s} \le 0\right\}$$

(2.14b) $$\mathcal{P} = \left\{(s,\alpha) \,|\, F(s,\alpha) = Z_0 \text{ and } \frac{\partial F}{\partial s}\dot{s} > 0\right\}$$

Obviously, operator A is now defined by (2.13) and (2.14) for all $e(t)$ which are continuous and piecewise differentiable. (We assume $e(0) = \alpha(0) = s(0) = 0$).

The family of the constitutive laws (2.12)-(2.14) based on the yield surface $F(s,\alpha)$ satisfying (2.6), (2.8) will be called an *admissible plastic law* or briefly an *admissible law*.

Now we have

Theorem 2.1. Let A be admissible. Then the problem (2.1) has unique solution with $\varepsilon(x,t)$, $\sigma(x,t)$ piecewise analytic in t for every x and $L^2(I)$ for every t.

Theorem 2.2. Let A be admissible. Then for every mesh and linear element the finite element solution of the problem (2.1) exists is unique and is piecewise analytic in t. Moreover, if we let the mesh size tend to zero, then the sequence of the finite element solution will weakly* converge to the solution of the problem (2.1).

Theorems 2.1 and 2.2 have been proven in a general form in two dimensions (under weaker conditions than we mentioned) in [8]. See also [9].

3) One-dimensional engineering constitutive laws

In engineering the constitutive laws of plasticity are formulated in different ways, usually in an incremental form using the notion of reversals. Nevertheless these laws can be cast in a form of differential equation. This form will be called the *standard form*.

Let us mention some of these laws.

a) *Bilinear kinematic hardening law.* Here we have one internal parameter $(n = 1)$ and

$$\begin{aligned} &(3.1a) \qquad \dot{s} = E\dot{e} \\ &(3.1b) \qquad \dot{\alpha} = 0 \end{aligned} \qquad \text{for} \begin{cases} |s-k^{-1}\alpha| < s_0 \\ \text{or} \quad s - k^{-1}\alpha = s_0 \text{ and } \dot{s} \le 0 \\ \text{or} \quad -s + k^{-1}\alpha = s_0 \text{ and } \dot{s} \ge 0 \end{cases}$$

$$\begin{aligned} &(3.2a) \qquad \dot{s} = E_p\dot{e} \\ &(3.2b) \qquad \dot{\alpha} = E_pk\dot{e} \end{aligned} \qquad \text{for} \begin{cases} s - k^{-1}\alpha = s_0 \text{ and } \dot{s} > 0 \\ -s + k^{-1}\alpha = s_0 \text{ and } \dot{s} < 0 \end{cases}$$

The operator A based on (3.1) and (3.2) is independent of k. It depends on the model of elasticity E, the model of plastic flow E_p and the yield stress s_0. Of course $\alpha(t)$ depends on k. Usually $k = 1$ is used. Then, in a physical interpretation, this law keeps the shape and size of the yield surface fixed, and the surface only shifts by the hardening. The parameter α is the position of the center of the yield surface.

The problem is whether we can cast (3.1), (3.2) in a form (2.12)-(2.14) with $F(s,\alpha)$ satisfying (2.6)-(2.8). Selecting $k = \left(\frac{E-E_p}{EE_p}\right)^{1/2}$, (3.1), (3.2) will coincide with (2.12)-(2.14) with

$$F(s,\alpha) = [\max F_1(s,\alpha), F_2(s,\alpha)]^* \tag{3.3}$$

where $[\cdot,\cdot]^*$ means a smoothening operator in the neighborhood $Q = \{(s,\alpha) | F_1(s,\alpha) = F_2(s,\alpha)\}$. (This smoothening is only formal to achieve C^1 continuity of $F(s,\alpha)$). In (3.3) we have

$$F_1(s,\alpha) = s - \left(\frac{EE_p}{E-E_p}\right)^{1/2} \alpha \tag{3.4a}$$

$$F_2(s,\alpha) = -s + \left(\frac{EE_p}{E-E_p}\right)^{1/2} \alpha. \tag{3.4b}$$

We mention that the value $k^2 = (E-E_p)/EE_p$ is the only value which allows to cast (3.1), (3.2) into the form (2.12)-(2.14).

b) *Bilinear isotropic hardening law.* Here the standard form is as follows

$$\begin{aligned} &(3.5a) \quad \dot{s} = E\dot{e} \\ &(3.5b) \quad \dot{\alpha} = 0 \end{aligned} \quad \text{for} \begin{cases} & |s| - k^{-1}\alpha < s_0 \\ \text{or} & s - k^{-1}\alpha = s_0 \text{ and } \dot{s} \le 0 \\ \text{or} & -s - k^{-1}\alpha = s_0 \text{ and } \dot{s} \ge 0 \end{cases}$$

$$\begin{aligned}&(3.6a) && \dot{s} = E_p\dot{e} \\ &(3.6b) && \dot{\alpha} = E_p k\dot{e}\end{aligned} \quad \text{for} \quad s - k^{-1}\alpha = s_0 \quad \text{and} \quad \dot{s} > 0$$

$$\begin{aligned}&(3.7a) && \dot{s} = E_p\dot{e} \\ &(3.7b) && \dot{\alpha} = E_p k\dot{e}\end{aligned} \quad \text{for} \begin{cases} -s - k^{-1}\alpha = s_0 \quad \text{and} \quad \dot{s} < 0. \end{cases}$$

Once more, the operator A here does not depend on k, while $\alpha(t)$ does.

When $k = 1$ then an analogous physical interpretation as before holds. The yield function keeps shape and its center fixed and the hardening increases its size. Once more the proper selection of k, namely $k = \left(\frac{E-E_p}{EE_p}\right)^{1/2}$ allows to cast this law in the form (2.12), (2.14) with

$$\hat{F}(s,\alpha) = [\max \hat{F}_1(s,\alpha), \hat{F}_2(s,\alpha)]^* \tag{3.8}$$

$$\hat{F}_1(s,\alpha) = s - \left(\frac{E-E_p}{E-E_p}\right)^{1/2} \alpha \tag{3.9a}$$

$$F_2(s,\alpha) = -s - \left(\frac{E-E_p}{E-E_p}\right)^{1/2} \alpha. \tag{3.9b}$$

We get the operator A which coincides with the one defined by (3.5)-(3.7).

Functions $F(s,\alpha)$ and $\hat{F}(s,\alpha)$ satisfy the condition (2.6)-(2.8).

c) *Chaboche law [2]*. This law has become recently popular and (see Sect.4) approximates the experimental results relatively well. Although it is formulated in [2] in a different form, it is possible to write it in the standard form.

$$
\begin{array}{ll}
(3.10a) & \dot{s} = E\dot{e} \\
(3.10b) & \dot{\chi} = 0 \\
(3.10c) & \dot{R} = 0 \\
(3.10d) & {}_h = 0 \\
(3.10e) & {}_\ell = 0
\end{array}
\quad \text{for} \quad
\left\{
\begin{array}{l}
e_\ell < e < e_h \quad \text{or} \\
e = e_h \quad \text{and} \quad e \le 0 \\
e = e_\ell \quad \text{and} \quad e \ge 0
\end{array}
\right.
$$

$$
\begin{array}{ll}
(3.11a) & \dot{s} = \dfrac{E[c(a-\chi) + b(Q-R)]}{c(a-\chi) + b(Q-R) + E}\,\dot{e} \\
(3.11b) & \dot{\chi} = \dfrac{Ec(a-\chi)}{c(a-\chi) + b(Q-R) + E}\,\dot{e} \\
(3.11c) & \dot{R} = \dfrac{Eb(Q-R)}{c(a-\chi) + b(Q-R) + E}\,\dot{e} \\
(3.11d) & \dot{e}_h = \dot{e} \\
(3.11e) & \dot{e}_\ell = \dot{e} - 2\,\dfrac{R}{E}
\end{array}
\quad \text{for} \quad
\left\{
\begin{array}{c}
e_h = e \\
\text{and} \\
\dot{e} > 0
\end{array}
\right.
$$

$$
\begin{array}{ll}
(3.12a) & \dot{s} = \dfrac{E[c(a+\chi) + b(Q-R)]}{c(a+\chi) + b(Q-R) + E}\,\dot{e} \\
(3.12b) & \dot{\chi} = \dfrac{Ec(a+\chi)}{c(a+\chi) + b(Q-R) + E}\,\dot{e} \\
(3.12c) & \dot{R} = \dfrac{-Eb(Q-R)}{c(a+\chi) + b(Q-R) + E}\,\dot{e} \\
(3.12d) & \dot{e}_h = \dot{e} + 2\,\dfrac{R}{E} \\
(3.12e) & \dot{e}_\ell = \dot{e}
\end{array}
\quad \text{for} \quad
\left\{
\begin{array}{c}
e = e_\ell \\
\text{and} \\
\dot{e} < 0
\end{array}
\right.
$$

with the initial conditions $s(0) = \chi(0) = R(0) = 0$,

$$e_h(0) = e_0 \,, \qquad e_\ell(0) = -e_0 \tag{3.13}$$

In [2] the law is derived using certain physical heuristic arguments. The following coefficients are input in the Chaboche law.

a: kinematic coefficient

c: kinematic exponents

Q: isotropic coefficient

b: isotropic exponent

e_0: yield strain

E: elastic modulus

In contrast to the kinematic and isotropic laws mentioned earlier, the Chaboche law cannot be cast to the form of an admissible law as simple as before. Nevertheless, an admissible law which is close to the Chaboche law can be formulated. In [10] such law (with n = 2) was proposed analyzed in detail.

d) *The Babuska-Li law*

We have

$$\left.\begin{aligned} &\dot{s} = E\dot{e} && \text{(3.14a)} \\ &\dot{\alpha} = 0 && \text{(3.14b)} \\ &\dot{\beta} = 0 && \text{(3.14c)} \end{aligned}\right\} \quad \text{for} \quad (s,\alpha,\beta) \in \mathcal{E}$$

For $(s, \alpha, \beta) \in \mathcal{P}_+$:

(3.15a) $$\dot{s} = \frac{E\left[\left(\frac{c}{2}\alpha - \sqrt{ac}\right)^2 + \left(\frac{b}{2}\beta - \sqrt{bQ}\right)^2\right]}{\left(\frac{c}{2}\alpha - \sqrt{ac}\right)^2 + \left(\frac{b}{2}\beta - \sqrt{bQ}\right)^2 + E}\,\dot{e}$$

(3.15b) $$\dot{\alpha} = \frac{-E\left(\frac{c}{2}\alpha - \sqrt{ac}\right)}{\left(\frac{c}{2}\alpha - \sqrt{ac}\right)^2 + \left(\frac{b}{2}\beta - \sqrt{bQ}\right)^2 + E}\,\dot{e} \quad \text{for } (s, \alpha, \beta) \in \mathcal{P}_+$$

(3.15c) $$\dot{\beta} = \frac{-E\left(\frac{b}{2}\beta - \sqrt{bQ}\right)}{\left(\frac{c}{2}\alpha - \sqrt{ac}\right)^2 + \left(\frac{b}{2}\beta - \sqrt{bQ}\right)^2 + E}\,\dot{e}$$

(3.16a) $$\dot{s} = \frac{E\left[\left(\frac{c}{2}\alpha + \sqrt{ac}\right)^2 + \left(\frac{b}{2}\beta - \sqrt{bQ}\right)^2\right]}{\left(\frac{c}{2}\alpha + \sqrt{ac}\right)^2 + \left(\frac{b}{2}\beta - \sqrt{bQ}\right)^2 + E}\,\dot{e}$$

(3.16b) $$\dot{\alpha} = \frac{E\left(\frac{c}{2}\alpha + \sqrt{ac}\right)}{\left(\frac{c}{2}\alpha + \sqrt{ac}\right)^2 + \left(\frac{b}{2}\beta - \sqrt{bQ}\right)^2 + E}\,\dot{e} \quad \text{for } (s, \alpha, \beta) \in \mathcal{P}_-$$

(3.16c) $$\dot{\beta} = \frac{E\left(\frac{b}{2}\beta - \sqrt{bQ}\right)}{\left(\frac{c}{2}\alpha + \sqrt{ac}\right)^2 + \left(\frac{b}{2}\beta - \sqrt{bQ}\right)^2 + E}\,\dot{e}$$

where

(3.17a) $$\mathcal{E} = \Big\{(s, \alpha, \beta) \mid F(s, \alpha, \beta) < Z_0 \text{ or } F_1(s, \alpha, \beta) = Z_0 \text{ and } \dot{e} \le 0 \text{ or } F_2(s, \alpha, \beta) = Z_0 \text{ and } \dot{e} \ge 0\Big\}$$

(3.17b) $$\mathcal{P}_+ = \left\{(s,\alpha,\beta) \mid F_1(s,\alpha,\beta) = Z_0 \text{ and } \dot{e} > 0\right\}$$

(3.17c) $$\mathcal{P}_- = \left\{(s,\alpha,\beta) \mid F_2(s,\alpha,\beta) = Z_0 \text{ and } \dot{e} < 0\right\}$$

and $Z_0 = s_y$ is initial yield stress.

Here

(3.18a) $$F_1(s,\alpha,\beta) = \frac{c}{4}\alpha^2 - \sqrt{ac}\,\alpha + \frac{b}{4}\beta^2 - \sqrt{bQ}\,\beta + s$$

(3.18b) $$F_2(s,\alpha,\beta) = \frac{c}{4}\alpha^2 + \sqrt{ac}\,\alpha + \frac{b}{4}\beta^2 - \sqrt{bQ}\,\beta - s$$

and

(3.19) $$F(s,\alpha,\beta) = [\max(F_1(s,\alpha,\beta),\ F_2(s,\alpha,\beta)]^*.$$

We see that this law has two internal parameters $(n = 2)$ and depends on the constants a, b, c, Q, E, Z_0.

It is easy to see that F defined by (3.19) satisfies (2.6) and (2.7). The property (2.8) is more complicated and needs some relations among coefficients of input data. For the detailed analysis of this law we refer to [10]. In [10] also other laws are proposed and analyzed in detail.

4) Experimental data

It is obvious that the constitutive law influences essentially any computed data (for example, data computed by the finite element method). Hence it is essential to analyze the reliability of the assumed material behavior under realistic conditions. By this, it is meant the behavior of commercial material (without laboratory treatment) under various random strain conditions. The experiments related to the constitutive law have to address the following:

a) Reproducibility of the stress response for the imputed strain.

b) Selection and quantitative assessment of an constitutive law.

For our experimental program we selected 5454 aluminum alloy in the H32 condition. The major alloying elements in the workhardening material are magnesium (2.7%), manganese (0.8%) and chromium (0.12%). This particular alloy is widely used in marine applications and is generally available in a variety of structural shapes as sheet, plate, tube, wire, rod and bar. The normal mechanical properties are tensile strength of 40 ksi, yield strength of 30 ksi and elongation 10%. Samples of the material were obtained from two suppliers.

For the testing program three different classes of strain were used: a triangular (or constant magnitude) strain functions and two different random strain functions (using random number generator for different means and standard deviation). For these cases seven tests were conducted with each test having a constant mean strain level ranging from -0.6% to 0.6%. All tests had a strain range of 1.0% and constant strain rate 0.001/sec. Replicate tests were concluded for each strain, thus total of 84 data sets were created. A typical data set contained approximately 40,000 data points. For two samples of measured data σ the functions φ_i^σ, Ψ_{ui}^σ

$$\varphi_i^\sigma = \sigma_1(t_i) - \sigma_2(t_i)$$

$$\Psi_i^\sigma = \frac{\sigma_1(t_i) + \sigma_2(t_i)}{2}$$

were computed together with the norms

$$\|\varphi^\sigma\|_\infty = \max_{1\le i\le N} |\varphi_i^\sigma|$$

$$\|\Psi^\sigma\|_\infty = \max_{1\le i\le N} |\Psi_i^\sigma|$$

As an error measure of the difference the ratio

$$B^{\sigma} = \|\varphi^{\sigma}\|_{\infty} / \|\Psi^{\sigma}\|_{\infty}$$

is used.

For the measure of strain reproducibility the ratio B^{e} was computed (i.e., two identical random strains were compared) and $B^{e} < 1.8\%$ was obtained.

The constants for the constitutive laws were computed for every sample by minimization of the (relative) difference between the experimental and computed data leading to minimal error measure. Typical results are given in tables 1 and 2. Here the results are for 28 periodic and 28 random strains for 10 loading loops. In Table 1 the error measure B^{σ} is reported for the group of periodic and random strain. Here the values of the constants in the constitutive laws are the averages of the best fit constants for every class separately. We report the smallest (best) and largest (worst) error measure of the difference between the samples and the law used as well as the average error measure. In table 2 we report analogous data for the combined set of periodic and random history (56 samples). We use the average constants of the best fits for all (56) samples and the average of the best fit constants for the random strain (28 samples) only. Tables 1 and 2 show clearly that the periodic and random strain lead to different results. They also show the importance of the selection of the values of the constants in the constitutive laws.

Table 3 shows the results for the error measure B^{σ} for one sample (FC), the span of 10, 20, 50 and 100 loops for *the random strain*. The constants in the constitutive laws are the average values of 28 samples of *random strain* and 10 loops. We report the error measure of the difference in stresses between the experiment (sample FC) (mean strain 0.006) and the stresses computed from the law.

Table 1

	28 CYCLIC PERIODIC LOAD HISTORY			28 RANDOM LOAD HISTORY		
	BEST	AVERAGE	WORST	BEST	AVERAGE	WORST
ISOTROPIC	17.5678%	26.1709%	34.8327%	25.9049%	36.6324%	54.3445%
KINEMATIC	13.9703%	21.9894%	29.6227%	24.9118%	30.3313%	37.9495%
CHABOCHE	10.0580%	13.4342%	19.7299%	9.7581%	15.1567%	21.6382%
BABUSKA-LI	8.7208%	13.0558%	17.6285%	11.1595%	17.7304%	23.6274%

Table 2

ALL 28 CYCLIC PERIODIC LOAD AND 28 RANDOM LOAD HISTORY TOGETHER

	MEAN OF 56 REAL DATAS			MEAN OF 28 RANDOM LOAD DATAS		
	BEST	AVERAGE	WORST	BEST	AVERAGE	WORST
ISOTROPIC	20.2992%	32.9285%	55.9337%	20.2992%	36.6324%	32.2566%
KINEMATIC	15.7110%	36.6296%	55.7103%	19.5086%	29.0444%	37.9495%
CHABOCHE	9.1939%	14.3332%	21.9578%	9.0194%	14.2968%	21.6382%
BABUSKA-LI	9.5089%	19.3730%	36.2526%	8.9993%	16.5079%	23.6274%

Table 3

	10 LOOPS	20 LOOPS	50 LOOPS	100 LOOPS
FC - HH	18.6020%	18.1069%	17.6902%	19.0026%
FC - ISOTROPIC	41.9331%	45.4107%	44.1721%	44.1659%
FC - KINEMATIC	27.7171%	33.2344%	32.8032%	32.5333%
FC - CHABOCHE	21.4401%	21.1359%	24.1460%	26.6967%
FC - BABUSKA-LI	21.3331%	21.7366%	26.3657%	24.7862%

Table 4

	10 LOOPS	20 LOOPS	50 LOOPS	100 LOOPS
GL - ES	10.2216%	10.2265%	10.6647%	10.6341%
GL - ISOTROPIC	29.1335%	31.9397%	34.6639%	36.8122%
GL - KINEMATIC	24.5334%	27.4146%	30.2475%	32.4730%
GL - CHABOCHE	12.7273%	12.4393%	11.9824%	11.5788%
GL - BABUSKA-LI	12.3832%	12.0705%	11.5576%	11.1017%

The first row reports the error measure of the stresses for two samples (FC and HH) (for the same random strain). This data characterize the reproducibility of the response and are related to the uncertainty of the material properties.

Table 4 shows analogous results *for periodic strain* (once more the constants in the constitutive laws were computed from the random strain for 10 loops and 28 samples). The error measure for the difference between two samples (GS and ES) (mean strain is -0.004) is reported too. We see once more the difference of the responses for the random and periodic load. Periodic load response is obviously more predictable. This clearly shows the need to make experiments not only for periodic strain.

Figure 1 shows the difference in stresses for the sample FC and different constitutive laws for the random strain. Figure 2 shows analogous results for the periodic strain (for the sample GL). Figure 3 shows the strain-stress relation for the sample FC and random constitutive laws.

We refer to [11] for additional experimental results, statistical analysis of the obtained data and also comparison with other laws. See [11] also for details of the experimental procedures. Obviously unreliability (uncertainty) in the material responses is large and hence reliability (uncertainty) in computed (finite element) data is expected to be large too. We will briefly address this problem in Section 6.

5) Two dimensional constitutive law

The problem is *in principle* the same as in one dimension. Here instead the scalar strain and stress we have to consider strain and stress tensors. Also the internal parameter space is larger.

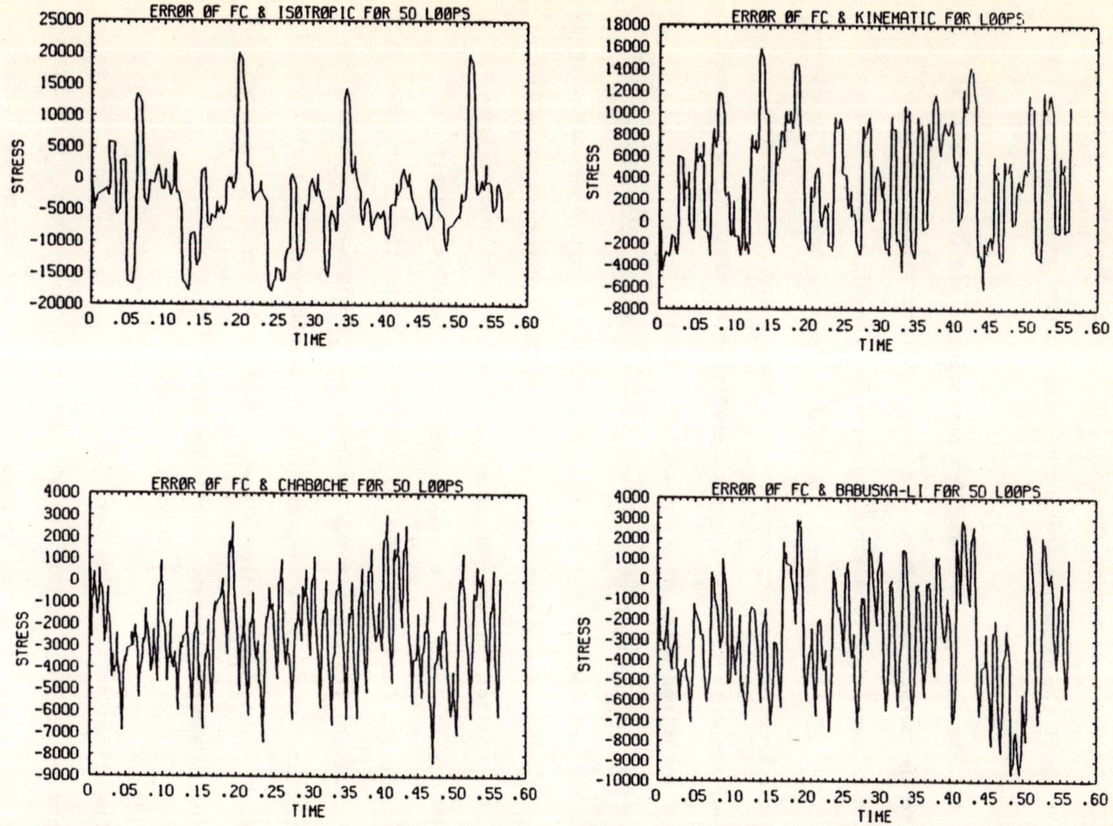

Figure 1. The difference in stresses of the experiment and constitutive law for random strain.

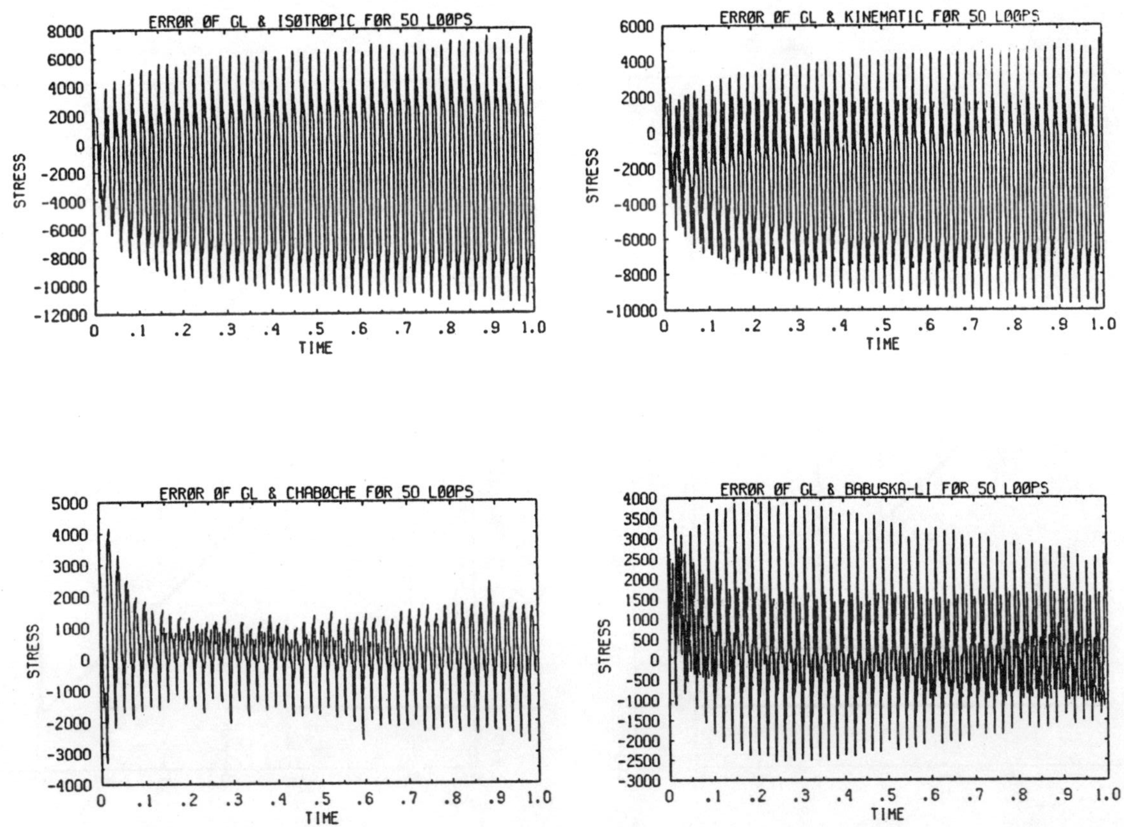

Figure 2. The difference in stresses of the experiment and constitutive law for periodic strain.

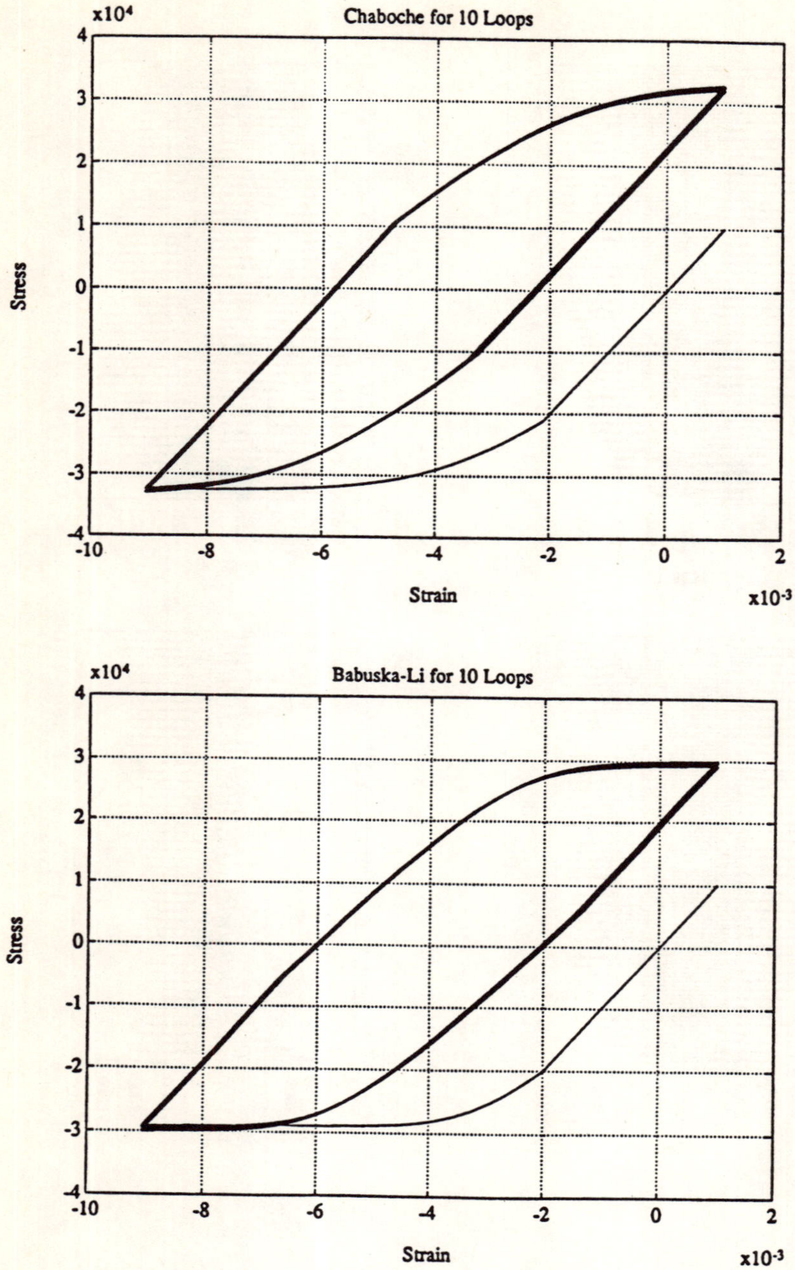

Figure 3. The stress strain relation for the experiment and for various laws and periodic strain.

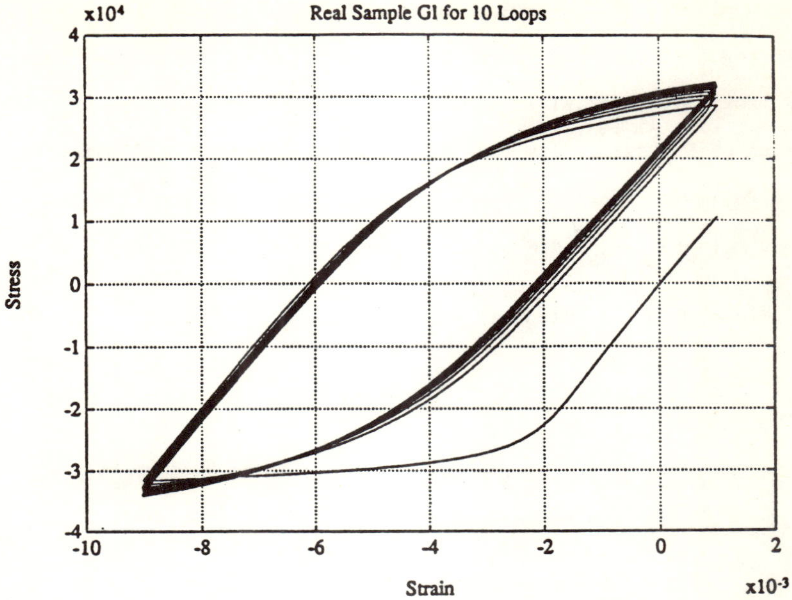

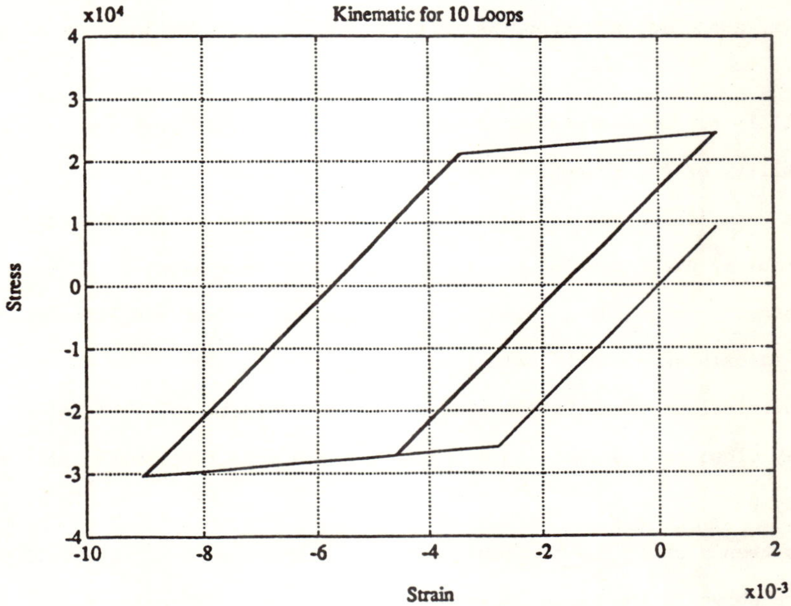

Figure 3. The stress strain relation for the experiment and for various laws and periodic strain.

Because of complexity of the experimentation, one dimensional laws are usually used together with additional hypothesis such as von Mises, etc. In the case of isotropic material we need rotational symmetry of the law. The generalization of Babuska-Li law (see [10] preserving rotational symmetry leads to the following function $F(s,\alpha,\beta)$ where s is tensor with the components $s = (s_x, s_y, s_{xy}) \in \mathbb{R}^3$, $\alpha = [\alpha_1, \alpha_2, \alpha_3] \in \mathbb{R}^3$ and $\beta \in \mathbb{R}^1$.

$$F(s,\alpha,\beta) = f(s - \sqrt{ac}\sqrt{R}\alpha) + \frac{c}{4}f^2(\sqrt{R}\alpha) + \frac{b}{4}\beta^2 - \sqrt{bQ}\,\beta$$

where
$$f^2(s) = s_x^2 + s_y^2 - s_x s_y + 3s_{xy}^2$$

$$R = \begin{bmatrix} 4/3 & 2/3 & 0 \\ 2/3 & 4/3 & 0 \\ 0 & 0 & 1/3 \end{bmatrix}$$

In the constitutive law, the strain e has to be in the form $\varepsilon_{xx}, \varepsilon_{yy}, \gamma_{xy}$. For details, analysis, and proof we refer to [10] where additional results are presented.

6) Reliability of the computational analyses of plasticity problems

Reliability of the computational analyses depends on

a) Mathematical modelling restricting the characteristic features of the physical problem, available information and aims of the analysis.

b) Correctness of the mathematical formulation guaranteeing the basic properties as existence of the solution, etc.

c) Numerical treatment which produces an approximate solution which is sufficiently close to the exact solution in a sense associated with the aims of the analyses.

As has been seen in the previous section, an important characteristic of plasticity problem is the large uncertainty in available information, especially in the constitutive law. In 2 and 3 dimensional problems, the uncertainty described by the error measure B is of order 30%-40% or more. (In the *one dimensional* setting we have seen in the best cases uncertainty of order 20-25%.)

The following problems arise:

1) How much the of uncertainty in the constitutive law is projected in the uncertainty of the data of interest?

2) Is it possible to formulate and solve the problem to determine the optimal bounds for the data of interest depending on uncertainties in available information?

3) How can one formulate and solve the problem in a stochastic sense? Here, of course, various probability fields or at least correlations of the experimental material properties would be needed and are practically unavailable today.

In [8] an attempt to derive bounds depending on the uncertainties in the constitutive law of the worst scenario was made. The idea is to assume that all uncertainty in constitutive law is in the unknown past. The problem is then to consider the "worst" possible past (in a set of admissible function) leading to the largest bound of data of interest.

There are other possibilities to address: the influence of the uncertainties analyzed by Monte Carlo methods and approaches of the type used in optimal design, etc.

It seems that any finite element consideration in plasticity problems using constitutive law without addressing the effects of uncertainties has to be interpreted and accepted in a very conservative way.

The problem of the correct mathematical formulation is a crucial one. The formulation of the (direct) plasticity problem guaranteeing well posedness can be achieved without any adverse effects of approximating experimental data as has been seen in Section 4.

The numerical treatment should lead to the approximate solution with acceptable accuracy in the sense needed for application. Today's available weak convergence theory seems to be far from what is needed.

Summarizing we see that there are major problems and questions related to the reliability of the computational analysis of the plasticity problems.

References

[1] Zuckowski, M: *Combined loading in the Theory of Plasticity.* P.W.N. Polish Scientific Publishers, Warsaw 1981.

[2] Chaboche, J.L.: Time Independent Constitutive Theories for Cyclic Plasticity, *Internal, J. of Plasticity* 2(1986) 149-188.

[3] Dafalis, Y.F., Popov, E.P.: Plastic Internal Variables Formalism of Cyclic Plasticity, *J. of Applied Mechanics* 98(1976) 645-657.

[4] Johnson, W., Mellor, P.B.: *Engineering Plasticity*, Ellis Harwood Ltd., Chichester, England 1983.

[5] Krempe, E.: Models of Viscoplasticity--Some Comments on Equilibrium (Back) Stress and Drag Stress. *Acta Mechanica* 69(1987) 25-42.

[6] Mroz, Z.: On the Description of Anistropic Work Hardening, *J. Mech. Phys. Solids* 15 (1967) 163-175.

[7] Valanis, K.C., Lee, C.F.: Endochronic theory of Cyclic Plasticity with Applications, *J. of Applied Mechanics* 51 (1984) 367-374.

[8] Bonnetier, E.: *Mathematical treatment of uncertainties appearing in the formulation of some models for plasticity*, Ph.D. thesis 1988, University of Maryland, College Park, MD 20742.

[9] Miyoshi, T.: *Foundations of the numerical analysis of plasticity*, North-Holland, Mathematics Studies Lecture Notes in Numerical and Applied Analysis, Vol. 7, 1985, eds. H. Fukita, M. Yamaguti.

[10] Li, Y.: *Theoretical and numerical problems of constitutive laws of plasticity and their applications*, Ph.D. Theses, University of Maryland, College Park, MD 20742. In preparation.

[11] Jerina, K.L., Babuska, I.: Mathematical modelling of Physical Systems with consideration of uncertainties. To appear.

Shear Band Formation in Impact Problems

O.T. Bruhns

Lehrstuhl für Mechanik I
Ruhr-Universität Bochum, D-4630 Bochum

ABSTRACT

The behaviour of metals under dynamic loading is described by introducing a continuum damage theory, which is also valid for nonisothermal processes. Microstructural changes are modelled by internal variables, e.g. damage due to shear bands is described with a second order tensor. Dislocation induced viscoplastic deformations are modelled using a flow rule of the overstress-type. This theory is applicable for strain rates of 10^{-4} s^{-1} up to $10^5 s^{-1}$.

1 INTRODUCTION

In many methods for the analysis of impact loaded structures S, a near field S_n in the vicinity of the point of impact and a far field S_f are distinguished, which are separated by the boundary ∂S_{nf}. In general, S_n is deformed inelastically, the state of stress is three dimensional and propagation phenomena are of importance. On the contrary, elastic behaviour prevails in S_f; concerning the state of stress the approximations of beam, plate or shell theory are valid and the characteristic times of the structural behaviour are large compared with L/c. Here L denotes a typical length of the structure and c is the velocity of propagation of acceleration waves. Hence the solution in S_f can be expanded in eigenfunctions corresponding to the lower eigenfrequencies, provided plausible assumptions are made about the stresses on ∂S_{nf}.

Structural failure under dynamic loading, however, frequently depends on the behaviour in S_n. There are exceptions of this observation, e.g. when a focussing of waves leads to failure at points of S_f , but nevertheless we are convinced that constitutive modelling and numerical simulation of the behaviour of S_n deserves more attention if one aims to predict structural failure. According to what has been said about the state of stress in S_n, a complete continuum mechanics description is required.

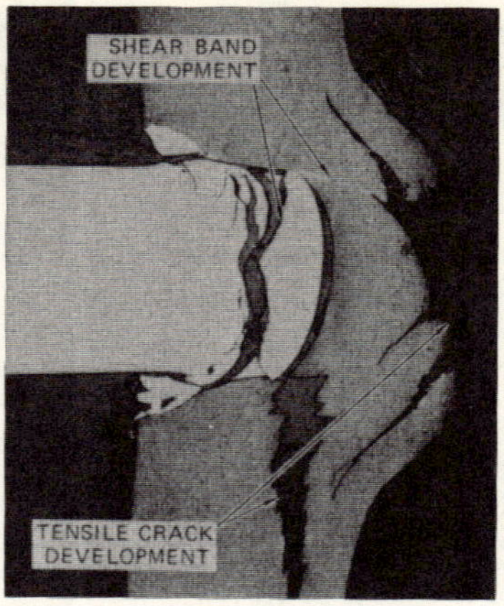

Fig. 1. Failure mechanisms during vertical impact (from Shockey et al. [1])

Fig. 1 shows a vertical impact of a blunt steel projectile against a steel target. The target plate fails by "plugging" due to shear band formation and by "spalling" due to cracks, which are nucleated by the tensile wave which is reflected on the rear side. Further investigations from Curran et al. [2] have shown that two types of mechanisms contribute to the inelastic behaviour of metals under dynamic loading, i.e. dislocation processes and (with a larger characteristic length) damage processes. The latter comprise nucleation, growth and coalescence of defects like microvoids, microcracks or micro shear bands. In vertical impact problems, shear band formation has been found the most important of these damage mechanisms.

In this context a defect will be called a microdefect if its extension is small compared with the linear dimension of a representative volume element ΔV of S; otherwise it will be named a macrodefect. During a typical loading process, the nucleation and the growth of these microdefects may finally result in the nucleation of a macrodefect at some instant time $t = t^*$. The time interval $[0, t^*]$, where $t = 0$ marks the beginning of the process, will be called phase 1. Only this phase can be described within the frame of continuum mechanics.

For phase 2., i.e. for $t > t^*$, we have to deal with the growth of macrodefects (either macrocracks or macro shear bands) in a damaged solid and thus elements of fracture mechanics have to be included in order to complete the constitutive relations. However, even for phase 2 an accurate description of phase 1 is necessary since otherwise all failure criteria (fracture, localization etc.) would be based on physically unrealistic values of Cauchy stress σ, temperature Θ etc.

In this paper we therefore will concentrate on the description of phase 1 processes. To this end in chapter 2 a continuum damage theory for moderate and high strain rate deformations is developed, which takes into account damage due to shear bands [3, 4]. This theory will be valid for strain rates between 10^{-4} sec^{-1} and 10^{5} sec^{-1} and for arbitrarily large deformations and rotations. The model has been implemented into 1-D [4] and 3-D axisymmetric [5] finite element codes. Some numerical results for vertical impact problems are given in chapter 3.

A description of the material behaviour during phase 2 would surpass the frame of this paper and is therefore reserved for a separate one. For some details we here refer to Diehl [4] where first fundamental results have been reported.

2 A CONTINUUM DAMAGE THEORY

2.1 General considerations

Consider a representative volume element ΔV as a polycrystal containing a finite number of defects like stress concentrators etc. The motion of this element may be defined by averaging values over ΔV, e.g.

$$\mathbf{D} := \text{averaging strain rates} \tag{2.1}$$

the rate of deformation tensor is the average effect of the different mechanisms that contribute to the motion in the disproportionately stressed grains of ΔV.

The continuum damage theory is now characterized by the additive decomposition of this rate of deformation tensor $\mathbf{D}$ according to

$$\left.\begin{array}{l} \mathbf{D} = \mathbf{D}_r + \mathbf{D}_i \qquad \text{with} \\ \\ \mathbf{D}_i = \mathbf{D}_{da} + \mathbf{D}_{dt} + \mathbf{D}_s = \mathbf{D}_d + \mathbf{D}_s \end{array}\right\} \tag{2.2}$$

where $\mathbf{D}_r$ and $\mathbf{D}_i$ are reversible and irreversible (inelastic) parts, respectively. $\mathbf{D}_i$ contains contributions $\mathbf{D}_{da}$ of athermal dislocation motion, $\mathbf{D}_{dt}$ of thermally activated dislocation motion and $\mathbf{D}_s$ of micro shear band formation. Additional variables are introduced in order to describe microstructural changes and to formulate necessary yield and loading conditions. The identification of these variables is based on an approximate homogenization method which is explained in [3]. We here simply refer to the results of this method. Let

$$\mathcal{Q} = \{\mathbf{X}, \varkappa, \boldsymbol{\tau}, \vartheta, \mathbf{Z}, \boldsymbol{\alpha}_q\} \tag{2.3}$$

be the set of internal variables and

$$\bar{\mathcal{Q}} = \{u, \mathbf{A}_{dl}\} \tag{2.4}$$

the set of process variables, where $\mathbf{X}$ is a kinematic hardening tensor and $\varkappa$ is an isotropic hardening parameter related to the total dislocation density N. The micro-stress concentration tensor $\boldsymbol{\tau}$ represents an average of local stress concentrations in the vicinity of microdefects. ϑ is an average of local temperature concentrations, $\mathbf{Z}$ is a symmetric positive definite damage tensor, and $u = N_{mob}/N$ is the quotient of mobile and total dislocation densities. $\mathbf{A}_{dl}$ in (2.4) is an average of local dislocation induced deformations in the vicinity of microdefects and $\boldsymbol{\alpha}_q$ is a vector-valued thermal variable which is related to heat flux vector $\mathbf{q}$.

For the description of nonisothermal processes the specific free enthalpy g is introduced as a function

$$g = g(\sigma, \Theta; \mathcal{Q}) \, , \tag{2.5}$$

while the process variables $\bar{\mathcal{Q}}$ do not alter the energy of the solid.

2.2 The constitutive law

The constitutive law comprehends:

i) the state function for the specific free enthalpy, from which the reversible strains and the specific entropy are deduced,
ii) evolution laws for the internal as well as the process variables $\mathcal{Q}$ and $\bar{\mathcal{Q}}$, respectively, and finally
iii) flow rules for $\mathbf{D}_{da}$, $\mathbf{D}_{dt}$, $\mathbf{D}_s$ and the inelastic spin $\mathbf{W}_i$.

According to the nonlinear inelastic behaviour of the material some of the flow rules and evolution laws are subjected to yield and/or loading conditions.

The decomposition (2.2) is motivated by the introduction of a so-called intermediate configuration $\hat{\chi}_t$ (cf. e.g. Lehmann [6]) with the following properties

$$\left.\begin{array}{l} \mathbf{F} = \mathbf{F}_r \mathbf{F}_i \\[2ex] \mathbf{F}_i : \chi_o \longrightarrow \hat{\chi}_t, \quad \mathbf{F}_r : \hat{\chi}_t \longrightarrow \chi_t \end{array}\right\} \tag{2.6}$$

of the reversible and the irreversible parts, respectively, of the deformation gradient $\mathbf{F}$, where χ_o denotes the initial configuration and χ_t the actual configuration of the considered body in motion.

Contrary to Lehmann [6] we here adopt the concept of isoclinic intermediate configurations with the special choice

$$\mathbf{F}_r = \mathbf{F}_r^T, \quad \mathbf{V}_r = \mathbf{F}_r \tag{2.7}$$

with $\mathbf{V}_r$ the reversible (left) stretch tensor. This concept has the advantage that a linear relation between $\mathbf{D}_r$ and an objective rate of the reversible Green's strain tensor

$$\mathbf{E}_r := \frac{1}{2}(\mathbf{F}_r^T\mathbf{F}_r - \mathbf{1}) \tag{2.8}$$

can be formulated irrespective of any isotropy condition, namely

$$\mathbf{D}_r := \mathrm{sym}(\overset{\circ}{\mathbf{V}}_r\mathbf{V}_r^{-1}) = \mathbf{V}_r^{-1}\overset{\circ}{\mathbf{E}}_r\mathbf{V}_r^{-1} \tag{2.9}$$

and where the objective time derivative (Jaumann-type rate), e.g.

$$\overset{\circ}{\mathbf{A}} := \dot{\mathbf{A}} - (\mathbf{W}-\mathbf{W}_i)\mathbf{A} + \mathbf{A}(\mathbf{W}-\mathbf{W}_i) \tag{2.10}$$

which has been introduced in the intermediate configuration has the (chain rule) property

$$\frac{\partial\varphi}{\partial\mathbf{A}}\cdot\dot{\mathbf{A}} = \frac{\partial\varphi}{\partial\mathbf{A}}\cdot\overset{\circ}{\mathbf{A}} \quad . \tag{2.11}$$

The consequence on the other hand, however, is that an additional constitutive law is needed for the inelastic spin $\mathbf{W}_i$ (cf. Dafalias [7, 8]).

The evolution laws and flow rules are first formulated for variables with respect to this intermediate configuration and thereafter transformed to the actual configuration. This conversion includes multiplications with $\mathbf{F}_r$ and is simplified in the present case since in metals only small reversible changes of the shape occur

$$\mathbf{F}_r \approx \Delta\mathbf{1}\,, \quad \Delta = \det(\mathbf{F}_r) \;:\; \hat{\chi}_t \longrightarrow \chi_t \;. \tag{2.12}$$

The formulation of constitutive laws with respect to the actual configuration is particularly helpful for an updated Lagrangian formulation of the finite element code.

In this paper the constitutive laws are listed in a somewhat simplified form compared with Diehl [4]; this means. e.g. that only for simplicity certain recovery terms in the hardening laws for $\mathbf{X}$ and $\varkappa$ and moreover some coupling effects of minor interest have been neglected.

i) Reversible part of deformations

Contrary to what is usually adopted in the formulation of elastic-plastic constitutive relations in the present case we no longer maintain the assumption that the inelastic deformations will not influence the reversible material behaviour. This is especially due to the fact that the damage will influence the material properties (stiffness of the material) in the elastic domain.

Under the aforementioned conditions the reversible strains follow from enthalpy function $g(\cdot)$

$$\mathbf{E}_r = \frac{f(\zeta)}{2\mu\Delta}\left\{\sigma - \frac{\nu}{1+\nu}\operatorname{tr}(\sigma)\,\mathbf{1}\right\} + \alpha_t(\Theta - \Theta_o)\mathbf{1} \tag{2.13}$$

and thus the rate of deformation tensor becomes (2.9)

$$\mathbf{D}_r = \frac{f(\zeta)}{2\mu\Delta^3}\left\{\overset{\circ}{\sigma} - \frac{\nu}{1+\nu}\operatorname{tr}(\overset{\circ}{\sigma})\,\mathbf{K}(\cdot)\right\} + \frac{1}{\Delta^2}\,\alpha_t\dot{\Theta}\,\mathbf{B}(\cdot) \tag{2.14}$$

for an instantaneously isotropic material. Herein μ (shear modulus), ν (Poisson's ratio) and α_t (thermal expansion coefficient) are the material parameters of the undamaged elastic material. The influence of the damage tensor $\mathbf{Z}$ on the elastic response is introduced by functions

$$\Delta = \Delta(\sigma, \Theta, \mathbf{Z}), \qquad f = f(\zeta) \tag{2.15}$$

wherein $\zeta := \|\mathbf{Z}\|$ and

$$\mathbf{K} = \mathbf{K}(\sigma, \Theta, \mathbf{Z}), \qquad \mathbf{B} = \mathbf{B}(\sigma, \Theta, \mathbf{Z}). \tag{2.16}$$

are material tensors.

Neglecting now in both functions of (2.16) some higher order terms which for processes with moderate velocities remain small compared with one we finally can arrive at $\mathbf{K} = \mathbf{1}$ and $\mathbf{B} = \mathbf{1}$, where $\mathbf{1}$ is the unit tensor. Thus $\mathbf{K}(\cdot)$ and $\mathbf{B}(\cdot)$ in (2.14) describe (higher order) anisotropy effects due to damage of the material. These effects will be disregarded in the following.

ii) Flow rules - part 1: Dislocation induced inelastic deformations

Neglecting for moderate up to high strain rates the contribution of the athermal dislocation motion the dislocation induced inelastic rate of deformation tensor reads

$$\left.\begin{array}{l} \mathbf{D}_d = \mathbf{D}_{dt} = \gamma_t(u,\dots)\left[\Phi_t\right]_t \mathbf{N}_t \qquad \text{if} \\ \\ F_t = \dfrac{1}{\Delta^2}(\sigma' - \mathbf{X}')\cdot(\sigma' - \mathbf{X}') - g_t(x, \Theta, u) \geq 0 \,, \end{array}\right\} \tag{2.17}$$

where

$$\mathbf{N}_t = \frac{\sigma' - \mathbf{X}}{\|\sigma' - \mathbf{X}\|} \tag{2.18}$$

is the normal to the surface $F_t = \text{const.}$, $F_t \geq 0$ is the yield condition and the square brackets here and in the following mean

$$[x]_y = \begin{cases} x & \text{if } F_y \geq 0, \\ 0 & \text{otherwise} \end{cases} \tag{2.19}$$

$F_y \geq 0$ is a yield condition for changes of a variable $y \in (\mathcal{J} \cup \bar{\mathcal{J}})$. The prime $(\;)'$ in (2.17) as usual denotes the deviatoric part of a tensor. Here and in what follows functions γ_y, Φ_y and g_y have been introduced as in usual viscoplasticity except an additional dependency of some of these functions on the process variable u, which has been taken to allow for the necessary transitions of $\mathbf{D}_i$ according to the rate dependency of the different contributions. Moreover different functions have been introduced for the description of the different mechanisms that contribute to the motion.

Equation (2.17) means that an associated flow rule is used. It has been found that this is a sufficient condition for compatibility with the second law of thermodynamics.

iii) Flow rules - part 2: Shear band induced processes

For the shear band induced motion we find in agreement with the second law of thermodynamics

$$\mathbf{D}_s = \left[\Phi_s\right]_s \langle\langle\langle LC_s \rangle\rangle\rangle \frac{\sigma'}{\|\sigma'\|} \qquad \text{if} \tag{2.20}$$

$$\left.\begin{aligned} F_s &= \frac{1}{\Delta^2}\,\tau' \cdot \tau' - \frac{\nu_s}{\Delta}\,\mathrm{tr}\,\tau - g_s(\zeta) \geq 0 \quad \text{and} \\ LC_s &= \frac{1}{\Delta}\left(\frac{2}{\Delta}\,\tau' - \nu_s \mathbf{1}\right) \cdot \left(\overset{\#}{\tau} + m\tau\right) > 0. \end{aligned}\right\} \qquad (2.21)$$

where $\langle\langle\langle \cdot \rangle\rangle\rangle$ denotes a Maccauly bracket with respect to the loading condition $LC_s > 0$. We note that herein and in what follows the necessary yield and loading conditions for damage processes are formulated in (τ, ϑ)-space rather than in (σ, Θ)-space as in the case of the dislocation processes in (2.17), since damage is connected with local stress and temperature concentrations at weak spots within the material. Moreover, damage processes depend on the local hydrostatic pressure through $\mathrm{tr}\,\tau$. Furthermore

$$\overset{\#}{\tau} := \overset{\circ}{\tau} - \tau\,\mathrm{tr}(\mathbf{D}) \qquad (2.22)$$

is a second objective time derivative which has to be introduced in addition to (2.10) in connection with the transformation to the actual configuration.

iv) Flow rules - part 3: Inelastic spin

Following some ideas of Mandel [9], Dafalias [7, 8] has introduced the concept of plastic spin where

$$\mathbf{W} := \boldsymbol{\Omega} + \mathbf{W}_i \qquad (2.23)$$

$\mathbf{W}$ is the total spin, $\mathbf{W}_i$ is the inelastic spin defined as antisymmetric part of the inelastic velocity gradient and $\boldsymbol{\Omega}$ is explained as rigid body spin of the substructure, which is different from the material spin $\mathbf{W}$. We note that herein the reversible part of the total spin has been neglected for the assumed small reversible distortions. According to this concept the inelastic spin or $\boldsymbol{\Omega}$, respectively, have to be given by a proper constitutive law. We note further that to our mind there is no necessity for explaining $\boldsymbol{\Omega}$ as mentioned above, since $\boldsymbol{\Omega}$ is defined through (2.23) once a constitutive relation for the inelastic spin has been formulated.

Therefore similar as in (2.2) $\mathbf{W}_i$ has to be decomposed into

$$\mathbf{W}_i = \mathbf{W}_{dt} + \mathbf{W}_s \qquad (2.24)$$

the different parts that contribute to the motion, where here the simplification of (2.17) has already been taken into consideration. Since the contribution of the damage to the motion can be assumed small compared with the thermally activated dislocation motion $\mathbf{W}_s$ will be neglected and thus

$$\mathbf{W}_i = \gamma_w(u)\left[\Phi_w\right]_t \frac{(\mathbf{X}\sigma' - \sigma'\mathbf{X})}{\|\sigma'\|\,\|\mathbf{X}\|} \;; \qquad \text{if } F_t \geq 0 \,. \tag{2.25}$$

Here $\mathbf{W}_i$ is introduced in a form of the same type as (2.17). We note that this form differs somewhat from the relations of Paulun and Pecherski [10], where with the use of the representation theory different tensor functions were given. Furthermore similar as has been mentioned by Dafalias [11] the overstress-function Φ of (2.25) will also be taken from (2.17), i.e. $\Phi_w = \Phi_t$.

v) Evolution laws - part 1: Dislocation variables

The evolution of the hardening of the material is here described in an analogous way as in usual viscoplasticity (cf. Perzyna and Wojno [12], Chaboche [13] and Bruhns [14])

$$\left.\begin{aligned} \overset{\#}{\mathbf{X}} &= \Delta c_t(\varkappa)\, \mathbf{D}_d \\ \dot{\varkappa} &= k(\varkappa)\, \sigma \cdot \mathbf{D}_d \end{aligned}\right\} \tag{2.26}$$

where $c_t(\cdot)$ and $k(\cdot)$ are functions of the average dislocation density of ΔV.

vi) Evolution laws - part 2: Damage variables

It is known from the theory of (elastic) composites that the local stress concentrations τ can be related to the overall stresses σ e.g. by

$$\overset{\#}{\tau} = \mathbb{D}\, \overset{\#}{\sigma} \tag{2.27}$$

(Dvorak [15]) where $\mathbb{D}$ is a positive definite material tensor. This relation here will be generalized in the form

$$\overset{\#}{\tau} = \mathbb{D}\, \overset{\#}{\sigma} + d_4 \dot{\varkappa}\, \tau - m\tau \tag{2.28}$$

where

$$\mathbb{D}\, \overset{\#}{\sigma} = d_1 \overset{\#}{\sigma} + \frac{d_2}{\Delta}(\tau \overset{\#}{\sigma} + \overset{\#}{\sigma} \tau) + \frac{d_3}{\Delta}(\mathbf{Z} \overset{\#}{\sigma} + \overset{\#}{\sigma} \mathbf{Z}) \tag{2.29}$$

and $m(\cdot)$ and $d_1(\cdot)$ to $d_4(\cdot)$ are yet undetermined functions of $\|\tau\|$ and the process variables u and $\|\mathbf{A}_{dl}\|$, respectively.

For the evolution of the local temperature concentrations we adopt a relation

$$\dot{\vartheta} = \left[\Phi_\Theta\right]_V \quad \text{if } F_V = \frac{1}{\Delta^2}\,\tau' \cdot \tau' - g_V \geq 0 \tag{2.30}$$

i.e we here neglect possible contributions of damage induced deformations to the evolution of the temperature concentration. The evolution of the damage tensor **Z** is characterized by nucleation and growth of the damage. We thus state

$$\overset{\#}{\mathbf{Z}} = \left[\Phi_Z\right]_Z \langle\langle LC_Z \rangle\rangle \,(\mathbf{N}_n + \mathbf{N}_g) \quad \text{if} \tag{2.31}$$

$$\left.\begin{aligned} F_Z &= \frac{1}{\Delta^2}\,\tau' \cdot \tau' - \frac{\nu_Z}{\Delta}\,\mathrm{tr}\,\tau - g_Z(\vartheta) \geq 0 \quad \text{and} \\ LC_Z &= \frac{1}{\Delta}\left(\frac{2}{\Delta}\,\tau' - \nu_Z \mathbf{1}\right) \cdot \left(\overset{\#}{\tau} + m\tau\right) > 0. \end{aligned}\right\} \tag{2.32}$$

where now $\langle\langle \cdot \rangle\rangle$ is a second Maccauly bracket with respect to the loading condition $LC_Z > 0$. Herein in a specific form

$$\mathbf{N}_n = \frac{\mu_1}{\Delta^2}\,\sigma'^2, \qquad \mathbf{N}_g = \frac{1}{\Delta}(\mu_2 + \omega(\zeta, \sigma))\mathbf{Z} \tag{2.33}$$

describe the changes of **Z** due to shear band nucleation and shear band growth, respectively. **Z** itself contains informations about average shear band size and average shear band orientation and can be used in formulating the failure criteria.

vii) Evolution laws - part 3: Thermal variables
We introduce

$$\overset{\circ}{\boldsymbol{\alpha}}_q = \Delta\,\mathrm{grad}\,\Theta - \frac{1}{\tau_q}\,\boldsymbol{\alpha}_q\,, \tag{2.34}$$

where the objective rate $(\overset{\circ}{\cdot})$ is according to (2.10) defined as

$$\overset{\circ}{\boldsymbol{\alpha}}_q = \dot{\boldsymbol{\alpha}}_q - (\mathbf{W} - \mathbf{W}_i)\,\boldsymbol{\alpha}_q \tag{2.35}$$

for vector-valued variables. With an appropriate choice of the specific free enthalpy function g, we can derive $\mathbf{q} = -(\boldsymbol{k}/\tau_q)\boldsymbol{\alpha}_q$ from the entropy inequality (cf. Kosinski [16]) and thus find

$$\tau_q\,\overset{\circ}{\mathbf{q}} + \mathbf{q} = -\,\Delta\,\boldsymbol{k}\,\mathrm{grad}\,\Theta. \tag{2.36}$$

where k is the heat conductivity. (2.36) is known as the Maxwell-Cattaneo law of heat conduction.

viii) Evolution laws - part 4: Process variables $\bar{\mathcal{Q}}$

The evolution law for the process variables u has to be adopted in the form of a second order differential equation

$$\ddot{u} + c\dot{u} + \frac{c^2}{4}(u - u_{eq}) = 0 ; \qquad \text{with } c = c_0(\alpha + v_p) \tag{2.37}$$

to fulfill the above mentioned transition conditions. Here $v_p := \max\{0, \|\mathbf{D}\| + \chi\dot{\Theta}\}$ is a sufficient measure to describe the process velocity and $u_{eq}(\cdot)$ is the equilibrium value of u which for different values of v_p can be determined from experiments (e.g. Shioiri and Sakino [17]).

Finally the influence of the local dislocation motion on the deformations is described by a most simple rate dependent relation of the overstress-type

$$\overset{\ast\ast}{\mathbf{A}}_{dl} = \Delta^2 \gamma_{dl} \left[\Phi_{dl} \right]_v \frac{\tau'}{\|\tau'\|} + (\mathrm{tr}\mathbf{D})\mathbf{A}_{dl} : \quad \text{if } F_v \geq 0 . \tag{2.38}$$

2.3 Failure criteria

Two different criteria for failure of the structure can be formulated

i) failure due to micro shear band formation during phase 1 $t \leq t^*(\mathbf{x})$, and

ii) nucleation of a macro shear band at $t^*(\mathbf{x})$.

Both criteria depend on the magnitude of the representative volume element ΔV that has been selected according to the grain size of the material and the characteristic length of the loading of the specific problem.

A useful indicator of micro shear band damage is the relative part ξ of the shear band area of a reference plane $\Delta\mathbf{A} = \Delta A\,\mathbf{n}$ of ΔV

$$\xi(\mathbf{x}, \mathbf{n}) := \sum \frac{\delta A_\alpha}{\Delta A} \, |\mathbf{e}_\alpha \cdot \mathbf{n}| \tag{2.39}$$

where δA_α is the area of a typical micro shear band with direction $\mathbf{e}_\alpha$ and the summation runs over all micro shear bands that are intersected by the reference plane ΔA. If ξ exceeds a threshold value ξ_{crit} no further shear stress can be transmitted. We therefore postulate that damage will be not dangerous as long as

$$\xi(\mathbf{x}, \mathbf{n}) \leq \xi_{crit}, \quad \forall\, \mathbf{x}, \mathbf{n} . \tag{2.40}$$

ξ itself is not a material property since it depends on $\mathbf{n}$. However, bounds of ξ can be expressed in terms of damage tensor $\mathbf{Z}$ [4]. We thus can formulate

$$\xi(\mathbf{x}, \mathbf{n}) \leq c_1 \, \| \mathbf{Z} \|^3 \tag{2.41}$$

and with this result criterion (2.40) can be formulated somewhat more rigorously

$$c_1 \, \| \mathbf{Z} \|^3 \leq \xi_{crit}, \quad \forall \mathbf{x} \, . \tag{2.42}$$

Since, however, a few rather crude estimates are involved in the derivation of (2.42) one might ask for a phenomenological generalization. This leads us to the demand that a monotonically increasing function of $\| \mathbf{Z} \|$ must be bounded

$$\| \mathbf{Z} \| \leq Z_{crit} \tag{2.43}$$

where Z_{crit} can be determined empirically.

As to the criterion for the nucleation of a macro shear band we propose the following. Let $\mathcal{M}^* = \mathcal{M}(t^*)$ be the initial configuration of this macro shear band at time $t = t^*$. Let furthermore $z_i \geq 0$, $(i = 1, 2, 3)$ be the (not identical) eigenvalues of $\mathbf{Z}$ at t^* and $\mathbf{e}^z{}_i$ the corresponding eigenvectors. If then the maximum of the eigenvalues z_{max} attains the value

$$z_{max} = \triangle V^{-1/3} \tag{2.44}$$

then a macro shear band $\mathcal{M}^*$ is nucleated at $\mathbf{x}$ with

$$A^* = \triangle V^{2/3} \, , \quad \mathbf{n}^* = \mathbf{e}^z{}_{max} \tag{2.45}$$

initial area A^* and orientation $\mathbf{n}^*$, respectively.

2.4 Identification of material parameters

A strategy for the determination of the material parameters was first discussed by Diehl and Fornefeld [18] and later on completed by Fornefeld [5]. Since experiments with high strain rate deformations are characterized by inhomogeneous states of stress and strain, the interpretation of experiments is difficult and fitting the parameters to experimental curves requires the numerical solution of an initial and boundary value problem. Therefore the identification process represents an iterative procedure where an initial set $\mathcal{C}$ of parameters is chosen such that the simulation of fictitious homogeneous deformations gives plausible results. $\mathcal{C}$ is modified until the numerical simulation coincides with experimental results.

A useful mathematical tool for this iteration is the evolutionary strategy. This is an statistical optimization method for engineering systems, which imitates the principles of biological evolution like mutation and selection and which was introduced by Rechenberg [19]. The evolutionary strategy easily circumvents relative minima of the cost function, where classical deterministic methods are "trapped".

The complete set of experimental data necessary for the identification process has yet not been performed for a real material. We therefore have defined a so-called reference material, whose behaviour is representative for a wide class of metals and alloys. This reference material is defined by prescription of σ-ε-curves at different constant strain rates, creep curves, relaxation curves etc. The investigation of material data for an armour steel is currently under progress.

2.5 Some results for homogeneous uniaxial states of stress

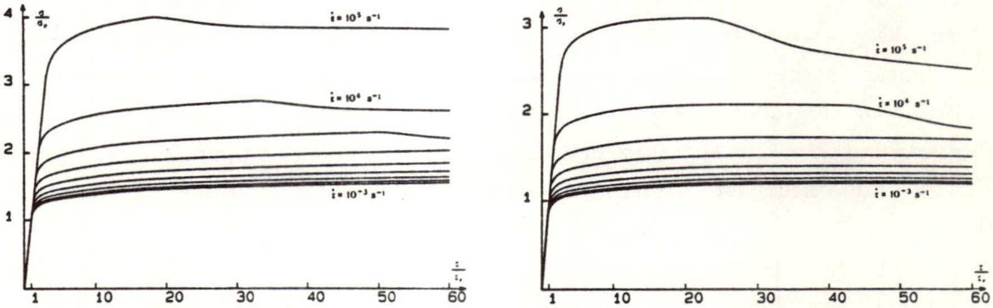

Fig. 2. Stress-strain curves for different strain rates and initial temperatures, σ/σ_F versus $\varepsilon/\varepsilon_F$, where σ_F is the flow stress for extremely slow processes. $\varepsilon_F = \sigma_F/E$ (E is Young's modulus), a) Θ_{init} = 300 K (left), b) Θ_{init} = 600 K (right)

The main features of the constitutive model of equations (2.13) to (2.38) can be illustrated in the case of small deformations and uniaxial states of stress. Fig. 2 shows two families of σ-ε-curves for different $\dot\varepsilon$. Figs. 2.a and 2.b differ with respect to the initial temperature. We observe

- an increasing stress level with increasing $\dot\varepsilon$,
- an increasing flow stress with increasing $\dot\varepsilon$,
- a decreasing stress level with icreasing Θ.

- a decreasing flow stress with icreasing Θ,
- damage induced softening at larger strains only for sufficiently large $\dot{\varepsilon}$.

These observations are in agreement with experimental results. Further predictions of our model in the case of cyclic tests, relaxation tests, strain rate jump tests etc. can be found in [4] together with predictions for combined tension-torsion tests. Several time integration methods were used for the integration of the rather stiff system (2.13)-(2.38). With this respect the explicit methods of Runge-Kutta and of Fehlberg as well as the implicit trapezoidal rule have been compared. The Fehlberg algorithm was found to be the most efficient one [5].

3. NUMERICAL SIMULATION OF VERTICAL IMPACTS

3.1 Uniaxial states of stress

As a preliminary examination the impact of a slender rod against a rigid wall with an impact velocity $v_0 \lesssim 50$ m/sec is studied. We assume that effects of radial inertia can be neglected and thus a uniaxial state of stress prevails. 25 simplex elements for the discretization of a rod of length $l_0 = 5 \cdot 10^{-2}$m and diameter $d_0 = 6 \cdot 10^{-3}$m were used. Furthermore an explicit formulation of the balance equations of momentum and energy (in the terminology of Bathe [20]) and also explicit time integration of the incremental constitutive law were used. The use of conditionally stable explicit time integration procedures is justified in impact problems since very small time steps are already required for reasons of accuracy (cf. Belytschko [21]).

For $v_0 \lesssim 10$ m/sec elastic behaviour was observed. This could be expected from the classical result $\sigma = \rho c v$, which holds along characteristics during uniaxial elastic wave propagation.

For $v_0 = 100$ m/sec the following results were obtained [4]:
i) Even for very high rates of strain. the assumption of adiabatic behaviour should be inspected carefully. Calculations with $\boldsymbol{k} = 0$ and $\boldsymbol{k} \neq 0$, respectively, gave temperature rises which differ by about 50 %. This discrepancy has to be explained with the extremely large temperature gradients in the rod. Thus the assumption of adiabatic conditions in processes with high rates of strain frequently used in the literature has to be questioned when large temperatute gradients are involved.
ii) At $t \approx 9.7\,\mu$sec the incident wave arrives at the end of the rod. Then at the head of the rod the following values of variables can be observed

$$|\varepsilon|_{max} = 14.1\ \%$$
$$\Delta\Theta_{max} = 16.8\ \text{K}$$
$$N_{max} = 2.34 \cdot 10^5\ \text{m}^{-2}$$

The damage accumulation is concentrated in the first three elements of the discretized rod with a maximum value of $Z_{max} = 1.53\ \text{m}^{-1}$. If one element is now regarded as representative volume element we find $\Delta V = 5.65 \cdot 10^{-8}\ \text{m}^3$ and thus from (2.44) that

$$Z_{max} < \Delta V^{-1/3} = 260.6\ \text{m}^{-1}$$

and therefore damage is far from a critical value. The damage, however, causes softening of the material which is observed in the stress-curves.

iii) About 80 % of the CPU time were consumed during the time integration of the constitutive law.

3.2 Three dimensional axisymmetric states of stress

The calculation of adiabatic axisymmetric high strain rate deformations is based on the principle of virtual work for the configuration at t $+\Delta t$ (i.e. an implicit formulation was used) and on an updated Lagrangian formulation. We note that according to our aforementioned results the assumption of adiabatic conditions might be questionable since it could lead to slight modifications of the results. Therefore a careful examination of the temperature field seems to be necessary. Moreover future considerations will take care about this point by introducing the heat flux into the calculation. The resulting nonlinear equations of motion

$$\underline{\underline{M}}\ {}^{t+\Delta t}\underline{\ddot{U}} = {}^{t+\Delta t}\underline{R} - {}^{t+\Delta t}\underline{F} \tag{3.1}$$

($\underline{\underline{M}}$ = consistent mass matrix, $\underline{U}$ = displacement vector of global degrees of freedom, $\underline{R}$ = vector of nodal forces due to external loads, $\underline{F}$ = vector of nodal forces due to internal stresses) are solved iteratively with the modified Newton-Raphson method:

$$\underline{\underline{M}}\,{}^{t+\Delta t}\underline{\ddot{U}}^{[k]} + {}^{t}\underline{\underline{K}}\ \Delta\underline{U}^{[k]} = {}^{t+\Delta t}\underline{R} - {}^{t+\Delta t}\underline{F}^{[k-1]} - {}^{t}\underline{\underline{Q}}\,\delta_{k1}, \tag{3.2}$$

$$ {}^{t+\Delta t}\underline{U}^{[k]} = {}^{t+\Delta t}\underline{U}^{[k-1]} + \Delta\underline{U}^{[k]}\ , \tag{3.3}$$

for k = 1, 2,.... This equation differs from that given by Bathe [20] inasmuch as only an elastic "tangential" stiffness matrix $\underline{\underline{K}}$ can be derived in the presence of rate dependent constitutive relations. On the contrary, an additional load vector $\underline{Q}$ appears, which contains $\mathbf{D}_i$ as well as the rate of temperature.

The iterative search for dynamic equilibrium in (3.2) has to be combined with the time integration of the equations of motion. Here the Newmark-method (with δ = 1/2, α = 1/4) has been compared with the Wilson-Θ-method (with the recommended value of Θ = 1.4). Both methods are unconditionally stable, however, the Newmark-method was superior to the Wilson-Θ-method due to the fact that the latter has overestimated the velocity of wave propagation. After each step of the iteration, the constitutive laws (2.13)-(2.44) and the energy balance (which is reduced to an "evolution law" for temperature during adiabatic processes) were integrated explicitly.

We used isoparametric 4-, 8- and 12-node elements. Higher order elements were by far superior to elements of low order. 9 point Gaussian quadrature was used for the 8-node elements.

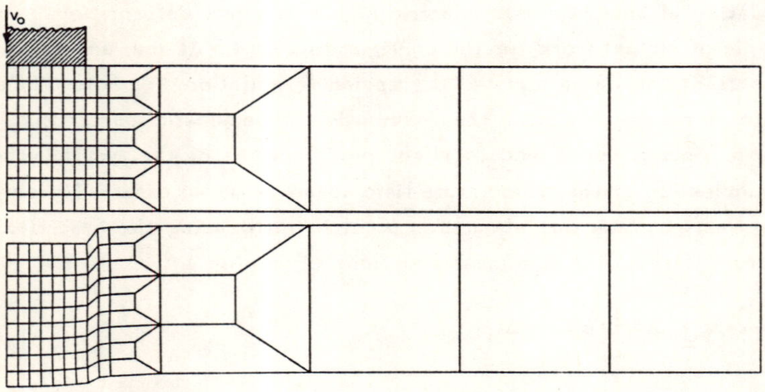

Fig. 3. FE discretization (upper part) and deformations at t = 20 µsec (lower part) of a circular plate

As an example of application the problem of a vertical impact of a cylindrical (rigid) projectile (d_o = 30 mm, v_o = 100 m/s) on a circular plate of dimensions: thickness 30 mm and diameter 300 mm has been calculated. The discretization of the plate as well as the deformations of the structure after 20 µsec are depicted with Fig. 3. Furthermore with Figs. 4 to 6 the spatial variations of the Z_{zz}-components of the

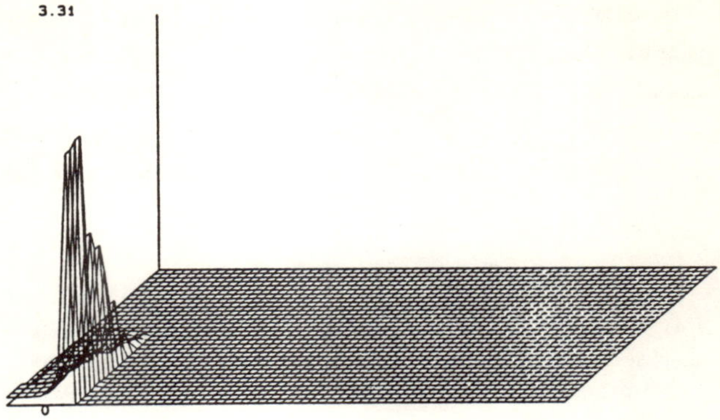

Fig. 4. Z_{zz} distribution at t = 8 μsec

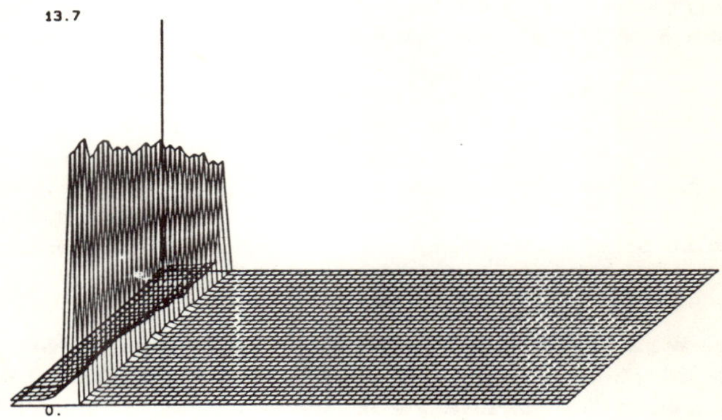

Fig. 5. Z_{zz} distribution at t = 20 μsec

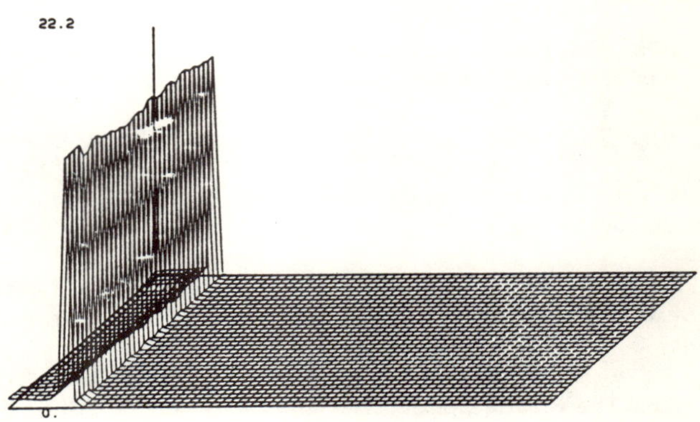

Fig. 6. Z_{zz} distribution at t = 40 μsec

damage tensor are shown at different times t = 8, 20, 40 μsec. The maximum values of these components were found to be $Z_{zz\,max}$ = 3.31. 13.7, 22.2 m^{-1}, which here indicate micro (and macro) shear band formation in a failure zone of cylindrical geometry. Together with a concentrated heat generation in this zone this is a distinct indication for plugging of the plate due to adiabatic shear band formation.

References

1. Shockey, D.A., Seaman, L. and Curran, D.R.; Microfailure Models and Their Application to Nonlinear Dynamic Fracture Problems, in Perrone, N. and Atluri, S.N. (eds.), Nonlinear and Dynamic Fracture Mechanics, ASME AMD-Vol. 35 (1979), 79-105

2. Curran, D.R., Seaman, L. and Shockey, D.A.; Linking Dynamic Fracture to Microstructural Processes, in Meyers, M.A. and Murr, L.E. (eds.), Shock Waves and High Strain Rate Phenomena in Metals, New York: Plenum Press, 1981

3. Bruhns, O.T. and Diehl, H.; An Internal Variable Theory of Inelastic Behaviour at High Rates of Strain", Arch. Mech. **41** (1989), 427-460

4. Diehl, H.; Ein Materialmodell zur Berechnung von Hochgeschwindigkeitsdeformationen metallischer Werkstoffe unter besonderer Berücksichtigung der Schädigung durch Scherbänder, Report No. 66, Institute of Mechanics, RU Bochum 1989

5. Fornefeld, W.; Zur Parameteridentifikation und Berechnung von Hochgeschwindigkeitsdeformationen metallischer Werkstoffe anhand eines Kontinuums-Damage -Modells, Report No. 73, Institute of Mechanics, RU Bochum 1990

6. Lehmann, T., The Constitutive Law in Thermoplasticity, CISM Courses and Lectures No. 281, Springer-Verlag 1984

7. Dafalias, Y.N.: The Plastic Spin Concept and a Simple Illustration of its Role in Finite Plastic Transformations, Mech. Materials **3** (1984), 223-233

8. Dafalias, Y.N.: The Plastic Spin, J. Appl. Mech. **52** (1985), 865-871

9. Mandel, J.; Plasticitè classique et viscoplasticitè, CISM Courses and Lectures No. 97, Springer-Verlag 1971

10. Paulun, J.E. and Pecherski, R.B.; On the Application of the Plastic Spin Concept for the Description of Anisotropic Hardening in Finite Deformation Plasticity, J. Plasticity **3** (1987) 303-314

11. Dafalias, Y.F.; The Plastic Spin in Viscoplasticity, Int. J. Solids Struct. **26** (1990), 149-163

12. Perzyna, P. and Wojno, W.; Unified Constitutive Equations for Elastic-Viscoplastic Material, Bull. Acad. Polon. Sci., Sèr. Sci. Techn. **24** (1976), 85-94

13. Chaboche, J.L.; Constitutive Equations for Cyclic Plasticity and Cyclic Viscoplasticity, Int. J. Plasticity **5** (1989), 247-302

14. Bruhns, O.T.; On the Constitutive Relations of Austenitic Stainless Steels, in Desai, C.S., Krempl, E., Kiousis, P.D. and Kundu, T. (eds.), Constitutive Laws for Engineering Materials - Theory and Applications, Elsevier 1987

15. Dvorak, G.J.; Mechanical Properties of Composites. in Boehler, J.P. (ed.), Mechanical Behaviour of Anisotropic Solids, M. Nijhoff Publ. 1982

16. Kosinski, W.; Thermal Waves in Inelastic Bodies, Arch. Mech. **27** (1975), 733-748

17. Shioiri, J. and Sakino, K.; Comparison of Flow Stress of Aluminum at High Rates of Strain Calculated on the Basis of Ultrasonic Study of Dislocation Behavior with Directly Measured Ones, in Chiem, C.Y., Kunze, H.D. and Meyer, L. W. (eds.), Impact Loading and Dynamic Behaviour of Materials, DGM 1988

18. Diehl, H. und Fornefeld, W.; Werkstoffverhalten bei hochdynamischer Belastung - FE-Rechnungen zu Phase 1, theoretische Beschreibung von Phase 2, Forschungsbericht Institut für Mechanik, RU Bochum 1990

19. Rechenberg, I.; Evolutionsstrategie, Stuttgart: F. Frommann Verlag 1973

20. Bathe, K.J., Finite-Elemente-Methoden, Berlin: Springer-Verlag 1986

21. Belytschko, T.; An Overview of Semidiscretization and Time Integration Procedures, in Belytschko, T. and Hughes, T.J.R. (eds.), Computational Methodes for Transient Analysis, North-Holland 1983

22. Wriggers, P.; Zur Berechnung von Stoß- uund Kontaktproblemen mit Hilfe der Finite-Element Methode, Bericht Nr. F81/1, Lehrgebiet Baumechanik, U Hannover 1981

On the Improvement of Fracture Toughness due to Transformation Induced Plasticity

H. Okada, T. Tamura, N. Ramakrishnan[1], S. N. Atluri and J. S. Epstein[2]
Computational Mechanics Center
Georgia Institute of Technology
Atlanta, Georgia 30332-0356, U.S.A.

ABSTRACT

An analysis on transformation toughening of magnesia partially stabilized zirconia (Mg-PSZ) is presented in this paper. Transformation toughening of Mg-PSZ is attributed to the stress induced phase transformation of tetragonal zirconia to monoclinic structure at the neighborhood of crack tip. A rate form constitutive model [13, 14], using a micromechanics approach, is employed. Stationary and stably propagating crack problems are analyzed, using finite element method. The results of finite element analysis are compared to those of experimental study of Perry et al. [12]. In the comparison, it is found that displacement field and toughness enhancement during stable crack propagation predicted by the finite element analysis, are very analogous to those obtained in the experimental study.

1 Introduction

Stress induced phase transformation of zirconia in ceramic materials has been found to improve the fracture toughness. Tetragonal phase zirconia contained in certain ceramic materials such as zirconia toughened alumina (ZTA) and partially stabilized zirconia (PSZ), as their second phase particles, undergo phase transformation to monoclinic structure under an applied stress. This stress induced phase transformation results in a nonlinear stress-strain relationship at macroscopic level. According to Evans and Cannon [7], this phase transformation of zirconia is associated with 4–6 % of dilatation and about 16 % of shear without constraint from surrounding material. When a macroscopic crack is present in the material, high stress field around the crack tip induces the zirconia phase transformation, thereby resulting in a significant modification of stress field around the crack tip. Inside the transformed zone, fully and partially transformed zones are considered [2]. In the fully transformed zone, all the transformable tetragonal zirconia particles are transformed, and in the partially transformed zone, a portion of the transformable zirconia particles are transformed. The fully transformed zone is embedded in the partially transformed zone. In the fully transformed zone, the material behaves linearly, and in the partially transformed zone, the material exhibits nonlinear stress-strain behavior. Therefore, if the fully transformed zone exists, the stress field immediately near the crack tip has an $1/\sqrt{r}$

[1]on leave from DMRL. Kanchanbagh. Hyderabad. India-500258

[2]Currently at Fracture and Dynamic Behavior Group. Idaho National Engineering Laboratory. P.O. BOX 1625. IRC-2218. Idaho Falls. Idaho 83415-2218. U.S.A.

singularity, characterized by the *crack tip stress intensity factor* K_I^{tip}. And stress field is characterized by the *far field stress intensity factor* K_I^{far} outside the transformed zone. The stress intensity at the crack tip is reduced in certain conditions.

In this paper, a constitutive model of Okada [13, 14], based on a micromechanics approach, is used to predict the toughness enhancement effect. The constitutive model is implemented in a nonlinear finite element code. In order to evaluate toughness enhancement in magnesia partially stabilized zirconia (Mg-PSZ), a stably propagating crack problem is analyzed. A uniaxial stress-strain curve presented by Marshall [9] is used to determine the behavior of the phase transformation of tetragonal zirconia in Mg-PSZ. By using constitutive model of Okada [13, 14], microscopic and macroscopic behaviors of the material can be revealed, such as the size of transformed zone, the distribution of the transformed zirconia and displacement field. Furthermore, the result of moire interferometry experiment presented in Perry et al. [12] is compared to the result of present finite element analysis. Trends in the displacement fields match each other. The wake zone height behind the crack tip is predicted from both the results.

2 Constitutive equation

In this section, a rate (incremental) form of constitutive equation for dilatational transformation plasticity ([13, 14]) is summarized. Self-consistent method is used to develop a constitutive model, considering three different material constituents, matrix, transformed and untransformed tetragonal zirconia particles, as shown in Fig. 1. It is noted here that the matrix and the untransformed tetragonal zirconia are considered to be identically the same material, in this study. However, for the sake of discussions, the matrix and the untransformed tetragonal zirconia particles are treated differently. In developing the constitutive equation, several assumptions, which make the derivation possible, have been introduced. One of those is that amount of transformed zirconia is related to stress level in untransformed tetragonal zirconia, as stated earlier. Also, this assumption forms a mathematical basis of developing the constitutive model. Since the long range effect of the shear transformation strain can be considered to be small, because of kinking and twinning taking place inside the transformed zirconia particles (Dravid et al. [5]) and the randomness of the distribution of zirconia particles, we consider only the dilatational part of the transformation strain. For the simplicity, tetragonal zirconia particles are assumed to be spherical in shape[3].

<u>Notations</u>:

K:	Bulk modulus
$\bar{\sigma}_{ij}$:	Macroscopic effective stress
$\bar{\varepsilon}_{ij}$:	Macroscopic effective strain
ε_{kk}^T:	Dilatational transformation strain associated with the phase transformation of zirconia from tetragonal to monoclinic structure
$\bar{\varepsilon}_{kk}^T$:	Macroscopic effective transformation strain

[3]The effect of shape of the zirconia particles is discussed in Ramakrishnan et al. [15].

ε^u_{kk}: Dilatational strain in untransformed tetragonal zirconia
σ^u_{kk}: Hydrostatic tensile component of stress in untransformed tetragonal zirconia
$S_{ilk\ell}$: Eshelby's tensor (Eshelby [6])
C_T: Volume fraction of transformed zirconia

The macroscopic stress-strain relationship for its hydrostatic tensile-dilatational components can be written, as:

$$\frac{1}{3}\sigma_{kk} = K(\varepsilon_{kk} - \varepsilon^T_{kk}) \tag{1}$$

A relationship between the stress in the untransformed second phase particle and the volume fraction of the transformed zirconia particles is assumed to exist (see Fig. 2). The slope of the curve in Fig. 2 gives the rate of the phase transformation.

$$\begin{aligned} \dot{C}_T &= h\ \{\tfrac{1}{3}\dot{\sigma}^u_{kk}\} \quad \ldots\ \tfrac{1}{3}\dot{\sigma}^u_{kk} > 0 \\ \dot{C}_T &= 0 \quad \ldots\ \tfrac{1}{3}\dot{\sigma}^u_{kk} \leq 0 \end{aligned} \tag{2}$$

and,

$$h = h(\frac{1}{3}\sigma^u_{kk}, \text{etc}) \tag{3}$$

where, $(\dot{\ })$ denotes the time derivative. h is the slope of the curve depicted in Fig. 2. Eqs. (2) and (3) form the basis of the constitutive equation developed in this study.

The rate form constitutive equation is written, as:

$$\dot{\sigma}_{ij} = \delta_{ij}K(1 - \frac{\varepsilon^T_{mm}hK}{1 + \frac{1}{3}KhS_{iijj}\varepsilon^T_{kk}})\dot{\varepsilon}_{kk} + 2\bar{\mu}\dot{\varepsilon}'_{ij} \tag{4}$$

and,

$$\dot{\varepsilon}^T_{kk} = \dot{C}_T\varepsilon^T_{kk} = \frac{\varepsilon^T_{mm}hK\dot{\varepsilon}_{kk}}{1 + \frac{1}{3}KhS_{iijj}\varepsilon^T_{kk}}$$

when, $\frac{1}{3}\dot{\sigma}^u_{kk} \geq 0$

And,

$$\dot{\sigma}_{ij} = \delta_{ij}K\dot{\varepsilon}_{kk} + 2\bar{\mu}\dot{\varepsilon}'_{ij} \tag{5}$$

when, $\frac{1}{3}\dot{\sigma}^u_{kk} < 0$

where, $\dot{\varepsilon}'_{ij}$ is the rate of effective deviatoric strain.

This constitutive model is employed to evaluate the toughness enhancement of Mg-PSZ during a stable crack propagation.

3 Finite element analysis on the toughness enhancement of Mg-PSZ

In this section, the enhancement of the fracture toughness is discussed using the constitutive model developed. The stress intensity at the crack tip is characterized by the crack tip stress intensity factor K_I^{tip} and that of outside the transformed zone, the stress field is governed by the far-field stress intensity factor K_I^{far} (Budiansky et. al. [2], and McMeeking and Evans [8]). The toughness increment is shown by the difference between K_I^{tip} and K_I^{far}. It is assumed here that the critical value of the crack tip stress intensity factor K_{IC}^{tip} is a material property, and is unchanged by the transformation of zirconia particles.

In order to evaluate the enhancement of the fracture toughness using the finite element analysis, the path independent integrals J and T^* are employed. Both the J integral [16] and the T^* [1] integral have their basis on the energy release per unit crack extension. The J integral is limited to the case of self-similar crack extension in linear or nonlinear elastic solid under static loading, whereas the T^* integral can be applied to the case of any general nonlinear solids under dynamic loading [1]. Since these two integrals compute the energy release rate, one can show the relations between the crack tip stress intensity factor and the far field stress intensity factor to be,

$$J = \frac{1-\nu^2}{E}(K_I^{far})^2 \tag{6}$$

$$T^* = \frac{1-\nu^2}{E}(K_I^{tip})^2 \tag{7}$$

Since K_I^{tip} takes a constant value, which is also equal to the value of far field stress intensity factor at the initiation of crack propagation, toughness enhancement during the stable crack propagation can be defined, as:

$$\text{toughness enhancement effect} = \frac{K_I^{far}}{K_I^{far}(\text{Initiation})} \tag{8}$$

In the current analysis, the stress intensity factor at the initiation $K_{IC}(=K_I^{tip})$ is assumed to be 4.5 (MPa $M^{\frac{1}{2}}$), which is a value obtained in the experimental study of Perry et al. [12].

3.1 Material properties of Mg-PSZ

The present dilatational transformation plasticity model is applied to an analysis on magnesia partially stabilized zirconia (Pg-PSZ). The elastic properties are available in [11]. The uniaxial stress-strain curve (Marshall [9]) is used to determine the transformation behavior. Stationary and stably propagating cracks are analyzed using finite element method, in which the present constitutive model is implemented. The uniaxial stress strain curve [9] can be empirically expressed.

$$\begin{aligned}
\sigma &= 208000\ \varepsilon && \text{(MPa)} \quad \ldots \sigma < 280\text{(MPa)} \\
\sigma &= 208000\ (\varepsilon - 0.99717 \times 10^{3.5\times10^{-4}}) - 0.2071 \times 10^6 && \text{(MPa)} \quad \ldots \sigma \geq 280\text{(MPa)}
\end{aligned} \tag{9}$$

From this expression, the relationship between the volume fraction of transformed zirconia and the stress in untransformed zirconia particles can be obtained. Although the uniaxial stress-strain curve given by Marshall [9] terminates without the linear portion of full transformation, it is assumed in the present analysis that the linear portion of full transformation exists from its terminal point. Poisson's ratio ν is assumed to be 0.23, according to [11]. The transformation strain ε^T_{kk} is taken to be 0.04.

3.2 Stable crack propagation analysis

For stable crack propagation analysis, a problem of semi-circular region, to which far field stress intensity factor K_I^{far} dominant field is imposed, is chosen. At the boundary of the semi-circular region, a displacement prescribed boundary condition, is imposed. The finite element mesh discretization in given in Fig. 3. Four nodded isoparametric element is used and the analysis is carried out under the plane strain condition. The crack tip is initially at point (i) in Fig. 3-(c), and is moved to point (ii), using a nodal release technique. Only one node is released at a time. Associated with the nodal release procedure, the external load is chosen in each nodal release step such that the T^* integral remains constant.

3.3 Result and discussion

The toughness enhancement during the stable crack growth (R-Curve) with constant crack tip stress intensity factor [4] is shown in Fig. 4. A toughness enhancement of 40 percent is seen in the steady state. In the experimental study of Perry et al. [12], nearly the same order of the toughness enhancement in the steady state is reported and is shown in Fig. 5. Although, there is a small discrepancy between the result of the present analysis and that of the experimental study, it can be said that the transformation induced plasticity is the major source of the toughness enhancement in the case of the partially stabilized zirconia. A common trend seen in Figs. 4 and 5 is that most of the fracture toughness enhancement effect is obtained within a very small length of crack propagation. The growth of the partially transformed zone is depicted in Fig. 6. After the initiation, the transformed zone height rapidly increases approaching the steady state transformed zone height. From Figs. 4 and 6, it is seen that while the far field stress intensity factor is increasing, the height of the transformation also increases. The transformed zone height is estimated to be about 0.16 (mm) at the initiation of the crack propagation. The height of transformed zone in the steady state is approximately twice of that at the initiation, as seen in Fig. 6. The steady state transformation zone height is estimated to be about 0.33 (mm).

Iso-y-displacement contour generated from the result of the finite element analysis is shown in Fig. 7 and it is compared to the experimental observation of moire interferometry ([12]), which is shown in Fig. 8 and also shows iso-y-displacement curves. Near the free surface behind the crack tip, the iso-y-displacement contour shifts towards the opposite direction of the crack propagation and this is the effect of the irreversible strain produced by the zirconia phase transformation. In Fig. 7, the distance from the crack plane at which the iso-y-displacement curve starts shifting is estimated to be about 0.16 (mm) using the finite element analysis. And according to the experimental observation, it is about 0.12 (mm). The patterns of iso-y-displacement curves of finite element analysis agree with those of the experimental study excellently.

---

[4]Here, it is assumed that the fully transformed zone exists around the crack tip, so that there is no toughness enhancement for the case of stationary crack.

However, as seen in recent paper of Dadkhah [4], the displacement field in wake zone is not under plane strain nor under plane stress condition. The displacement field is very three dimensional, which is also shown in Cox et al. [3], because of a surface uplifting due to the dilatational transformation strain in wake zone. This surface uplifting can be captured only by three dimensional analysis. The mechanics of fracture in such a case has not fully explored yet. Although Dadkhah et al. [4] analyzed the surface uplifting in steady state wake zone, three dimensional effects on the phenomenon of toughness enhancement have not been described. The authors strongly feel a need for a complete three dimensional analysis on the mechanisms of transformation toughening.

Acknowledgement

The results reported herein were obtained during the course of investigation supported by the U.S. Army office. This support is gratefully acknowledged.

References

[1] S. N. Atluri, *in Computational Methods in the Mechanics of Fracture*, chapter 5 (1986), North-Holland

[2] B. Budiansky, J. W. Hutchinson and J. C. Lambropoulos, "Continuum theory of dilatant transformation in ceramics", *Int. J. Solids Structures*, vol. 19(1983), pp. 337-355

[3] B. N. Cox, D. B. Marshall, D. Kouris, and T. Mura, "Surface Displacement analysis of the transformed zone in magnesia partially stabilized zirconia", *Journal of Engineering Materials and Technology*, vol. 110, (1988), pp. 105-109

[4] M. S. Dadkhah, D. B. Marshall, W. L. Morris and B. N. Cox, "Direct Measurement of transformation zone strains in toughened zirconia", *Tech. Rep.*, Rockwell Science Center, Thousand Oaks, California, (1990)

[5] V. P. Dravid, M. R. Notis, and C. E. Lyman, "Twinning and microcracking associated with monoclinic zirconia in the eutectic system zirconia-mullite", *J. Am. Ceram. Soc.*, vol. 71, no. 4 (1988), pp c-219-c-221

[6] J. D. Eshelby, "The determination of the elastic field of an ellipsoidal inclusion, and related problems", *Proc. Roy. Soc.* A 241(1957), pp376-396

[7] A. G. Evans and R. M. Cannon, "Toughening of brittle solids by martensitic transformations", *Acta Metall.*, vol. 34, no. 5(1986), pp. 761-800

[8] R. M. McMeeking and A. G. Evans, "Mechanics of transformation-toughening in brittle materials", *J. Am. Cera. Soc.*, vol. 65, no. 5(1982), pp 242-246

[9] D. B. Marshall, "Strength Characteristics of transformation-toughened Zirconia", *Am. Cera. Soc.*, vol. 69, no. 3 (1986), pp. 173-180

[10] Y. Mori, Y. Kitao, A. Ishitani, and T. Masaki, "X-ray determination of transformed zone size in toughened zirconia ceramics", *J. Am. Ceram. Soc.*, vol. 71, no. 7, (1988), pp. C-322-C-324

[11] Nilsen (U.S.A.) Inc., Technical data of Nilsen's PSZ

[12] K. E. Perry, J. S. Epstein, H. Okada and S. N. Atluri "R curve behavior in partially stabilized zirconia using moire interferometry", Submitted to ICM 6.

[13] H. Okada, Ph.D. Thesis, Georgia Institute of Technology, Atlanta Georgia (August 1990)

[14] H. Okada, N. Ramakrishnan and S. N. Atluri, ' Constitutive modeling of a solid exhibiting transformation induced plasticity", in proceedings of *Third Intenational Conference on Constitutive Laws for Engineering Materials*, edt. C.S.Desai

[15] N. Ramakrishnan, H. Okada and S. N. Atluri, "Computer simulation of transformation plasticity using finite element method", *Acta Metall.* (to appear) , (1990)

[16] J. R. Rice, " A path independent and the approximate analysis of strain concentration by notches and cracks", *J. App. Mech.*, (1968), pp. 379-386

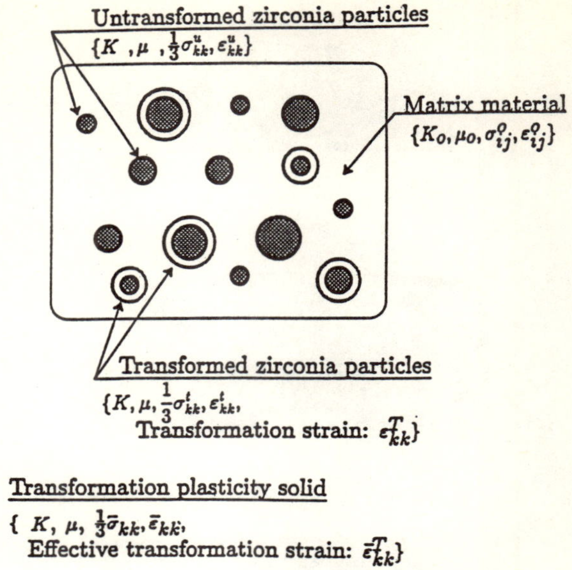

Fig. 1 Dilatational transformation induced plasticity model for Mg-PSZ

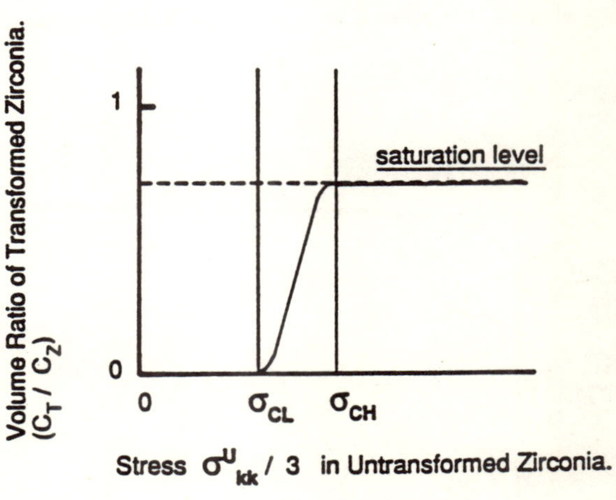

Fig. 2 Relationship between the volume fraction of transformed zirconia and the stress in untransformed tetragonal zirconia particles

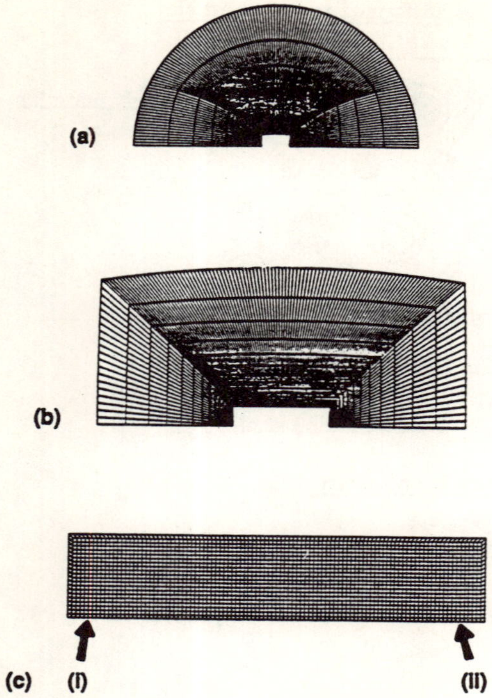

Fig. 3 Finite element mesh discretizartion for the stably propagating crack problem

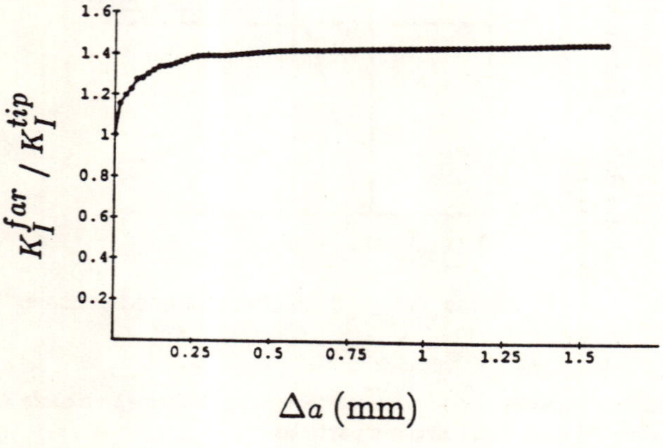

Fig. 4 Toughness enhancement during stable crack propagation (R- curve)

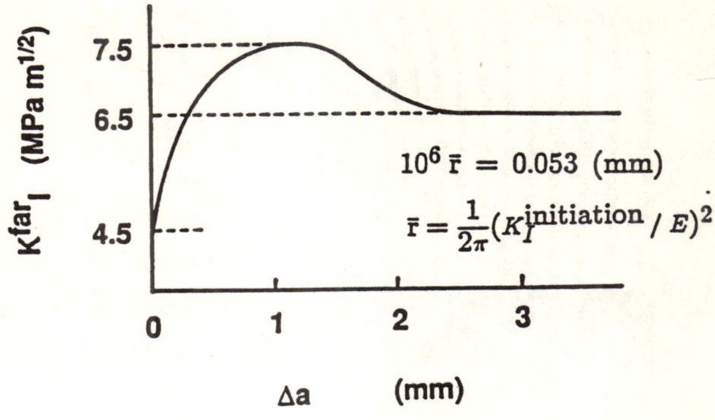

Fig. 5 R-Curve of Perry et al. [12]

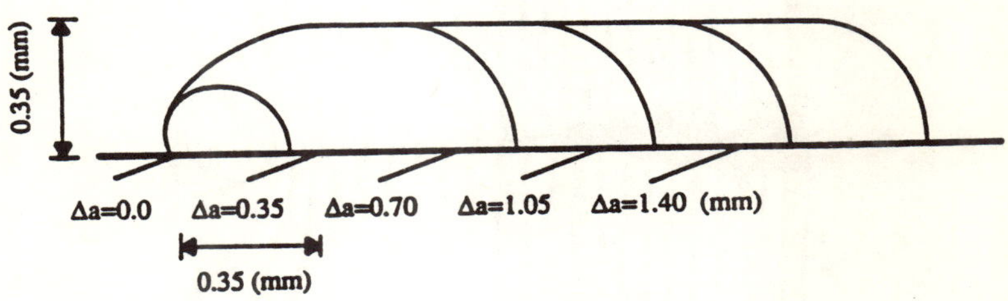

Fig. 6 Growth of Partially transformed zone

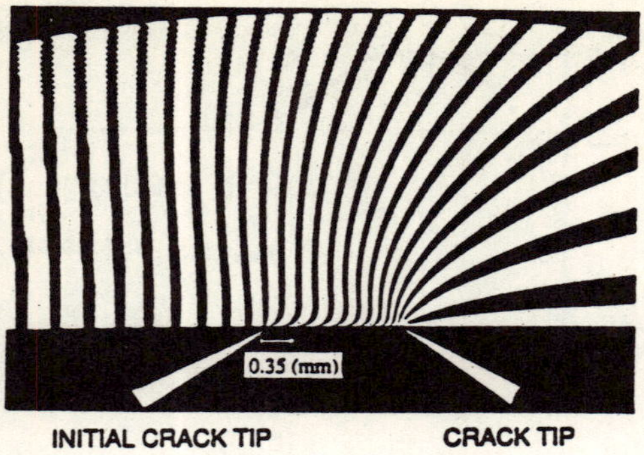

Fig. 7 Iso-y-displacement curve of the result of finite element analysis

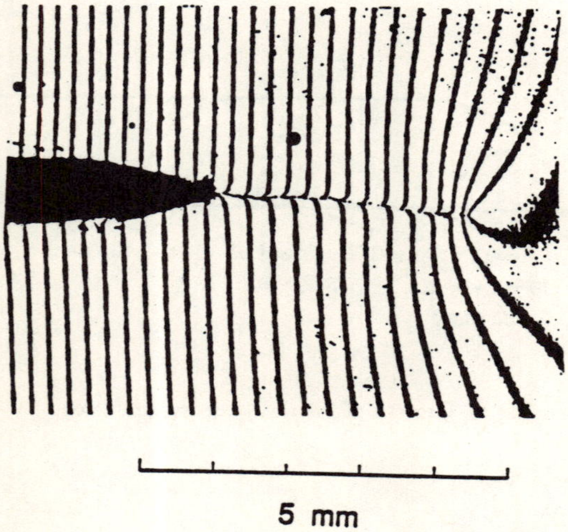

Fig. 8 Moire fringe pattern of Perry et al. [12] (Iso-y-displacement curve)

Extremum Properties of Finite-Step Solutions in Elastoplasticity with Nonlinear Mixed Hardening

C. Comi, A. Corigliano and G. Maier
Department of Structural Engineering, Technical University (Politecnico) of Milan, Piazza Leonardo da Vinci 32, 20133 Milan, Italy

(*Received* 20 *May* 1989; *in revised form* 28 *March* 1990)

Abstract—The class of elastic–plastic constitutive laws assumed herein can be described as follows. We envisage $y \geqslant 1$ yield criteria (or "modes") which define the current yield surface; each yield function ϕ_α consists of two addends: an effective stress Φ_α and a yield limit Y_α ($\alpha = 1, \ldots, y$). The former is an order-one, positively homogeneous function of the difference between the stresses σ_{ij} and reference stresses α_{ij}, which are generally nonlinear functionals of the plastic strain history, so that kinematic hardening is accounted for. The yield limit Y_α is generally a nonlinear function of y nondecreasing internal variables λ_β, so that another hardening mechanism (which may reduce to isotropic hardening) is accommodated in the model. The rates of these variables play the role of plastic multipliers in the flow rule.

A finite step in the geometrically linear evolutive analysis of such solids is defined according to the backward-difference (or "stepwise holonomic") strategy for approximate time integration. For the "finite-step" boundary value problem thus arising, various extremum characterizations of solutions are established and the underlying constitutive restrictions are pointed out and discussed.

1. INTRODUCTION

Extremum theorems for the incremental (rate) solution to boundary value problems in elastoplasticity, and for the path-independent solution in the "deformation theory" of plasticity, have been the subjects of a fairly abundant literature. The earlier part of this is synthesized in treatises such as Martin's (1975), the more recent part is still disseminated in research papers, while broad mathematical foundations and alternative characterizations of solutions (in terms of variational inequalities) can be found in comprehensive works by Panagiotopoulos (1985) and Sewell (1987).

Some of the early contributions to extremum properties in plasticity, precisely those concerning constitutive models with linear yield functions and linear hardening (Maier, 1969, 1970a), have had fruitful implications on the computational aspects of inelastic analysis. In fact, for those models, rate problems, path-independent ("holonomic", single-step) analysis and finite-step (stepwise–holonomic) problems within marching solution methods, all become amenable to and solvable by quadratic programming (after suitable space discretization). The stepwise–holonomic method (now more frequently called "backward difference" method) for approximate time integration in elastic–plastic analysis was proposed and developed by De Donato and Maier (1972, 1973) in such a context of plasticity with piecewise-linearized yield criteria. Mechanically interpreted, this procedure basically rests on two provisions: within each load step the elastic–plastic, path-dependent (nonholonomic), stress–strain law is replaced by a path-independent (holonomic) law generated from it. Between two successive steps, all quantities which depend on the yielding history are updated, thus accounting for the intrinsic irreversibility of plasticity. In the context of piecewise-linearized plasticity, stepwise holonomic elastic–plastic analysis exhibits the appealing feature of being exposed only to errors entailed by the constitutive idealization, and by possible local unloading within a single load step (no error accumulation, unconditional stability). Surveys of the literature on this area of plasticity can be found in Cohn *et al.* (1979) and Maier *et al.* (1982). These surveys, however, do not cover more recent contributions such as those proposed by Dittmer *et al.* (1985), Martin *et al.* (1987), Franchi and Genna (1987) and Zouain *et al.* (1988).

In the last few years, methods for the approximate time integration of the (nonlinear, differential) relationships of elastoplasticity have become a fashionable research topic in computational mechanics. Valuable contributions are due to Ortiz and Popov (1985), Simo and Taylor (1986), Matthies and Strang (1979), Simo *et al.* (1988), Casciaro and Mancuso (1988) and to other authors. The stepwise–holonomic or backward difference time integration procedure is actively investigated, *in primis* by J. B. Martin and his coworkers.

In this paper, the finite-step boundary value problem arising in stepwise–holonomic elastoplastic analysis is addressed with reference to a category of material models (described in Section 2), endowed with mixed nonlinear hardening. Namely, as a consequence of yielding processes the elastic domain translates in the stress space (kinematic hardening) and the yield limits change (isotropic hardening, called so here as it may describe changes in size but not in shape of the yield domain). For the above boundary value problem in finite increments (but in the small deformations, geometrically linear range), we derive in Sections 3–4 extremum properties of solutions. Constitutive hypotheses found to be sufficient for the validity of some of them are discussed in Section 2. In a parallel paper (Comi and Maier, 1990), one of the present results is shown to provide a convergence criterion for a popular iterative solution scheme, thus substantiating their potential use in computational plasticity.

In general terms, whenever suitable, space discretizations (e.g. finite element models) are adopted, the extremum theorems derived in what follows make the step problem amenable to a problem in nonlinear programming.

The basic purpose of the present paper is to extend to nonlinear multiple yield criteria and nonlinear mixed hardening, the aforementioned earlier work on piecewise-linearized plastic models.

Such an extension appears to be supplementary to the contributions so far available on the subject, to the authors' knowledge, such as those proposed by Franchi and Genna (1984), Feijòo and Zouain (1988), Reddy and Griffin (1986) and Maier and Novati (1990).

As for the notation and analytical dress, the developments in what follows are carried out by the conventional formalism of classical plasticity (rather than by the more elegant but also more abstract mathematical formalism of nonsmooth mechanics or convex analysis).

The usual Cartesian tensor description is used, with the index summation convention (when this does not apply, repeated indices are avoided and a single subscript is used for the parentheses enclosing the factors). A factor in parentheses is preceded by a dot denoting product when it might be confused with the argument of a functional dependence.

2. PROBLEM FORMULATIONS: INCREMENTAL VS STEPWISE HOLONOMIC CONSTITUTIVE LAWS

2.1. *A class of material models*

The most general kind of elastoplastic constitutions considered herein, are specified by the following relation set, in an orthogonal reference x_i $(i = 1, 2, 3)$:

$$\sigma_{ij} = E_{ijrs}\varepsilon^{e}_{rs} = E_{ijrs}(\varepsilon_{rs} - \varepsilon^{p}_{rs} - \theta_{rs}); \qquad \text{(1a, b)}$$

$$\phi_\alpha = \Phi_\alpha(\sigma_{rs} - \alpha_{rs}(\varepsilon^{p}_{\ell m})) - Y_\alpha(\lambda_\beta) \leqslant 0 \quad (\alpha, \beta = 1, \ldots, y); \qquad \text{(2a, b)}$$

$$\dot{\varepsilon}^{p}_{ij} = \frac{\partial \psi_\alpha}{\partial \sigma_{ij}}(\sigma_{rs} - \alpha_{rs}(\varepsilon^{p}_{\ell m}))\dot{\lambda}_\alpha, \quad \dot{\lambda}_\alpha \geqslant 0; \qquad \text{(3a, b)}$$

$$\phi_\alpha \dot{\lambda}_\alpha = 0 \quad \text{or} \quad (\phi\dot{\lambda})_\alpha = 0. \qquad \text{(4a, b)}$$

Here, and henceforth, dots mark time derivatives; the customary properties (positive definiteness and symmetries) of the moduli tensor of linear elasticity E_{ijrs} are assumed. Equation (1b) expresses the additivity of elastic (ε^{e}_{rs}), plastic (ε^{p}_{rs}) and imposed (e.g. thermal, θ_{rs}) strains. Equations (2) define the current elastic domain (and the current yield surface as its boundary) in the stress space by y yield functions ϕ_α, each of which is expressed as the sum of an "effective stress" Φ_α, and a yield limit Y_α. The former is seen to depend on plastic strains so that a translation of the yield surface occurs at yielding and is defined by the "reference stresses" α_{ij} (kinematic hardening). The dependence of the latter addend Y_α on the variables λ_β will be called "isotropic hardening", although the shape of the elastic domain is preserved only in special cases, e.g. for a single mode $(y = 1)$ and if Φ_α is a first-order homogeneous function. The role of the nondecreasing variables λ_α is specified by eqns (2)–(4): their time derivatives act as "plastic multipliers" in eqn (3a) and are related to the yield functions by the complementarity equation (4). By virtue of the sign constraints on the variables involved, eqn (4) holds also component-wise for each α, and implies that:

$$\dot{\phi}_\alpha \dot{\lambda}_\alpha = 0 \quad \text{or} \quad (\dot{\phi}\dot{\lambda})_\alpha = 0, \qquad \text{(5a, b)}$$

which expresses Prager's consistency rule for "loading" (yielding) and "unloading" processes. The variables λ_α are thought of as measures, at the phenomenological macroscale level, of irreversible rearrangements occurring at the microscale level inside the material and, hence, will be called "internal variables" in what follows, e.g. Martin (1981). The above set of relations is homogeneous in time t: this implies the time-independent ("inviscid") character of the material behaviour described and thus reduce t to an event-ordering variable.

2.2. *Stepwise holonomic constitutive laws*

Assume a situation in which all variables are known (and will be marked by barred symbols). Starting from this situation, consider finite increments (denoted by Δ; finite in contrast to infinitesimal or rates) and relate them by the following relation set:

$$\Delta\sigma_{ij} = E_{ijrs} \cdot (\Delta\varepsilon_{rs} - \Delta\varepsilon^{p}_{rs} - \Delta\theta_{rs}); \qquad \text{(6)}$$

$$\phi_\alpha = \Phi_\alpha(\bar{\sigma}_{rs}+\Delta\sigma_{rs}-\alpha_{rs}(\bar{\varepsilon}^p_{\ell m}+\Delta\varepsilon^p_{\ell m}))-Y_\alpha(\bar{\lambda}_\beta+\Delta\lambda_\beta) \leqslant 0; \tag{7a, b}$$

$$\Delta\varepsilon^p_{ij} = \frac{\partial\psi_\alpha}{\partial\sigma_{ij}}(\bar{\sigma}_{rs}+\Delta\sigma_{rs}-\alpha_{rs}(\bar{\varepsilon}^p_{\ell m}+\Delta\varepsilon^p_{\ell m}))\Delta\lambda_\alpha, \quad \Delta\lambda_\alpha \geqslant 0; \tag{8a, b}$$

$$\phi_\alpha\Delta\lambda_\alpha = 0 \quad (\alpha, \beta = 1, \ldots, y). \tag{9}$$

Equations (6)–(9) will be called the stepwise–holonomic laws (path-independent within the step), generated by the elastoplastic incremental (nonholonomic, path-dependent) constitution (1)–(4), to which they are easily seen to reduce for $\Delta t \to 0 (\Delta \to d)$. Clearly, these relations are algebraic, nonlinear, while eqns (1)–(4) are differential nonlinear.

2.3. *Finite-step boundary value problems*

Consider a solid which occupies a region V (in R^3), with a smooth boundary S, where n_i denotes the outward normal direction. The evolutive response is sought to a given history of external actions: body forces $b_i(t)$ and imposed strains $\theta_{ij}(t)$ in V; tractions $T_i(t)$ on S_T; imposed displacements $\bar{u}_i(t)$ on S_u (S_T and S_u being complementary parts of S). The compatibility and equilibrium equations are assumed to be unaffected by geometry changes. These equations associated with the constitutive eqns (1)–(4) give rise to the set of governing relations, i.e. to the b.v. problem in rates. An approximate marching solution of this nonlinear differential equation set is usually based on a preliminary sub-division of the external action history into the sequence of loading conditions at instants $t_0 = 0$, t_1, $t_2, \ldots, t_n = t_{n-1}+\Delta t_n$. These are chosen in such a way that the external action variation over each time interval Δt, is not far from linear.

Suppose that all the variable fields are known throughout V at instant $\bar{t} \equiv t_{n-1}$ (their values will be represented by barred symbols). Denoting by Δ increments over $\Delta t \equiv \Delta t_n$ of both input data and response variables, let us associate the stepwise–holonomic constitutive laws (6)–(9) to the geometric compatibility and equilibrium equations:

$$\Delta\varepsilon_{ij} = \tfrac{1}{2}(\Delta u_{i,j}+\Delta u_{j,i}) \quad \text{in } V, \quad \Delta u_i = \Delta\bar{u}_i \quad \text{on } S_u, \tag{10a, b}$$

$$\Delta\sigma_{ij,i}+\Delta\bar{b}_j = 0 \quad \text{in } V, \quad \Delta\sigma_{ij}n_i = \Delta\bar{T}_j \quad \text{on } S_T. \tag{11a, b}$$

The finite-step b.v. problem governed by eqns (6)–(11) arises when the approximate time integration of the original b.v. problem in rates, is formulated according to an Euler, backward difference, fully implicit scheme. The procedure for each step defines all state-dependent variables and all derivatives in the unknown state at the step end, and enforces the algebraic, plastic relations there [eqns (7)–(9)]. Allowance for the irreversible, history-dependent nature of plasticity is made from step to step, by updating the internal variables λ_α with sign-constrained increments.

As mentioned in Section 1, stepwise–holonomic (or backward difference) time-integration in the above sense, leading to the b.v. problem (6)–(11), is the subject of abundant recent and nonrecent literature. Here, this finite-step problem will be studied (in Sections 3, 4, 5) as for the extremum properties of solutions, under some of the restrictions to be considered next.

2.4. *Constitutive restrictions*

The following hypotheses which specialize the plastic constitutive law (2)–(4), have a significant role in the subsequent discussions.

(a) The effective stresses are differentiable convex functions of their argument (denoted by τ_{ij}), namely:

$$\Phi_\alpha(\tau_{ij}) \geqslant \Phi_\alpha(\tau'_{ij}) + \frac{\partial \Phi_\alpha}{\partial \tau_{rs}}(\tau'_{ij}) \cdot (\tau_{rs} - \tau'_{rs}), \quad \text{for any } \tau_{ij}, \tau'_{ij}. \tag{12}$$

This hypothesis is equivalent to the convexity of the yield functions ϕ_α, with respect to stresses and hence, entails the convexity of all current elastic domains, which is one of the requirements of Drucker's stability postulate (Drucker, 1960; 1988).

(b) The effective stresses are positively homogeneous functions of order one of their argument τ_{ij}, i.e.

$$\Phi_\alpha(c\tau_{ij}) = c\Phi_\alpha(\tau_{ij}), \quad \text{for any } c \geqslant 0. \tag{13}$$

By Euler's theorem and by an easily proved consequence of it, property (13) implies that

$$\Phi_\alpha(\tau_{ij}) = \frac{\partial \Phi_\alpha}{\partial \tau_{rs}}(\tau_{ij})\tau_{rs}; \quad \frac{\partial^2 \Phi_\alpha}{\partial \tau_{rs}\, \partial \tau_{hk}}(\tau_{ij})\tau_{hk} = 0. \tag{14a, b}$$

Note that eqn (14b) is satisfied whenever the argument τ_{ij} coincides with the factor τ_{hk}. Clearly, the latter tensor τ_{hk} can be interpreted as a solution of the homogeneous linear equation system whose coefficient matrix is the Hessian of Φ_α, evaluated for any fixed τ_{ij}. Most plastic constitutive laws in use comply with this hypothesis, or can be re-formulated so that they comply with it.

(c) The yield functions ϕ_α and the plastic potentials ψ_α are identical, namely:

$$\psi_\alpha = \phi_\alpha. \tag{15}$$

This association of the flow rule with the yield criterion or normality property, required by Drucker's stability postulate, is known to be violated by important categories of materials such as geomaterials and concrete.

(d) The plastic work (per unit volume) is a functional of the plastic strain history:

$$W^{\mathrm{p}}(t) = \int_0^t \sigma_{ij}(t')\dot{\varepsilon}^{\mathrm{p}}_{ij}(t')\,\mathrm{d}t'. \tag{16}$$

When the constitutive restrictions (b) and (c) are satisfied, i.e. eqns (13) and (15) hold, then through eqns (3a) and (4), expression (16) for W^{p} can be given the form:

$$W^{\mathrm{p}}(t) = \Pi(t) + \Gamma(t), \tag{17}$$

$$\Pi(t) \equiv \int_0^t Y_\alpha(t')\dot{\lambda}_\alpha(t')\,\mathrm{d}t'; \quad \Gamma(t) \equiv \int_0^t \alpha_{ij}(t')\dot{\varepsilon}^{\mathrm{p}}_{ij}(t')\,\mathrm{d}t'. \tag{18a, b}$$

The hypothesis, called henceforth "reciprocal hardening", means that:

$$\frac{\partial Y_\alpha}{\partial \lambda_\beta} = \frac{\partial Y_\beta}{\partial \lambda_\alpha}; \quad \frac{\partial \alpha_{ij}}{\partial \varepsilon^{\mathrm{p}}_{rs}} = \frac{\partial \alpha_{rs}}{\partial \varepsilon^{\mathrm{p}}_{ij}}. \tag{19a, b}$$

This symmetry of Jacobian matrices is a necessary and sufficient condition for the path-independence of the two plastic work addends (18): the former, associated to the so-called isotropic hardening, becomes a function of the internal variables only $\Pi(\lambda_\alpha)$; the latter, associated to the kinematic hardening, becomes a function of the plastic strains only $\Gamma(\varepsilon^{\mathrm{p}}_{ij})$. Such reciprocity in the interaction between yield modes is violated by some material models (e.g. by the three-mode model due to Resende and Martin, 1985).

(e) When the preceding hypothesis is complied with, a further restriction is the convexity of functions $\Pi(\lambda_\alpha)$ and $\Gamma(\varepsilon^{\mathrm{p}}_{ij})$, i.e. the positive, semi-definiteness of their Hessian matrices which, clearly, coincide with the (symmetric) Jacobian matrices (19). Using eqns (18) and (19), the second-order plastic work can be expressed in the form:

$$\delta^{(2)} W^{\mathrm{p}} = \frac{1}{2}\delta Y_\alpha \delta\lambda_\alpha + \frac{1}{2}\delta\alpha_{ij}\delta\varepsilon^{\mathrm{p}}_{ij} = \frac{1}{2}\delta\lambda_\alpha \frac{\partial Y_\alpha}{\partial \lambda_\beta}\delta\lambda_\beta + \frac{1}{2}\delta\varepsilon^{\mathrm{p}}_{ij}\frac{\partial \alpha_{ij}}{\partial \varepsilon^{\mathrm{p}}_{rs}}\delta\varepsilon^{\mathrm{p}}_{rs}. \tag{20}$$

Equation (20) shows that by the present hypothesis, only non-negative work can be done by an external agency, causing an infinitesimal perturbation while preserving equilibrium. This means material stability by Hill's criterion (Hill, 1957); it is one of the requirements of Drucker's postulate and rules out softening behaviour.

3. MINIMUM PROPERTIES OF SOLUTIONS IN THE FINITE INCREMENTS OF KINEMATIC VARIABLES

In this section, we address the b.v. problem of Section 2.3, defined by eqns (6)–(11) and prove two extremum characterizations of its solution on the basis of all the constitutive restrictions (a)–(e), discussed in Section 2.4.

3.1. *Kinematic (potential energy) theorem*

The following statement can be regarded as a manifestation of the potential energy principle in the present context of incremental elastoplasticity, discretized in time in the sense of Sections 2.2 and 2.3.

Proposition 1. Assuming the constitutive restrictions (a)–(e) of Section 2.4, consider the optimization problem:

$$\min \Big\{ \Omega(\Delta u_i, \Delta\lambda_\alpha, \Delta\varepsilon^{\mathrm{e}}_{ij}, \Delta\varepsilon^{\mathrm{p}}_{ij}) \equiv \frac{1}{2}\int_V \Delta\varepsilon^{\mathrm{e}}_{ij} E_{ijhk} \Delta\varepsilon^{\mathrm{e}}_{hk}\,\mathrm{d}V + \int_V \Pi(\bar{\lambda}_\alpha + \Delta\lambda_\alpha)\,\mathrm{d}V$$
$$+ \int_V \Gamma(\bar{\varepsilon}^{\mathrm{p}}_{ij} + \Delta\varepsilon^{\mathrm{p}}_{ij})\,\mathrm{d}V + \int_V \bar{\sigma}_{ij}\Delta\varepsilon^{\mathrm{e}}_{ij}\,\mathrm{d}V - \int_V (\bar{b}_i + \Delta\bar{b}_i)\Delta u_i\,\mathrm{d}V - \int_{S_T} (\bar{T}_i + \Delta\bar{T}_i)\Delta u_i\,\mathrm{d}S \Big\}. \tag{21}$$

subject to the relations ("constraints"):

$$\Delta\varepsilon^{e}_{ij} = \tfrac{1}{2}(\Delta u_{j,i}+\Delta u_{i,j})-\Delta\varepsilon^{p}_{ij}-\Delta\bar{\theta}_{ij} \quad \text{in } V; \quad \Delta u_i = \Delta\bar{u}_i \quad \text{on } S_u; \tag{22a, b}$$

$$\Delta\varepsilon^{p}_{ij} = \frac{\partial\Phi_\alpha}{\partial\sigma_{ij}}(E_{hk\ell m}\cdot(\bar{\varepsilon}^{e}_{\ell m}+\Delta\varepsilon^{e}_{\ell m})-\alpha_{hk}(\bar{\varepsilon}^{p}_{\ell m}+\Delta\varepsilon^{p}_{\ell m}))\Delta\lambda_\alpha, \quad \Delta\lambda_\alpha \geqslant 0 \quad \text{in } V. \tag{23a, b}$$

The fact that a field Δu_i, $\Delta\lambda_\alpha$, $\Delta\varepsilon^{e}_{ij}$ minimizes the functional Ω (21), subject to the constraints (22)–(23), is a sufficient and necessary condition for solving the finite-step problem defined by eqns (6)–(11); for the given external action increments $\Delta\bar{b}_i$, $\Delta\bar{u}_i$, $\Delta\bar{\theta}_{ij}$, starting from a known situation (barred symbols).

Proof of sufficiency. Consider the Lagrangian functional:

$$L(\Delta u_i, \Delta\lambda_\alpha, \Delta\varepsilon^{e}_{ij}, \Delta\varepsilon^{p}_{ij}, \ell_\alpha; \xi_{ij}, \beta_{ij}, \eta_\alpha) \equiv \Omega+\int_V \xi_{ij}\cdot[\Delta\varepsilon^{e}_{ij}-\text{r.h.s. (22a)}]\,\mathrm{d}V$$

$$+\int_V \beta_{ij}\cdot[\Delta\varepsilon^{p}_{ij}-\text{r.h.s. (23a)}]\,\mathrm{d}V+\int_V \eta_\alpha\cdot(\Delta\lambda_\alpha-\ell^2_\alpha)\,\mathrm{d}V; \tag{24}$$

where ξ_{ij}, β_{ij} and η_α are Lagrange multipliers, r.h.s. (·) means right-hand side of eqn (·) for brevity, and eqn (23b) has been interpreted as $\Delta\lambda_\alpha = \ell^2_\alpha$.

Mark by a tilde a solution of the optimization problem considered (and, for brevity, also the cumulative values; e.g. $\tilde{\lambda}_\alpha = \bar{\lambda}_\alpha+\Delta\tilde{\lambda}_\alpha$), calculate the first variation of L around this solution and, setting:

$$\tau_{hk} \equiv E_{hkrs}\varepsilon^{e}_{rs}-\alpha_{hk}(\varepsilon^{p}_{rs}) = \sigma_{hk}-\alpha_{hk}(\varepsilon^{p}_{rs}), \tag{25}$$

express its vanishing for all variations of the independent variables listed in eqn (24):

$$\delta^{(1)}L = \delta^{(1)}\Omega+\int_V \xi_{ij}\{\delta\varepsilon^{e}_{ij}-\delta[\text{r.h.s. (22a)}]\}\,\mathrm{d}V$$

$$+\int_V \beta_{ij}\cdot\left[\delta\varepsilon^{p}_{ij}-\frac{\partial\Phi_\alpha}{\partial\sigma_{ij}}(\tilde{\tau}_{hk})\delta\lambda_\alpha-\Delta\tilde{\lambda}_\alpha\frac{\partial^2\phi_\alpha}{\partial\sigma_{ij}\,\partial\sigma_{rs}}(\tilde{\tau}_{hk})\delta\tau_{rs}\right]\mathrm{d}V+\int_V \eta_\alpha\cdot[\delta\lambda_\alpha-2(\tilde{\ell}\delta\ell)_\alpha])\,\mathrm{d}V = 0. \tag{26}$$

The stationarity of L with respect to the Lagrangian multipliers is another way of enforcing the constraints (22)–(23), being understood that the boundary compatibility (22b) is tacitly required as $\delta u_i = 0$ on S_u. The stationarity (26) with respect to the variations δu_i, $\delta\varepsilon^{e}_{ij}$, $\delta\varepsilon^{p}_{ij}$, $\delta\lambda_\alpha$ and $\delta\ell_\alpha$ in turn, gives rise to the following equations, respectively:

$$\int_V(\bar{b}_i+\Delta\bar{b}_i)\delta u_i\,\mathrm{d}V+\int_{S_T}(\bar{T}_i+\Delta\bar{T}_i)\delta u_i\,\mathrm{d}S = \int_V -\xi_{ij}\frac{1}{2}\delta(u_{i,j}+u_{j,i})\,\mathrm{d}V. \tag{27}$$

$$\xi_{ij} = \beta_{hk}\Delta\tilde{\lambda}_\alpha \frac{\partial^2\Phi_\alpha}{\partial\sigma_{hk}\,\partial\sigma_{rs}}(\tilde{\tau}_{lm})E_{rsij} - \Delta\tilde{\varepsilon}^e_{hk}E_{ijhk} - \tilde{\sigma}_{ij}, \tag{28}$$

$$\alpha_{ij}(\tilde{\varepsilon}^p_{lm}) + \xi_{ij} + \beta_{ij} + \beta_{hk}\Delta\tilde{\lambda}_\alpha \frac{\partial^2\Phi_\alpha}{\partial\sigma_{hk}\,\partial\sigma_{rs}}(\tilde{\tau}_{lm})\frac{\partial\alpha_{rs}}{\partial\varepsilon^p_{ij}}(\tilde{\varepsilon}^p_{lm}) = 0, \tag{29}$$

$$\eta_\alpha = \beta_{ij}\frac{\partial\Phi_\alpha}{\partial\sigma_{ij}}(\tilde{\tau}_{lm}) - Y_\alpha(\tilde{\lambda}_\beta), \tag{30}$$

$$2(\eta\tilde{\ell})_\alpha = 0. \tag{31}$$

Equation (27), by the virtual work principle, expresses the equilibrium of the tensor field $-\xi_{ij}$ (if interpreted as stresses) under the loads at the end of the step.

Now substitute eqn (28) into (29) and rearrange; denoting by I_{ijhk} the identity tensor:

$$\left[I_{ijhk} + \Delta\tilde{\lambda}_\alpha \frac{\partial^2\Phi_\alpha}{\partial\sigma_{hk}\,\partial\sigma_{rs}}(\tilde{\tau}_{\ell m})\cdot\left(E_{rsij} + \frac{\partial\alpha_{rs}}{\partial\varepsilon^p_{ij}}(\tilde{\varepsilon}^p_{lm})\right)\right]\beta_{hk} = \tilde{\sigma}_{ij} + E_{ijhk}\Delta\tilde{\varepsilon}^e_{hk} - \alpha_{ij}(\tilde{\varepsilon}^p_{lm}). \tag{32}$$

Interpret eqn (32) as a linear system in the unknowns β_{hk}. Noting, by eqn (25), that the r.h.s. is $\tilde{\tau}_{ij}$ and using the identity (14b) due to the hypothesis (b) Section 2.4, one realizes that this system admits the solution:

$$\beta_{ij} = \tilde{\tau}_{ij} = \tilde{\sigma}_{ij} + E_{ijhk}\Delta\tilde{\varepsilon}^e_{hk} - \alpha_{ij}(\tilde{\varepsilon}^p_{lm}). \tag{33}$$

This solution is unique. In fact, the identity tensor and the elastic tensor are positive definite; the Jacobian of α_{rs} (ε^p_{hk}) is positive semi-definite because of the convexity hypothesis (e) in Section 2.4; the Hessian of Φ_α too is so, because of the convexity hypothesis (a). Therefore, pre-multiplying both sides by the inverse of the positive definite tensor in round brackets, eqn (32) is transformed into an equivalent system, whose coefficient matrix is clearly nonsingular.

Substituting (33) into eqn (28), by virtue of identity (14b) again, we obtain:

$$\xi_{ij} = -\tilde{\sigma}_{ij} - E_{ijhk}\Delta\tilde{\varepsilon}^e_{hk} = -\tilde{\sigma}_{ij}. \tag{34}$$

Substitute (34) into eqn (27) which holds for any δu_i: this turns out to express, by the virtual work principle, the equilibrium of the stress field $\tilde{\sigma}_{ij}$ with the load at step end, namely the fulfilment of eqn (11).

Taking into account (33) in eqn (30), η_α is recognized as the yield function ϕ_α, eqn (7a), in view of the homogeneity hypothesis (b) and of the consequent eqn (14a).

As a consequence, eqn (31) becomes equivalent to the complementarity equation (9), since the constraint (23b), expressed as $\Delta\lambda_\alpha = \ell^2_\alpha$, holds at the minimum ($\sim$) as well.

At the minimum, the second variation $\delta^{(2)}L$ of functional L, is non-negative which implies, in particular:

$$-\int_V [\eta(\delta\ell)^2]_\alpha \,\mathrm{d}V \geqslant 0. \tag{35}$$

Therefore, $\eta_\alpha \leqslant 0$ and hence, $\phi_\alpha \leqslant 0$.

At this stage, it can be concluded that the constraints (22)–(23), which include geometric compatibility, together with the constrained minimization of Ω imply the fulfilment of all the relations (6)–(11) governing the finite-step problem, subject to the restrictions of Section 2.4.

Proof of necessity. Capped symbols will denote quantities pertaining to a solution of the finite-step elastoplastic problem (6)–(11); primed symbols will mark quantities pertaining to fields which satisfy the constraints (22)–(23) ("feasible solution" for the optimization problem). Denoting by unbarred symbols values at the end of the step (e.g. $\lambda' = \bar{\lambda}_\alpha + \Delta\lambda'_\alpha$), after some trivial algebra we can write:

$$\Omega' - \hat{\Omega} = \int_V \frac{1}{2}(\Delta\varepsilon^{e'}_{ij} - \Delta\hat{\varepsilon}^{e}_{ij})E_{ijhk} \cdot (\Delta\varepsilon^{e'}_{hk} - \Delta\hat{\varepsilon}^{e}_{hk})\,\mathrm{d}V + \int_V [\Pi(\lambda'_\alpha) - \Pi(\hat{\lambda}_\alpha)]\,\mathrm{d}V$$

$$+ \int_V [\Gamma(\varepsilon^{p'}_{ij}) - \Gamma(\hat{\varepsilon}^{p}_{ij})]\,\mathrm{d}V + \int_V \Delta\hat{\varepsilon}^{e}_{ij} E_{ijhk} \cdot (\Delta\varepsilon^{e'}_{hk} - \Delta\hat{\varepsilon}^{e}_{hk})\,\mathrm{d}V + \int_V \bar{\sigma}_{ij} \cdot (\Delta\varepsilon^{e'}_{hk} - \Delta\hat{\varepsilon}^{e}_{hk})\,\mathrm{d}V$$

$$- \int_V b_i \cdot (\Delta u'_i - \Delta\hat{u}_i)\,\mathrm{d}V - \int_{S_T} T_i \cdot (\Delta u'_i - \Delta\hat{u}_i)\,\mathrm{d}S. \quad (36)$$

Note now that: the first integrand is non-negative; because of the convexity of Π and Γ, respectively, the second and third integrands are bounded below by the linear term of their Taylor expansions around the solution, in view of an inequality similar to (12); the algebraic sum of the last four integrals can be replaced by a single integral making use of a virtual work equation (since the actual static quantities at the step end are equilibrated and the differences of kinematic quantities are compatible if plastic strain increments are added). Therefore, we may write:

$$\Omega' - \hat{\Omega} \geqslant \int_V Y_\alpha(\hat{\lambda}_\beta) \cdot (\Delta\lambda'_\alpha - \Delta\hat{\lambda}_\alpha)\,\mathrm{d}V + \int_V \alpha_{ij}(\hat{\varepsilon}^{p}_{rs}) \cdot (\Delta\varepsilon'^{p}_{ij} - \Delta\hat{\varepsilon}^{p}_{ij})\,\mathrm{d}V - \int_V \hat{\sigma}_{ij} \cdot (\Delta\varepsilon'^{p}_{ij} - \Delta\hat{\varepsilon}^{p}_{ij})\,\mathrm{d}V. \quad (37)$$

Substituting the usual expressions for plastic strain increments, using position (25) and rearranging, the inequality (37) becomes:

$$\Omega' - \hat{\Omega} \geqslant -\int_V \frac{\partial\Phi_\alpha}{\partial\sigma_{ij}}(\tau'_{rs})\hat{\tau}_{ij}\Delta\lambda'_\alpha\,\mathrm{d}V + \int_V Y_\alpha(\hat{\lambda}_\beta)\Delta\lambda'_\alpha\,\mathrm{d}V + \int_V \left[-Y_\alpha(\hat{\lambda}_\beta) + \frac{\partial\Phi_\alpha}{\partial\sigma_{ij}}(\hat{\tau}_{rs})\hat{\tau}_{ij}\right]\Delta\hat{\lambda}_\alpha\,\mathrm{d}V. \quad (38)$$

The last integrand vanishes because of the complementarity eqn (9), since the factor in brackets is recognized as $\hat{\phi}_\alpha$. The first integral can be replaced by its lower bound, provided by eqn (12) and evaluated with $\tau_{ij} = \hat{\tau}_{ij}$ if ϕ_α is convex (hyp. a), taking into account eqn (14a) if it is also homogeneous of first-order (hyp. b). Then we have:

$$\Omega' - \hat{\Omega} \geqslant \int_V [Y_\alpha(\hat{\lambda}_\beta) - \Phi_\alpha(\hat{\sigma}_{ij} - \alpha_{ij}(\hat{\varepsilon}^{p}_{hk}))]\Delta\lambda'_\alpha\,\mathrm{d}V. \quad (39)$$

The factor in brackets equals $-\hat{\phi}_\alpha \geqslant 0$, while $\Delta\lambda'_\alpha \geqslant 0$. Therefore, $\Omega' \geqslant \hat{\Omega}$, namely the value attained by the functional Ω at the finite-step solution, turns out to be the (absolute) minimum over the feasible domain defined by constraints (22)–(23).

3.2. *A theorem in the increments of plastic strains and internal variables*

The linear elastic stress response $\Delta\bar{\sigma}^e_{ij}$ to the given external action changes over Δt is now assumed to be preliminarily computed. For the remaining computations in the step, $\Delta\bar{\sigma}^e_{ij}$ represents data containing all information on the load increments. We denote by $\Delta\sigma^p_{ij}$ and $\Delta\varepsilon^{ep}_{ij}$ the (self-equilibrated) stresses and corresponding elastic strains generated in the body by the (unknown) plastic strains $\Delta\varepsilon^p_{ij}$, should they act as imposed strains on the body assumed as linear elastic in the absence of any external action (hence, with homogeneous boundary conditions). In principle, one can write:

$$\Delta\sigma^p_{ij} = \Delta\sigma^p_{ij}(\Delta\varepsilon^p_{rs}) = \int_V Z_{ijhk}(x_r, \xi_s)\Delta\varepsilon^p_{hk}(\xi_s)\,\mathrm{d}V. \tag{40}$$

The tensor-valued Green's function in (40) is analytically known or obtainable in few cases, e.g. for the isotropic homogeneous elastic domain (for which it can be derived from Kelvin's fundamental solution of the Navier equation, see Bui, 1978). However, conventional discretizations in space reduce the linear integral operator in (40) to a matrix. Its discrete version by, say, compatible finite element models, preserves the symmetry and negative semi-definiteness which are the essential features of the above operator (and can be easily proved by the virtual work principle, see e.g. Maier, 1970b). These facts confer computational interest to the following statement, which will be proved as a consequence of the kinematic theorem of Section 3.1.

Proposition 2. Under the constitutive restrictions (a)–(e) of Section 2.4, the finite-step b.v. problem (6)–(11) is equivalent to the optimization problem:

$$\min\Big\{\bar{\Omega}(\Delta\varepsilon^p_{ij}, \Delta\lambda_\alpha) \equiv -\frac{1}{2}\int_V\int_V \Delta\varepsilon^p_{ij}(x_r)Z_{ijhk}(x_r, \xi_s)\Delta\varepsilon^p_{hk}(\xi_s)\,\mathrm{d}V_x\,\mathrm{d}V_\xi$$
$$+\int_V [\Pi(\bar{\lambda}_\alpha + \Delta\lambda_\alpha) + \Gamma(\bar{\varepsilon}^p_{ij} + \Delta\varepsilon^p_{ij})]\,\mathrm{d}V - \int_V (\bar{\sigma}^e_{ij} + \Delta\bar{\sigma}^e_{ij})\Delta\varepsilon^p_{ij}\,\mathrm{d}V - \int_V \bar{\sigma}^p_{ij}\Delta\varepsilon^p_{ij}\,\mathrm{d}V\Big\} \tag{41}$$

subject to:

$$\Delta\varepsilon^p_{ij} = \frac{\partial\Phi_\alpha}{\partial\tau_{ij}}(\tau_{hk})\Delta\lambda_\alpha, \quad \Delta\lambda_\alpha \geqslant 0 \tag{42a, b}$$

where

$$\tau_{ij} \equiv \bar{\sigma}_{ij} + \Delta\bar{\sigma}^e_{ij} + \int_V Z_{ijhk}(x_r, \xi_s)\Delta\varepsilon^p_{hk}(\xi_s)\,\mathrm{d}V_\xi - \alpha_{ij}(\bar{\varepsilon}^p_{rs} + \Delta\varepsilon^p_{rs}). \tag{43}$$

Proof. Substitute into the functional of Proposition 1, the new interpretation of increments:

$$\Delta\varepsilon^{e}_{ij} = \Delta\varepsilon^{ee}_{ij} + \Delta\varepsilon^{ep}_{ij} \equiv C_{ijhk}(\Delta\bar{\sigma}^{e}_{hk} + \Delta\sigma^{p}_{hk}). \tag{44}$$

where C_{ijhk} is the elastic compliance tensor.

Write virtual work equations which exploit equilibrium and compatibility of the two fictitious elastic responses (to external action and plastic strain increments), such as:

$$\int_V b_i \Delta u_i \,\mathrm{d}V + \int_{S_T} T_i \Delta u_i \,\mathrm{d}S + \int_{S_u} (\bar{\sigma}^{e}_{ij} + \Delta\bar{\sigma}^{e}_{ij}) n_j \Delta\bar{u}_i \,\mathrm{d}S$$

$$= \int_V (\bar{\sigma}^{e}_{ij} + \Delta\bar{\sigma}^{e}_{ij})(\Delta\varepsilon^{ee}_{ij} + \Delta\varepsilon^{ep}_{ij} + \Delta\varepsilon^{p}_{ij} + \Delta\bar{\theta}_{ij}) \,\mathrm{d}V; \tag{45}$$

$$\int_V \sigma^{p}_{ij} \cdot (\Delta\varepsilon^{ep}_{ij} + \Delta\varepsilon^{p}_{ij}) \,\mathrm{d}V = 0. \tag{46}$$

Using these equations and eqn (40) and dropping constant terms, functional (21) reduces to (41). The compatibility constraints (22) no longer intervene as they are fulfilled by the very definitions of the two addends which appear in (44). This implies that displacements are no longer among the variables.

As for the constraints (23a) and (23b), they remain unaltered except that eqn (40) is taken into account in the argument τ_{ij}.

4. DUAL EXTREMUM PROPERTIES OF FINITE-STEP HOLONOMIC SOLUTIONS

4.1. *Static (complementary energy) theorem*

4.1.1. Referring again to the step–holonomic b.v. problem (6)–(11), we prove below the following statement.

Proposition 3. Consider the optimization problem:

$$\min \Big\{ \Omega_c(\Delta\sigma_{ij}, \Delta\lambda_\alpha, \Delta\varepsilon^{p}_{ij}) \equiv \frac{1}{2}\int_V \Delta\sigma_{ij} C_{ijhk} \Delta\sigma_{hk} \,\mathrm{d}V + \int_V \Pi_c(\bar{\lambda}_\alpha + \Delta\lambda_\alpha) \,\mathrm{d}V + \int_V \Gamma_c(\bar{\varepsilon}^{p}_{ij} + \Delta\varepsilon^{p}_{ij}) \,\mathrm{d}V$$

$$- \int_V Y_\alpha(\bar{\lambda}_\beta + \Delta\lambda_\beta)\bar{\lambda}_\alpha \,\mathrm{d}V - \int_V \alpha_{hk}(\bar{\varepsilon}^{p}_{ij} + \Delta\varepsilon^{p}_{ij})\bar{\varepsilon}^{p}_{hk} \,\mathrm{d}V + \int_V \Delta\sigma_{ij}\Delta\bar{\theta}_{ij} \,\mathrm{d}V - \int_{S_u} \Delta\sigma_{ij} n_i \Delta\bar{u}_j \,\mathrm{d}S \Big\}, \tag{47}$$

where Π_c and Γ_c denote the following functions:

$$\Pi_c \equiv Y_\alpha(\lambda_\beta)\lambda_\alpha - \Pi; \quad \Gamma_c \equiv \alpha_{ij}(\varepsilon^{p}_{rs})\varepsilon^{p}_{ij} - \Gamma \tag{48}$$

subject to the relations:

$$\Delta\sigma_{ij,i} + \Delta\bar{b}_j = 0 \quad \text{in } V, \quad n_i\Delta\sigma_{ij} = \Delta\bar{T}_j \quad \text{on } S_T \tag{49a, b}$$

$$\phi_\alpha = \Phi_\alpha(\bar{\sigma}_{ij} + \Delta\sigma_{ij} - \alpha_{ij}(\bar{\varepsilon}^{p}_{rs} + \Delta\varepsilon^{p}_{rs})) - Y_\alpha(\bar{\lambda}_\beta + \Delta\lambda_\beta) \leqslant 0 \quad \text{in } V. \tag{50}$$

When the constitutive restrictions (a)–(e) of Section 2.4 hold, any solution to the b.v. problem (6)–(11) minimizes the functional Ω_c, subject to the constraints (49)–(50). Also the converse is true (sufficient condition) when hypotheses (a)–(e) are fulfilled with strict convexity of Π and Γ.

Proof of sufficiency. Being understood that equilibrium (49) is implicitly complied with by any stress field considered (so that $\delta\sigma_{hk}$ will be self-equilibrated), we write the augmented functional:

$$L_c = \Omega_c + \int_V \eta_\alpha \cdot (\phi_\alpha + \ell_\alpha^2)\,\mathrm{d}V. \tag{51}$$

Compute and set to zero its first variation due to variations $\delta\sigma_{ij}$, $\delta\lambda_\alpha$, $\delta\varepsilon_{ij}^p$, $\delta\ell_\alpha$ and $\delta\eta_\alpha$ around the minimum marked by $\sim$; thereafter, take into account eqn (50) written as $\phi_\alpha + \ell_\alpha^2 = 0$, and, finally rearrange to obtain:

$$\begin{aligned}\delta^{(1)}L_c = &\int_V \left[C_{ijhk}\Delta\tilde{\sigma}_{ij} + \Delta\bar{\theta}_{hk} + \eta_\alpha \frac{\partial\Phi_\alpha}{\partial\sigma_{hk}}(\tilde{\tau}_{ij})\right]\delta\sigma_{hk}\,\mathrm{d}V \\ &- \int_{S_u} \Delta\bar{u}_h n_k \delta\sigma_{hk}\,\mathrm{d}S + \int_V \left[\frac{\partial Y_\alpha}{\partial\lambda_\beta}(\tilde{\lambda}_\gamma)(\tilde{\lambda}_\beta - \bar{\lambda}_\beta - \eta_\beta)\right]\delta\lambda_\alpha\,\mathrm{d}V \\ &+ \int_V \left[\frac{\partial\alpha_{ij}}{\partial\varepsilon_{hk}^p}(\tilde{\varepsilon}_{rs}^p)\left(\tilde{\varepsilon}_{hk}^p - \bar{\varepsilon}_{hk}^p - \frac{\partial\Phi_\alpha}{\partial\sigma_{hk}}(\tilde{\tau}_{rs})\eta_\alpha\right)\right]\delta\varepsilon_{ij}^p\,\mathrm{d}V + 2\int_V [\eta\ell]_\alpha \delta\ell_\alpha\,\mathrm{d}V = 0.\end{aligned} \tag{52}$$

The stationarity of L_c at the minimum entails the following consequences:

(a) The kinematic quantities involved in the first two integrals are geometrically compatible, due to the virtual work principle because $\delta\sigma_{hk}$ is self-equilibrated, but otherwise arbitrary.

(b) The expressions in brackets in the last three integrals vanish everywhere in V. This implies, respectively [subject to: strict convexity of Π, i.e. nonsingularity of its Hessian, as for (53); strict convexity of Γ as for (54)]:

$$\eta_\beta = \Delta\tilde{\lambda}_\beta \tag{53}$$

$$\Delta\tilde{\varepsilon}_{ij}^p = \frac{\partial\Phi_\alpha}{\partial\sigma_{ij}}\Delta\tilde{\lambda}_\alpha \tag{54}$$

$$[\tilde{\phi}\Delta\tilde{\lambda}]_\alpha = 0. \tag{55}$$

The non-negativeness of the second variation $\delta^{(2)}L_c$, of the Lagrangian implies, in particular, that

$$\eta_\beta \geqslant 0. \tag{56}$$

Thus, it can be concluded that the constrained minimization of Ω_c leads to the fulfilment of all those relations governing the b.v. problem, which are not expressed by the constraints (49) and (50).

Proof of necessity. Capped symbols will mark quantities which solve the step problem (6)–(11), starred symbols quantities which are feasible with respect to constraints (49)–(50); barred and unbarred symbols for cumulative quantities refer to the starting and the end instant of the step, respectively (e.g. $\lambda_\alpha^* = \bar{\lambda}_\alpha + \Delta\lambda_\alpha^*$). Using eqns (47), after some easy manipulations, we may write:

$$\begin{aligned}\Omega_c^* - \hat{\Omega}_c = {} & \int_V \frac{1}{2}(\Delta\sigma_{ij}^* - \Delta\hat{\sigma}_{ij}) C_{ijhk} \cdot (\Delta\sigma_{hk}^* - \Delta\hat{\sigma}_{hk})\,\mathrm{d}V + \int_V [\Pi(\hat{\lambda}_\alpha) - \Pi(\lambda_\alpha^*)]\,\mathrm{d}V \\ & + \int_V [\Gamma(\hat{\varepsilon}_{ij}^p) - \Gamma(\varepsilon_{ij}^{p*})]\,\mathrm{d}V + \int_V [Y_\alpha(\lambda_\beta^*)(\lambda_\alpha^* - \bar{\lambda}_\alpha) - Y_\alpha(\hat{\lambda}_\beta)(\hat{\lambda}_\alpha - \bar{\lambda}_\alpha)]\,\mathrm{d}V \\ & + \int_V [\alpha_{ij}(\varepsilon_{rs}^{p*})(\varepsilon_{ij}^{p*} - \bar{\varepsilon}_{ij}^p) - \alpha_{ij}(\hat{\varepsilon}_{rs}^p)(\hat{\varepsilon}_{ij}^p - \bar{\varepsilon}_{ij}^p)]\,\mathrm{d}V \\ & + \int_V (\Delta\hat{\sigma}_{ij} C_{ijhk} + \Delta\bar{\theta}_{hk})(\Delta\sigma_{hk}^* - \Delta\hat{\sigma}_{hk})\,\mathrm{d}V - \int_{S_u} (\Delta\sigma_{hk}^* - \Delta\hat{\sigma}_{hk}) n_h \Delta\bar{u}_k\,\mathrm{d}S. \qquad (57)\end{aligned}$$

The following circumstances should be noted in eqn (57): the first integrand is non-negative; because of the assumed convexity of $\Pi(\lambda_\alpha)$ and $\Gamma(\varepsilon_{ij}^p)$ (which does not imply convexity of Π_c and Γ_c), the second and third integrands are not less than the linear terms of their Taylor expansions around λ_α^* and ε_{ij}^{p*}, respectively. Since $\Delta\sigma_{ij}^* - \Delta\hat{\sigma}_{ij}$ represents a self-equilibrated field and $\Delta\hat{\varepsilon}_{hk}$ and $\Delta\hat{u}_i$ are compatible, a virtual work equation permits one to replace the two last integrals by a single integral involving $\Delta\hat{\varepsilon}_{ij}^p$. Therefore, eqn (57) gives rise to the inequality:

$$\begin{aligned}\Omega_c^* - \hat{\Omega}_c \geqslant {} & \int_V Y_\beta(\lambda_\alpha^*)(\hat{\lambda}_\beta - \lambda_\beta^*)\,\mathrm{d}V + \int_V \alpha_{ij}(\varepsilon_{rs}^{p*})(\hat{\varepsilon}_{ij}^p - \varepsilon_{ij}^{p*})\,\mathrm{d}V + \int_V [Y_\alpha(\lambda_\beta^*)\Delta\lambda_\alpha^* - Y_\alpha(\hat{\lambda}_\beta)\Delta\hat{\lambda}_\alpha]\,\mathrm{d}V \\ & + \int_V [\alpha_{ij}(\varepsilon_{rs}^{p*})\Delta\varepsilon_{ij}^{p*} - \alpha_{ij}(\hat{\varepsilon}_{rs}^p)\Delta\hat{\varepsilon}_{ij}^p]\,\mathrm{d}V - \int_V (\Delta\sigma_{hk}^* - \Delta\hat{\sigma}_{hk})\Delta\hat{\varepsilon}_{hk}^p\,\mathrm{d}V. \qquad (58)\end{aligned}$$

By rearranging, expressing $\Delta\hat{\varepsilon}_{hk}^p$ in the form (8a) with normality ($\psi_\alpha = \phi_\alpha$), and taking into account eqns (12) and (14a) as in the necessity proof of Proposition 1, one obtains successively:

$$\begin{aligned}\Omega_c^* - \hat{\Omega}_c \geqslant {} & -\int_V [Y_\beta(\hat{\lambda}_\alpha) - Y_\beta(\lambda_\alpha^*)]\Delta\hat{\lambda}_\beta\,\mathrm{d}V - \int_V [\alpha_{ij}(\hat{\varepsilon}_{rs}^p) - \alpha_{ij}(\varepsilon_{rs}^{p*})]\Delta\hat{\varepsilon}_{ij}^p\,\mathrm{d}V \\ & + \int_V (\hat{\sigma}_{ij} - \sigma_{ij}^*)\Delta\hat{\varepsilon}_{ij}^p\,\mathrm{d}V \geqslant \int_V \left\{[\hat{\sigma}_{ij} - \alpha_{ij}(\hat{\varepsilon}_{rs}^p)]\frac{\partial\Phi_\beta}{\partial\tau_{ij}}(\hat{\tau}_{rs}) - Y_\beta(\hat{\lambda}_\alpha)\right\}\Delta\hat{\lambda}_\beta\,\mathrm{d}V \\ & - \int_V \left\{[\sigma_{ij}^* - \alpha_{ij}(\varepsilon_{rs}^{p*})]\frac{\partial\Phi_\beta}{\partial\tau_{ij}}(\tau_{rs}^*) - Y_\beta(\lambda_\alpha^*)\right\}\Delta\hat{\lambda}_\beta\,\mathrm{d}V. \qquad (59a, b)\end{aligned}$$

In (59b), the expressions in { } are recognized as the yield function $\hat{\phi}_\beta$ in the solution (former integral) and as ϕ^*_β for a feasible field (latter integral). Therefore, the former integral vanishes and the latter is nonpositive in view of eqn (50), since $\Delta\hat{\lambda}_\beta \geqslant 0$.

4.1.2. The restriction mentioned at the end of Proposition 3 for the validity of the sufficient condition, rules out the important case of ideal plasticity. This is characterized by the constitutive features: Y_α = constant and $\alpha_{ij} = 0$, which imply $\Pi_c = 0$ and $\Gamma_c = 0$. By virtue of such specializations, Proposition 3 can be rephrased as follows.

Proposition 3*. The optimization problem

$$\min\left\{\Omega_c(\Delta\sigma_{ij}) = \frac{1}{2}\int_V \Delta\sigma_{ij}C_{ijhk}\Delta\sigma_{hk}\,\mathrm{d}V + \int_V \Delta\sigma_{ij}\Delta\theta_{ij}\,\mathrm{d}V - \int_{S_u}\Delta\sigma_{ij}n_i\Delta\bar{u}_j\,\mathrm{d}S\right\}, \tag{60}$$

subject to:

$$\Delta\sigma_{ij,i} + \Delta\bar{b}_j = 0 \quad \text{in } V; \quad n_i\Delta\sigma_{ij} = \Delta\bar{T}_j \quad \text{on } S_T \tag{61a, b}$$

$$\phi_\alpha = \Phi_\alpha(\bar{\sigma}_{ij} + \Delta\sigma_{ij}) - Y_\alpha \leqslant 0 \quad \text{in } V, \tag{62}$$

is equivalent to the b.v. problem (6)–(11) specialized to ideal plasticity.

Proof of sufficiency. Let us follow once again the path of reasoning which led to the sufficiency part of Proposition 3: note that $\Delta\lambda_\alpha$ and $\Delta\varepsilon^p_{ij}$ do not show up in problems (60)–(62) and hence, in the first variation (52) of the specialized Lagrangian functional, the third and fourth addends (volume integrals) are missing. Therefore, the requirements that Hessians of $\Pi(\lambda_\alpha)$ and $\Gamma(\varepsilon^p_{ij})$ contained in them be nonsingular (i.e. Π and Γ strictly convex) are unnecessary. However, eqns (53) and (54) are still legitimate and the whole set of the relations governing the b.v. problem is still recovered. In fact, the compatibility arises again from the first and second addends in eqn (52) by the virtual work principle; the complementarity (55) from the last addend in (52); the sign constraints (56) from the non-negativeness of the second variations as before. Interpreting the Lagrangian multipliers η_α as plastic multiplier increments $\Delta\lambda_\alpha$, eqn (53), we can write eqn (54).

Proof of necessity holds unaltered with self-evident specializations.

Remarks. It is worth stressing that the main difference between the proofs of sufficiency in Proposition 3 and Proposition 3* rests on the fact that in the latter, ideal plasticity case, stresses only (not also $\Delta\varepsilon^p_{ij}$) appear in the argument τ_{ij} in eqn (52).

Proposition 3* can be recognized as an extension to the finite increment problem of Haar–Kàrmàn theorem (Martin, 1975), to which it reduces if the step starts from the unstressed state ($\bar{\sigma}_{ij} = 0$).

4.2. *A dual extremum property of plastic strains and internal variables*

Interpreting again the elastic–plastic response $\Delta\sigma_{ij}$ as the sum of the linear–elastic stress response $\Delta\bar{\sigma}^e_{ij}$ and the self-equilibrated stresses $\Delta\sigma^p_{ij}$ generated by plastic strains $\Delta\varepsilon^p_{ij}$, we prove below the following extremum theorem dual to Proposition 2.

Proposition 4. Consider the problem:

$$\min\left\{\Omega_c(\Delta\varepsilon^p_{ij},\Delta\lambda_\alpha)\equiv-\frac{1}{2}\int_V\int_V\Delta\varepsilon^p_{ij}(x_r)Z_{ijhk}(x_r,\xi_s)\Delta\varepsilon^p_{hk}(\xi_s)\,\mathrm{d}V_x\,\mathrm{d}V_\xi+\int_V[\Pi_c(\bar\lambda_\alpha+\Delta\lambda_\alpha)\right.$$

$$\left.+\Gamma_c(\bar\varepsilon^p_{ij}+\Delta\varepsilon^p_{ij})]\,\mathrm{d}V-\int_V Y_\beta(\bar\lambda_\alpha+\Delta\lambda_\alpha)\bar\lambda_\beta\,\mathrm{d}V-\int_V\alpha_{hk}(\bar\varepsilon^p_{ij}+\Delta\varepsilon^p_{ij})\bar\varepsilon^p_{hk}\,\mathrm{d}V\right\},\quad(63)$$

subject to the relation:

$$\phi_\alpha=\Phi_\alpha(\tau_{ij})-Y_\alpha(\bar\lambda_\beta+\Delta\lambda_\beta)\leqslant 0,\qquad(64)$$

where τ_{ij} is defined by eqn (43) and contains the data of the external action step through the pre-calculated elastic stresses $\Delta\bar\sigma^e_{ij}$.

Under the constitutive restrictions (a)–(e) any solution to the boundary value problem (6)–(11) solves the minimization problem (63)–(64), and the converse is true if strict convexity of Π and Γ is assumed.

Proof. Splitting again the actual response into two addends, as in eqn (43), use eqn (45) again and the further virtual work equation

$$\int_V\Delta\sigma^p_{ij}(C_{ijrs}\Delta\bar\sigma^e_{rs}+\Delta\bar\theta_{ij})\,\mathrm{d}V=\int_{S_u}\Delta\sigma^p_{ij}n_i\Delta\bar u_j\,\mathrm{d}S.\qquad(65)$$

Thus, taking into account eqn (40) and dropping constant terms, the functional (47) of Proposition 3 is easily transformed into (63). The equilibrium constraints (49) are *a priori* fulfilled by both fictitious linear responses (to external action and plastic strain increments) in view of their very definition; therefore these constraints can be dropped. Condition (50) becomes constraint (64) with the integral expression (40) of self-stresses induced by plastic strains.

Remark. The complementary plastic work can be given the following expression, if the constitutive restrictions (b)–(d) hold and eqns (48) are used:

$$W^p_c(\bar t+\Delta t)\equiv\int_0^{\bar t}\varepsilon^p_{ij}(t')\dot\sigma_{ij}(t')\,\mathrm{d}t'+\int_{\bar t}^{\bar t+\Delta t}\varepsilon^p_{ij}(t')\dot\sigma_{ij}(t')\,\mathrm{d}t'=\Pi_c(\bar\lambda_\alpha+\Delta\lambda_\alpha)$$

$$+\Gamma_c(\bar\varepsilon^p_{ij}+\Delta\varepsilon^p_{ij})+\sigma_{ij}\bar\varepsilon^p_{ij}-Y_\alpha(\bar\lambda_\beta+\Delta\lambda_\beta)\bar\lambda_\alpha-\alpha_{ij}(\bar\varepsilon^p_{rs}+\Delta\varepsilon^p_{rs})\bar\varepsilon^p_{ij}.\quad(66)$$

This shows that the functions Π_c and Γ_c defined by eqns (48) represent the two addends into which the complementary plastic work can be split (as Π and Γ for W^p), provided the starting situation for the step coincides with the original undeformed state ($\bar\varepsilon^p_{ij}=0$, $\bar\lambda_\alpha=0$).

5. GENERAL EXTREMUM STATEMENTS FOR FINITE-STEP PROBLEMS

We will keep in what follows, only two of the constitutive restrictions of Section 2.4, namely: (b) the effective stresses are positively homogeneous functions of the order of one of their argument; (c) the flow rule is associative, i.e. the plastic potentials coincide with the yield functions. Note that no condition is imposed on the hardening rule and, in particular, softening is now admitted. The two statements given below rest on this weaker hypothesis basis.

Proposition 5. Under restrictions (b) and (c), the finite-step b.v. problem (6)–(11) is equivalent to the following constrained optimization, provided the minimum is zero (otherwise the b.v. problem has no solution):

$$\min \Big\{ \bar{\bar{\Omega}}(\Delta u_i, \Delta\lambda_\alpha, \Delta\varepsilon^p_{ij}, \Delta\sigma_{ij}) \equiv \frac{1}{2}\int_V \Delta\varepsilon^e_{ij} E_{ijhk} \Delta\varepsilon^e_{hk}\, \mathrm{d}V + \int_V \bar{\sigma}_{ij}\Delta\varepsilon^e_{ij}\, \mathrm{d}V$$

$$-\int_V (\bar{b}_i + \Delta\bar{b}_i)\Delta u_i\, \mathrm{d}V - \int_{S_T} (\bar{T}_i + \Delta\bar{T}_i)\Delta u_i\, \mathrm{d}S + \frac{1}{2}\int_V \Delta\sigma_{ij} C_{ijhk} \Delta\sigma_{hk}\, \mathrm{d}V$$

$$+\int_V \Delta\varepsilon^p_{ij}\alpha_{ij}(\bar{\varepsilon}^p_{hk} + \Delta\varepsilon^p_{hk})\, \mathrm{d}V + \int_V \Delta\lambda_\alpha Y_\alpha(\bar{\lambda}_\beta + \Delta\lambda_\beta)\, \mathrm{d}V - \int_V \Delta\bar{\theta}\Delta\sigma_{ij}\, \mathrm{d}V$$

$$-\int_{S_u} \Delta\sigma_{ij} n_j \Delta\bar{u}_i\, \mathrm{d}S - \int_V \Delta\bar{\theta}_{ij}\bar{\sigma}_{ij}\, \mathrm{d}V - \int_{S_u} \bar{\sigma}_{ij} n_j \Delta\bar{u}_i\, \mathrm{d}S \Big\} = 0, \tag{67}$$

subject to:

$$\Delta\varepsilon^p_{ij} = \frac{\partial\Phi_\alpha}{\partial\sigma_{ij}}\Delta\lambda_\alpha; \quad \Delta\lambda_\alpha \geqslant 0; \quad \Phi_\alpha(\bar{\sigma}_{ij} + \Delta\sigma_{ij} - \alpha_{ij}(\bar{\varepsilon}^p_{hk} + \Delta\varepsilon^p_{hk})) - Y_\alpha(\bar{\lambda}_\beta + \Delta\lambda_\beta) \leqslant 0 \tag{68a, b, c}$$

$$\Delta\sigma_{ij} = E_{ijhk} \cdot (\tfrac{1}{2}(\Delta u_{h,k} + \Delta u_{k,h}) - \Delta\varepsilon^p_{hk} - \Delta\bar{\theta}_{hk}) \quad \text{in } V; \quad \Delta u_i = \Delta\bar{u}_i \quad \text{on } S_u \tag{69a, b}$$

$$\Delta\sigma_{ij,i} + \Delta\bar{b}_j = 0 \quad \text{in } V; \quad \Delta\sigma_{ij} n_i = \Delta\bar{T}_j \quad \text{on } S_T. \tag{70a, b}$$

Proof. Using hypotheses (b) and subsequently (c), we can write, taking into account eqns (7a) and (25):

$$\Delta\lambda_\alpha\phi_\alpha = \Delta\lambda_\alpha \frac{\partial\Phi_\alpha}{\partial\tau_{ij}}(\tau_{rs})\tau_{ij} - \Delta\lambda_\alpha Y_\alpha(\bar{\lambda}_\beta + \Delta\lambda_\beta)$$

$$= \Delta\varepsilon^p_{ij} \cdot [\bar{\sigma}_{ij} + \Delta\sigma_{ij} - \alpha_{ij}(\bar{\varepsilon}^p_{rs} + \Delta\varepsilon^p_{rs})] - \Delta\lambda_\alpha Y_\alpha(\bar{\lambda}_\beta + \Delta\lambda_\beta). \tag{71a, b}$$

Rearranging and using (6), one obtains:

$$\Delta\lambda_\alpha\phi_\alpha = -\Delta\varepsilon^e_{ij} E_{ijrs}\Delta\varepsilon^e_{rs} - \Delta\varepsilon^p_{ij}\alpha_{ij}(\varepsilon^p_{hk}) - (\bar{\sigma}_{ij} + \Delta\sigma_{ij})\Delta\theta_{ij}$$

$$-\Delta\varepsilon^e_{ij}\bar{\sigma}_{ij} - \Delta\lambda_\alpha Y_\alpha(\lambda_\beta) + (\bar{\sigma}_{ij} + \Delta\sigma_{ij})\Delta\varepsilon_{ij}. \tag{72}$$

Integration of the r.h.s. of (72) over the volume V and a virtual work equation to transform the last addend, leads to functional (67) with a reversed sign. Therefore, its minimization under constraints (68)–(70) means minimizing the integral over V of the non-negative integrand $-\phi_\alpha \Delta\lambda_\alpha$. If the minimum is zero, then the complementary condition (9) is satisfied everywhere and, since all other governing relations are expressed by the constraints, the b.v. problem is solved. Clearly, if the minimum is not zero, the set of governing relations (6)–(11) turns out to be incompatible. Conversely, any solution of (6)–(11) is clearly feasible for optimization (67)–(70) and makes its objective zero [and, hence, minimum, in view of the above noted meaning of functional (67)].

Proposition 6. When the elastic stress increments $\Delta\bar{\sigma}^e_{ij}$ have been determined, under the constitutive restrictions (b) and (d), the b.v. problem (6)–(11) is equivalent to the following minimization problem, if the minimum is zero (otherwise the b.v. problem is not solvable):

$$\min \Big\{ \bar{\Omega}(\Delta\varepsilon^p_{ij}, \Delta\lambda_\alpha) \equiv -\int_V \int_V \Delta\varepsilon^p_{ij}(x_r) Z_{ijhk}(x_r, \xi_s) \Delta\varepsilon^p_{hk}(\xi_s) \, \mathrm{d}V_x \, \mathrm{d}V_\xi - \int_V (\bar{\sigma}_{ij} + \Delta\bar{\sigma}^e_{ij}) \Delta\varepsilon^p_{ij} \, \mathrm{d}V$$

$$+ \int_V \Delta\varepsilon^p_{ij} \alpha_{ij} (\bar{\varepsilon}^p_{hk} + \Delta\varepsilon^p_{hk}) \, \mathrm{d}V + \int_V \Delta\lambda_\alpha Y_\alpha (\bar{\lambda}_\beta + \Delta\lambda_\beta) \, \mathrm{d}V \Big\}, \quad (73)$$

subject to [eqn (43) defines τ_{ij}]:

$$\Delta\varepsilon^p_{ij} = \frac{\partial \Phi_\alpha}{\partial \tau_{ij}} (\tau_{hk}) \Delta\lambda_\alpha, \quad \Delta\lambda_\alpha \geqslant 0, \quad \Phi_\alpha(\tau_{ij}) \leqslant Y_\alpha. \quad (74a, b, c)$$

Proof. As in Propositions 2 and 4, the superposition of the elastic responses to external actions (pre-determined as $\Delta\bar{\sigma}^e_{ij}$) and to plastic strain increments [through the integral operator (40) and relevant Green's kernel Z] makes compatibility and equilibrium *a priori* satisfied. The governing relations of the b.v. problem are imposed as constraints (74), except the complementarity condition (9); in view of the sign constraints of both factors, this can be enforced through the constrained minimization of:

$$\int_V -\phi_\alpha(\Delta\varepsilon^p_{ij}, \Delta\lambda_\beta) \Delta\lambda_\alpha \, \mathrm{d}V. \quad (75)$$

This integral is easily transformed into the functional (73), using the constitutive hypotheses (b) and (c) alone, as done in the proof of Proposition 5.

Remark 1. The optimization problem concerned by Proposition 5 is easily seen to be related to those in Propositions 1 and 2 by means of two noteworthy links: (i) the functional $\bar{\Omega}$ is the sum of the two functionals Ω and Ω_c, and of a term which is constant within the step [last two integrals in (67)]; (ii) the feasible region defined by the constraints (68)–(70) is the intersection of the two regions defined by (22)–(23) and by (49)–(50), respectively. Similar connections relate Proposition 6 to Propositions 2 and 4. In other terms, it might be said that, when all the constitutive restrictions of Section 2.4 hold, Proposition 5 "splits" into the pair of dual, much simpler extremum characterizations stated in Propositions 1 and 3. Similarly, Proposition 6 "splits" into Propositions 2 and 4.

Remark 2. None of the constitutive restrictions of Section 2.4 are needed for the extremum characterization trivially achievable for the b.v. problem by minimizing to an integral like (75) subject to all governing relations except complementarity. Restrictions (b) and (c), i.e. normality and effective stress homogeneity of the order of one, as seen above by proving Propositions 5 and 6, confer to the functional to minimize a favorable, mechanically interpretable form and give rise to the links pointed out in the preceding remark.

Remark 3. The functionals envisaged in Propositions 5 and 6 are bounded below over the relevant feasible sets and hence, if these are not empty, their minimizations always have solutions. The fact that the minimum fails to vanish and hence, the b.v. problem has no solution, means that the body is incapable of carrying the current step of external actions. Clearly, such situations of collapse (whose identification will usually be improved by reducing the step amplitude) are characterized by lack of (bounded) solution in the context of the other Propositions 1–4.

6. CLOSING REMARKS

As a conclusion we briefly point out below the computational potentialities of the present results, their connections with earlier results via specialization and, finally, their possible extensions.

(a) If a suitable space discretization (e.g. by finite element modelling) is adopted, each one of the propositions pointed out in what precedes, reduces the finite-step problem arising from a backward–difference (or stepwise–holonomic) strategy for the approximate time integrations of the (nonlinear differential) relations governing elastic–plastic processes, to nonlinear programming. The nonlinear programming problem equivalent to the b.v. problem is generally nonconvex and not easy to solve. However, extremum characterizations of the solution to the finite-step elastic–plastic problem can lead to rigorous sufficient conditions for the convergence of iterative algorithms, devised to solve numerically that problem. With reference to Proposition 1, this circumstance is demonstrated in a parallel paper (Comi and Maier, 1990).

(b) A first kind of specializations of the present findings is obtained by assuming yield functions linear in the stresses and linear hardening. This constitutive piecewise-linearization reduces Propositions 1–6 to earlier extremum theorems (Maier, 1969, 1970a) which involve quadratic functionals and linear constraints and, in discrete versions, quadratic programming. A second way of specializing the developments expounded herein is to consider infinitesimal increments only ($\Delta t = \delta t \to 0$). This reduces again to quadratic–linear optimizations (or programming) and leads back to earlier rate theorems, precisely to statements proposed by Capurso (Proposition 1), Ceradini (Proposition 2), Capurso–Maier (Proposition 3) and Maier (Propositions 4, 5, 6). Surveys of such and other antecedents can be found e.g. in Martin (1975), Panagiotopoulos (1985), Cohn *et al.* (1979) and Maier *et al.* (1982). For perfectly-plastic materials and a single-step (starting from an unstressed state), Proposition 3 reduces to classical Haar–Kàrmàn theorem (see Martin, 1975). Recent contributions due to Martin and Reddy (1988), Reddy and Griffin (1986) and to Feijòo and Zouain (1988) seem to be related to Propositions 1 and 2, respectively, although they differ in approach, constitutive basis and path of reasoning.

(c) A relatively straightforward generalization of the present study would be to a broader class of elastic–plastic material models (primarily a class where the rates of the internal variables are distinct from the plastic multipliers intervening in the flow rules). No difficulty is expected in allowing for geometric effects on equilibrium, provided these are expressed by terms linearized in the displacement increments and containing the stresses at the beginning of the step. This would permit one to capture the influence of possible combined constitutive and geometrical instabilizing effects on the overall finite-step response (as on the rate response studied by Maier, 1971). A promising and a more challenging current development of this study is its extension to damage material models and to other constitutive laws not included in traditional elastoplasticity.

Acknowledgements—A grant from MPI (40%) is gratefully acknowledged.

REFERENCES

Bui, H. J. L. (1978). Some remarks about the formulation of three-dimensional thermoelastoplastic problems by integral equations. *Int. J. Solids Structures* **14,** 935–939.

Casciaro, R. and Mancuso, M. (1988). Un approccio numerico al problema dell'elastoplasticità incrementale. Omaggio a Giulio Ceradini, University of Rome, 177–187.

Cohn, M. Z., Maier, G. and Grierson, D. E. (1979). *Engineering Plasticity by Mathematical Programming.* Pergamon Press, Oxford.

Comi, C. and Maier, G. (1990). Extremum theorem and convergence criterion for an iterative solution to the finite-step problem in elastoplasticity with mixed nonlinear hardening. *Eur. J. Mech.*, A/Solids, **9**(6), to appear.

De Donato, O. and Maier, G. (1972). Mathematical programming methods for the inelastic analysis of reinforced concrete frames allowing for limited rotation capacity. *Int. J. Num. Meth. Engng* **4,** 307–329.

De Donato, O. and Maier, G. (1973). Finite element elastoplastic analysis by quadratic programming: the multistage method. *2nd International Conference on Structural Mechanics in Nuclear Technology* SMiRT, Berlin, V, Part M.

Dittmer, C. T., Griffin, T. B. and Martin, J. B. (1985). A programming approach to the numerical analysis of elastoplastic continua. *Engng Fract. Mech.* (Olszak Memorial Volume) **21,** 817–825.

Drucker, D. C. (1960). Plasticity. In *Structural Mechanics* (Edited by J. N. Goodier and N. J. Hoff), pp. 407–455. Pergamon Press, New York.

Drucker, D. C. (1988). Conventional and unconventional plastic response and representation. *Appl. Mech. Rev.* **41**(4), 151–167.

Feijòo, R. A. and Zouain, N. (1988). Formulations in rates and increments for elastic–plastic analysis. *Int. J. Num. Meth. Engng* **26**(9), 2031–2048.

Franchi, A. and Genna, F. (1984). Minimum principles and initial stress method in elastic–plastic analysis. *Engng Struct.* 65–69.

Franchi, A. and Genna, F. (1987). A numerical scheme for integrating the rate plasticity equations with an 'a priori' error control. *Comp. Meth. Appl. Mech. Engng* **60**(3), 317–342.

Hill, R. (1957). A general theory of uniqueness and stability in elastic–plastic solids. *J. Mech. Phys. Solids* **5,** 229–241.

Maier, G. (1969). Teoremi di minimo in termini finiti per continui elastoplastici con leggi costitutive linearizzate a tratti. *Rendiconti dell'Istituto Lombardo di Scienze e Lettere* **103,** 1066–1080.

Maier, G. (1970a). A matrix structural theory of piecewise–linear plasticity with interacting yield planes. *Meccanica* **5**(1), 55–66.

Maier, G. (1970b). A minimum principle for incremental elastoplasticity with nonassociated flow laws. *J. Mech. Phys. Solids* **18,** 319–330.

Maier, G. (1971). Incremental plastic analysis in the presence of large displacements and physical instabilizing effects. *Int. J. Solids Structures* **7,** 345–372.

Maier, G., Munro, J. and Lloyd-Smith, D. (1982). Mathematical programming applications to engineering plastic analysis. *Appl. Mech. Rev.* **35,** 1631–1643. *Appl. Mech.* Update (1986) (Edited by C. R. Steele and G. S. Springer), pp. 377–383. ASME, New York.

Maier, G. and Novati, G. (1990). Extremum theorems for finite-step backward-difference analysis of elastic–plastic nonlinearly hardening solids. *Int. J. Plast.* **6,** 1–10.

Martin, J. B. *Plasticity.* (1975). MIT Press, Cambridge, MA.

Martin, J. B. (1981). An internal variable approach to the formulation of finite element problems in plasticity. In *Physical Nonlinearities in Structural Analysis* (Edited by J. Hult and J. Lemaitre), pp. 165–176. Springer, Berlin.

Martin, J. B. and Reddy, B. D. (1988). Variational principles and solution algorithms for internal variable formulations of problems in plasticity. Omaggio a Giulio Ceradini, University of Rome, 465–477.

Martin, J. B., Reddy, B. D., Griffin, T. B. and Bird, W. W. (1987). Applications of mathematical programming concepts to incremental elastic–plastic analysis. *Engng Struct.* **9,** 171–176.

Matthies, H. and Strang, G. (1979). The solution of nonlinear finite element equations. *Int. J. Num. Meth. Engng* **14**(11), 1613–1626.

Ortiz, M. and Popov, E. P. (1985). Accuracy and stability of integration algorithms for elastoplastic constitutive relations. *Int. J. Num. Meth. Engng* **21,** 1561–1576.

Panagiotopoulos, P. D. (1985). *Inequality Problems in Mechanics and Applications*. Birkhäuser, Boston.

Perego, U. (1988). Explicit backward difference operators and consistent predictors for linear hardening elastic–plastic constitutive laws. *Sol. Mech. Arch.* **13**(2), 65–102.

Reddy, B. D. and Griffin, T. B. (1986). Variational principles and convergence of finite element approximations of a holonomic elastic–plastic problem. AMRU Report No. 83, University of Cape Town.

Resende, L. and Martin, J. B. (1985). Formulation of Drucker–Prager cap model. *ASCE J. Engng Mech.* **111**(7), 855–881.

Sewell, M. J. (1987). *Maximum and Minimum Principles*. Cambridge University Press, Cambridge.

Simo, J. C., Kennedy, J. G. and Govindjee, S. (1988). Non-smooth multisurface plasticity and viscoplasticity loading/unloading conditions and numerical algorithms. *Int. J. Num. Meth. Engng* **26**, 2161–2185.

Simo, J. C. and Taylor, R. L. (1986). A return mapping algorithm for plane stress elastoplasticity. *Int. J. Num. Meth. Engng* **22**, 649–670.

Zouain, P. N., Borges, L. A. and Hecke, M. B. (1988). A force method for elastic–plastic analysis of frames by quadratic optimization. *Int. J. Solids Structures* **24**, 211–221.

Computational Model for Large Elastic-Plastic Deformations at Moderate Elastic Strains

D.Perić†, and D.R.J.Owen

Department of Civil Engineering, University College of Swansea, Swansea SA2 8PP, Wales, U.K.

Summary — *Issues related to the constitutive modelling and computational treatment of finite deformation elasto-plasticity are examined employing logarithmic stretches as strain measures. A strain-energy function for isotropic elastic materials is proposed, which in conjunction with notions of multiplicative decomposition of the deformation gradient and unstressed configuration provides a computationally effective framework. It is pointed out that in this context, an algorithm could be proposed for rate-independent finite strain elasto-plasticity, which is exact for elastic processes. Large elasto-plastic deformations at moderate elastic strains are examined within the approximation theory and displacement based finite element formulation of the boundary value problem proposed. Numerical examples are provided and results are compared with available experimental evidence and existing results.*

1. Introduction

In a situation where the theoretical framework for large strain elastoplasticity still suffers from lack of consistency with regards to various theories and experimental evidence, and is still the subject of vehement debate often resulting in disagreement between different approaches (see e.g [1,15-16,24-25,40]), the computational scene has provided a series of unexpected and physically unacceptable results. Dienes [5] 1979 observed, for the unrestricted homogenous simple shear deformations of a purely elastic material, that the Jaumann rate used in the hypoelastic constitutive equation leads to oscillatory Cauchy stresses. In their intriguing paper [22], Nagtegaal and de Jong 1982, observed oscillatory Cauchy stresses for the Jaumann rate used in the hypoelastic constitutive equation in connection with kinematic hardening plasticity, where elastic deformations were restricted by the plastic flow. These observations triggered numerical experiments with various stress rates exchanging the Jaumann rate of Kirchhoff (or Cauchy) stress $\overset{\triangledown}{\boldsymbol{\tau}}$ in the hypoelastic constitutive equation $\overset{\triangledown}{\boldsymbol{\tau}} = \mathbf{h} : (\boldsymbol{d} - \boldsymbol{d}^p)$ with the 'proper' objective stress rate. However, it was pointed out by Hill [8], and later shown by Nemat-Nasser [24], that all stress rates are equivalent in the sense that any other could be used in the constitutive equation with an appropriate change of constitutive moduli. Şimo and Pister [38] have even proved that the usual assumption of constant and isotropic spatial elasticities in finite strain regimes conflicts with elasticity. In [30] this observation has

† On leave from *Gradjevinski institut, Fakultet gradjevinskih znanosti, Zagreb, Yugoslavia*

been confirmed explicitly showing that a unique stress rate consistent with hyperelasticity can be defined for every conjugate pair through the rate form of the associated hyperelastic constitutive equation and deriving an anisotropic spatial elasticity tensor associated with the conjugate pair $\{\boldsymbol{T}, \ln \boldsymbol{U}\}$. Unexpected phenomena observed for the kinematic hardening of elasto-plastic materials identified the need for reconsideration of the constitutive theory of finite strain elasto-plasticity, particularly for anisotropic materials and brought into focus notions of the plastic spin and plastically induced rotation, directors etc. (see [1,4,16,24,27]).

In computational circles, effort has been directed at the formulation of an algorithm for integration of the constitutive equations that would maintain incremental objectivity. Various schemes have been proposed , e.g. [11-12,36]. Some of these use the rotationally invariant part of the deformation gradient in the rate constitutive equation based on the natural approach [2] or employ logarithmic stretches as strain measures [23,35].

In this paper it is shown that, motivated by Rice [34] and Hill [9] and following Rolph and Bathe [35], an algorithm based on logarithmic stretches and their conjugate stresses could be proposed for rate-independent finite strain elasto-plasticity, which is exact for elastic processes.

Following Ogden [26], in Section 3 a new isotropic strain energy function is introduced and applied to the example of homogenous simple shear deformations. Approximations to the rate form of the hyperelastic constitutive equation are derived for the small elastic strain case.

In Section 4, a framework for finite strain elasto-plasticity is proposed, based on multiplicative decomposition of the deformation gradient as postulated by Lee [15], namely $\boldsymbol{F} = \boldsymbol{F}^e \boldsymbol{F}^p$, and following Nemat-Nasser [24], additive strain rate decomposition is adopted, i.e.

$$(\ln \boldsymbol{U})^{\cdot} = (\ln \boldsymbol{U}^e)^{\cdot} + (\ln \boldsymbol{U}^p)^{\cdot}$$

As proposed by Lee [15], we postulate $\boldsymbol{F}^e = \boldsymbol{V}^e$, which is justified, at least, for both isotropic elastic and plastic responses (see [4,40]). In such a situation, the elasto-plastic constitutive equations are simply transformed onto the intermediate configuration. Restriction to isotropic elasto-plasticity, i.e. the situation where the plastic deformation rate tensor is coaxial with the stress tensor, allows formulation of large strain elasto-plasticity on the continuum level without introduction of a substructure.

Numerical integration of the constitutive equation relying on operator split methodology is presented in Section 5. The elastic predictor - plastic corrector procedure is generalized for the finite strain case, employing the simplicity of treatment of small strain elasto-plasticity.

In Sections 6-7, we present formulation of the boundary value problem in terms of logarithmic strain and its conjugate stress and the displacement based finite element formulation, exploiting approximation of the rate form of the hyperelastic constitutive equation given in Section 3.

Numerical examples are provided in Section 8. For the example of necking and failure of thin sheet in uniaxial tension, close agreement with experiment is shown. Finally, failure analysis of the Scordelis-Lo roof is presented and capabilities are foreseen for the modelling of highly complex physical phenomena such as shear band formation and structural failure analysis.

2. Preliminaries

2.1 Notation

We shall use the notation and terminology which is already standard in continuum mechanics (see e.g. Marsden and Hughes [17] or Ogden [26]):

$\chi : \Omega \subset \mathbb{R}^3 \to \mathbb{R}^3$ ... *deformation* from the *reference* configuration $\Omega \subset \mathbb{R}^3$ of the body B (time dependence omitted)

$x = \chi(X)$ for each $X \in \Omega$... *current* or *deformed* configuration

$E_I(e_i), (I, i = 1,2,3)$... bases in the reference (current) configuration

$F = \frac{\partial \chi(X)}{\partial X} \equiv \frac{\partial x}{\partial X}$... *deformation gradient*

$J = \det F$... Jacobian of the deformation mapping

$F = RU = VR$... polar decomposition of F

$U(V)$... right (left) *stretch* tensor

R... *rotation* tensor

$\lambda_{(I)} = \lambda_{(i)} (I, i = 1,2,3)$... *principal stretches* – eigenvalues of $U(V)$

$N^{(I)}(n^{(i)}), (I, i = 1,2,3)$... *Lagrangian (Eulerian) triad* – eigenvectors of $U(V)$

$C = F^T F = U^2 = \sum_{I=1}^{3} \lambda_{(I)}^2 N^{(I)} \otimes N^{(I)}$... *right Cauchy-Green* tensor

$b = FF^T = V^2 = \sum_{i=1}^{3} \lambda_{(i)}^2 n^{(i)} \otimes n^{(i)}$... *left Cauchy-Green* tensor

$v = v(x)$... *spatial velocity*

$l = \nabla v(x) = \frac{\partial v(x)}{\partial x}$... *velocity gradient*

$d = \operatorname{sym} l = (l + l^T)$... *rate of deformation* tensor

$w = \operatorname{asym} l = (l - l^T)$... *spin* tensor

σ... *Cauchy* stress tensor

$\tau = J\sigma$... *Kirchhoff* stress tensor

$\rho_{\mathrm{Ref}}(\rho)$... reference (current) mass densities

$\dot{w} = \frac{1}{\rho_{\mathrm{Ref}}}(\tau : d) = \frac{1}{\rho}(\sigma : d)$... rate of stress work per unit mass

$S = F^{-1} \tau F^{-T}$... *second Piola-Kirchhoff* stress tensor

$P = \tau F^{-T} = F S$... *first Piola-Kirchhoff* stress tensor

$\Omega^{(L)} = \sum_{I,J=1}^{3} \Omega_{IJ}^{(L)} N^{(I)} \otimes N^{(J)}$... spin of the Lagrangian triad

$\omega^{(E)} = \sum_{i,j=1}^{3} \omega_{ij}^{(E)} n^{(i)} \otimes n^{(j)}$... spin of the Eulerian triad

$\omega^{(R)} = \dot{R} R^T$... spin of the Eulerian triad relative to the Lagrangian triad †

2.2 Geometry and kinematical relations

Following Hill [8-9] and Ogden [26], we consider the general Lagrangian strain

† Note that $\omega^{(R)}$ is the spatial tensor

tensors

$$E = \left\{ \begin{array}{ll} \frac{1}{m}(U^m - 1), & m \neq 0 \\ \ln U, & m = 0 \end{array} \right\} \tag{2.1}$$

with eigenvalues in the Lagrangian triad

$$f(\lambda_{(I)}) = \left\{ \begin{array}{ll} \frac{1}{m}(\lambda_{(I)}^m - 1), & m \neq 0 \\ \ln\lambda_{(I)}, & m = 0 \end{array} \right\} \tag{2.2}$$

$(I = 1,2,3)$, where $\lambda_{(I)}$ are principal stretches, and f is a real continuous monotonic increasing mapping such that $\{f: (0,\infty) \to \mathbb{R} \mid f(1) = 0,\ f'(1) = 1\}$. Members of this family of strain measures in standard use are: Green-Lagrange $E^{(2)} = \frac{1}{2}(U^2 - 1)$ for $m = 2$; Almansi $E^{(-2)} = \frac{1}{2}(1 - U^{-2})$ for $m = -2$; Biot $E^{(1)} = (U - 1)$ for $m = 1$; Hencky $E^{(0)} = \ln U$ for $m = 0$ †.

Material-time differentiation of the general Lagrangian strain tensors (2.1) defines the general class of strain rate tensors. The general *Lagrangian strain rate* tensor,

$$\dot{E} = \sum_{I,J=1}^{3} \dot{E}_{IJ} N^{(I)} \otimes N^{(J)} \tag{2.3}$$

with components

$$\left. \begin{array}{ll} \dot{E}_{II} = f'(\lambda_{(I)})\dot{\lambda}_{(I)}, & I = 1,2,3 \\ \dot{E}_{IJ} = \Omega_{IJ}^{(L)}\{f(\lambda_{(J)}) - f(\lambda_{(I)})\}, & I \neq J \end{array} \right\} \tag{2.4}$$

being itself objective, as opposed to a general *Eulerian strain rate* tensor which includes the spin tensor w , is of special interest.

Writing the rate of deformation and the rotated rate of deformation tensor in the Eulerian triad and Lagrangian triad, respectively,

$$d = \sum_{i,j=1}^{3} d_{ij} n^{(i)} \otimes n^{(j)}, \qquad D = \sum_{I,J=1}^{3} D_{IJ} N^{(I)} \otimes N^{(J)} \tag{2.5}$$

with,

$$D = R^T d\, R \tag{2.6}$$

and using the connection derived by Hill (see [9])

$$\left. \begin{array}{ll} D_{II} = \dfrac{\dot{\lambda}_{(I)}}{\lambda_{(I)}}, & I = 1,2,3 \\ D_{IJ} = \Omega_{IJ}^{(L)} \dfrac{\lambda_{(J)}^2 - \lambda_{(I)}^2}{2\lambda_{(I)}\lambda_{(J)}}, & I \neq J \end{array} \right\} \tag{2.7}$$

† Observe that corresponding relations are plausible in the Eulerian triad.

a useful relationship between the general Lagrangian strain rate and rate of deformation can be obtained, (see Hill [9], Nemat-Nasser [24], or Ogden [26]) which is expressed (in component form) as

$$\left.\begin{aligned} \dot{E}_{II} &= \lambda_{(I)} f'(\lambda_{(I)}) D_{II} && I = 1,2,3 \\ \dot{E}_{IJ} &= \frac{f(\lambda_{(J)}) - f(\lambda_{(I)})}{\lambda_{(J)}^2 - \lambda_{(I)}^2} 2\lambda_{(I)}\lambda_{(J)} D_{IJ}, && I \neq J, \lambda_{(I)} \neq \lambda_{(J)}, \end{aligned}\right\} \tag{2.8}$$

where no summation over indices is implied.

2.3 Work conjugacy and stress measures

Exploiting the fundamental notion of work conjugacy introduced by Hill [8] stress measures may be obtained independently of any constitutive law. In particular, we are interested in stress measures conjugate to general Lagrangian strain measures (2.1).

Indeed, if with $\boldsymbol{\Sigma}$ we denote the general stress in the Lagrangian triad

$$\boldsymbol{\Sigma} = \sum_{I,J=1}^{3} \Sigma_{IJ} \boldsymbol{N}^{(I)} \otimes \boldsymbol{N}^{(J)} \tag{2.9}$$

and write the Kirchhoff stress and rotated Kirchhoff stress in the Eulerian triad and Lagrangian triad, respectively,

$$\boldsymbol{\tau} = \sum_{i,j=1}^{3} \tau_{ij} \boldsymbol{n}^{(i)} \otimes \boldsymbol{n}^{(j)}, \qquad \boldsymbol{T} = \sum_{I,J=1}^{3} T_{IJ} \boldsymbol{N}^{(I)} \otimes \boldsymbol{N}^{(J)} \tag{2.10}$$

with,

$$\boldsymbol{T} = \boldsymbol{R}^T \boldsymbol{\tau} \boldsymbol{R} \tag{2.11}$$

then, using the identity

$$\frac{1}{\rho_{\text{Ref}}}(\boldsymbol{\tau} : \boldsymbol{d}) = \frac{1}{\rho_{\text{Ref}}}(\boldsymbol{\Sigma} : \dot{\boldsymbol{E}}) \tag{2.12}$$

it can be shown (see Hill [9] or Nemat-Nasser [24]) that the stress $\boldsymbol{\Sigma}$ conjugate to the general strain measure $\boldsymbol{E}$ may be written (in component form)

$$\left.\begin{aligned} \Sigma_{II} &= \frac{T_{II}}{\lambda_{(I)} f'(\lambda_{(I)})}, && I = 1,2,3 \\ \Sigma_{IJ} &= \frac{\lambda_{(J)}^2 - \lambda_{(I)}^2}{f(\lambda_{(J)}) - f(\lambda_{(I)})} \frac{T_{IJ}}{2\lambda_{(I)}\lambda_{(J)}}, && I \neq J, \lambda_{(I)} \neq \lambda_{(J)}, \end{aligned}\right\} \tag{2.13}$$

where no summation over indices is implied.

EXAMPLE 2.1 If the logarithmic strain $\ln \boldsymbol{U}$ is employed as a Lagrangian strain measure relationship (2.1) reads

$$\left.\begin{aligned} (\ln \boldsymbol{U})\dot{}_{II} &= D_{II}, && I = 1,2,3 \\ (\ln \boldsymbol{U})\dot{}_{IJ} &= \frac{\ln(\lambda_{(I)}/\lambda_{(J)})}{\lambda_{(I)}^2 - \lambda_{(J)}^2} 2\lambda_{(I)}\lambda_{(J)} D_{IJ}, && I \neq J, \lambda_{(I)} \neq \lambda_{(J)} \end{aligned}\right\} \tag{2.14}$$

(no summation on indices).

The stress conjugate to the logarithmic strain $\ln U$ can be expressed

$$\left.\begin{aligned} \Sigma^{(0)}_{II} &= T_{II}, && I = 1,2,3 \\ \Sigma^{(0)}_{IJ} &= \frac{\lambda^2_{(I)} - \lambda^2_{(J)}}{\ln(\lambda_{(I)}/\lambda_{(J)})} \frac{T_{IJ}}{2\lambda_{(I)}\lambda_{(J)}}, && I \neq J, \lambda_{(I)} \neq \lambda_{(J)}, \end{aligned}\right\} \tag{2.15}$$

(no summation on indices).

For an isotropic elastic material the principal axes of the Kirchhoff stress tensor coincide with the Eulerian triad and equation (2.15) takes the simple form

$$\Sigma^{(0)}_{II} = T_{II}, \qquad I = 1,2,3 \tag{2.16}$$

that is,

$$\boldsymbol{\Sigma}^{(0)} = \boldsymbol{T} = \boldsymbol{R}^T \boldsymbol{\tau} \boldsymbol{R} \tag{2.17}$$

For a further discussion on stress conjugate to logarithmic strain we refer to Hoger [10].

3. Isotropic elastic solids

3.1 Strain-energy function

At the outset we assume the existence of a strain energy function in the form of a scalar symmetric function of its principal stretches $\lambda^e_{(i)}, (i = 1,2,3)$

$$W^{(0)}(\lambda^e_{(1)}, \lambda^e_{(2)}, \lambda^e_{(3)}) = \mu\ ((\ln\lambda^e_{(1)})^2 + (\ln\lambda^e_{(2)})^2 + (\ln\lambda^e_{(3)})^2) + \tfrac{1}{2}\lambda\ (\ln J^e)^2 \tag{3.1}$$

where μ, λ are non-zero material constants and

$$J^e = \lambda^e_{(1)}\lambda^e_{(2)}\lambda^e_{(3)} \tag{3.2}$$

is defined in (2.4). Introducing $I^{(0)e}$ as the first principal invariant of $(\ln \boldsymbol{U}^e)^2$ (and of $(\ln \boldsymbol{V}^e)^2$), (3.1) reads

$$W^{(0)}(I^{(0)e}, J^e) = \mu I^{(0)e} + \tfrac{1}{2}\lambda(\ln J^e)^2 \tag{3.3}$$

3.2 Hyperelastic constitutive model

Material version. Standard procedure leads to a constitutive model (see [30-31] for details) with rotated stress tensor $\boldsymbol{T}$ and material version of the elasticity tensor $\mathsf{H}^{(C)e}$ given, respectively, as

$$T_{IJ} = 2\mu(\ln \boldsymbol{U}^e)_{IJ} + \lambda(\operatorname{tr}(\ln \boldsymbol{U}^e))\delta_{IJ} \tag{3.4}$$

$$\mathsf{H}^{(0)e}_{IJKL} = \mu\,(\delta_{IK}\delta_{JL} + \delta_{IL}\delta_{JK}) + \lambda\delta_{IJ}\delta_{KL} \tag{3.5}$$

Note that (3.5) defines a constant and isotropic fourth order tensor.

Spatial version. Following the procedure described in [30], we rephrase the constitutive model (3.4-5) in spatial description, with the Kirchhoff stress and spatial elasticity tensor given in Eulerian triad as

$$\tau_{ij} = 2\mu(\ln V^e)_{ij} + \lambda(\operatorname{tr}(\ln V^e))\delta_{ij} \tag{3.6}$$

$$\mathsf{h}^{(0)e}_{ijkl} = \left\{ \begin{array}{ll} (J^e)^{-1}\,\widehat{\mathsf{h}}^{(0)e}_{ijkl}, & k = l \\ (J^e)^{-1}\,\widehat{\mathsf{h}}^{(0)e}_{ijkl}\dfrac{\ln(\lambda^e_{(l)}/\lambda^e_{(k)})}{\lambda^{e2}_{(l)} - \lambda^{e2}_{(k)}}2\lambda^e_{(k)}\lambda^e_{(l)}, & k \neq l \end{array} \right\} \tag{3.7}$$

(no summation over indices is implied) where

$$\widehat{\mathsf{h}}^{(0)e}_{ijkl} = \mu\,(\delta_{ik}\delta_{jl} + \delta_{il}\delta_{jk}) + \lambda\delta_{ij}\delta_{kl} \tag{3.8}$$

Observe that (3.7-8) defines a fourth order tensor which is in general *non-constant* and *anisotropic* (or more precisely *orthotropic*), even for an isotropic material.

Due to isotropy of $\widehat{\mathsf{h}}^{(0)e}$, the spatial elasticity tensor $\mathsf{h}^{(0)e}$ has major symmetries, namely $\mathsf{h}^{(0)e}_{ijkl} = \mathsf{h}^{(0)e}_{klij}$, which guarantees the existence of the rate potential (Hill [7]).

3.3 Rate forms and approximation

In material description, the rate form of hyperelastic constitutive equation (3.4) reads

$$\dot{T} = \mathsf{H}^{(0)e} : (\ln U^e)^{\cdot} \tag{3.9}$$

which, rewritten alternatively in spatial description, results in

$$\overset{\square}{\tau} = J^e \mathsf{h}^{(0)e} : d \tag{3.10}$$

where

$$\overset{\square}{\tau} = \dot{\tau} - \omega^{(R)e}\tau + \tau\omega^{(R)e} \tag{3.11}$$

is the *Green-Naghdi rate* of Kirchhoff stress and $\mathsf{h}^{(0)e}$ is a spatial elasticity tensor.

Considering the representation of the spatial elasticity tensor $\mathsf{h}^{(0)}$ in the Eulerian triad (3.7-8), it can be shown that (see Hill [9])

$$\frac{\ln(\lambda_{(l)}/\lambda_{(k)})}{\lambda^2_{(l)} - \lambda^2_{(k)}}2\lambda_{(k)}\lambda_{(l)} = 1 - \frac{1}{6}\left(\frac{\lambda_{(l)}}{\lambda_{(k)}} - 1\right)^2 + \ldots \tag{3.12}$$

(no summation over indices is implied) so that within second order approximation we may write

$$\mathsf{h}^{(0)} \approx \widehat{\mathsf{h}}^{(0)} \tag{3.13}$$

with strict equality satisfied only when the same fibers remain principal ($d_{kl} = 0$ if $k \neq l$). Whenever $0.75 \leq \lambda_{(l)}/\lambda_{(k)} \leq 1.25$ the relative error of the componentwise approximation is less than 1 per cent.

Furthermore, choosing the Eulerian triad as a particular basis we may write (see [30] for details), within second order approximation,

$$\overset{\square}{\tau} = \dot{\tau} - \omega^{(R)e}\tau + \tau\omega^{(R)e}$$
$$\approx \dot{\tau} - w^e\tau + \tau w^e + 2\mu \sum_{i \neq j} \frac{1}{4}\left(\frac{\lambda^e_{(j)}}{\lambda^e_{(i)}} - 1\right)^2 d_{ij} n^{(i)} \otimes n^{(j)} \tag{3.14}$$

so that, within the same order of approximation, the rate form (3.10) reads

$$\overset{\nabla}{\tau} \approx J^e \mathsf{h}^{(0)e} : d \tag{3.15}$$

where

$$\overset{\nabla}{\tau} = \dot{\tau} - w^e\tau + \tau w^e \tag{3.16}$$

is the *Jaumann rate* of Kirchhoff stress. As long as $0.80 \leq \lambda^e_{(j)}/\lambda^e_{(i)} \leq 1.20$ relative error of the componentwise approximation is less than 1 per cent.

Following the same argument, but now within first order approximation (see [30] for details), the spatial version of the *first elasticity* tensor associated with the conjugate pair $\{T, \ln U\}$ (with standard meaning, see [17,30]), may be written

$$\mathsf{a}^{(0)}_{ijkl} \approx \mathsf{h}^{(0)e}_{ijkl} - \mathsf{m}_{ijkl} + \delta_{ik}\sigma_{jl} \tag{3.17}$$

where

$$\mathsf{m}_{ijkl} = \tfrac{1}{2}(\sigma_{ik}\delta_{jl} + \sigma_{il}\delta_{jk} + \sigma_{jk}\delta_{il} + \sigma_{jl}\delta_{ik}) \tag{3.18}$$

which allows a simple description of the linearized form of balance principles in terms of the logarithmic strain and rotated stress. In the approximation (3.17) the componentwise error introduced is less than 1 per cent of the corresponding component of the tensor m if $0.98 \leq \lambda_{(j)}/\lambda_{(i)} \leq 1.02$.

REMARK 3.1 We point out that the chosen strain-energy function is similar to the class of strain-energy functions for isotropic elastic materials proposed by Ogden [26], where the commonly used invariants $I^{(2)e}$ of $b^e = (V^e)^2$ are substituted with invariants $I^{(0)e}$ of $(\ln V^e)^2$. It is in fact equivalent to the model of the *Hadamard material* which is a simple element of that class, and was proposed by Simo and Pister [38] as the simplest possible hyperelastic material that has an isotropic tangent elasticity tensor. Conditions $W^{(0)}(1,1,1) = 0$, and $W^{(0)}(\lambda^e_{(1)}, \lambda^e_{(2)}, \lambda^e_{(3)}) \to \infty$ as $\lambda^e_{(i)} \to \infty (i = 1,2,3)$ as well as $W^{(0)}(\lambda^e_{(1)}, \lambda^e_{(2)}, \lambda^e_{(3)}) \to \infty$ as $J^e \to \infty$ or $J^e \to 0$, are naturally satisfied.

REMARK 3.2 The Jaumann rate of the Kirchhoff stress coincides with the rate of stress conjugate to the logarithmic strain $\ln U$ when the current configuration is the reference one, as shown by Hill [8-9], and recently pointed out by Nemat-Nasser [24].

4. Isotropic finite strain elasto-plasticity

Through multiplicative decomposition of the deformation gradient as postulated by Lee [15]

$$F = F^e F^p \tag{4.1}$$

we introduce the intermediate configuration, not necessarily physically attainable, as the set of all unloaded infinitessimal cells.

Following the arguments of Nemat-Nasser [24], additive strain rate decomposition is adopted, i.e.

$$(\ln \boldsymbol{U})^{\cdot} = (\ln \boldsymbol{U}^{e})^{\cdot} + (\ln \boldsymbol{U}^{p})^{\cdot} \tag{4.2}$$

which has already been frequently utilized in computational finite strain plasticity with different strain measures involved [40].

Further, it is assumed that elastic deformations are not influenced by plastic flow i.e. are independent of the history. This has a physical counter-argument, specially for large plastic flow (see Nemat-Nasser [24] for discussion), but this effect is not considered here.

As proposed by Lee [15], we assume that local rotation does not affect the elastic response, i.e. $\boldsymbol{F}^{e} = \boldsymbol{V}^{e}$, which is justified, at least for both isotropic elastic and plastic responses (see Simo [40]). This gives

$$\boldsymbol{F} = \boldsymbol{V}^{e} \boldsymbol{F}^{p} \tag{4.3}$$

With these assumptions introduced, the usual elasto-plastic constitutive equations are simply transformed onto the intermediate configuration. Here, we confine ourselves to isotropic elasto-plasticity and review only the basic equations. (For further discussions see [12,19,29,39])

$$\left.\begin{aligned}
(\ln \boldsymbol{U})^{\cdot} &= (\ln \boldsymbol{U}^{e})^{\cdot} + (\ln \boldsymbol{U}^{p})^{\cdot} \\
\boldsymbol{T} &= \frac{\partial W^{(0)}}{\partial (\ln \boldsymbol{U}^{e})} \\
&= 2\mu (\ln \boldsymbol{U}^{e}) + \lambda \left(\operatorname{tr} (\ln \boldsymbol{U}^{e}) \right) \mathbf{1} \\
(\ln \boldsymbol{U}^{p})^{\cdot} &= \dot{\gamma} \frac{\partial \Phi (\boldsymbol{T}, K)}{\partial \boldsymbol{T}} \\
\dot{K} &= \dot{\gamma} H (\ln \boldsymbol{U}^{p}, K)
\end{aligned}\right\} \tag{4.4}$$

Here $\Phi(\boldsymbol{T}, K)$ is the yield surface and function $H(\ln \boldsymbol{U}^{p}, K)$ defines the hardening law. Finally, loading/unloading conditions may be formulated in the standard Kuhn-Tucker form

$$\Phi \leq 0, \qquad \dot{\gamma} \geq 0, \qquad \dot{\gamma}\Phi = 0, \tag{4.5}$$

For convenience, the basic equations governing the material version of the constitutive model for J_2-flow theory of elastoplasticity have been summarized in Table 1.

Standard procedure leads to the rate constitutive equation

$$\dot{\boldsymbol{T}} = \mathsf{H}^{(0)ep} : (\ln \boldsymbol{U})^{\cdot} \tag{4.6}$$

where $\mathsf{H}^{(0)ep}$ is the elastoplastic moduli tensor

$$\mathsf{H}^{(0)ep} = \mathsf{H}^{(0)e} - \frac{\left(\mathsf{H}^{(0)e} : \widehat{\boldsymbol{N}}\right) \otimes \left(\mathsf{H}^{(0)e} : \widehat{\boldsymbol{N}}\right)}{H + \widehat{\boldsymbol{N}} : \mathsf{H}^{(0)e} : \widehat{\boldsymbol{N}}} \tag{4.7}$$

Table 1
Constitutive model for J_2-*flow theory* of elastoplasticity — *Material version*

(i) Additive decomposition of the rate of deformation tensor

$$(\ln U)^{\cdot} = (\ln U^e)^{\cdot} + (\ln U^p)^{\cdot}$$

(ii) Linear isotropic constitutive equation

$$\dot{T} = 2\mu(\ln U^e)^{\cdot} + \lambda\,(\operatorname{tr}(\ln U^e)^{\cdot})\,\mathbf{1}$$

(iii) Yield surface and hardening law

$$\Phi(T, K) = \| \operatorname{dev} T \| - \sqrt{\tfrac{2}{3}}K$$

$$\dot{K} = \dot{\gamma} H(\ln U^p, K)$$

(iv) Flow rule

$$(\ln U^p)^{\cdot} = \dot{\gamma}\hat{N}, \qquad \hat{N} = \frac{\operatorname{dev} T}{\| \operatorname{dev} T \|}$$

(v) Kuhn-Tucker loading/unloading conditions

$$\Phi \le 0, \qquad \dot{\gamma} \ge 0, \qquad \dot{\gamma}\Phi = 0$$

---

and

$$\hat{N} = \frac{\partial \Phi}{\partial T} \tag{4.8}$$

defines the direction of plastic flow.

Making use of the transformation described in Section 3.2 (details may be found in [30]), we rewrite the elastoplastic moduli (4.7) in spatial description, as

$$\mathsf{h}^{(0)ep}_{ijkl} = \left\{ \begin{array}{ll} (J^e)^{-1}\,\hat{\mathsf{h}}^{(0)ep}_{ijkl}\,, & k = l \\ (J^e)^{-1}\,\hat{\mathsf{h}}^{(0)ep}_{ijkl}\,\dfrac{\ln(\lambda^e_{(l)}/\lambda^e_{(k)})}{\lambda^{e\,2}_{(l)} - \lambda^{e\,2}_{(k)}}\,2\lambda^e_{(k)}\lambda^e_{(l)}, & k \ne l \end{array} \right\} \tag{4.9}$$

(no summation over indices is implied) where

$$\hat{\mathsf{h}}^{(0)ep}_{ijkl} = R_{iI}R_{jJ}R_{kK}R_{lL}\mathsf{H}^{(0)ep}_{IJKL} \tag{4.10}$$

Again, (4.9-10) define a fourth order tensor which is in general *non-constant* and *anisotropic* under the set of all proper orthogonal spatial transformation $Q \subset \mathrm{SO}(3)$. Observe that anisotropy in (4.9-10) is strain-induced.

As noted by Mandel [16], if isotropy is maintained in the intermediate configuration, the elastic spin equals the continuum spin, (or in other words, the plastic strain rate tensor is coaxial with the stress tensor, Nemat-Nasser [24]), and, for the case of small elastic deformations with deviatoric plasticity, following the arguments of Section 3.3, the spatial rate form of (4.6) may be written

$$\overset{\nabla}{\tau} \approx J\mathsf{h}^{(0)ep} : d \tag{4.11}$$

where

$$\overset{\nabla}{\tau} = \dot{\tau} - w\tau + \tau w \tag{4.12}$$

is the Jaumann rate of Kirchhoff stress, and

$$\mathsf{h}^{(0)ep} \approx J^{-1}\mathsf{H}^{(0)ep} \tag{4.13}$$

REMARK 4.1 Logarithmic stretches as a 'convenient approximation in the constitutive description of plastic response'(Hill [9]) were recommended by Rice [34], and later confirmed by Hill [9], with an imposed interval of moderate stretches where the model is applicable. Introduction of an intermediate configuration in conjunction with the isotropic elastic response removes such a restriction.

5. Numerical integration of the constitutive equations

From the computational standpoint the problem consists of formulating numerical integration procedures for updating the known state variables $\{ F^e_n, F^p_n, T_n, K_n \}$ at the configuration $\chi_n(X)$, to obtain $\{ F^e_{n+1}, F^p_{n+1}, T_{n+1}, K_{n+1} \}$ at the configuration $\chi_{n+1}(X)$, where the incremental displacement $u(x_n)$ is assumed given and the updated configuration and updated deformation gradient can be evaluated through

$$\begin{aligned} \chi_{n+1} &= \chi_n + u \\ F_{n+1} &= F_u F_n \end{aligned} \tag{5.1}$$

where

$$F_u = \mathbf{1} + \operatorname{Grad} u \tag{5.2}$$

These must be consistent with constitutive equations (4.4-5).

Particularly suitable for implementation are procedures connected with general operator split methodology, where the original problem of evolution is solved through composition, applying first the elastic and then the plastic algorithm.

• *Elastic predictor*

Calculating F_{n+1} from (5.1), and because

$$F^{p\,\text{trial}}_{n+1} = F^p_n \tag{5.3}$$

we easily obtain

$$F^{e\,\text{trial}}_{n+1} = F_{n+1}(F^p_n)^{-1} \tag{5.4}$$

and with logarithmic stretches computed using the algorithm in **Box 1** we evaluate the elastic trial stress. This leads to exact solution of the elastic problem governed by strain-energy function (3.1).

Box 1. *Computation of logarithmic stretches*

- **Evaluate the right Cauchy-Green tensor and its spectral representation**

$$C = F^T F$$

$$C = \sum_{I=1}^{3} C_{(I)} N^{(I)} \otimes N^{(I)}$$

- **Evaluate the principal stretches and obtain the logarithmic stretches**

$$\lambda_{(I)} = (C_{(I)})^{\frac{1}{2}}$$

$$\ln U = \sum_{I=1}^{3} \ln \lambda_{(I)} N^{(I)} \otimes N^{(I)}$$

- *Plastic corrector*

The equations describing plastic response (4.4-7) we solve on the fixed intermediate configuration employing standard numerical procedures. We refer to Simo and Hughes [39] and Mitchell and Owen [19], for discussion on the recent advances and details of implementation. For completeness, the equations governing a one step Euler backward scheme are summarized

$$\left.\begin{aligned} \Delta(\ln U) &= \Delta(\ln U^e) + \Delta(\ln U^p) \\ T_{n+1} &= \left.\frac{\partial W^{(0)}}{\partial(\ln U^e)}\right|_{n+1} \\ &= 2\mu(\ln U^e)_{n+1} + \lambda(\operatorname{tr}(\ln U^e)_{n+1})\mathbf{1} \\ (\ln U^p)_{n+1} &= (\ln U^p)_n + \Delta\gamma \frac{\partial \Phi(T_{n+1}, K_{n+1})}{\partial T_{n+1}} \\ K_{n+1} &= K_n + \Delta\gamma H((\ln U^p)_{n+1}, K_{n+1}) \end{aligned}\right\} \tag{5.3}$$

In addition, the constraint condition

$$\Phi(T_{n+1}, K_{n+1}) = 0$$

must be satisfied.

The algorithm is summarized in Box 2 for von Mises elasto-plastic constitutive equations.

6. Formulation of the boundary value problem: moderate elastic strains

The formulation of the general boundary value problem in terms of logarithmic strain and rotated stress $\{T, \ln U\}$, due to the explicit appearance of R and U in the

Box 2. *Finite strain extension of the stress integration procedure for deviatoric elasto-plasticity*

- Initialization

$$\chi_{n+1}^{(0)} = \chi_n \,, \qquad F_{n+1}^{(0)} = F_n$$

- Update configuration

$$\chi_{n+1} = \chi_{n+1}^{(i)} + u$$

- Compute current deformation gradient

$$F_{n+1} = F_u F_{n+1}^{(i)} \,, \qquad F_u = 1 + \operatorname{Grad} u$$

- Start elastic predictor phase

$$F_{n+1}^{p\,\mathrm{trial}} = F_n^p \,, \qquad F_{n+1}^{e\,\mathrm{trial}} = F_{n+1}(F_n^p)^{-1}$$

- Evaluate logarithmic strains
  See **Box 1**
- Elastic stress

$$T_{n+1}^{\mathrm{trial}} = 2\mu \operatorname{Dev}(\ln U^e)_{n+1} + \kappa\,(\operatorname{tr}(\ln U^e)_{n+1})\,1$$

- Check for yielding

$$\text{IF} \;\; \Phi_{n+1}^{\mathrm{trial}} = \| \operatorname{Dev} T_{n+1}^{\mathrm{trial}} \| - \sqrt{\frac{2}{3}}\, K\,(\ln \bar{U}^p)_n \le 0$$

$$\text{Set } (\bullet)_{n+1} = (\bullet)_{n+1}^{\mathrm{trial}} \text{ then EXIT}$$

ELSE

- Plastic corrector phase

$$\hat{N}_{n+1} = \frac{\operatorname{Dev} T_{n+1}^{\mathrm{trial}}}{\| \operatorname{Dev} T_{n+1}^{\mathrm{trial}} \|}$$

$$(\ln \bar{U}^p)_{n+1} = (\ln \bar{U}^p)_n + \sqrt{\frac{2}{3}}\,\Delta\gamma$$

$$\operatorname{Dev} T_{n+1} = \operatorname{Dev} T_{n+1}^{\mathrm{trial}} - 2\mu\,\Delta\gamma\,\hat{N}_{n+1}$$

- Update intermediate configuration

$$V_{n+1}^e = R\,U_{n+1}^e\,R^T \,, \qquad F_{n+1}^p = (V_{n+1}^e)^{-1} F_{n+1}$$

balance principles, leads naturally to multifield variational principles. Our main goal is development of a displacement type finite element formulation. Henceforth, introducing approximations from Sections 3-4., we formulate the classical weak form of momentum balance (see Marsden and Hughes [17]), convenient for the finite element discretization. Doing so, we restrict attention to the static case.

6.1 Weak form of momentum balance

To generate *admissible variations* we introduce the space

$$\mathcal{V} = \{\boldsymbol{\eta} : \chi(\Omega) \rightarrow \mathbb{R}^3 \mid \boldsymbol{\eta} = 0 \quad \text{on} \quad \chi(\partial\Omega_u)\} \tag{6.1}$$

The *strong form* of the boundary value problem, written in *material description*

$$\operatorname{Div} \boldsymbol{P} + \rho_{\text{Ref}}\bar{\boldsymbol{b}} = 0 \qquad \text{in} \quad \Omega \tag{6.2}$$

with boundary conditions

$$\left.\begin{aligned} \boldsymbol{P}\cdot\boldsymbol{N} &= \bar{\boldsymbol{t}}_{\text{Ref}} \quad \text{on} \quad \partial\Omega_\sigma \\ \chi &= \bar{\chi} \quad \text{on} \quad \partial\Omega_u \end{aligned}\right\} \tag{6.3}$$

takes the classical weak form (see Marsden and Hughes [17])

$$G_0(\chi(\boldsymbol{X}), \boldsymbol{\eta}_0) = \int_\Omega (\boldsymbol{P} : \operatorname{Grad}\boldsymbol{\eta}_0 - \rho_{\text{Ref}}\bar{\boldsymbol{b}}\cdot\boldsymbol{\eta}_0)dV - \int_{\partial\Omega_\sigma} \bar{\boldsymbol{t}}_{\text{Ref}}\cdot\boldsymbol{\eta}_0 dA \tag{6.4}$$

where $\boldsymbol{\eta}_0 = \boldsymbol{\eta}\circ\chi$.

Alternatively, the spatial version of the weak form (6.4) reads

$$G(\chi, \boldsymbol{\eta}) = \int_{\chi(\Omega)} (\boldsymbol{\sigma} : \nabla\boldsymbol{\eta} - \rho\bar{\boldsymbol{b}}\cdot\boldsymbol{\eta})dv - \int_{\chi(\partial\Omega_\sigma)} \bar{\boldsymbol{t}}\cdot\boldsymbol{\eta}\, da \tag{6.5}$$

6.2 Consistent linearization

Exploiting rate forms derived in Sections 3-4. with appropriate linearization we generate incremental forms. Straightforward procedure leads to a linearized form of the functional $G(\chi, \boldsymbol{\eta})$ which in spatial description reads

$$\int_{\chi(\Omega)} \nabla\boldsymbol{u} : \mathbf{a}^{(0)} : \nabla\boldsymbol{\eta}\, dv = -G(\widehat{\chi}, \boldsymbol{\eta}) \tag{6.6}$$

The term $G(\widehat{\chi}, \boldsymbol{\eta})$ has standard interpretation as an *unbalanced force* at the configuration $\widehat{\chi}$, while the term $\int_{\chi(\Omega)} \nabla\boldsymbol{u} : \mathbf{a}^{(0)} : \nabla\boldsymbol{\eta}\, dv$, linear in $\boldsymbol{u}$, provides the *tangent stiffness*.

Using (3.17-18) and (4.11-13) we recover the standard *material* and *geometrical* parts of the tangent stiffness, respectively,

$$ {}^{\mathrm{M}}K = \int_{\chi(\Omega)} \nabla^s \boldsymbol{u} : (\mathsf{h}^{(0)ep} - \mathsf{m}) : \nabla^s \boldsymbol{\eta} dv, \tag{6.7} $$

$$ {}^{\mathrm{G}}K = \int_{\chi(\Omega)} \nabla \boldsymbol{u} : (\boldsymbol{\sigma} \otimes \mathbf{1}) : \nabla \boldsymbol{\eta} dv, \tag{6.8} $$

where ∇^s denotes the symmetric part of ∇. Analogously, one obtains *residual forces* as,

$$ \widehat{G} = \int_{\chi(\Omega)} \widehat{\boldsymbol{\sigma}} : \nabla^s \boldsymbol{\eta} \, dv - {}^{\mathrm{EXT}}\widehat{G} \tag{6.9} $$

where ${}^{\mathrm{EXT}}\widehat{G}$ is the virtual work of external loading

$$ {}^{\mathrm{EXT}}\widehat{G} = \int_{\chi(\Omega)} \rho \bar{\boldsymbol{b}} \cdot \boldsymbol{\eta} dv + \int_{\chi(\partial\Omega_\sigma)} \bar{\boldsymbol{t}} \cdot \boldsymbol{\eta} da \tag{6.10} $$

7. Finite element formulation

Following standard methodology (see Zienkiewicz [42] or Hughes [13]), we introduce the space $\mathcal{V}^h \subset \mathcal{V}$ as a finite dimensional approximation to $\mathcal{V}$ described in (6.1). Furthermore, let subspace $\mathcal{V}^h$ be generated through spatial finite element discretization in the current configuration $\chi(\Omega) = \bigcup_{e=1}^{\mathrm{NEL}} \chi(\Omega_e)$, with $\chi(\Omega_a) \bigcap \chi(\Omega_b) = \emptyset$ if $a \neq b$, so that over the typical element $\chi(\Omega_e)$ one has the interpolation

$$ \boldsymbol{u}^h \Big|_{\chi(\Omega_e)} = \sum_{a=1}^{\mathrm{NEN}} N_a(\boldsymbol{x}) \boldsymbol{u}_a \tag{7.1} $$

where N_a are the standard shape functions.

Tangent stiffness matrix. Representing the discrete gradient operators $\nabla^s \boldsymbol{u}^h$ and $\nabla \boldsymbol{u}^h$ in the form

$$ \nabla^s \boldsymbol{u}^h \Big|_{\chi(\Omega_e)} = \sum_{a=1}^{\mathrm{NEN}} B_a \boldsymbol{u}_a, \qquad \nabla \boldsymbol{u}^h \Big|_{\chi(\Omega_e)} = \sum_{a=1}^{\mathrm{NEN}} G_a \boldsymbol{u}_a, \tag{7.2} $$

we may write the tangent operator (6.7-8) in the *matrix format*

$$ \boldsymbol{k}^e = {}^{\mathrm{M}}\boldsymbol{k}^e + {}^{\mathrm{G}}\boldsymbol{k}^e \tag{7.3} $$

where

$$ {}^{\mathrm{M}}\boldsymbol{k}^e = \left[{}^{\mathrm{M}}\boldsymbol{k}^e_{ab} \right] = \int_{\chi(\Omega_e)} B_a^T \left[\mathsf{h}^{(0)ep} - \mathsf{m} \right] B_b \, dv, \tag{7.4} $$

$$ {}^{\mathrm{G}}\boldsymbol{k}^e = \left[{}^{\mathrm{G}}\boldsymbol{k}^e_{ab} \right] = \int_{\chi(\Omega_e)} G_a^T \left[\widehat{\boldsymbol{\sigma}} \right] G_b \, dv \tag{7.5} $$

are the standard *material* and *geometric* element stiffness matrices, respectively, and

$$K = \underset{e=1}{\overset{\text{NEL}}{\mathsf{A}}} (k^e) \tag{7.6}$$

is the global stiffness matrix obtained applying the finite element assembly operator [13].

Residual vector. Discrete version of (6.9-10) supplies the element residual force vector

$$f^e = {}^{\text{INT}}f^e - {}^{\text{EXT}}f^e \tag{7.7}$$

where

$${}^{\text{INT}}f^e = \{{}^{\text{INT}}f^e_a\} = \int_{\chi(\Omega_e)} B^T_a \{\hat{\sigma}\}\, dv, \tag{7.8}$$

$${}^{\text{EXT}}f^e = \{{}^{\text{EXT}}f^e_a\} = \int_{\chi(\Omega_e)} N_a\{\rho\bar{b}\}dv + \int_{\chi(\partial\Omega_e)} N_a\{\bar{t}\}da \tag{7.9}$$

are the so-called element *internal* and *external* force vectors, respectively, and

$$F = F_{\text{nodal}} + \underset{e=1}{\overset{\text{NEL}}{\mathsf{A}}} (f^e) \tag{7.10}$$

is the global *residual vector*.

8. Numerical examples

The algorithm presented herein has interesting properties in relation to the choice of reference configuration, as has been discussed in [31]. We present results for the case when the current configuration is the reference one, i.e. the incremental deformation gradient is used in polar decomposition setting $F^{(0)}_{n+1} = \mathbf{1}$ in **Box 2**.

In all examples a full Newton-Raphson method is employed, with equlibrium iterations in each loading increment. Use of the consistent tangent moduli (see [19,39]) is of utmost importance in achieving the quadratic rate of convergence of the Newton-Raphson method, which substantially improves computational effectivness.

EXAMPLE 8.1 Necking and failure of thin sheet in uniaxial tension test † The geometry and material characterization for this example are taken from [6] and shown in Fig.8.1. The analysis is performed for quarter of the sheet, restricting deformations to be symmetric along the lines $X_1 = 0$ and $X_2 = 0$. The tensile axis is aligned with the X_2 direction, so that the boundary conditions can be expressed in the material description as

$$\left.\begin{array}{llll} U_1 = 0, & T_2 = 0, & \text{at} & X_1 = 0, \\ U_2 = 0, & T_1 = 0, & \text{at} & X_2 = 0, \\ & T_1 = T_2 = 0, & \text{at} & X_1 = B_\circ, \\ U_2 = \overline{U}_2, & T_1 = 0, & \text{at} & X_2 = L_\circ, \end{array}\right\} \tag{8.1}$$

† To simplify notation, in Examples 8.1 and 8.2, we identify $T \equiv t_{\text{Ref}}$ as a traction vector over the reference area

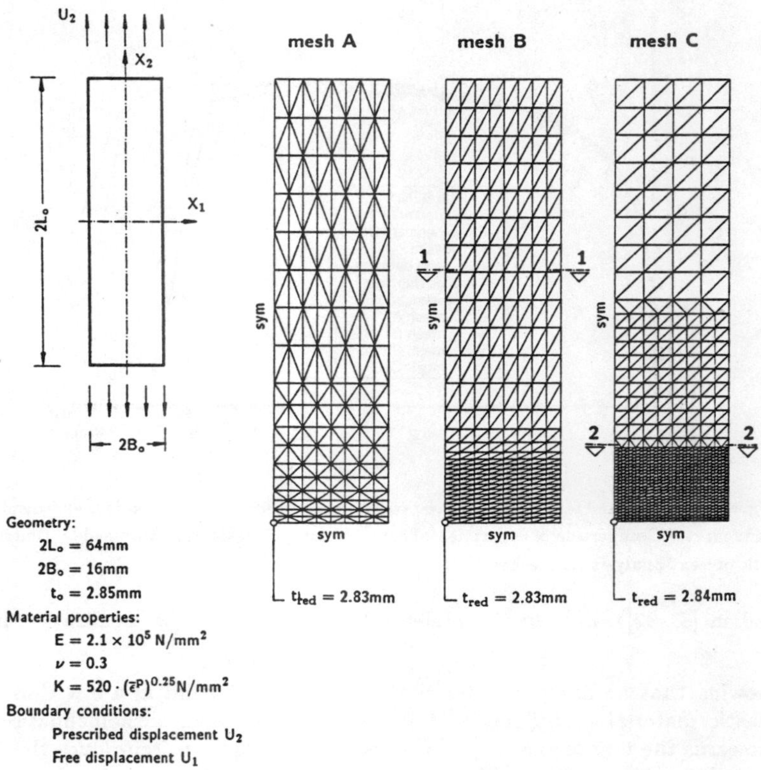

Fig.8.1: Necking and failure of thin sheet in uniaxial tension test: Geometry, material characteristics and finite element meshes employed

where U_1, U_2 are displacement components and T_1, T_2 are components of the traction vector in the X_1, X_2 directions, respectively. Three different meshes, denoted A,B and C are considered consisting of 288, 464 and 974 constant strain triangular finite elements, respectively. To trigger the failure we introduce a geometrical perturbation by thinning the sheet at the central node to $t_{red} = 2.83$mm for meshes A and B, and $t_{red} = 2.84$mm for mesh C.

Results obtained using the computational model based on logarithmic strains (details may be found in [32]) for meshes A, B and C are compared in Fig.8.2 with experimental data, and numerical solutions presented by Doltsinis [6].

Up to the point A, all numerical solutions closely reproduce the experimental data. From that point results for the mesh A show localization in tensile mode of failure in the first row of finite elements, and closely follow the results obtained in [6]. A good agreement with the experimental data for mesh A may be obtained by introducing damage parameters, as shown by Doltsinis [6], or a larger geometrical imperfection (as

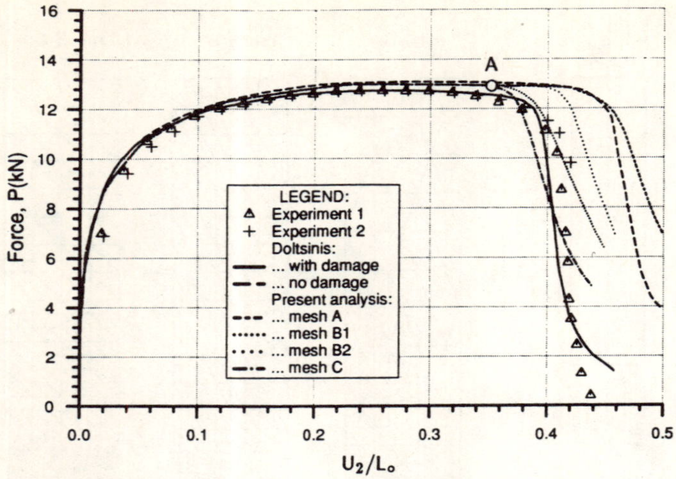

Fig.8.2: Necking and failure of thin sheet in uniaxial tension test: Load versus end-displace ment curves. Comparison of experimental data and numerical solution obtained by Doltsinis, with present analysis for meshes A, B and C

discussed in [31-32]), but the finite element mesh again shows localization in tensile mode.

Knowing that an analytical description of a Lüders band in a flat thin sheet for rigid plastic material (see Thomas [41]) gives a shear band with inclination of $\theta = 54°44'$ towards the tensile axis, mesh B1 is designed in order to reproduce this idealized behaviour, by setting in the initial configuration longer sides of triangles in the first 13 rows of elements with inclination of $\Theta = 65°8'$ towards the tensile axis. Due to the nonuniformity introduced by the geometrical imperfection and diffuse necking, for a strain hardening elasto-plastic material this can only be done in an approximate way. In the early stages of localisation (identified with $U_2/L_\circ = 0.36 - 0.38$) when the deformation pattern deviates slightly from the homogenous solution, results for mesh B1 closely follow experimentally obtained values in the load versus end-displacement diagram (see Fig.8.2). Then, with increasing deformation ($U_2/L_\circ > 0.39$) due to an improper finite element mesh alignment, the localisation shows a diffusive pattern, which is reflected as a flattening of the load versus end displacement curve (see Fig.8.2).

By setting in the initial configuration longer sides of triangles in the first 13 rows of elements with inclination of $\Theta = 70°36'$ towards the tensile axis, mesh B2 is designed to produce a possible shear band inclined at $\theta \approx 60°$ towards the tensile axis, which is reported as an experimentally observed value (see Thomas [41]). Initiation of localisation, which is numerically detected by a sudden increase in the number of equilibrium iterations required, for $U_2/L_\circ = 0.42$ shows a slight delay compared to experimental results. A sharp drop in the load versus end displacement curve for $U_2/L_\circ = 0.43 - 0.44$ is associated with the formation and development of the shear band depicted in Fig.8.3.

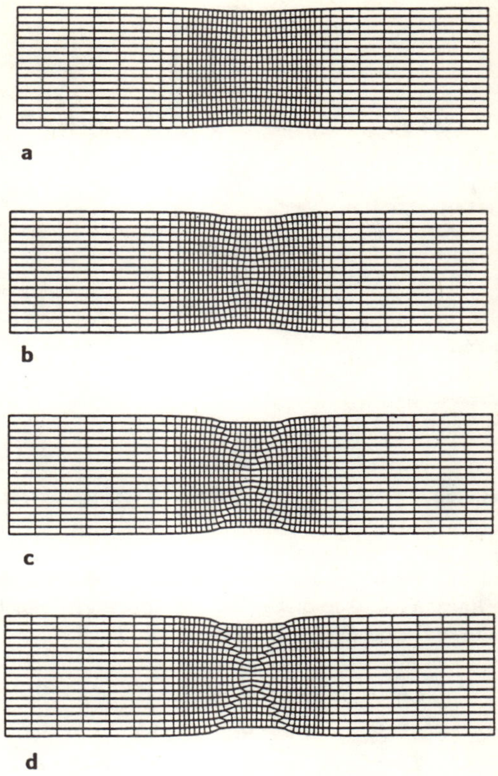

Fig.8.3: Necking and failure of thin sheet in uniaxial tension test: Deformed finite element mesh B2 at various stages of shear band formation. Only central part 1—1 of the mesh B2 is shown. Each quadrilateral consists of two triangles. (a) $U_2/L_o = 0.41$, (b) $U_2/L_o = 0.42$, (c) $U_2/L_o = 0.43$, (d) $U_2/L_o = 0.44$.

Finally, we employ the mesh C with an increased number of elements in the central part 2—2 of the model, where a non-uniform deformation pattern is initiated by the introduction of a geometrical imperfection. Homogenous deformation is followed by the gradual development of diffuse necking for loading beyond the limit point which corresponds to $U_2/L_o = 0.25$. Then, at $U_2/L_o = 0.36$, with an abrupt change of equilibrium path detected by an increased number of equilibrium iterations and a necessity for small load increments, the approximately uniform deformation pattern gives way to highly localised deformation in the form of a shear band.

EXAMPLE 8.2 Failure analysis of Scordelis-Lo roof The geometry, material characteristics and loading conditions for this example are shown in Fig.8.4 (see also [3,14]).

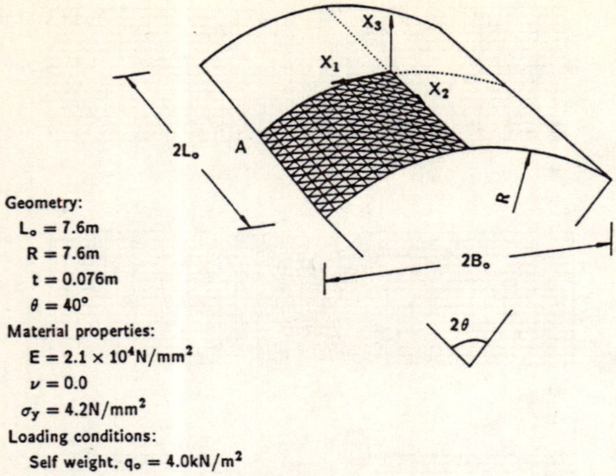

Fig.8.4: Failure analysis of Scordelis-Lo roof: Geometry, material characteristics and loading conditions

The analysis is performed for one quarter of the shell, restricting deformations to be symmetric along the lines $X_1 = 0$ and $X_2 = 0$. Accordingly, the boundary conditions can be expressed in the material description as

$$\left.\begin{aligned} U_1 = 0, \quad & \varphi_t = 0, && \text{at} \quad X_1 = 0, \\ U_2 = 0, \quad & \varphi_t = 0, && \text{at} \quad X_2 = 0, \\ T_1 = T_2 = & T_3 = 0, && \text{at} \quad X_1 = B_\circ, \\ U_1 = U_3 = 0, \quad & \varphi_n = 0, && \text{at} \quad X_2 = L_\circ, \end{aligned}\right\} \tag{8.2}$$

where $U_I, (I = 1,2,3)$ are displacement components, φ_t, φ_n are rotations around the tangential and normal vector of the middle surface, T_I are components of the traction vector in the X_I directions, respectively. Three different meshes, A (8×8), B (12×12) and C (16×16) of constant strain – constant curvature triangular thin shell finite elements are considered (so called Morley shell element, see Morley [21] and Perić and Owen [33] for recent contributions in large deformation problems).

Computation is performed employing standard arc-length control, and gravitational force versus vertical displacement of point A curves are obtained for the three different meshes and presented in Fig.8.5. All meshes give virtually the same results for deformation prior to localisation and compare well with available numerical results for this problem. The highly localised mode of failure which arises through formation of a plastic hinge around axis X_2 in the central part of the shell which is evident in Fig.8.6b, leads to a sharp drop in the force-displacement diagram which is captured with mesh C. It is less evident for mesh B and smoothed for the coarsest mesh A, which corresponds

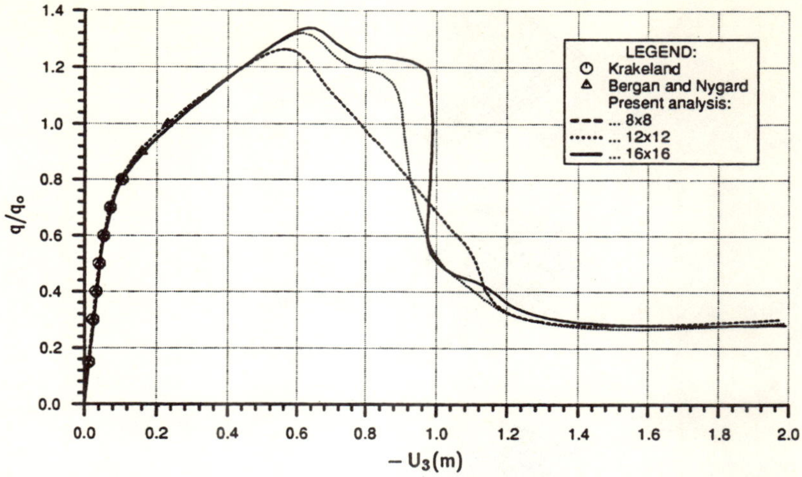

Fig.8.5: Failure analysis of Scordelis-Lo roof: Load versus displacement curves for point A. Comparison of numerical solutions obtained for various finite element meshes

to a diffuse mode of failure. The second failure mode that follows formation of the first plastic hinge in the central part arises through plastic hinge formation around axis X_1 across the full middle section of the shell (see Fig.8.6c-d) and is captured by all three finite element meshes giving almost identical force-displacement curves.

9. Summary and conclusions

A model for large strain elasto-plasticity based on logarithmic strains and multiplicative decomposition of the deformation gradient is presented, which extends the applicability of the algorithm proposed in [35]. Approximation of the rate form of the hyperelastic constitutive equation is discussed, and a finite element procedure formulated suitable for large scale computations. We point out that the variational formulation and finite element discretization presented in Sections 6-7. results in a matrix format equivalent to that proposed by McMeeking and Rice [18].

Restriction to isotropic elasto-plastic response allows modelling of elasto-plastic materials at finite strains under monotonic and proportional loading conditions. A general constitutive model that extends to elasto-plastic materials at finite strains accounting for anisotropy and texture effects, feasible for large scale computations, is yet to emerge. We refer to [1,4,20] for the present state of development. The lack of experimental evidence under complex loading conditions renders constitutive description more difficult and makes it, at present, essentially an academic issue.

The strong influence of the spatial finite element discretization (mesh size and orientation) on the initiation and pattern of the localisation was exemplified by the emergence of shear bands in the thin sheet under uniaxial tension and localised failure

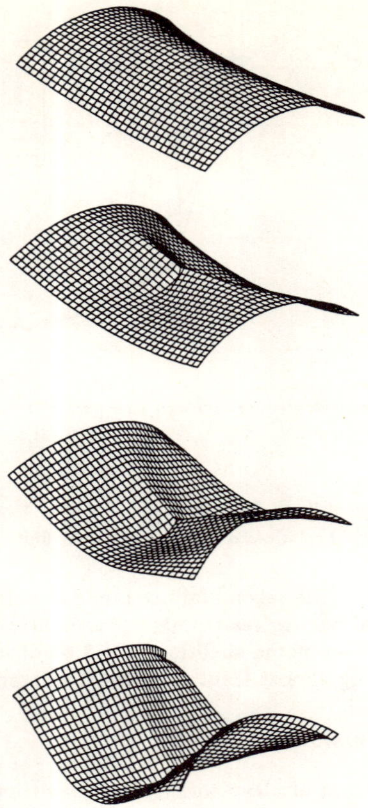

a

b

c

d

Fig.8.6: Failure analysis of Scordelis-Lo roof: Deformed finite element mesh (16× 16) at various stages of loading. Each quadrilateral consists of two triangles. (a) $q/q_o = 1.335$, (b) $q/q_o = 0.505$, (c) $q/q_o = 0.280$, (d) $q/q_o = 0.351$.

mode in the thin shell problem, confirming the necessity for special techniques (mesh adaptivity, shape function enrichment) for the general finite element analysis of localisation problems (see Ortiz and Quigley [28] for recent review).

Acknowledgment

This work was partially funded through the *BRITE* initiative of the *European Community* (Project P-2029) and this support is gratefully acknowledged.

References

[1] A.AGAH-TEHRANI, E.H.LEE, R.L.MALLET and E.T.ONAT, ' The theory of elastic-plastic deformation at finite strain with induced anisotropy modeled as combined isotropic-kinematic hardening ', *J. Mech. Phys. Solids,* **35**, 519–539 (1987)

[2] J.H.ARGYRIS and M.KLEIBER, ' Incremental formulation in nonlinear mechanics and large strain elasto-plasticity — Natural approach. Part I and Part II ', *Comp. Meth. Appl. Mech. Engng.*, **11**, 215–247 (1977) and **14**, 259–294 (1978)

[3] P.G.BERGAN and M.K.NYGÅRD, ' Nonlinear shell analysis using free formulation finite elements ', In *Finite Element Methods for Nonlinear Problems,* (Edited by P.G.Bergan et al.), 317–338, Springer-Verlag, Berlin (1986)

[4] Y.F.DAFALIAS, ' Issues on the constitutive formulation at large elastoplastic deformations: Part I. Kinematics and Part II. Kinetics ', *Acta Mech.*, **69**, 119–138 (1987) and **73**, 121–146 (1988)

[5] J.K.DIENES, ' On the analysis of rotation and stress rates in deforming bodies ', *Acta Mech.,* **32**, 217–232 (1979)

[6] I.St.DOLTSINIS, ' Nonlinear concepts in the analysis of solids and structures ', In *Advances in Computational Nonlinear Mechanics*, (Edited by I.St.Doltsinis), 1–80, CISM *Courses and Lectures* No.300, Springer-Verlag, Wien (1989)

[7] R.HILL, A general theory of uniqueness and stability in elastic-plastic solids, *J. Mech. Phys. Solids,* **6**, 236–249 (1958)

[8] R.HILL, ' On constitutive inequalities for simple materials. I and II ', *J. Mech. Phys. Solids,* **16**, 229–242 and 315–322 (1968)

[9] R.HILL, ' Aspects of invariance in solid mechanics ', *Adv. Appl. Mech.,* **18**, 1–75 (1978)

[10] A.HOGER, ' The stress conjugate to logarithmic strain ', *Int. J. Solids Struct.,* **23**, 1645–1656 (1987)

[11] T.J.R.HUGHES and J.WINGET, ' Finite rotations effects in numerical integration of rate constitutive equations arising in large-deformation analysis ', *Int. J. Num. Meth. Engng,* **15**, 1413–1418 (1980)

[12] T.J.R.HUGHES, ' Numerical implementation of constitutive models: rate-independent deviatoric plasticity ', In *Theoretical Foundations for Large Scale Computations of Nonlinear Material Behavior,* (Edited by S.Nemat-Nasser et al.), 29–57, Martinus Nijhoff Publishers, Dodrecht (1984)

[13] T.J.R.HUGHES, *The Finite Element Method: Linear Static and Dynamic Finite Element Analysis,* Prentice-Hall, Englewood Cliffs, New Jersey (1987)

[14] B.KRÅKELAND, ' Nonlinear analysis of shells using degenerate isoparametric elements ', In *Proc. Int. Conf. on Finite Elements in Nonlinear Mechanics,* 265–284, Geilo, Norway (1977), Tapir (1978)

[15] E.H.LEE, ' Elastic-plastic deformations at finite strains ', *J. Appl. Mech.,* **36**, 1–6 (1969)

[16] J.MANDEL, ' Equations constitutives et directeurs dans les milieux plastiques et viscoplastiques ', *Int. J. Solids Struct.,* **9**, 725–740 (1973)

[17] J.E.MARSDEN and T.J.R.HUGHES, *Mathematical Foundations of Elasticity,* Prentice-Hall, Englewood Cliffs, New Jersey (1983)

[18] R.M.MCMEEKING and J.R.RICE, ' Finite-element formulations for problems of large elastic-plastic deformation ', *Int. J. Solids Structures,* **11**, 601–616 (1975)

[19] G.P.MITCHELL and D.R.J.OWEN, ' Numerical solution for elastic-plastic problems ', *Eng. Comp.,* **5**, 274–284 (1988)

[20] B.MORAN, M.ORTIZ and C.F.SHIH, ' Formulation of implicit finite element methods for multiplicative finite deformation plasticity ', *Int. J. Num. Meth. Engng.,* **29**, 483–514 (1990)

[21] L.S.D.MORLEY, ' The constant-moment plate-bending element ', *J. Strain Anal.,* **6**, 20–24 (1971)

[22] J.C.NAGTEGAAL and J.E.DE JONG, ' Some aspects of non-isotropic work-hardening in finite strain plasticity ', In *Plasticity of Metals at Finite Strain: Theory, Experiment and Computation,* (Edited by E.H.Lee and R.L.Mallet), 65–102, Div.Appl.Mech., Stanford Univ. and Dept.Mech.Eng., R.P.I.,Troy, (1982)

[23] J.C.NAGTEGAAL and F.E.VELDPAUS, ' On the implementation of finite strain plasticity equation in a numerical model ', In *Numerical Analysis of Forming Processes,* (Edited by J.F.T.Pittman et al.), 351–371, Wiley, Chicester (1984)

[24] S.NEMAT-NASSER, ' On finite strain elasto-plasticity ', *Int. J. Solids Struct.,* **18**, 857–872 (1982)

[25] S.NEMAT-NASSER, ' On finite plastic flow of crystalline solids and geomaterials ', *J. Appl. Mech.*, **50**, 1114–1126 (1983)

[26] R.W.OGDEN, *Non-linear Elastic Deformations*, Ellis Horwood, Chicester (1984)

[27] E.T.ONAT, ' Represenation of inelastic behaviour in the presence of anisotropy and of finite deformations', In *Recent Advances in Creep and Fracture of Engineering Materials and Structures*, (Edited by B.Wilshire and D.R.J.Owen), 231–264, Pineridge Press, Swansea (1982)

[28] M.ORTIZ and J.J.QUIGLEY IV, ' Element design and adaptive meshing in strain localization problems ', In *Computational Plasticity II: Models, Software and Applications*, (edited by D.R.J.Owen et al.), 213–236, Pineridge Press, Swansea (1989)

[29] D.R.J.OWEN and E.HINTON, *Finite Elements in Plasticity*, Pineridge Press, Swansea (1980)

[30] D.PERIĆ, ' Consistent stress rates in solid mechanics: computational implications ', (to appear), (1990)

[31] D.PERIĆ, D.R.J.OWEN and M.E.HONNOR, ' A model for finite strain elasto-plasticity based on logarithmic strains: computational issues ', (submitted for publication) (1990)

[32] D.PERIĆ, G.P.MITCHELL and D.R.J.OWEN, ' Some aspects of strain localisation problems for rate independent solids ', *Int. J. Num. Meth. Engng.*, (in press) (1991)

[33] D.PERIĆ and D.R.J.OWEN, ' The Morley thin shell element for large deformation problems ', (in preparation) (1991)

[34] J.R.RICE, ' Continuum mechanics and thermodynamics of plasticity in relation to microscale deformation mechanisms ', In *Constitutive Equations in Plasticity*, (Edited by A.S.Argon), 23–75, MIT Press, Cambridge, Massachusetts (1975)

[35] W.D.ROLPH III and K.J.BATHE, ' On a large strain finite element formulation for elasto-plastic analysis ', In *Constitutive Equations: Macro and Computational Aspects*, (Edited by K.J.Willam), Winter Annual Meeting, 131–147, ASME (1984)

[36] R.RUBINSTEIN and S.N.ATLURI, ' Objectivity of incremental constitutive relations over finite time steps in computational finite deformation analysis ', *Comp. Meth. Appl. Mech. Engng.*, **36**, 277–290 (1983)

[37] A.C.SCORDELIS and K.S.LO, ' Computer analysis of cylindrical shells ', *J. Am. Concr. Inst.*, **61**, 539–561 (1969)

[38] J.C.SIMO and K.S.PISTER, ' Remarks on rate constitutive equations for finite deformations problems: computational implications ', *Comp. Meth. Appl. Mech. Engng.*, **46**, 201–215 (1984)

[39] J.C.SIMO and T.J.R.HUGHES, ' General return mapping algorithms for rate-independent plasticity ', In *Constitutive Laws for Engineering Materials: Theory and Applications*, (Edited by C.S.Desai et al.), 221–231, Elsevier Science Publ., New York (1987)

[40] J.C.SIMO, ' A framework for finite strain elastoplasticity based on maximum plastic dissipation and multiplicative decomposition: Part I. Continuum formulation and Part II. Computational aspects', *Comp. Meth. Appl. Mech. Engng.*, **66**, 199–219 and **68**, 1–31 (1988)

[41] T.Y.THOMAS, *Plastic Flow and Fracture of Solids*, Academic Press, New York, (1961)

[42] O.C.ZIENKIEWICZ, *The Finite Element Method*, McGraw-Hill, London, 3-rd.ed. (1977)

A 3D Finite Element Approach for Predicting the Deformation of an Anisotropic Elasto-Plastic Membrane

E.MASSONI J.L. CHENOT M. BELLET

Ecole Nationale Supèrieure des Mines De Paris, CEMEF
Sophia Antipolis, 06560 Valbonne FRANCE

Summary

A finite element method (F.E.M.) for the solution of non-linear elasto-plastic plate problems is presented. A membrane mechanical formulation is used, associated with a Lagrangian description of motion and an incremental implicit form of the flow rule of the material. The flow theory used in this work is based on quadratic Hill's theory for anisotropy. The expression of the yield criterion and related equations are given in the surface curvilinear coordinates system. The non-linear equilibrium equation of the deformed configuration is solved by a complete Newton-Raphson method. The formulation is tested for the bulging of circular diaphragms. The numerical results are compared with an analytical solution and with experimental results. Contact problems at interfaces between tools and sheets are solved, taking friction into account. The method is applied to the problem of stretch forming of a blank using arbitrarily shaped punches and dies. Results from a numerical study of punch stretching with hemispherical and elliptical punches are presented with experimental comparisons in the case of hemispherical stretching. Deep drawing is simulated by introduction of restraining forces, and is tested with a flat-headed punch.

1. Introduction

Thin sheet forming processes may be divided into three major classes: hydraulic bulging, punch stretching and deep drawing. During these processes the plate material is exposed to large plastic deformation. The governing equations of such a process contain therefore non-linearities of both materials and geometries. Generally authors use the F.E.M. for the solution of this non-linear elastoplastic problem [1,2]. In order to formulate the incremental F.E.M. we use a convected coordinate system embedded in the body and then obtain relations between incremental quantities [3]. The thinness of the sheet allows the use of the membrane assumption in writing the governing equations of motion. This paper presents such a formulation in terms of an updated Lagrangian formulation. The material is assumed to obey HILL's quadratic anisotropic law of which the Von Mises criterion is subset [4]. Local incremental constitutive laws, based upon an implicit integration scheme of the plastic strain rate [5], are written with convected coordinates [6]. Numerical results are compared with analytical and experimental data in the case of a bulge test. Contact and friction problems during punch stretching and deep drawing processes are studied. Comparisons with experimental data are made in the case of stretch forming over an hemispherical punch. The transition from stretch to draw forming would require changing boundary conditions: for stretch forming we have enforced a no-motion condition on the periphery of the sheet whereas for deep drawing this condition is no longer valid [7]. The influence of drawbeads on the plastic flow is simulated by the introduction of restraining forces applied to the membrane boundary. Numerical results are presented for various constant values of these forces.

2. Mechanical formulation

2.1 Description of membrane kinematics

The deformed membrane is assumed to be a geometric surface, with material points identified by two convected coordinates (θ^1, θ^2) which move with the solid in such a way that the coordinates for a given material point do not change with time t. The position vector $\mathbf{r}$ denotes the position of the particle at time t with respect to a fixed orthogonal Cartesian frame $(0; \mathbf{I}_1, \mathbf{I}_2, \mathbf{I}_3)$. The covariant basis vectors $\mathbf{g}_i$ and the metric tensor g_{ij} at time t are given by:

$$\mathbf{g}_i = \frac{\partial \mathbf{r}}{\partial \theta^i} \qquad i=1,2 \tag{1}$$

$$g_{ij} = \mathbf{g}_i \cdot \mathbf{g}_j \tag{2}$$

The displacement $\mathbf{u}$ between two configurations can be decomposed as follows:

$$\mathbf{u} = u^i \, \mathbf{g}_i = u_i \, \mathbf{g}^i \tag{3}$$

where $\mathbf{g}^i$ denote the reciprocal basis vectors to $\mathbf{g}_i$. Subscripts and superscripts denote covariant and contravariant components respectively and the repeated indices signify the usual summation convention:

$$g^{ik}\, g_{kj} = \delta^i{}_j \tag{4}$$

$$\mathbf{g}^i = g^{ij}\, \mathbf{g}_j \tag{5}$$

Convected derivatives of a vector are expressed as follows:

$$\frac{\partial \mathbf{u}}{\partial \theta^j} = u^i_{/j}\, \mathbf{g}_i$$

$$u^i_{/j} = \frac{\partial \mathbf{u}}{\partial \theta^j}\, \mathbf{g}^i \tag{6}$$

2.2 Strain and stress measurement

If we denote the length of an infinitesimal material line segment in the initial configuration by ds_0 and in the current one by ds, the measure of strain between these two configurations is given by:

$$ds^2 - ds_0{}^2 \tag{7}$$

In convected coordinates we have:

$$ds_0{}^2 = g_{ij(t=0)}\, d\theta^i\, d\theta^j \tag{8}$$

$$ds^2 = g_{ij(t)}\, d\theta^i\, d\theta^j \tag{9}$$

So we have:

$$ds^2 - ds_0{}^2 = (\, g_{ij(t)} - g_{ij(t=0)}\,)\, d\theta^i\, d\theta^j = 2\, \varepsilon_{ij}\, d\theta^i\, d\theta^j$$

where ε_{ij} are the covariant components of the Lagrangian strain tensor. Using the covariant derivatives of the displacement between the two configurations we can write:

$$2\varepsilon_{ij} = u_{i/j} + u_{j/i} + u^k{}_{/i}\, u_{k/j} \tag{10}$$

The sheet is assumed to have a thickness, h, much smaller than the other dimensions. Membrane equations are then used to study the evolution of the sheet during the process:

$$h = h\, (\theta^1, \theta^2) \tag{11}$$

$$\mathbf{r} = \mathbf{r}\, (\theta^1, \theta^2) \tag{12}$$

Tangent vectors are given by (1): a three-dimensional local basis is obtained by appending to (1) the unit normal vector $\mathbf{g}_3$, given by:

$$\mathbf{g}_3 = \frac{\mathbf{g}_1 \wedge \mathbf{g}_2}{|\mathbf{g}_1 \wedge \mathbf{g}_2|} \tag{13}$$

Equation (2) gives $g_{33} = g^{33} = 1$ and $g_{3i} = g^{i3} = 0$ for $i = 1,2$.
The strain rate tensor $\dot{\boldsymbol{\varepsilon}}$ is written in terms of its covariant components $\dot{\varepsilon}_{ij}$.

$$\dot{\boldsymbol{\varepsilon}} = \dot{\varepsilon}_{ij}\, \mathbf{g}^i \otimes \mathbf{g}^j \tag{14}$$

$$\dot{\varepsilon}_{ij} = \frac{1}{2}\left(V_{i/j} + V_{j/i}\right) \tag{15}$$

V is a velocity field:

$$\mathbf{V} = V_i\, \mathbf{g}^i = V^i\, \mathbf{g}_i$$

(/i denote covariant differentiation with respect to convected coordinates θ^i given by (6))

We use the covariant components σ^{ij} of the Cauchy stress tensor σ for its expression:

$$\sigma = \sigma^{ij}\, \mathbf{g}_i \otimes \mathbf{g}_j \tag{16}$$

2.3 Mechanical equilibrium

The principle of virtual work is used:

$$\int_{\Omega} \sigma : \dot{\varepsilon}^* \, dv - \int_{\Omega} \mathbf{F}\, \mathbf{V}^* \, dv - \int_{\partial\Omega} \mathbf{T}\, \mathbf{V}^* \, ds \quad = 0 \tag{17}$$

for any arbitrary velocity field $\mathbf{V}^*$. σ is the Cauchy stress tensor, $\dot{\varepsilon}^*$ is the virtual strain rate tensor associated with the virtual velocity field $\mathbf{V}^*$ by relation (15), **T** is the external applied surface forces, and **F** the external applied forces on the boundary $\partial\Omega$ of the deformed surface Ω.
The tensor σ and $\dot{\varepsilon}^*$ are symmetric, thus we have:

$$\sigma : \dot{\varepsilon}^* = \sigma^{ij}\, \dot{\varepsilon}^*_{ij} = \sigma^{ij}\, V^*_{j/i}$$

Equation (17) becomes:

$$\int_{\Omega} \sigma^{ij}\, V^*_{i/j}\, dv - \int_{\Omega} \mathbf{F}\, \mathbf{V}^* \, dv - \int_{\partial\Omega} \mathbf{T}\, \mathbf{V}^* \, ds \quad = 0 \qquad i,j = 1,3 \tag{18}$$

for any $\mathbf{V}^*$.
The membrane assumption states that the virtual work of normal forces is negligible with respect to the virtual work of in-plane forces, or that the dissipated bending energy is far smaller than stretching energy. So we can consider that the

equilibrium equation (18) is rewritten with the plane stress hypothesis:

$$\int_{\Omega} \sigma^{ij} V^{*}_{i/j}\, dv - \int_{\Omega} \mathbf{F} \mathbf{V}^{*}\, dv - \int_{\partial\Omega} \mathbf{T} \mathbf{V}^{*}\, ds = 0 \qquad i,j = 1,2 \tag{19}$$

3.<u>Constitutive equation</u>

3.1 Plasticity equations

We assume that the material follows classical flow plasticity, which can be expressed as follows:

a/ The total strain rate $\dot{\varepsilon}$ is decomposed into elastic and plastic contributions.

$$\dot{\varepsilon} = \dot{\varepsilon}^{(el)} + \dot{\varepsilon}^{(p)} \tag{20}$$

b/ The linear elasticity is written as:

$$\dot{\sigma} = D_{(el)}\, \dot{\varepsilon}^{(el)}$$

or

$$\dot{\varepsilon}^{(el)} = D^{-1}_{(el)}\, \dot{\sigma}$$

$\dot{\sigma}$ is the stress rate tensor, $D_{(el)}$ represents the tensor of elastic constants for an isotropic elastic material

c/ The plastic strain rate $\dot{\varepsilon}^{(p)}$ is given by the normality rule:

$$\dot{\varepsilon}^{(p)} = \dot{\lambda}^{(p)} \frac{\partial f}{\partial \sigma} \tag{21}$$

$\dot{\lambda}^{(p)}$ is the plastic multiplier, and f represents the yield surface:

$$f(\sigma, \bar{\varepsilon}^{(p)}) = 0 \tag{22}$$

in which $\bar{\varepsilon}^{(p)}$ is the equivalent plastic strain computed by:

$$\bar{\varepsilon}^{(p)} = \int_0^t \dot{\bar{\varepsilon}}\, dt \tag{23}$$

It is important to note that we have choosen a quadratic criterion to determine the threshold of plasticity (22):

$$f(\sigma, \bar{\varepsilon}^{(p)}) = \sigma^2_{eq} - \sigma^2_0(\bar{\varepsilon}^{(p)}) \tag{24}$$

where $\sigma^2_0(\bar{\varepsilon}^{(p)})$ represents the yield stress squared, $\sigma^2_{eq} = \sigma\tilde{A}\sigma$ is the equivalent stress and $\tilde{A}$ a tensor of rank 4, which is called the criterion tensor. The quadratic form of f gives:

$$\frac{\partial f}{\partial \sigma} = 2\tilde{A}\sigma = A\sigma \tag{25}$$

So (21) gives :

$$\dot{\varepsilon}^{(p)} = \dot{\lambda}^{(p)} A\sigma \tag{26}$$

The inverse of equation (26) gives:

$$\sigma = \frac{1}{\dot{\lambda}^{(p)}} C \ \dot{\varepsilon}^{(p)} \tag{27}$$

where $C = A^{-1}$.

So the equivalent strain rate $\dot{\bar{\varepsilon}}$ of relation (23) is defined by :

then
$$\dot{\bar{\varepsilon}} = \sqrt{\dot{\varepsilon}^{(p)} C \ \dot{\varepsilon}^{(p)}} \tag{28}$$

We can summarize the local elastic-plastic flow rules :

if $f(\sigma, \bar{\varepsilon}^{(p)}) < 0$ or $f = 0$ and $\frac{\partial f}{\partial \sigma} : \dot{\sigma} < 0$

then $\dot{\varepsilon} = D^{-1}_{(el)} \dot{\sigma}$ and $\dot{\varepsilon}^{(p)} = 0$

if $f(\sigma, \bar{\varepsilon}^{(p)}) = 0$ and $\frac{\partial f}{\partial \sigma} : \dot{\sigma} \geq 0$

then $\dot{\varepsilon} = D^{-1}_{(el)} \dot{\sigma} + \dot{\varepsilon}^{(p)}$ and $\dot{\varepsilon}^{(p)} = \dot{\lambda}^{(p)} \frac{\partial f}{\partial \sigma}$

3.2 Different forms of the equivalent stress

The intrinsic form of the equivalent stress, $\sigma^2_{eq} = \sigma A \sigma$, permits the use of any coordinate system to write the criterion equation; in particular we can express the criterion equation in the convected coordinate system.

- *Isotropic equation*

The Von Mises yield criterion is:

$$f(\sigma, \bar{\varepsilon}^{(p)}) = \sigma^2_{eq} - \sigma^2_0 (\bar{\varepsilon}^{(p)}) \tag{29}$$

where:

$$\sigma^2_{eq} = \sigma A \sigma = \frac{3}{2} \text{trace}(\sigma^2) - \frac{1}{2} (\text{trace}\, \sigma)^2 \tag{30}$$

and we have:

$$A_{ijkl} = \frac{3}{2} g_{ik} g_{jl} - \frac{1}{2} g_{ij} g_{kl} \tag{31}$$

in the convected coordinates system, g_{ij} is the metric tensor and σ is given in equation (16).

- *Normal anisotropic criterion*

Hill's normal anisotropic criterion is given by :

$$\sigma_{eq}^2 = \frac{1+2r}{1+r} \operatorname{trace}(\sigma^2) - \frac{r}{1+r} (\operatorname{trace} \sigma)^2 \tag{32}$$

where r is the normal plastic anisotropy parameter.

This equation means that the anisotropy is only through the normal direction to the sheet and the behavior is isotropic in the plane. In this case we have:

$$A_{ijkl} = \frac{1+2r}{1+r} g_{ik}\, g_{jl} - \frac{r}{1+r} g_{ij}\, g_{kl} \tag{33}$$

in the system of convected coordinates.

- *Planar anisotropy*

The Hill's planar anisotropy criterion assumes that the material anisotropy has three mutually orthogonal symmetry planes at every point. The intersections of these planes are known as the principal axes of anisotropy. In this case, it is more difficult to write the criterion in intrinsic form with the equivalent stress given by

$$\sigma_{eq}^2 = \frac{1}{G+H}\left[(G+H)\sigma^{\circ 2}_{xx} - 2H\sigma^{\circ}_{xx}\sigma^{\circ}_{yy} + (H+F)\sigma^{\circ 2}_{yy} + 2N\tau^{\circ 2}_{xy}\right] \tag{34}$$

Equation (34) is written such that the principal axes of anisotropy are the coordinate axes. The reference axes are choosen such that:

x is the rolling direction

y is the transverse direction in the plane.

F,G,H,N are anisotropic parameters. The equivalent stress σ_{eq} is chosen equal to the yield stress σ_0 measured in a tensile test performed in the rolling direction. This choice is convenient to linking the flow rule to experimental results.

If we assume that the axes of reference, x and y, are the same as the initial base vectors $g_{i(t=0)} = g^{\circ}_i$, the equation (34) can be written in the following intrinsic form

$$f\left(\sigma^0, \bar{\varepsilon}^{(p)}\right) = \sigma^0\ A^0\ \sigma^0 - \sigma_0^2(\bar{\varepsilon}^{(p)}) \tag{35}$$

First approach

In order to take into account the effect of geometry change, the rotation of the initial axes (x,y) is studied; we introduce an intermediate configuration (figure 1) represented by the local basis $\mathbf{g}'$(in Ω') deduced from the initial base $\mathbf{g}^\circ$ (in Ω°) by a pure rotation:

$$\mathbf{g}' = \mathbf{R}\mathbf{g}^\circ \tag{36}$$

from the deformation gradient tensor $\mathbf{F}$:

$$\mathbf{F} = 1 + \text{grad}\,\mathbf{u} \tag{37}$$

which is decomposed according to:

$$\mathbf{F} = \mathbf{V}\mathbf{R}, \tag{38}$$

where $\mathbf{V}$ is the displacement tensor and $\mathbf{R}$ the pure rotation tensor.

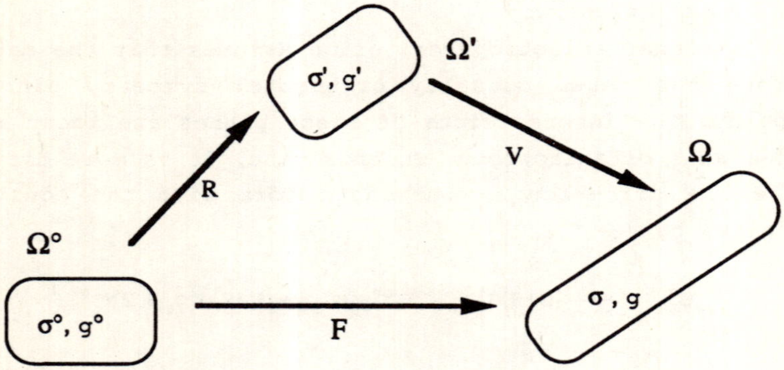

Figure 1: Polar decomposition of deformation gradient tensor F.

We have:

$$\mathbf{g} = \mathbf{F}\mathbf{g}^\circ = \mathbf{V}\mathbf{R}\mathbf{g}^\circ = \mathbf{V}\mathbf{g}' \tag{39}$$

In terms of displacement $\mathbf{u}$, a first-order approximation can be written:

$$\mathbf{R} = 1 + (\,\text{grad}\,\mathbf{u} - {}^{T}\text{grad}\,\mathbf{u}\,)/2 \tag{40}$$

The stress tensor can be defined by:

$$\sigma = \sigma^{\circ ij}\, \mathbf{g}^\circ{}_i \otimes \mathbf{g}^\circ{}_j = \sigma'^{ij}\, \mathbf{g}'_i \otimes \mathbf{g}'_j \quad i,j = 1,2 \tag{41}$$

By substitution of the decomposition (36) of $\mathbf{g}^\circ$ in the local basis $\mathbf{g}'$ we can write:

$$\sigma^{\circ ij} = R_{ki}\, \sigma'^{kl}\, R_{lj} \tag{42}$$

we can write substituting equation (42) in (35):

$$f = \sigma' A' \sigma - \sigma_0^2 \tag{43}$$

with: $$\sigma^2_{eq} = \sigma' A' \sigma' = \sigma'^{ij} A'_{ijkl} \sigma'^{kl} \tag{44}$$

in which: $$A'_{ijkl} = R_{i\alpha} R_{j\beta} A^\circ_{\alpha\beta\gamma\delta} R_{k\gamma} R_{l\delta} \tag{45}$$

It is important to note that in the equation (45) the anisotropy parameter (in A°) are known from experimental tests (as a first approximation it is assumed that they remain constant along the deformation), so at each step of the deformation the equivalent stress σ_{eq} is known when we can determine the stress tensor.

Second approach

Let $\mathbf{g}_i$ be a base vector along a co-ordinate line and $\mathbf{e}_i$ be unit vectors along the anisotropic axes, given as the x and y axes in equation (34) (Figure 2).

Then the following decomposition can be made:

$$\mathbf{g}_i = P_{ij}\, \mathbf{e}_j$$

$$\mathbf{e}_i = \frac{\mathbf{g}_i}{|\mathbf{g}_i|}$$

with:

If $\sigma^{*\alpha\beta}$ are the component of Cauchy stress in the local rectangular frame and σ^{ij} the contravariant component of Cauchy stress in the convective co-ordinate system we have:

$$\sigma^{*\alpha\beta} = P_{i\alpha}\ \sigma^{ij}\ P_{j\beta}$$

The equivalent stress can be rewritten in the surface curvilinear co-ordinate system:

$$\sigma^2_{eq} = \sigma^{ij} A_{ijkl} \sigma^{kl}$$

Where $$A_{ijkl} = P_{i\alpha} P_{j\beta} A^*_{\alpha\beta\gamma\delta} P_{k\gamma} P_{l\delta}$$

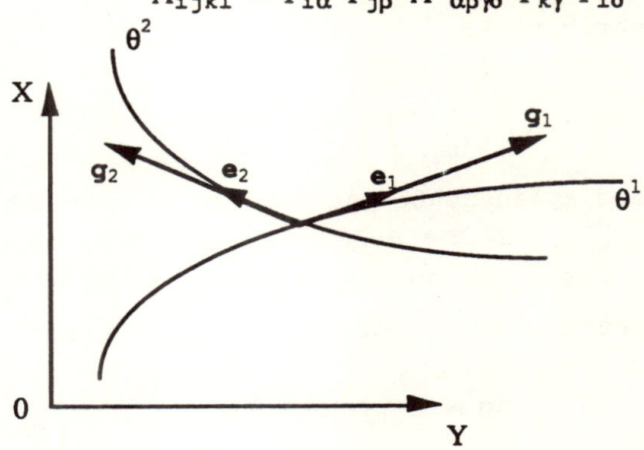

Figure 2: Co-ordinate systems used in the 2nd analysis ; axes of anisotropy.

4. Numerical solution

4.1 Time and space discretization

Incremental loading procedure

The external load is applied increment by increment. If t is the time (or parameter characterisying the extent of deformation), Δt represents an increment of the variable between t and $t+\Delta t$. The motion of the deformed membrane is thus discretized in time according to the following incremental formulation, an updated Lagrangian one:

Considering the known equilibrium configuration at time t, the problem consists of calculating the unknown equilibrium state at time $t+\Delta t$. By application of equation (18) at $t+\Delta t$ we obtain:

$$\int_{\Omega(t+\Delta t)} (\sigma^{ij}+\Delta\sigma^{ij})\, V^{*}_{i/j}(h+\Delta h)\, ds = \int_{\Omega(t+\Delta t)} \mathbf{T}(\mathbf{u}+\Delta\mathbf{u}) . \mathbf{V}^{*} ds + \int_{\partial\Omega(t+\Delta t)} \mathbf{F}(\mathbf{u}+\Delta\mathbf{u}) . \mathbf{V}^{*} dl$$

$\forall\ \mathbf{V}^{*}$

This equation is solved in terms of incremental displacements $\Delta\mathbf{u}$ during Δt. For this purpose, it is necessary to calculate the incremental stress $\Delta\sigma$ as a function of the incremental Lagrangian strain tensor $\Delta\varepsilon$. The load increment can be fully elastic-plastic or mixed with an elastic->elastic-plastic transition between t and $t+\Delta t$. The basic idea is a semi-implicit integration scheme of the plastic deformation rate. In order to consider elastic-plastic loading, we decompose $\Delta\varepsilon$ into elastic and plastic contributions:

$$\Delta\varepsilon = \Delta\varepsilon^{(el)} + \Delta\varepsilon^{(p)} \tag{46}$$

in which
$$\Delta\varepsilon = \Delta\varepsilon_{ij}\ \mathbf{g}^{i} \otimes \mathbf{g}^{j} \tag{47}$$

with
$$2\ \Delta\varepsilon_{ij} = g_{ij} - g^{0}_{ij} \tag{48}$$

and in terms of displacements the nonzero components are:

$$2\ \Delta\varepsilon = \Delta u_{\alpha/\beta} + \Delta u_{\beta/\alpha} + \Delta u^{\gamma}{}_{/\alpha}\ \Delta u_{\gamma/\beta} \tag{49}$$

$$2\Delta\varepsilon = \mathrm{grad}\ \Delta\mathbf{u} + {}^{T}\mathrm{grad}\ \Delta\mathbf{u} + \mathrm{grad}\ \Delta\mathbf{u}\ {}^{T}\mathrm{grad}\ \Delta\mathbf{u} \tag{50}$$

The stress increment is known in terms of the elastic strain increment:

$$\Delta\sigma = \mathbf{D}_{(el)}\ \Delta\varepsilon^{(el)}\ , \tag{51}$$

while the plastic strain increment is given by:

$$\Delta\varepsilon^{(p)} = \int_{t}^{t+\Delta t} \dot{\lambda}^{p} \frac{\partial f}{\partial \sigma} dt \tag{52}$$

The semi-implicit integration scheme [5] is chosen to extrate equation (52):

$$\Delta\varepsilon^{p} = \Delta\lambda^{p}\left[(1-\eta)\frac{\partial f(\sigma,\bar{\varepsilon})}{\partial\sigma} + \eta\frac{\partial f(\sigma+\Delta\sigma,\bar{\varepsilon}+\Delta\bar{\varepsilon})}{\partial\sigma}\right] \quad (53)$$

in which $\Delta\bar{\varepsilon}$ is the equivalent incremental plastic strain defined by:

$$\Delta\bar{\varepsilon} = \sqrt{\Delta\varepsilon^{p}_{ij}\ C\ \Delta\varepsilon^{p}_{ij}} \quad (54)$$

η is an arbitrary fixed parameter in [0,1] which varies the integration scheme between fully explicit and fully implicit, respectively. In general, the value 0.5 is chosen.
The criterion equation is satisfied at the end of the increment:

$$f\left(\sigma+\Delta\sigma,\bar{\varepsilon}+\Delta\bar{\varepsilon}\right)=0 \quad (55)$$

This set of equations is solved locally for $\sigma+\Delta\sigma$ by a Newton-Raphson method.

Spatial discretization

The membrane surface Ω is meshed with finite isoparametric plate elements with three degrees of freedom at each node, which are the Cartesian components of the displacement. All integrals are calculated with the Gauss technique. We use a vectorial notation adjusted to the F.E.M.: a tensor is represented by a vector calculated at each integration point of the mesh. The velocity and the coordinates at each node N of the mesh are noted $\mathbf{V}_N$ and X_N. The fundamental relation between covariant derivatives and nodal cartesian components for any virtual velocity field is written:

$$V^{*}{}_{i/j} = B_{ijNk}\ V^{*}{}_{Nk} \quad (56)$$

for j=1,3; i=1,2;k=1,3; N=1,Nel

(Nel is the number of nodes)

with:

$$B_{ijNk} = \frac{\partial\psi_N}{\partial\theta^i}\ \mathbf{I}_k\ \mathbf{g}_j \quad (57)$$

ψ_N is the value of interpolation function at node N and $\mathbf{I}_k$ are the Cartesian base vectors.

Therefore, the discretized form of the equilibrium equation at time t+Δt is:

$$\int_{\Omega(t+\Delta t)} (\sigma^{ij}+\Delta\sigma^{ij})\, B_{ijNk} V^*_{Nk} (h+\Delta h)\, ds =$$

$$\int_{\Omega(t+\Delta t)} \psi_N T^k (\Delta \mathbf{u})\, . V^*_{Nk} ds + \int_{\partial\Omega(t+\Delta t)} \psi_N F^k (\Delta \mathbf{u})\, . V^*_{Nk} dl \qquad (58)$$

for any component V^*_{Nk} of the discretized virtual velocity field with N=1,Nel; (i,j)=1,2; k=1,3.

It is important to note that all quantities which appear in this equation are written on the configuration at time t+Δt, and depend on $\Delta \mathbf{u}$ which is the incremental displacement vector between configuration at time t and configuration at time t+Δt.

Equation (58) must be verified for any virtual velocity field V^*_{Nk}; and the final discretized equilibrium is, for any degree of freedom (N,k):

$$R_{Nk}(\Delta \mathbf{u}) = \int_{\Omega(t+\Delta t)} (\sigma^{ij}+\Delta\sigma^{ij})\, B_{ijNk} (h+\Delta h)\, ds \; -$$

$$\int_{\Omega(t+\Delta t)} \psi_N T^k (\Delta \mathbf{u})\, ds - \int_{\partial\Omega(t+\Delta t)} \psi_N F^k (\Delta \mathbf{u})\, dl \quad = \quad 0 \qquad (59)$$

The problem consists in solving the non linear system, equation (59), in which the nonlinearities are both geometric and material. This necessitates the use of an iterative method for solving.

4.2 Solution algorithm

The Newton-Raphson algorithm is used:

start with $\Delta \mathbf{u}_0 = 0$

then if $\Delta \mathbf{u}_i$ is the incremental displacement field at iteration number i,

at iteration i+1, $\Delta \mathbf{u}_{i+1} = \Delta \mathbf{u}_i + \delta\Delta \mathbf{u}_{i+1}$

with
$$\left[\frac{\partial R(\Delta \mathbf{u}_i)}{\partial \Delta \mathbf{u}}\right] \delta\Delta \mathbf{u}_{i+1} = - R(\Delta \mathbf{u}_i) \qquad (60)$$

The loop is to be continued until convergence is reached, i.e. ||R|| is small enough. The expression of the tangent stiffness matrix, including the geometrical contribution, can be found in reference [8].

4.3 Contact equations

During a sheet forming process there is an evolution of the contact between the sheet metal and the tools. This supposes a finite element code with changing incremental boundary conditions, in order to describe correctly the slip of membrane nodal points along the surface of tools, and possibly the onset or the loss of contact.

The punch displacement is assumed to be vertical, along the z direction. The contact condition may be expressed by the incremental displacement relation:

$$\Delta \mathbf{u}_{Nr} . \mathbf{n}_N = (\Delta \mathbf{u}_N - \Delta \mathbf{W}_p) . \mathbf{n}_N = 0 \qquad (61)$$

where:

$\Delta \mathbf{W}_p$ is the incremental punch displacement vector;

$\Delta \mathbf{u}_N$ is the incremental displacement vector of node N;

$\Delta \mathbf{u}_{Nr}$ is the relative displacement vector of node N, which must be tangential to the punch surface;

$\mathbf{n}_N$ is a normal unit vector to the punch surface at node N location.

Therefore three different problems are to be tackled:

a/ the analysis of nodal relative locations, with respect to punch surface;

b/ the introduction of the contact condition, equation (61), in the solution of the equilibrium equation;

c/ the computation of the friction stresses during slip at the interface between punch and sheet.

5. Numerical results

5.1 Bulging test of a circular diaphragm

The chosen test for this study is the bulging by one-sided fluid pressure of a circular diaphragm clamped on its edge.

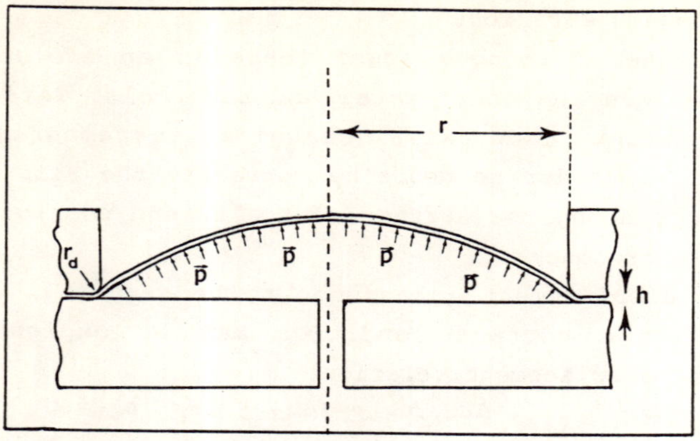

Figure 3: Experimental device.

The F.E.M. results are compared with an analytical solution (not included friction) for a rigid-plastic membrane. The initial geometric characteristics of the diaphragm are:

radius $r = 100$ mm

thickness $h_0 = 1$ mm

The yield stress of the material σ_0 has a constant value of 400MPa (no strain hardening).

a/Comparaison with a rigid-plastic analytical solution

Rigid-plastic analytical solution

The diaphragm is considered as a thin shell with a plane stress state. The deformed dome is assumed to be part of a sphere, the thickness of which is supposed to be homogeneous. Then the Laplace equation gives :

$$R = \frac{2h}{p}\sigma_0$$

where R is the radius of curvature

h is the thickness of the deformed shell

p is the applied pressure

The radius of curvature and the maximum deflection w at pole are related by:

$$w = \sqrt{R^2 - r^2} - R$$

and the thickness can be calculated by writing the volume conservation.

F.E.M. results

Numerical results were obtained with the elasto-plastic code. The mesh is composed of triangular elements and is drawn in figure 4. The same material characteristics are assumed.

We have verified that a change in elasticity coefficients didn't affect the displacement values, so that they may be compared with the rigid-plastic analytical solution. The hoop stress remains very nearly equal to the yield stress σ_0 throughout the dome (and the dome itself remains a part of the sphere), which is consistent with the analytical solution.

The agreement about maximum deflection is quite good, although some disparity appears as the applied pressure increases. The computed maximum deflection is greater than the analytical solution (by about 4% for p = 4MPa)and this may be explained by the fact that the analytical calculation does not take into account the thinning at the pole of the membrane.

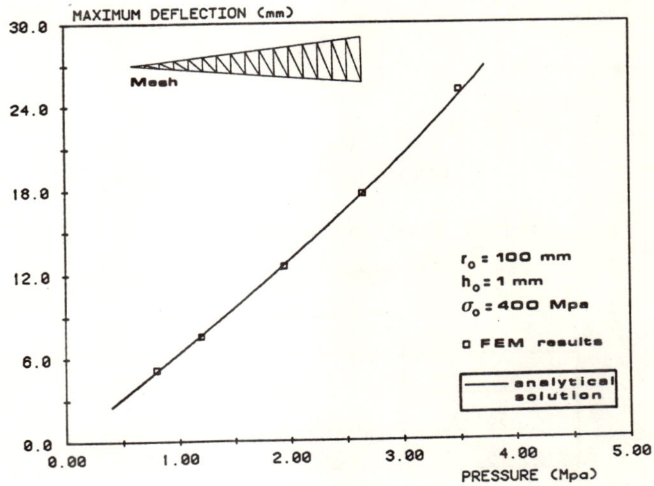

Figure 4: Bulging of an elastic-plastic diaphragm. Comparison of the F.E.M. results with the rigid-plastic analytical solution.

b/comparison with experimental results

The geometric data are

-clamping radius r = 75 mm

-initial sheet thickness h_0 = 0.7 mm

The material data are:

-Young's modulus E = 208 000 MPa

-Poisson's coefficient ν = 0.3

The experimental test have been made for two steels. The first is a low carbon steel(steel 1); the other is a HSLA steel (Nb alloyed)(steel 2).

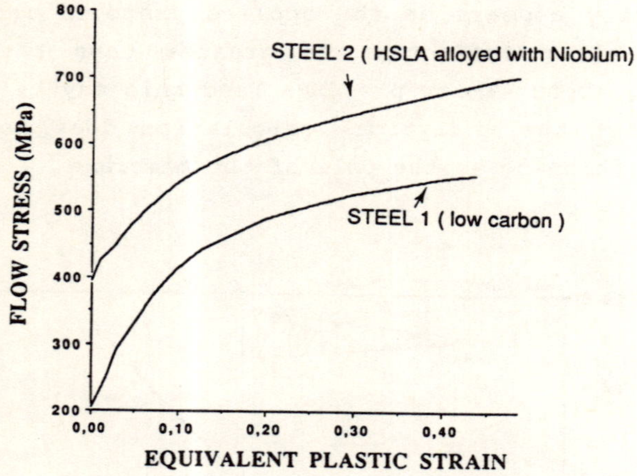

Figure 5: Material work-hardening curve.

	steel 1	steel 2
R_0	1.88	0.863
R_{90}	2.18	1.15
R_{45}	1.37	1.27

Figure 6: Anisotropy parameter.

$R\alpha$ is an anisotropic parameter or strain ratio, α is the angle to the rolling direction of sheet. We have:

$$R_0 = \frac{H}{G} \; ; \; R_{90} = \frac{H}{F} \; ; \quad R_{45} = \frac{N}{F+G} - \frac{1}{2}$$

F,G,H,N are the anisotropy coefficients of equation (34).
For the numerical test we have meshed a quarter of the sample using isoparametric elements with three nodes. The sheet is clamped on its edge and a uniform pressure is applied.

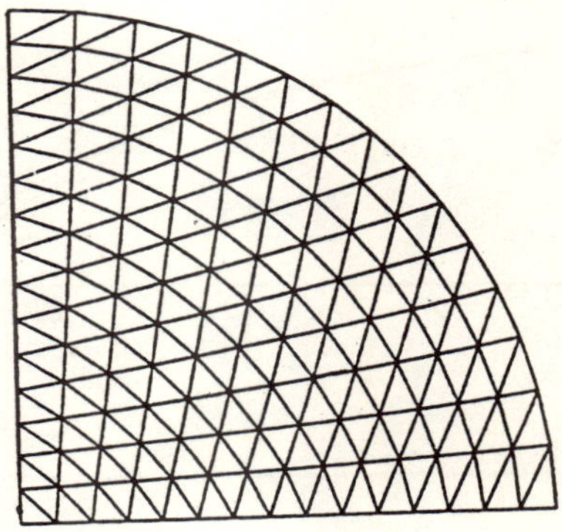

Figure 7: Finite element mesh.

Numerical result is obtained for an isotropic elastic-plastic material. The work hardening laws represented in figure 5 are introduced in the computation, but without anisotropy, meaning that the equivalent stress is given by the isotropic relation (30). The comparison has been made for two final pressures for each steel.

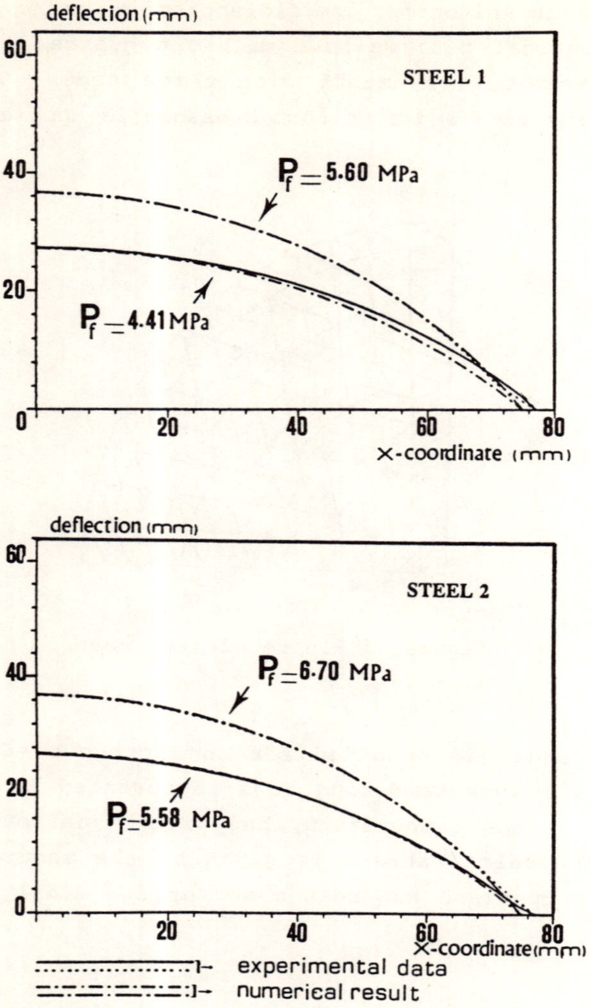

Figure 8: Radial section of the bulged sheet: comparison between experimental and numerical results.

The agreement between experimental and numerical results is good although the computational formulation does not take into account the influence of the anisotropy parameter.

5.2 Hemispherical punch stretching

Experiments have been carried out for HSLA Nb-alloyed steel(steel 2). The material parameters were given previously. Figure 9 shows the punch-die geometry.

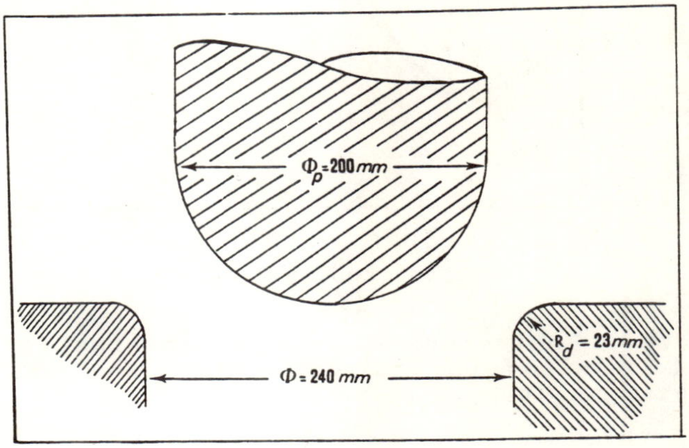

Figure 9: Geometric data used for the FEM simulation.

The sheet is 0.7 mm thick and is clamped firmly around a circle with radius 130mm. Experimental results are available for a punch travel of 45mm. For the numerical simulation, both die and punch surfaces have been discretized into triangular elements. The sheet material is considered as a planar anisotropic elastic-plastic work-hardening material (equation (43), (44), (45)). Because of axial symmetry, only a narrow sector is studied: it is meshed with 29 triangular elements and a no-displacement condition is assumed at the boundary (r=130mm). Concerning friction, experimental data gives a Coulomb's coefficient μ=0.15 during the process. A short approximate calculation of the friction stresses values leads to the equivalent range for the value of the Tresca law coeficient 0.007. The thickness distribution is plotted on figure 10 with experimental data. Comparison of curves shows that the general trend is qualitatively good, and the quantitative agreement is satisfactory.

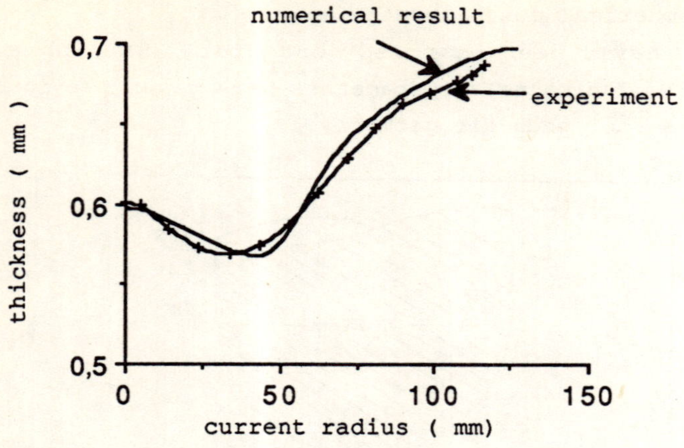

Figure 10: Hemispherical punch stretching: thickness distribution along the initial radius and comparison with experimental result.

5.3 Elliptical Punch Stretching

The elliptical punch dimensions are:

-X axis : 200 mm

-Y axis : 140 mm

-Z axis : 100 mm

By reason of symmetry, only a quarter of the sheet is meshed (positive X and Y coordinates). The dimensions of the studied zone are 150mm x 100mm. A regular mesh is used: it is made of 126 elements with four nodes and four integration points each. A clamping condition is assigned to the domain boundary and there is no die. In this calculation, the same material parameters as for modelling the hemispherical punch stretching case are used with no friction.

The corresponding thickness distribution along the major(x) and the minor(y) axis is given in figure 11 for a 52mm punch travel. The minimum value is found on the minor axis. The general tendency is in fair agreement with the experimental results reported in [3].

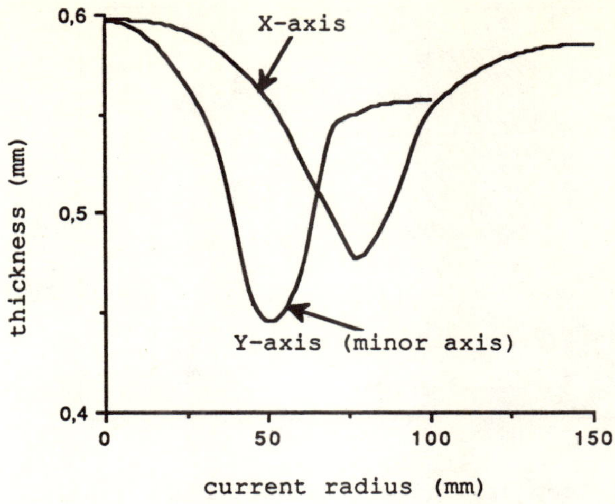

Figure 11: Elliptical punch stretching. Thickness distribution along major and minor axis.

5.4 Deep drawing: introduction of restraining forces

In order to model the influence of drawbeads on the plastic flow, we have implemented in the model restraining forces at the boundary of the finite element mesh. The value of these drawbead forces should be calculated by a detailed analysis of the evolution of the sheet in the drawbead region, where the membrane assumption is no longer valid because of the small radii of curvature. The results are presented for an implementation of forces of constant value, all around a circular sheet, drawn by a flat headed punch in a cylindrical die.

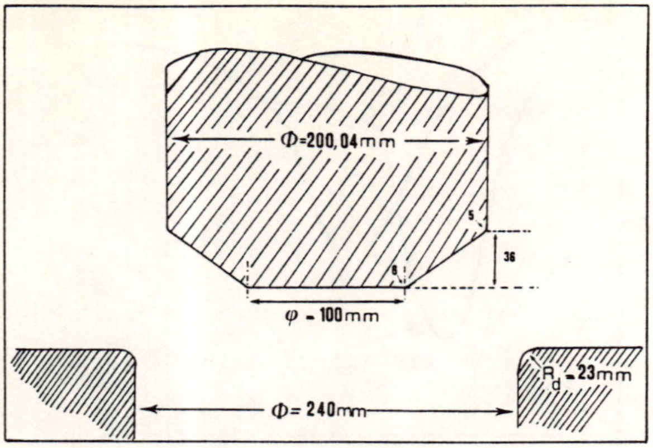

Figure 12: Deep drawing with a flat headed punch.

The initial sheet diameter is 300mm and the material parameters are the same as for hemispherical and elliptical punch stretching. There is no friction at sheet-punch and sheet-die interfaces. Figure 13 shows the thickness profile for a punch travel of 40mm and for two different values of restraining forces: 50N/mm and 100N/mm.

It is clear that as the value of the force decreases, the deformation is less and less concentrated at the edge of the punch (r=50mm), and consequently, a localized necking in this region can be eliminated. It is to be noticed that for the lower value of the restraining force, in-plane compressive state is found near the edge of the die, with increasing thickness, which could lead to wrinkling.

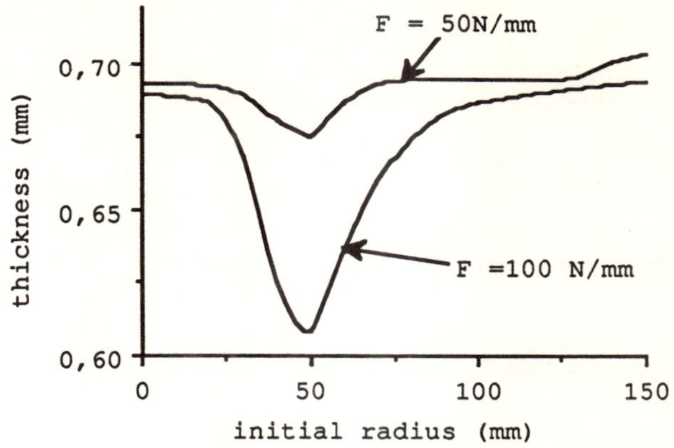

Figure 13: Deep-drawing: influence of restraining forces on the thickness distribution.

6. Conclusion

An incremental elastic-plastic approach for solving thin sheet forming has been presented. The general deformation of an isotropic and a planar anisotropic sheet metal has been considered by using a convected coordinate system. The formulation incorporates the rotation of anisotropic axes as well as the effect of geometric change for each step of deformation. The F.E.M. program has been developed for the analysis of large flange deformation in the deep drawing process, as well as for the analysis of hydrostatic bulging of sheet metal. In these two cases the formulation is in good agreement with experimental results. Friction effects and contact sheet-tool are also studied with this model. Deep drawing modelling has been shown to be possible with the introduction of restraining forces at the edge of the forming sheet. To improve the calculations further, consideration might be given to the calculation of these restraining forces by means of a detailed analysis of the plastic flow near drawbeads.

<u>references</u>

1. Wang, N.M.; Budiansky, B.: Analysis of Sheet Metal Stamping by a Finite Element Method. J. of Ap. Mech. 45 (1978) 73-82

2. Tang, S.C.: Large Elastoplastic Strain Analysis Flanged Hole Forming. Comp. & Structures 13 (1981) 363

3. Kitagawa, H. ; Tomita, Y.: An Incremental Finite Element Analysis of two dimensional Large Strain and Large Displacement Problems for Elastoplastic Material. Proc. 21st Japan National Congress for Applied Mechanics. (1971)

4. Hill, R.; The Mathematical Theory of Plasticity, Clarendon Press, Oxford (1950)

5. Braudel, H.J.; Abouaf, M.; Chenot, J.L.: An Implicit and an Incremental Formulation for the solution of Elastic-plastic Problems by the Finite Element Method. Comp & Struct. 22 (1986)

6. Yang, D.Y.; A Rigid Plastic Finite Element Formulation for the Analysis of General Deformation of Planar Anisotropic Sheet Metals and its Applications. Int. J. Mech.Sci. 28 (1986) 825

7. Andersen, B.S. : A Numerical Study of the Deep Drawing Process. Num. Meth. in Ind. Form. Proc. eds Pineridge Press (1982) 709

8. Massoni, E.; Bellet, M.; Abouaf, M.; Chenot, J.L.: Large Displacements Numerical Calculation of 3-dimensional Elastic plastic and Elastoviscoplastic Membrane by the Finite Element Method. Numerical Methods for Non linear Problems. Eds Pineridge Press (1986) 480

9. Wang, N.M. : A Rigid Plastic Rate Sensitive Finite Element Method for Modelling Sheet Metal Forming Processes. John Wiley & sons Ltd (1984)

10. Nakamachi, E.; Takezono, S.; Sowerby, R.: A numerical Analysis of the hydraulic Bulging of Circular Disks into Axisymmetric Dies. J. of Ap. Mech. 49 (1982) 501

11. Kobayashi, S.; Kim, J.H.: Deformation Analysis of Axisymmetric Sheet Metal Forming Processes by the Rigid Plastic Finite Element Method. Proceedings of a Symposium on Mechanics of Sheet Metal Forming,Eds. Koistinen, D.P.; Wang, N.M.;Plenum press (1978)

12. Matiasson, K.; Melander, A.: Numerical Simulation of stretch forming processes. International Workshop on Simulation of Metal Forming Processes by the Finite Element Method, Ed. Springer Verlag (1985)

14. Massoni, E.; Lene, F.; Duvaut, G.; Chenot, J.L.: An Elastic plastic Membrane Mechanical Approach: Application to Thin Sheet Forming Analysis. Computational plasticity, Pinerige Press 2 (1989) 827

Models of Plasticity in Strain Space

L.G. POPOV

Chair of Higher Mathematics
Moscow Power Engineering Institute,
Krasnokazarmennaya, 14, Moscow E250, USSR.

J.R. WHITEMAN

BICOM, Institute of Computational Mathematics
Department of Mathematics and Statistics
Brunel University, Uxbridge, England.

SUMMARY
The constitutive equations of plasticity are formulated for yield surfaces, that are either smooth or contain vertices, in strain space in terms of strain and strain rates, which are treated as independent variables. The starting point is the Drucker Prager postulate and use is made of the Rabotnov model of plasticity in terms of yield planes which form an envelope of the yield surface. Calculated results obtained in this way are compared with results of traditional models of plasticity (isotropic and kinematic flow theory models and deformation theory) and results of one-dimensional experiments.

1. Introduction

The constitutive equations of plasticity have traditionally been formulated in such a way that the current stress and the stress increments are independent variables and with these the corresponding strain increments are then computed. Using this approach to solve solid mechanics problems one is faced with two main difficulties see [1]. First, the material behaviour at any point of the solid depends on the very quantaties which are being sought. As a result, when using finite element methods to approximate the deformation, it is necessary for each load increment to check throughout the solid for elastic or plastic response. Secondly, in every step, in order to obtain a stiffness matrix one has to invert a matrix which has been obtained from the constitutive equation for the particular loading increment. These difficulties become more pronounced in the case of a yield surface with vertices, when the current stress state lies at the intersection of two or more loading surfaces.

It is therefore more convenient in any load increment for the constitutive equation to be formulated in such a manner that the current strain and the strain increment are independent variables, so that the corresponding stress increment can be calculated from these. Attempts to reformulate the constitutive equations in such a manner have been undertaken repeatedly (see e.g. [1] - [3]), but the theory is as yet incomplete. The purpose of this paper is to provide such constitutive equations for certain types of plastic deformation and to compare results calculated from these with those of traditional models.

2. Constitutive equations of plasticity based on flow theory in strain space

2.1 A general equation of associated flow theory at smooth point of yield surface

We consider a form of the stress-strain relations of plasticity based on the following assumptions:

(i) the strain tensor ϵ_{ij} is represented as the sum of elastic ϵ^e_{ij} and plastic ϵ^p_{ij} parts:

$$\epsilon_{ij} = \epsilon^e_{ij} + \epsilon^p_{ij} \tag{2.1}$$

where the elastic part is related to the stress tensor σ_{ij} by Hooke's law

$$\sigma_{ij} = D_{ijk\ell}(\epsilon_{k\ell} - \epsilon^p_{k\ell}) \tag{2.2}$$

and the constant elastic moduli satisfy $D_{ijk\ell} = D_{jik\ell} = D_{ij\ell k} = D_{k\ell ij}$.

(ii) the yield surface in strain space exists and its equation is

$$\phi(\epsilon_{ij}, \epsilon^p_{ij}, K_n) = 0 \tag{2.3}$$

where the K_n are parameters depending on the loading history. This surface limits the elastic behaviour and current strain is always located inside or on this surface. Thus any strain increment which is directed from a point on this surface outwards causes (as the surface changes) a new plastic strain increment.

(iii) The Drucker postulate [4] is satisfied, i.e. in a stress cycle the work produced by change of stress from σ^0_{ij} to σ_{ij} is non-negative, i.e.

$$\oint_\sigma (\sigma_{ij} - \sigma^0_{ij}) d\epsilon_{ij} \geq 0 \tag{2.4}$$

where σ^0_{ij} is the original stress and σ_{ij} is the current stress at any point inside or on the yield surface. For the cycle with increment $d\epsilon_{ij}$ directed out of the yield surface inequality (2.4) leads to the following inequalities:

$$D_{ijk\ell}(\epsilon_{k\ell} - \epsilon^0_{k\ell})d\epsilon^p_{ij} \geq 0 \tag{2.5}$$

$$D_{ijk\ell}(d\epsilon_{k\ell} - d\epsilon^p_{k\ell})d\epsilon^p_{ij} \geq 0 \tag{2.6}$$

where $d\epsilon^p_{ij}$ is the corresponding plastic strain increment.

Let us consider the product $D_{ijk\ell}d\epsilon^p_{ij}$ in (2.6). It can be seen that

$$D_{ijk\ell}\, d\epsilon_{k\ell}\, d\epsilon^p_{ij} \geq D_{ijk\ell}\, d\epsilon^p_{k\ell}\, d\epsilon^p_{ij} \geq 0 \tag{2.7}$$

where the last inequality follows from the fact that $D_{ijk\ell}$ is symmetric and positive definite producing a positive definite quadratic form for the middle term. As a result of (2.7) and (2.5) it follows that the yield surface (2.3) is not concave and thus the tensor $D_{ijk\ell}\, d\epsilon^p_{k\ell}$ is directed along the outward normal to the yield surface at every smooth point. At such a point the law of plastic flow may be written as

$$D_{ijk\ell}\, d\epsilon^p_{k\ell} = cg\, n_{ij}(n_{pq}\, d\epsilon_{pq}) \tag{2.8}$$

where the function $g \equiv g(\epsilon_{ij},\epsilon^p_{ij},K_n)$,

$$n_{ij} = \frac{\partial\phi}{\partial\epsilon_{ij}}\left(\frac{\partial\phi}{\partial\epsilon_{k\ell}}\cdot\frac{\partial\phi}{\partial\epsilon_{k\ell}}\right)^{-1/2} \tag{2.9}$$

is a tensor corresponding to the unit normal to the yield surface in strain space and the parameter c is

$$c = \begin{cases} 1 \text{ if } \phi = 0\,,\ n_{pq}\, d\epsilon_{pq} \geq 0 \\ \\ 0 \text{ if } \phi < 0 \text{ or } \phi = 0 \text{ but } n_{pq}\, d\epsilon_{pq} < 0\,. \end{cases} \tag{2.10}$$

Because of (2.6) it is necessary in the case of the Drucker postulate for the strain hardening that g satisfies

$$0 < g < \inf\left(D_{ijk\ell}\, n_{ij}\, n_{k\ell} \,\middle|\, n_{ij}\, n_{ij} = 1\right) \tag{2.11}$$

where the inf in fact produces the smallest eigenvalue of matrix $D_{ijk\ell}$.

As a result of (2.2) and (2.8) the relation between $d\epsilon_{ij}$ and $d\sigma_{ij}$ may be written as

$$d\sigma_{ij} = D_{ijk\ell}\, d\epsilon_{k\ell} - cgn_{ij}(n_{pq}\, d\epsilon_{pq})\,. \tag{2.12}$$

Let us add to assumptions (i) - (iii) the assumptions (iv) that we have isotropic linear elasticity, i.e.

$$D_{ijk\ell} = \frac{E}{1+\nu}\left(\delta_{ik}\delta_{j\ell} + \frac{\nu}{1-2\nu}\delta_{ij}\delta_{k\ell}\right) \tag{2.13}$$

holds, where E is Young's modulus, ν is Poisson's ratio and δ_{ij} is Kronecker delta and (v) that the first invariants of the stress and strain tensors are related linearily, i.e. the loading of the hydrostatic pressure (volume deformation) is reversible:

$$\sigma_{ii} = \frac{E}{1-2\nu}\epsilon_{ii} \,.$$

Therefore the arguments of the function ϕ in (2.3) will be deviators and equation (2.12) may be written as

$$ds_{ij} = 2G\left(de_{ij} - cq\, n_{ij}(n_{pq}\, de_{pq})\right) \tag{2.14}$$

where s_{ij}, e_{ij} are deviators of the stress and strain tensors respectively, $G \equiv E/(2(1+\nu))$ is shear modulus and from (2.11) the function $q \equiv g/2G$ satisfies

$$0 < q < 1 \,. \tag{2.15}$$

For every case where the function ϕ in (2.3) is known the values of q and n_{ij} may be computed.

2.2 Some simple examples: isotropic and kinematic hardening models in strain space

Let us consider the case where the yield surface in strain space is a sphere which may undergo translation and distortion. Let the equation of this surface be

$$\phi \equiv (e_{ij}-c(\chi^P)e^P_{ij})(e_{ij}-c(\chi^P)e^P_{ij}) - k^2(\chi^P) = 0 \tag{2.16}$$

where

$$\chi^P = \int (2de^P_{ij}\, de^P_{ij})^{1/2}$$

is the Odkvist parameter and the functions c and k are determined by the material properties [5].

If we suppose that $c(\chi^P) \equiv 1$, $k(\chi^P) = a_0\chi^P+b_0$ then we have that equation (2.14) will be the same as that which arises from the *isotropic* hardening model. The constants

a_0 and b_0 are determined from pure shear experiments as $a_0 = G'/(\sqrt{2}(G-G'))$ and $b_0 = \gamma_0/\sqrt{2}$, where G' is the tangent shear modulus and γ_0 is the elastic shear strain limit. Here and below for simplicity we consider a linear hardening material (G' = const). The parameters q and n_{ij} in (2.14) for this case are

$$q = 1 - G'/G\,,\quad n_{ij} = (e_{ij} - e^p_{ij})/((e_{k\ell} - e^p_{k\ell})(e_{k\ell} - e^p_{k\ell}))^{1/2}\,. \tag{2.17}$$

If we suppose in equation (2.16) that $c(\chi^p) = c_0 > 1$ and $k(\chi^p) = b_0$, we find that equation (2.14) now becomes that of the *kinematic* hardening model. In this case we have

$$q = 1/c_0 = 1 - G'/G\,,\quad n_{ij} = (e_{ij} - e^p_{ij})/b_0\,. \tag{2.18}$$

During "proportional" deformation where $e_{ij} = \lambda e^0_{ij}$ with e^0_{ij} any constant and λ an increasing parameter, equation (2.14) may be integrated and we get the equation of *deformation theory*

$$s_{ij} = 2G_s(\Gamma)e_{ij} \tag{2.19}$$

where $G_s(\Gamma) \equiv T/\Gamma$, $T \equiv (1/2\ s_{ij}s_{ij})^{1/2}$, $\Gamma \equiv (2\ e_{ij}e_{ij})^{1/2}$. For the case considered here of linear hardening we have

$$G_s(\Gamma) = \begin{cases} G \text{ if } \Gamma \leq \gamma_0 \\ \\ G + (G - G')\gamma_0/\Gamma \text{ if } \Gamma > \gamma_0\,. \end{cases} \tag{2.20}$$

Despite the limitation that it is in fact based on nonlinear elasticity, deformation theory is often used in modelling plasticity because of its simplicity. The possibility of using equation (2.19) for non-proportional deformation and the relation of this theory to other models of plasticity is discussed in [6].

2.3 A yield surface with singular point

It may be seen from inequality (2.5) that for a singular (conical) point of the yield surface the tensor $D_{ijk\ell}\, de^p_{k\ell}$ may only lie inside or on the boundary of the cone of the normals. Equation (2.8) is in this case not applicable. The most suitable way to satisfy inequality (2.5) in this case is based on the Koiter principle of superposition [7]. In strain space this principle may be formulated in the following way. The yield surface at every instant is an envelope of a finite or infinite number of *regular* yield surfaces ϕ_α (α = 1,2,...,N). Each of these surfaces has all the properties of a yield surface and for each the elementary increment of plastic strain is given by (2.8) for ϕ_α. Thus the total plastic strain increment is the average of the sum of elementary increments so that

$$de^p_{ij} \equiv \frac{1}{N}\sum_{\alpha=1}^{N} C_\alpha q_\alpha n^\alpha_{ij}(n^\alpha_{pq}\, de_{pq}) \, , \tag{2.21}$$

where n^α_{ij} is a tensor corresponding to the normal to the yield surface ϕ_α and the function q_α, which is determined by the condition that ϕ_α changes continuously, satisfies the inequality $0 < q_\alpha < 1$.

It can be shown that, with de^p_{ij} defined in this way, inequality (2.6) which may be written as

$$(de_{ij} - de^p_{ij})de^p_{ij} \geq 0 \tag{2.22}$$

is also satisfied.

Finally, the constitutive equation at a conical point of the yield surface may be written as

$$ds_{ij} = 2G\Big(de_{ij} - \frac{1}{N}\sum_{\alpha=1}^{N} C_\alpha q_\alpha n^\alpha_{ij}(n^\alpha_{pq}\, de_{pq})\Big) \, . \tag{2.23}$$

In the case of an infinite number of yield surfaces as $N \to \infty$ equation (2.23), omitting the tensor indices, may be written asymptotically in the form

$$ds = 2G\Big(de - k\int_{\Omega_1} qn(nde)d\Omega\Big) \tag{2.24}$$

where k is a constant and the integral is taken over all angles of orientation of the normals $\mathbf{n}$ for which the current value $\mathbf{e}$ is located on the corresponding yield surface and the condition of "active" deformation: $\delta h \equiv \mathbf{n} \cdot de \geq 0$, takes place, where h is the deviation of the loading plane.

We next consider another important case where the regular surfaces are planes. This was considered by Sanders [8] in stress space and it was shown that the constitutive equation in increments may be integrated along a stress path to produce stress-strain relations in a form which partially resembles deformation theory. Using an analogous technique in strain space with q = q(h) one can write

$$\mathbf{s} = 2G\left(\mathbf{e} - k \int_{\Omega_2} Q(\dot{h})\mathbf{n}\, d\Omega\right) \tag{2.25}$$

where Q = Q(h) is a function obtained by integration.

In the next section we show how equation (2.25) may be obtained from the Rabotnov model as in [9]. At the same time this approach indicates how the various quantities in (2.25) are produced.

3. Constitutive equation based on the Rabotnov model of plasticity

3.1 Vector space of deviators

The method described is located in five-dimensional Euclidean space $\mathbb{R}^5$ with orthonormal base vectors $\mathbf{e}_k$, k = 1,2,...,5, see [10]. The strain deviator e_{ij} and the stress deviator s_{ij} correspond to the *strain vector* $\xi = \xi_k \mathbf{e}_k$ and *stress vector* $\beta = \beta_k \mathbf{e}_k$, which are determined by their projections

$$\xi_1 = e_{11}\,, \quad \xi_2 = (e_{22} - e_{33})/\sqrt{3}\,, \quad \xi_3 = 2e_{12}/\sqrt{3}\,,$$

$$\xi_4 = 2e_{23}/\sqrt{3}\,, \quad \xi_5 = 2e_{31}/\sqrt{3}$$

and (3.1)

$$\beta_1 = 3s_{11}/2\,, \quad \beta_2 = \sqrt{3}\,(s_{22} - s_{33})/2\,, \quad \beta_3 = \sqrt{3}\, s_{12}\,,$$

$$\beta_4 = \sqrt{3}\, s_{23}\,, \quad \beta_5 = \sqrt{3}\, s_{31}\,.$$

The absolute values of these vectors correspond to the second invariants of the deviators

$$|\xi| = (\xi_k \xi_k)^{1/2} = (2/3\; e_{ij} e_{ij})^{1/2}\,, \quad |\beta| = (s_k s_k)^{1/2} = (3/2\; s_{ij} s_{ij})^{1/2}\,. \tag{3.2}$$

The reason for introducing $\mathbf{R}^5$ is that only five of the six components of these deviators are independent. Furthermore, as was shown in [10], every tensor relation between any two second order deviators which includes only three invariant operations such as addition, scalar multiplication and inner product, may be represented in $\mathbf{R}^5$, where the tensors are replaced by the corresponding vectors and the inner products by scalar products.

The process of change of the strain deviator is represented in $\mathbf{R}^5$ as a *strain trajectory* . The radius-vector of the point of the strain trajectory is the strain vector $\boldsymbol{\xi}$. The constitutive equation of plasticity has to define a rule by means of which one can, for each point of the strain trajectory, find the corresponding stress vector $\boldsymbol{\beta}$. The hodograph of the vector $\boldsymbol{\beta}$ represents the *stress trajectory*.

The relation between two vectors remains under every orthogonal transformation of coordinates. Thus, if for a strain trajectory the corresponding stress trajectory is found by the constitutive equation, then for any subsequent strain trajectory which may be produced from the first one by a rotation and a mirror type transformation the corresponding stress trajectory may be re-established using the same rotation and transformation. The reverse of this is also true and has been called the *postulate of isotropy* [10]. If every trajectory in the vector space $\mathbf{R}^5$ satisfies this feature then the corresponding tensor equation can include only the three above operations of addition, scalar multiplication and inner product.

3.2 Generalization of the Rabotnov model to the 5D space of deviators

Rabotnov proposed his two-dimensional model in the context of plastic deformation of a circular thin-walled tube under the action of complex pure bending [9] (see Appendix). We here generalize the model in the context of a material which is initially isotropic, plastic incompressible and with constant elastic modulus.

In the initial (unstressed) state the yield surface in strain space is the sphere with radius ξ_0: $x_k x_k = \xi_0^2$ where ξ_0 is the elastic strain limit. This sphere is an envelope of the *"planes of plasticity"* which play the role of the above regular yield surfaces. The equation of the plane in the initial state is $x_k n_k = \xi_0$, where $\mathbf{n} = n_k e_k$ is the unit vector of the normal to the plane.

Each plane in the envelope may move only parallel to its original position. The distance between the planes of plasticity, whose normal vectors are in opposing directions (we shall call these a *pair of planes*) is constant and equal to $2\xi_0$. The increase in the strain vector ξ causes the planes of plasticity connected with it to move in the normal direction away from the origin of coordinates. The movement of each of the planes causes an equal displacement of the second of the pair. The elementary plastic strain ξ_n^p, which is caused by displacement of each plane of plasticity, is directed along its normal $\mathbf{n}$ and depends on the value of the deviation $h - \xi_0$ of this plane from its initial position. The value of ξ_n^p is

$$\xi_n^p = q(h - \xi_0)\mathbf{n} \tag{3.3}$$

where $q = q(h-\xi_0)$ is some odd function, which is now determined.

The complete plastic strain is equal to the sum of all the elementary plastic strains arising from the planes of plasticity, i.e.

$$\xi^p = k\int_\Omega q(h-\xi_0)\mathbf{n}\, d\Omega , \tag{3.4}$$

where $d\Omega$ is an element of solid angle in the space $\mathbf{R}^5$, Ω is the complete solid angle of the unit sphere with centre at the origin of coordinates and k is a constant ($k = 15/8\pi^{-2}$). Hence the stress-strain relation is

$$\beta = E(\xi - k\int_\Omega q(h-\xi_0)\mathbf{n}\, d\Omega) \tag{3.5}$$

where E is Young's modulus.

It is convenient for calculating the integrals in (3.5) to use unified polar angles ψ_i ($i = 1,\ldots,4$). The components of the unit normal vector $\mathbf{n}$ of the plane of plasticity can be written as

$$\begin{aligned} n_1 = \cos\psi_1 , \quad n_2 = \sin\psi_1\cos\psi_2 , \quad n_3 &= \sin\psi_1\sin\psi_2\cos\psi_3 , \\ n_4 &= \sin\psi_1\sin\psi_2\sin\psi_3\cos\psi_4 \\ n_5 &= \sin\psi_1\sin\psi_2\sin\psi_3\sin\psi_4 \end{aligned} \tag{3.6}$$

where

$$0 \le \psi_i \le \pi , \quad i = 1,2,3 , \quad 0 \le \psi_4 \le 2\pi .$$

Taking into account that $d\Omega = \sin^3\psi_1 \sin^2\psi_2 \sin\psi_3 d\phi_1 d\psi_2 d\psi_3 d\psi_4$ we can write equation (3.5) as

$$\boldsymbol{\beta} = E\Big(\boldsymbol{\xi} - 2k \int_0^\pi\!\!\int_0^\pi\!\!\int_0^\pi\!\!\int_0^\pi q(h-\xi_0)\mathbf{n} \sin^3\psi_1 \sin^2\psi_2 \sin\psi_3 d\psi_1 d\psi_2 d\psi_3 d\psi_4\Big) . \tag{3.7}$$

Here we may see that the stress vector $\boldsymbol{\beta}$ is a functional of the strain trajectory because the value h of the deviation of the plane depends on the strain trajectory. The rule for determining the value of h is the same as that of the Rabotnov model (see Appendix), but the roles of $\boldsymbol{\varkappa}$, $\mathbf{r}$, e_0 are played respectively by $\boldsymbol{\xi}$, $\mathbf{n}$, ξ_0 and the scalar product is considered in $\mathbf{R}^5$.

3.3 Definition of material constants and numerical procedure

In the simplest case we may consider the function $q = q(h-\xi_0)$ in equation (3.7) as a linear function and take it as a product of a constant g_0 and $(h-\xi_0)$ so that

$$q(h-\xi_0) = g_0 \cdot (h-\xi_0) . \tag{3.8}$$

To define g_0 let us consider a deformation along the axis $0\xi_1$. From the above we can write

$$h = \begin{cases} \xi_0 , & \text{if } -\xi_0/\xi_1 \le \cos\psi_1 \le \xi_0/\xi_1 \\ \xi_1\cos\psi_1 , & \text{if } \xi_0/\xi_1 \le \cos\psi_1 \le 1 \\ -\xi_1\cos\psi_1 + 2\xi_0 , & \text{if } -1 \le \cos\psi_1 < -\xi_0/\xi_1 . \end{cases} \tag{3.9}$$

Substitution of (3.9) into (3.7) gives, after integration,

$$\begin{aligned} \beta_1 &= E\xi_0\Big((1-g_0)\frac{\xi_1}{\xi_0} + \frac{15}{8}g_0 - \frac{5}{4}g_0\Big(\frac{\xi_0}{\xi_1}\Big)^2 + \frac{3}{8}g_0\Big(\frac{\xi_0}{\xi_1}\Big)^4\Big) \\ \beta_2 &= \beta_3 = \beta_4 = \beta_5 = 0 . \end{aligned} \tag{3.10}$$

The same equation can be obtained for uniaxial deformation along each of the axes $0\xi_i$, $i = 2,3,4,5$, because equation (3.7) satisfies the postulate of isotropy.

From (3.10) since $\xi_1 \gg \xi_0$ we see that asymptotically

$$\beta_1 = E(1-g_0)\xi_1 + \frac{15}{8}g_0\, E\xi_0 . \tag{3.11}$$

Thus g_0 and ξ_0 can be found from the uniaxial stress-strain diagram.

Equation (3.10) gives a rather satisfactory description of the uniaxial diagram of a natural strain hardening material, when $g_0 = 1$ this diagram has a horizontal asymptote, whilst when $0 < g_0 < 1$ it has an inclined asymptote. Equation (3.7) makes possible the description of a material with any uniaxial smooth diagram; for example, with a yield plateau and further strain hardening. We may choose the function $q = q(h-\xi_0)$ in such a form that the relation between β_1 and ξ_1 from (3.7) is the same as that found experimentally and represented as $\beta_1 = \beta_1(\xi_1)$.

Using (3.9) and (3.7) in the case of uniaxial deformation we have

$$\beta_1 = E(\xi_1 - \frac{15}{2}\int_0^{\omega} q(\xi_1\cos\psi_1 - \xi_0)\sin^3\psi_1\cos\psi_1 d\psi_1) \tag{3.12}$$

where $\omega = \arccos(\xi_0/\xi_1)$.

Let the value β_1 from (3.12) equal the experimental value $\beta_1(\xi_1)$. By using the following notation

$$\begin{gathered} t = \cos\psi_1\,, \quad x = \xi_1/\xi_0\,, \quad h_1 = h/\xi_0 \\ q_1(h_1 - 1) = q(h-\xi_0)/\xi_0\,, \quad f(x) = x - \beta_1(\xi_1)/(E\xi_0) \end{gathered} \tag{3.13}$$

we have

$$\int_{1/x}^{1} q_1(xt-1)(1-t^2)t\,dt = \frac{2}{15}f(x)\,. \tag{3.14}$$

The introduction of the new variable $z = xt$ enables equation (3.14) to be written as

$$\int_1^x q_1(z-1)(x^2-z^2)z\,dz = \frac{2}{15}x^4 f(x)\,. \tag{3.15}$$

After differentiation of (3.15) with respect to x and multiplication of both parts of the resulting equation by $x/2$ we have

$$\int_1^x q_1(z-1)x^2 z\,dz = \frac{1}{15}x\frac{d}{dx}(x^4 f(x))\,. \tag{3.16}$$

If we subtract (3.15) from (3.16) we have

$$\int_1^x q_1(z-1)z^3dz = \frac{1}{15}\left(x\frac{d}{dx}(x^4f(x)) - 2x^4f(x)\right). \tag{3.17}$$

Differentiation of (3.17) with respect to x gives

$$q_1(x-1) = \frac{1}{15x^3}\left\{x\frac{d^2}{dx^2}\left(x^4f(x)\right) - \frac{d}{dx}\left(x^4f(x)\right)\right\} \tag{3.18}$$

or, taking into account that $f(x) = x - \beta^*(x)$ where $\beta^*(x) = \beta_1(\xi_1)/(E\xi_0)$, we obtain

$$q_1(x-1) = \frac{1}{15}\left(15x - 8\beta^*(x) - 7x\frac{d\beta^*}{dx} - x^2\frac{d^2\beta^*}{dx^2}\right). \tag{3.19}$$

From (3.19) it is possible to find the unknown function q_1 and hence $q = q(h-\xi_0)$ from (3.13). From now on we use only the simplest form of q as in (3.8). We note that in this case the coefficients of more traditional models of plasticity may be related to those of the Rabotnov model as follows

$$G = E/3\ ,\ \ G' = G(1-g_0)\ ,\ \ \gamma_0 = 15\sqrt{3}\,\xi_0/8\ . \tag{3.20}$$

In order to use formula (3.7) we have to calculate the multiple integral over the sphere Ω. Let $\Omega^+ = \{x_1^2 + ... + x_5^2 = 1, x_5 \geq 0\}$ be the top half of the unit sphere in the space $\mathbb{R}^5$:

$$I = \int_{\Omega^+} F(x)d\Omega\ . \tag{3.21}$$

(Note that the bottom half Ω^- is associated with the second plane of the pair of planes of plasticity.) Designating the new variables $t_k = \cos\psi_k$, $k = 1,...,4$, where the ψ_k are unified polar angles, integral (3.21) may be written as

$$I = \int_{-1}^1\int_{-1}^1\int_{-1}^1\int_{-1}^1 F(t_1,t_2,t_3,t_4)(1-t_1^2)(1-t_2^2)^{1/2}(1-t_4^2)^{-1/2}dt_1dt_2dt_3dt_4\ . \tag{3.22}$$

The calculation of this integral may be represented as four-times the calculation of an integral of the type

$$\int_{-1}^1 \psi(t)(1-t^2)^{m/2}\ dt\ ,\ \ m = -1,0,1,2\ . \tag{3.23}$$

For evaluating these one may use the well-known Gaussian quadrature rule with weight function $\psi_m = (1-t^2)^{m/2}$ so that

$$\int_{-1}^1 \psi(t)(1-t^2)^{m/2}dt \doteq \sum_{j=1}^{\ell} A_j^{(m)}\psi(t_j^{(m)})\ . \tag{3.24}$$

The values of the coefficients $A_j^{(m)}$ and nodes $t_j^{(m)}$ are given in tables (see e.g. [11]). The calculation procedure for finding the value of the stress vector β on a given strain trajectory is the following:

Using the values $t_j^{(m)}$ in formulae (3.24) we find the values of the unified polar angles ψ_k and using these in formulae (3.6) we determine the components of the vectors of the normals $\mathbf{n}$ of the planes of plasticity, the contribution of which we take into account in the formulae of integration when calculating the integral in (3.4). For each of these planes we determine the values of the deviation h from the origin of coordinates at the relevant moment of deformation. Then, by means of formulae of integration (3.24) we calculate the values of the stress vector $\beta(\tau)$ at the corresponding points of the strain trajectory $\xi = \xi(\tau)$.

4. One-dimensional experiments - Baushinger effect

We now show how results of 1D experiments can be reproduced using the above models. These experiments consist of loading, unloading and further loading during pure shear, so that the strain trajectory has three segment branches which are located along the axis $0\xi_3$. The parametric equation of this trajectory is

$$\xi_3 = \begin{cases} s \ \text{ if } \ 0 \le s \le s_1 \\ 2s_1 - s \ \text{ if } \ s_1 < s \le s_1 + s_2 \\ -2s_2 + 3 \ \text{ if } \ s_1 + s_2 < s_1 \le s_1 + s_2 + s_3 \end{cases} \tag{4.1}$$

$$\xi_k = 0\,, \quad k = 1,2,4,5$$

where s is the current arclength of strain trajectory, and s_1, s_2, s_3 are the lengths of the first, second and third branches respectively.

For shear strain γ and shear stress τ by (3.1) we have

$$\gamma = 2e_{12} = \sqrt{3}\,\xi_3\,, \quad \tau = s_{12} = \beta_3/\sqrt{3}$$

and the values of shear strain at the moments when the direction of loading is changing are

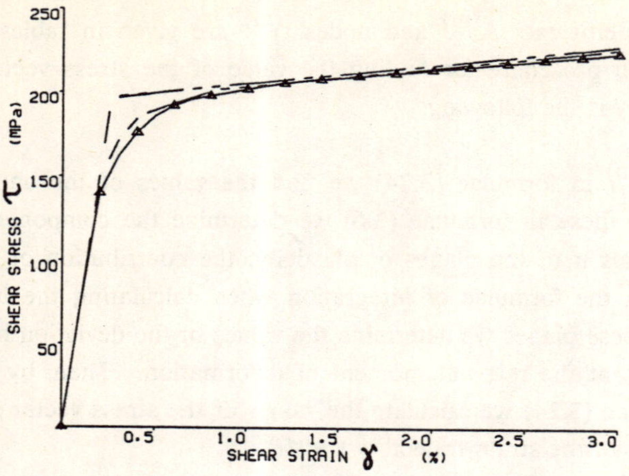

Fig 1 Stress-strain diagram of pure shear

Fig 2 (a) - (c) Stress strain loops

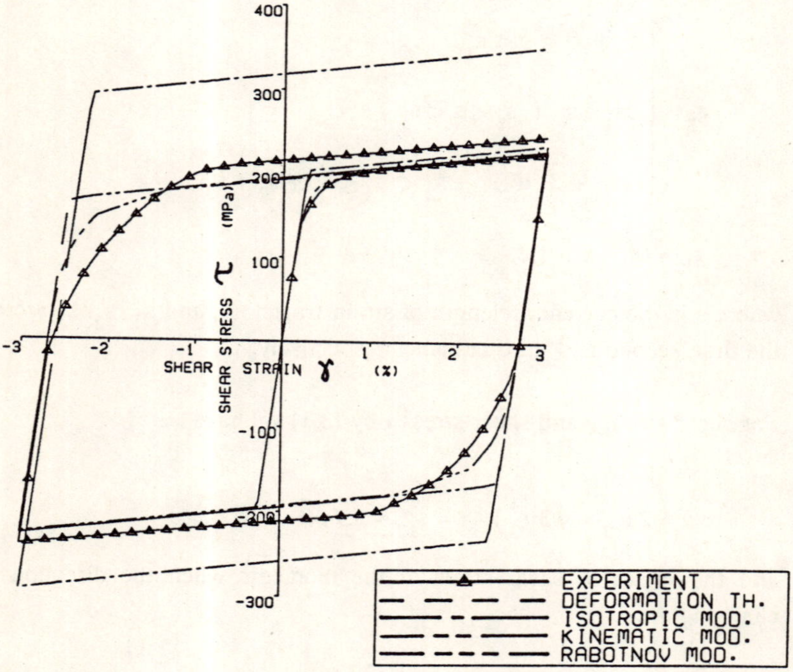

Fig 2(a)

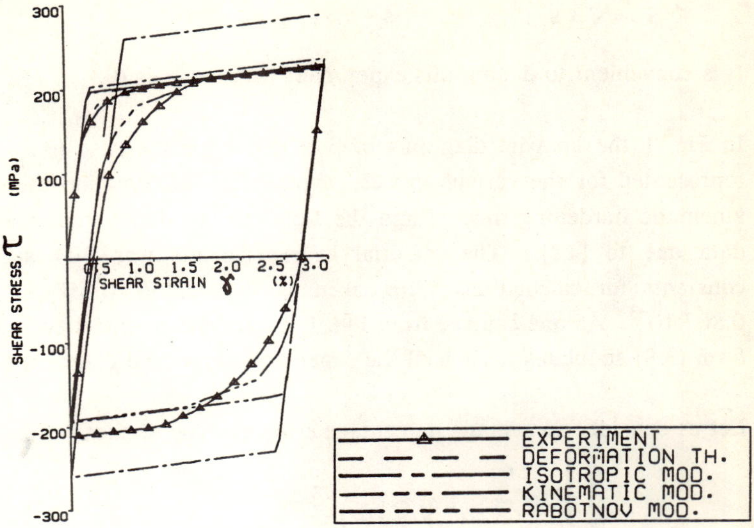
SHEAR STRESS τ (MPa)
SHEAR STRAIN γ (%)
EXPERIMENT
DEFORMATION TH.
ISOTROPIC MOD.
KINEMATIC MOD.
RABOTNOV MOD.

Fig 2(b)

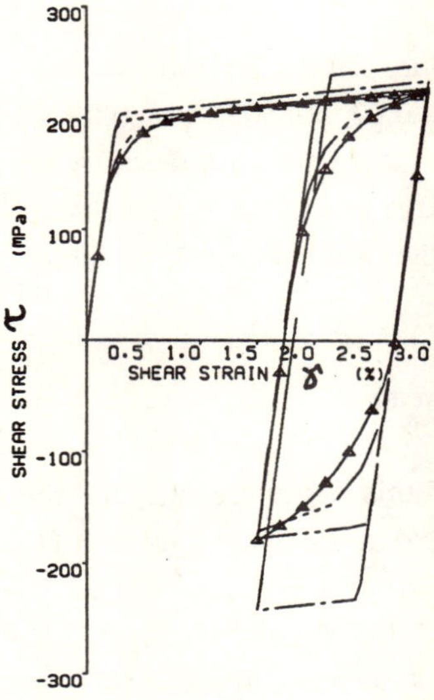
SHEAR STRESS τ (MPa)
SHEAR STRAIN γ (%)

Fig 2(c)

$$\gamma_1 = \sqrt{3}\, s_1\,, \quad \gamma_2 = \sqrt{3}\,(s_1 - s_2)\,, \quad \gamma_3 = \sqrt{3}\,(s_1 - s_2 + s_3)\,. \tag{4.2}$$

It is convenient to denote this experiment with symbols $(0,\gamma_1,\gamma_2,\gamma_3)$.

In Fig. 1 the uniaxial diagrams of pure shear for the case $(0,\gamma_1) = (0.0,0.03)$ are represented for the various models of plasticity (deformation theory, isotropic and kinematic hardening models and the Rabotnov model) together with experimental data due to [12]. The material in experiments was 304 stainless steel, the constants for calculations were taken as $E = 2.256.10^5$MPa, $g_0 = 0.895$, $\xi_0 = 0.807.10^{-3}$. As one can see from Fig. 1 the Rabotnov model even with the simplest form (3.8) simulates fairly well the experimental uniaxial diagram.

Let us consider an average percentage error of value of shear stress

$$\epsilon_1 = \left(\frac{1}{\gamma}\int_0^{\gamma_1}\left(\left(\tau^{ex} - \tau^{th}\right)/\tau^{ex}\right)^2 d\gamma\right)^{1/2} \tag{4.3}$$

where τ^{ex} and τ^{th} are respectively the shear stress as a function of shear strain γ obtained both experimentally and theoretically.

For bilinear approximation as used in traditional models the value $\epsilon_1 = 5.4\%$ and for the Rabotnov model this value is 1.6%.

In Fig. 2(a) - 2(c) analogous results are shown for the following case of uniaxial deformation: (0.0,0.03,-0.03,0.03), (0.0,0.03,0.0,0.03), (0.0,0.03,0.015,0.03) respectively. It can be seen that deformation theory does not describe the hysteresis loops, the isotropic hardening model gives large errors but the kinematic hardening model and the Rabotnov model describe the ideal Baushinger effect with linear and nonlinear hardening respectively.

5. Further work and acknowledgements

It is acknowledged that the results of this paper are only for the one-dimensional case. Further results for two and three dimensions are given in [13].

Part of this work was done whilst L.G. Popov visited BICOM for the period February - November 1990 under the auspices of the British Council. The support for this visit is gratefully acknowledged.

References

1. Yoder, P.J., Iwan, W.D., On the formulation of strain-space plasticity with multiple loading surfaces. ASME, J. of Appl. Mechanics **48** (1981) 773-778.
2. Naghdi, P.M., Trapp, J.A., On the nature of normality of plastic strain rate and convexity of yield surfaces in plasticity. ASME, J. of Appl. Mechanics **42** (1975) 61-66.
3. Gasey, J., Naghdi, P.M., On the nonequivalence of the stress space and strain space formulation of plasticity theory. ASME, J. of Appl. Mechanics **50** (1983) 350-354.
4. Drucker, D.C., Some implications of work hardening and ideal plasticity. Q. Appl. Math. **7** (1950) 411-418.
5. Klushnikov, V.D., Mathematical Theory of Plasticity. Moscow University Press 1979 (in Russian).
6. Klushnikov, V.D., Some new concepts in plasticity and deformation theory. Applied Mathematics and Mechanics **23** (1959) 722-731 (in Russian).
7. Koiter, W.T., Stress-strain relations, uniqueness and variational theorems for elastic plastic material with singular yield surface. Q. Appl. Math. **11** (1953) 350-354.
8. Sanders, J.L., Plastic stress-strain relations based on linear loading functions. p.455-460, Proc. Second U.S. Congress of Applied Mechanics 1954.
9. Rabotnov, Yu.N., Model illustrating some properties of hardening plastic solids. Applied Mathematics and Mechanics **23** (1959) 164-168 (in Russian).
10. Ilyushin, A.A., Plasticity, Foundations of General Plastic Theory. Izdat. Acad. Nauk USSR, Moscow (1963) (in Russian).
11. Stroud,A.H., Secrest, D., Gaussian Quadrature Formulas. Prentice Hall, Englewood Cliffs, N.J. (1966).
12. Ohno, N., Kachi, Y., A constitutive model of cyclic plasticity for nonlinear hardening materials. ASME, J. of Appl. Mechanics **53** (1986) 395-403.
13. Popov, L.G., Constitutive models of plasticity in strain space. Comparison of theoretical predictions with 2D and 3D experiments of complex loading. Technical Report BICOM 90/7, Institute of Computational Mathematics, Brunel University 1990.

Appendix

Two-dimensional model of plasticity by Yu.N. Rabotnov

To explain the effects of elastic-plastic deformation Rabotnov, [9], proposed a simple mechanical model - a round thin-walled tube under a complex pure bend (Fig. A1). Let us consider this model in more detail, supposing that the tube's material is elasto-plastic, with symmetric bilinear stress-strain diagram, which is the same during tension and compression (Fig. A2). According to the kinematic assumption of bend we have

$$e_{33} = (\mathbf{x}.\boldsymbol{\tau}) \tag{A1}$$

where e_{33} is the axial deformation of a "fibre" running along the tube wall, $\boldsymbol{\varkappa} = \varkappa_1 \mathbf{i}_1 + \varkappa_2 \mathbf{i}_2$ is the vector of curvature, $\varkappa_1$, $\varkappa_2$ are curvatures in the planes $0x_1x_2$ and $0x_2x_3$, $\mathbf{i}_1$,$\mathbf{i}_2$ are orthonormal base vectors in the plane $0x_1x_2$, which is perpendicular to the axis of symmetry $0x_3$ of the tube.

For the bending moments, taking into account the thinness of the tube wall we can write

$$\mathbf{M} = \mathrm{H} \oint \sigma_{33}\, \mathbf{r}\, ds$$

where $\mathbf{M} = \mathrm{M}_1 \mathbf{i}_1 + \mathrm{M}_2 \mathbf{i}_2$ is the vector of moments, M_1, M_2 are bending moments in the planes $0x_1x_2$ and $0x_2x_3$, $\mathbf{r} = \mathrm{R}(\cos\eta \mathbf{i}_1 + \sin\eta \mathbf{i}_2)$ is the vector of location of a "fibre", σ_{33} is the axial stress of the fibre, H is the thickness of the wall, s is the length of arc, measured in positive direction from axis $0x_1$: $s = \left|\mathbf{r}\right| . \eta$, where η is the angle of fibre in polar coordinates. From now on it is assumed that the radius R of the tube is equal to 1.

Let us consider the behaviour of a fibre characterized by angle η. During the loading from unstressed state the first plastic deformation will appear when the absolute value of e_{33} is equal to proportionality limit e_0:

$$\left| \boldsymbol{\varkappa} . \mathbf{r} \right| = e_0 \, . \tag{A3}$$

On the plane $0x_1x_2$ it will determine a pair of parallel lines. We shall call these the "lines of plasticity". When the vector $\boldsymbol{\varkappa}$ touches one of these lines for the first time, the plastic deformation of this fibre will begin. If further deformation takes the vector $\boldsymbol{\varkappa}$ out of this strip, the plastic deformation of this fibre will continue, but the

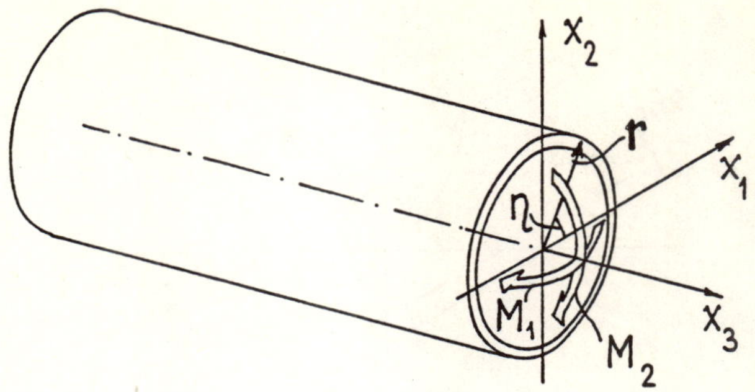

Fig A1 Rabotnov's mechanical model - a round thin walled tube under a complex pure bend

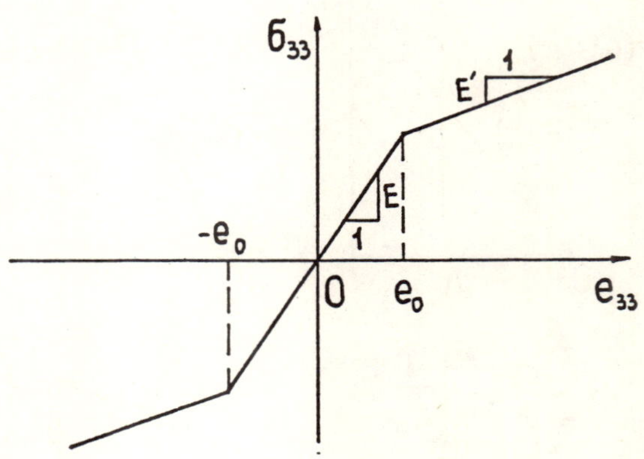

Fig A2 Stress-strain diagram of tube's material

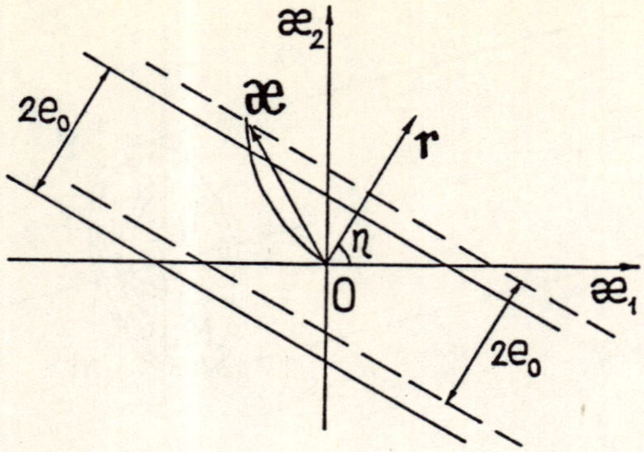

Fig A3 The initial and current locations of the pair of "lines of plasticity" with vector of hormal r

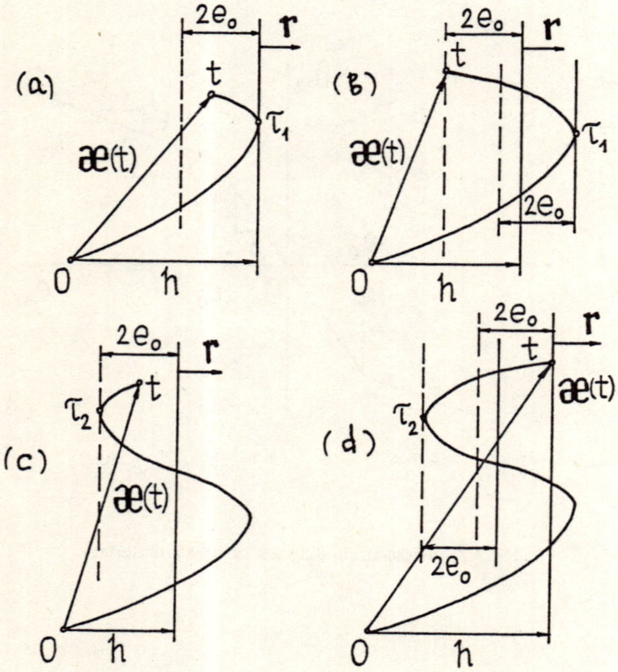

Fig A4 The rule to determine value h

strip will move after the vector $\varkappa$ as a unit and the later positions of the strip will always be parallel to its original position, see Fig. A3. If the vector $\varkappa$ finds itself inside the strip, then its change will not cause plastic deformation of this fibre. The same situation (only with opposite sign) takes place for the diametrically opposite fibre, and therefore the stresses σ_{33} in these two fibres contribute equally to the vector of moments **M**.

For the value of stress σ_{33} in the fibre we have

$$\sigma_{33} = E(e_{33} - (1-E'/E)(h-e_0)) \tag{A4}$$

where E is the modulus of elasticity, E′ is the tangent modulus in the stress-strain diagram, $h-e_0$ is the deviation of the straight line with equation $\varkappa \cdot \mathbf{r} = h$ from its position in the unstressed state. The value of h depends on the whole history of the deformation (change of the vector $\varkappa$).

Let us consider the rule for determining the value h. Assume that the deformation of the tube is determined by the curve $\mathbf{x} = \mathbf{x}(\tau)$, where $\tau \in [0,t]$ is the time of any increase in the deformation parameter. At time t one of the following three outcomes will occur:

1. $h = e_0$, if $\max\limits_{\tau\in[0,t]} (|\varkappa(\tau).\mathbf{r}| \leq e_0)$.
2. $h = \min(\varkappa(\tau_1).\mathbf{r},\ \varkappa(t).\mathbf{z} + 2e_0)$.
3. $h = \max(\varkappa(\tau_2).\mathbf{r} + 2e_0,\ \varkappa(t).\mathbf{r})$.

This rule is shown in Fig. A4 for the case $\eta = 0$, i.e. $\mathbf{r} = \mathbf{i}_1$.

In the plane $0\varkappa_1\varkappa_2$ every fibre with angle $\eta \in [0,\pi]$ has its own strip. Their mutual part determines the field of elastic deformation. In the initial state this field is a circle with radius e_0.

As a result of (A1), (A2), (A4) and the above discussion, we can write

$$M = 2EH \int_0^{\pi} (\varkappa.\mathbf{r})\mathbf{r}\, d\eta - 2EHQ \int_0^{\pi} (h-e_0)\mathbf{r}\, d\eta$$

where $Q = (1 - E'/E)$ or

$$M = \pi EH\left(\varkappa - \frac{2Q}{\pi} \int_0^{\pi} (h-e_0)\mathbf{r}\, d\eta\right) . \tag{A5}$$

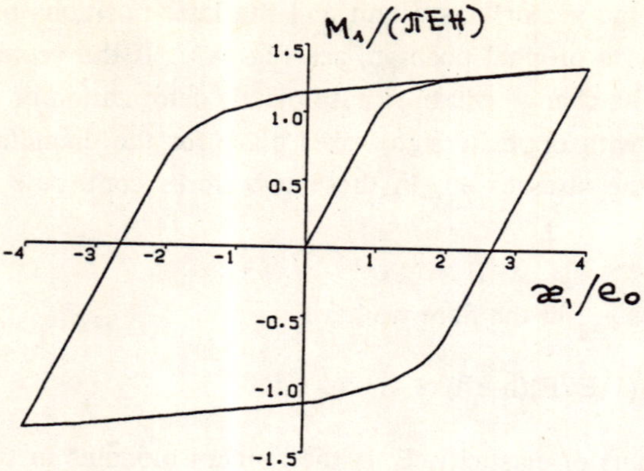

Fig A5 The dependence of value M_1 on value $\varkappa_1$ for uniaxial cyclic bend $[0,\ 4e_0,\ -4e_0,\ 4e_0]$ for the case Q = 0.97

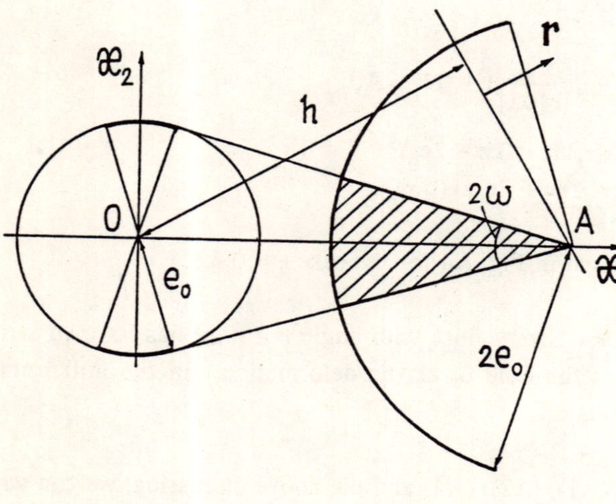

Fig A6 The location of the "lines of plasticity" after the uniaxial bend for the state A

Hence we obtain

$$\mathbf{M} = \pi EH(\boldsymbol{\varkappa} - \boldsymbol{\varkappa}^P) \tag{A6}$$

where we designate

$$\boldsymbol{\varkappa}^P = \frac{2Q}{\pi}\int_0^{\pi}(h-e_0)\mathbf{r}\, d\eta\ . \tag{A7}$$

The vector $\boldsymbol{\varkappa}^P$ may be considered as a measure of the plastic deformation of the whole tube and equation (A6) is analogous to the usual separation of the deformation into elastic and plastic parts.

As a simple example of using equation (A5) let us consider the uniaxial bend with moment $\mathbf{M} = M_1\mathbf{i}_1$, i.e. deformation along axis $0\varkappa_1$. According to the above discussion we can write

$$h = \begin{cases} e_0 & \text{if} \quad -e_0/\varkappa_1 \le \cos\eta \le e_0/\varkappa_1 \\ \varkappa_1\cos\eta & \text{if} \quad e_0/\varkappa_1 < \cos\eta < 1\ . \\ \varkappa_1\cos\eta + 2e_0 & \text{if} \quad -1 \le \cos\eta < -e_0/\varkappa_1 \end{cases} \tag{A8}$$

If we substitute h from (A8) in (A5) we shall have

$$M_1 = \pi EH(\varkappa_1 - \frac{4Q}{\pi}\int_0^{\pi/2-\omega}(\varkappa_1\cos\eta - e_0)\cos\eta\, d\eta)\ .$$

By integrating we find that

$$M_1 = \pi EH\varkappa_1\left(1 - Q + \frac{2Q}{\pi}(\omega + \frac{1}{2}\sin 2\omega)\right) \tag{A9}$$

where $\sin\omega = e_0/\varkappa_1$.

If we consider cyclical uniaxial loading it may be easily shown that equation (A5) describes the ideal Baushinger effect, as is shown in Fig. A5.

The location of the "lines of plasticity" is shown in Fig. A6, where the thick lines are envelopes of the "lines of plasticity" for the state A and the field of elastic deformation is shaded. It is important to notice that the field of elastic behaviour in the plane $0\varkappa_1\varkappa_2$ has a singularity (corner) at the point A.

Sheet Metal Forming Modelled as a Quasi-Dynamic Process

Lars Bernspång, Kjell Mattiasson, Alf Samuelsson

Chalmers University of Technology
S–412 96 Göteborg, Sweden

ABSTRACT

Metal sheet stretch forming and deep drawing are large strain elastoplastic problems. The stamping process is quasistatic, with typically 0.1s duration, so the inertia effects are negligible. By solving the problem as a dynamic one with an explicit time stepping, the solution of a large number of sets of FEM equations can be avoided. The price is that, for stability and accuracy, a large number of time steps is needed. We show by comparison with quasistatic solutions and experiments that the solution for the given problem can be obtained from the solution of a fictive problem with a duration of 0.001s – 0.01s. Solution of the fictive problem requires only a small fraction of the CPU–time required for the quasistatic solution.

1. INTRODUCTION

Sheet metal forming by stamping is certainly a very important manufacturing process particularly in the automotive industry. It happens also to be a very interesting and challenging problem in large deformation mechanics. The strains are really high, regularly 30% and more, there is friction between the sheet and the punch, the blank holder and the die, and wrinkles may appear in the sheet. The problem is an elasto-plastic quasistatic example of shell analysis. Finally, in the so-called deep-drawing process material flows into the original region of the plate under consideration, thus complicating the analysis.

The main reason from the industrial point of view for the design of an efficient computer code is the demand for reduction of the time required to design the stamping tools. Without a code based on mathematical analysis the design is a costly process with repeated steps of design, manufacturing and tryout.

2. THE QUASISTATIC APPROACH

The first comprehensive analysis of the stamping problem was published by Wang and Budiansky [1] in 1978. Inspired by this large-strain elastoplastic approach a code has been developed at our department. It is based on Hutchinson's generalization of the J2-theory of elastoplasticity to large strain conditions. Anisotropy in the perpendicular direction to the plane of the plate was taken into account by Hill's approach and the Jaumann type of objective stress rate was used.

In the finite element model only membrane stresses were taken into account in such a way that the shell was modelled as a facet surface by plane constant strain triangles. Since the sheet is plane before deformation a total Lagrange formulation is suitable and was thus chosen for the evolution process.

The constraint condition in the elastoplasticity was treated by a variational inequality approach including an implicit time stepping with an iterative procedure in each step involving the repeated calculation of a fictive elastic stress and a projection in complementary energy onto the convex surface of the admissible region. The contact problem was treated by the Coulomb friction theory with slip or stick at each node.

The code has been checked against an extensive series of experiments under displacement control on stretching and deep-drawing of sheets in steel, brass, stainless steel and aluminium. By cylindrical stamps, cups were produced according to Figs. 1. and 3. The experiments have been performed at the Swedish Institute for Metals Research.

Comparison between calculated and experimentally obtained measures as strains, punch force and diameter reduction at deep drawing showed reasonable agreement, see [2], [3], [4]. Similar codes have been used in a routine way for industrial applications. Still there are drawbacks:

* The computing time is large (often too large) for many car body details. This is mainly due to the large number of equilibrium iterations and solutions of sets of equations in each time step.

* Numerical instability is difficult to avoid completely giving problems for the user.

A way to avoid the first of these problems is to use an Euler, viscosity dominated, flow formulation as advised for general industrial forming problems by Zienkiewicz and coworkers [5] by which the computing time can be considerably reduced. However, for sheet forming this approach that assumes pure plasticity without an elastic phase, is not generally applicable. Obviously, springback can not be modelled. For sheet forming in superplastic materials it is, however, very efficient [6].

Another way to overcome the drawbacks with the quasistatic elastoplastic approach is to use a quasidynamic elastoplastic formulation which will be described here. It has earlier been discussed in [7].

3. THE QUASIDYNAMIC APPROACH

Stamping is a relatively slow process – about 0.1s – so the inertia forces can generally be neglected. However, inclusion of the inertia forces can simplify the solution considerably. By lumping the mass matrix and solving the separated ordinary differential equations in time by an explicit method no systems of equations have to be solved. This also implies that the storage requirement is reduced and that a vector facility speeds up the execution considerably. The price for this is that much smaller time steps than for the quasistatic implicit method have to be chosen in order to ensure numerical stability and accuracy. Also, the stability limit is proportional to the smallest element size and depends upon the regularity of the mesh. The question to be answered thus is if it is possible to get reliable solutions to stamping problems with this approach within a reasonable CPU–time.

By courtesy of John O Hallquist we have got access to his DYNA 3D–code which is a finite element program for the analysis of dynamic problems including large deformations and nonlinear material behavior. It uses the central difference integration method and chooses automatically stable time steps. Among large strain shell elements in the code we have chosen a four–node Reissner–Mindlin, bilinear one given by Belytschko and Tsay [8]. In order to reduce the number of variables it is integrated with one single integration point in the plane. The serious stability problem that can be expected is prevented by addition of fictitious viscosity or stiffness.

Further, DYNA3D provides contact and friction algorithms of penalty type. We have augmented the code with anisotropy out of the plane according to Hill, steered by the parameter R defined from a uniaxial tension test as the ratio between the strain out of the plane and the strain in the plane perpendicular to the load. The flow properties of the materials were described by the Ludwik–Hollomon two parameters expression

$$\sigma = K\epsilon^n \tag{1}$$

which gives a good description for the carbon steel DD used in deep-drawing but also reasonably well describes other materials used for stamping.

3. STRETCH FORMING

At stretch forming, Fig. 1, the blank is firmly clamped by serrated dies in order to avoid drawing-in from the blankholder during stamping.

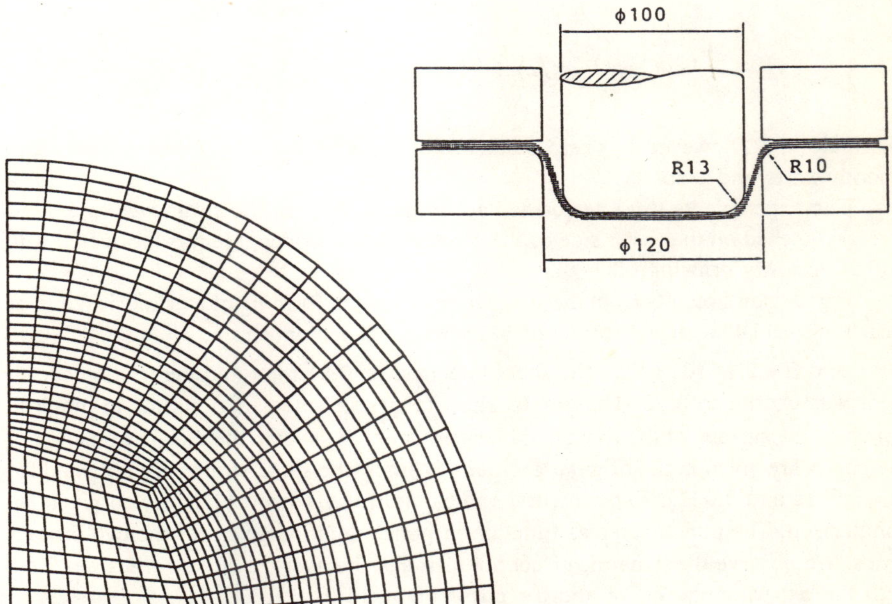

Fig. 1. Stretching by cylindrical punch and FEM–mesh

Contact with friction is then obtained only between the punch and the sheet. We have assumed a constant Coulomb friction coefficient here. In our comparisons with experiments the coefficient has been chosen to such a value that the calculated and measured strains coincide on the top of the punch.

In order to study the influence of the inertia effects the process time has been varied between 0.001s and 0.002s corresponding to 9000 respectively 18000 time steps. Compared to the quasistatic analysis a substantial reduction of the CPU–time was obtained.

The punch is displacement controlled so a velocity function has to be chosen. We have chosen for $0 < t < T$

$$v_1 = \sin(\pi t/T) \tag{2}$$

which, however, gives a non–zero acceleration at $t = 0$ so the punch force calculated as mass times acceleration can be expected to be overestimated. The function

$$v_2 = \frac{1}{2}[\sin(2\pi t/T - \pi/2) + 1] \tag{3}$$

for $0 < t < T$, however, has zero acceleration at $t = 0$ and $t = T$ corresponding to smooth starts and stops.

The experiments were performed according to Fig. 1. The strain was measured by a grid etched on the lower side of the sheet while the upper side was lubricated with a high viscosity drawing oil.

Fig. 2 shows results from the simulation of stretch forming of a cylindrical cup of a carbon steel DD with yield stress of 173 MPa at 0.2 percent strain, $n = 0.22$, $K = 513$ MPa, and $E = 2.1 \cdot 10^5$ MPa. The sheet thickness is 0.81 mm and the friction coefficient was chosen to 0.12. The graphs show tangential and radial strains in material points on the surface of the sheet from the center of the sheet to the die. The position of the points are given in the original geometry. Almost the same strains are obtained for T = 0.001s and 0.002s. Experimental values and results from [2] are also given for comparison. The punch force as function of punch depth is given for the two process times. We observe the dynamic effect but also that the average curve agrees well both with the experimental curve and the curve obtained in [2].

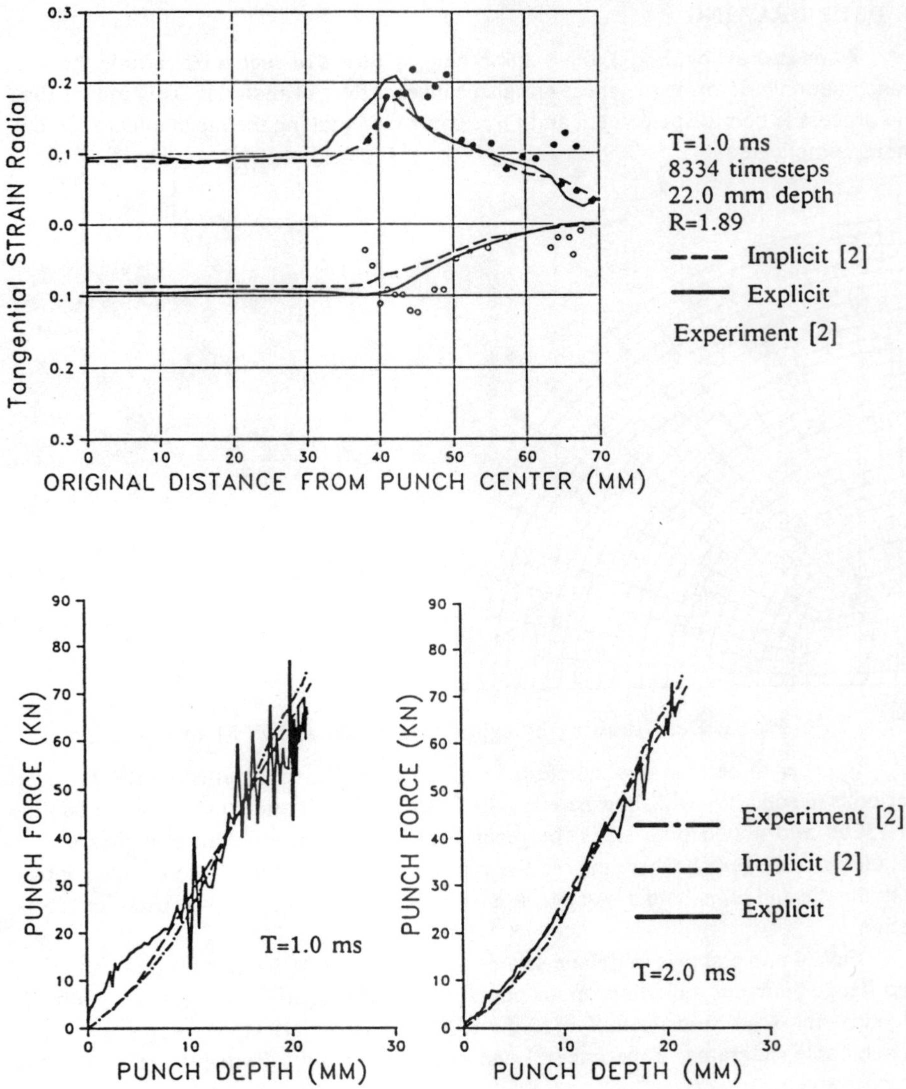

Fig. 2. Strains and punch forces at stretch forming

4. DEEP DRAWING

At deep drawing, Fig. 3, the blank is held by forces of such a magnitude that the punch not only deforms the sheet but also causes inflow of material. This means that the process is both displacement and force controlled making the mechanical problem more complicated.

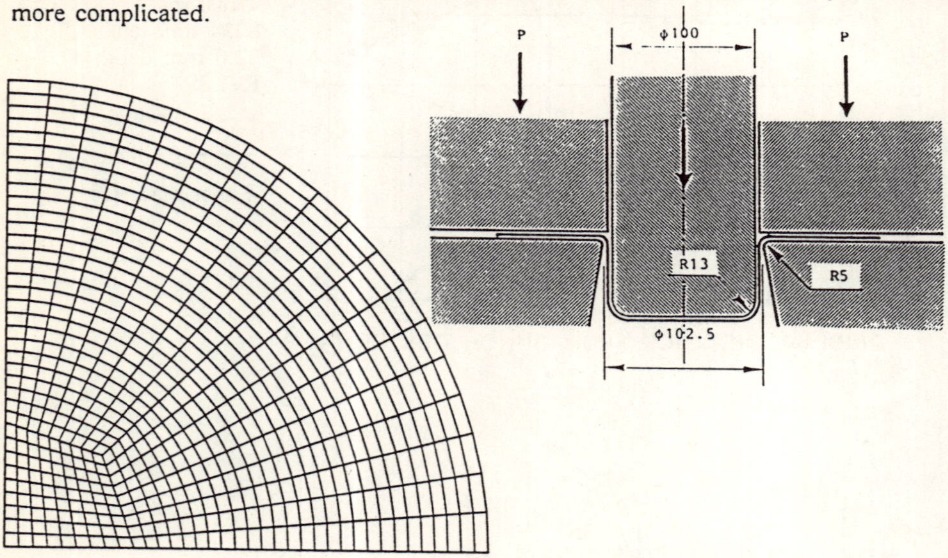

Fig. 3. Deep drawing by cylindrical punch and FEM–mesh

In order to be able to simulate the process by a fictive dynamic model a longer period is needed. We will show here results from T = 0.004s and 0.010s corresponding to 15000 and 40000 time steps. In general, the friction coefficient is higher at the contact with the punch where the relative movements are very small than at the contact with the die and blankholder. Also here a high viscosity drawing oil was used for lubrication.

Figs. 4 and 5 shows midplane strains distribution at 31.5 mm depth, punch force and flange diameter reduction versus punch depth. The calculated strains were almost identical for T = 4.0 ms and T = 10.0 ms. The dynamic effects are apparent at the punch force diagrams. Experimental and numerical results from [3] are shown for comparison.

The material is a carbon steel DD with yield stress of 220 MPa at 0.2 percent strain, n = 0.18, K = 547 MPa, and $E = 2.1 \cdot 10^5$ MPa. The sheet thickness is 0.70 mm and the friction coefficient was chosen to 0.14 at the flanges and 0.40 at the punch.

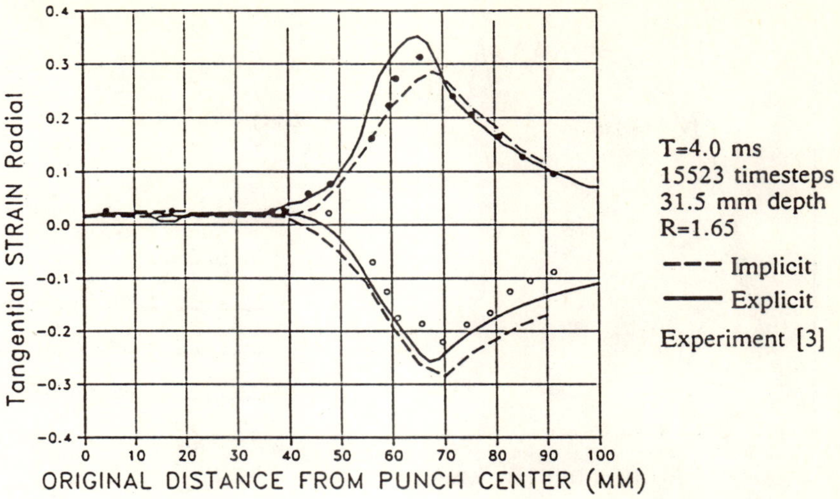

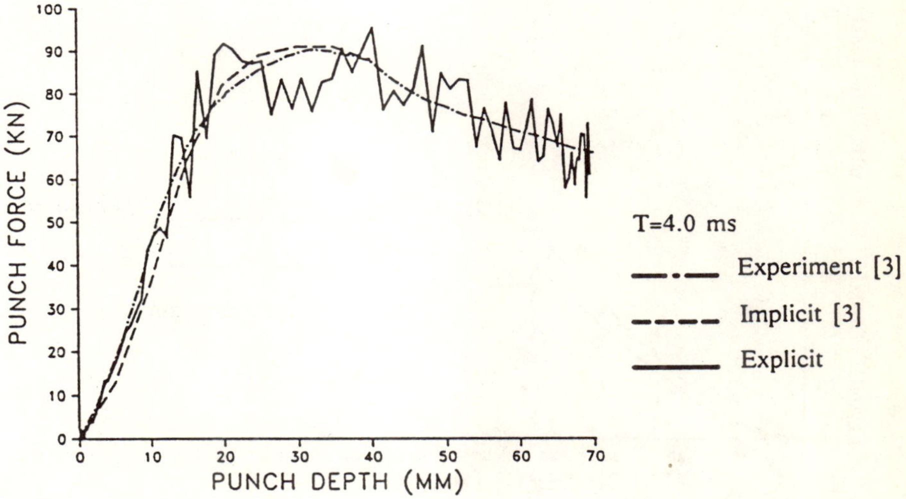

Fig. 4. Strains and punch forces for T = 4.0 ms at deep drawing

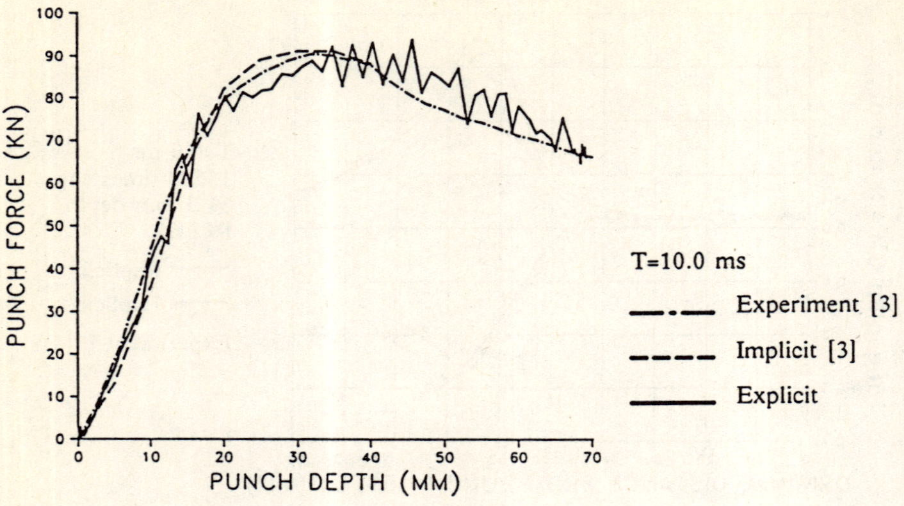

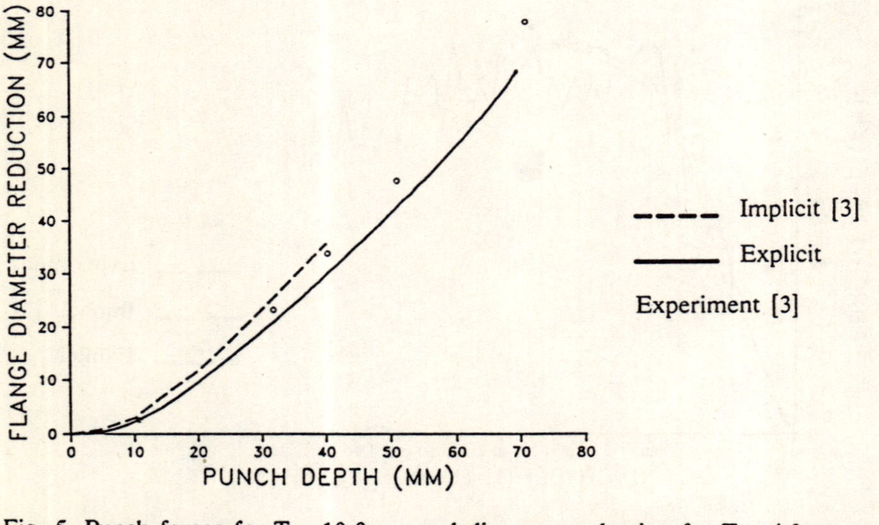

Fig. 5. Punch forces for T = 10.0 ms and diameter reduction for T = 4.0 ms at deep drawing

CONCLUSIONS

Our investigations show that it is possible to solve sheet stretching and deep drawing problems in an economic and reliable way with an explicit dynamic code. As seen from Figs. 4 and 5, the punch force is overestimated for high punch depths. The main reason for that is probably that our 2 parameter stress–strain law is not valid for the very high strains met here.

The investigation is part of a joint project between groups in Belgium, Greece, Italy and Sweden under the auspices of ECSC.

REFERENCES

1. Wang, N.M., Budiansky, B.: Analysis of sheet metal stamping by the finite element method, J. appl. mechanics, 45, 73–82, 1978.
2. Saran, M., Schedin, E., Samuelsson, A., Melander, A., Gustafsson, C.: Computational and experimental modelling of stretch forming operations, J. mater. shaping techn. 8, 31–38, 1990.
3. Saran, M., Schedin, E., Samuelsson, A., Melander, A., Gustafsson, C.: Numerical and experimental investigations of deep drawing of metal sheets, J. eng. indust. ASME, 112, 272–277, 1990.
4. Saran, M., Samuelsson, A.: Elastic–viscoplastic implicit formulation for finite element simulation of complex sheet forming processes, Int. j. numer. methods eng. 30, 1675–1697, 1990.
5. Zienkiewicz, O.C., Onate, E., Heinrich, J.C.: Plastic flow in metal forming. Applications of numerical methods in forming processes, Proc. Winter Annual meeting ASME, AMD 28, 107–120. 1978.
6. Bonet, J., Wood, R.D., Wargadipura, H.S.: Numerical simulation of the superplastic forming of thin sheet components using the finite element method, Int. j. numer. methods eng. 30, 1718–1737, 1990.
7. Honecker, A., Mattiasson, K.: Finite element procedures for 3D sheet forming simulation, Proc. NUMIFORM 89, Balkema 1989.
8. Belytschko, T., Tsay, C.S.: Explicit algorithms for the nonlinear dynamics of shells, AMD, 48, ASME, 209–231, 1981.

Some Thoughts about Metal Forming Simulation Models and Numerical Performance

J. Appeltauer, O. Mahrenholtz
Technical University Hamburg-Harburg

1 Introduction

Numerical treatment of sophisticated mechanical problems became nowadays a very large, manifold topic. This is true even for the particular case of various metal forming problems, due to multitude of physical models, correlated mathematical models, and correspondent numerical procedures. From this mass we will shortly depict the engineering aspect of very simple, efficient, problem oriented simulation programs, and insist upon essential correlations between physical and mathematical model versus required numerical procedures. As an illustration of thoughts, threedimensional free-forging is selected.

2 Some reflections on simplification modalities

A simple simulation program of forming processes can from the beginning give most important answers only, and this problem oriented for a certain category of forming processes. Engineering use of such simulation programs involves generally knowledge of time-variable velocity and stress fields, to control secondary deformations like spreading and bulging, dynamical interaction between workpiece and forming tool, or material behaviour till crack initiation, respectively. Residual stresses are from this point of view out of interest. The forming process of its part will be considered as quasi-static and isothermal. Then temperature affects material yield limit only. The deformed material will retain its homogeneity and isotropy.

Numerical efficiency of a simulation program means small storage place demand and short computing time, in order to be performed on a rather small computer, without loss of technically required accuracy. To accomplish these partial contradictory exigencies, one has to do on three different levels, i.e. modeling, setting up the simulation algorithm, and actual programming, in the course of which cross connections between levels must be observed.

On the other hand, all numerical procedures must be as simple as possible. That means numerical time integration (called time discretization also), spatial discretization in finite elements, numerical integration of element characteristics, and iterative solution of substitutive algebraic nonlinear equations system. For all that homogeneous simplifications must be considered, since inopportune isolated accuracy remains generally labour loss, or worst, can have an unfavourable influence on the whole calculation. Such a coordination must exist notably between time and space discretization and iterative solution of equations, since these routines are very extensive.

Interconnection between the mentioned action levels is due primarily to the fact that some simplification at one level will be possible only with corresponding simplification at another one. Significant examples will be treated subsequently. Secondly, rough time and space discretization softens the whole simulation process, and than a tight formulation of the mathematical model can be required.

3 Modeling operations

Time integration of essentially nonsteady forming processes can be opportune subdivided in many quasi-steady steps. Then, integration scheme will be very simple, if material behaviour can be expressed incrementally. This is elementary possible for a so-called rigid-plastic material behaviour by the Saint Venant-Lévy-Mises flow rule

$$\mathbf{T}' = \frac{k}{\sqrt{II_D}}\mathbf{D} , \tag{1}$$

where $\mathbf{T}'$ denotes the deviator of the Cauchy stress tensor in a Eulerian description, favourable for such incremental snapshots, $\mathbf{D}$ the rate of deformation tensor, while II_D is its second invariant, and k the shear yield stress. Relation (1) can be accepted as valid for cold forming with normal velocities, when elastic and viscous deformation components are negligible with respect to plastic deformation. Since this plastic deformation becomes in forming very large, there is another restrictive condition for applying relation (1). Indeed, plastic rotation is not contained in (1), which means that coaxiality must be postulated between principal directions of strain and stress tensors respectively. As deformations are less homogeneous, this assumption will be more rough. Although (1) is rather a state equation, it can be applied successively according to evolution character of the isotropic hardening function

$$k = k(\mathbf{B}^{-1}, \Theta) , \tag{2}$$

where $\mathbf{B}^{-1}$ denotes the left Cauchy-Green tensor and Θ the constant temperature [1]. Another property of the flow rule (1) is very disadvantageous, namely it implies plastic incompressibility as a secondary condition. Thus, an important correlation of physical model to simulation algorithm has been developed.

Another essential correlation of physical model to simulation algorithm consists in the difficult definition of frictional contact. Omiting nonlocal and nonlinear friction laws, very simple Amonton-Coulomb

$$\tau = -\mu\sigma_n \frac{\Delta \mathbf{v}}{|\Delta \mathbf{v}|} \tag{3}$$

respectively Siebel

$$\tau = -mk \frac{\Delta \mathbf{v}}{|\Delta \mathbf{v}|} \tag{4}$$

friction stresses are available. Herein τ denotes the friction stress vector, σ_n the normal pressure on contact surfaces, $\Delta \mathbf{v}$ the vector of relative velocity between workpiece and forming tool, while μ and m are friction coefficients. The last is included into the interval [0; 1]. Thus, friction stress value does not depend on velocity. It will be seen in the scope of mathematical model that Amonton-Coulomb definition (3) is noncompatible for rigid-plastic material behaviour. In planar or axially symmetrical forming processes there is no difficulty with Siebel definition (4), since velocities are enclosed in the significant plane. Differently in three-dimensional problems: here velocity directions are not known initially and so friction stress directions remain generally unknown also. This can lead to troublesome difficulties in simulation algorithm, namely on the occasion of iterative solution of equation system. It will be seen at this topic that by-passing possibilities exist, at least for important particular cases. However, a significant extension of mathematical model is required for this reason, capable also of smoothing mentioned rough approximations.

With assumptions as mentioned, mathematical model describes the control volume by equilibrium equations

$$\nabla \cdot \mathbf{T} = \mathbf{O} \tag{5}$$

and first order kinematics

$$\mathbf{D} = \frac{1}{2}(\mathbf{L} + \mathbf{L}^T)\,, \tag{6}$$

where

$$\mathbf{L} = \nabla \mathbf{v} \tag{7}$$

is the tensor of spatial gradient of velocity, $\mathbf{v}$ the velocity vector, and ∇ the Nabla operator. Constitutive equation (1) renders (5), (6) as field equations. With surface boundary conditions, in velocities, stresses, or mixed definition, a boundary value problem is generally attained, of such a kind in planar or axially symmetrical problems at least. In threedimensional deformations, directionally unknown friction stresses, as surface stresses, break the common boundary value problem, which must be extended.

However, velocity stringing by secondary incompressibility condition call for a variational form of the boundary value problem, or its extended version, with embeded incompressibility condition. For proper boundary value problems this can be obtained with the principle of virtual velocities in the weak (Galerkin) form

$$\int_V \nabla \cdot \mathbf{T} \delta \mathbf{v} dV = 0 \,, \tag{8}$$

which leads for symmetrical tensors, according to the Gauss theorem, to

$$\delta_{\mathbf{v}} [\int_V (\nabla \cdot \mathbf{T}' + \sigma_m \nabla \cdot \mathbf{v}) dV - \int_S \mathbf{n T v} dS] = 0 \,, \tag{9}$$

where σ_m denotes the mean normal (hydrostatic) stress, $\mathbf{n}$ the normal unit vector in the running surface point, and $\delta_{\mathbf{v}}$ variation with respect to velocity. Since direct variation with respect to σ_m does not appear, this mean normal stress will be interpreted as the penalty function (or part of it)

$$\sigma_m = K \nabla \cdot \mathbf{v} \,, \tag{10}$$

where K is obviously a large numerical factor, regardless of its physical meaning [2], or better [1]

$$K = c \frac{k}{2\sqrt{II_D}} \,, \tag{11}$$

with c a real large numerical factor. For extended boundary value problems, the expression between square brackets of (9) will be interpreted as a Markov functional Π, and its stationary value will be forced by

$$\frac{\partial \Pi}{\partial v} = 0 \,, \tag{12}$$

whereby the derivative must be extended on the stress factor of frictional power too [3]. The penalty function method is maintained for its numerical tightness with regard to the Lagrange parameter method. It is an important fact, that factor c will depend only on word length of employed computer. According to this Markov functional is demonstrated in [4], that Amonton-Coulomb friction law (3) does not satisfy the natural boundary conditions of coupled boundary value problem.

4 Algorithmical aspects

To begin with time integration, incremental version of forming process allows application of the gradiental Euler method. Simplifying the prototype of differential equation as

$$y'(x) = f[x, y(x)] \; , \tag{13}$$

integration scheme can be written as

$$y_{k+1} = y_k + hf(x_k, y_k) \; , \tag{14}$$

where h is generally the step width between equidistant integration points, time for present case. In the mentioned frame of simple numerical procedures, predictor-corrector integration methods, like explicit Heun method [5], must not be taken into consideration. Stresses and upsetting force are less susceptible to approximation errors than displacement fields. Thus, if a good description of deformations is desirable, a relative small die velocity must be considered, for instance 0,5% of investigated height. Otherwise 2% of investigated height can be a satisfactory value.

An essential part of simplification effort is represented by spatial discretization in finite elements. A rather rough discretization is envisaged. This means as few elements as actually feasible, respectively as simple as possible. Only hexahedral elements with eight nodes, consequently linear shape functions, and isoparametric velocity distributions were considered. Naturally, no large performance pretentions can be postulated. Nevertheless, in upsetting cases excellent results were obtained. In stretching cases, with local die application, some performance limits of few isoparametric hexahedrons can be attained. Namely, contact pressure jump from compressed part to free part of workpiece surface cannot be reproduced without some irregularities. To avoid these partially, not only number but sizes also of finite elements were found as significant. Under the die, equal narrow elements are preferable in longitudinal directions, while outside, width has to increase continuously till the decay length is reached. This arrangement can certain avoid some oscillations of iteration of nonlinear equation system

$$\mathbf{C}\mathbf{V} + \mathbf{F} = 0 \; , \tag{15}$$

reached by finite element discretization. Herein, $\mathbf{C}$ is the system matrix (according to extended variational formulation, no stiffness matrix), $\mathbf{V}$ the vector of nodal velocities, and $\mathbf{F}$ the free friction term. Workpiece height will also be uniformly subdivided. A rough division accelerates the pressure decay, but residual compression as 17% of upsetting force remains on the frontal area; in contrast to this finer division leads to a more slow decay, but practically without front compression. It is important to mention, that corresponding deformations are practically identical. Foregoing is a noticeable example for algorithmical routines interconnection.

Rough subdivision of workpiece height can lead to some boundary irregularities even in uniform upsetting cases. To avoid these, frictional density must be canceled at perpendicular edges, best linearly from centre of mass of boundary elements. This time, we have a connection between modeling and algorithmical level.

Numerical integration of first term in (9) is a troublesome, very often repeated operation. It is of interest, that Gauss-Legendre method can be equivalently substituted by the Irons procedure [1], which reduces the integration points from eight so six, very simple situated in the six surface centres.

A last but most important topic, since very calculation intensive, is represented by the iterative solution of system (15) for each time step. Choice of solution procedure is strongly related not only to allowable error norm, but also to the rough finite element discretization. According to the first point of view, velocity and stress fields prove to be not very sensitive with respect to iteration error level. Thus, a size of 10^{-2} is sufficient with respect to velocity error norm, proved about equations residua and final control of stress boundary conditions, too. Consequently, simple fixed-point iteration of successive approximations is available. This is a substantial simplification because, according to the second point of view, rough discretization may cause serious difficulties with the generally used Newton-Raphson iteration procedure. Apart from its more complicated algorithmical structure, application of Newton-Raphson method suffers from nonuniform distribution of equations residua, for instance in upsetting direction or cross to it, respectively. On the other hand the gradiental relaxation factor α, appearing in

$$\mathbf{V}_{i+1} = \mathbf{V}_i + \alpha_i \Delta \mathbf{V}_i \, , \tag{16}$$

is not at all easy to be determined, and little values of α can be misleading, because veloctiy norm gets better by worse residuum norm.

However, iteration must begin with a good approximation of unknown veloctiy field. It is the great advantage of extended variational formulation that iteration can begin with the homogeneous deformation of friction lack, and this as well for upsetting as for stretching. To iterate according to the classical boundary value problem formulation, a very good distribution of directionally unknown friction stresses must be established at the beginning. This is possible mainly for simple workpiece forms. Therefore, tightening of mathematical model is always preferable.

5 Actual programming

Structure and linking of subroutines, logical organization of frequently repeated computations, advantageous management of storage volume contribute on their part to make a program efficient.

Applying all these points of view the efficient simulation program FARF (Finite-Element Analysis of Rigid-Plastic Free-Forging) was obtained, capable of pursue upsetting and stretching processes with satisfactory engineering accuracy.

References

[1] Appeltauer, J.; Dung, N.L.; Mahrenholtz, O.: Some theoretical and numerical aspects of large deformations in three-dimensional metal forming. NUMIFORM 89. Proc. 3rd Int. Conf. on Numerical Methods in Industrial Forming Processes. Fort Collins, Colorado. A.A. Balkema 1989 pp. 523–527

[2] Park, J.J.; Kobayashi, S.: Three-dimensional finite element analysis of block compression. Int. J. Mech. Sci. 26 (1984) 165–176

[3] Appeltauer, J.; Dung, N.L.; Mahrenholtz, O.: Untersuchung des dreidimensionalen Freiformschmiedens mit der Finite-Element-Methode. Präzisionsumformtechnik. Springer-Verlag Berlin etc. 1990 pp. 274–285

[4] Besdo, D.: Numerik. In: Plastomechanische und metallkundliche Grundlagen der Umformtechnik. To be published by VDEh

[5] Schwarz, H.R.: Numerische Mathematik. B.G. Teubner Stuttgart 1986

On Thermomechanical Shakedown Analysis using Incremental Procedures

T. NIEZGODA

Military Academy of Technology
ul. Kaliskiego 2, 01-489 Warsaw, Poland

M. KLEIBER and J.A. KÖNIG

Institute of Fundamental Technological Research
ul. Świętokrzyska 21, 00-049 Warsaw, Poland

Abstract

A method has been previously proposed by two of the present authors to bound shakedown loads by means of procedures typical of the thermo-elastic-plastic incremental structural analysis. This paper aims at showing numerical effectiveness of the approach by considering a complex problem of a thick - walled cylindrical shell subjected to variable repeated mechanical and temperature loads. The results are critically assessed and the numerical accuracy of the analysis is discussed in detail. Some modyfications to the original algorithm are proposed. The conclusion is that the method lends itself well to the numerical analysis of complex thermo-mechanical shakedown problems of practical significance.

1. Introduction

It is generally accepted that safety of elastic-plastic structures exposed to variable repeated loads and/or temperature fields has to be assessed by means of the theory of shakedown. The theory accounts for such effects as a possible accumulation of plastic strains leading to excessive total structural deformations as well as for the possibility of the low-cycle fatique due to reversals of plastic strain increments. The numerical shakedown analysis has been mainly based on a direct application of the fundamental shakedown theorems which leads to a dual pair of constrained optimization problems, cf. sec. 2.

However, in the case of more complex systems the size of the resulting optimization problems grows up exponentially with the number of finite elements employed. Therefore, it was proposed in [9] to bound shakedown loads by means of procedures typical of the elastic-plastic incremental analysis. It can be expected that for very large systems the computational effort (which does not increase so drastically with the number of elements in this case) will become at least of the same order as in the former case.

Some simple examples have been satisfactorily analyzed in this way, cf. [3], also for the case of thermal loads, cf. [7]. For the simple structures considered in those papers the amount of computational work was higher than this needed by the classical approach using optimization methods but they were intended merely to check the general applicability of the new approach.

The present paper is devoted to the analysis of a more complex example and to the discussion of the effectiveness and accuracy of the 'incremental' approach in such a case.

2. Shakedown analysis of structures

As mentioned above, the majority of existing solutions to shakedown problems has been obtained by a more or less direct use of the fundamental shakedown theorems which are natural extensions of the well-known theorems of the plastic limit analysis to variable repeated loads and temperature fields, cf. [11], for instance.

When employing these theorems one has first to determine the envelope of elastic or thermoelastic stresses, i.e. stresses calculated under the assumption of unlimited elastic material response. Next, one has to either

- use the statical approach in which one looks for an optimum distribution of residual stresses (i.e. stresses being in equilibrium with the vanishing load); or
- use the kinematical approach in which one looks for an optimum distribution of plastic strain increments.

Both the approaches result in corresponding optimization problems for the safety factor against inadaptation. These optimization problems constitute a dual pair, i.e. their solutions provide the same magnitude of the safety factor.

As mentioned above, it has been shown in [9] that bounds to the actual shakedown load limits can be obtained by analyzing the response of a given structure under properly selected strictly cyclic load paths. Namely, if shakedown takes place for such a path then it will do so in any load path (not necessarily a cyclic one) contained within the smallest convex set containing the cyclic path. This result can be extended to the case of thermal actions as shown in [7].

Practical application of the method proposed consists in performing the incremental elastic-plastic analysis along a cyclic path and recording the plastic strain increments. If they cease to exist after a number of cycles, a state of shakedown has been attained. This implies in turn that the load/temperature limits of the cyclic path are below shakedown limits for any path contained within the same limits. This being the case the load/temperature limits can be expanded and the whole procedure repeated.

On the other hand, if at a certain stage the plastic strain increments do not tend to vanish, this means that shakedown limits have been exceeded. The analysis of cyclic processes provides us in such a way with a lower and an upper bound to shakedown load/temperature limits.

When it comes to practical computations one has to realize that systems of practical siqnificance can be analyzed only via certain discretized methods. Then, a serious problem arises with respect to the equivalence between the response of original (continuous) structure and its model having a finite number of degrees of freedom. Duality of the statical versus kinematical approach is preserved since the fundamental shakedown theorems hold also for discrete systems. However, the magnitude of the safety factor obtained for the discrete model may differ from that of the original structure. Clearly, the same applies to any method using incremental procedures which are based on the time discretization of the deformation process. Errors due to approximate discretization procedures lead in particular to such undesirable effects as:

- inexact determination of elastic and residual stresses
- neglecting of some kinamatically admissible plastic strain increment fields (a more detailled analysis of possible errors of the kinematical method was given in [10]).

3. Finite element analysis of thermo-elastic-plastic problems

The no-hardening isotropic thermo-elastic-plastic material is described by adopting the following basic assumptions [1], [8]:

- additivity of elastic, plastic and thermal strain contributions

$$\epsilon_{ij} = \epsilon_{ij}^{(e)} + \epsilon_{ij}^{(p)} + \epsilon_{ij}^{(\Theta)} \tag{1}$$

- linear behaviour in the elastic range

$$\sigma_{ij} = C_{ijkl}\left(\epsilon_{kl} - \epsilon_{kl}^{(p)} - \epsilon_{kl}^{(\Theta)}\right) \tag{2}$$

$$C_{ijkl} = \lambda\,\delta_{ij}\,\delta_{kl} + G\,(\delta_{ik}\,\delta_{jl} + \delta_{il}\,\delta_{jk}) \tag{3}$$

- linear thermal expansion law

$$\epsilon_{ij}^{(\Theta)} = \alpha\Theta\delta_{ij}\ , \qquad \Theta = T - T_0 \tag{4}$$

  where T_0 is reference temperature and α is the thermal expansion coefficient,

- Huber - Mises yield condition

$$\sqrt{\frac{3}{2}\,\sigma_{ij}^{D}\,\sigma_{ij}^{D}} - \sigma_y = 0 \tag{5}$$

  where σ_y =const, and

$$\sigma_{ij}^{D} = \sigma_{ij} - \frac{1}{3}\,\sigma_{kk}\,\delta_{ij} \tag{6}$$

  (σ_y is assumed independent of temperature only to facilitate comparisons with analytical solutions obtained in sec. 4).

The standard derivation (presented for a more general situation in [6,8], for instance) leads to the rate - form of the constitutive equation for thermo - elastic - plastic material in the form.

$$\dot{\sigma}_{ij} = C_{ijkl}^{(e-p)}\left(\dot{\epsilon}_{kl} - \alpha\dot{\Theta}\,\delta_{ij}\right) + \sigma_{ij}^{*} + \sigma_{ij}^{**} \tag{7}$$

where

$$C_{ijkl}^{(e-p)} = C_{ijkl} - 3G\,\frac{\sigma_{ij}^{D}\,\sigma_{kl}^{D}}{\sigma_y^2} \tag{8}$$

$$\sigma_{ij}^{*} = \dot{C}_{ijkl}\left(\epsilon_{kl} - \epsilon_{kl}^{(p)} - \alpha\Theta\delta_{kl}\right) \tag{9}$$

---

[1] For the sake of discussion clarity the cartesian components of vectors and tensors are considered only.

$$\sigma_{ij}^{**} = -\frac{3}{2}\,\frac{\sigma_{ij}^{D}\,\sigma_{kl}^{D}\,\sigma_{kl}^{*}}{\sigma_y^2} \tag{10}$$

Assuming that in the temperature range of interest for the present study the elestic characteristics of the material λ and G are independent of temperature we clearly have $\sigma_{ij}^{*} = 0$, $\sigma_{ij}^{**} = 0$.
The yield condition is adopted in the form

$$\sqrt{\frac{3}{2}\,\sigma_{ij}^{D}\,\sigma_{ij}^{D}} - \sigma_y = 0 \tag{11}$$

with the local loading condition given by

$$n_{ij}\,\dot{\sigma}_{ij}^{(tr)} > 0 \tag{12}$$

$$\dot{\sigma}_{ij}^{(tr)} = C_{ijkl}\left(\dot{\epsilon}_{kl} - \alpha\dot{\Theta}\delta_{kl}\right) \tag{13}$$

which is computationally consistent and corresponds to the known strain space formulation of the theory of plasticity, [6].

The fundamental equations describing the continuing equilibrium of discretized system have the form, [1,5]

$$\mathbf{K}_T\,\dot{\mathbf{r}} = \dot{\mathbf{R}} \tag{14}$$

Eq. (14) represents a system of N simultaneous ordinary differential equations of the first order with N unknown functions $\mathbf{r} = \mathbf{r}(t)$, N being the total number degrees of freedom. The structure is assumed to be subjected to a slowly varying load $\mathbf{R} = \mathbf{R}(t)$.
The build-up of the tangent stiffness matrix $\mathbf{K}_T$ is a standard issue in the finite element methodology and refers directly to the rate form of the material constitutive law (7).

The system (14) has to be solved with appropriate initial conditions of the form

$$\mathbf{r}(0) = \hat{\mathbf{r}}_0 \tag{15}$$

Any kind of a numerical integration scheme known in the vast literature can be employed to solve eqs. (14), (15).
In this study the full Newton-Raphson iterative scheme was adopted as described in [1,5], for instance.
The only aspect of the procedure which requires additional comments in this paper is the integration of the constitutive low, which is the central issue in the algorithm.
The widely accepted and effective algorithm for integration of isothermal elastic-plastic constitutive laws is known as the radial return scheme, [1,5]. It is based on the idea that the normal vector at the subsequent iteration is approximated in terms of the stress $\sigma^{(tr)D}$.
The generalization of the algorithm valid in the thermo-elastic-plastic case goes as follows, (parameter $s = 1, 2, \ldots$ numbers iterations at a given load level).

Step 1. Obtain the s-th approximation to incrmental strain $\Delta\epsilon_{ij}^{(s)}$ by solving eq. (14) for the s-th correction to Δr and using the explicit relation between generalized nodal displacements r and elemental strains ϵ_{ij}.

Step 2. Calculate the $(s+1)$-th approximation to the trial stress increment

$$\Delta\sigma_{ij}^{(tr)(s+1)} = C_{ijkl}\left(\Delta\epsilon_{kl}^{(s)} - \alpha\dot{\Theta}\delta_{kl}\right)$$

Step 3. Calculate the trial stresses

$$\sigma_{ij}^{(tr)(s+1)} = \sigma_{ij}^{(tr)(s)} + \Delta\sigma_{ij}^{(tr)(s+1)}$$

Step 4. Calculate the deviatoric part of $\sigma^{(tr)(s+1)}$

$$\sigma_{ij}^{(tr)D(s+1)} = \sigma_{ij}^{(tr)(s+1)} - \frac{1}{3}\sigma_{kk}^{(tr)(s+1)}\delta_{ij}$$

Step 5. Calculate the product of $\sigma_{ij}^{(tr)D(s+1)}$ by itself

$$A = \sigma_{ij}^{(tr)D(s+1)}\sigma_{ij}^{(tr)D(s+1)}$$

Step 6. Check the yield condition:

$$\text{if} \quad \frac{3}{2}A \leq \sigma_y^{(s)^2} \quad \text{then} \quad \begin{aligned} \sigma_{ij}^{(s+1)} &= \sigma^{(tr)(s+1)} \\ \sigma_y^{(s+1)} &= \sigma_y^{(s)} \end{aligned}$$

(elastic process)

go to step 1

$$\text{if} \quad \frac{3}{2}A > \sigma_y^{(s)^2} \quad \text{then} \quad \text{go to step 7} \qquad \text{(plastic process)}$$

Step 7. Calculate the approximation to the unit normal

$$n_{ij} = \sqrt{\frac{3}{2}}\frac{\sigma_{ij}^{(tr)D(s+1)}}{\sigma_y^{(s)}}$$

Step 8. Calculate the effective plastic strain increment:

$$\Delta\bar{\epsilon}^{(pl)} = -\frac{1}{3G+\xi}\left(\sigma_{ij}^{(tr)D(s+1)} - \sigma_y^{(s)}\right)$$

Step 9. Update

$$\begin{aligned} \sigma_{ij}^{(s+1)} &= \sigma_{ij}^{(tr)(s+1)} - 2G\Delta\bar{\epsilon}^{(pl)}n_{ij} \\ \sigma_y^{(s+1)} &= \sigma_y^{(s)} \\ \epsilon_{ij}^{(pl)(s+1)} &= \epsilon_{ij}^{(pl)(s)} + \Delta\bar{\epsilon}^{(pl)}n_{ij} \\ \bar{\epsilon}^{(pl)(s+1)} &= \bar{\epsilon}^{(pl)(s)} + \Delta\bar{\epsilon}^{(pl)} \end{aligned}$$

4. Example

Let us analyze a thick-walled circular cylindrical tube in the plane strain state (axial deformation neglected), with internal and external radii equal to a and b, respectively, subjected to a variable repeated internal pressure p_0 and to a variable repeated (independently of the pressure) internal temperature Θ_0, external temperature being zero. Material of the tube obeys the Huber - Mises yield condition:

$$(\sigma_r - \sigma_y)^2 + (\sigma_y - \sigma_z)^2 + (\sigma_r - \sigma_z)^2 \leq 2\,\sigma_y^2 \tag{16}$$

where σ_r, σ_y, σ_z denote the radial, circumferencial and axial stress, respectively, and σ_y is the yield-point stress in tension-compression. To make the analytical solution feasible the yield-point stress has been assumed to be independent of temperature.

The internal pressure p_0 and the internal temperature Θ_0 may vary independently of each other within the following limits

$$0 \leq p_0 \leq \bar{p}, \qquad 0 \leq \theta_0 \leq \bar{\Theta} \tag{17}$$

$\bar{p}$ and $\bar{\Theta}$ being constants.

The example has been selected, because of its practical importance. On the other hand, the problem allows to obtain the exact analytical solution, which is necessans to estimate the accuracy of numerical results. In this way we shall be able to assess the practical applicability of the shakedown analysis using incremental procedures.

4.1. Analytical solution

The elastic stress solution to the problem may be presented as follows:

$$\begin{aligned}
\sigma_r^E &= p_0\, \tfrac{a^2}{b^2-a^2} \left(1 - \tfrac{b^2}{r^2}\right) + \Theta_0\, \tfrac{E\alpha}{2(1-\nu)\ln\frac{a}{b}} \left[\tfrac{a^2}{b^2-a^2}\left(\tfrac{b^2}{r^2} - 1\right)\ln\tfrac{a}{b} - \ln\tfrac{r}{b}\right] \\
\sigma_\phi^E &= p_0\, \tfrac{a^2}{b^2-a^2} \left(1 + \tfrac{b^2}{r^2}\right) - \Theta_0\, \tfrac{E\alpha}{2(1-\nu)\ln\frac{a}{b}} \left[\tfrac{a^2}{b^2-a^2}\left(\tfrac{b^2}{r^2} + 1\right)\ln\tfrac{a}{b} + \ln\tfrac{r}{b} + 1\right] \\
\sigma_z^E &= p_0\, \tfrac{2\,\nu\, a^2}{b^2-a^2} - \Theta_0\, \tfrac{E\alpha}{2(1-\nu)\ln\frac{a}{b}} \left[\tfrac{2a^2}{b^2-a^2}\ln\tfrac{a}{b} + \tfrac{2}{\nu}\ln\tfrac{r}{b} + 1\right]
\end{aligned} \tag{18}$$

where E is the Young modulus and ν denotes the Poisson ratio.
The thermoelastic part of the solution results from the assumed logarithmic distribution of the temperature along the tube radius

$$\Theta(r) = \Theta_0\, \frac{\ln\frac{b}{r}}{\ln\frac{b}{a}} \tag{19}$$

In the course of the analysis the following problem characteristics have been determined in the $\bar{p} - \bar{\Theta}$ plane, cf. Fig. 4-1:

a) Region of perfectly elastic response of the tube, curve (a),

b) Limits of $\bar{p}$, $\bar{\Theta}$ allowing for shakedown. They result from two conditions: safety with respect to incremental collapse and safety against alternating plasticity,curve (d-b),

c) Limit load of the tube (which is independent of Θ), straight line (e).

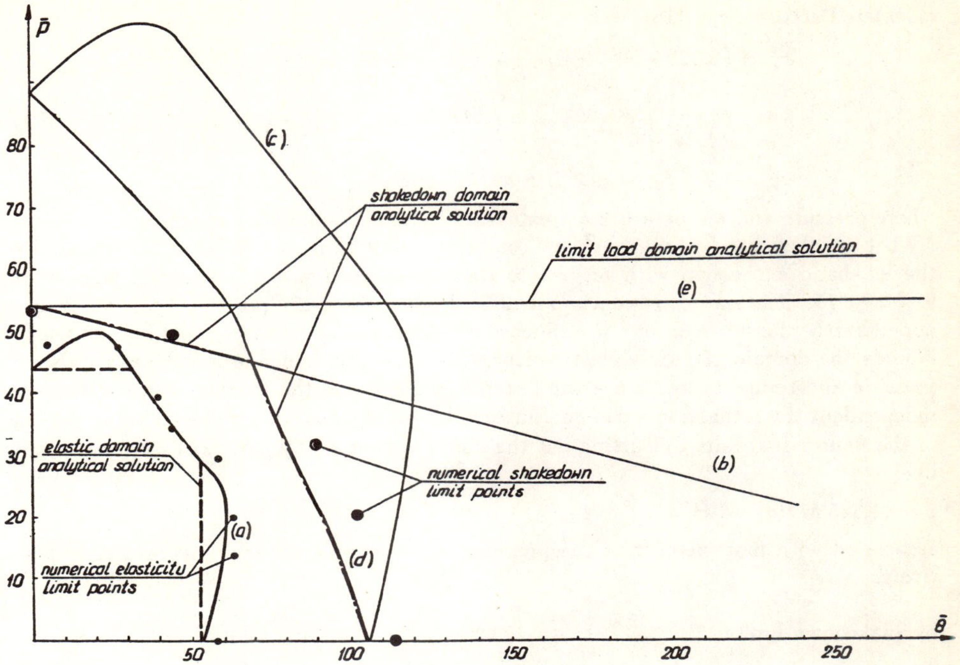

Fig. 4-1 Analytical and numerical limits of the elastic response of the tube, of the shakedown domain as well as the ultimate load:

- - - elastic response limit (analytical)

– – – shakedown limit (analytical)

o elastic response limit (numerical)

⊙ shakedown limit (numerical)

Calculation were performed for the following values of the tube dimensions and material parameters:

$$\begin{array}{lll} a = 9\,mm & b = 11\,mm & E = 208 \cdot 10^3 M\,Pa \\ v = 0.3 & \alpha = 2 \cdot 10^{-5}/K & \sigma_y = 235\ M\,Pa \end{array} \tag{20}$$

Details of the calculations are as follows:

a) The elasticity limit

For the assumed values of material and geometrical parameters (20) the elastic stress calculated according to (18) become

$$\begin{aligned}\sigma_r^E &= \left(2.025 - \tfrac{245.025}{r^2}\right) p_0 + \left(\tfrac{728.07}{r^2} + 14.81 \cdot \ln r - 41.52\right) \Theta_0 \\ \sigma_y^E &= \left(2.025 + \tfrac{245.025}{r^2}\right) p_0 + \left(-\tfrac{728.07}{r^2} + 14.81 \cdot \ln r - 26.72\right) \Theta_0 \\ \sigma_z^E &= 1.215\, p_0 + (29.615 \ln r - 70.18)\Theta_0\end{aligned} \tag{21}$$

where pressure and temperature are expressed in $[MPa]$ and $[K]$, respectively.

Let us substitute (21) into the yield condition (16) with $p_0 = \bar{p}$, $\Theta_0 = \bar{\Theta}$, and maximize the left-hand expression with respect to the radial coordinate r keeping in mind that $9 \leq r \leq 11$. The maximization is not difficult since for every ratio $\bar{p}/\bar{\Theta}$ the maximum appears either for $r = a$ or for $r = b$. Such a procedure yields the curve (a) of Fig. 4-1 which bounds the domain of purely elastic response of the tube under proportionally growing pressure and temperature. Since the internal pressure and the internal temperature are independent the actual domain of guaranteed elastic response is slightly smaller as marked in the figure. It results by cutting out the domain bounded by curve (a) by the straight lines

$\bar{p} = 44.325\, MPa$ and $\bar{\Theta} = 52.58\, K$

associated with first yielding in the presence of either pressure or temperature, respectively.

b) Shakedown limits

There exist two modes of inadaptation in the form of the incremental callapse and alternating plasticity. The safety conditions with respect to each one of them can be considered separately, cf. [10].

Let us begin with the incremental collapse. Since (16) implies plastic incompressibility, there exists only one possible mode of accumulation of plastic strains which defined by

$$\bar{\mu} = C/r\,, \quad \Delta\bar{\epsilon}_r = -C/r^2\,, \quad \Delta\bar{\epsilon}_\varphi = C/r^2\,, \quad C > 0 \tag{22}$$

where $\bar{\mu}$ is the increment of radial displacement in a kinematically admissible cycle of plastic strains. To establish the safety condition with respect to this mode of incremental collapse we can employ the ready-to-use formula (8.20) of [11]. In our case it assumes, the following form

$$\int_a^b a_p(r)\, J_p(r) r\,dr + \int_a^b a_\Theta(r)\, J_\Theta(r) r\,dr \leq \int_a^b D\left(\Delta\bar{\epsilon}_r, \Delta\bar{\epsilon}_\varphi\right) r\,dr \tag{23}$$

where $D\left(\Delta\bar{\epsilon}_r, \Delta\bar{\epsilon}_\varphi\right)$ denotes the plastic dissipation associated with the plastic strain increments $\Delta\bar{\epsilon}_r$, $\Delta\bar{\epsilon}_\varphi$ via the flow rule associated with the yield condition (16), whereas

$$\begin{aligned}J_p(r) &= \sigma_r^{Ep} \Delta\bar{\epsilon}_r + \sigma_\varphi^{Ep} \Delta\bar{\epsilon}_\varphi \\ J_\Theta(r) &= \sigma_r^{E\Theta} \Delta\bar{\epsilon}_r + \sigma_\varphi^{E\Theta} \Delta\bar{\epsilon}_\varphi\end{aligned} \tag{24}$$

σ_r^{Ep}, σ_φ^{Ep}, $\sigma_r^{E\Theta}$, $\sigma_\varphi^{E\Theta}$ being the elastic and thermoelastic stresses for $p_0 = 1$, $\Theta_0 = 0$ and $p_0 = 0$, $\Theta_0 = 1$, respectively. Moreover

$$a_p(r) = \begin{cases} \bar{p} & \text{if} \quad J_p(r) > 0 \\ 0 & \text{if} \quad J_p(r) < 0 \end{cases}, \quad a_\Theta(r) = \begin{cases} \bar{\Theta} & \text{if} \quad J_\Theta(r) > 0 \\ 0 & \text{if} \quad J_\Theta(r) < 0 \end{cases} \tag{25}$$

By substituting the proper quantities into (24) one obtains

$$J_p(r) = \frac{C}{r^4}\,490.5 \;, \qquad J_\Theta(r) = \frac{C}{r^2}\left[14.81 - \frac{1456.15}{r^2}\right] \tag{26}$$

Obviously, $J_p(r) > 0$ everywhere $J_\Theta(r) > 0$ for $r_0 < r$ where

$$r_0 = 9.9166 \tag{27}$$

In the case of plane strain plastic flow and the Huber - Mises yield condition the plastic dissipation is given by

$$D\left(\Delta\bar{\epsilon}_r,\, \Delta\bar{\epsilon}_r\right) = \frac{2\,\sigma_y}{\sqrt{3}}\frac{C}{r^2} \tag{28}$$

Finally, by substituting all the necessary quantities into (23) and performing the prescribed integrations we arrive at the following condition of safety with respect to incremental collapse as defined by the mode (22)

$$\bar{p} + 0.1487\,\bar{\Theta} \le 54.45 \tag{29}$$

cf. line (b) in Fig. 4-1.

To obtain the safety condition with respect to alternating plasticity we employ the ready-to-use formula (4.39) of [11] valid for even yield functions. This formula can be explained briefly as follows. Consider yield conditionssatisfying

$$f\left(\sigma_{ij}\right) = f\left(-\,\sigma_{ij}\right) \le 2\,\sigma_y^2 \tag{30}$$

Then, the safety domain with respect to alternating plasticity is the intersection of all domains defined by

$$f\left[\frac{1}{2}\left(\pm\bar{p}\,\sigma_{ij}^{Ep} \pm \bar{\Theta}\,\sigma_{ij}^{E\Theta}\right)\right] \le 2\,\sigma_y^2 \tag{31}$$

where all the combinations of the signs must be accounted for, σ_{ij}^{Ep}, σ_{ij}^{Ep} possessing the same meaning as in (24). In our case there are only two essentially different combinations of signs yielding the following two conditions

$$\begin{aligned}&\left(\bar{p}\,\sigma_r^{Ep} + \bar{\Theta}\,\sigma_r^{E\Theta} - \bar{p}\,\sigma_\varphi^{Ep} - \bar{\Theta}\,\sigma_\varphi^{E\Theta}\right) + \left(\bar{p}\,\sigma_r^{Ep} + \bar{\Theta}\,\sigma_r^{E\Theta} - \bar{p}\,\sigma_z^{Ep} - \bar{\Theta}\,\sigma_z^{E\Theta}\right)^2 + \\ &\left(\bar{p}\,\sigma_\varphi^{Ep} + \bar{\Theta}\,\sigma_\varphi^{E\Theta} - \bar{p}\,\sigma_z^{Ep} - \bar{\Theta}\,\sigma_z^{E\Theta}\right)^2 \le 8\,\sigma_y^2\end{aligned} \tag{32}$$

and

$$\begin{aligned}&\left(\bar{p}\,\sigma_r^{Ep} - \bar{\Theta}\,\sigma_r^{E\Theta} - \bar{p}\,\sigma_\varphi^{Ep} + \bar{\Theta}\,\sigma_\varphi^{E\Theta}\right) + \left(\bar{p}\,\sigma_r^{Ep} - \bar{\Theta}\,\sigma_r^{E\Theta} - \bar{p}\,\sigma_z^{Ep} + \bar{\Theta}\,\sigma_z^{E\Theta}\right)^2 + \\ &\left(\bar{p}\,\sigma_\varphi^{Ep} - \bar{\Theta}\,\sigma_\varphi^{E\Theta} - \bar{p}\,\sigma_z^{Ep} + \bar{\Theta}\,\sigma_z^{E\Theta}\right)^2 \le 8\,\sigma_y^2\end{aligned} \tag{33}$$

They have to hold for every r satisfying $a \le r \le b$. Practically, one has to determine for every $\bar{p}/\bar{\Theta}$ ratio the maximum of the left-hand side of either (32) or (33) with respect to r. It is not particulary involving in our case since the maxima occur again only at $r = a$ or $r = b$. Let us note that (32), cf. curve (c) in Fig. 4-1, bounds a domain being similar to that bounded by curve (a) but twice as large. However, the curve (d) associated with the condition (33) turns out to be more stringent in our case. Finally, the shakedown domain is bounded by lines (b) and (d).

c) Ultimate load

Since there is no temperature dependence of the yield-point stress σ_y, the state of instantaneous collapse of the tube takes place at

$$\bar{p} = p_\mu = \frac{2\sigma_y}{\sqrt{3}} \ln \frac{b}{a} = 54.45 \tag{34}$$

as shown by line (e) in Fig. 4-1, according to the well-known solution of the plastic limit analysis.

4.2. Numerical shakedown analysis of the tub by means of an elastic-plastic incremental procedure

Finding the elastic response range is simple; one has to perform the incremental procedure along various straight lines p_0/Θ_0 =const stopping at the first yield. In this way points on the boundary of the elastic domain are determined. Black points in Fig. 4-1 signify a number of such points.

The shakedown analysis is much more involving. According to the theormes presented in [9] and [7], the response of a given structure to an arbitrary load/temperature path contained within prescribed limits is equivalent to its response in a cyclic load/temperature path bounded by the same limits. The equivalence means here that, depending on the limits, shakedown or inadaptation takes place in both the cases. The computations suggested by this consideration were performed in the following way.

A pair of magnitudes $\bar{p}_1$, $\bar{\Theta}_1$, presumably belonging to the shakedown domain is selected and the tube is subjected to the load/temperature path $0A_1\ 0B_1\ 0C_1\ 0$, cf. Fig. 4-2, the plastic strain increments being recorded. Such a cycle is repeated several times until no new plastic strain inerement would appear in a whole subsequent cycle. This implies that the point $\bar{p}_1$, $\bar{\Theta}_1$ belongs to the actual shakedown domain.

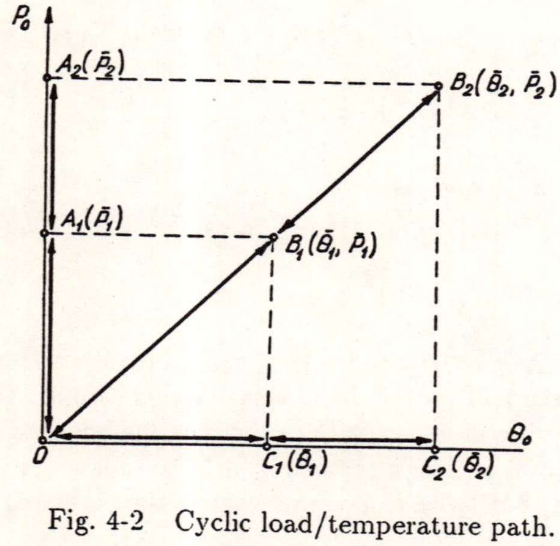

Fig. 4-2 Cyclic load/temperature path.

In the next step the magnitudes of $\bar{p}_1$ and $\bar{\Theta}_1$ are increased proportionally to each other up the walue of, say, $\bar{p}_2$, $\bar{\Theta}_2$ and the whole procedure is repeated.

If, at a certain step, say for a pair $\bar{p}_k$, $\bar{\Theta}_k$, the tube does not tend to shakedown any more, i.e. if non - decreasing increments of plastic strains appear during a single load/temperature cycle in at least one of the finite element integration points, then one can state that the pair $\bar{p}_k$, $\bar{\Theta}_k$ defines a point which does not belong to the shakedown domain.

The two points $\bar{p}_{k-1}$, $\bar{\Theta}_{k-1}$ and $\bar{p}_k$, $\bar{\Theta}_k$ provide us with a lower and an upper bound to the actual shakedown limits. Accuracy of the result can be improved by diminishing the distance between the points which provids us with the lower and the upper bounds. A few bounds to shakedown limits obtained in such way for the tube are shown in Fig. 4-1 with small encireled points.

It must be stressed that all the plastic strain increments must be recorded. The total increments over a whole cycle may become zero even though there is no shakedown. This occurs in the case of alternating plasticity.

In the course of the computations we have observed, that the rate of convergence of plastic strains can be improved if the cyclic loading is performed in a slightly different way; namely, by applying higher load/temperature amplitudes durring the first cycle (a sort of overloading) and only then applying the proper ones. After such operation plastiç strains usually converged much faster than in the case of using the same amplitudes all the time. The single overloading does not influence the final result but accelerates the generation of residual stresses required for the tube to shake down.

5. Conclusions

In view of the fact that the finite element model of the tube was not very sophisticated, the computations perfomed do not look discouraging. The numerically obtained points suggest both the elastic region as well as the shakedown domain to be slightly bigger than those obtained on the basis of the exact analytical solution. This should be expected since we have modelled the tube with a system of a lower number of degrees of freedom. Another source of possible errors is the temperature distribution within the tube wall which was assumed linear in the numerical computations.

The overall conclusion is even a moderate number of the finite elements employed allows to obtain a quite reliable result.

References

[1] Bathe K.J., Finite element procedures in engineering analysis, Prentice-Hall, New Jersey, 1982.

[2] Belytschko T., Plane stress shakedown analysis by finite elements, Int. J. Mech. Sci., 14, 619 - 625, 1972.

[3] Borkowski A., Kleiber M. , On a numerical approach to shakedown analysis of structures, Comp. Meth. Appl. Mech. Eng., 22, 101 - 119, 1980.

[4] Corradi L., Zavelani A., A linear programming approach to shakedown analysis of structures, Comp. Meth. Mech. Eng., 3, 37 - 53, 1974.

[5] Kleiber M., Incremental finite element modelling in nonlinear solid mechanics, Ellis Horwood/PWN, 1989.

[6] Kleiber M., Computational coupled non-associative thermo- plasticity, Comp. Meths. Appl. Mech. Eng., in print.

[7] Kleiber M., König J.A., Incremental shakedown analysis in the case of thermal effects, Int. J. Num. Meth. Eng., 20, 1567 - 1573, 1986.

[8] Kleiber M., Niezgoda T., Numerical analysis of thermo- elastic-plastic, Rozprawy Inzynierskie, 645-660, 1988.

[9] König J.A., Kleiber M., On a new method of shakedown analysis, Bull. Ac. Pol. Sci., Ser. Sci. Techn., 26, 165 - 171, 1978.

[10] König J.A., On exactness of the kinematical approach in the structural shakedown and limit analysis, Eng. - Archiv, 52, 421 - 428, 1982.

[11] König J.A., Shakedown of Elastic - Plastic Structures, Elsevier/PWN Warsaw, 1987.

[12] Maier G., A shakedown matrix theory allowing for work-hardening and second order geometric effects, In "Foundations of Plasticity", A. Sawczuk editor, Noordhoff, Leiden, 417 - 433, 1972.

[13] Moss W.C., On the computational significance of the strain space formulation of plasticity theory, Int. J. Num. Meths. Eng. 20, 1703-1709, 1984.

[14] Nguyen Dang Hung, König J.A., A finite element formulation for shakedown problems using a yield criterion of the mean, Comp. Meth. Appl. Mech.Eng., 8, 179 - 192, 1976.

[15] Nguyen Dang Hung, Palgen L., Shakedown analysis by displacement method and aquilibrium finite element, Fifth SMiRT Conf., Berlin, paper L 3/3, 1979.

Computational Aspects of a Rate Dependent Yield Hinge Theory

H. Rothert, N. Gebbeken, V. Brandt

Institut für Statik (Institute of Structural Analysis), Universität Hannover, Deutschland

1 Introduction

The numerical analysis of steel frame structures exposed to extreme loads is subject to specific conditions. In these specific cases, it is not the usability of frame structures that has to be guaranteed, but their safety. Since extreme loads, such as earthquake, impact, wind gust and explosion, are purely dynamic actions, a dynamic analysis is indispensable. Even constructions calculated according to the theory of elasticity contain unpredictable plastifications, for example in connections or notches. However, construction components, which are plasticized and subject to a dynamic loading, can tolerate only a limited number of load cycles less than 100 ([4]). Consequently, the load-carrying behaviour of dynamically loaded structures can only be analyzed realistically by means of a geometrical and material non-linear dynamic numerical analysis. The static simulation of dynamic loads determined with the help of dynamical calculations makes only sense in connection with a practice-related method of analysis. At the moment, the economical (as far as the effort of discretization, CPU time and evaluation of the results are concerned) and dynamical calculation of the limit loads of entire steel-frame structures is only feasible on the basis of the yield hinge theory, of which a detailed description concerning static forces is given for example in [10] and [11].

The reliability of the yield hinge theory depends predominantly firstly on the realistic modeling of the actual behaviour of the material, parts of the structure and the entire structure by mathematical and mechanical models, and secondly on the reliability and stability of the corresponding algorithms. This unity of model and analysis, the so-called "computational mechanics", is a relatively young field of research.

Based on the methods of computational mechanics, this paper deals mainly with a reliable method of physical modeling for the dynamic calculation of the limit load of steel frame structures.

2 Rod theory

The rod theory applied here is an extended geometrical non-linear theory of the second order, as described in [8], for example.

The rod theory is based on the following assumptions:
- Bernoulli-hypothesis,
- in the elastic stress range the material is linear elastic, homogenous and isotropic,
- straight beams with a constant and symmetrical cross section,
- ratio of height h to length l: $\frac{h}{l} < 0.1$,
- elastic strain: $\varepsilon < 10^{-3}$.

The displacements of the actual configuration at time t are subdivided into rigid body motions and stress relevant displacements; the latter cause only small strains.

Applying the principle of virtual work in the actual configuration at time t

$$\delta W = \int_0^L {}^t N\, \delta(\Delta^t u') + {}^t M\, \delta(\Delta^t \varphi') + m\, {}^t\ddot{v}\, \delta\, {}^t v \; dx \quad , \tag{1}$$

with the mass coverage m, the relation between internal forces, displacements and accelerations is given in matrix notation by

$${}^t\underline{s} = {}^t\underline{k}(v)\, {}^t\underline{v} + {}^t\underline{m}\, {}^t\ddot{\underline{v}} \quad , \tag{2}$$

with the stiffness matrix ${}^t\underline{k}(v)$, the consistent mass matrix ${}^t\underline{m}$, the vector of internal forces ${}^t\underline{s}$, the nodal displacement vector ${}^t\underline{v}$ and the nodal acceleration vector ${}^t\ddot{\underline{v}}$. Introducing the equilibrium between external forces ${}^\circ\underline{P}$ and internal forces ${}^\circ\underline{S}$, the equation of motion is

$${}^\circ\underline{P} = {}^\circ\underline{K}(V)\, {}^\circ\underline{V} + {}^\circ\underline{M}(V)\, {}^\circ\ddot{\underline{V}} \quad . \tag{3}$$

This equation does not contain any damping terms, as there are no reliable input parameters available at the present. As the influences of the damping of the soil, connections and the elastic as well as the plastic material deformations on the load-carrying behaviour have to be examined more closely in further investigations, the assumed data for a global damping will not be discussed here.

3 Strain rate dependent material model

One of the main parameters in a calculation which is based on the theory of plasticity is the yield stress $_Y\sigma$. It is normally found by an uniaxial tension test under quasi static loading. In dynamic loading cases the yield stress depends on the loading velocity.

Manjoine [13] carried out uniaxial tension tests at unalloyed and untreated steel. The strain rates varied from $\dot{\varepsilon} = 10^{-6}\,s^{-1}$ to $\dot{\varepsilon} = 10^{3}\,s^{-1}$. The Young's modulus E was found to be independent of the strain rates. The situation in the plastic zone is characterized by the lowest yield stress $_Y\sigma$, the failure stress $_U\sigma$, and the failure strain $_U\varepsilon$ in relation to the strain rates as shown in Fig. 1. Both, yield stress and failure stress rise monotonously with increasing strain rates.

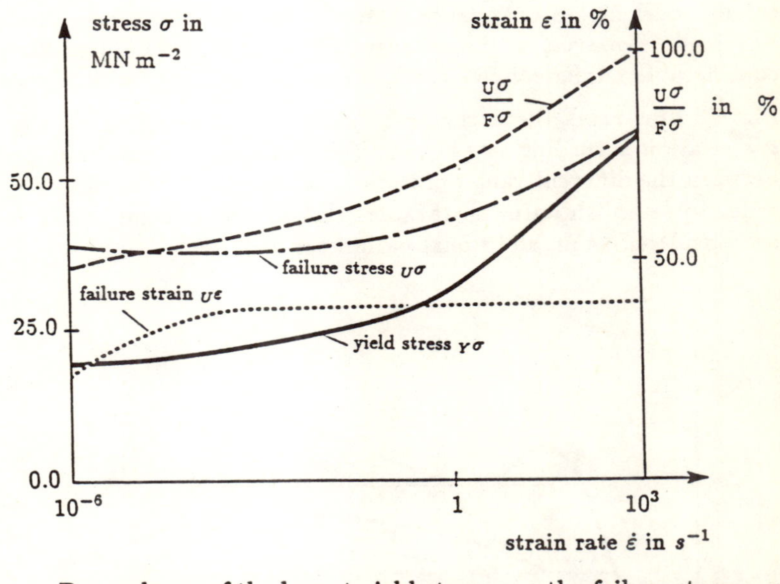

Fig. 1 Dependence of the lowest yield stress $_Y\sigma$, the failure stress $_U\sigma$ and the failure strain $_U\varepsilon$ on the strain rates $\dot{\varepsilon}$

Cowper and Symonds [9] developed a formula for the relation between the dynamic yield stress $_Y\sigma$, the quasi-static yield stress $_{Y,s}\sigma$ and the strain rate $\dot{\varepsilon}$ (4 b). The "overstrain" tolerated by the material is consequently an exponential function of the strain rate, with quantity D formally representing a reciprocal viscosity constant. It can such be interpreted that if $\dot{\varepsilon} = D$, the yield stress increases by twice the amount of $_{Y,s}\sigma$. The dependency of the rates, as shown in form (4b), is regarded as certain concerning unalloyed and untreated steel (Bodner [3], Brandes/Limberger[5]).

The determination of the material parameters D and p will be dealt with in chapter 5. The following holds for the stress-strain relations depending on the strain rate $\dot{\varepsilon}$:

$$\varepsilon < {}_{Y,s}\varepsilon \left[1 + \left(\frac{|\dot{\varepsilon}|}{D}\right)^{\frac{1}{p}}\right] \quad \longrightarrow \quad \sigma = E\varepsilon \quad , \tag{4a}$$

$$\varepsilon \geq {}_{Y,s}\varepsilon \left[1 + \left(\frac{|\dot{\varepsilon}|}{D}\right)^{\frac{1}{p}}\right] \quad \longrightarrow \quad {}_{Y}\sigma = {}_{Y,s}\sigma \left[1 + \left(\frac{|\dot{\varepsilon}|}{D}\right)^{\frac{1}{p}}\right] \tag{4b}$$

with

${}_{Y,s}\varepsilon$: yield strain under quasi-static loading.

A linearization of the above-mentioned relations analogous to Prandtl's material model makes only sense after the realistic σ-ε-curves have been determined. However, as the analysis of these relations is feasible without a considerable numerical effort, a linearization will not be considered here.

Fig. 2 illustrates the actual σ-ε-relation in comparison to the idealized σ-ε-relation according to (4a) and (4b). There is no well-defined relation between the different values of the strain and the corresponding stresses. In order to establish an unmistakable relation, the current strain rate has to be considered as an additional parameter.

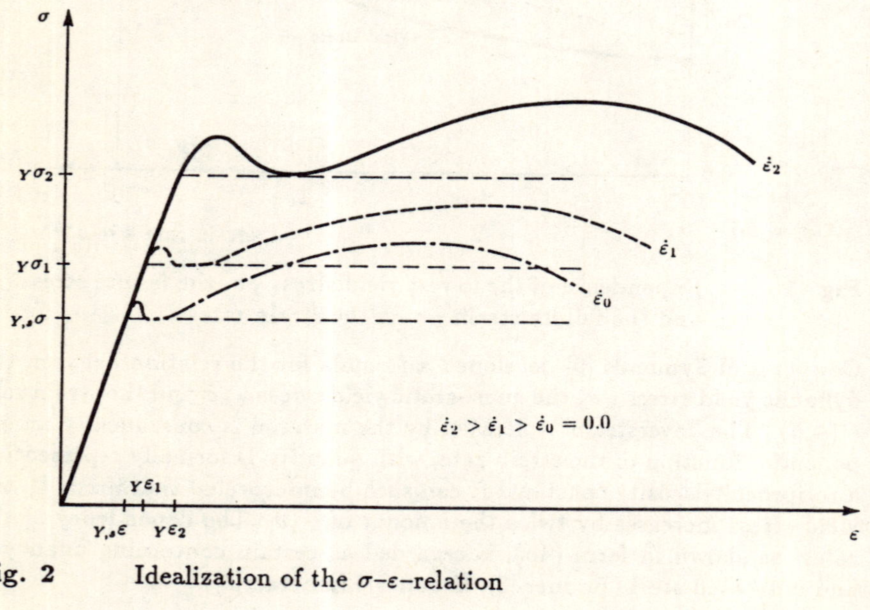

Fig. 2 Idealization of the σ–ε–relation

In uniaxial states of stress yielding occurs in any given material point if the stress σ equals the yield stress ${}_{Y}\sigma$. For three-dimensional states of stress, the beginning of yielding is determined by means of a yield condition. The yield condition according to Huber, v. Mises and Hencky which has been applied here represents a criterion to decide whether the material behaviour in a given material point is elastic or plastic.

In order to take the rate dependence of the material rule into consideration, the scalar parameter k resulting from the uniaxial tension test is described as a function of the strain rates $\dot{\varepsilon}$. From this the following yield condition f results for a viscous material:

$$f = f\left[\sigma_{ik}, k(\dot{\varepsilon})\right] \begin{cases} > 0 & \text{not possible ,} \\ = 0 & \text{ideal plastic ,} \\ < 0 & \text{elastic .} \end{cases} \tag{5}$$

Thus, the following statements are assumed:

- in the elastic stress range the material behaves linear elastic,
- in the plastic stress range the material behaves ideal plastic,
- the yield condition depends on the strain rates.

Hereby, the material behaviour has been defined by various hypotheses which lead to an efficient material model.

4 Strain rate dependent interaction relations

For the yield hinge theory the interaction relations in the space of internal forces have to be determined. Here, Bernoulli's hypothesis concerning cross sections remaining plane also in the deformed state is valid in the same way as it is known in the theory of elasticity. The cross sections are double symmetrical and the plane of symmetry is identic with the loading plane. Since data concerning the rate dependent shear stress are not available, a shear force interaction will not be given. Assuming that the cross sections remain plane, a linear distribution of the strain and the strain rates with respect to the cross section is given. With the help of equations (4a,b) the corresponding stress distributions can be determined, which yield the rate dependent interaction relations if integrated over the cross section. To the authors' knowledge, these relations were first determined by Brandt for I-rectangular- and box-type-sections. The derivation of these relations can be found in [7].

Fig. 3 illustrates the analysis of interaction relations for rolled sections of a HE 200 B, with regard to various strain rates ${}_{s}\dot{\varepsilon}$ in the centroidal axis and various curvature rates $\dot{\kappa}$.

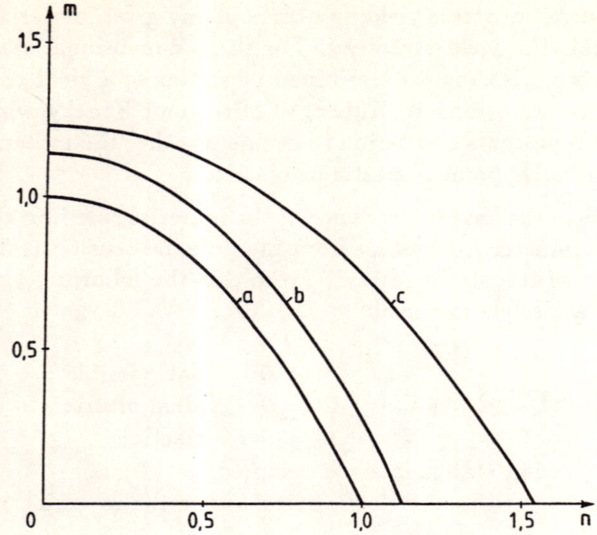

Fig. 3 Bending moment - normal force interaction of a HE 200 B, with
$D = 75\ s^{-1}$ and $p = 3.3$ with regard to
a) $s\dot{\varepsilon} = 0.\ s^{-1}$ and $\dot{\kappa} = 0.\ m^{-1}s^{-1}$,
b) $s\dot{\varepsilon} = 0.1\ s^{-1}$ and $\dot{\kappa} = 1.\ m^{-1}s^{-1}$,
c) $s\dot{\varepsilon} = 10.\ s^{-1}$ and $\dot{\kappa} = 5.\ m^{-1}s^{-1}$.

5 Material parameters D and p

The material parameters D and p (cf. chapter 2) can be obtained by fitting the numerical results to those which found experimentically.

Cowper and Symonds [9] obtained a good agreement between Manjoines's experimental results and their own analytical results by using a parameter combination $D = 40s^{-1}$ and $p = 5$.

Recently, Brandes and Limberger [5] were able to confirm the validity of the results of Cowper and Symonds' for uniaxial tension tests. Experimental results concerning the dependence of the tolerable bending moment on the curvature rate are hardly available. Aspden and Campbell [2] examined beams which are fixed at both ends and have a rectangular cross-section. Fig. 4 illustrates the M–κ–curves which Aspden and Campbell obtained for various curvature rates. The fully-plastic bending moments determined on the basis of the interaction relations for the parameter combination suggested by Cowper and Symonds are marked by straight broken lines.

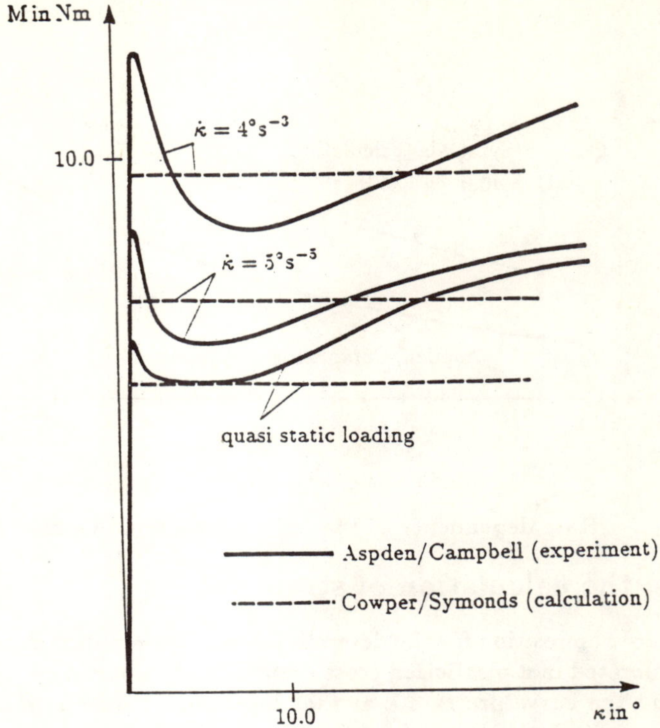

Fig. 4 Bending moment - curvature relations according to Cowper/ Symonds and Aspden/Campbell

The values for the fully-plastic bending moment, calculated according to the parameter combination of Cowper and Symonds, lie above the lower experimental values for the yield moment, i.e. they are on the "unsafe side". Therefore, D and p have been adapted to the rate dependence of the fully-plastic bending moment. For our own analysis we found a combination of $D = 75\,s^{-1}$ and $p = 3.3$, which is consistent with the numerically analyzed fully-plastic moment and the experimental results for the lower yield moment which have been published by Aspden and Campbell (Fig. 5).

With $D = 75\,s^{-1}$ and $p = 3.3$ the lower yield stress is 12 to 18 % below the experimental results for the uniaxial tension test which have been obtained by Manjoine as well as by Brandes and Limberger. However, the fully plastic normal force available cannot be completely taken advantage of by the material parameters D and p. Yet, the dynamic loads computed on the basis of the above-mentioned assumptions are on the "safe side".

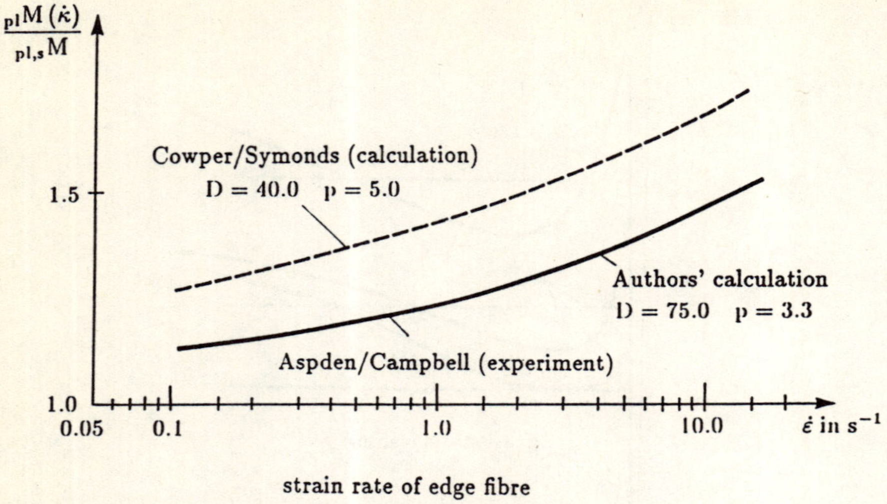

Fig. 5 Rate dependence of the fully-plastic bending moment

6 On the calculation of strain rates

A necessary supposition for the determination of those internal forces which can be tolerated in a plasticized cross section, is the knowledge of the strain rate $\dot{\varepsilon}$ and the curvature rate $\dot{\kappa}$ at that location. It is impossible to carry out an exact analysis in the plasticized cross section by using a yield hinge model.

The strain rate and the curvature rate consist of an elastic and a plastic part in a plasticized cross section:

$$\dot{\varepsilon} = {}_{el}\dot{\varepsilon} + {}_{pl}\dot{\varepsilon} \quad , \tag{6}$$

$$\dot{\kappa} = {}_{el}\dot{\kappa} + {}_{pl}\dot{\kappa} \quad . \tag{7}$$

On the basis of these assumptions and approximations the elastic strain rates ${}_{el}\dot{\varepsilon}$ and curvature rates ${}_{el}\dot{\kappa}$ can be exactly calculated for every cross sections with the help of a finite element formulation using the form functions $\underline{H}$ and the elastic node velocity ${}_{el}\underline{\dot{v}}$. For the mechanical model it is assumed that the zone of plastification of a rod is locally concentrated in only one individual cross section. Therefore, it is impossible to determine the plastic rate. A reference length necessary for the calculation of the strains is not existent. Accordingly, neither the strains nor the strain rates in the yield hinge can be determined.

One possible solution to the problem of determining the plastic strain rates is the assumption of plastic zones with respect to the axis of the rod; feasible for example is the so-called yield zone theory. However, this method leads to a considerable increase in the number of elements per rod. It is therefore not very well applicable to the numerical analysis of entire frame structures.

The determination of the size of the plastic zones by means of parametrization proves to be a reasonable but time consuming approach, since the form and the size of the plastic zones depend on too many parameters, such as type of load, geometry of the system etc. Therefore, a method has been chosen here, which, although being uneconomical as far as the design of the structure is concerned – due to the calculated rates being smaller than the actual ones – is on the "safe side". In the numerical analysis of strains the plastic nodal displacements are treated just like elastic displacements, they can be added. The plastic strains are generally assigned to an area which is much larger than the actually plasticized area of the rod, i.e., the computed plastic strains and the corresponding rates turn out to be smaller than the actual values.

The influence of the length of the element on the strain rates can be shown by comparative calculations which are demonstrated in chapter 8.

Possible errors have to be examined with respect to both the internal forces and the deformations. The results of the numerical analysis of the one-bay, one-story frame (chapter 8) depend on the chosen discretization, i.e. the strain rates had a maximum error of 8 % with respect to the internal forces and of 17 % concerning the rotations and displacements. Thus it can be concluded that these maximum errors are tolerable. On the average they will be much less.

7 Physical modeling of plasticized cross sections

A realistic representation of plasticized cross sections on the basis of the yield hinge theory is of essential importance for the correct description of the plastic load-carrying behaviour. Interaction relations are used to determine only the ultimate load capacity of the cross sections. In extension to this, yield hinge models are not only used for the computation of plastic deformation rates and the energy dissipation in the yield hinges, but also to establish criteria that indicate the situation when a yield hinge is unloaded making that cross section once again elastic. Physical modeling of plasticized cross sections has to be carried out with regard to these aspects.

The widely used mechanical model for the analysis of plasticized rod cross sections is the yield hinge. Kazinczy (1914) introduced the so-called centric yield hinge, while Onat and Prager (1963) were the first to describe the interaction of a plane bending moment and a normal force. For further historical reference see [10].

In a centric yield hinge, the dissipated mechanical energy can only be determined for those internal forces for which the corresponding degree of freedom has been released. This is true because the plastic portion can only be determined for the corresponding deformation component. The thus calculated dissipation energy is merely an adequate approximation of the actual dissipation energy and describes only the behaviour due to plane bending or due to only a normal force respectively shear force acting. Although being an inconsistent combination from the theoretical point of view, the introduction of centric yield hinges in combination with interaction relations offers comparatively satisfactory results for the limit load.

A consistent model can be established by the combination of interaction relations and eccentric yield hinges. Using eccentric yield hinges the kinematic relations of hinge movements can be interpreted as an analysis of shifting the internal forces along the hypersurface of the interaction relations. By using consistent analytical non-linear interaction conditions and by assuming small load increments (no iteration on the interaction surface), a comprehensive description can be obtained for the movements of the yield hinge and the energy that dissipates in the yield hinge. The algorithmic disadvantage of the concept of eccentric yield hinges becomes obvious when the element matrices have to be derived for a second time. In this case, a mere modification of the elastic matrices is not sufficient (Gebbeken [10]).

Moreover, non-symmetrical mass matrices appear (Brandt [6]) when a static condensation of plastic displacements is carried out in the domain of the elements.

Therefore, in addition to the model of centric yield hinges another, more advantageous numerical model will be presented which requires less numerical and algorithmical effort. It is assumed for the above mentioned model that all hypotheses of the theory of plasticity which are valid for the stress-strain-space are also applicable for the internal force-deformation-space. Thus, a theoretically consistent formulation is obtained. However, a kinematic model can generally not be realized in this way. The plastic deformations of plasticized cross sections are determined using Drucker's postulate of material stability (Ueda et al. [14], Argyris et al. [1]).

The following holds for the internal forces in the current configuration at time t:

$$ {}^{t}\underline{s}(t) = {}^{t}\underline{k}(v(t))\, {}^{t}_{el}\underline{v}(t) + {}^{t}_{1}\underline{s}(v(t)) + {}^{t}\underline{m}\, {}^{t}\ddot{\underline{v}}(t) \quad . \tag{8} $$

If a cross section of a rod is fully plasticized, the interaction relation

$$f\{{}^{t}\underline{s}(t), \dot{\varepsilon}(t)\} = 0 \tag{9}$$

must be fulfilled at time t. Under the above mentioned conditions the end point of the vector of the internal forces must lie on the surface of interaction, in the case that an unloading of the cross section does not occur. At time $t + \Delta t$ either

$$f\{{}^{t+\Delta t}\underline{s}(t+\Delta t), \dot{\varepsilon}(t+\Delta t)\} = 0 \tag{10}$$

or

$$\Delta f(t) = f(t+\Delta t) - f(t) = 0 \tag{11}$$

has to be fulfilled for the interaction relation (Δ marks the increment and should not be understood as the Laplace operator). The development of the increment of the interaction relation into a Taylor's series which is approximated by neglecting higher than linear terms yields

$$\begin{aligned} \Delta f(t) &= \frac{\partial f\{{}^{t}\underline{s}(t), \dot{\varepsilon}(t)\}}{\partial\, {}^{t}\underline{s}(t)} \Delta\, {}^{t}\underline{s}(t) \\ &= \operatorname{Grad} f\{{}^{t}\underline{s}(t), \dot{\varepsilon}(t)\}\ \Delta\, {}^{t}\underline{s}(t) \\ &= {}^{t}\underline{\Phi}^{T}(t)\, \Delta\, {}^{t}\underline{s}(t) \end{aligned} \tag{12}$$

with

$$\Delta\, {}^{t}\underline{s}(t) = {}^{t}\underline{s}(t+\Delta t) - {}^{t}\underline{s}(t) \quad . \tag{13}$$

The increment of internal forces reads in the tangential formulation with respect to the metric of the current configuration at time t

$$\Delta\, {}^{t}\underline{s}(t) = {}^{t}_{T}\underline{k}(v(t))\, \Delta\, {}^{t}_{el}\underline{v}(t) + {}^{t}\,\underline{m}\, \Delta\, {}^{t}\underline{\ddot{v}}(t), \tag{14}$$

with

$$\Delta\, {}^{t}_{el}\underline{v}(t) = {}^{t}_{el}\underline{v}(t+\Delta t) - {}^{t}_{el}\underline{v}(t) \quad , \tag{15}$$

$$\Delta\, {}^{t}\underline{\ddot{v}}(t) = {}^{t}\underline{\ddot{v}}(t+\Delta t) - {}^{t}\underline{\ddot{v}}(t) \quad . \tag{16}$$

The total displacements ${}^{t}\underline{v}$ consist of the remaining plastic displacements ${}^{t}_{pl,bl}\underline{v}$, the elastic displacements ${}^{t}_{el}\underline{v}$ and the current plastic displacements in the yield hinge ${}^{t}_{pl,ak}\underline{v}$:

$${}^{t}\underline{v}(t) = {}^{t}_{el}\underline{v}(t) + {}^{t}_{pl,bl}\underline{v}(t) + {}^{t}_{pl,ak}\underline{v}(t) \quad . \tag{17}$$

Drucker's postulate of stability requires that the material behaviour of isothermic deformations is stable. Given this postulate and the condition of convex interaction surfaces in the space of the internal forces, the current plastic deformations - based on Prandtl-Reuss' associated yield rule - generally change to

$$\triangle\, {}^{\;t}_{pl,ak}\underline{v}(t) = \lambda \frac{\partial f(t)}{\partial\, {}^t\underline{s}(t)} = \lambda\, {}^t\underline{\Phi}(t) \quad , \tag{18}$$

or with respect to the nodes i and j of an element read

$$\begin{bmatrix} \triangle\, {}^{\;t}_{pl,ak}\underline{v}_i(t) \\ \\ \triangle\, {}^{\;t}_{pl,ak}\underline{v}_j(t) \end{bmatrix} = \begin{bmatrix} \lambda_i & 0 \\ \\ 0 & \lambda_j \end{bmatrix} \begin{bmatrix} {}^t\underline{\Phi}_i(t) \\ \\ {}^t\underline{\Phi}_j(t) \end{bmatrix} \quad , \tag{19}$$

with proportionality constants λ, or λ_i and λ_j. For λ only non-negative values are allowed. At $\lambda < 0$, an unloading of the cross section takes place and it behaves elastically again. At $\lambda \geq 0$ the cross section remains plasticized and a alteration of internal forces within the frame structure takes place.

Equation (18) means that the vector of the plastic displacement increments has to coincide with the normal vector of the interaction curve in the space of the internal forces.

The remaining plastic displacements do not change in a plasticized cross section within a given time increment Δt. Accordingly, (14) with (17) can be expressed by the following equation:

$$\begin{bmatrix} \triangle\, {}^t\underline{s}_i(t) \\ \\ \triangle\, {}^t\underline{s}_j(t) \end{bmatrix} = \begin{bmatrix} {}^t_T\underline{k}_{ii}(v(t)) & {}^t_T\underline{k}_{ij}(v(t)) \\ \\ {}^t_T\underline{k}_{ji}(v(t)) & {}^t_T\underline{k}_{jj}(v(t)) \end{bmatrix} \begin{bmatrix} \triangle\, {}^t\underline{v}_i(t) - \triangle\, {}^{\;t}_{pl,ak}\underline{v}_i(t) \\ \\ \triangle\, {}^t\underline{v}_j(t) - \triangle\, {}^{\;t}_{pl,ak}\underline{v}_j(t) \end{bmatrix}$$
$$+ \begin{bmatrix} {}^t m_{ii} & {}^t m_{ij} \\ \\ {}^t m_{ji} & {}^t m_{jj} \end{bmatrix} \begin{bmatrix} \triangle\ddot{\underline{v}}_i(t) \\ \\ \triangle\ddot{\underline{v}}_j(t) \end{bmatrix} \quad . \tag{20}$$

By inserting (18) into (20), the incremental alteration of internal forces in the end nodes of a rod element with plasticized cross sections becomes

$$\begin{bmatrix} \triangle\, {}^t\underline{s}_i(t) \\ \\ \triangle\, {}^t\underline{s}_j(t) \end{bmatrix} = \begin{bmatrix} {}^t_T\underline{k}_{ii}(v(t)) & {}^t_T\underline{k}_{ij}(v(t)) \\ \\ {}^t_T\underline{k}_{ji}(v(t)) & {}^t_T\underline{k}_{jj}(v(t)) \end{bmatrix} \begin{bmatrix} \triangle\, {}^t\underline{v}_i(t) - \lambda_i\, {}^t\underline{\Phi}_i(t) \\ \\ \triangle\, {}^t\underline{v}_j(t) - \lambda_j\, {}^t\underline{\Phi}_j(t) \end{bmatrix}$$
$$+ \begin{bmatrix} {}^t m_{ii} & {}^t m_{ij} \\ \\ {}^t m_{ji} & {}^t m_{jj} \end{bmatrix} \begin{bmatrix} \triangle\ddot{\underline{v}}_i(t) \\ \\ \triangle\ddot{\underline{v}}_j(t) \end{bmatrix} \quad . \tag{21}$$

Simultaneously, the increment of the internal forces has to meet the requirements of the alteration of the interaction equation (11):

$$\begin{bmatrix} \Delta f_i(t) \\ \Delta f_j(t) \end{bmatrix} = \begin{bmatrix} {}^t\underline{\Phi}_i^T(t) & 0 \\ 0 & {}^t\underline{\Phi}_j^T(t) \end{bmatrix} \begin{bmatrix} \Delta\, {}^t\underline{s}_i(t) \\ \Delta\, {}^t\underline{s}_j(t) \end{bmatrix} \stackrel{!}{=} \begin{bmatrix} 0 \\ 0 \end{bmatrix} \quad . \tag{22}$$

The constants λ_i and λ_j can be determined by (21) and (22). Inserting (21) into (22) one gets

$$\begin{aligned} 0 = {} & {}^t\underline{\Phi}_i^T(t)\, {}^t_T\underline{k}_{ii}(v(t))\Delta\, {}^t\underline{v}_i(t) - \lambda_i c_{ii} + \\ & {}^t\underline{\Phi}_i^T(t)\, {}^t_T\underline{k}_{ij}(v(t))\Delta\, {}^t\underline{v}_j(t) - \lambda_j c_{ij} + \\ & {}^t\underline{\Phi}_i^T(t)\, {}^t m_{ii}\, \Delta\underline{\ddot{v}}_i(t) + {}^t\underline{\Phi}_i^T(t)\, {}^t m_{ij}\, \Delta\underline{\ddot{v}}_j(t) \quad , \end{aligned} \tag{23 a}$$

$$\begin{aligned} 0 = {} & {}^t\underline{\Phi}_j^T(t)\, {}^t_T\underline{k}_{ji}(v(t))\Delta\, {}^t\underline{v}_i(t) - \lambda_i c_{ji} + \\ & {}^t\underline{\Phi}_j^T(t)\, {}^t_T\underline{k}_{jj}(v(t))\Delta\, {}^t\underline{v}_j(t) - \lambda_j c_{jj} + \\ & {}^t\underline{\Phi}_j^T(t)\, {}^t m_{ji}\, \Delta\underline{\ddot{v}}_i(t) + {}^t\underline{\Phi}_j^T(t)\, {}^t m_{jj}\, \Delta\underline{\ddot{v}}_j(t) \quad , \end{aligned} \tag{23 b}$$

with the scalars

$$c_{ii} = {}^t\underline{\Phi}_i^T(t)\, {}^t_T\underline{k}_{ii}(v(t))\, {}^t\underline{\Phi}_i(t) \quad , \tag{24 a}$$
$$c_{ij} = {}^t\underline{\Phi}_i^T(t)\, {}^t_T\underline{k}_{ij}(v(t))\, {}^t\underline{\Phi}_j(t) \quad , \tag{24 b}$$
$$c_{ji} = {}^t\underline{\Phi}_j^T(t)\, {}^t_T\underline{k}_{ji}(v(t))\, {}^t\underline{\Phi}_i(t) \quad , \tag{24 c}$$
$$c_{jj} = {}^t\underline{\Phi}_j^T(t)\, {}^t_T\underline{k}_{jj}(v(t))\, {}^t\underline{\Phi}_j(t) \quad . \tag{24 d}$$

Equations (23 a,b) can be solved with respect to λ_i and λ_j

$$\begin{aligned} \lambda_i = {} & \left\{ c_4\, {}^t\underline{\Phi}_i^T(t)\, {}^t_T\underline{k}_{ii}(v(t)) - c_2\, {}^t\underline{\Phi}_j^T(t)\, {}^t_T\underline{k}_{ji}(v(t)) \right\} \Delta\, {}^t\underline{v}_i(t) + \\ & \left\{ c_4\, {}^t\underline{\Phi}_i^T(t)\, {}^t_T\underline{k}_{ij}(v(t)) - c_2\, {}^t\underline{\Phi}_j^T(t)\, {}^t_T\underline{k}_{jj}(v(t)) \right\} \Delta\, {}^t\underline{v}_j(t) + \\ & \left\{ c_4\, {}^t\underline{\Phi}_i^T(t)\, {}^t\underline{m}_{ii} - c_2\, {}^t\underline{\Phi}_j^T(t)\, {}^t\underline{m}_{ji} \right\} \Delta\, {}^t\underline{\ddot{v}}_i(t) + \\ & \left\{ c_4\, {}^t\underline{\Phi}_i^T(t)\, {}^t\underline{m}_{ij} - c_2\, {}^t\underline{\Phi}_j^T(t)\, {}^t\underline{m}_{jj} \right\} \Delta\, {}^t\underline{\ddot{v}}_j(t) \quad , \end{aligned} \tag{25 a}$$

$$\begin{aligned} \lambda_j = {} & \left\{ c_3\, {}^t\underline{\Phi}_i^T(t)\, {}^t_T\underline{k}_{ii}(v(t)) - c_1\, {}^t\underline{\Phi}_j^T(t)\, {}^t_T\underline{k}_{ji}(v(t)) \right\} \Delta\, {}^t\underline{v}_i(t) + \\ & \left\{ c_3\, {}^t\underline{\Phi}_i^T(t)\, {}^t_T\underline{k}_{ij}(v(t)) - c_1\, {}^t\underline{\Phi}_j^T(t)\, {}^t_T\underline{k}_{jj}(v(t)) \right\} \Delta\, {}^t\underline{v}_j(t) + \\ & \left\{ c_3\, {}^t\underline{\Phi}_i^T(t)\, {}^t\underline{m}_{ii} - c_1\, {}^t\underline{\Phi}_j^T(t)\, {}^t\underline{m}_{ji} \right\} \Delta\, {}^t\underline{\ddot{v}}_i(t) + \\ & \left\{ c_3\, {}^t\underline{\Phi}_i^T(t)\, {}^t\underline{m}_{ij} - c_1\, {}^t\underline{\Phi}_j^T(t)\, {}^t\underline{m}_{jj} \right\} \Delta\, {}^t\underline{\ddot{v}}_j(t) \quad , \end{aligned} \tag{25 b}$$

with the factors

$$c_1 = \frac{c_{ii}}{c_{ij}\ c_{ji} - c_{ii}\ c_{jj}} \quad , \tag{26 a}$$

$$c_2 = \frac{c_{ij}}{c_{ii}\ c_{jj} - c_{ij}\ c_{ji}} \quad , \tag{26 b}$$

$$c_3 = \frac{c_{ji}}{c_{ij}\ c_{ji} - c_{ii}\ c_{jj}} \quad , \tag{26 c}$$

$$c_4 = \frac{c_{jj}}{c_{ii}\ c_{jj} - c_{ij}\ c_{ji}} \quad . \tag{26 d}$$

The following relation for the increment of the internal forces can be obtained by inserting (25 a,b) into (21)

$$\begin{bmatrix} \triangle\, {}^t\underline{s}_i(t) \\ \triangle\, {}^t\underline{s}_j(t) \end{bmatrix} = \begin{bmatrix} {}^{\ t}_{T,pl}\underline{k}_{ii}(v(t)) & {}^{\ t}_{T,pl}\underline{k}_{ij}(v(t)) \\ {}^{\ t}_{T,pl}\underline{k}_{ji}(v(t)) & {}^{\ t}_{T,pl}\underline{k}_{jj}(v(t)) \end{bmatrix} \begin{bmatrix} \triangle\, {}^t\underline{v}_i(t) \\ \triangle\, {}^t\underline{v}_j(t) \end{bmatrix} + \begin{bmatrix} {}^{\ t}_{T,pl}m_{ii}(t) & {}^{\ t}_{T,pl}m_{ij}(t) \\ {}^{\ t}_{T,pl}m_{ji}(t) & {}^{\ t}_{T,pl}m_{jj}(t) \end{bmatrix} \begin{bmatrix} \triangle\ddot{\underline{v}}_i(t) \\ \triangle\ddot{\underline{v}}_j(t) \end{bmatrix} \quad . \tag{27}$$

This is equivalent to the notation

$$\triangle\, {}^t\underline{s}(t) = {}^{\ t}_{T,pl}\underline{k}(v(t))\ \triangle\, {}^t\underline{v}(t) + {}^{\ t}_{T,pl}\underline{m}(t)\ \triangle\, {}^t\ddot{\underline{v}}(t) \tag{28}$$

with

$$\begin{aligned} {}^{\ t}_{T,pl}\underline{k}_{ii}(v) = {} & {}^t_T\underline{k}_{ii}(v) - \\ & {}^t_T\underline{k}_{ii}(v)\ {}^t\underline{\Phi}_i\ \left\{ c_4\ {}^t\underline{\Phi}_i^T\ {}^t_T\underline{k}_{ii}(v) - c_2\ {}^t\underline{\Phi}_j^T\ {}^t_T\underline{k}_{ji}(v) \right\} - \\ & {}^t_T\underline{k}_{ij}(v)\ {}^t\underline{\Phi}_j\ \left\{ c_3\ {}^t\underline{\Phi}_i^T\ {}^t_T\underline{k}_{ii}(v) - c_1\ {}^t\underline{\Phi}_j^T\ {}^t_T\underline{k}_{ji}(v) \right\} \quad , \end{aligned} \tag{29 a}$$

$$\begin{aligned} {}^{\ t}_{T,pl}\underline{k}_{ij}(v) = {} & {}^t_T\underline{k}_{ij}(v) - \\ & {}^t_T\underline{k}_{ii}(v)\ {}^t\underline{\Phi}_i\ \left\{ c_4\ {}^t\underline{\Phi}_i^T\ {}^t_T\underline{k}_{ij}(v) - c_2\ {}^t\underline{\Phi}_j^T\ {}^t_T\underline{k}_{jj}(v) \right\} - \\ & {}^t_T\underline{k}_{ij}(v)\ {}^t\underline{\Phi}_j\ \left\{ c_3\ {}^t\underline{\Phi}_i^T\ {}^t_T\underline{k}_{ij}(v) - c_1\ {}^t\underline{\Phi}_j^T\ {}^t_T\underline{k}_{jj}(v) \right\} \quad , \end{aligned} \tag{29 b}$$

$$\begin{aligned} {}^{\ t}_{T,pl}\underline{k}_{ji}(v) = {} & {}^t_T\underline{k}_{ji}(v) - \\ & {}^t_T\underline{k}_{ji}(v)\ {}^t\underline{\Phi}_i\ \left\{ c_4\ {}^t\underline{\Phi}_i^T\ {}^t_T\underline{k}_{ii}(v) - c_2\ {}^t\underline{\Phi}_j^T\ {}^t_T\underline{k}_{ji}(v) \right\} - \end{aligned} \tag{29 c}$$

$$ {}^{t}_{T}\underline{k}_{jj}(v)\, {}^{t}\underline{\Phi}_j \left\{ c_3\, {}^{t}\underline{\Phi}_i^T\, {}^{t}_{T}\underline{k}_{ii}(v) - c_1\, {}^{t}\underline{\Phi}_j^T\, {}^{t}_{T}\underline{k}_{ji}(v) \right\} \quad , $$

$$ \begin{aligned} {}^{\;t}_{T,pl}\underline{k}_{jj}(v) = {} & {}^{t}_{T}\underline{k}_{jj}(v) - \\ & {}^{t}_{T}\underline{k}_{ji}(v)\, {}^{t}\underline{\Phi}_i \left\{ c_4\, {}^{t}\underline{\Phi}_i^T\, {}^{t}_{T}\underline{k}_{ij}(v) - c_2\, {}^{t}\underline{\Phi}_j^T\, {}^{t}_{T}\underline{k}_{jj}(v) \right\} - \\ & {}^{t}_{T}\underline{k}_{jj}(v)\, {}^{t}\underline{\Phi}_j \left\{ c_3\, {}^{t}\underline{\Phi}_i^T\, {}^{t}_{T}\underline{k}_{ij}(v) - c_1\, {}^{t}\underline{\Phi}_j^T\, {}^{t}_{T}\underline{k}_{jj}(v) \right\} \quad , \end{aligned} \tag{29 d} $$

and

$$ \begin{aligned} {}^{\;t}_{T,pl}\underline{m}_{ii} = {} & {}^{t}\underline{m}_{ii} - \\ & {}^{t}_{T}\underline{k}_{ii}(v)\, {}^{t}\underline{\Phi}_i \left\{ c_4\, {}^{t}\underline{\Phi}_i^T\, {}^{t}\underline{m}_{ii} - c_2\, {}^{t}\underline{\Phi}_j^T\, {}^{t}\underline{m}_{ji} \right\} - \\ & {}^{t}_{T}\underline{k}_{ij}(v)\, {}^{t}\underline{\Phi}_j \left\{ c_3\, {}^{t}\underline{\Phi}_i^T\, {}^{t}\underline{m}_{ii} - c_1\, {}^{t}\underline{\Phi}_j^T\, {}^{t}\underline{m}_{ji} \right\} \quad , \end{aligned} \tag{30 a} $$

$$ \begin{aligned} {}^{\;t}_{T,pl}\underline{m}_{ij} = {} & {}^{t}\underline{m}_{ij} - \\ & {}^{t}_{T}\underline{k}_{ii}(v)\, {}^{t}\underline{\Phi}_i \left\{ c_4\, {}^{t}\underline{\Phi}_i^T\, {}^{t}\underline{m}_{ij} - c_2\, {}^{t}\underline{\Phi}_j^T\, {}^{t}\underline{m}_{jj} \right\} - \\ & {}^{t}_{T}\underline{k}_{ij}(v)\, {}^{t}\underline{\Phi}_j \left\{ c_3\, {}^{t}\underline{\Phi}_i^T\, {}^{t}\underline{m}_{ij} - c_1\, {}^{t}\underline{\Phi}_j^T\, {}^{t}\underline{m}_{jj} \right\} \quad , \end{aligned} \tag{30 b} $$

$$ \begin{aligned} {}^{\;t}_{T,pl}\underline{m}_{ji} = {} & {}^{t}\underline{m}_{ji} - \\ & {}^{t}_{T}\underline{k}_{ji}(v)\, {}^{t}\underline{\Phi}_i \left\{ c_4\, {}^{t}\underline{\Phi}_i^T\, {}^{t}\underline{m}_{ii} - c_2\, {}^{t}\underline{\Phi}_j^T\, {}^{t}\underline{m}_{ji} \right\} - \\ & {}^{t}_{T}\underline{k}_{jj}(v)\, {}^{t}\underline{\Phi}_j \left\{ c_3\, {}^{t}\underline{\Phi}_i^T\, {}^{t}\underline{m}_{ii} - c_1\, {}^{t}\underline{\Phi}_j^T\, {}^{t}\underline{m}_{ji} \right\} \quad , \end{aligned} \tag{30 c} $$

$$ \begin{aligned} {}^{\;t}_{T,pl}\underline{m}_{jj} = {} & {}^{t}\underline{m}_{jj} - \\ & {}^{t}_{T}\underline{k}_{ji}(v)\, {}^{t}\underline{\Phi}_i \left\{ c_4\, {}^{t}\underline{\Phi}_i^T\, {}^{t}\underline{m}_{ij} - c_2\, {}^{t}\underline{\Phi}_j^T\, {}^{t}\underline{m}_{jj} \right\} - \\ & {}^{t}_{T}\underline{k}_{jj}(v)\, {}^{t}\underline{\Phi}_j \left\{ c_3\, {}^{t}\underline{\Phi}_i^T\, {}^{t}\underline{m}_{ij} - c_1\, {}^{t}\underline{\Phi}_j^T\, {}^{t}\underline{m}_{jj} \right\} \quad . \end{aligned} \tag{30 d} $$

The dependence of each individual quantity on the time (cf. (25 a,b) and (21)) is not shown in the equations (29 a-d, 30 a-d, 34 a-d and 37 a,b).

Those cases in which the cross section is only plasticized in either node i or node j are described as special cases. ${}^{t}\underline{\Phi}_j(t)$ does not exist when the cross section is only plasticized in node i. This fact can be considered formally by

$$ {}^{t}\underline{\Phi}_j(t) = \underline{0} \tag{31} $$

in (21) and (22). The parameter λ_i can be derived in the same manner:

$$\lambda_i = \frac{1}{c_{ii}}\, {}^t\underline{\Phi}_i^T(t)\, {}^t_T\underline{k}_{ii}(v(t))\, \triangle\, {}^t\underline{v}_i(t) + \frac{1}{c_{ii}}\, {}^t\underline{\Phi}_i^T(t)\, {}^t_T\underline{k}_{ij}(v(t))\, \triangle\, {}^t\underline{v}_j(t) + \frac{1}{c_{ii}}\, {}^t\underline{\Phi}_i^T(t)\, {}^t\underline{m}_{ii} - \triangle\, {}^t\ddot{\underline{v}}_i(t) + \frac{1}{c_{ii}}\, {}^t\underline{\Phi}_i^T(t)\, {}^t\underline{m}_{ij} - \triangle\, {}^t\ddot{\underline{v}}_j(t) \quad . \tag{32}$$

The components of ${}^t_{T,pl}\underline{k}(v(t))$ and ${}^t_{T,pl}\underline{m}$ become

$${}^t_{T,pl}\underline{k}_{ii}(v) = {}^t_T\underline{k}_{ii}(v) - \frac{1}{c_{ii}}\, {}^t_T\underline{k}_{ii}(v)\, {}^t\underline{\Phi}_i\, {}^t\underline{\Phi}_i^T\, {}^t_T\underline{k}_{ii}(v) \quad , \tag{33 a}$$

$${}^t_{T,pl}\underline{k}_{ij}(v) = {}^t_T\underline{k}_{ij}(v) - \frac{1}{c_{ii}}\, {}^t_T\underline{k}_{ii}(v)\, {}^t\underline{\Phi}_i\, {}^t\underline{\Phi}_i^T\, {}^t_T\underline{k}_{ij}(v) \quad , \tag{33 b}$$

$${}^t_{T,pl}\underline{k}_{ji}(v) = {}^t_T\underline{k}_{ji}(v) - \frac{1}{c_{ii}}\, {}^t_T\underline{k}_{ji}(v)\, {}^t\underline{\Phi}_i\, {}^t\underline{\Phi}_i^T\, {}^t_T\underline{k}_{ii}(v) \quad , \tag{33 c}$$

$${}^t_{T,pl}\underline{k}_{jj}(v) = {}^t_T\underline{k}_{jj}(v) - \frac{1}{c_{ii}}\, {}^t_T\underline{k}_{ji}(v)\, {}^t\underline{\Phi}_i\, {}^t\underline{\Phi}_i^T\, {}^t_T\underline{k}_{ij}(v) \quad , \tag{33 d}$$

$${}^t_{T,pl}\underline{m}_{ii} = {}^t\underline{m}_{ii} - \frac{1}{c_{ii}}\, {}^t_T\underline{k}_{ii}(v)\, {}^t\underline{\Phi}_i\, {}^t\underline{\Phi}_i^T\, {}^t\underline{m}_{ii} \quad , \tag{34 a}$$

$${}^t_{T,pl}\underline{m}_{ij} = {}^t\underline{m}_{ij} - \frac{1}{c_{ii}}\, {}^t_T\underline{k}_{ii}(v)\, {}^t\underline{\Phi}_i\, {}^t\underline{\Phi}_i^T\, {}^t\underline{m}_{ij} \quad , \tag{34 b}$$

$${}^t_{T,pl}\underline{m}_{ji} = {}^t\underline{m}_{ji} - \frac{1}{c_{ii}}\, {}^t_T\underline{k}_{ji}(v)\, {}^t\underline{\Phi}_i\, {}^t\underline{\Phi}_i^T\, {}^t\underline{m}_{ii} \quad , \tag{34 c}$$

$${}^t_{T,pl}\underline{m}_{jj} = {}^t\underline{m}_{jj} - \frac{1}{c_{ii}}\, {}^t_T\underline{k}_{ji}(v)\, {}^t\underline{\Phi}_i\, {}^t\underline{\Phi}_i^T\, {}^t\underline{m}_{ij} \quad . \tag{34 d}$$

If only the cross section at node j is plasticized, the following holds accordingly:

$${}^t\underline{\Phi}_i(t) = \underline{0} \quad . \tag{35}$$

In equation (28) the tangential formulation for the alteration of internal forces with respect to the element is considered. If secant matrices are used, the equation is

$$\triangle\, {}^{t}\underline{s}(t) \;=\; {}^{t}_{pl}\underline{k}(v(t))\, \triangle\, {}^{t}\underline{v}(t) \;+\; {}^{\;\;t}_{1,pl}\underline{s}(v(t)) \;+\; {}^{t}_{pl}\underline{m}(t)\, \triangle\, {}^{t}\ddot{\underline{v}}(t) \qquad (36)$$

with

$$\begin{aligned} {}^{\;\;t}_{1,pl}\underline{s}_i(v) \;=\;& {}^{t}_{1}\underline{s}_i(v) \;- \qquad (37\,a)\\ & {}^{t}\underline{k}_{ii}(v)\, {}^{t}\underline{\Phi}_i \left\{c_4\, {}^{t}\underline{\Phi}_i^T - c_2\, {}^{t}\underline{\Phi}_j^T\right\} \left\{{}^{t}_{1}\underline{s}_i(v) + {}^{t}_{1}\underline{s}_j(v)\right\} \;-\\ & {}^{t}\underline{k}_{ij}(v)\, {}^{t}\underline{\Phi}_j \left\{c_3\, {}^{t}\underline{\Phi}_i^T - c_1\, {}^{t}\underline{\Phi}_j^T\right\} \left\{{}^{t}_{1}\underline{s}_i(v) + {}^{t}_{1}\underline{s}_j(v)\right\} \quad , \end{aligned}$$

$$\begin{aligned} {}^{\;\;t}_{1,pl}\underline{s}_j(v) \;=\;& {}^{t}_{1}\underline{s}_j(v) \;- \qquad (37\,b)\\ & {}^{t}\underline{k}_{ji}(v)\, {}^{t}\underline{\Phi}_i \left\{c_4\, {}^{t}\underline{\Phi}_i^T - c_2\, {}^{t}\underline{\Phi}_j^T\right\} \left\{{}^{t}_{1}\underline{s}_i(v) + {}^{t}_{1}\underline{s}_j(v)\right\} \;-\\ & {}^{t}\underline{k}_{jj}(v)\, {}^{t}\underline{\Phi}_j \left\{c_3\, {}^{t}\underline{\Phi}_i^T - c_1\, {}^{t}\underline{\Phi}_j^T\right\} \left\{{}^{t}_{1}\underline{s}_i(v) + {}^{t}_{1}\underline{s}_j(v)\right\} \quad . \end{aligned}$$

If the components of ${}^{t}_{pl}\underline{k}(v(t))$ are calculated with respect to (29 a-d) or (23 a-d), ${}^{t}_{T}\underline{k}(v)$ is to be replaced by ${}^{t}\underline{k}(v)$. The same is true for ${}^{t}_{pl}\underline{m}(t)$. The tangential incremental equation of motion of the entire system is

$$\triangle\, \underline{P}\,(t) \;=\; {}_{T,pl}\underline{K}\,(v(t))\, \triangle\, \underline{V}\,(t) \;+\; {}_{T,pl}\underline{M}\,(t)\, \triangle\, \ddot{\underline{V}}\,(t) \qquad (38)$$

and, following the application of the Newmark-procedure it is

$$\triangle\, \hat{\underline{P}}\,(t + \triangle t) \;=\; {}_{pl,T}\hat{\underline{K}}\,(v(t + \triangle t))\, \triangle\, \underline{V}\,(t + \triangle t) \quad . \qquad (39)$$

The effective load increment $\triangle \hat{\underline{P}}\,(t + \triangle t)$ of $\triangle t$ is

$$\triangle\, \hat{\underline{P}}\,(t + \triangle t) \;=\; \hat{\underline{P}}\,(t + \triangle t) \;-\; \hat{\underline{P}}\,(t) \qquad (40)$$

with

$$\hat{\underline{P}}(t{+}\triangle t) \;=\; \underline{P}(t{+}\triangle t) + {}_{pl}\underline{M}\,(t{+}\triangle t) \left[a_1\, \underline{V}(t) \;+\; a_2\, \dot{\underline{V}}(t) \;+\; a_3\, \ddot{\underline{V}}(t)\right] \quad . \qquad (41)$$

The same is true for $\hat{\underline{P}}(t)$. The resulting effective plastic tangential stiffness matrix finally becomes:

$${}_{pl,T}\hat{\underline{K}}\,(v\,(t + \triangle t)) \;=\; {}_{pl,T}\underline{K}\,(v\,(t + \triangle t)) \;+\; a_1\; {}_{T,pl}\underline{M}\,(t + \triangle t) \quad . \qquad (42)$$

The scheme of calculation is as follows:

1) Given the values

$$\underline{V}(t),\ \dot{\underline{V}}(t),\ \ddot{\underline{V}}(t) \text{ and } {}_{bl,pl}\underline{V}(t),\ {}_{pl}\underline{K}(v(t)),\ {}_{pl}\underline{M}(t),$$
$${}_{pl,T}\underline{K}(v(t)),\ {}_{pl,T}\underline{M}(t),\ {}_{1,pl}\underline{S}(v(t)) \text{ and } \hat{\underline{P}}(t),$$

at time t one is able to carry out the

2) calculation of the effective load increment vector of the time interval Δt

$$\Delta \hat{\underline{P}}(t+\Delta t) = \hat{\underline{P}}(t+\Delta t) - \hat{\underline{P}}(t) \quad , \tag{40}$$

3) the calculation of the effective plastic tangential stiffness matrix

$${}_{pl,T}\hat{\underline{K}}(v(t)) = {}_{pl,T}\underline{K}(v(t)) + a_1\ {}_{T,pl}\underline{M}(t) \quad , \tag{42}$$

4) solution of the equation system

$$\Delta \hat{\underline{P}}(t+\Delta t) = {}_{pl,T}\hat{\underline{K}}(v(t))\, \Delta \underline{V}_n(t+\Delta t) \quad , \tag{43}$$

5) calculation of internal forces and displacements at time $t+\Delta t$

$$\underline{V}_n(t+\Delta t), \dot{\underline{V}}_n(t+\Delta t) \text{ and } \ddot{\underline{V}}_n(t+\Delta t),$$

6) calculation of the ratio of the Euclidean norm of the increments of displacements

$$\text{QuotiV} = \frac{||\Delta \underline{V}_n(t+\Delta t)||}{||\underline{V}_n(t+\Delta t)||} \quad , \tag{44}$$

7) calculation of system matrices at time $t+\Delta t$

$${}_{pl}\underline{K}_n(v(t+\Delta t)),{}_{pl}\underline{M}_n(t+\Delta t) \text{ and } {}_{1,pl}\underline{S}_n(v(t+\Delta t)),$$

8) calculation of the residuum

$$\begin{aligned}\hat{\underline{\Psi}}_n(t+\Delta t) &= {}_{pl}\underline{K}_n(v(t+\Delta t))\ \underline{V}_n(t+\Delta t) + \\ &\quad {}_{pl}\underline{M}_n(t+\Delta t)\ \ddot{\underline{V}}_n(t+\Delta t) + \\ &\quad {}_{1,pl}\underline{S}_n(v(t+\Delta t)) - \underline{P}(t+\Delta t),\end{aligned} \tag{45}$$

9) calculation of the ratio of the Euclidean norm of the residuum

$$\text{QuotiK} = \frac{||\underline{\Psi}_n(t+\Delta t)||}{||\underline{P}(t+\Delta t)||} \quad . \tag{46}$$

10) In case QuotiV or QuotiK are larger than the stipulated upper limits ${}_v\eta$ respectively ${}_k\eta$, equation

$$\hat{\underline{\Psi}}_n(t+\Delta t) = {}_{pl,T}\hat{\underline{K}}(v(t))\,\Delta\underline{\dot{V}}_{n+1}(t+\Delta t) \tag{47}$$

has to be solved. The procedure continues with 5).

11) Check of the interaction relations at the nodes of the rod ends.
If

$$f\{{}^t\underline{s}(t), \dot{\varepsilon}(t)\} = 0 \quad \wedge \quad f\{{}^{t+\Delta t}\underline{s}_n(t+\Delta t), \dot{\varepsilon}_n(t+\Delta t)\} \geq {}_f\eta \tag{9}$$

the internal forces of the corresponding rod ends are reduced to the interaction curve, and QuotiV is assumed to be larger than ${}_v\eta$. The procedure continues with 7).
If

$$f\{{}^t\underline{s}(t), \dot{\varepsilon}(t)\} < 0 \quad \wedge \quad f\{{}^{t+\Delta t}\underline{s}_n(t+\Delta t), \dot{\varepsilon}_n(t+\Delta t)\} \geq 0 \tag{9}$$

the plastification of the cross sections is considered, and QuotiV has to be chosen larger than ${}_v\eta$. The procedure continues with 7).

12) Calculation of the increment of the dissipation energy ΔD in the plasticized cross sections within time interval Δt:

$$\Delta D = \frac{{}^{t+\Delta t}_{pl}\underline{s}(t) + {}^{t+\Delta t}_{pl}\underline{s}(t+\Delta t)}{2}\left[{}^{t+\Delta t}_{pl,ak}\underline{v}(t+\Delta t) - {}^{t+\Delta t}_{pl,ak}\underline{v}(t)\right] . \tag{48}$$

If $\Delta D < 0$ in one yield hinge, this yield hinge has become elastic again. After the calculation of ${}_{bl,pl}\underline{V}(t+\Delta t)$, it has to be continued with 7).

13) The calculated state of equilibrium meets the requirements of the interaction conditions in all plasticized cross sections. The dissipation energy is in all yield hinges non-negative. Consequently, the iterative calculation of the internal forces and the displacements at time $t+\Delta t$ has been completed.

Annotations:

To the authors' knowledge the model presented in this paper which incorporates the associated yield rule is best suited for the numerical analysis of the load-carrying behaviour of plasticized cross sections in steel frame structures. Being consistent in its derivation, this model requires only a little numerical effort as compared to the eccentric yield hinges which have been derived in [10]. Moreover, with λ_i or λ_j this model offers a reliable criterion to determine the unloaded state of plasticized cross sections and allows the calculation of all components of the plastic deformation rates with the help of the associated yield rule.

8 Numerical results

In [6] various rod structures have been examined which were exposed to earthquake and impact loads. The comparative calculations in this paper are based on the frame illustrated in Fig. 6a which is subjected to a load-time function as shown in Fig. 6b. Damping effects were not considered in order to make sure that the analysis remains free of further influences. The material law is linear elastic-visco plastic as described in chapters 3 and 4.

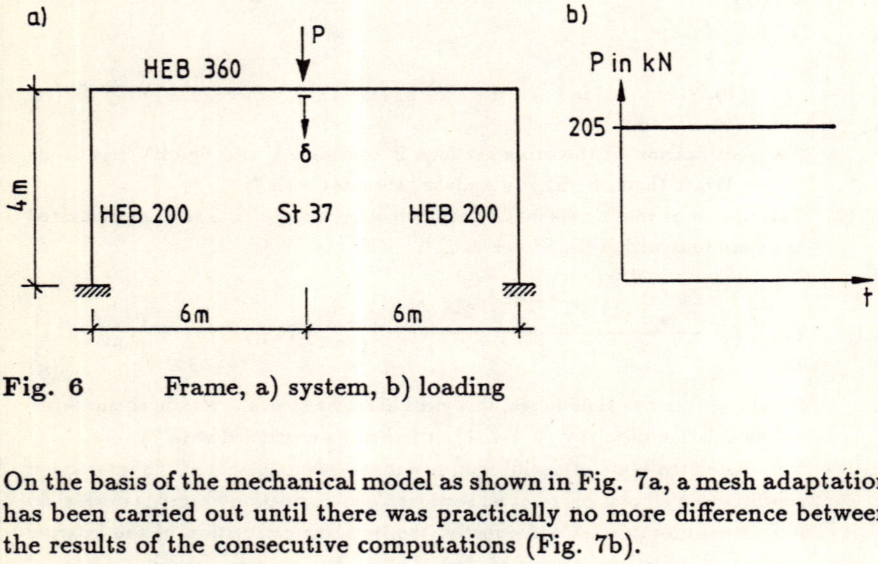

Fig. 6 Frame, a) system, b) loading

On the basis of the mechanical model as shown in Fig. 7a, a mesh adaptation has been carried out until there was practically no more difference between the results of the consecutive computations (Fig. 7b).

The parameters needed for the calculation were chosen as

$$_{v}\eta = {}_{k}\eta = 0.0001 \quad , \quad {}_{f}\eta = 0.005 \quad ,$$

$$D = 75.0 \quad \text{and} \quad p = 3.3,$$

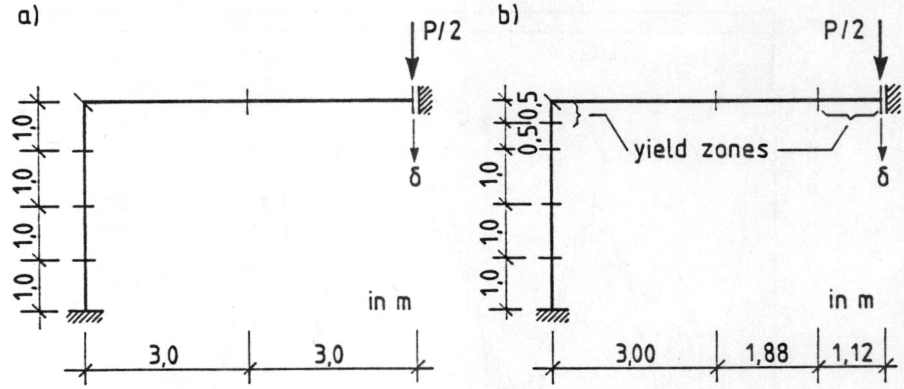

Fig. 7 Frame, discretization, a) initial mesh, b) mesh after an element adaptation

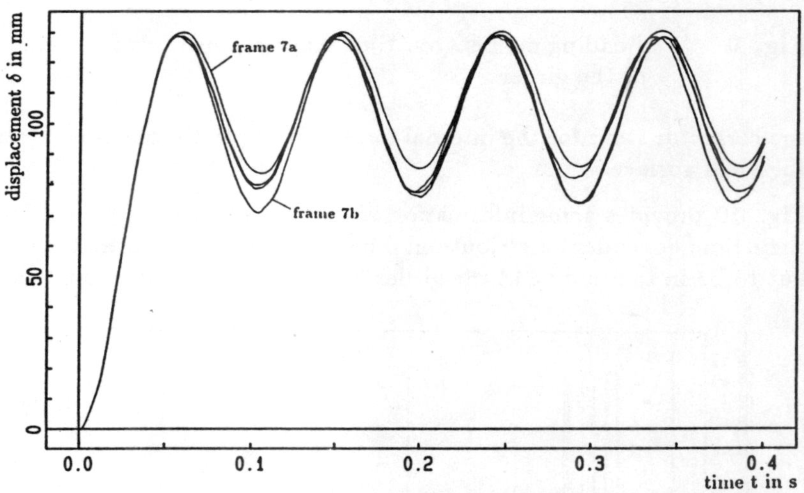

Fig. 8 Displacement vs. time diagram with respect to δ in the middle of the girder

while the time increment Δt was taken as 1 ms.

Fig. 8 shows the displacement vs. time diagrams with respect to the middle of the girder. The maximum error of the vertical displacement δ amounts to 17 %. Two additional graphs demonstrate the results of different states of the adaptation procedure.

The maximum error of the internal forces was found for the bending moment in the middle of the girder and amounts to 8 %. The corresponding bending moment vs. time diagram is given in Fig. 9. The little peaks in the curves concerning frame 7b result from the required accuracy ${}_F\eta$ while computing

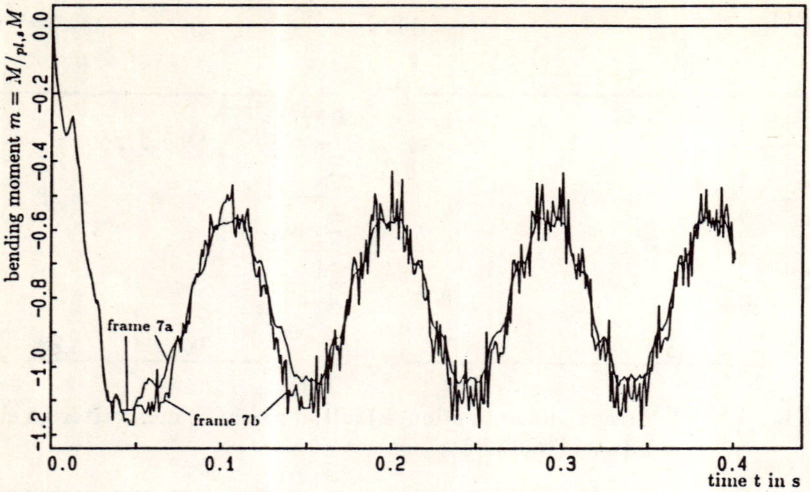

Fig. 9 Bending moment vs. time diagram with respect to the middle of the girder

the corrector steps for the internal forces of the plasticized cross sections on the yield surface.

Fig. 10 provides some information about the number of yield hinges and their time dependent distribution. The location of the plastic hinges turned out to be in the middle of the girder and in the top of the columns.

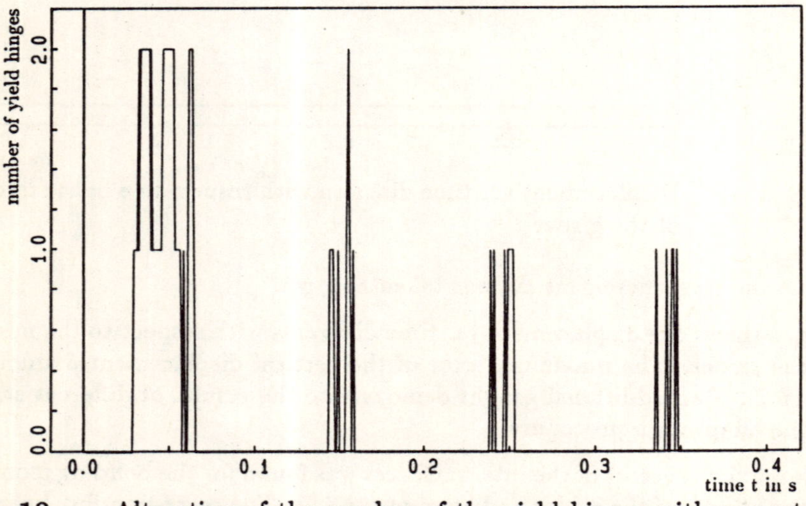

Fig. 10 Alteration of the number of the yield hinges with respect to the response of the system

9 Conclusions

Every realistic simulation of the reality by means of models in the range of computational mechanics requires assumptions and hypotheses which have to be checked. For this reason it is necessary to demonstrate and quantify the possible errors. When using the very clearly-structured yield hinge theory, the physical modeling of plasticized cross sections presents the most important problem, especially if strain-rate dependent material laws are to be taken into consideration. A quantification of the resulting errors has been determined by comparative calculations that are based on real structures.

10 References

[1] Argyris, J. et al.: Finite element analysis of two- and three-dimensional elastoplastic frames – the natural approach. Comp. Meth. Appl. Mech. Eng., 35 (1982) 221-248

[2] Aspden, R.J.; Campbell, J.D.: The effect of loading rate on the elasto-plastic flexure of steel beams. Proc. of the Royal Society, A290, London (1966) 266-285

[3] Bodner, S.R.: Strain rate effects in dynamic loading of structures. In: Behavior of materials under dynamic loading, Ed.: Huffington, N.J., ASME, New York (1965) 93-105

[4] Brandes, K.: Niedrigzyklisches Versagen von Stahlträgern unter Erdbebenbeanspruchung. In: Festschrift Roik, Institut für Konstruktiven Ingenieurbau, Ruhr-Universität Bochum, Mitteilung Nr. 84-3, Bochum (1984) 186-199

[5] Brandes, K.; Limberger, E.: Zur Beeinflussung der Festigkeitswerte von Betonstahl durch die Dehngeschwindigkeit. Beton- und Stahlbetonbau, 80 (1985) 90-94, 128-133

[6] Brandt, V.: Zur Berechnung ebener Rahmentragwerke mit dynamischen Einwirkungen bei verzerrungsratenabhängigem Materialverhalten, Institut für Statik, Universität Hannover, Mitteilung Nr. 38/90, Hannover, 1990

[7] Brandt, V.; Rothert, H.: Interaktionsbeziehungen für viskoses Materialverhalten. ZAMM 71 (1991)

[8] Brandt, V.; Gebbeken, N.; Rothert, H.: Dynamic inelastic analysis of plane steel structures taking into account the effect of rate-dependent behaviour. European Conference on Structural Dynamics (EURODYN '90), in: Eds. W.-B. Krätzig et al.: Structural Dynamics, Rotterdamm, Brookfield: A.A. Balkema (1991) 1023-1029

[9] Cowper, G.R.; Symonds, P.S.: Strain hardening and strain rate effects in the impact loading of cantilever beams. Tech. rep. 28, Brown Univ., Prov., 1957

[10] Gebbeken, N.: Eine Fließgelenktheorie höherer Ordnung für räumliche Stabtragwerke (zugleich ein Beitrag zur historischen Entwicklung), Institut für Statik, Universität Hannover, Mitteilung Nr. 32/88, Hannover, 1988

[11] Gebbeken, N.: Eine Fließgelenktheorie höherer Ordnung für räumliche Stabtragwerke. Stahlbau 57, H.12 (1988) 365-372

[12] Gebbeken, N.; Rothert, H.: Ein räumliches exzentrisches Fließgelenk für die Traglastberechnung räumlicher Stabwerke. ZAMM 60 (1989) T346-T348

[13] Manjoine, M.J.: Influence of rate of strain and temperature on yield stress of mild steel. Journal of Applied Mechanics 11 (1944) A211-A218

[14] Ueda, T. et al.: A new theory on elastic-plastic analysis of framed structures. Technology reports of the Osaka University, 19, No. 875 (1969) 263-276

Computation of Constitutive Response

Hermann G. Matthies
Germanischer Lloyd
Hamburg

Summary: **Deformation-path dependent irreversible material behaviour is increasingly modelled with internal variables. The evolution laws for these internal variables, derived from thermodynamics, display a mathematical structure which is analogous to the well known case of plasticity. This analogy supplies the basis for a derivation of numerical methods for the computation of constitutive response for general materials with internal variables.**

1 Introduction

With ever increasing complexity in material modelling, the formalism of using internal variables to describe deformation-path dependent irreversible material behaviour is steadily gaining ground. Irreversible behaviour often means that structural changes take place in the material. These changes are the memory of the material, its state depending on the treatment experienced so far. The evolution of these internal variables is derived on a thermodynamic basis. The mathematical structure is analogous to the familiar cases of classical inviscid plasticity and visco-plasticity. In fact, these two model cases shall guide our construction of numerical methods. By exploiting this analogy, numerical methods valid for any set of internal variables are derived, leading to a generalisation of the familiar projection or return mapping methods of plasticity.

In general, the processes taking place are thermo-mechanical, but for the sake of simplicity we shall limit ourselves here to purely mechanical behaviour; therefore exclude thermal processes from consideration and only be concerned with deformation processes, in order to exhibit more clearly the essence of the numerical procedure.

Classical plasticity is a prime example of a theory with internal variables—in the conceptually simplest case of perfect plasticity this internal variable is the accumulated plastic strain. Extensions to more complex material models which include visco-elastic (delayed elastic), visco-plastic (creep), hardening or damage mechanisms are often performed on an ad-hoc basis.

However a phenomenological constitutive law should conform with the laws of thermodynamics, obey the principle of material frame indifference and be invariant under the symmetry group actions appropriate for the particular material under consideration. The use of irreversible thermodynamics as a starting point not only identifies the appropriate thermodynamic fluxes and forces, but it also provides a basis to construct constitutive relations satisfying these requirements.

Often the evolution of some internal variable is assumed as rate-independent, in analogy to classical plasticity. The mathematical structure therefore turns out to be very similar, with the possibility of identifying "yield surfaces" and "flow rules". Rate dependent behaviour on the other hand leads to a structure analogous to visco-plasticity. Hence, by imagining a material with perfect, inviscid, rate-independent plasticity combined with visco-plasticity, we have a model with the same mathematical structure as the general case with arbitrary internal variables.

In the second section we shall introduce the general setting, where it will be assumed that the state of the material is characterised by the current values of strain and internal variables. These internal variables will have to change in general during the deformation if they are to be of any use in describing the memory of the material. The stress and the rates of change of the internal variables as functions of the present state of the material will be specified in the third section; these evolution laws for the internal variables are one of the key ingredients for constitutive relations—the other main one being of course the judicious choice of internal variables.

In most cases the evolution of the state of a material point cannot be computed analytically and some numerical approximation is necessary. In the fourth section it will be shown that the thermodynamic considerations not only guide the development of constitutive relations but also lend themselves to a consistent numerical approximation of the rate problem for the evolution of internal parameters. We will arrive at a generalisation of the familiar projection or return mapping methods of plasticity.

In essence, the incremental rate problem is solved by computing the constrained minimum of a function of the dual internal variables at each time step. This in itself is a non-trivial problem and will be discussed in the fifth section.

2 Preliminaries and General Setting

2.1 Placement, Displacement and Strain

The material body under consideration may be represented by a differentiable manifold, with its placement given by an invertible and smooth mapping from this manifold into the Euclidian observer space. An intrinsic description would be based on quantities defined on the manifold [1, 2], but since it will make no difference to our subject, we may as well choose a fixed reference configuration and recover the total Lagrangian description.

From the displacement $\boldsymbol{u}_t$ at time t, we obtain the deformation gradient at $\boldsymbol{x}$ by $\boldsymbol{F}_t(\boldsymbol{x}) = \nabla \boldsymbol{u}_t(\boldsymbol{x})$, where ∇ is the Nabla operator. The right Cauchy-Green deformation tensor is defined as usually by $\boldsymbol{C} = \boldsymbol{F}^T \cdot \boldsymbol{F}$, where we have dropped the index t and argument $\boldsymbol{x}$ for notational convenience and $\boldsymbol{F}^T$ is the transpose of $\boldsymbol{F}$.

The main variable describing the externally observable state of our material in this purely mechanical theory is the Green-Lagrange strain $\boldsymbol{E} = \frac{1}{2}(\boldsymbol{C} - \boldsymbol{I})$, where $\boldsymbol{I}$ denotes the identity tensor.

2.2 Forces and Momentum Balance

Given some time dependent forces $\boldsymbol{f}_t(\boldsymbol{x})$ acting on the body, we wish determine the evolution of the placement by computing the material velocity (the superimposed dot denotes the time derivative) $\boldsymbol{v}_t(\boldsymbol{x}) = \dot{\boldsymbol{u}}_t(\boldsymbol{x})$. The velocity may be found by considering the momentum balance or equilibrium equation, conveniently stated in the form of the principle of virtual work

$$\int \varrho \boldsymbol{w} \cdot \dot{\boldsymbol{v}}_t \, d\boldsymbol{x} = \int \varrho \boldsymbol{w} \cdot \boldsymbol{f}_t \, d\boldsymbol{x} + \int \boldsymbol{w} \cdot \nabla(\boldsymbol{F}_t \cdot \boldsymbol{S}_t) \, d\boldsymbol{x}. \tag{2.1}$$

Here ϱ is the density, $\boldsymbol{w}$ a virtual displacement and $\boldsymbol{S}$ the second Piola-Kirchhoff stress tensor.

Therefore, given the stress $\boldsymbol{S}_t(\boldsymbol{x})$, the acceleration and hence the placement evolution could be computed. But as is well known, kinematic and balance principles alone are not sufficient to determine the placement, different material bodies behave differently. The problem therefore is to determine $\boldsymbol{S}_t(\boldsymbol{x})$, when given $\boldsymbol{u}_s(\boldsymbol{x})$ and hence $\boldsymbol{E}_s(\boldsymbol{x})$ for all times s from 0 to t.

2.3 State of the Material Point

We shall assume that we are dealing with a simple material [1], so that the stress $\boldsymbol{S}_t(\boldsymbol{x})$ at the material point $\boldsymbol{x}$ depends only on quantities defined at that point. Instead of considering the whole history of $\boldsymbol{E}_s(\boldsymbol{x})$ for $0 \leq s \leq t$, we shall further assume that the present state of the material is completely specified by the present value of the strain $\boldsymbol{E}_t(\boldsymbol{x})$ and a set of internal variables or parameters which will collectively be denoted by $\boldsymbol{Q}_t(\boldsymbol{x})$. These internal variables are "hidden" and cannot be observed, but represent the memory of the material point, i.e. sufficient information to compute the stress $\boldsymbol{S}_t(\boldsymbol{x})$. This as well as the evolution of $\boldsymbol{Q}_t(\boldsymbol{x})$ with time will be addressed in the third section.

2.4 Examples for Internal Variables

The internal variables will usually consist of a set of tensors of even order. For the sake of simplicity we shall assume that these components are elements of some vector space, i.e. not a general manifold, as would be required for example if orthogonal tensors describing orientation were one of the components of $\boldsymbol{Q}$. The evolution then would of course have to take place on that manifold, introducing additional aspects which would distract from our subject.

In the following we list some examples of material models which implicitly or explicitly use internal variables, further examples may be found in refs. [3, 4, 5]:

- *Perfectly elasto-plastic material.* The internal variable is the plastic strain $\boldsymbol{E}_p$. Knowing $\boldsymbol{E}_p$, we may compute the elastic strain $\boldsymbol{E}_e$ and therefore the stress $\boldsymbol{S}$ (e.g. cf. [6]).
- *Elasto-plastic material with hardening.* The internal variable is $(\boldsymbol{E}_p, r, \boldsymbol{\alpha})$, where $\boldsymbol{E}_p$ is the plastic strain as before and r and $\boldsymbol{\alpha}$ are associated with the size (isotropic hardening) and the position (kinematic hardening) of the yield surface (e.g. cf. [3, 4, 5, 6]).

- *Material with plasticity, damage mechanisms and ageing.* For each of these mechanisms a suitable internal variable may be found (e.g. cf. [5, 6, 7, 8]).
- *Visco-plastic material combined with inviscid plasticity.* The internal variable is the inelastic strain $\boldsymbol{E}_i$, further decomposed into the visco-plastic strain $\boldsymbol{E}_v$, the plastic strain $\boldsymbol{E}_p$ plus eventually some hardening parameters (e.g. cf. [4]).

This last example will be our main vehicle to show the analogy to the general case with arbitrary internal variables. The visco-plastic strain will be the model for those internal variables which are rate dependent, and the inviscid strain will stand for those which are rate-independent.

To make things even simpler, in this model case we shall look only at the small strain/displacement case, so we shall denote here the Cauchy stress by σ and the total strain by ε. The total strain is additevly decomposed into elastic strain ε_e and the internal variable, the inelastic strain ε_i; which is further decomposed into visco-plastic strain ε_v and inviscid plastic strain ε_p. The elastic strain is defined as usually by $\sigma = \mathcal{A} : \varepsilon_e$, where $\mathcal{A}$ is the fourth order positive definite elasticity tensor. We therefore have the well known strain decomposition

$$\varepsilon = \varepsilon_e + \varepsilon_i = \varepsilon_e + \varepsilon_v + \varepsilon_p. \tag{2.2}$$

In order to facilitate the comparison, we shall display the devolpment of the general case side by side with the model case.

3 Constitutive Response

3.1 Thermodynamic Considerations

As stated already in the introduction, we shall require that the laws of thermodynamics be satisfied for the constitutive relations to be developed. More specifically, from the energy balance, expressing the first law of thermodynamics, and the Clausius-Duhem inequality, which we take to express the second law, one may deduce the following reduced dissipation inequality (e.g. cf. [3, 5]):

$$N\dot{T} + \dot{\Phi} - \boldsymbol{S} : \dot{\boldsymbol{E}} + \frac{1}{T}\boldsymbol{q} \cdot \nabla T \leq 0. \tag{3.1}$$

By N we denote the entropy per unit volume, Φ is the free energy per unit volume, T the absolute temperature and $\boldsymbol{q}$ the heat flux vector. In our model case, the free energy per unit volume is simply

$$\Phi(\varepsilon, \varepsilon_i) = \frac{1}{2}(\varepsilon - \varepsilon_i) : \mathcal{A} : (\varepsilon - \varepsilon_i) = \frac{1}{2}\varepsilon_e : \mathcal{A} : \varepsilon_e. \tag{3.2}$$

Since we want to consider only isothermal processes, the first and last terms of eq. (3.1) may be dropped from consideration, and we are left with:

$$\boldsymbol{S} : \dot{\boldsymbol{E}} - \dot{\Phi} \geq 0; \qquad \begin{cases} \sigma : \dot{\varepsilon} - \dot{\Phi} \geq 0 \quad \Rightarrow \\ \sigma : \dot{\varepsilon} - (\varepsilon - \varepsilon_i) : \mathcal{A} : (\dot{\varepsilon} - \dot{\varepsilon}_i) \geq 0 \quad \Rightarrow \\ \sigma : \dot{\varepsilon} - \sigma : (\dot{\varepsilon} - \dot{\varepsilon}_i) \geq 0 \quad \Rightarrow \\ \sigma : \dot{\varepsilon}_i \geq 0 \end{cases} \tag{3.3}$$

Hence this inequality requires the intrinsic dissipation to be always non-negative. It is a direct consequence of the second law.

3.2 Constitutive Model

The free energy Φ is assumed to be a smooth function of the strain E and the internal variables Q. Using the chain rule, we arrive at:

$$(S - d_E\Phi) : \dot{E} - \langle d_Q\Phi, \dot{Q}\rangle \geq 0; \qquad \begin{cases} \sigma : \dot{\varepsilon} - \dot{\Phi} \geq 0 & \Rightarrow \\ -(-\mathcal{A} : (\varepsilon - \varepsilon_i)) : \dot{\varepsilon}_i \geq 0 & \Rightarrow \\ \sigma : \dot{\varepsilon}_i \geq 0 & \end{cases} \tag{3.4}$$

Here $\langle \cdot, \cdot \rangle$ is a suitable duality pairing depending on the number and type of internal variables and $d_G F$ the partial Gâteaux derivative of F with respect to G.

We shall assume that Q is independent of E, so that we deduce from eq. (3.4):

$$\begin{array}{lll} & S = d_E\Phi(E, Q), & \sigma = \mathcal{A} : (\varepsilon - \varepsilon_i), \\ \text{and} & \langle P, \dot{Q}\rangle \geq 0, & p : \dot{\varepsilon}_i \geq 0; \\ \text{where} & P = -d_Q\Phi(E, Q), & p = -(-\mathcal{A} : (\varepsilon - \varepsilon_i)) = \sigma. \end{array} \tag{3.5}$$

The P's are the thermodynamic forces associated to the fluxes $\dot{Q}$. In our model case this turns out to be just the Cauchy stress. The second law has reduced to the inequality on the pairing between thermodynamic forces P and fluxes $\dot{Q}$ — the intrinsic dissipation. In our model case this again repeats the previous inequalities (3.3) and (3.4). In order to satisfy this inequality 'a priori' and in view of the Onsager relations (e.g. cf. [4, 9, 10]), this suggest the existence of a convex conjugate dissipation 'potential' $\Omega(P)$ (cf. also [4–10]), in extionsion of the linear thermodynamic case.

This is equivalent with the assumption that the intrinsic dissipation is maximal as we shall see in the next subsection. This has been termed one of the extremal principles of irreversible thermodynamics by Ziegler [9].

The conjugate dissipation potential will detemine the evolution of the internal parameters, classically by setting $\dot{Q} = d\,\Omega(P)$ (e.g. cf. [10]). But when we deal with rate-independent behaviour, Ω is not smooth, not even continuous and problems will arise here. This is due to the fact that the conjugate function of Ω in the sense of convex analysis (e.g. cf. [11, 12]), the dissipation potential

$$\Omega^*(\dot{Q}) = \sup_P(\langle P, \dot{Q}\rangle - \Omega(P)), \tag{3.6}$$

has to be homogeneous of degree one if the evolution of Q is to be rate-independent. How this leads to a non-smooth Ω and in turn to "yield surfaces" was shown by Lubliner in a series of papers [13–16]. This may also be seen directly here, because if we want the relation $P = d\,\Omega^*(\dot{Q})$ (this is 'dual' to $\dot{Q} = d\,\Omega(P)$) to give the same value for $\dot{Q}$ arbitrarily positively scaled (this is the meaning of rate-independence), then Ω^* has to be homogeneous of degree one for the derivative to be constant on rays from the origin.

Convex analysis then tells us—and this again is easily seen from the definition of Ω in terms of Ω^* by interchanging Ω and Ω^* and taking the supremum over the $\dot{Q}$'s in eq. (3.6)—that Ω is either 0 or $+\infty$, depending on whether $P = d\,\Omega^*(\dot{Q})$ or not. Such a function Ω is called an indicator function of the set of P's where it vanishes. In plasticity, this set B is called the "elastic domain" (for P in the interior of B there is no dissipation) and its boundary the "yield surface". We see how these concepts arise naturally from the assumption of rate-independence.

Such a function Ω is clearly not differentiable at the "yield surface", so in generalisation of the classical concept of derivative we have to use the concept of subdifferential from convex analysis, defined by

$$\dot{Q} \in \partial \Omega(P) \quad \Leftrightarrow \quad \forall R\colon \langle R - P, \dot{Q} \rangle + \Omega(P) \leq \Omega(R). \tag{3.7}$$

This condition is satisfied by all hyperplanes lying entirely underneath the graph of Ω and touching it at P. Obviously, if Ω is differentiable at P, there is only one such hyperplane, but in general there may be more. Therefore, the subdifferential is a (convex) set, and its elements are called subgradients. The subdifferential is a generalisation of the notion of Gâteaux differential, and indeed if Ω is Gâteaux differentiable, then the Gâteaux differential is the only element in the subdifferential.

In the general case we have both rate-dependent and rate-independent components in Q, and we shall assume Ω to be a sum of smooth functions and such indicator functions; and the evolution law for internal parameters can be described by the relation eq. (3.7). Additionally, we shall assume that Ω will be such that it achieves its minimum for $P = 0$, i.e. if there is no force then there is no flux—in other words equilibrium.

3.3 Variational Inequalities

The constitutive relation eq. (3.7) is termed a differential inclusion or variational inequality in convex analysis, and in our mechanical context it may be called a "flow rule". Taking $R = 0$ in eq. (3.7), we see that $\langle P, \dot{Q} \rangle \geq \Omega(P) - \Omega(0)$. Since the minimum of Ω occurs at $P = 0$, the right hand side is not negative and the dissipation inequality eq. (3.5) is satisfied.

But not only is the intrinsic dissipation non-negative, it is also maximal. Because for any other choice of rate for the internal variables, say Q', not satisfying the relation eq. (3.7), we can—by multiplying the relation for Q' with -1, adding it to the one for $\dot{Q}$ and setting $R = 0$ —deduce that $\langle P, \dot{Q} \rangle \geq \langle P, Q' \rangle$.

We want to come back to what was said above, that in general we want to assume Ω to be the sum of a smooth function Ω_1 for the rate-dependent components and an indicator function Ω_2 for the rate-independent components of Q. It then follows from eq. (3.7) that

$$\begin{aligned} &\dot{Q} - d\,\Omega_1(P) \in \partial\, \Omega_2(P) \quad \Leftrightarrow \quad \forall R\colon \langle R - P, \dot{Q} - d\,\Omega_1(P) \rangle \leq 0, \\ &\qquad\qquad\qquad\qquad\qquad \text{or} \\ &\dot{Q}_1 = d\,\Omega_1(P) \quad \text{and} \quad \dot{Q}_2 \in \partial\, \Omega_2(P) \quad \Leftrightarrow \quad \forall R\colon \langle R - P, \dot{Q}_2 \rangle \leq 0. \end{aligned} \tag{3.8}$$

Here $Q = Q_1 + Q_2$ has been decomposed into rate-dependent and rate-independent parts.

The convex set B where Ω_2 vanishes is usually given by functions f_i describing the boundary—the "yield (hyper)-surface"—in the form $B = \{R \,|\, \forall\, i\colon f_i(R) \leq 0\}$, where the functions f_i are convex. From now on we shall assume that B is described in such a manner. If the functions f_i are differentiable and the derivatives $d f_i$ linearly independent, the "flow rule" eq. (3.8) may be written as

$$\begin{aligned} &\dot{Q}_1 = d\Omega_1(P), \qquad \dot{Q}_2 = \sum_i \lambda_i \, d f_i(P), \\ &\forall\, i\colon\ \lambda_i \geq 0, \qquad f_i(P) \leq 0, \qquad \lambda_i f_i(P) = 0; \end{aligned} \tag{3.9}$$

where we have employed Lagrange-multipliers λ_i.

We have now finally arrived at a formal analogue to the classical theory of multi-surface visco-plasticity (e.g. cf. [4]). One may think of Ω_1 as a visco-plastic flow potential, and of B as the set where there is no rate-independent flow if P is in the interior.

Let us remark that we have chosen the strain $\boldsymbol{E}$ and the internal parameters Q as "primitive variables". As is common in thermodynamics, by using Legendre transforms—or Fenchel transforms in the case of non-differentiable functions—other variables could have been chosen. For numerical computations the present ones seem the most common and efficient.

It is well known [17] that the conditions in eq. (3.9) are the Kuhn-Tucker conditions of a minimisation problem, and we shall use a discrete analogue of this for the numerical procedure to be described in the next section.

3.4 The Model Case

It remains to translate these last developments to our model case. Let us define an 'equivalent' stress $\sigma_{eq}^2 = \frac{3}{2}\sigma^D : \sigma^D$ (this is the usual von Mises equivalent stress), $\sigma^D = \sigma - \frac{1}{3}\mathrm{tr}(\sigma)I$ being the stress deviator. For the inviscid part we set $f(\sigma) = \frac{1}{3}(\sigma_{eq}^2 - \sigma_y^2)$, where σ_y is the yield stress for this perfectly plastic part. Hence $B = \{\sigma \,|\, f(\sigma) \leq 0\}$ is the set where there is no inviscid plastic flow in the interior. But since we want to combine this perfectly plastic behaviour for our model case with visco-plasticity, we may imagine that there is actually some inelastic deformation before the stress reaches the boundary of B (the yield surface), namely creep with the potential $\Omega_1(\sigma) = \frac{2c}{3(n+1)}\langle\sigma_{eq}^2/\sigma_0^2 - 1\rangle^{(n+1)/2}$ (a variant of Norton's law), where c and σ_0 are constants, and n is the power law creep exponent, and $\langle x\rangle = \frac{1}{2}(x + |x|)$ is the positive part of x. Writing $\dot{\varepsilon}_i = \dot{\varepsilon}_v + \dot{\varepsilon}_p$, eq. (3.8) is translated to

$$\begin{aligned} &\dot{\varepsilon}_v = d\Omega_1(\sigma) = c\left\langle \frac{\sigma_{eq}^2}{\sigma_0^2} - 1 \right\rangle^{\frac{n-1}{2}} \sigma^D, \\ &(\tau - \sigma) : \dot{\varepsilon}_p \leq 0 \qquad \forall\, \tau \in B \end{aligned} \tag{3.10}$$

The second relation, which is the classical Prandtl-Reuß flow rule, may be re-written with the help of a Lagrange (or plastic) multiplier λ to the more familiar form

$$\begin{aligned} &\dot{\varepsilon}_p = \begin{cases} \lambda\sigma^D, & \text{if } f(\sigma) = 0; \\ 0, & \text{otherwise.} \end{cases} \\ &\lambda \geq 0, \qquad f(\sigma) \leq 0, \qquad \lambda f(\sigma) = 0; \end{aligned} \tag{3.11}$$

In case $f(\sigma) = 0$, the plastic multiplier λ is determined from the consistency condition $d f(\sigma)/dt = 0$.

For the stress rising from zero our model case shows the following phases: first the behaviour is purely elastic up to the point when σ_{eq} reaches σ_0, from here on Ω_1 is non-zero and we have power law creep. When the stress rises so high that σ_{eq} reaches σ_y, inviscid, perfectly plastic behaviour sets in limiting the stress to stay inside B.

4 Numerical Approximation

4.1 Spatial Discretisation

We shall only briefly touch this subject, which is concerned with constructing a finite dimensional approximation to the balance equation (2.1), e.g. with finite elements, as it is fairly standard. The displacement is assumed as a finite linear combination of basis functions and forces are discretised by duality with the principle of virtual work.

But now we also have to introduce a discretisation of the field of internal variables $Q_t(\boldsymbol{x})$, since it is not possible to store it at infinitely many material points in the computer. Usually the internal variables are only stored at the integration points. This implies also an discretisation of the stress field $S_t(\boldsymbol{x})$, and we see immediately that in problems with internal variables we always have at least implicitly a mixed interpolation.

4.2 Temporal Descretisation

We still have to solve at each material point—actually normally only integration points are considered in the computation—the rate problem or variational evolution inequality (3.8) and (3.9). The 'outer' iteration, which tries to establish global equlibrium or solve the finite dimensional analogue of eq. (2.1), will produce trial displacement increments, and hence for our material model we are presented with a strain increment ΔE (model case: $\Delta\varepsilon$), and we have to compute the increment of the internal variables ΔQ (model case: $\Delta\varepsilon_i$). Then the stress at the new state is given by eq. (3.5). When increments instead of rates are written in the evolution inequality for the internal variables (3.8/3.9), there is still a choice at which point in time we want the inequality to be satisfied.

Probably the simplest is at the beginning of the time step, which has been used a great deal in the past (e.g. cf. [18–20], or at the end of the step, which leads to the usual projection or return mapping methods (e.g. cf. [21–25]). But is is also possible to use an intermediate point in time, as for example in Simo and Govindjee [26]. For reasons of stability and for the sake of simplicity of exposition, we choose a fully implicit scheme and want the inequality satisfied at the end of the increment.

Let $\hat{E}, \hat{Q}, \hat{P}$ (model case: $\hat{\varepsilon}, \hat{\varepsilon}_i, \hat{\sigma}$) be the values at the start of the step, and E, Q, P (model case: $\varepsilon, \varepsilon_i, \sigma$) at the end where E (model case: ε) is given (at least as a trial strain from the outer iteration), and Q and P have to be found (model case: ε_i and σ), and $\Delta Q = Q - \hat{Q}$ (model case: $\Delta\varepsilon_i = \varepsilon_i - \hat{\varepsilon}_i$), then

$$
\begin{aligned}
&\frac{\Delta Q_1}{\Delta t} = d\,\Omega_1(P), && \frac{\Delta\varepsilon_v}{\Delta t} = d\,\Omega_1(\sigma) = c\left\langle \frac{\sigma_{eq}^2}{\sigma_0^2} - 1\right\rangle^{\frac{n-1}{2}} \sigma^D, \\
\text{and}\quad &\forall R\text{: } \langle R - P, \Delta Q_2\rangle \le 0; && \forall\sigma\text{: } (\tau - \sigma) : \Delta\varepsilon_p \le 0; \\
\text{or}\quad &\Delta Q_2 = \sum_i \lambda_i f_i(P), && \Delta\varepsilon_p = \lambda\sigma^D, \\
\text{and}\quad &\forall i\text{: } \lambda_i \ge 0,\ f_i(P) \le 0, && \lambda \ge 0,\ f(\sigma) \le 0, \\
&\lambda_i f_i(P) = 0, && \lambda f(\sigma) = 0.
\end{aligned}
\tag{4.1}
$$

Here $\Delta Q/\Delta t$ (model case: $\Delta\varepsilon_i/\Delta t$) is the finite difference quotient approximation to $\dot{Q}$ (model case: $\dot{\varepsilon}_i$). This inequality has to be solved at every time step in each outer iteration at each integration point.

4.3 The Minimisation Problem

We want to show that the solution of the inequality (4.1) is at the same time the solution of a minimisation problem (e.g. cf. [21–24]), as these relations are the Kuhn-Tucker conditions of a minimisation problem. To this end we need the partial Legendre transform of the free energy $\Phi(E, Q)$ with respect to the internal variables:

$$
\Phi^{*Q}(E, P) = -\langle P, Q\rangle - \Phi(E, Q) \qquad \Phi^{*\varepsilon_p}(\varepsilon, \sigma) = \frac{1}{2}\sigma : \mathcal{A}^{-1} : \sigma - \sigma : \varepsilon \tag{4.2}
$$

Here P resp. σ is such that the relations (3.5) are satisfied. Not always is it so easy as in our model case to actually compute Φ^{*Q}, and we may have to resort to the following procedure: Given E and P, find Q such that eq. (3.5) is satisfied, i.e. $P = -d_Q\Phi(E, Q)$. With Q so determined, return to equation (4.2) and compute Φ^{*Q}.

Now consider the problem to minimise the functional

$$
\begin{aligned}
H(P) &:= \Phi^{*Q}(E, P) + \langle P, \hat{Q}\rangle + \Delta t\,\Omega_1(P), \\
H(\sigma) &:= \frac{1}{2}\sigma : \left(\mathcal{A}^{-1} : \sigma - \varepsilon + \hat{\varepsilon}_i\right) + \Delta t\,\frac{2c}{3(n+1)}\left\langle \frac{\sigma_{eq}^2}{\sigma_0^2} - 1\right\rangle^{\frac{n+1}{2}},
\end{aligned}
\tag{4.3}
$$

for those P (model case: σ), such that $P \in B$ (model case: $f(\sigma) \le 0$). In some sense this is the dual problem to the one whose Kuhn-Tucker conditions are stated in eq. (4.1).

4.4 Generalised Projection or Return Mapping Method

In the description of the computational procedure it will be necessary to distinguish several cases. We shall try to explain these distinctions again by refering to our model case.

The computational steps may be described in the following order:

(1) Given $E = \hat{E} + \Delta E$ (model case: $\varepsilon = \hat{\varepsilon} + \Delta\varepsilon$), assume $\Delta Q \equiv 0$ (model case: $\Delta\varepsilon_i \equiv 0$) and compute $\tilde{P} = -d_Q\Phi(E, Q)$ (model case: $\tilde{\sigma} = -(-\mathcal{A} : (\varepsilon - \varepsilon_i))$). This is the so-called elastic predictor, because here we assume that the internal parameters do not change, i.e. in our model case there is no inelastic strain increment.

(2) Check that $\tilde{P} \in B$ (model case: $\tilde{\sigma}_{eq} \leq \sigma_y$)—i.e. we have no rate-independent "flow"—and $d\Omega(\tilde{P}) = 0$ (model case: $\tilde{\sigma}_{eq} \leq \sigma_0$)—also no rate-dependent "flow". If this is the case, relation (4.1) is satisfied and the final value is $P = \tilde{P}$ (model case: $\sigma = \tilde{\sigma}$) and $\Delta Q = 0$ (model case: $\Delta\varepsilon_i = 0$), i.e. $Q = \hat{Q}$ (model case: $\varepsilon = \hat{\varepsilon}$), so that this step is elastic.

(3) On the other hand, if $\tilde{P} \in B$ but $d\Omega(\tilde{P}) \neq 0$ ($\tilde{\sigma}_{eq} \leq \sigma_y$, but $\tilde{\sigma}_{eq} > \sigma_0$)—i.e. we have only rate-dependent "flow"—solve the following equation for a new value of $\tilde{P}$ (model case: $\tilde{\sigma}$): (This is the usual step in visco-plasticity with no constraint set B, and the equation is the condition for an unconstrained minimum of $H(P)$ resp. $H(\sigma)$)

$$
\begin{aligned}
d_P\Phi^{*Q}(E, \tilde{P}) + \hat{Q} + \Delta t\, d\Omega_1(\tilde{P}) &= 0 \\
\mathcal{A}^{-1} : \tilde{\sigma} - \varepsilon + \hat{\varepsilon}_i + \Delta t\, c\left\langle \frac{\tilde{\sigma}_{eq}^2}{\sigma_0^2} - 1 \right\rangle^{\frac{n-1}{2}} \tilde{\sigma}^D &= 0
\end{aligned}
\tag{4.4}
$$

(4) If $\tilde{P}$ (model case: $\tilde{\sigma}$) from step (3) satisfies $\tilde{P} \in B$ (model case: $f(\tilde{\sigma}) \leq 0$), we are again finished, and the final value is $P = \tilde{P}$ and $\Delta Q = \Delta t\, d\Omega_1(P)$, or more precisely $\Delta Q_1 = \Delta Q$, $\Delta Q_2 = 0$ (model case: $\sigma = \tilde{\sigma}$ and $\Delta\varepsilon_i = \Delta t\, c\langle \tilde{\sigma}_{eq}^2/\sigma_0^2 - 1\rangle^{\frac{n-1}{2}} \tilde{\sigma}^D$, or more precisely $\Delta\varepsilon_v = \Delta\varepsilon_i$, $\Delta\varepsilon_p = 0$).

(5) Otherwise find the point $P \in B$ (model case: σ such that $f(\sigma) \leq 0$), where the functional $H(P)$ resp. $H(\sigma)$ in eq. (4.3) assumes its minimum. Since eq. (4.4) is the condition $dH(\tilde{P}) = 0$ (model case: $dH(\tilde{\sigma}) = 0$) and $\tilde{P}$ was not in B (model case: $\tilde{\sigma}$ did not satisfy $f(\tilde{\sigma}) \leq 0$), we know that the point P where $H(P)$ is minimal with $P \in B$ (model case: σ where $H(\sigma)$ is minimal with $f(\sigma) \leq 0$) must be on the boundary of B (model case: must satisfy $f(\sigma) = 0$). Here we have the generalised projection or return mapping method: The point $\tilde{P}$ from eq. (4.4) outside of B (model case: $\tilde{\sigma}$ with $f(\tilde{\sigma}) > 0$) is 'projected' onto the boundary to give the final point P resp. σ. The increment in internal varibales is given by $\Delta Q = -d_P\Phi^{*Q}(E, P) - \hat{Q}$ and $\Delta Q_1 = \Delta t d\Omega_1(P)$, $\Delta Q_2 = \Delta Q - \Delta Q_1$ (model case: $\Delta\varepsilon_i = -(\mathcal{A}^{-1} : \sigma - \varepsilon) - \hat{\varepsilon}_i = \varepsilon_i - \hat{\varepsilon}_i$, and $\Delta\varepsilon_v = \Delta t c\langle \tilde{\sigma}_{eq}^2/\sigma_0^2 - 1\rangle^{\frac{n-1}{2}} \tilde{\sigma}^D$, $\Delta\varepsilon_p = \Delta\varepsilon_i - \Delta\varepsilon_v$), and we are finished.

The steps (1) to (4) involve no particular difficulty except possibly for the equation solving in step (3), which may be achieved by standard procedures such as Newton's method. Step (5) is much harder and will be discussed in the next section.

5 Solving the Minimisation Problem

5.1 Special Cases

These special cases arise generally in a situation where there is only one "yield"-surface given by a function $f(P)$ (like in our model case). The constrained minimisation problem may then be formulated with a single Lagrange multiplier:

$$\begin{aligned} d\,H(P) + \lambda d\,f(P) &= 0, \\ \text{and} \quad f(P) &= 0, \end{aligned} \tag{5.1}$$

since we are in step (5) and know that the constraint $f(P) \leq 0$ is active.

The implicit function theorem allows us to solve these two equations for P (model case: σ) as a function of λ, $P = P(\lambda)$. Inserting this into the second of eq. (5.1), we are left with solving $f(P(\lambda)) = 0$, one equation in one unknown. Having found λ such that this eqation is satisfied, we compute $P(\lambda)$ and are finished. The main problem here is finding an explicit representation for $P(\lambda)$. This is possible for special forms of the dissipation potential Ω. We shal not dwell further on these matters and refer to refs. [8, 23] and the references therein.

Another special case is when essentially only one component of P is responsible for the changes in internal parameters. An example for this situation is the familiar setting where the Lévy - von Mises theory is used and only the magnitude of deviatoric stress is important. This has been called the 'effective stress function' algorithm by Kojić and Bathe [24]. Equation (5.1) is esentially only one-dimensional when viewed in the right co-ordinates. In the same manner the plane stress problem for visco-plasticity is approached in ref. [25].

5.2 General Algorithm

We now go back to the general problem as given in eq. (4.3) with $B = \{R \,|\, \forall i\colon f_i(R) \leq 0\}$. When we formulate the problem with Lagrangian $L(P, \lambda_i) = H(P) + \sum_i \lambda_i f_i(P)$, the optimality conditions are

$$\begin{aligned} &dH(P) + \sum_i \lambda_i d\,f_i(P) = 0, \\ \text{and} \quad &\forall i\colon \lambda_i \geq 0, \qquad f_i(P) \leq 0, \qquad \lambda_i f_i(P) = 0, \end{aligned} \tag{5.2}$$

but for numerical stability and effectiveness it is usually better to use a modified or augmented Lagrangian; in the simplest case this may be taken as

$$L_r(P, \lambda_i) = H(P) + \sum_i \left(\lambda_i f_i(P) + \frac{1}{2} r^2 \langle f_i(P) \rangle^2\right) \tag{5.3}$$

The last term may be viewed as a penalty with parameter r^2, enforcing the constraints. It may be shown that every saddle point of eq. (5.3) is a saddle point of eq. (5.2). The algorithm we want to use to find a saddle point of eq. (5.3) is called Uzawa's algorithm [11], and it proceeds by first fixing the Lagrange multipliers, and then minimising with respect to P. In the last step the Lagrange multipliers are updated:

(1) Select $\lambda_i \geq 0 \quad \forall i$.

(2) Find $P = \arg\min L_r(P, \lambda_i)$, i.e. solve $d_P L_r(P, \lambda_i) = 0$.

(3) Set $\lambda_i := \langle \lambda_i + r^2 f_i(P) \rangle \quad \forall i$.

(4) Check for convergence. If the algorithm has converged, then stop. Otherwise go to step (2).

The minimisation in step (2) is unconstrained, $L_r(P, \lambda_i)$ is convex in P and differentiable and so may be solved by standard algorithms like Newton's method. The step (3) may be viewed as a projection (onto the set $\lambda_i \geq 0$) of the gradient of the dual problem, and it changes the Lagrange multipliers by an amount such that the penalty should vanish.

For our model case, the modified Lagrangian function can be easily written down explicitly:

$$\begin{aligned} L_r(\sigma, \lambda) = \sigma : \left(\mathcal{A}^{-1} : \sigma - \varepsilon + \varepsilon_i\right) + \Delta t \frac{2c}{3(n+1)} \left\langle \frac{\sigma_{eq}^2}{\sigma_0^2} - 1 \right\rangle^{\frac{n+1}{2}} \\ + \lambda \left(\frac{1}{3}(\sigma_{eq}^2 - \sigma_y^2)\right) + \frac{1}{2} r^2 \left\langle \frac{1}{3}(\sigma_{eq}^2 - \sigma_y^2)\right\rangle^2 \end{aligned} \quad (5.4)$$

It is now a simple differentiation to derive the equation which has to be solved in step (2).

6 Conclusion

Concluding the general description, the combination of the procedure given in section 4.4 together with Uzawa's algorithm completely specifies the computation of the increment in internal variables for a given strain increment. We see that it may be given without reference to any particular material behaviour, and that its mathematical structure is analogous to multi-surface visco-plasticity.

7 References

1 Noll, W. — A new mathematical theory of simple materials, Arch. Rat. Mech. Anal., **48**, p.1, 1972

2 Marsden, J.E. & Hughes, T.J.R. — **Mathematical Foundations of Elasticity**, Prentice Hall, Englewood Cliffs N.J., 1983

3 Lemâitre, J. & Marquis, D. — 'Modelling elasto- plasticity, damage and ageing as coupled behaviours in engineering materials', **Applied Solid Mechanics-2**, Eds. Tooth, A.S., & Spence, J., Elsevier Applied Science, London, 1988, p.277

4 Halphen, B. & Nguyen, Q.S. — Sur les matériaux standards généralisés, J. de Mécanique, **14**, p.39, 1975

5 Lemâitre, J. — Coupled elasto-plasticity and damage constitutive equations, Comp. Meth. Appl. Mech. Eng.,**51**, p.31, 1985

6 Lubliner, J. — A simple theory of plasticity, Int. Jnl. Solids & Structures, **10**, p.313, 1974

7 Krajčinović, D. - Continuous damage mechanics revisited: Basic concepts and definitions, J. Appl. Mech., **52**, p.829, 1985

8 Matthies, H.G. — 'The rate problem for complex material behaviour with internal variables', **Computational Plasticity—COMPLAS II**, Eds. Owen, D.R.J., Hinton, E. and Oñate, E., Pineridge Press, Swansea, 1989, p.27

9 Ziegler, H. — Zwei Extremalprinzipien der irreversiblen Thermodynamik, Ing. Archiv, **30**, p.410, 1961

10 de Groot,S.R. & Masur, P. — **Non-Equilibrium Thermodynamics**, Dover, New York, 1984

11 Ekeland, I. & Temam, R. — **Convex Analysis and Variational Problems**, North Holland, Amsterdam, 1976

12 Oden, J.T. — **Qualitative Methods in Nonlinear Mechanics**, Prentice Hall, Englewood Cliffs N.J., 1986

13 Lubliner, J. — On loading, yield and quasi- yield hypersurfaces in plasticity theory, Int. J. Solids & Structures, **11**, p.1011, 1975

14 Lubliner, J. — Non-smooth dissipation functions and yield criteria, Acta Mechanica, **22**, p.289, 1975

15 Lubliner, J. — A thermodynamic yield criterion in visco-plasticity, Acta Mechanica, **30**, p.165, 1978

16 Lubliner, J. — On thermodynamic yield criteria, Acta Mechanica, **37**, p.259, 1980

17 Koiter, W.G. — 'General theorems for elasto-plastic bodies', **Progress in Solid Mechanics**, Eds. Sneddon, S,. & Hill, R., North-Holland, Amsterdam, 1960, p.165

18 Zienkiewicz, O.C. — **The Finite Element Method**, Mc Graw Hill, London, 1977

19 Bathe, K.-J. — **Finite Element Procedures in Engineering Analysis**, Prentice Hall, Englewood Cliffs N.J., 1982

20 Owen, D.R.J. & Hinton, E. — **Finite Elements in Plasticity**, Pineridge Press, Swansea, 1980

21 Nguyen, Q.S. — On the elastic-plastic initial boundary value problem and its numerical integration, Int. J. Num. Eng., **11**, p.817, 1977

22 Matthies, H. — Finite element approximations in thermo-plasticity, Numer. Funct. Anal. and Optimiz., **1**, p.145, 1979

23 Matthies, H.G. — A decomposition method for the integration of the elastic-plastic rate problem, Int. J. Num. Meth. Eng., **28**, p.1, 1989

24 Kojić, M. & Bathe, K.-J. — The 'effective stress function' algorithm for thermo-elasto-plasticity and creep, Int. J. Num. Meth. Eng., **24**, p.1509, 1987

25 Simo, J.C. & Govindjee, S. — Exact closed-form solution of the return mapping algorithm in plane stress elasto-visco-plasticity, Eng. Comput., **5**, p.254, 1988

26 Simo, J.C. & Govindjee, S. — Nonlinear B-stability and symmetry preserving return mapping algorithms for plasticity and viscoplasticity, submitted to Int. J. Num. Meth. Eng.

A Critical Assessment of FE-Performance in Localized Failure Analysis

Paul Steinmann and Kaspar Willam
Institute of Mechanics, University of Karlsruhe, D-7500 Karlsruhe 1
CEAE-Department, University of Colorado at Boulder, CO 80309-0428

Abstract

The performance of different finite element models is evaluated when applied to elasto-plastic failure computations. The weak form of localization proposed by STEINMANN & WILLAM [10] is utilized to check the ability of several element enrichements to reproduce localization within a single element. Several variations of four noded elements are studied within the frame of elasto-plastic constitutive models which exhibit localization in the form of discontinuous bifurcation. The results of J_2-plasticity are presented in detail for plane stress conditions. These findings lead to the development of the singularized finite element formulation **QS** advanced by STEINMANN & WILLAM [11], which reproduces discontinuous strain fields in the weak sense. The importance of localization capturing element formulations is emphasized for homogeneous bifurcation- as well as for nonhomogeneous deformation problems. In this context the uniaxial tension of a flat metal sheet is examined and compared with experimental results.

1 Introduction

Localization theory signals the condition for which the assumption of a smooth strain field fails to hold, so that discontinuous bifurcation at the constitutive level may develop. This localization condition describes the formation of a discontinuity in the field of displacement- or rather velocity gradients across a surface with orientation **N**. Discontinuous bifurcation is signalled by a singularity of the localization tensor, i.e. when $\det(\mathbf{Q}^{ep}) = \det(\mathbf{N} \cdot \mathbf{E}^{ep} \cdot \mathbf{N}) = 0$ where $\mathbf{E}^{ep}$ denotes the tangential material operator. For numerical computations the underlying singularity of $\mathbf{Q}^{ep}$ is responsible for the loss of ellipticity which causes severe mesh-dependence. Thereby the post-peak behaviour depends crucially on the mesh design and the choice of the element formulation. If the fixed mesh approach is utilized, the finite element grid has to transmit the local singularity throughout the structural system without mesh-adaption to capture the particular failure mode.

2 Localized Failure : Constitutive Level

Let us assume that the homogeneously deformed solid is subjected to quasi-static increments of deformation. While the velocity field is assumed to be C^0-continuous, i.e.

$$[[\dot{\mathbf{u}}]] = \dot{\mathbf{u}}^+ - \dot{\mathbf{u}}^- = \mathbf{0} \tag{1}$$

the velocity gradient field may exhibit a jump across the discontinuity surface defined by the normal vector **N**

$$[[\nabla\dot{\mathbf{u}}]] = \nabla\dot{\mathbf{u}}^+ - \nabla\dot{\mathbf{u}}^- \neq \mathbf{0} \tag{2}$$

Here the superindices '+' and '−' denote the quantities at the plus and minus side of the discontinuity surface. Maxwell's compatibility conditions require that the jump of the velocity gradient is of the form

$$[[\nabla \dot{\mathbf{u}}]] = \dot{\gamma}\mathbf{M} \otimes \mathbf{N} \tag{3}$$

Analogous to the wave propagation argument the vector $\mathbf{M}$ denotes the polarization direction and the localization amplitude $\dot{\gamma}$ is related to the wave speed. The corresponding discontinuity in the strain rate is defined by the tensor product of the unit vectors $\mathbf{N}$ and $\mathbf{M}$

$$[[\dot{\epsilon}]] = [[\nabla^s \dot{\mathbf{u}}]] = \frac{1}{2}\dot{\gamma}(\mathbf{M} \otimes \mathbf{N} + \mathbf{N} \otimes \mathbf{M}) = \dot{\gamma}[\mathbf{M} \otimes \mathbf{N}]^s \tag{4}$$

Introducing the jump in the stress rate with the aid of the elasto-plastic material operator $\mathbf{E}^{ep}$, and requiring balance of linear momentum across the plane of discontinuity leads to the localization condition

$$\dot{\mathbf{t}} = \dot{\gamma}\mathbf{Q}^{ep} \cdot \mathbf{M} \doteq 0 \quad \text{or} \quad \mathbf{M} \cdot \mathbf{Q}^{ep} \cdot \mathbf{M} = \rho c^2 \doteq 0 \tag{5}$$

Localization initiates when this condition is satisfied by some directions $\mathbf{N}$ and $\mathbf{M}$ at the first time in the deformation history, see RUDNICKI & RICE [8]. Thereby, the acoustic, or in the static case, the localization tensor is defined as

$$\mathbf{Q}^{ep} = \mathbf{N} \cdot \mathbf{E}^{ep} \cdot \mathbf{N} \tag{6}$$

Integrating the jump of the velocity gradient renders the difference of the velocity field at the positive and negative side of the discontinuity surface where $\dot{\mathbf{u}}^+ = \dot{\mathbf{u}}^-$ along $(\mathbf{N}\cdot\mathbf{x}) = 0$.

$$\dot{\mathbf{u}}^+ - \dot{\mathbf{u}}^- = \dot{\gamma}(\mathbf{M} \otimes \mathbf{N}) \cdot \mathbf{x} \tag{7}$$

Thus the bifurcated velocity field is defined by

$$\dot{\mathbf{u}}^b = \dot{\mathbf{u}} + \dot{\gamma}\Lambda(\mathbf{x})(\mathbf{M} \otimes \mathbf{N}) \cdot \mathbf{x}; \quad \Lambda = \begin{cases} \lambda & \text{if } (\mathbf{N} \cdot \mathbf{x}) > 0 \\ (\lambda - 1) & \text{if } (\mathbf{N} \cdot \mathbf{x}) < 0 \end{cases} \tag{8}$$

3 Singularities and Dimensional Constraints

To examine the implications of dimensional constraints let us briefly review the rate forms of elasto-plastic constitutive laws, $\dot{\sigma} = \mathbf{E}^{ep} : \dot{\epsilon}$ where the compliance relation $\dot{\epsilon} = \mathbf{C}^{ep} : \dot{\sigma}$ is readily found with the aid of the SHERMAN / MORRISON formula.

$$\mathbf{E}^{ep} = \mathbf{E} - \frac{\mathbf{E} : \mathbf{m} \otimes \mathbf{n} : \mathbf{E}}{E_p + \mathbf{n} : \mathbf{E} : \mathbf{m}}; \qquad \mathbf{C}^{ep} = \mathbf{C} + \frac{\mathbf{m} \otimes \mathbf{n}}{E_p} \tag{9}$$

Here $\mathbf{m}$ and $\mathbf{n}$ denote the gradients of the plastic potential and the yield surface. For isotropic conditions the linear-elastic operators are defined as :

$$\mathbf{E} = 2\mu \mathbf{1}_4 + \lambda \mathbf{I} \otimes \mathbf{I}; \qquad \mathbf{C} = \frac{1}{2\mu}\mathbf{1}_4 - \frac{\lambda}{2\mu(2\mu + 3\lambda)}\mathbf{I} \otimes \mathbf{I} \tag{10}$$

where μ, λ denote the LAMÉ-constants. For the investigation of the elasto-plastic tangential operator under dimensional constraints it is useful to define the following tensor space of symmetric second order tensors $V^{3D} := \{\mathbf{T} \mid T_{ij} = T_{ji},\ i, j = 1, 3\}$ $\quad \text{Dim}\, V^{3D} = 6$, and its 2-dimensional subspace $V^{2D} := \{\mathbf{T} \in V^{3D} \,|\, T_{13} \equiv T_{23} \equiv T_{33} \equiv 0\}$ $\text{Dim}\, V^{2D} = 3$. In our notation upper case indices designate the 2-dimensional subspace $\mathbf{T} \in V^{2D} \rightarrow T_{IJ} = T_{JI},\ I, J = 1, 2$, while the axisymmetric reduction is indicated by lower case indices $V^{AX} := \{\mathbf{T} \in V^{3D} \,|\, T_{13} \equiv T_{23} \equiv 0, T_{33} \neq 0\}$ $\text{Dim}\, V^{AX} = 4$

(i) Plane Strain Constraint :

For plane strain conditions the strain tensor is restricted to $\boldsymbol{\epsilon} \in V^{2D}$, whereas the stress tensor belongs to $\boldsymbol{\sigma} \in V^{AX}$. Thus the gradients of the yield surface and plastic potential are evaluated in $\mathbf{n}, \mathbf{m} \in V^{AX}$. Consequently the elasto-plastic operator expands into :

$$\dot{\sigma}_{IJ} = E^{ep,\epsilon}_{IJKL} \dot{\epsilon}_{KL}; \qquad E^{ep,\epsilon}_{IJKL} = E_{IJKL} - \frac{E_{IJmn} m_{mn}\, n_{op} E_{opKL}}{E_p + n_{qr} E_{qrst} m_{st}}$$

$$\mathbf{E}^{ep,\epsilon} = \mathbf{E}^{2D,2D} - \frac{\mathbf{E}^{2D,AX} : \mathbf{m}^{AX} \otimes \mathbf{n}^{AX} : \mathbf{E}^{AX,2D}}{E_p + \mathbf{n}^{AX} : \mathbf{E}^{AX,AX} : \mathbf{m}^{AX}} \tag{11}$$

(ii) Plane Stress Constraint :

For plane stress conditions the stress tensor is restricted to $\boldsymbol{\sigma} \in V^{2D}$, whereas the strain tensor is in $\boldsymbol{\epsilon} \in V^{AX}$. Thus the gradients of the yield surface and plastic flow potential are evaluated in $\mathbf{n}, \mathbf{m} \in V^{2D}$. The elasto-plastic compliance is in this case :

$$\dot{\epsilon}_{KL} = C^{ep,\sigma}_{IJKL} \dot{\sigma}_{IJ}; \qquad C^{ep,\sigma}_{IJKL} = C_{IJKL} + \frac{m_{IJ}\, n_{KL}}{E_p} \tag{12}$$

The corresponding stiffness relation $\mathbf{E}^{ep,\sigma} = (\mathbf{C}^{ep,\sigma})^{-1}$ is obtained utilizing the SHERMAN / MORRISON formula.

$$\mathbf{E}^{ep,\sigma} = \mathbf{E}^{\sigma} - \frac{\mathbf{E}^{\sigma} : \mathbf{m}^{2D} \otimes \mathbf{n}^{2D} : \mathbf{E}^{\sigma}}{E_p + \mathbf{n}^{2D} : \mathbf{E}^{\sigma} : \mathbf{m}^{2D}} \tag{13}$$

Here $\mathbf{E}^{\sigma}$ designates the plane elastic operator of isotropic linear elasticity.

$$\mathbf{E}^{\sigma} = (\mathbf{C}^{\sigma})^{-1} = 2\mu \mathbf{1}^{2D,2D}_{4} + \frac{2\mu\lambda}{2\mu + \lambda} \mathbf{I}^{2D} \otimes \mathbf{I}^{2D} \tag{14}$$

(iii) Axisymmetric Constraint :

The elasto-plastic tangent operator for axisymmetric conditions is derived in an analogous manner with $\boldsymbol{\sigma} \in V^{AX}, \boldsymbol{\epsilon} \in V^{AX}$ and $\mathbf{n}, \mathbf{m} \in V^{AX}$.

$$\mathbf{E}^{ep,AX} = \mathbf{E}^{AX,AX} - \frac{\mathbf{E}^{AX,AX} : \mathbf{m}^{AX} \otimes \mathbf{n}^{AX} : \mathbf{E}^{AX,AX}}{E_p + \mathbf{n}^{AX} : \mathbf{E}^{AX,AX} : \mathbf{m}^{AX}} \tag{15}$$

For ideal plasticity, when the elasto-plastic modulus is zero $E_p = 0$, the singularities of the tangent operator can be readily evaluated associated with the left and right critical eigendirections. In the plane strain case the rank 1 modification singularizes the tangent operator as long as the gradients $\mathbf{n}, \mathbf{m}$ are in V^{2D}. This is a very crucial restriction, since in general $\mathbf{n}, \mathbf{m}$ are in V^{AX} constraining the singularity usually associated with ideal plasticity.

$$\dot{\epsilon}^{cr,l} : \mathbf{E}^{ep,\epsilon} = \dot{\epsilon}^{cr,l} : \mathbf{E}^{2D,2D} - \Big(\frac{\dot{\epsilon}^{cr,l} : \mathbf{E}^{2D,AX} : \mathbf{m}^{AX}}{\mathbf{n}^{AX} : \mathbf{E}^{AX,AX} : \mathbf{m}^{AX}}\Big)\mathbf{n}^{AX} : \mathbf{E}^{AX,2D} \rightarrow \dot{\epsilon}^{cr,l} = \mathbf{n}^{2D} \tag{16}$$

$$\mathbf{E}^{ep,\epsilon} : \dot{\epsilon}^{cr,r} = \mathbf{E}^{2D,2D} : \dot{\epsilon}^{cr,r} - \mathbf{E}^{2D,AX} : \mathbf{m}^{AX}\Big(\frac{\mathbf{n}^{AX} : \mathbf{E}^{AX,2D} : \dot{\epsilon}^{cr,r}}{\mathbf{n}^{AX} : \mathbf{E}^{AX,AX} : \mathbf{m}^{AX}}\Big) \rightarrow \dot{\epsilon}^{cr,r} = \mathbf{m}^{2D} \tag{17}$$

In contrast, for plane stress the singularity of the tangent operator can be activated for $\mathbf{n}, \mathbf{m} \in V^{2D}$. It is this condition which is responsible for the wide range of instabilities met under plane stress assumptions.

$$\dot{\epsilon}^{cr,l} : \mathbf{E}^{ep,\sigma} = \dot{\epsilon}^{cr,l} : \mathbf{E}^{\sigma} - (\frac{\dot{\epsilon}^{cr,l} : \mathbf{E}^{\sigma} : \mathbf{m}^{2D}}{\mathbf{n}^{2D} : \mathbf{E}^{\sigma} : \mathbf{m}^{2D}})\mathbf{n}^{2D} : \mathbf{E}^{\sigma} \rightarrow \dot{\epsilon}^{cr,l} = \mathbf{n}^{2D} \tag{18}$$

$$\mathbf{E}^{ep,\sigma} : \dot{\epsilon}^{cr,r} = \mathbf{E}^{\sigma} : \dot{\epsilon}^{cr,r} - \mathbf{E}^{\sigma} : \mathbf{m}^{2D}(\frac{\mathbf{n}^{2D} : \mathbf{E}^{\sigma} : \dot{\epsilon}^{cr,r}}{\mathbf{n}^{2D} : \mathbf{E}^{\sigma} : \mathbf{m}^{2D}}) \rightarrow \dot{\epsilon}^{cr,r} = \mathbf{m}^{2D} \tag{19}$$

In analogy the elasto-plastic tangent operator for axisymmetric constraints is rank deficient if $\mathbf{n}, \mathbf{m} \in V^{AX}$.

$$\dot{\epsilon}^{cr,l} : \mathbf{E}^{ep,AX} = \dot{\epsilon}^{cr,l} : \mathbf{E}^{AX,AX} - (\frac{\dot{\epsilon}^{cr,l} : \mathbf{E}^{AX,AX} : \mathbf{m}^{AX}}{\mathbf{n}^{AX} : \mathbf{E}^{AX,AX} : \mathbf{m}^{AX}})\mathbf{n}^{AX} : \mathbf{E}^{AX,2D} \rightarrow \dot{\epsilon}^{cr,l} = \mathbf{n}^{AX} \tag{20}$$

$$\mathbf{E}^{ep,AX} : \dot{\epsilon}^{cr,r} = \mathbf{E}^{AX,AX} : \dot{\epsilon}^{cr,r} - \mathbf{E}^{AX,AX} : \mathbf{m}^{AX}(\frac{\mathbf{n}^{AX} : \mathbf{E}^{AX,AX} : \dot{\epsilon}^{cr,r}}{\mathbf{n}^{AX} : \mathbf{E}^{AX,AX} : \mathbf{m}^{AX}}) \rightarrow \dot{\epsilon}^{cr,r} = \mathbf{m}^{AX} \tag{21}$$

The singularity of the tangent operator is normally responsible for 'diffuse' failure. In contrast, 'localized' failure is associated with discontinuous bifurcation of the strain rate field. The jump of the strain rate $[[\dot{\epsilon}]] = \dot{\gamma}\frac{1}{2}[\mathbf{M}\otimes\mathbf{N}+\mathbf{N}\otimes\mathbf{M}]$ is restricted to the subspace V^{R2} with

$$V^{R2} := \{\mathbf{T} \,|\, \mathrm{rank}\,(\mathbf{T}) \leq 2\} \tag{22}$$

With this preliminaries at hand the quadratic form of the discontinuous bifurcation condition is related to the jump of the strain rate.

$$\rho c^2 = (\mathbf{M}\otimes\mathbf{N}) : \mathbf{E}^{ep} : (\mathbf{N}\otimes\mathbf{M}) = [[\dot{\epsilon}]]^t : \mathbf{E}^{ep} : [[\dot{\epsilon}]] = 0 \tag{23}$$

Clearly, this loss of ellipticity which is synonymous with the zero wave speed argument when $c = 0$ is not nessecarily associated with the loss of positive definitness of the tangential material operator if non-symmetric models are considered. For the sake of simplicity let us consider symmetric operators for 2 dimensional problems in what follows. In this case the neccesary condition for bifurcation reads

$$\dot{\epsilon}^{cr} : \mathbf{E}^{ep} : \dot{\epsilon}^{cr} = 0 \quad \text{for} \quad \dot{\epsilon}^{cr} = \mathbf{n} \in V^{2D} \tag{24}$$

Recall that this quadratic form provides a sufficient condition for loss of ellipticity as shown in the analysis of non-symmetric operators where discontinuous bifurcation is not precluded by positive spectral properties due to the BROMWICH bounds [4], see WILLAM & ETSE [13]. Since $[[\dot{\epsilon}]]$ belongs to the rank one space V^{R2}, the sufficient condition for discontinuous bifurcation is simply

$$\dot{\epsilon}^{cr} \quad \text{coaxial} \quad [\mathbf{M}\otimes\mathbf{N}+\mathbf{N}\otimes\mathbf{M}] \in V^{R2} \tag{25}$$

For the 2-dimensional case this expression expands into

$$\dot{\epsilon}^{cr} = \begin{pmatrix} n_{11} & n_{12} \\ n_{21} & n_{22} \end{pmatrix} = \gamma \begin{pmatrix} M_1 N_1 & \frac{1}{2}[M_1 N_2 + M_2 N_1] \\ \frac{1}{2}[M_2 N_1 + M_1 N_2] & M_2 N_2 \end{pmatrix} \tag{26}$$

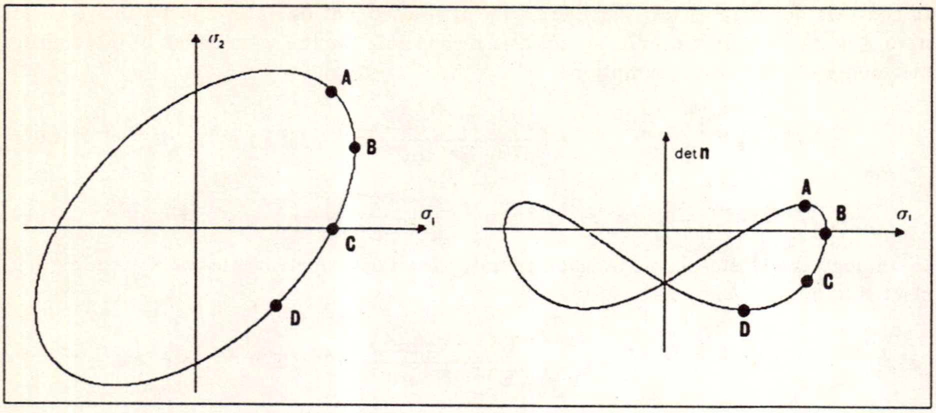

Figure 1: Plane Stress J_2 Yield Surface and Localization Condition

subjected to the subsidiary condition $|\mathbf{M}| = 1 = |\mathbf{N}|$. Basic algebraic manipulations reveal that this condition is synonymous with $\det \mathbf{n} = n_{11}n_{22} - n_{12}n_{21} \leq 0$. This condition is studied for the v.MISES yield ellipse in plane stress, see Fig.1. The variation of the eigenspectrum of the localization tensor $\mathbf{Q}^{ep}$ in form of phase velocity plots and its determinant are depicted in Fig.2[14]. Remarkably there is no singularity of $\mathbf{Q}$ under equibiaxial tension at point (A), the first loss of ellipticity is met under biaxial tension at point (B) with one discontinuity aligned with the 90 deg direction. Uniaxial tension at point (C) leads to two discontinuity surfaces with normals inclined under ±35.26 deg. For pure shear at point (D) the classical ±45 deg orientations of shear bands are predicted by the $\mathbf{Q}$-analysis.

4 Element Distribution of Localized Discontinuity

Let us consider localization under the assumption of uniform stress- and state variables within the element domain. When a singularity in the localization tensor is detected at the element centroid a discontinuity develops with the normal vector $\mathbf{N}$ which divides the element area A into subareas A^+ and A^- at the positive and negative side of the discontinuity. Without loss of generality we set $\lambda = 1$ and add the velocity field $\dot{\mathbf{u}} = \dot{\gamma}\Lambda(\mathbf{M} \otimes \mathbf{N}) \cdot \hat{\mathbf{x}}$ with $\hat{\mathbf{x}} = (\mathbf{x} - \mathbf{x}_0)$ to the standard finite element approximation $\dot{\mathbf{u}}^h$. The nodal velocities are then $\dot{\mathbf{d}}^b = \dot{\mathbf{d}} + \dot{\mathbf{d}}_{loc}$ where the nodal velocities for localization $\dot{\mathbf{d}}_{loc}$ are determined as

$$\dot{\mathbf{u}}_{loc} = \dot{\gamma} \begin{pmatrix} \Lambda M_1(N_1\hat{x}_1 + N_2\hat{y}_1) \\ \Lambda M_1(N_1\hat{x}_2 + N_2\hat{y}_2) \\ \Lambda M_1(N_1\hat{x}_3 + N_2\hat{y}_3) \\ \Lambda M_1(N_1\hat{x}_4 + N_2\hat{y}_4) \end{pmatrix} \quad \text{and} \quad \dot{\mathbf{v}}_{loc} = \dot{\gamma} \begin{pmatrix} \Lambda M_2(N_1\hat{x}_1 + N_2\hat{y}_1) \\ \Lambda M_2(N_1\hat{x}_2 + N_2\hat{y}_2) \\ \Lambda M_2(N_1\hat{x}_3 + N_2\hat{y}_3) \\ \Lambda M_2(N_1\hat{x}_4 + N_2\hat{y}_4) \end{pmatrix} \tag{27}$$

These nodal degrees of freedom are equivalent to discontinuous bifurcation within the element domain, see Fig.3. However, since discontinuous fields are usually not permitted within the element domain, the discontinuity is redistributed by the shape functions and the $\mathbf{B}$-operator respectively. This leads to the smeared form of localization depicted in

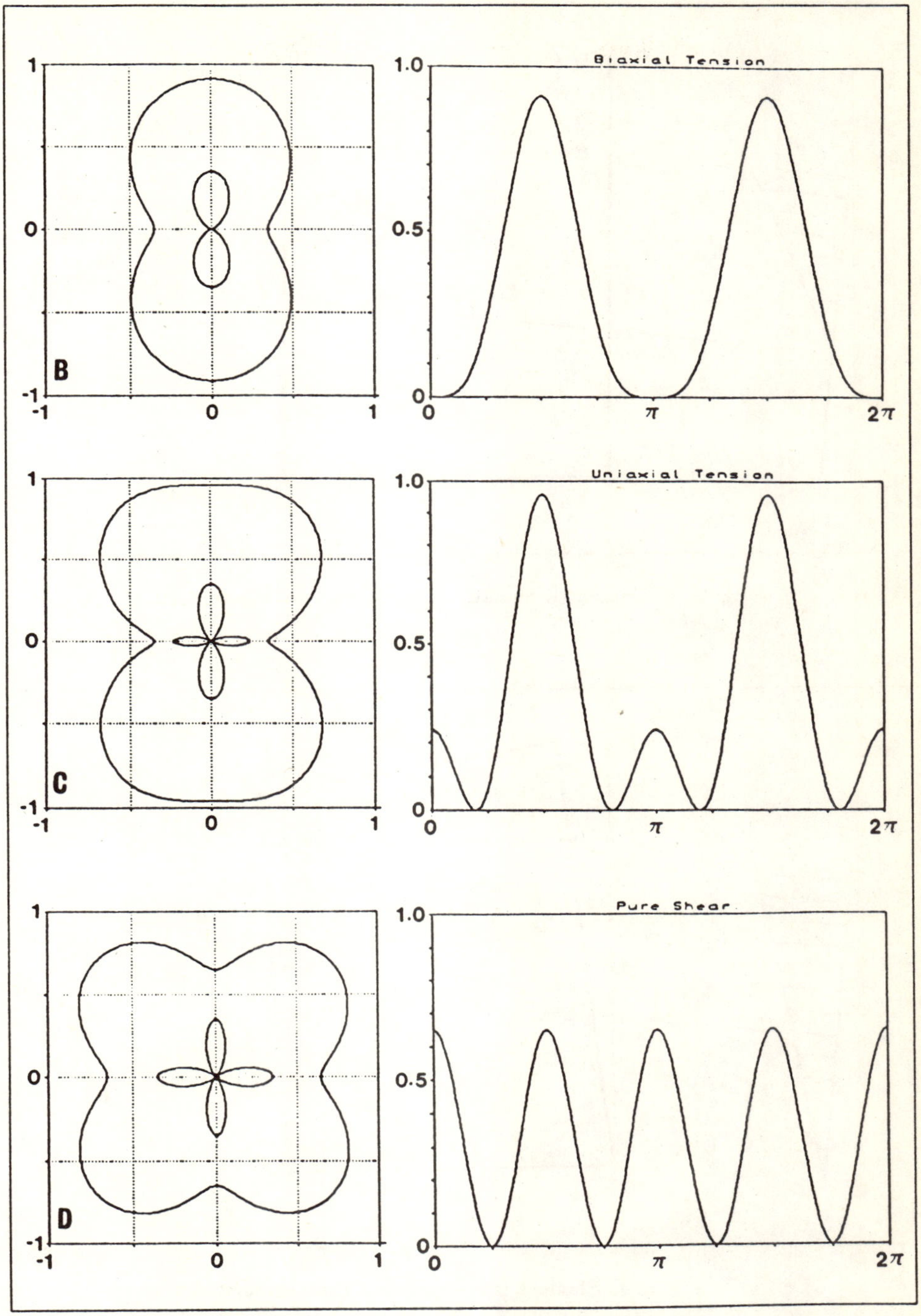

Figure 2: Phase velocity and Variation of det **Q** : Plane Stress J_2 Plasticity

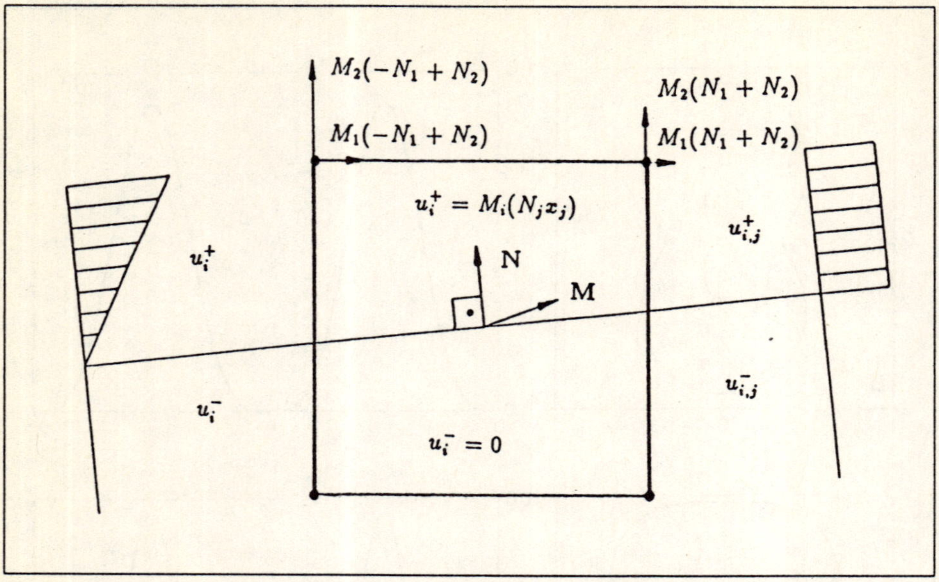

Figure 3: Element Kinematics of Discontinous Bifurcation

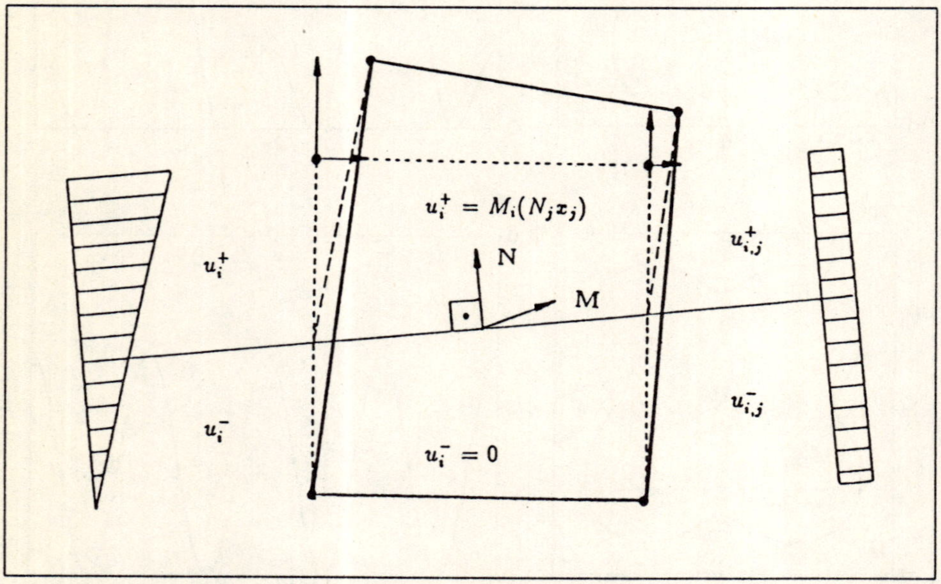

Figure 4: Element Distribution of Discontinuity

Fig. 4 which is energetically equivalent to the formation of the localized discontinuity.

$$\dot{\mathbf{u}}^{h,b}_{dis} = \mathbf{N}\dot{\mathbf{d}} + \mathbf{N}\dot{\mathbf{d}}_{loc} \tag{28}$$

$$\dot{\epsilon}^{h,b}_{dis} = \mathbf{B}\dot{\mathbf{d}} + \mathbf{B}\dot{\mathbf{d}}_{loc} = \mathbf{B}\dot{\mathbf{d}} + [[\dot{\epsilon}]]_{dis} \tag{29}$$

5 Weak Form of Localization

The zero value of the quadratic form $\rho c^2 = \mathbf{M} \cdot \mathbf{Q}^{ep} \cdot \mathbf{M}$ indicates a vanishing wave speed which is synonymous with the localization condition for forming a discontinuity surface. Thereby, the quadratic form is a sufficient but not a necessary condition for localization. Because of the minor symmetry of the material operator this jump condition can be restated in terms of the pointwise localization condition

$$\rho c^2 = \mathbf{M} \otimes \mathbf{N} : \mathbf{E}^{ep} : \mathbf{N} \otimes \mathbf{M} = [\mathbf{M} \otimes \mathbf{N}]^s : \mathbf{E}^{ep} : [\mathbf{N} \otimes \mathbf{M}]^s = \dot{\gamma}^{-2}[[\dot{\epsilon}]]^t : \mathbf{E}^{ep} : [[\dot{\epsilon}]] = 0 \tag{30}$$

Distribution of the discontinuous strain rate within the element domain infers the following approximation of the jump condition

$$[[\dot{\epsilon}]]_{dis} = \mathbf{B}_1\dot{\mathbf{d}}^{loc}_1 + \mathbf{B}_2\dot{\mathbf{d}}^{loc}_2 = (\mathbf{B}_1 - \mathbf{B}_2\mathbf{k}^{-1}_{22}\mathbf{k}_{21})\dot{\mathbf{d}}^{loc}_1 \neq \dot{\gamma}[\mathbf{N} \otimes \mathbf{M}]^s \tag{31}$$

using condensation of the internal degrees of freedom $\dot{\mathbf{d}}^{loc}_2 = -\mathbf{k}^{-1}_{22}\mathbf{k}_{21}\dot{\mathbf{d}}^{loc}_1$. In general the distributed approximation of the strain rate jump violates pointwise localization, i.e.

$$[[\dot{\epsilon}]]^t_{dis} : \mathbf{E}^{ep} : [[\dot{\epsilon}]]_{dis} \neq 0 \tag{32}$$

However, $[[\dot{\epsilon}]]_{dis}$ satisfies the localization condition in the integral sense if the element description captures discontinuous bifurcation. The weak form is obtained by integrating the localization condition over the element domain

$$\mathrm{d}^2W_{loc} = \int_{V^e} \rho c^2 \mathrm{d}V = \int_{V^e} [[\dot{\epsilon}]]^t_{dis}\mathbf{E}^{ep}[[\dot{\epsilon}]]_{dis}\mathrm{d}V \doteq 0 \tag{33}$$

In terms of the external nodal degrees of freedom the weak localization criterion reduces to

$$\mathrm{d}^2W_{loc} = \dot{\mathbf{d}}^{loc}_1(\mathbf{k}_{11} - \mathbf{k}_{12}\mathbf{k}^{-1}_{22}\mathbf{k}_{21})\dot{\mathbf{d}}^{loc}_1 \doteq 0 \tag{34}$$

Alternatively, in terms of the external as well as internal degrees of freedom this condition expands into

$$\mathrm{d}^2W_{loc} = \dot{\mathbf{d}}^{loc}_1\mathbf{k}_{11}\dot{\mathbf{d}}^{loc}_1 - \dot{\mathbf{d}}^{loc}_2\mathbf{k}_{22}\dot{\mathbf{d}}^{loc}_2 \doteq 0 \tag{35}$$

In other terms, the singularity $\det(\mathbf{Q}^{ep}) = 0$ of the localization tensor is reproduced at the element level, if the localization kinematics above is associated with zero energy. Analysis of the localization tensor determines the kinematics of discontinuous bifurcation and furnishes the nodal values $\dot{\mathbf{d}}_{loc} = [\dot{\mathbf{u}}_{loc}, \dot{\mathbf{v}}_{loc}]$. For each constitutive model, the localization kinematics $\dot{\mathbf{d}}_{loc}$ identifies those eigenmodes which are active and which participate in the formation of discontinuity surface during failure. In that case, all eigenvalues associated with those eigenmodes must vanish in order to reproduce discontinuous bifurcation at the element level without numerical dissipation.

	$Q4$	$QR4$	$CST4$	$QM6$	$QE5$	$Q\bar{B}$	$QLOC$	QS	OI	QBI	FIB	OB
(B) :		•		†	†			•				
(C) :		•		†	†			•				
(D) :		•		†	†			•				

Table 1: Localization Properties of 4-Node Elements : Plane Stress J_2-Model

6 Localization Properties of Different Elements

In the following, different four noded elements are examined with regard to their ability to capture localized failure. Thereby a square element domain is considered which is subjected to plane stress conditions. In J_2 theory of ideal elasto-plasticity different failure modes emerge. Depending on the position of the yield stress on the elliptic yield locus tensile splitting as well as shear banding modes emerge at points B, C and D of Fig.1. However no localization takes place at yielding under equibiaxial tension at point A.

In order to verify 'isotropy' or 'directional bias' we assume a homogeneous distribution of all state quantities and rotate the natural $\xi - \eta$ co-ordinate system of the element with respect to the fixed global $x - y$ co-ordinate system. The weak localization test is performed continuously to check the ability of the element formulation to capture localization and to check frame invariance. Fig.5 illustrates the normalized energy dissipation during localization for a 2π rotation of the natural co-ordinate system. The localization properties of the different quadrilateral elements are summarized in Tab.1. Elements marked with a bullet • capture localization for all orientations of the natural co-ordinate system, those marked with a cross † fail numerically in certain cases and those without any mark do not satisfy the weak localization test in general.

There are only some elements which satisfy the weak localization test throughout. The one-point underintegrated **QR4**-element [1], the incompatible **QM6**-element by TAYLOR, BERESFORD & WILSON [12] which was recently reformulated by SIMO & RIFAI [9] in the context of mixed enhanced strain models, the **QE5**-element proposed by the same authors and the singularized **QS**-element of STEINMANN & WILLAM [11]. Nevertheless the two mixed enhanced strain formulations fail numerically for certain orientations of the natural co-ordinate system due to a singular substiffness $\mathbf{k}_{22}$. The standard **Q4**-element, the incompressible $\mathbf{Q\bar{B}}$-element of HUGHES [3], the assembly of four constant strain triangles **CST4**, different stabilization techniques advocated by BELYTSCHKO & BINDEMAN [2] **OI, QBI, FIB, OB** and the enriched **QLOC**-element by ORTIZ, LEROY & NEEDLEMAN [7] fail the weak localization test for almost all orientations of the natural co-ordinate system with respect to the direction of the discontinuity surface. Note that all elements render the correct zero-energy dissipation during localization as long as the element edges are aligned with the orientation of the discontinuity surface. The **CST4**-element has been considered particularly suitable for capturing localization in J_2-computations, see e.g. [14]. Since the **CST4**-element is enriched by additional internal element edges it satisfies the weak localization test for more orientations then the other unsatisfactory elements, see Fig.5, but it does not capture localization in general. The reason for this sometimes unexpected behaviour is explained in detail in the next section.

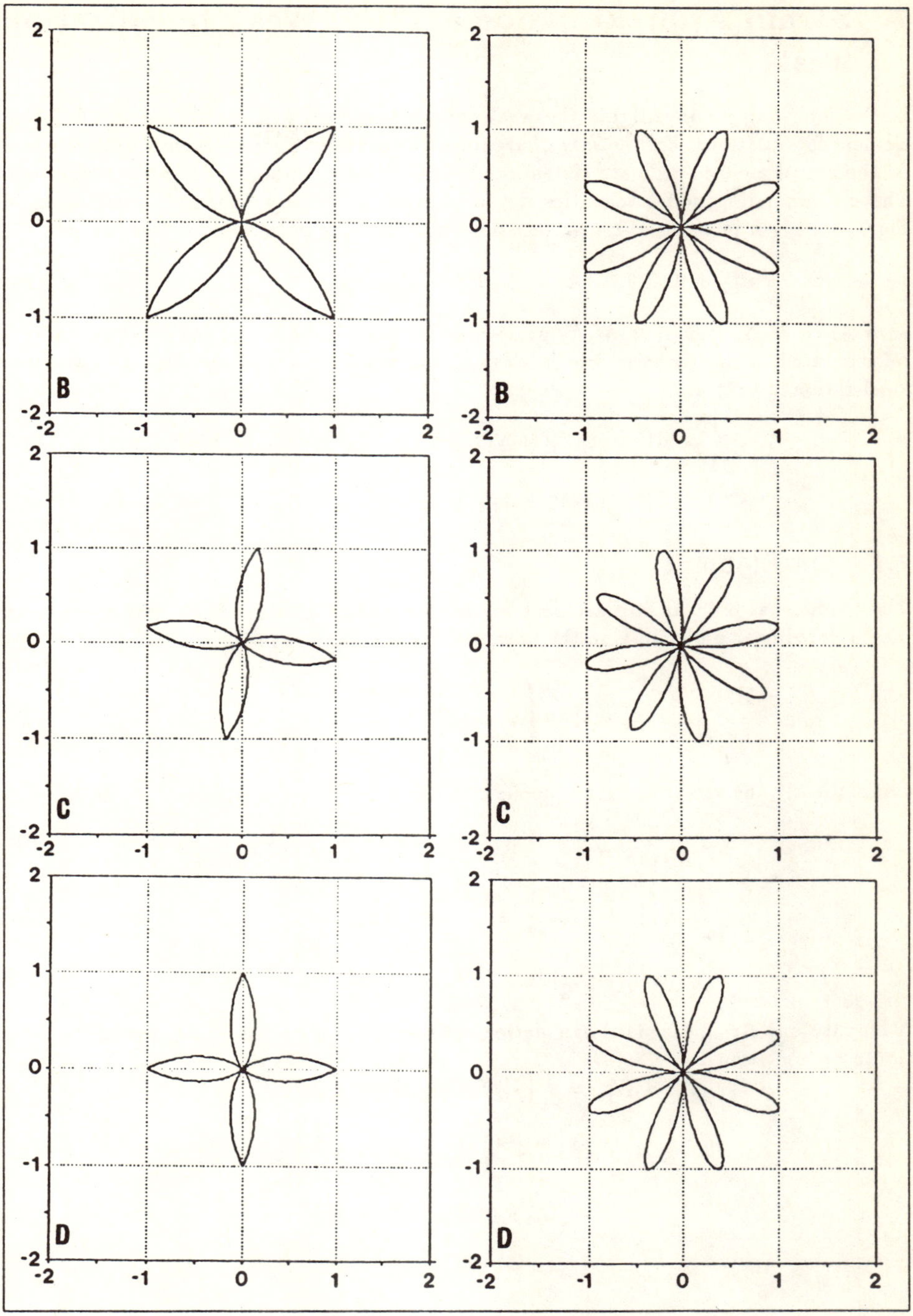

Figure 5: Weak Localization Test : **Q4, CST4**

7 Strain Approximation and the Weak Localization Test

In the following we investigate the weak localization test analytically for the different element formulations. For the sake of argument we consider a square element domain and let the global $x-y$ co-ordinate system coincide with the natural $\xi-\eta$ co-ordinate system. The corresponding nodal velocities $\dot{\mathbf{d}}_{loc}$ associated with the localization kinematics in Fig.3. are given in terms of the surface normal $\mathbf{N}$ and the polarization direction $\mathbf{M}$.

$$\dot{\mathbf{d}}_{loc}^t = [0,0,0,0,(\mathcal{X}_1+\mathcal{X}_2),(\mathcal{Y}_1+\mathcal{Y}_2),(\mathcal{X}_2-\mathcal{X}_1),(\mathcal{Y}_2-\mathcal{Y}_1)] \tag{36}$$

with $\mathcal{X}_i = M_1 N_i$ and $\mathcal{Y}_i = M_2 N_i$ while $\dot{\gamma} = 1$. The elasto-plastic material operator is reformulated in matrix form which leads to the following expression for the quadratic localization indicator.

$$\int_V \rho c^2 \mathrm{d}V = \mathcal{X}_1^2 E_{11}^{ep} + \mathcal{Y}_2^2 E_{22}^{ep} + (\mathcal{X}_2^2 + \mathcal{Y}_1^2 + 2\mathcal{X}_2\mathcal{Y}_1)E_{33}^{ep}] \tag{37}$$

$$+2[\mathcal{X}_1\mathcal{Y}_2 E_{12}^{ep} + (\mathcal{Y}_1\mathcal{Y}_2 + \mathcal{X}_2\mathcal{Y}_2)E_{23}^{ep} + (\mathcal{X}_1\mathcal{X}_2 + \mathcal{X}_1\mathcal{Y}_1)E_{31}^{ep}] = 0$$

7.1 Q4-Element

The traditional bilinear displacement expansion renders a strain field, which spans an incomplete linear polynomial in the natural ξ-η co-ordinates.

$$\epsilon^h \in \text{span}\begin{pmatrix} 1 & 0 & 0 & \eta & 0 & 0 & 0 \\ 0 & 1 & 0 & 0 & \xi & 0 & 0 \\ 0 & 0 & 1 & 0 & 0 & \xi & \eta \end{pmatrix} \tag{38}$$

Consequently the smeared discontinuous strain rate field results with eq.(36) in

$$\begin{aligned} [[\dot{\epsilon}_{11}]] &= \frac{1}{2}\mathcal{X}_1 && && + \frac{1}{2}\mathcal{X}_1\eta \\ [[\dot{\epsilon}_{22}]] &= \frac{1}{2}\mathcal{Y}_2 && + \frac{1}{2}\mathcal{Y}_1\xi \\ [[\dot{\epsilon}_{12}]] &= \frac{1}{2}[\mathcal{X}_2+\mathcal{Y}_1] && + \frac{1}{2}\mathcal{X}_1\xi && + \frac{1}{2}\mathcal{Y}_1\eta \end{aligned} \tag{39}$$

A lenghty but straightforward calculation renders the energy dissipated during localizationto be non-zero.

$$\mathrm{d}^2 W_{loc} = \int_V [[\dot{\epsilon}]]^t \mathbf{E}^{ep}[[\dot{\epsilon}]]\mathrm{d}V = \int_V \rho c^2 \mathrm{d}V$$

$$+\frac{1}{3}[\mathcal{X}_1^2 E_{11}^{ep} + \mathcal{Y}_1^2 E_{22}^{ep} + (\mathcal{X}_1^2 + \mathcal{Y}_1^2)E_{33}^{ep}] \tag{40}$$

$$+\frac{2}{3}[\mathcal{X}_1\mathcal{Y}_1(E_{23}^{ep} + E_{31}^{ep})] \neq 0$$

7.2 QR4-Element

The QR4-element results from underintegration of the incomplete linear strain field which corresponds to an uniform strain approximation in the element domain.

$$\epsilon^h \in \text{span}\begin{pmatrix} 1 & 0 & 0 \\ 0 & 1 & 0 \\ 0 & 0 & 1 \end{pmatrix}. \tag{41}$$

The smeared discontinuous strain rate field reads

$$\begin{aligned} [[\dot{\epsilon}_{11}]] &= \frac{1}{2}\mathcal{X}_1 \\ [[\dot{\epsilon}_{22}]] &= \frac{1}{2}\mathcal{Y}_2 \\ [[\dot{\epsilon}_{12}]] &= \frac{1}{2}[\mathcal{X}_2 + \mathcal{Y}_1] \end{aligned} \tag{42}$$

It is easily verified that the weak localization test is satisfied identically under all conditions.

$$\mathrm{d}^2 W_{loc} = \int_V [[\dot{\epsilon}]]^t \mathbf{E}^{ep} [[\dot{\epsilon}]] \mathrm{d}V = \int_V \rho c^2 \mathrm{d}V = 0 \tag{43}$$

7.3 $\bar{\mathbf{B}}$-Element

The $\bar{\mathbf{B}}$-approach was proposed by HUGHES [3] to improve the isochoric performance of the 4-node quadrilateral. It consists of replacing the compatible strain-displacement operator $\mathbf{B}_1$ by the modified $\bar{\mathbf{B}}_1$-operator. To this end the $\mathbf{B}_1$-operator is split into $\mathbf{B}_1^{dev}$ and $\mathbf{B}_1^{dil}$ to separate the deviatoric and dilatational components of the strain tensor when multiplied with $\mathbf{d}_1$. The $\mathbf{B}_1^{dil}$- operator is replaced by an improved dilatational contribution which is evaluated best in case of the bilinear quadrilateral at the centroid of the element, as suggested originally by NAGTEGAAL, PARKS & RICE [6].

$$\bar{\mathbf{B}}_1(\xi,\eta) = \mathbf{B}_1^{dev}(\xi,\eta) + \mathbf{B}_1^{dil}(\xi_0,\eta_0) \tag{44}$$

Thus the locking behavior of the traditional bilinear element is reduced for nearly incompressible problems.

$$\epsilon^h \in \text{span}\begin{pmatrix} 1 & 0 & 0 & -\frac{1}{3}\xi & +\frac{2}{3}\eta & 0 & 0 \\ 0 & 1 & 0 & +\frac{2}{3}\xi & -\frac{1}{3}\eta & 0 & 0 \\ 0 & 0 & 1 & 0 & 0 & \xi & \eta \end{pmatrix} \tag{45}$$

Even though this formulation looses its genuine meaning under plane stress conditions, the same element description is used for the sake of comparison. The smeared discontinuous strain rate field results in

$$\begin{aligned} [[\dot{\epsilon}_{11}]] &= \frac{1}{2}\mathcal{X}_1 - \frac{1}{6}\mathcal{Y}_1\xi + \frac{2}{6}\mathcal{X}_1\eta \\ [[\dot{\epsilon}_{22}]] &= \frac{1}{2}\mathcal{Y}_2 + \frac{2}{6}\mathcal{Y}_1\xi - \frac{1}{6}\mathcal{X}_1\eta \\ [[\dot{\epsilon}_{12}]] &= \frac{1}{2}[\mathcal{X}_2 + \mathcal{Y}_1] + \frac{1}{2}\mathcal{X}_1\xi + \frac{1}{2}\mathcal{Y}_1\eta \end{aligned} \tag{46}$$

This leads to energy dissipation during localization, which is revealed by considering the weak localization test.

$$\mathrm{d}^2W_{loc} = \int_V [[\dot{\epsilon}]]^t \mathbf{E}^{ep}[[\dot{\epsilon}]]\mathrm{d}V = \int_V \rho c^2 \mathrm{d}V$$

$$+\frac{1}{27}[(\mathcal{Y}_1^2 + 4\mathcal{X}_1^2)E_{11}^{ep} + (4\mathcal{Y}_1^2 + \mathcal{X}_1^2)E_{22}^{ep} + 9(\mathcal{X}_1^2 + \mathcal{Y}_1^2)E_{33}^{ep}] \tag{47}$$

$$+\frac{2}{27}[3\mathcal{X}_1\mathcal{Y}_1(E_{23}^{ep} + E_{31}^{ep}) - 2(\mathcal{X}_1^2 + \mathcal{Y}_1^2)E_{12}^{ep}] \neq 0$$

7.4 QM6-Element

The incompatible **QM6**-element by TAYLOR, BERESFORD & WILSON [12] was reformulated by SIMO & RIFAI [9] within the mixed formulation of an equivalent enhanced strain field. Let us consider stress states which permit decoupling of the volumetric-deviatoric representation of the elasto-plastic material operator. After lenghty but straightforward calculations the internal degrees of freedom $\dot{\mathbf{d}}_2^{loc} = -\mathbf{k}_{22}^{-1}\mathbf{k}_{21}\dot{\mathbf{d}}_1^{loc}$ associated with the localization kinematic $\dot{\mathbf{d}}_1^{loc}$ can be expressed as

$$\dot{\mathbf{d}}_2^{loc,t} = -\frac{1}{2}\left[\frac{\det \hat{\mathbf{E}}_{21}^{ep}}{\det \hat{\mathbf{E}}_{22}^{ep}}\mathcal{Y}_1, \frac{\det \hat{\mathbf{E}}_{12}^{ep}}{\det \hat{\mathbf{E}}_{11}^{ep}}\mathcal{X}_1, \mathcal{X}_1, \mathcal{Y}_1\right] \tag{48}$$

with $\hat{\mathbf{E}}_{ij}^{ep}$ denoting the matrix which is obtained by deleting the i-th row and j-th column of $\mathbf{E}^{ep}$. This leads to an approximation of the normal strain components with linear polynomials which are complete, while the shear strain is reduced to uniform variation.

$$\epsilon^h \in \mathrm{span}\begin{pmatrix} 1 & 0 & 0 & \xi & \eta & 0 & 0 \\ 0 & 1 & 0 & 0 & 0 & \xi & \eta \\ 0 & 0 & 1 & 0 & 0 & 0 & 0 \end{pmatrix} \tag{49}$$

To satisfy the weak localization test the internal degrees of freedom have to eliminate the energy dissipation produced by the external part of the tangential stiffness matrix $\mathrm{d}^2W_{loc} = \dot{\mathbf{d}}_1^{loc}\mathbf{k}_{11}\dot{\mathbf{d}}_1^{loc} - \dot{\mathbf{d}}_2^{loc}\mathbf{k}_{22}\dot{\mathbf{d}}_2^{loc} \doteq 0$. The equivalent smeared discontinuous strain rate field reads

$$\begin{aligned}
[[\dot{\epsilon}_{11}]] &= \frac{1}{2}\mathcal{X}_1 - \frac{1}{2}\frac{\det \hat{\mathbf{E}}_{21}^{ep}}{\det \hat{\mathbf{E}}_{22}^{ep}}\mathcal{Y}_1\xi + \frac{1}{2}\mathcal{X}_1\eta \\
[[\dot{\epsilon}_{22}]] &= \frac{1}{2}\mathcal{Y}_2 + \frac{1}{2}\mathcal{Y}_1\xi - \frac{1}{2}\frac{\det \hat{\mathbf{E}}_{12}^{ep}}{\det \hat{\mathbf{E}}_{11}^{ep}}\mathcal{X}_1\eta \\
[[\dot{\epsilon}_{12}]] &= \frac{1}{2}[\mathcal{X}_2 + \mathcal{Y}_1]
\end{aligned} \tag{50}$$

which yields the localization energy of the internal degrees of freedom

$$\begin{aligned}
\mathrm{d}^2W_{loc}^{int} &= \dot{\mathbf{d}}_2^{loc}\mathbf{k}_{22}\dot{\mathbf{d}}_2^{loc} \\
&= \frac{1}{3}\left[\left(\frac{\det \hat{\mathbf{E}}_{21}^{ep}}{\det \hat{\mathbf{E}}_{22}^{ep}}\mathcal{Y}_1\right)^2 E_{11}^{ep} + \left(\frac{\det \hat{\mathbf{E}}_{12}^{ep}}{\det \hat{\mathbf{E}}_{11}^{ep}}\mathcal{X}_1\right)^2 E_{22}^{ep} + (\mathcal{X}_1^2 + \mathcal{Y}_1^2)E_{33}^{ep}\right] \\
&\doteq \dot{\mathbf{d}}_1^{loc}\mathbf{k}_{11}\dot{\mathbf{d}}_1^{loc} = \mathrm{d}^2W_{loc}^{ext}
\end{aligned} \tag{51}$$

This is satisfied under the above mentioned assumptions and is synonymous with $\det \mathbf{E}^{ep} = \det \hat{\mathbf{E}}_{33}^{ep} = 0$.

7.5 QLOC-Element

The enrichement strategy by ORTIZ, LEROY & NEEDLEMAN [7] for constructing the augmented stiffness **QLOC** leads to the $\mathbf{B}_2$-operator in the augmented stiffness $\mathbf{k}_{22} = \int \mathbf{B_2}^t \mathbf{E}^{ep} \mathbf{B_2} dV$ which is based on the vectors $\mathbf{N}, \mathbf{M}$ of localization analysis.

$$\mathbf{B}_2 = \Lambda[\mathbf{M} \otimes \mathbf{N} + \mathbf{N} \otimes \mathbf{M}] = 2\Lambda[\mathbf{M} \otimes \mathbf{N}]^s \tag{52}$$

Therefore, the augmented stiffness reduces to

$$\mathbf{k}_{22} = 4 \int_{V^e} \Lambda^2 [\mathbf{M} \otimes \mathbf{N}]^s : \mathbf{E}^{ep} : [\mathbf{N} \otimes \mathbf{M}]^s dV \tag{53}$$

which is the weak form of localization with vanishing stiffness contributions, i.e.

$$\mathbf{k}_{22} = 4 \int_{V^e} \Lambda^2 \rho c^2 dV = 0 \tag{54}$$

This observation leads to the counterintuitive result that the localization kinematics activates only the standard stiffness contribution $\mathbf{k}_{11}$ of the enriched element expansion, i.e.

$$d^2 W_{loc} \doteq \dot{\mathbf{d}}_1^{loc} \mathbf{k}_{11} \dot{\mathbf{d}}_1^{loc} - \dot{\mathbf{d}}_2^{loc} \mathbf{k}_{22} \dot{\mathbf{d}}_2^{loc} = \dot{\mathbf{d}}_1^{loc} \mathbf{k}_{11} \dot{\mathbf{d}}_1^{loc} \tag{55}$$

and does not improve the weak localization properties of the element since $\mathbf{k}_{22} = \mathbf{0}$.

7.6 QS-Element

Recently an alternative enrichement strategy was proposed by STEINMANN & WILLAM [11]. The weak form of localization is introduced directly into the element formulation, thus leading to a singularized stiffness. For the sake of argument we adopt here the Q4-element as reference stiffness and denote the singularized element by **QS**. In that case, the reference strain field is enhanced by the smeared form of localization which was derived as $\mathbf{B}\dot{\mathbf{d}}_{loc}$. Since two discontinuities are found in the hyperbolic range this expression is utilized directly to define the augmented strain field in terms of the generalized coordinates $\dot{d}_{21}$ and $\dot{d}_{22}$

$$\dot{\epsilon}^{h,b} = \nabla^s \dot{\mathbf{u}}^h + \dot{\tilde{\epsilon}}^h = \mathbf{B}_1 \dot{\mathbf{d}}_1 + \mathbf{B}_{21} \dot{d}_{21} + \mathbf{B}_{22} \dot{d}_{22} \quad \text{with} \quad \mathbf{B}_{21} = \mathbf{B}_1 \dot{\mathbf{d}}^1_{loc}, \mathbf{B}_{22} = \mathbf{B}_1 \dot{\mathbf{d}}^2_{loc} \tag{56}$$

Since the enhanced strain field is contained in the original strain field such that $\dot{\tilde{\epsilon}}^h \in \nabla^s \dot{\mathbf{u}}^h$, this strategy introduces the desired singularity in the tangent stiffness below

$$\begin{pmatrix} \mathbf{k}_{11} & \mathbf{k}_{11}\dot{\mathbf{d}}^1_{loc} & \mathbf{k}_{11}\dot{\mathbf{d}}^2_{loc} \\ \dot{\mathbf{d}}^{1,t}_{loc}\mathbf{k}_{11} & \dot{\mathbf{d}}^{1,t}_{loc}\mathbf{k}_{11}\dot{\mathbf{d}}^1_{loc} & \dot{\mathbf{d}}^{1,t}_{loc}\mathbf{k}_{11}\dot{\mathbf{d}}^2_{loc} \\ \dot{\mathbf{d}}^{2,t}_{loc}\mathbf{k}_{11} & \dot{\mathbf{d}}^{2,t}_{loc}\mathbf{k}_{11}\dot{\mathbf{d}}^1_{loc} & \dot{\mathbf{d}}^{2,t}_{loc}\mathbf{k}_{11}\dot{\mathbf{d}}^2_{loc} \end{pmatrix} \begin{pmatrix} \Delta\dot{\mathbf{d}}_1 \\ \Delta\dot{d}_{21} \\ \Delta\dot{d}_{22} \end{pmatrix} = \begin{pmatrix} -\mathbf{R}_1 \\ -R_{21} \\ -R_{22} \end{pmatrix} \tag{57}$$

where the residuals are defined as

$$\mathbf{R}_1 = \int_B \mathbf{B}_1^t \Sigma dV - \mathbf{F}_{ext}; \qquad R_{21} = \dot{\mathbf{d}}^{1,t}_{loc} \int_B \mathbf{B}_1^t \Sigma dV; \qquad R_{22} = \dot{\mathbf{d}}^{2,t}_{loc} \int_B \mathbf{B}_1^t \Sigma dV \tag{58}$$

Satisfaction of the patch test requires that the element formulation consistently reproduces constant strain states. In the context of mixed enhanced strain fields this condition requires that

$$\int_B \mathbf{B}_2^{(1/2)} dV = \int_B \mathbf{B}_1 \dot{\mathbf{d}}^{(1/2)}_{loc} dV \doteq \mathbf{0} \tag{59}$$

The definition of the $\dot{\mathbf{d}}_{loc}^{(1/2)}$ contains the undeterminate parameter λ which distributes the bifurcated displacement field to the plus and minus side of the discontinuity surface. By setting $\lambda = A^-/A = 1/2$ the patch test is satisfied exactly for square element domains. If we consider the condensed format of the tangent stiffness we observe the singularities associated with the two rank one modifications of the reference stiffness $\mathbf{k}_{11}$

$$\left[\mathbf{k}_{11} - \frac{(\mathbf{V}_1 W_{22} - \mathbf{V}_2 W_{21}) \otimes \mathbf{V}_1 + (\mathbf{V}_2 W_{11} - \mathbf{V}_1 W_{12}) \otimes \mathbf{V}_2}{W_{11} W_{22} - W_{21} W_{12}}\right] \Delta \dot{\mathbf{d}}_1 = \tag{60}$$

$$- \left[\mathbf{R}_1 - \frac{(\mathbf{V}_1 W_{22} - \mathbf{V}_2 W_{21}) R_{21} + (\mathbf{V}_2 W_{11} - \mathbf{V}_1 W_{12}) R_{22}}{W_{11} W_{22} - W_{21} W_{12}}\right]$$

where

$$\mathbf{V}_i = \mathbf{k}_{11} \dot{\mathbf{d}}_{loc}^i \quad \text{and} \quad W_{ij} = \dot{\mathbf{d}}_{loc}^{i,t} \mathbf{k}_{11} \dot{\mathbf{d}}_{loc}^j \tag{61}$$

For $\dot{\mathbf{d}}_{crit} = \dot{\mathbf{d}}_{loc}^{(1/2)}$ the singularized element is designed such that $\Delta \dot{\mathbf{d}}_1$ bifurcates if the external force is orthogonal to the orientation of localization, i.e.

$$\dot{\mathbf{d}}_{loc}^{t(1/2)} \mathbf{F}_{ext} = - \left[\dot{\mathbf{d}}_{loc}^{t(1/2)} \mathbf{R}_1 - \dot{\mathbf{d}}_{loc}^{t(1/2)} \frac{(\mathbf{V}_1 W_{22} - \mathbf{V}_2 W_{21}) R_{21} + (\mathbf{V}_2 W_{11} - \mathbf{V}_1 W_{12}) R_{22}}{W_{11} W_{22} - W_{21} W_{12}}\right] \doteq 0 \tag{62}$$

8 Bifurcation of the Homogeneous Uniaxial Tension Problem

The representation of discontinuity surfaces by the spatial FE-approximation and its connection to the weak localization test is demonstrated with the example of a rectangular metal sheet subjected to uniaxial tension. The analytical solution for the inclination of LÜDERS bands, which emerge as bifurcated failure mode, is in full agreement with experimental observations, see NADAI [5]. The assembly of elements demands satisfaction of the weak localization test on the system level which is obtained by applying the standard finite element assembly operator.

$$d^2 W_{loc} = \mathop{\mathrm{A}}_{e=1}^{Nel} d^2 W_{loc}^e = \mathop{\mathrm{A}}_{e=1}^{Nel} \dot{\mathbf{d}}_e^{loc} \mathbf{k}_e \dot{\mathbf{d}}_e^{loc} = \dot{\mathbf{d}}^{loc} \mathbf{K} \dot{\mathbf{d}}^{loc} \doteq 0 \tag{63}$$

The upper right quarter of the sheet is discretized with 10×20 four noded elements, thus precluding non-symmetric deformation patterns. The material behaviour is modelled by ideal J_2-flow theory in plane stress. Homogeneous deformations are prescribed at the top surface. After one elastic loadstep the yield surface is reached throughout the entire sheet. The eigenmodes associated with the lowest eigenvalue for this stress state are depicted in Fig.6 (a) ÷ (d) for several element formulations. The standard Q4-element in plot (a) is not able to trace the correct failure mode predicted by the analysis of the localization tensor. Underintegration, which happens to satisfy the weak localization test for a single element, leads to extensive hourglassing for the assemblage of the 200 elements in plot (d) if no stabilization is used. As anticipated by the single element test both the QS-element in plot (b) and the QM6-element in plot (c) capture the correct failure mode with an inclination of the localized band of approximately 35 deg. Thus these elements are not restricted to exact alignment of the element edges with the discontinuity surface. Clearly, this is the desirable feature that the fixed element discretization does capture localized failure in the context of the fixed mesh approach.

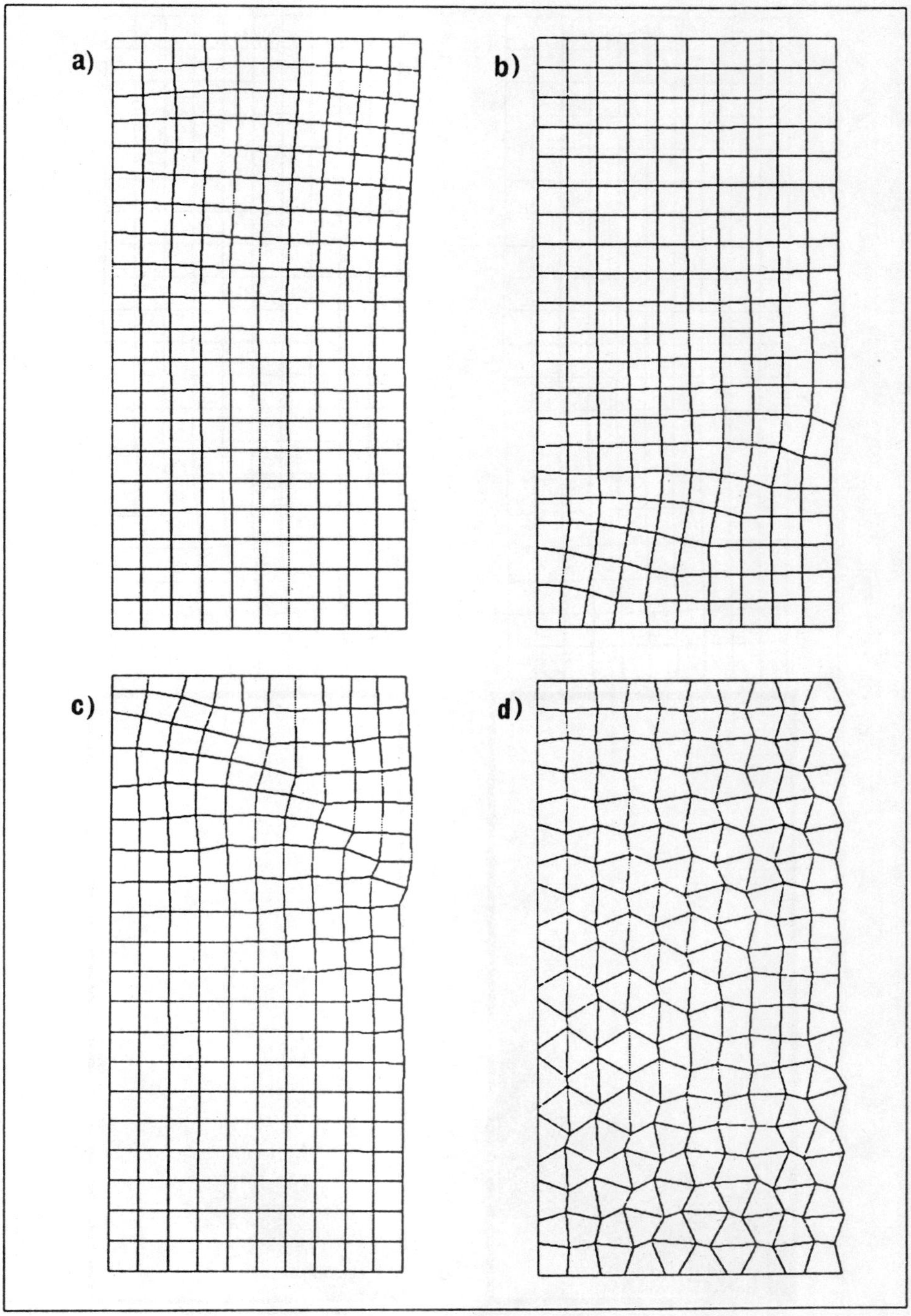

Figure 6: Eigenmodes for Homogeneous Tension Problem

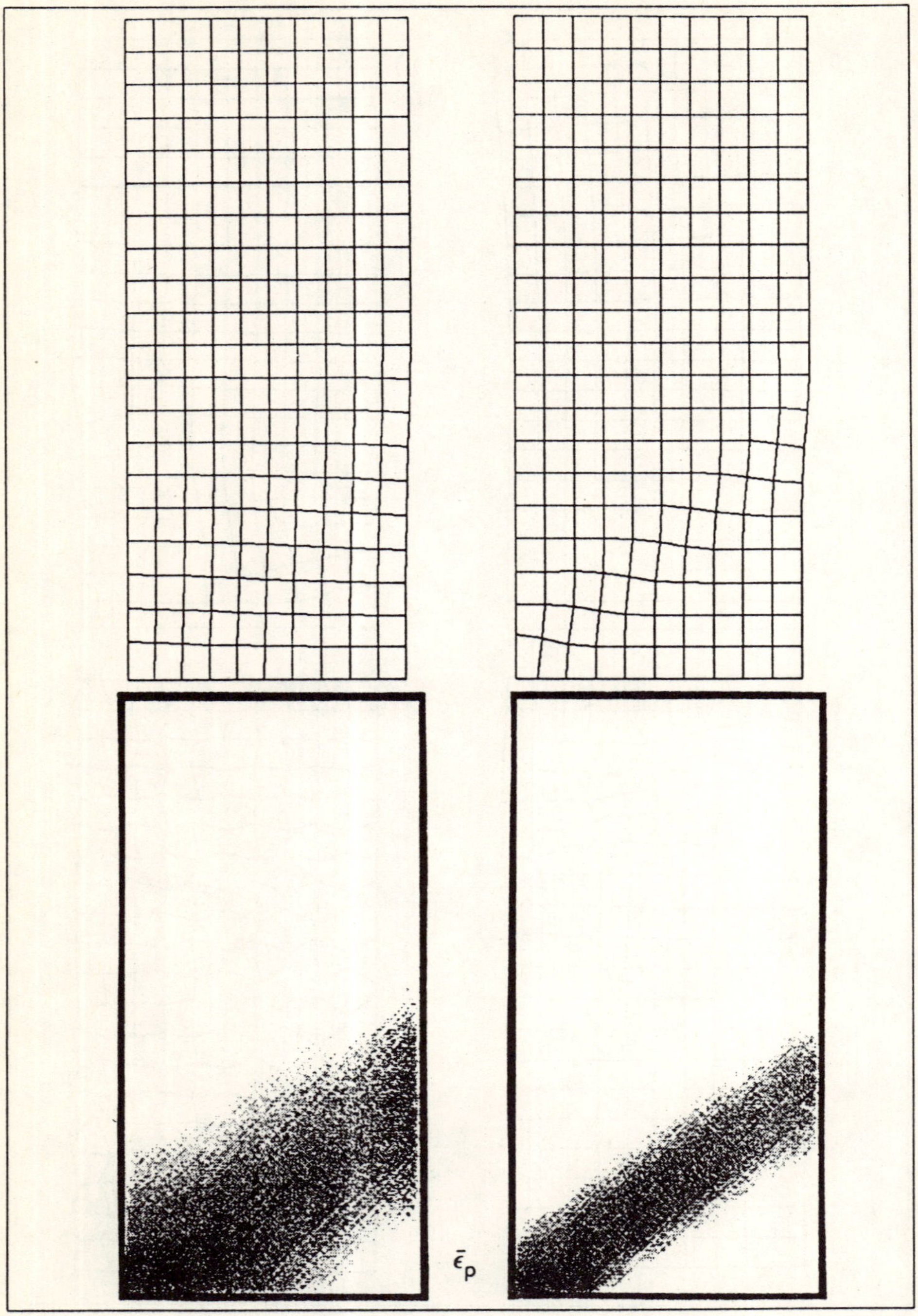

Figure 7: Nonhomogeneous Tension Problem : Q4, QS

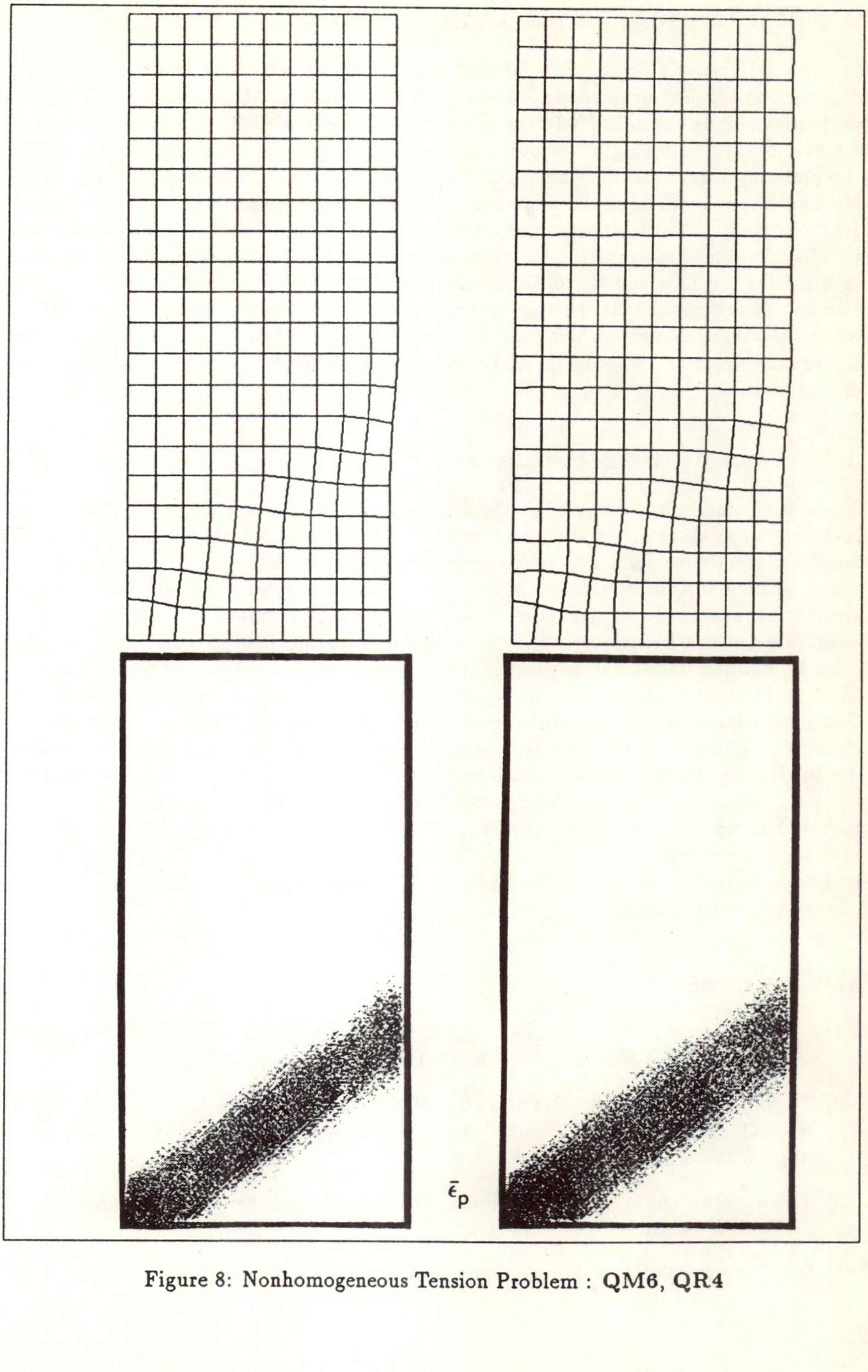

Figure 8: Nonhomogeneous Tension Problem : QM6, QR4

9 Nonhomogeneous Uniaxial Tension Problem

In the final example the bifurcation problem of the last section is transformed into a deformation problem by introduction of a small imperfection. All geometric and material properties remain the same, only the yield stress of the lower left element is reduced by a factor of 0.9. Softening behaviour would ease the capturing of localization, therefore the material is assumed to behave perfectly plastic. In the first elastic loadstep δ_0 the yield surface is reached in the weakened element. Then the specimen is streched up to $1.3125 \times \delta_0$ in 18 equal displacement-controlled load steps. The deformed meshes and the distribution of the equivalent plastic strain within the solution domain are depicted in Fig.7 for the **Q4**- and the **QS**-elements and in Fig.8 for the **QM6**- and the **QR4**-element. It is obvious that the standard **Q4**-element predicts a more distributed failure mode then those elements which pass the weak localization test. Since the deformations are equally magnified by a factor of 100 the hourglassing of the **QR4**-element remains almost invisible.

10 Concluding Remarks

The experimental observation of localization in terms of intense straining in narrow failure bands signals in classical continuum theory the loss of ellipticity. To capture this form of discontinuous bifurcation challenges numerical computations which are normally based on smooth approximations. In the framework of the fixed mesh approach enriched element expansions prove successful to reflect bifurcation at the constitutive level. The capability for this behaviour is checked with the weak localization test. On the other hand bifurcation from an homogeneous stress state is only one aspect of failure analysis. Most problems of practical importance are concerned with nonhomogeneous stress states, thus demanding solution of deformation rather than bifurcation problems. Therefore the element performance in the nonlinear responce regime should be taken into account besides the findings of the weak localization test. Nevertheless classical continuum theory is limited to describe the onset of localization but fails to describe the process of progressive failure. For more realistic descriptions non-classical constitutive models are a major research topic in computational mechanics. This should be noted particularly in the failure regime when realistic failure mechanisms are to be modelled and the results of numerical computations are assessed.

References

[1] J.H. ARGYRIS & K.J. WILLAM, [1974], "Some Considerations for the Evaluation of Finite Element Models", *Nucl. Engr. Design*, Vol. 28, pp. 76-96.

[2] T. BELYTSCHKO & L.P. BINDEMAN, [1990], "Assumed Strain Stabilization of the 4-Node Quadrilateral with 1-Point Quadrature for Nonlinear Problems", *Comp. Meth. Appl. Mech. Engr.*, to appear.

[3] T.J.R. HUGHES, [1987], *The Finite Element Method*, Prentice-Hall, Englewood Cliffs, New Jersey.

[4] L. MIRSKY, [1983], *An Introduction to Linear Algebra*, Dover Publ. Inc., New York.

[5] A. NADAI, [1931], *Plasticity*, McGraw-Hill.

[6] J.C. NAGTEGAAL, D.M. PARKS & J.R. RICE, [1974], "On Numerically Accurate Finite Element Solutions in the Fully Plastic Range", *Comp. Meth. Appl. Mech. Engr.*, Vol. 4, pp. 153-177.

[7] M. ORTIZ, Y. LEROY & A. NEEDLEMAN, [1987], "A Finite Element Method for Localization Failure Analysis", *Comp. Meth. Appl. Mech. Engr.*, Vol. 61, pp. 189-224.

[8] J.W. RUDNICKI & J.R. RICE, [1962], "Conditions for the Localization of Deformation in Pressure-Sensitive Dilatant Materials", *J. Mech. Phys. Solids*, Vol. 10, pp. 1-16.

[9] J.C. SIMO & M.S. RIFAI, [1990], "A Class of Mixed Assumed Strain Methods and the Method of Incompatible Modes", *Int. J. Num. Meth. Engr.*, Vol. 29.

[10] P. STEINMANN & K. WILLAM, [1990] "Finite Elements for Capturing Localized Failure", accepted for publication in *Ing. Archiv.*

[11] P. STEINMANN & K. WILLAM, [1990], "Finite Elements Performance in Localized Failure Computations", submitted for publication in *Comp. Meth. Appl. Mech. Engr.*.

[12] R.L. TAYLOR, P.J. BERESFORD & E.L. WILSON, [1976], "A Non-Conforming Element for Stress Analysis", *Int. J. Num. Meth. Engr.*, Vol. 10, No. 6, pp. 1211-1219.

[13] K. WILLAM & G. ETSE, [1990], "Assessment of the extended Leon model for plain concrete", *Proc. Int. Conf. Computer Aided Analysis and Design of Concrete Structures*, Zell am See, Austria, Vol II, pp. 851-870.

[14] K. WILLAM, N. SOBH & S. STURE, [1987], "Elastic-Plastic Tangent Operators: Localization Studies on the Constitutive and Finite Element Levels", *Advances in Inelastic Analysis*, ASME-WAM'87 Symp., Boston 1987, Editors S. Nakazawa, N. Rebelo and K. Willam, ASME, New York, Vol. 88, pp. 107-127.

3D Inelastic Constitutive Modeling for FE Analysis of Concrete Structures

G. Hofstetter, H.A. Mang, G. Meschke

Institute for Strength of Materials, Technical University of Vienna Vienna, Austria

Summary

In this paper two constitutive models for consideration of plastic deformations of concrete caused by 3D states of stress are presented. The first model is a modified cap model, serving as an example for nonsmooth multisurface plasticity. For this model a return mapping algorithm is presented. Tangent operators are derived, which are consistent with the integration algorithm. Hence, the quadratic rate of asymptotic convergence in the Newton solution procedure is preserved. The second constitutive model belongs to a special category of bounding surface models, characterized by the vanishing of the elastic range. This model is a so-called hypoplastic model. In contrast to ordinary plasticity, the tangent elasto-plastic material stiffness matrix also depends on the stress rate direction. In the theoretical part of the paper the mechanical fundamentals of the two material models are described. Subsequently, these constitutive models are applied to finite element analysis of a cylindrical concrete specimen subjected to a concentrated compressive force of the form of a steel punch. The results allow a comparison with experimental results and demonstrate advantages as well as disadvantages of the two constitutive models relative to one another.

Cap Model

Formulation of the cap model

The cap model, originally proposed by DiMaggio and Sandler [1], has proved to exhibit good predictive capabilities for both soil and rock [2] and concrete [3]. The yield surface of the two-invariant, rate-independent cap model consists of a failure envelope $f_1(\boldsymbol{\sigma})$, an elliptical cap $f_2(\boldsymbol{\sigma}, \kappa)$ and a tension cutoff region $f_3(\boldsymbol{\sigma})$, where $\boldsymbol{\sigma}$ and κ denote the stress tensor and the hardening parameter, respectively. In contrast to the strain hardening cap, the failure envelope and the tension cutoff region are fixed surfaces in the stress space. The original forms of the functions for these surfaces are (Fig.1):

$$f_1(\boldsymbol{\sigma}) = |\mathbf{s}| - F_e(I_1), \quad \text{for} \quad T \leq I_1 < \kappa, \tag{1}$$

$$f_2(\boldsymbol{\sigma}, \kappa) = |\mathbf{s}| - F_c(I_1, \kappa), \quad \text{for} \quad \kappa \leq I_1 \leq X(\kappa), \tag{2}$$

$$f_3(\boldsymbol{\sigma}) = T - I_1, \quad \text{for} \quad I_1 = T, \tag{3}$$

where

$$F_e(I_1) = \alpha - \lambda e^{-\beta I_1} + \theta I_1, \tag{4}$$

$$F_c(I_1, \kappa) = \frac{1}{R}\sqrt{[X(\kappa) - L(\kappa)]^2 - [I_1 - L(\kappa)]^2}. \tag{5}$$

$\alpha, \beta, \lambda, \theta, T$ and R are material parameters and I_1 and $|\mathbf{s}|$ are the first invariant of the stress tensor and the norm of the deviatoric stress tensor $\mathbf{s}$, respectively. $L(\kappa)$ is defined as

$$L(\kappa) = <\kappa> = \begin{cases} \kappa, & \text{if} \quad \kappa > 0, \\ 0, & \text{if} \quad \kappa \leq 0. \end{cases} \tag{6}$$

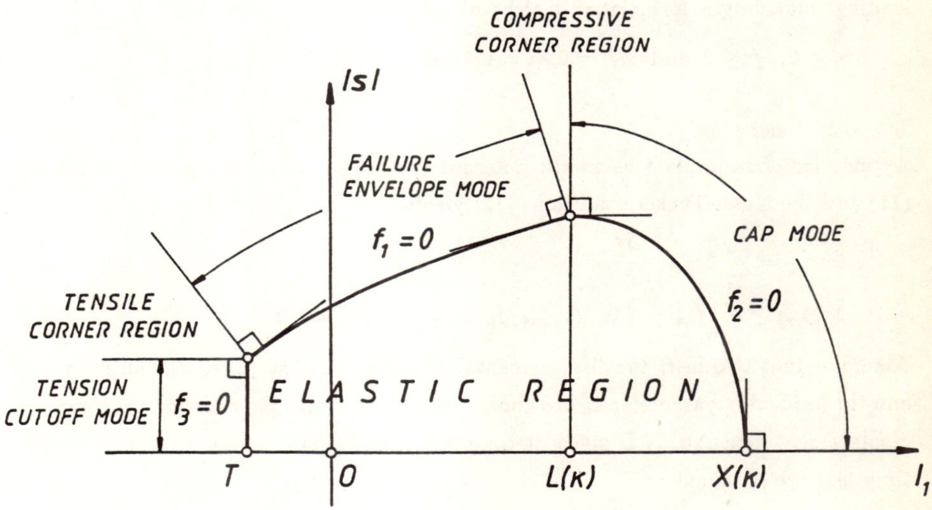

Fig. 1: Yield surface of the cap model

Following standard conventions in soil mechanics, compression and compaction are positive. The point of intersection of the cap with the I_1-axis is defined as

$$X(\kappa) = L(\kappa) + RF_e(\kappa). \tag{7}$$

The hardening parameter is implicitly defined as a functional of the plastic volumetric strain $tr(\boldsymbol{\varepsilon}^p)$ [1]. The qualitative shape of this functional is deduced from a hydrostatic test:

$$\bar{\varepsilon}_v^p = W\left(1 - e^{-DX(\kappa)}\right), \tag{8}$$

wherein W and D are material parameters and $\bar{\varepsilon}_v^p$ is the effective volumetric plastic strain, defined by Sandler and Rubin [4] by the rate equations

$$\dot{\bar{\varepsilon}}_v^p = \begin{cases} tr(\dot{\boldsymbol{\varepsilon}}^p), & \text{if } tr(\dot{\boldsymbol{\varepsilon}}^p) > 0 \text{ or } \kappa > 0 \text{ and } \kappa > I_1, \\ 0, & \text{otherwise}. \end{cases} \quad (9)$$

The range of the elastic domain is defined in terms of the three yield surfaces by

$$f_1(\boldsymbol{\sigma}) < 0 \ \wedge \ f_2(\boldsymbol{\sigma}, \kappa) < 0 \ \wedge \ f_3(\boldsymbol{\sigma}) < 0. \quad (10)$$

For multisurface plasticity an appropriate statement of an associative flow rule relies on Koiter's generalization [5], which in the present case takes on the form

$$\dot{\boldsymbol{\varepsilon}}^p = \sum_{i=1}^{3} \dot{\gamma}_i \frac{\partial f_i}{\partial \boldsymbol{\sigma}}, \quad (11)$$

where $\dot{\gamma}_i$, $i = 1,2,3$, are the plastic consistency parameters. Plastic loading or elastic loading/unloading is formulated in the Kuhn-Tucker form by means of the relations [6]

$$\dot{\gamma}_i \geq 0, \ \ f_i \leq 0 \ \text{ and } \ \dot{\gamma}_i f_i = 0, \ \ i = 1,2,3. \quad (12)$$

Integration algorithm

Application of an implicit backward Euler integration scheme to the generalized flow rule (11) and the Kuhn-Tucker conditions (12) yields

$$\Delta \boldsymbol{\varepsilon}_{n+1}^p = \sum_{i=1}^{3} \Delta\gamma_{i,n+1} \frac{\partial f_{i,n+1}}{\partial \boldsymbol{\sigma}_{n+1}} \quad (13)$$

$$\Delta\gamma_{i,n+1} \geq 0, \ \ f_{i,n+1} \leq 0, \ \ \Delta\gamma_{i,n+1} f_{i,n+1} = 0, \qquad i = 1,2,3. \quad (14)$$

Assuming that at time t_n the displacement field $\mathbf{u}_n$, the total strain $\boldsymbol{\varepsilon}_n$, the plastic strain $\boldsymbol{\varepsilon}_n^p$ and the hardening parameter κ_n are known and assuming that in addition the incremental displacement field $\Delta\mathbf{u}_{n+1}$ is given at time $n+1$, the plastic strain and stress tensor at time t_{n+1} are obtained as

$$\boldsymbol{\varepsilon}_{n+1}^p = \boldsymbol{\varepsilon}_n^p + \Delta\boldsymbol{\varepsilon}_{n+1}^p, \quad (15)$$

$$\boldsymbol{\sigma}_{n+1} = \mathbf{C} : (\boldsymbol{\varepsilon}_{n+1} - \boldsymbol{\varepsilon}_{n+1}^p), \quad (16)$$

where $\mathbf{C}$ denotes the elasticity tensor. It was shown by Simo and Hughes [7] that because of the assumption of convexity of f_i, plastic loading can be treated solely on the basis of an elastic trial stress. To this end, trial values are computed for the yield functions, freezing the plastic flow during the time step $n+1$:

$$f_{1,n+1}^{Trial} = f_{1,n+1}^{Trial}(|\mathbf{s}_{n+1}^{Trial}|, I_{1,n+1}^{Trial}) \quad (17)$$

$$f_{2,n+1}^{Trial} = f_{2,n+1}^{Trial}(|\mathbf{s}_{n+1}^{Trial}|, I_{1,n+1}^{Trial}, \kappa_n) \quad (18)$$

$$f_{3,n+1}^{Trial} = f_{3,n+1}^{Trial}(I_{1,n+1}^{Trial}), \quad (19)$$

where the elastic trial stress is

$$\mathbf{s}_{n+1}^{Trial} = 2G(\mathbf{e}_{n+1} - \mathbf{e}_n^p), \qquad I_{1,n+1}^{Trial} = 3K[tr(\boldsymbol{\varepsilon}_{n+1}) - tr(\boldsymbol{\varepsilon}_n^p)]. \tag{20}$$

In (20), G and K denote the shear modulus and the bulk modulus, respectively, and $\mathbf{e}$ is the deviatoric strain tensor.

If the trial value for I_1 is greater than $X(\kappa_n)$, then the original functional form for the cap (5), given in work by Sandler and Rubin [4] or Simo, Ju, Pister and Taylor [3], will lead to the square root of a negative number. Therefore, (2) together with (5) may not be used in the context with a return mapping algorithm. Reformulation of (2) and (5) by substituting (6) into (5) and inserting the so-obtained expression into (2), setting $f_2(\boldsymbol{\sigma}, \kappa) = 0$, yields

$$f_2(\boldsymbol{\sigma}, \kappa) = F_c(|\mathbf{s}|, I_1, \kappa) - F_e(\kappa), \qquad F_c(|\mathbf{s}|, I_1, \kappa) = \sqrt{|\mathbf{s}|^2 + \frac{1}{R^2}\left[I_1 - L(\kappa)\right]^2}. \tag{21}$$

(21) describes a cap with a similar functional form, avoiding, however, the possibility of a negative square root.

The process is elastic, if $f_{i,n+1}^{Trial} < 0$ for all $i \in [1,2,3]$, and plastic, if $f_{i,n+1}^{Trial} > 0$ for at least one $i \in [1,2,3]$. In the latter case it is necessary to determine $\Delta\boldsymbol{\varepsilon}_{n+1}^p$ and κ_{n+1}. Finally, the stresses are obtained from (16):

$$\mathbf{s}_{n+1} = \mathbf{s}_{n+1}^{Trial} - 2G\Delta\mathbf{e}_{n+1}^p \qquad I_{1,n+1} = I_{1,n+1}^{Trial} - 3K\Delta tr(\boldsymbol{\varepsilon}_{n+1}^p). \tag{22}$$

Loading for the various modes of the cap model

The loading conditions for the various modes of the cap model are obtained from (17) to (19) and from the loading conditions for the corner regions. From the hardening law (9) it follows that both the tensile corner region and the compressive corner region behave as in ideal plasticity. Therefore, the return point $\boldsymbol{\sigma}_{n+1}$ must lie at the intersection of the two surfaces, forming the respective corner region. For the tensile corner region, substitution of (1) and (3) into (13) yields

$$\Delta\mathbf{e}_{n+1}^p = \Delta\gamma_{1,n+1}\mathbf{n}_{n+1}, \qquad tr(\Delta\boldsymbol{\varepsilon}_{n+1}^p) = -3\left(\Delta\gamma_{1,n+1}\frac{dF_e(T)}{dI_1} + \Delta\gamma_{3,n+1}\right), \tag{23}$$

where (see Simo and Taylor [8])

$$\mathbf{n} = \frac{\mathbf{s}_{n+1}}{|\mathbf{s}_{n+1}|} = \frac{\mathbf{s}_{n+1}^{Trial}}{|\mathbf{s}_{n+1}^{Trial}|}. \tag{24}$$

The return point is the intersection of the failure envelope and the tension cutoff

$$I_{1,n+1} = T, \qquad |\mathbf{s}_{n+1}| = F_e(T). \tag{25}$$

Substitution of (23) and (25) into (22) and consideration of (24) gives

$$\Delta\gamma_{1,n+1} = \frac{|\mathbf{s}_{n+1}^{Trial}| - F_e(T)}{2G}, \qquad \Delta\gamma_{3,n+1} = \frac{T - I_{1,n+1}^{Trial}}{9K} - \Delta\gamma_{1,n+1}\frac{dF_e(T)}{dI_1}. \tag{26}$$

Loading in this corner region requires $\Delta\gamma_{1,n+1} > 0$, $\Delta\gamma_{2,n+1} = 0$ and $\Delta\gamma_{3,n+1} > 0$. If more than one yield criterion is active in multisurface plasticity, then $f_{i,n+1}^{Trial} > 0$ does not necessarily imply that $\Delta\gamma_{i,n+1} > 0$ (see Simo et al.[6]). Noting that for the tensile corner region $f_{3,n+1}^{Trial} > 0$ and enforcing $\Delta\gamma_{1,n+1} > 0$ and $\Delta\gamma_{3,n+1} > 0$, the inequalities characterizing loading within the tensile corner region are obtained as follows:

$$I_{1,n+1}^{Trial} < T, \qquad F_e(T) < |\mathbf{s}_{n+1}^{Trial}| < F_e(T) + \frac{2G}{9K}\frac{T - I_{1,n+1}^{Trial}}{dF_e(T)/dI_1}. \tag{27}$$

A similar argument provides the loading conditions for the compressive corner region. Loading within this corner region requires the return stress point to be located at the intersection of the cap and the failure envelope ($I_{1,n+1} = \kappa_n$, $|\mathbf{s}_{n+1}| = F_e(\kappa_n)$). Enforcing $\Delta\gamma_{1,n+1} > 0$ and $\Delta\gamma_{2,n+1} > 0$, yields

$$I_{1,n+1}^{Trial} < \kappa_n, \quad |\mathbf{s}_{n+1}^{trial}| > F_e(\kappa_n) + \frac{2G}{9K}\frac{\kappa_n - I_{1,n+1}^{Trial}}{dF_e(\kappa_n)/dI_1}. \tag{28}$$

From the equations (17), (18), (19) and the inequalities (27) and (28) the loading conditions for the various modes of the cap model are obtained.

Loading can easily be handled in the failure envelope mode, the tension cutoff mode or in one of the two corner regions, because loading in these modes is characterized by ideally plastic behavior. For this reason, the derivations in this paper are restricted to loading in the cap mode. Even for this mode the stress update can be reduced to the solution of one nonlinear scalar equation. Substitution of (2) together with (5) into (13) yields

$$\Delta\mathbf{e}_{n+1}^p = \Delta\gamma_{2,n+1}\frac{\partial f_{2,n+1}}{\partial \mathbf{s}_{n+1}} = \Delta\gamma_{2,n+1}\frac{\mathbf{s}_{n+1}}{F_c(|\mathbf{s}_{n+1}|, I_{1,n+1}, \kappa_{n+1})}, \tag{29}$$

$$\Delta tr(\boldsymbol{\varepsilon}_{n+1}^p) = 3\Delta\gamma_{2,n+1}\frac{\partial f_{2,n+1}}{\partial I_{1,n+1}} = \Delta\gamma_{2,n+1}\frac{3(I_{1,n+1} - \kappa_{n+1})}{R^2 F_c(|\mathbf{s}_{n+1}|, I_{1,n+1}, \kappa_{n+1})}. \tag{30}$$

Substituting (29)and (30) into (22), using (2), (5), (24) and the hardening law (8), yields

$$|\mathbf{s}_{n+1}| = |\mathbf{s}_{n+1}^{Trial}| - 2G\Delta\gamma_{2,n+1}\frac{|\mathbf{s}_{n+1}|}{F_c(|\mathbf{s}_{n+1}|, I_{1,n+1}, \kappa_{n+1})}, \tag{31}$$

$$I_{1,n+1} = I_{1,n+1}^{Trial} - 9K\Delta\gamma_{2,n+1}\frac{I_{1,n+1} - \kappa_{n+1}}{R^2 F_c(|\mathbf{s}_{n+1}|, I_{1,n+1}, \kappa_{n+1})}, \tag{32}$$

$$f_{2,n+1} = F_c(|\mathbf{s}_{n+1}|, I_{1,n+1}, \kappa_{n+1}) - F_e(\kappa_{n+1}) = 0, \tag{33}$$

$$tr(\Delta\boldsymbol{\varepsilon}_{n+1}^p) = W\left(e^{-DX(\kappa_n)} - e^{-DX(\kappa_{n+1})}\right) = H(\kappa_{n+1}). \tag{34}$$

Eqs.(30) to (33) provide four equations for the unknowns $|\mathbf{s}_{n+1}|$, $I_{1,n+1}$, $\Delta\gamma_{2,n+1}$ and κ_{n+1}. These equations can be reduced to one nonlinear scalar equation for κ_{n+1} [9] which can be solved by means of a Newton iteration technique.

Consistent tangent moduli for the cap mode
Differentiation of (16) and (15) yields, for the cap mode,

$$\begin{aligned} d\boldsymbol{\sigma}_{n+1} &= \mathbf{C} : (d\boldsymbol{\varepsilon}_{n+1} - d\boldsymbol{\varepsilon}^p_{n+1}), \\ d\boldsymbol{\varepsilon}^p_{n+1} &= d\Delta\gamma_{2,n+1}\frac{\partial f_{2,n+1}}{\partial \boldsymbol{\sigma}_{n+1}} + \Delta\gamma_{2,n+1}\left(\frac{\partial^2 f_{2,n+1}}{\partial \boldsymbol{\sigma}^2_{n+1}} : d\boldsymbol{\sigma}_{n+1} + \frac{\partial^2 f_{2,n+1}}{\partial \boldsymbol{\sigma}_{n+1}\partial \kappa_{n+1}} d\kappa_{n+1}\right). \end{aligned} \tag{35}$$

Inserting (35_2) into (35_1) yields

$$d\boldsymbol{\sigma}_{n+1} = \boldsymbol{\Xi}_{n+1} : d\boldsymbol{\varepsilon}_{n+1} - \boldsymbol{\Xi}_{n+1} : \frac{\partial f_{2,n+1}}{\partial \boldsymbol{\sigma}_{n+1}} d\Delta\gamma_{2,n+1} \tag{36}$$

$$- \Delta\gamma_{2,n+1}\boldsymbol{\Xi}_{n+1} : \frac{\partial^2 f_{2,n+1}}{\partial \boldsymbol{\sigma}_{n+1}\partial \kappa_{n+1}} d\kappa_{n+1}, \tag{37}$$

where

$$\boldsymbol{\Xi}_{n+1} = \left(\mathbf{C}^{-1} + \Delta\gamma_{2,n+1}\frac{\partial^2 f_{2,n+1}}{\partial \boldsymbol{\sigma}^2_{n+1}}\right)^{-1}. \tag{38}$$

Differentiation of $f_2(\boldsymbol{\sigma}_{n+1}, \kappa_{n+1}) = 0$ gives

$$\frac{\partial f_{2,n+1}}{\partial \boldsymbol{\sigma}_{n+1}} : d\boldsymbol{\sigma}_{n+1} + \frac{\partial f_{2,n+1}}{\partial \kappa_{n+1}} d\kappa_{n+1} = 0. \tag{39}$$

Combining (30) and (34) yields

$$\frac{1}{3}H(\kappa_{n+1}) - \Delta\gamma_{2,n+1}\frac{\partial f_{2,n+1}}{\partial I_{1,n+1}} = 0. \tag{40}$$

Differentiation of (40) gives

$$\begin{aligned} &\frac{1}{3}\frac{dH(\kappa_{n+1})}{d\kappa_{n+1}} d\kappa_{n+1} - d\Delta\gamma_{2,n+1}\frac{\partial f_{2,n+1}}{\partial I_{1,n+1}} \\ &-\Delta\gamma_{2,n+1}\left(\frac{\partial^2 f_{2,n+1}}{\partial I_{1,n+1}\partial \boldsymbol{\sigma}_{n+1}} : d\boldsymbol{\sigma}_{n+1} + \frac{\partial^2 f_{2,n+1}}{\partial I_{1,n+1}\partial \kappa_{n+1}} : d\kappa_{n+1}\right) = 0. \end{aligned} \tag{41}$$

Substituting (36) into (39) and (41) yields $d\Delta\gamma_{2,n+1}$ and $d\kappa_{n+1}$. Finally, substitution of the expressions for $d\Delta\gamma_{2,n+1}$ and $d\kappa_{n+1}$ into (36) gives the expression for the consistent tangent moduli for the cap mode

$$\frac{d\boldsymbol{\sigma}_{n+1}}{d\boldsymbol{\varepsilon}_{n+1}} = \boldsymbol{\Xi}_{n+1} - \sum_{i=1}^{2}\sum_{j=1}^{2} a^{-1}_{ij,n+1}\mathbf{N}_{i,n+1} \otimes \mathbf{N}_{j,n+1}, \tag{42}$$

where

$$\mathbf{N}_{1,n+1} = \boldsymbol{\Xi}_{n+1} : \frac{\partial f_{2,n+1}}{\partial \boldsymbol{\sigma}_{n+1}}, \qquad \mathbf{N}_{2,n+1} = \boldsymbol{\Xi}_{n+1} : \frac{\partial^2 f_{2,n+1}}{\partial \boldsymbol{\sigma}_{n+1}\partial \kappa_{n+1}}, \tag{43}$$

and

$$a_{11,n+1} = \frac{\partial f_{2,n+1}}{\partial \boldsymbol{\sigma}_{n+1}} : \Xi_{n+1} : \frac{\partial f_{2,n+1}}{\partial \boldsymbol{\sigma}_{n+1}}, \tag{44}$$

$$a_{12,n+1} = \frac{\partial f_{2,n+1}}{\partial \boldsymbol{\sigma}_{n+1}} : \Xi_{n+1} : \frac{\partial^2 f_{2,n+1}}{\partial \boldsymbol{\sigma}_{n+1} \partial \kappa_{n+1}} - \frac{1}{\Delta\gamma_{2,n+1}} \frac{\partial f_{2,n+1}}{\partial \kappa_{n+1}}, \tag{45}$$

$$a_{21,n+1} = \frac{\partial^2 f_{2,n+1}}{\partial \boldsymbol{\sigma}_{n+1} \partial \kappa_{n+1}} : \Xi_{n+1} : \frac{\partial f_{2,n+1}}{\partial \boldsymbol{\sigma}_{n+1}} + \frac{1}{\Delta\gamma_{2,n+1}} \frac{\partial f_{2,n+1}}{\partial I_{1,n+1}}, \tag{46}$$

$$a_{22,n+1} = \frac{\partial^2 f_{2,n+1}}{\partial \boldsymbol{\sigma}_{n+1} \partial \kappa_{n+1}} : \Xi_{n+1} : \frac{\partial^2 f_{2,n+1}}{\partial \boldsymbol{\sigma}_{n+1} \partial \kappa_{n+1}},$$

$$+ \frac{1}{\Delta\gamma_{2,n+1}} \frac{\partial^2 f_{2,n+1}}{\partial I_{1,n+1} \partial \kappa_{n+1}} - \frac{1}{3\Delta\gamma_{2,n+1}^2} \frac{dH(\kappa_{n+1})}{d\kappa_{n+1}}. \tag{47}$$

(48)

Note that because of $a_{12,n+1} \neq a_{21,n+1}$ the tensor of the consistent tangent moduli is not symmetric. The unsymmetry is caused by the fact that the cap model does not obey the principle of maximum plastic dissipation, which, as has been shown by Simo and Hughes [7], implies not only associativity of the flow rule but also of the hardening law. For the cap the latter condition would imply a rate equation for the hardening parameter κ of the form

$$\dot{\kappa} = -\dot{\gamma}_2 D_2(\kappa) \frac{\partial f_2}{\partial \kappa}, \tag{49}$$

where $D_2(\kappa)$ denotes a plastic modulus. Since the hardening law (49) is associated with the yield function, it is termed as associated hardening law. From (8) and (9) a rate form of the original hardening law for the cap is obtained as

$$\dot{\kappa} = \frac{1}{WD \frac{dX(\kappa)}{d\kappa} e^{-DX(\kappa)}} tr(\dot{\boldsymbol{\epsilon}}^p). \tag{50}$$

On the basis of an associated flow rule, it follows that

$$tr(\dot{\boldsymbol{\epsilon}}^p) = 3\dot{\gamma}_2 \frac{\partial f_2}{\partial I_1}. \tag{51}$$

Substitution of (51) into (50) yields

$$\dot{\kappa} = \dot{\gamma}_2 D_2(\kappa) \frac{\partial f_2}{\partial I_1}, \quad D_2(\kappa) = \frac{3}{WD \frac{dX(\kappa)}{d\kappa} e^{-DX(\kappa)}}. \tag{52}$$

Since, in general, $\partial f_2 / \partial \kappa \neq -\partial f_2 / \partial I_1$, comparison of (49) and (52) shows that the cap model does not obey the principle of maximum plastic dissipation. To overcome the drawback of a nonsymmetric consistent tangent in [9], an associated hardening law for the cap is proposed.

Hypoplastic Constitutive Model

Constitutive equations

In this Section strains and stresses will be referred to principal directions. The respective tensor components will be listed as vectors.

The vector of total strain rates, $d\boldsymbol{\epsilon}$, is decomposed as follows:

$$d\boldsymbol{\epsilon} = d\boldsymbol{\epsilon}^e + d\boldsymbol{\epsilon}^p \tag{53}$$

where $d\boldsymbol{\epsilon}^e$ and $d\boldsymbol{\epsilon}^p$ are the elastic and the plastic part of $d\boldsymbol{\epsilon}$, given as

$$d\boldsymbol{\epsilon}^e = (\mathbf{D}^e)^{-1} d\boldsymbol{\sigma} \tag{54}$$

and

$$d\boldsymbol{\epsilon}^p = (1/H^p) < L > \boldsymbol{\rho}, \tag{55}$$

respectively [10]. In (54), $\mathbf{D}^e$ is the matrix of the elastic stiffnesses, referred to principal directions, and $d\boldsymbol{\sigma}$ is the vector of stress rates. In (55), $\boldsymbol{\rho}$ is the direction vector, defined as

$$\boldsymbol{\rho} = d\boldsymbol{\epsilon}^p / [(d\boldsymbol{\epsilon}^p)^T d\boldsymbol{\epsilon}^p]^{1/2}, \tag{56}$$

$< L >$ is the loading function, defined as

$$\begin{aligned} < L > &= L = d\boldsymbol{\sigma}^T \mathbf{n}, \quad if \quad d\boldsymbol{\sigma}^T \mathbf{n} > 0, \\ < L > &= 0 \qquad\qquad\quad if \quad d\boldsymbol{\sigma}^T \mathbf{n} \leq 0, \end{aligned} \tag{57}$$

where $\mathbf{n}$ is the normal vector at the so-called image stress point $\boldsymbol{\sigma}_B$ on the bounding surface (Fig.2), and $H^p = H^p(\boldsymbol{\sigma}, \boldsymbol{\rho}, \mathbf{n})$ is the generalized plastic modulus.

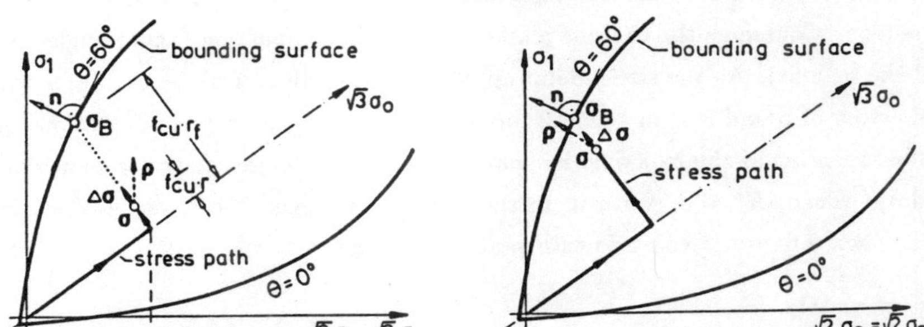

Fig. 2: Rotation of the direction vector $\boldsymbol{\rho}$ during deviatoric loading

In Fig.2, σ_0, f_{cu} and θ are the octahedral normal stress, the cylinder strength and the Lode angle, respectively; r and r_f are the deviatoric distances of the stress point $\boldsymbol{\sigma}$ and the image stress point σ_B from the hydrostatic axis, normalized with respect to f_{cu}. From (53), (54) and (55), the constitutive equations are obtained as [11]

$$d\sigma = \mathbf{D}^{ep} d\epsilon = [\mathbf{D}^e - (\mathbf{D}^e \rho \mathbf{D}^e \mathbf{n})/(\mathbf{n}^T \mathbf{D}^e \rho + H^p)] d\epsilon, \tag{58}$$

where $\mathbf{D}^{ep}$ is the unsymmetric elasto-plastic tangent material stiffness matrix.

Direction vector

The direction vector ρ controls the direction of plastic flow. ρ is determined from experimentally obtained results reported by Stankowski [12]. Fig.3 shows the rotation of an increment of the vector of deviatoric plastic strains, $\Delta \mathbf{e}^p$, from the direction of the first segment of the loading path in the deviatoric plane (σ_1-direction) to the direction normal to the bounding surface (fracture envelope). For a stress point, which is sufficiently

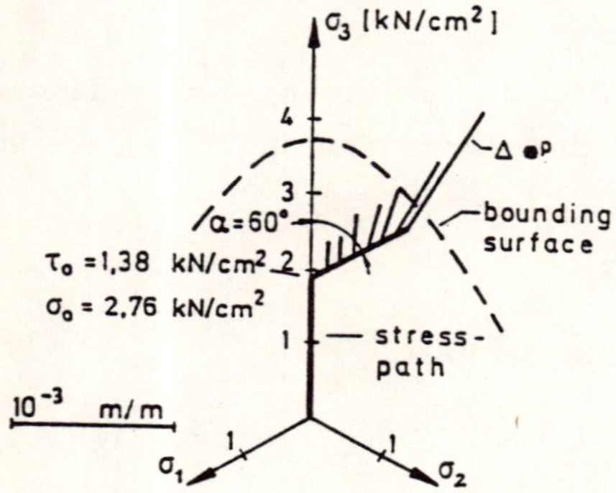

Fig. 3: Rotation of $\Delta \mathbf{e}^p$ during loading in the deviatoric plane [3]

far away from the failure envelope, the direction of the largest principal stress reached so far during loading controls the direction of ρ. Hence, in Fig.2a ρ is parallel to the σ_1-direction. Consequently, there is plastic volumetric compaction (acute angle between ρ and the σ_0-axis). As the stress point approaches the failure surface, ρ rotates towards the direction of $\mathbf{n}$ and σ_B. In general, this results in plastic deviatoric dilatation (obtuse angle between ρ and the σ_0-axis). For purely compressive stress states, ρ coincides with $\mathbf{n}$ and σ_B. Then, $H^p = 0$. With increasing tension, however, the rotation of ρ becomes slower. Then ρ does not coincide with $\mathbf{n}$ at σ_B, and $H^p > 0$.

ρ is given as [11]

$$\rho = f\mu + (1 - f)\mathbf{n}, \tag{59}$$

where μ is the unit vector in the direction of those stress components σ_i which exceed their previously attained maximum value, $\sigma_{i,max}$, and f is a pressure-sensitive weighting function, defined as [11]

$$f = (1 - f_0)(1 - D^c)^{1/2} + f_0, \tag{60}$$

where

$$D = r/r_f, \qquad 0 \leq D \leq 1, \tag{61}$$

is the distance parameter, $f_0 = f_0(\sigma_0)$ accounts for the diminution of the rotation of $\boldsymbol{\rho}$ with increasing tensile stress and $c(\sigma_0)$ refers to the zone between the onset of dilatation and the fracture envelope, which is shrinking with increasing hydrostatic stress. Expressions for c and f_0 are given in [11].

If $\sigma_{i,max}$ is exceeded, a sudden rotation of $\boldsymbol{\mu}$ and, hence, of $\boldsymbol{\rho}$ will occur. The extent of the corresponding sudden change of the direction of plastic flow is greater at low than at high levels of damage [12]. This is taken into account by the linear factor $f_r = f_r(\Delta\epsilon^p, \epsilon^{p,r})$ with $\Delta\epsilon^p = \epsilon^p - \epsilon^{p,r}$, where

$$\epsilon^p = \int d\epsilon^p = \int [(d\epsilon^p)^T d\epsilon^p]^{1/2} \tag{62}$$

is the effective plastic strain and $\epsilon^{p,r}$ represents ϵ^p when $\sigma_{i,max}$ is exceeded. $\boldsymbol{\mu}$ is then obtained as

$$\boldsymbol{\mu} = \boldsymbol{\mu}_0 + f_r(\boldsymbol{\mu}_n - \boldsymbol{\mu}_0), \tag{63}$$

where $\boldsymbol{\mu}_0$ and $\boldsymbol{\mu}_n$ denote the vector $\boldsymbol{\mu}$ before reaching and after having exceeded $\sigma_{i,max}$, respectively. Only strongly nonproportional loading paths are affected by f_r.

Generalized plastic modulus

The generalized plastic modulus H^p is given as [11]

$$H^p = H_b^p(\boldsymbol{\rho}, \mathbf{n}) + H_0^p(\sigma_0, D, D_{cr}), \tag{64}$$

where the plastic base modulus H_b^p governs the dilatant behaviour of concrete ($d\epsilon_0^p > 0$) and the plastic modulus H_0^p controls its ductile behavior, characterized by volume compaction ($d\epsilon_0^p < 0$). The basis for determination of H_b^p is the validity of the relation

$$W = \frac{1}{2} d\boldsymbol{\sigma}^T d\boldsymbol{\epsilon} \geq 0 \tag{65}$$

for any $d\boldsymbol{\epsilon}$, indicating material stability. In the expression for the second-order work density $W, d\boldsymbol{\sigma}$ is given as

$$d\boldsymbol{\sigma} = \mathbf{D}^e(d\boldsymbol{\epsilon} - d\lambda\boldsymbol{\rho}) \tag{66}$$

with $d\lambda$ following from (55). The solution of the present minimum-value problem [11], [13] provides a lower bound for H_b^p, as is required for material stability in the sense of (65):

$$H_b^p = -\frac{1}{2}(\mathbf{n}^T\mathbf{D}^e\boldsymbol{\rho}) + \frac{1}{2}(\boldsymbol{\rho}^T\mathbf{D}^e\mathbf{n})(\mathbf{n}^T\mathbf{D}^e\mathbf{n}). \tag{67}$$

For the ultimate compressive stress state, $\boldsymbol{\rho}$ and $\mathbf{n}$ coincide. Hence, for this case, $H_b^p = 0$, which is the lower bound of H^p for associated plasticity. The parameter f_0 in (60) prevents such a coincidence for loading in the mixed compression-tension material domain, allowing for a nonvanishing material stiffness at the ultimate stress level.

The transition from ductility to brittleness is characterized by the critical distance parameter D_{cr}, resulting from application of the condition

$$d\epsilon_0^p = 0 \tag{68}$$

to the expression for the volumetric component of $\boldsymbol{\rho}$, denoted as ρ_v. Analogous to $\boldsymbol{\rho}$ according to (59), ρ_v is obtained as

$$\rho_v = f_{cr}v_1 + (1 - f_{cr})v_2, \tag{69}$$

where r_1 and r_2 are the volumetric components of $\boldsymbol{\mu}$ and $\mathbf{n}$, respectively, and f_{cr} is the value of f related to D_{cr}. Following from (68) and (69),

$$f_{cr}v_1 + (1 - f_{cr})v_2 = 0. \tag{70}$$

For known v_1 and v_2, f_{cr} can be determined from (70). From (60) the following expression is obtained for the critical distance parameter [11]:

$$D_{cr} = \exp\{\ell n(1 - [(f_{cr}^I - f_0)/(1 - f_0)]^2)/c\}, \tag{71}$$

where the index "I" refers to the case $v_1 \leq 0$ and $v_2 \geq 0$, for which (71) yields physically relevant values of D_{cr}, ranging from 0.75 to 0.9. A negative volumetric component of $\mathbf{n}$ ($v_2 < 0$), however, means that for proportional loading up to failure there is only contraction. For this case a fictitious parameter $D_{cr} > 1$ is obtained from consideration of the bounding conditions $v_2 = -\sqrt{3}/3(\equiv d\gamma_0^{p\ell} = 0)$ for $D_{cr} \to \infty$ and $df_{cr}^I/dv_2 = df_{cr}^{II}/dv_2$ for $v_2 = 0$ as [11]:

$$D_{cr} = \exp\{\ell n(1 + [(f_{cr}^{II} - f_0)/(1 - f_0)]^2)/c\}, \tag{72}$$

where f_{cr}^{II} is given as

$$fcr^{II} = A(v_2 + \sqrt{3}/3)^{-2} + B(v_2 + \sqrt{3}/3)^{-1} \tag{73}$$

with

$$A = \sqrt{3}/9v_1, \qquad B = -1/3v_1. \tag{74}$$

Fig.4 shows $f_{cr} = f_{cr}(v_2)$ for several values of v_1.

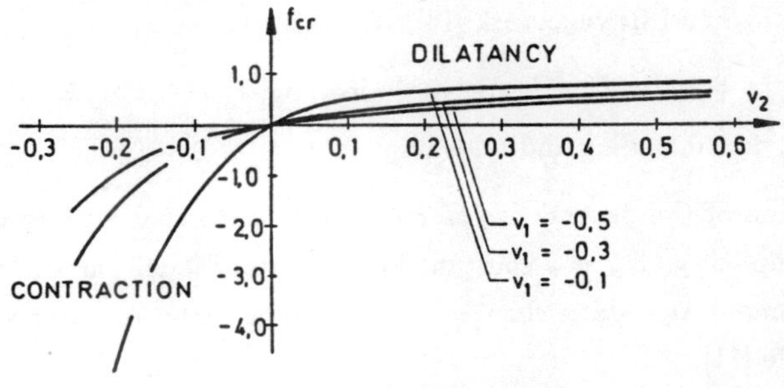

Fig. 4: $f_{cr}(v_2)$ for different values of v_1 [10]

Fig.5 illustrates $D_{cr} = D_{cr}(f_{cr})$ for different values of $c(\sigma_0)$ and for $f_0 = 0$.

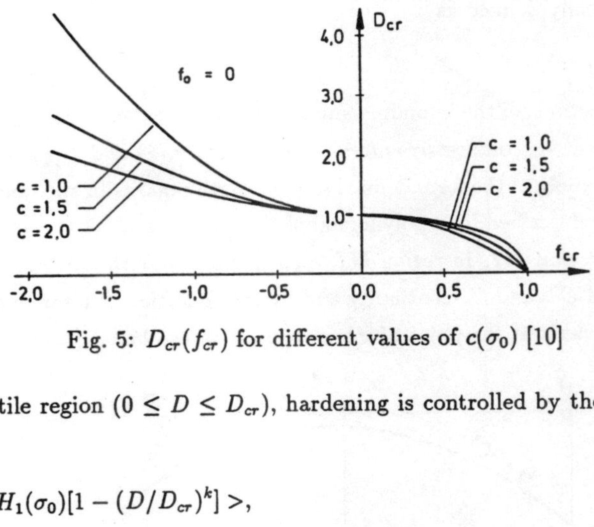

Fig. 5: $D_{cr}(f_{cr})$ for different values of $c(\sigma_0)$ [10]

In the ductile region ($0 \leq D \leq D_{cr}$), hardening is controlled by the superposition of H_b^p and of

$$H_0^p =< H_1(\sigma_0)[1 - (D/D_{cr})^k] >, \tag{75}$$

where the definition of the Macauley parameter is analogous to the one in (57). The term $1 - (D/D_{cr})^k$, where $k = 0.2 = const.$, allows consideration of damage resulting from frictional effects. $H_1(\sigma_0)$ accounts for the stiffness reduction following from pore collapse. Comparing the octahedral stress-strain relationship

$$d\sigma_0 = 3K_t d\epsilon_0, \tag{76}$$

where K_t is the tangent bulk modulus, with (58), specialized for purely hydrostatic loading, yields

$$H_1(\sigma_0) = (3K_tK_0)/(K_0 - K_t). \tag{77}$$

According to Chen and Buyukozturk [14], K_t is given as

$$K_t = K_0/(1 + k_1(\sigma_0/f_{cu})^{k_2}), \text{ for virgin loading; } k_1 = 0.358,\ k_2 = 1.5, \tag{78}$$

$$K_t = K_0, \text{ for unloading and reloading.} \tag{79}$$

For consideration of the dependence of H_0^p on D_{cr} the actual loading direction is taken into account implicitly. D_{cr} is a constant for the proportional parts of the loading path. D_{cr} is discontinuous at sudden changes of the loading direction. Hence, it is as discrete internal variable [11].

Bounding surface

The construction of the bounding surface is based on the failure surface of Willam and Warnke [15], formally defined as

$$F(\sigma) = r/r_f - 1 = 0. \tag{80}$$

The meridional sections of the bounding surface consist of 3 parts (Fig.6): In region I the tensile ($\theta = 0°$) and the compressive meridian ($\theta = 60°$), $r_t(\xi)$ and $r_c(\xi)$, with $\xi = \sigma_0/f_{cu}$, are taken as a parabola of the form defined in [15]. With regards to the tensile meridian in region II, a curve described by a polynomial of third order enables the smooth transition from region I to region III. In region III, a parabolic cap with $r_t = r_c$ is determined by the requirement of C^0- and C^1-continuity and by the condition of a tangent plane normal to the hydrostatic axis at the point of intersection with the latter [10].

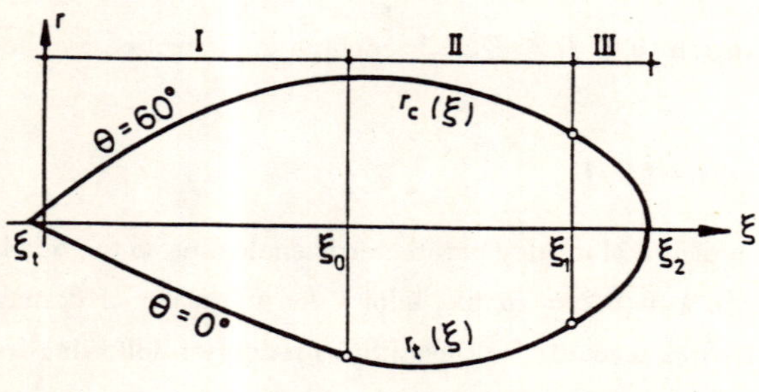

Fig. 6: Bounding surface

Loading criterion

The loading criterion of plasticity, which is based on the sign of the loading function in (55), fails for a degenerated loading surface [11], [16]. Thus, for determination of K_t in (78) and (79), the principal stress criterion is adopted [12]. Loading will occur, if the largest value of σ_i, $i = 1, 2, 3$, previously attained, is exceeded.

With respect to the distance parameter D, however, a criterion in terms of octahedral quantities is used for determination of a modified parameter for the case of hydrostatic and deviatoric unloading [11]:

$$D = (r_{in} + Lr)/(r_{in} + r_s), \quad \text{if} \quad \Delta\sigma_0 \geq 0 \quad \text{and} \quad \Delta\tau_0 \leq 0, \tag{81}$$

with

$$L = sign[(\mathbf{s}^T \mathbf{s}_B)/(|\mathbf{s}|\ |\mathbf{s}_B|)], \tag{82}$$

where $\mathbf{s}$ and $\mathbf{s}_B$ are the stress deviators corresponding to $\boldsymbol{\sigma}$ and $\boldsymbol{\sigma}_B$, respectively (Fig.7). r_{in} represents the deviatoric distance of the stress point, normalized with respect to f_{cu}, when unloading in the sense of (81) is signalled first. r_s is the deviatoric distance of the new image stress point $\boldsymbol{\sigma}_B$ after unloading. For $\Delta\sigma_0 < 0$ or $\Delta\tau_0 > 0, r_{in} = 0,\ r_s = r_f$. In this case, (81) degenerates to the expression for loading, given as $D = r/r_f$ [11].

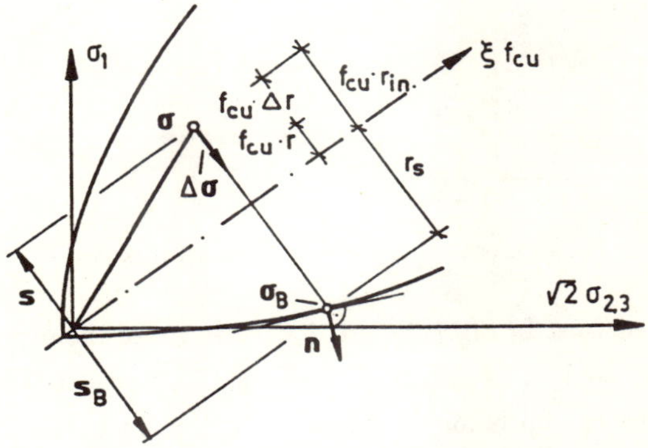

Fig. 7: Determination of D for the case of a significant change of the loading direction

Numerical investigation

The numerical study contains a finite element analysis of a cylindrical concrete specimen subjected to a concentrated compressive force by a steel punch (Fig.8). The numerical results, obtained by means of the two different constitutive models are compared with experimental results. The tests, conducted by Lieberum [17], are characterized by an extremely small loaded area. The steel punch and the cylindrical concrete specimen have diameters of 3.2 and 40.0 cm, respectively (Fig.8).

In [17], Young's modulus E_c, Poisson's ratio ν_c, the uniaxial compressive strength f_{cu} and the tensile strength f_{tu} are given as $E_c = 3000$ kN/cm^2, $\nu_c = 0.18$, $f_{cu} = 3.90$ kN/cm^2 and $f_{tu} = 0.30$ kN/cm^2. To account for the so-called wall effect [18], reduced Young's moduli are used in the small hatched area of the axial cross-section of the specimen (Fig.8),

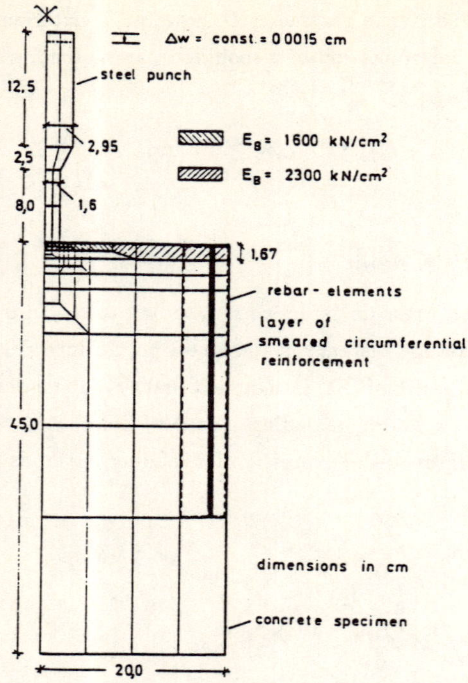

Fig. 8: Finite element model of the Lieberum test [17]

in the vicinity of its top surface. In Fig.9 the qualitative shape of the experimental load-penetration diagram is plotted. The numerical investigation is restricted to region I of Fig.9. Region II refers to the initiation of structural failure by radial splitting of the concrete cylinder. With increasing penetration of the steel punch, spalling of the concrete around the punch, accompanied by a sudden drop of the load-carrying capacity, is taking place (region III).

The cap model was implemented into the finite element program FEAP [19]. The hypoplastic model has been incorporated into the finite element program MARC. In FEAP, both, the concrete specimen and the steel punch are discretized by means of 9-node axisymmetric Lagrangian elements, and in MARC by means of 8-node axisymmetric elements. The test was run under displacement control. The numerical analysis was performed such that vertical displacement increments were applied at the top of the steel punch (0.01 cm for the cap model and 0.0015 cm for the hypoplastic model). The reason for the different size of the displacement increments is that only for the cap model consistent tangent moduli are available. Note that the use of consistent tangent moduli enables preserving the quadratic rate of asymptotic convergence in a Newton solution

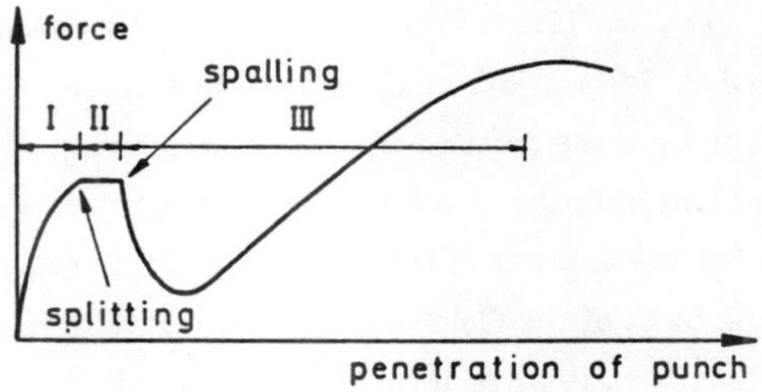

Fig. 9: Qualitative shape of the load-penetration diagram

procedure, yielding a considerable reduction of the number of iteration steps within one increment. For the cap model the criterion for termination of the Newton solution strategy is chosen in terms of an energy norm E, which is computed from the residual force and incremental displacement vector as

$$\Delta E_{n+1}^{(i)} \leq 10^{-10} \Delta E_{n+1}^{(1)}, \tag{83}$$

where the subscript $n+1$ denotes the displacement increment and the superscripts refer to the first and the i-th iteration step, respectively. The number of iteration steps to achieve this rather severe condition never exceeded 6. Values of the energy norm for typical displacement steps are given in Table 1.

Table 1. Energy norm for typical displacement steps

iteration step	increment 1	increment 15	increment 30
1	$0.211e+01$	$0.218e+01$	$0.218e+01$
2	$0.286e-03$	$0.246e-04$	$0.524e-05$
3	$0.257e-06$	$0.264e-08$	$0.501e-10$
4	$0.759e-12$	$0.174e-15$	$0.216e-19$

Model parameters for the modified cap model

Based on the Colorado Concrete Data [20] an extensive study of the predictive capabilities of the original cap model was performed by Simo et al. [3]. In [20], Young's modulus and Poisson's ratio are given as 2800 kN/cm^2 and 0.18, respectively; the average ultimate uniaxial compressive strength is given as 2.8 kN/cm^2.

In [3], the model parameters for the original cap model were computed by means of an optimization procedure. For the modified cap model the same tests described in [3] were employed for the purpose of parameter fitting, resulting in $\alpha = 3.03$ kN/cm^2, $\theta = 0.33, \lambda = 1.26$ kN/cm^2$, \beta = 0.77$ cm^2/kN, $R = 2.89, D = 0.0046$ cm^2/kN, $W = 0.42, T =$

-0.21 kN/cm^2 and $X(\kappa_0) = 11.2$ kN/cm^2. Based on the parameters for the modified cap model, the tests 1-2, 2-2 and 3-5 of the Colorado Concrete Data [20] were analyzed numerically. The results are identical to those obtained in [3]. With regards to the concrete of the Lieberum test only the elastic material parameters and the ultimate uniaxial strength are known. For the analysis of this test the model parameters of the modified cap model obtained on the basis of the Colorado Concrete Data are used. However, these parameters are scaled by the ratio or reciprocal ratio (according to the physical unit of the respective model parameter) of the respective ultimate uniaxial compressive strengths. Hence the model parameters are obtained as $\alpha = 4.22$ kN/cm^2, $\theta = 0.33$, $\lambda = 1.76$ kN/cm^2, $\beta = 0.55$ cm^2/kN, $R = 2.89$, $D = 0.0033$ cm^2/kN, $W = 0.42$, $T = -0.29$ kN/cm^2 and $X(\kappa_0) = 15.6$ kN/cm^2.

Material parameters for the hypoplastic model

The material parameters for this constitutive model are listed in Table 2.

Table 2. Material parameters for the hypoplastic model

Concrete			
Young's modulus:	$E_C = 3000$ kN/cm^2	Tensile strength:	$f_{tu} = 0.03$ kN/cm^2
Poisson's ratio:	$\nu = 0.18$	Softening modulus:	$E_S = 250$ kN/cm^2
Concrete Strength:	$f_{cu} = 3,90$ kN/cm^2	Shear retention factor:	SRF = const. = 0.2
Reinforcement			
Young's modulus:	$E = 19374$ kN/cm^2	Yield stress:	$\sigma_y = 37.4$ kN/cm^2
Steel punch			
Young's modulus:	$E = 20600$ kN/cm^2		

Discussion of the numerical results

Fig.10 shows a comparison of the computed load-penetration diagrams, based on the two different constitutive models, with the experimental one. Apart from the overestimation of the structural stiffness at the onset of the test, the experimental load-penetration diagram agrees reasonably well, up to a penetration of 0.30 cm, with the respective diagram obtained by means of the cap model. Although the computed structural stiffness is decreasing gradually with increasing penetration, the numerical calculation does not yield the failure plateau observed in the test.

For a given penetration of up to approximately 0.25 cm, the corresponding load acting on the steel punch is overestimated by the hypoplastic model. However, the experimental

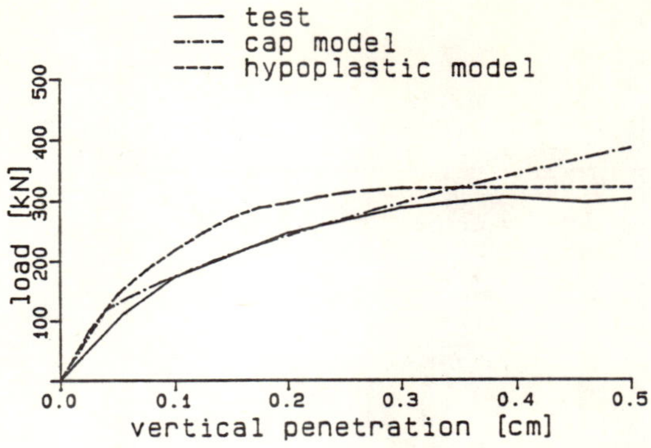

Fig. 10: Computed and experimental load-penetration diagram for region I

and the computed ultimate load agree quite well. The regions of cracked and plasticized concrete in the vincinity of the steel punch at penetrations of 0.03 cm and 0.3 cm, based on the calculation by means of the hypoplastic model, are shown in Fig.11 and Fig.12, respectively.

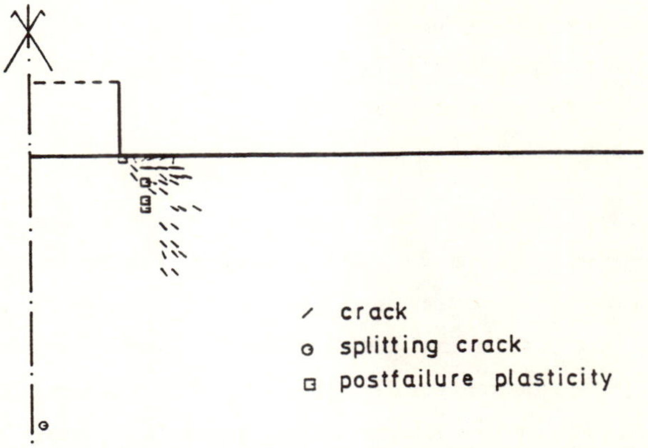

Fig. 11: Regions of cracked and plasticized concrete at a penetration of 0.03 cm

The reason why the cap model does not predict the failure plateau is that the material parameters for this model were determined on the basis of the Colorado Concrete Data. The range of these experiments is limited, in general, to strains up to 1 % to 1.5 %. The maximum strain does not exceed 3 %. At a penetration of 0.031 cm in the vicinity of the steel punch strains ranging from 1.7 % to 3.3 % are computed. Hence, the limit of the range of the experiments is reached already at this low penetration level. At a penetration of 0.3 cm in the vicinity of the steel punch the computed strains are about 7 % to 14 %. Thus, the limits of the experiments, which have served as the basis for the determination

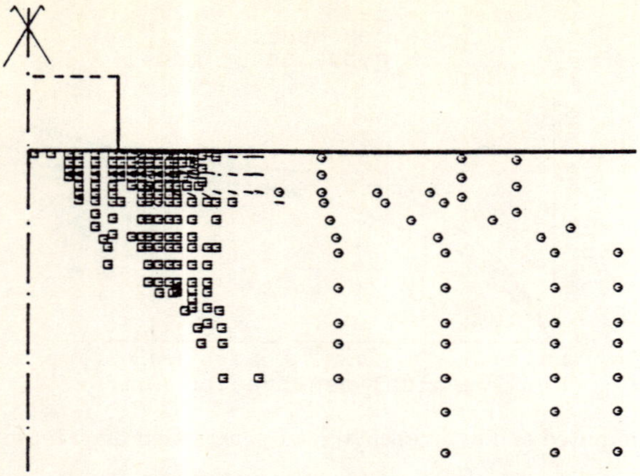

Fig. 12: Regions of cracked and plasticized concrete at a penetration of 0.3 cm

of the model parameters, are by far exceeded. Moreover, it is questionable, whether for such large strains an analysis based on the theory of small strains is at all appropriate. For the cap model an additional reason for deviations of the numerical results from the experimental results is the rough modeling of tensile failure, which is only considered by means of the tension cutoff surface. As far as the hypoplastic model is concerned, cracks are represented by a "smeared crack" approach.

Conclusions

The application of two constitutive models for consideration of plastic deformations of concrete to finite element analysis of a cylindrical concrete specimen subjected to a concentrated compressive force by a steel punch has shown the advantages and disadvantages of the two models. Apart from overestimation of the structural stiffness at the onset of the test, the load-penetration diagram computed by means of the cap model agrees reasonably well with the experimental load-displacement diagram up to a penetration level which is characterized by strains in the vicinity of the steel punch of about 10 %. The hypoplastic model predicts a structural behavior which is somewhat too stiff. The consistent formulation of the cap model within the framework of the return mapping algorithm guarantees a quadratic rate of asymptotic convergence in a Newton solution procedure. Compared with the classical continuum formulation, the size of the load (or displacement) increments may be increased considerably. At the same time, the number of iteration steps within an increment is reduced considerably, thus, yielding a higher degree of computational efficiency. The drawback of the cap model compared with the hypoplastic model is that the former, in contrast to the latter, does not predict the failure plateau.

References

1. DiMaggio, F.L.; Sandler, I.S.: Material models for granular soils. *J. of Engng. Mech.*, ASCE 97 (1971) 935-950.
2. Sandler, I.S.; DiMaggio F.L.; Baladi, G.Y.: Generalized cap model for geological materials. *J. of Geotech. Engng.*, ASCE 102 (1976) 683-699.
3. Simo, J.C.; Ju, J.W.; Pister, K.S.; Taylor, R.L.: An assessment of the cap model: Consistent return algorithms and rate-dependent extensions. *J. Eng. Mech.*, ASCE 114 (1988) 191-218.
4. Sandler, I.S.; Rubin, D.: An algorithm and a modular subroutine for the cap model. *Int. J. Numer. Analyt. Meth. Geomech.* 3 (1979) 173-186.
5. Koiter, W.T.: Stress strain relations, uniqueness and variational theorems for elastic-plastic materials with a singular yield surface. *Quart. J. Appl. Math.* 11 (1953) 350-354.
6. Simo, J.C.; Kennedy, J.G.; Godvindjee, S.: Unconditionally stable return mapping algorithms for non-smooth multi-surface plasticity amenable to exact linearization. *Int. J. Num. Meth. Engng.* 26 (1988) 2161 - 2185.
7. Simo, J.C.; Hughes, T.J.R.: *Elastoplasticity and viscoplasticity. Computational aspects.* In press.
8. Simo, J.C.; Taylor, R.L.: Consistent tangent operators for rate independent elastoplasticity, *Comp. Meth. Appl. Mech. Engng.* 48 (1985) 101-118.
9. Hofstetter, G.; Simo, J.C.; Taylor, R.L.: A modified cap model: consistent point solution algorithms, *Report No. UCB/SEMM-89/24*, Berkeley: University of California at Berkeley, Department of Civil Engineering 1989.
10. Meschke, G.; Mang, H.A.: Application of a new hypoplastic model to 3D finite element analyses of concrete structures. *Proc. 2nd Int. Conf. on Computational Plasticity, Models, Software and Applications* (Owen, D.J.R.; Hinton, E.; Onate, E.; eds.), Part II, 1219-1234, Swansea: Pineridge Press 1981.
11. Meschke, G.: Synthese aus konstitutivem Modellieren von Beton mittels dreiaxialer, elasto-plastischer Werkstoffmodelle und Finite-Elemente-Analysen dickwandiger Stahlbetonkonstruktionen. *Doctoral Dissertation*, Vienna: Technical University of Vienna 1989.
12. Stankowski, T.: Concrete under multiaxial load histories. *Doctoral Dissertation*, Boulder: University of Colorado 1983.
13. Maier, G.; Hueckel, T.: Nonassociated and coupled flow rules of elastoplasticity for rock-like materials. *Int. J. Rock Mech., Miner. Science and Geomech.* 16 (1978) 77-92.

14. Chen, E.S.; Buyukozturk, O.: Constitutive model for concrete in cyclic compression. *J. of Engng. Mech., ASCE* 191 (1985) 797-814.

15. Willam, K.J.; Warnke, E.P.: Constitutive model for the triaxial behavior of concrete. *Proc. Int. Assoc. for Bridge and Struct. Engng.* 19 (1975).

16. Dafalias, Y.F.; Popov, E.P.: Cyclic loading for materials with a vanishing elastic region. *Nuclear Engng. and Design* 41 (1977) 293-302.

17. Lieberum, K.H.: Das Tragverhalten von Beton bei extremer Teilflächenbelastung. *Doctoral Dissertation,* Darmstadt: Technical University of Darmstadt 1987.

18. Lieberum, K.H.; Reinhardt, H.W.: Strength of concrete on an extremely small bearing area. *ACI-Structural Journal* (1989) 67-76.

19. Taylor, R.L.: FEAP - A finite element analysis program, *Description and Users-Manual*, Berkeley: University of California at Berkeley, Department of Civil Engineering 1988.

20. Scavuzzo, R.; Stankowski, T.; Gerstle, K.H.; Ko, H.Y.: Stress-strain curves for concrete under multiaxial load histories. *NSF CME-80-01508,* Boulder: University of Colorado at Boulder, Department of Civil Engineering 1983.

Nonlinear Analysis of Cracked Concrete Beams under Bending and Torsion

W. WUNDERLICH, E. HEINS

Lehrstuhl für Statik
Technische Universität München
München, Deutschland

Summary

To investigate the nonlinear behaviour of slender beam structures of concrete, a theory of spatially loaded rods and the corresponding nonlinear analysis is presented. To include the realistic material description for the concrete including progressive cracks and post failure also in unloading and reloading a constitutive model is developed in which the damaging criterion is composed of several components each of them representing a special property of the material.

The nonlinear solution technique uses a Newton–Raphson technique in modified form in combination with a quasi–Newton–iteration for the inner equilibrium update of the material. The cross–sectional properties of the beam are obtained through integration over layers taking coupled bending and torsion into account.

The tangential stiffnesses of the finite element approach are obtained by direct numerical integration of the governing equations rather than through the use of and shape functions. Numerical results for lateral buckling of a concrete girder are compared to experimental ones.

1. Introduction

The use of concrete and steel with high strength in combination with modern construction techniques leads to increasing slender and complex structures. Already under usual load levels geometrically and material nonlinear behaviour may be observed. For a equally safe and economic design also the limit loads and the performance of the structures over the whole loading range has to be evaluated.

With the progress in computer development and numerical methods over the last decades it became possible to numerically simulate the nonlinear behaviour even of complicated structures realistically. The analysis may not only performed to investigate the load bearing capacity of specific structures but also to include certain influences through parametric studies, and possibly to calibrate the results by experiments.

This paper deals with a numerical method developed to investigate the nonlinear behaviour of slender beam structures made of concrete. The quite complex real three–dimensional behaviour is modelled by one–dimensional members under spatial loading and subject to bending and torsion.

In the first part the underlying nonlinear theory of arbitrary loaded beams is described in general. It may also be applied to investigate the stability behaviour of the structure.

For a realistic description of the load carrying capacity of concrete structures in addition to these geometric nonlinearities the material nonlinearities have to be included carefully. Even for smaller loads cracks in various directions may develop due to the low tensile strength of the concrete. Hence, the primary objective of this work in its secant part is to formulate a realistic material description for the concrete including progressive cracks and post failure behaviour of the damaged concrete also in unloading and reloading processes [1]. In contrast to other work in this field the in essence three-dimensional material model is reduced and implemented into a one-dimensional theory of rods including the influence of cracks on the stiffnesses of the beam directly.

In a third part some numerical results for the lateral buckling behaviour of slender beams made of concrete are presented.

2. Theory of beams

To meet the higher safety requirements which are necessary for a design using highly computational numerical analysis the formulation should predict the real behaviour quite accurately and not be restricted by many assumptions. In this respect the effect of coupling between bending and torsion as well as the influence of arbitrary cross sections in combination with a proper nonlinear formulation for the material have to be taken into account to realistically model the actual behaviour.

Following previous derivations [2,3] a nonlinear theory of rods was derived from the basic equations of the continuum [1]. It can be applied to initially straight beams under arbitrary cross sections and spatial loading in problems with large deflections, moderate rotations, and small strains. In the derivation the effects of average shear deformations and warping constraints are considered. The reference line may have an arbitrary position and not necessarily coincide with the centroid which is important for the changes due to the nonlinear material behaviour.

The derivation of theory was carried out consistently in the sense that all geometrically nonlinear terms are expanded into a Taylor-series up to the second order in formulating the incremental equations. In additionally restricting the rotations of the cross-section of the beam in the same manner, a theory of large deflections and moderate rotations is obtained which seems most appropriate for the concrete-structures under consideration.

The consequent and rather lengthy procedure of reduction to the one-dimensional beam problem finally leads to a generalized variational principle with seven kinematic variables and the same number of stress resultants as unknowns. This principle could be taken as the basis of a usual mixed-type finite element discretisation. For one-dimensional problems, however, a proven algorithm which is based on the integration of the corresponding Euler-equations, written in the form of a system of first-order ordinary differential equations [4], is more advantageous, because it results in highly accurate element stiffness matrices and corresponding load vectors. The whole structure can then be treated following the usual path of the finite element displacement formulation. The nonlinear solution is accomplished by a modified Newton–Raphson algorithm within each increment of the loading path. In this manner arbitrary systems of rods and beams may be analysed.

3. Constitutive modelling of concrete

3.1. General description

Due to the wide use of concrete for structural applications in civil engineering the realistic modelling of the behaviour of this material was investigated in numerous papers and research reports, see e.g. [6,7,8,9]. It is well-known that the theory of plasticity is the proper description for steel, and it seems obvious to use this theory also for concrete. As the real behaviour is by far more complex, however, no unified form for concrete exists to date, especially not for the general triaxial case. Mainly under hydrostatic compression the material can be characterized as ductile, under tension it behaves brittle and cracking changes the properties considerably.

It seems therefore not to be appropriate to characterize concrete by a single material law which is capable to describe its complex behaviour in different stress levels and under various loading histories. But its constitutive description may be composed by several special models which describe certain phenomena of the behaviour of concrete within given limits. Combining them to a general description in a consistent manner and observing the basic equations of the mechanics of continua may then lead to a quite general description, and gives the basis for applications within a practical range.

Following these general lines a triaxial material model was developed and applied for beams under combined bending and torsion [1]. The formulations are based on a generalized theory of plasticity with several components each of them representing a special property of the material [5]. In this way already existing and proven components may be included, e.g. those describing well–defined failure criteria or the strain hardening and strain softening under hydrostatic loading [10].

In a slender beam made of concrete and subject to bending and torsion specific stress states may occur for which the material exhibits very low tensile strength [11]. In these cases the first microcracks are already observed under low load levels. For these problems a special component for the constitutive modelling was developed with emphasis on the progressive crack developement and post–failure behaviour under arbitrary loading paths.

For the modelling of the cracks the smeared crack concept is used with multiple discrete and interacting and in general non–orthogonal crack directions. The concrete is assumed to be initially homogenous and isotropic and after cracking becoming a homogeneous and anisotropic continuum.

Several discrete cracks develop from initially arbitrary distributed microdefects. This process may also be considered as a general damaging of the material [12,13], connected with a global decrease of the elastic stiffness and strength. This change of elastic properties of the material is taken into account in the crack model under consideration. This is sometimes also called elasto-plastic coupling because the elastic stiffness is changing with the plastic part. In the classical theory of plasticity merely a superposition between elastic and inelastic deformations is assumed, and it is not connected with degradation of elastic stiffness.

3.2. Failure Criterion - Cracking of Concrete

If the largest principal tensile stress r_1 exceeds the tensile strength r^{tc}, the concrete is cracking. Assuming that the concrete is isotropic before the state of failure, the condition of failure may be formulated as:

$$f_B = r_1 - r^{tc} \leq 0 \quad . \tag{1}$$

It therefore may be concluded, that no other direction exists in the region with previous cracks, for which the tensile stress is higher than the cracking stress r^{tc}. This limiting tensile strength r^{tc} is a function of the multiaxial state of stress. It is assumed that normal to the direction of the principal tensile stress a crack is formed, and it follows that many different directions of cracking may develop in a material point, if the principal directions are changing. The cracks, however, do not change their direction in the progress of the loading history.

This strategy of cracking is described as "fixed crack approach", in contrast to the "rotating crack model". The numerical effort rises heavily by considering every new crack direction. In order to restrict the programming effort, a new crack shall be only formed, if the angle between the old and the new expected crack direction exceeds a defined measure (ca. 10°).

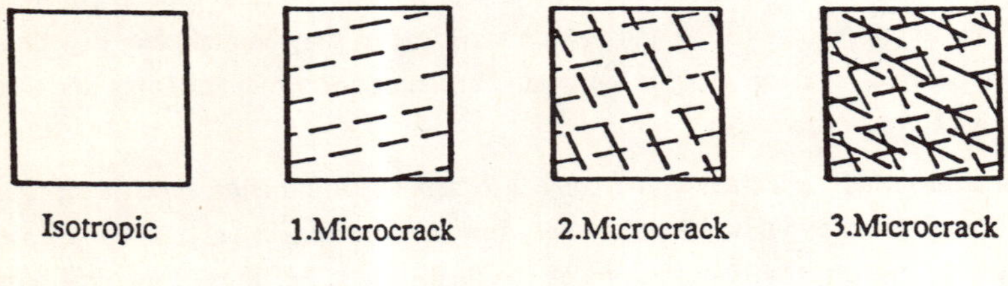

Fig. 1: Cracking and Microcracking for Rotating Principal Stress Directions

For the description of the postcracking behaviour, it is necessary for every plane of cracking to define a local orthonormal coordinate system of cracking $\mathbf{e}_\xi^{'}$ (Fig. 2). The base vector $\mathbf{e}_1^{'}$, has the direction of the principal tensile stress r_1, producing the crack, whereas $\mathbf{e}_2^{'}$ and $\mathbf{e}_3^{'}$

shall only lie in the plane of cracking. The transformation for the cartesian tensors is given by:

$$\text{Strains:} \quad h^{(x)}_{\xi\eta} = a^{(x)}_{\xi i}\, a^{(x)}_{\eta j}\, h_{ij} \quad , \tag{2}$$

$$\text{Stresses:} \quad r^{(x)}_{\xi\eta} = a^{(x)}_{\xi i}\, a^{(x)}_{\eta j}\, r_{ij} \quad . \tag{3}$$

The components $a^{(x)}_{\xi i}$ are the direction cosinus of crack number x:

$$a_{\xi i} =: \cos\left(\mathbf{e}'_{\xi} , \mathbf{e}_i \right) . \tag{4}$$

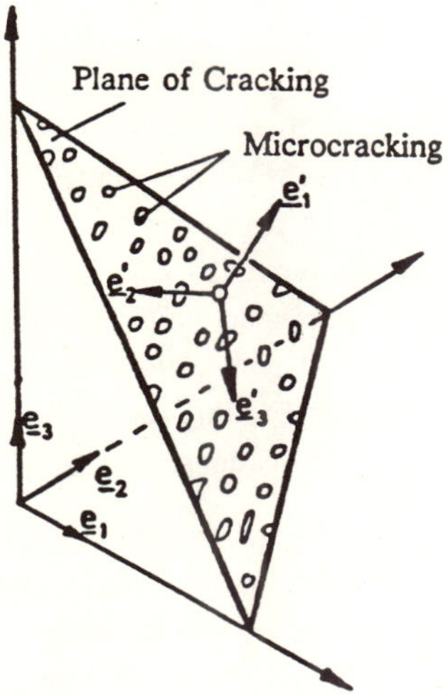

Fig. 2: Local Coordinate System for Cracking

3.3. Post-failure behaviour of the concrete

Under the phenomenological view point of solid mechanics the material behaviour is described by mathematical modelling in the frame work of the basic equations of thermomechanics. The models and parameters of a certain material like concrete have to be obtained by experiments. Investigating the microstructure of the concrete and the corresponding experimental results leads to a better understanding of the constitutive model and its assumptions.

In this respect some experimental results are considered. Fig. 3 shows a typical stress–strain–curve of a uniaxial tension test. Its post failure behaviour depends strongly on the deformations and on the external variables like stress and strains but is also a function of specific variables of the process which describe the variations in the structure of the material and the kind of the aggregate.

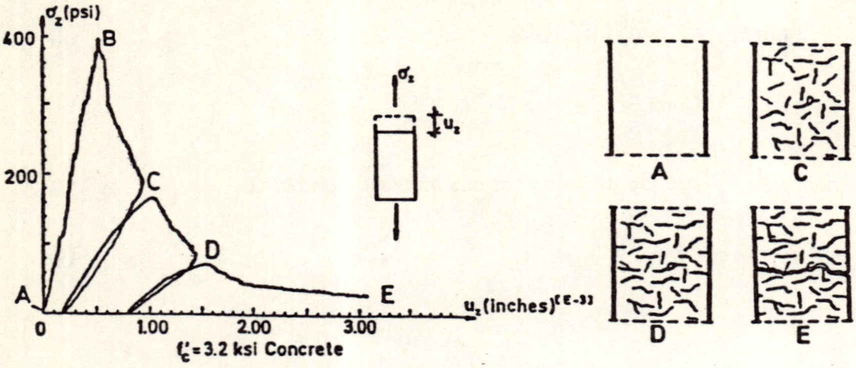

Fig. 3: Typical Strain Controled Uniaxial Tension Test of a Concrete Specimen

The range until reaching the ultimate strength will be defined in the following as initial state without damage based on the observation that the micro-cracks do not change even under slowly varying loads. Under unloading and rising strains discrete tensile cracks with local propagation and widening are observed. This effect leads to a reduction of the global tensile strength and a degradation of the elastic stiffness [13]. The comparison of the various paths of unloading and reloading shows that the global tensile strength and stiffness strongly depend on the maximum strain at the cracks during the loading path. For this reason they are used as a measuring variable for the damage in the following formulations.

Alternatively, the results of the tensile tests could be modelled numerically in detail on an aggregate level. Even for the uniaxial case it is quite difficult to measure the damaged concrete, and there are not enough reliable results available. The extension for the multiaxial case and softening creates obviously still higher testing problems. For the following formulations the idealized stress-strain-curve of Fig. 4 is used.

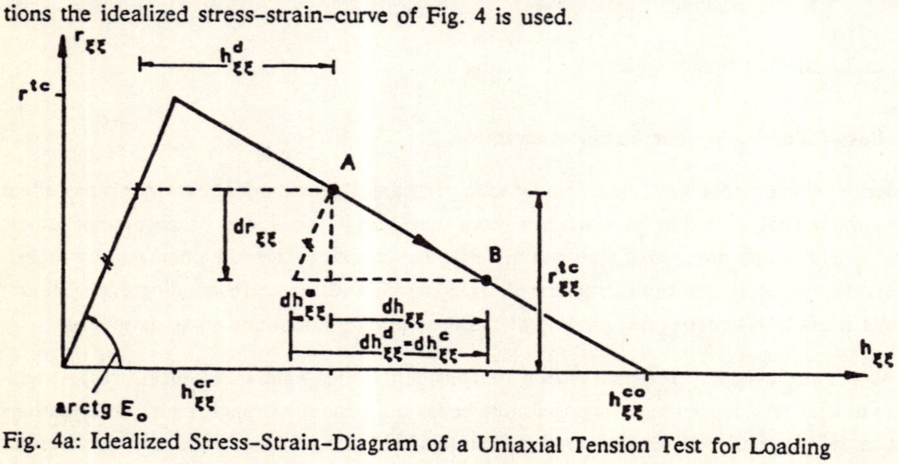

Fig. 4a: Idealized Stress-Strain-Diagram of a Uniaxial Tension Test for Loading

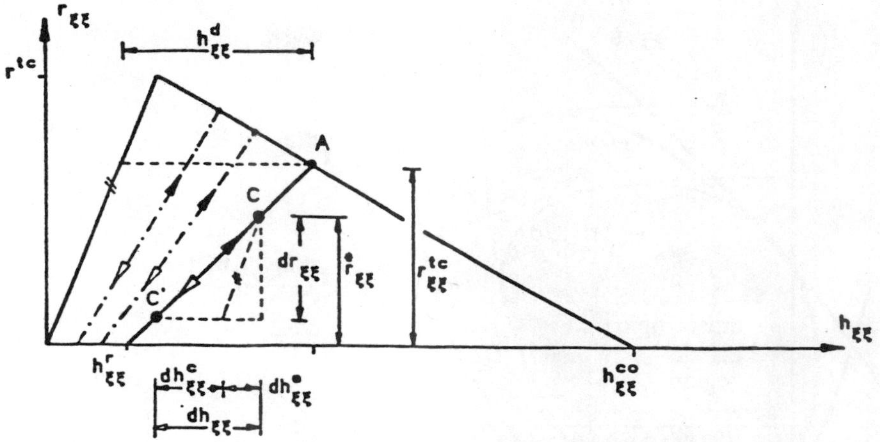

Fig. 4b: Idealized Stress–Strain–Diagram of a Uniaxial Tension Test for Unloading and Reloading

3.4. Formulation of the material model

For the description of the damaged material it is assumed (like in the theory of plasticity) that the total strain increments dh_{ij} are split up into an elastic part dh^e_{ij} and into a partially reversible strain increment dh^c_{ij}

$$dh_{ij} = dh^e_{ij} + dh^c_{ij} \quad . \tag{5}$$

The nondamaged part of the concrete is described by the elastic strain rates and renders a change of the stress rate by the elastic constitutive law

$$dr_{ij} = C_{ijkl}\, dh^e_{kl} \quad . \tag{6}$$

The tangential tensor of elasticity $C_{ijkl}\,(r_{mn})$ includes the parameters for the nondamaged component of the material. With the assumption of elastic and isotropic behaviour it depends on the current stresses (r_{mn}) only [14], and is not dependent on the loading path.

In analogy to the existence of the yield surface in the theory of plasticity it is assumed that the criterion for damage of the material has the form

$$F^d\,(r_{ij}\ ;\ K_{kl}) \leq 0 \quad . \tag{7}$$

It describes the changes in the structure of the material and depends on the stress state r_{ij} and through a damage–tensor K_{kl} also on the history of the damage development. The function F^d may also be geometrically visualized by a hyper–surface in the stress space and characterizes the begin of failure or a state of damage, see Fig. 5. It subdivides the space into two regions with different behaviour. In the inner region the material has purely elastic behaviour whereas irreversible and elastic reactions are described by points in the outer one.

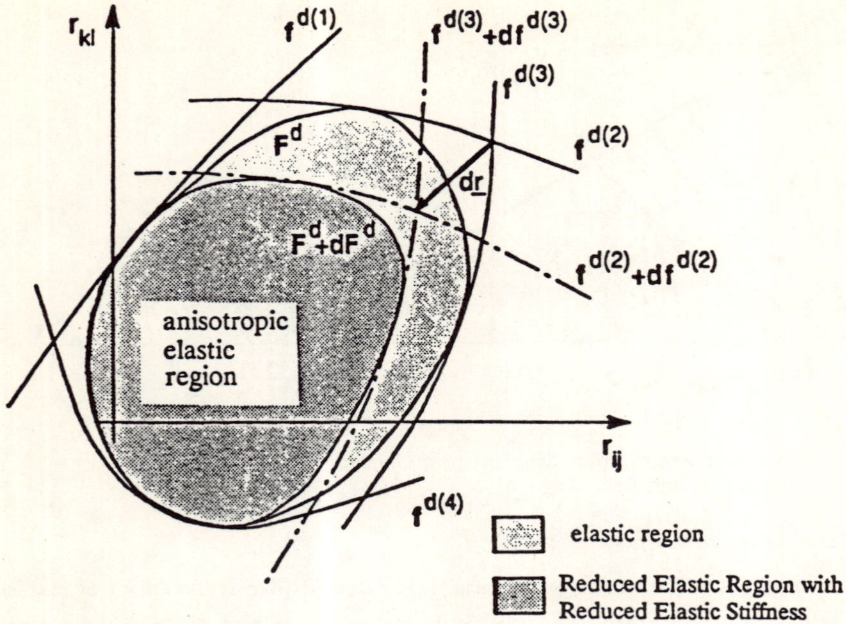

Fig. 5 : Approximation of the Damage Hypersurface F^d and the Increment $F^d + dF^d$ by n Partial Surfaces $f^{d(x)}$ and $f^{d(x)} + df^{d(x)}$

However, the complex anisotropic behaviour of the concrete with multiple cracks cannot be realistically modelled by one single function F^d. It is therefore approximately composed of several damaging surfaces $f^{d(x)}$ [3]. It is required that each is strictly

$$f^{d(x)} \leq 0 \qquad\qquad x = 1,2, \ldots, m \tag{8}$$

convex resulting in the convexity of the whole region, see Fig. 5. It may be seen that the structure of the material only changes, if at least one of the criteria for damage is satisfied. Each of the functions $f^{d(x)}$ represents a hardening or a softening part of the full material model and is determined in simple experimental tests, e.g. in an uniaxial tension or in a three-dimensional compression test.

In the formulations of this chapter the failure and post failure behaviour of concrete with cracking in n directions under tensile loads is emphasized. Thus, the constitutive model is composed of n components. It may be supplemented without difficulties by other components, e.g. describing elasto-plastic deformations [5].

it is characteristic for the brittle behaviour of the concrete that the global tensile strength of concrete is not suddenly reduced to zero after reaching the failure limit (1) which may be explained by the development of microcracks. The decrease of strength is governed by the n criteria for damage provided that the stress states satisfy the conditions:

$$f^{d(x)} = r_{\xi\xi} - r^{tc}_{\xi\xi} \; (I_1, h^d_{\xi\xi}) \leq 0 \; . \qquad x = 1,2, \; ...., \; n \tag{9}$$

The decrease of the tensile strength is oriented in the direction of ξ which is normal to the corresponding crack plane x, and n gives the number of actual crack directions. The criterion of damage is a function of the first stress invariant and mainly of the internal variables of the process $h^d_{\xi\xi}$, which denotes the maximum strains at n cracks during the loading history. As already described this variable is employed as a specific measure for the damage in direction ξ. It is also assumed that the process of damaging is irreversible; i.e. that the influence of the cracks cannot be reversed. This restriction is expressed through

$$dh^{d(x)}_{\xi\xi} \geq 0 \quad . \qquad x = 1,2, \; ...., \; n \tag{10}$$

As the hypersurface F^d is composed by n conditions of damage, the corresponding strain rates dh^c_{ij} of equ. (5) can be obtained by addition of n strains at the cracks

$$dh^c_{ij} = \sum_{x=1}^{n} dh^{c(x)}_{ij} \quad . \tag{11}$$

These cracking strain rates again consist of two parts: the strains $dh^c_{\xi\xi}$ normal to the crack and the shear strains $d\gamma^{c(x)}$ in the plane of the cracks see in Fig. 6. Both are transformed to a spatial strain state by equs. (2) and (4).

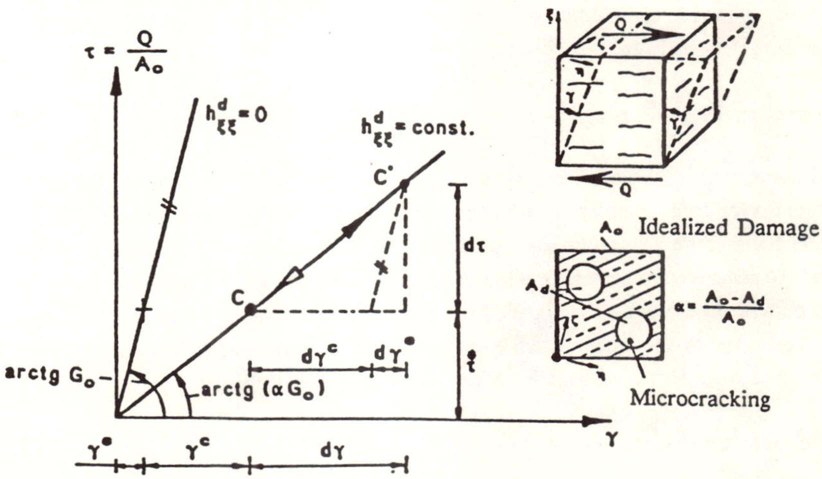

Fig. 6: Idealized Shear Stress-Strain-Diagram for Unloading and Reloading with Damage $dh^c_{\xi\xi}$ in one Direction

It is observed that with progressive damaging the global elastic stiffness decreases, see Fig. 4. In the following the anisotropic elastic behaviour shall be described such that the degradation in stiffness is coupled with an increase in damaging. With the development of the cracks not only the stiffness normal to the crack is decreasing but also the stiffness due to shear along the crack, see Fig. 6 [15]. Thus, for each crack orientation the degradation of the elastic stiffness renders two coupling conditions:

$$f^{\sigma(x)} = r_{\xi\xi} - \dot{r}_{\xi\xi} \; (I_1, h_{\xi\xi}, h^d_{\xi\xi}) = 0 \;, \qquad x = 1,2, \ldots, n \tag{12}$$

$$f^{\tau(x)} = \tau^{(x)} - \dot{\tau}^{(x)} \; (I_1, \gamma^{(x)}, h^d_{\xi\xi}) = 0 \,. \qquad x = 1,2, \ldots, n \tag{13}$$

These conditions are connected with the corresponding conditions of damage through the internal variable $h^d_{\xi\xi}$ of the process. This is also denoted as "elasto–plastic coupling".

In the example of Fig. 5 damaging increases only in two crack orientations accompanied by a decrease of the normal and stear stiffness in these two directions. In the other directions the mechanical properties are assumed to remain unchanged.

The basic equations (1) to (13) of the constitutive modelling now relate the incremental strain rates with their corresponding stress rates. The relations may be written symbolically as

$$d\underline{r} = \{\underline{I} + \underline{D}\} \; \underline{C} \; \underline{\Omega} \; d\underline{h} \tag{14}$$

in which the first part describes the elastic $(\underline{C})$, anisotropic $(\underline{\Omega})$ behaviour, and the second term $(\underline{D})$ the increase of damaging.

The steel of the reinforcement of the concrete mainly is under uniaxial tension. It is described here by an elasto–plastic constitutive law with isotropic hardening in a three parameter model due to Ramberg–Osgood [16].

4. Bond between steel and concrete

A detailed description of the nonlinear properties of the bond between the two materials steel and concrete is very complicated and cannot be used for the analysis of whole beam structures. On the other hand some general bond–slip–relation has to be included for a realistic description of cracked concrete structures. Due to the bond the concrete still carries a certain amount of tension between the cracks, the socalled tension stiffening. This effect will be considered indirectly analogous to the smeared crack concept by a modified stress–strain curve of the concrete such that its softening part is longer and less decreasing.

5. Numerical solution technique

For the numerical solution of the geometrical and material nonlinear beam problem with combined bending and torsion the development of a numerically stable algorithm plays an important role. One basis is the approach in which the system of first-order differential equations is integrated numerically over each element, and the resulting integral matrix is transformed to a stiffness matrix. It can then be treated in the same way as in the usual displacement finite element approach. This method is particularly well suited to the treatment of the one-dimensional rod problem and has been shown to give very accurate results. It has the additional advantage that a certain relative error-bound can be given a-priori for the element matrix. It also appears to better account for the rapidly decaying warping functions, and their sometimes significant effect on the load-carrying behaviour of slender beams, than stiffness matrices developed from a usual displacement formulation. For further details of the approach, which has been used successfully by the first author and his coworkers, see [2-4,17].

The other basis is the solution of a nonlinear system of the structure the incremental form of which is obtained by adding the element stiffnesses including the influence of initial deformations and initial stresses to the system matrix in the usual displacement formulation.

As the damage function is formulated for a certain point across the section of the beam the stress-resultants have to be obtained through integration over layers in which the stresses are assumed to be constant, see Fig. 7. A formulation of the yield function with the resultants directly seems not applicable here as the material behaviour may change considerably. At the stations across the thickness the strains are obtained from the displacements of the continuum by the usual kinematical assumptions for the beam and connected with the stresses by the material model. In this way the integration over the layers does not lead to an equilibrium state of the resultants, and the constitutive model of the concrete makes it necessary to consider the changes of the material iteratively in an inner loop of the incremental prodecure.

The nonlinear solution technique uses an Newton Raphson algorithm in modified form. Only at the begin of a new load increment the whole tangential system matrix is assembled. The inner equilibrium update for the material behaviour employes a Quasi-Newton iteration developed by Davidson, Fletcher and Powell (DFP-update) [18] and approximates the inverse tangential stiffness of the cross section by its secant. This has the advantage that the matrix of the cross sectional properties for the beams have to be evaluated by integration over the cross section only at the initial state, and the nonlinear variations are included in the incremental load vector. The resulting program was developed in FORTRAN 77, and it includes also routines provided by the software library DFGBIB [19].

6. Example: Lateral buckling of a concrete girder

The method described in the paper is applied to a slender concrete beam. The nonlinear behaviour exhibits sensitivity with respect to the lateral stability of the girder. Its numerical simulation is compared to testing results which were carried out in the testing laboratories in Darmstadt [20] by W. Pauli.

The load carrying capacity of the slender beam is very much dependent on the imposed imperfections. To separate the influence of the material nonlinearity the first simulation includes the geometric nonlinearities only, and it also takes the warping into account. Then the material model described above is applied, and the imperfections are taken as a parameter in the calculations. Fig. 8 shows the load-deflection-curves of the example obtained by the numerical simulation, which are in good agreement with the experimental results.

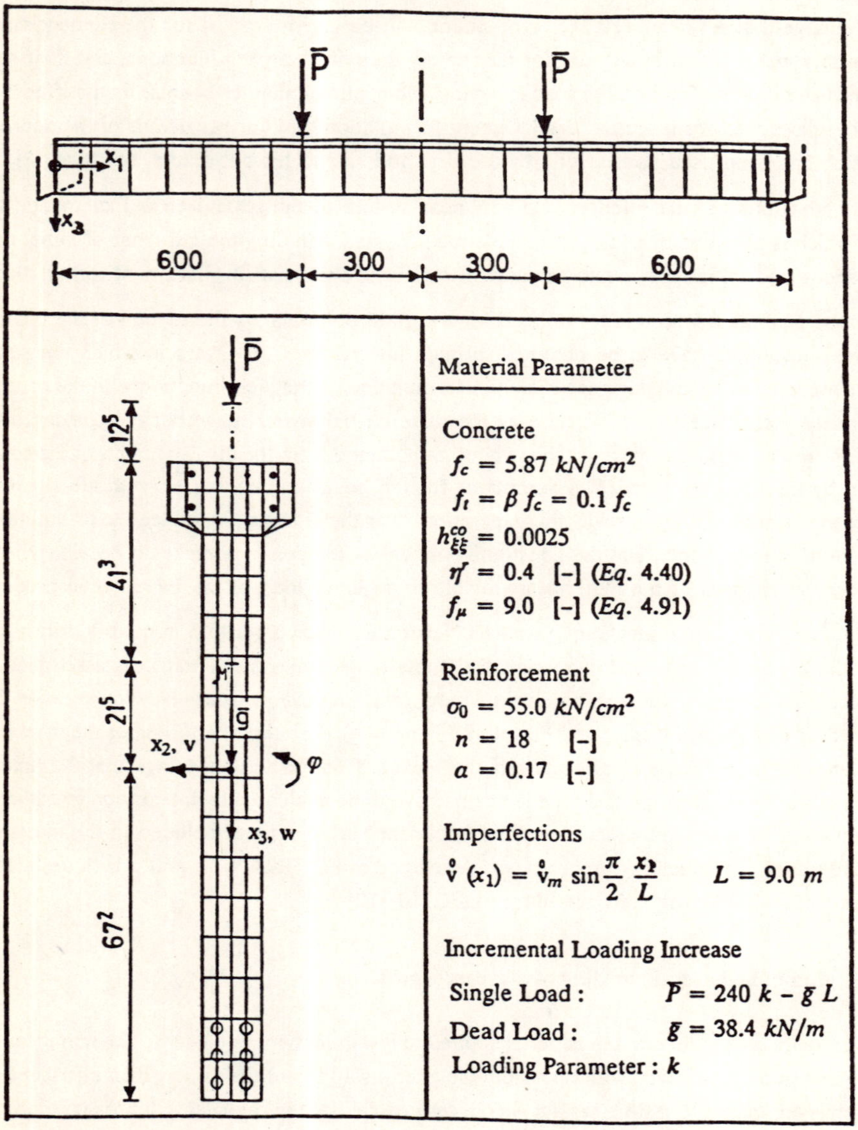

Fig. 7: Discretization of the System and the Cross Section, Imperfections, Material Parameters and Loading

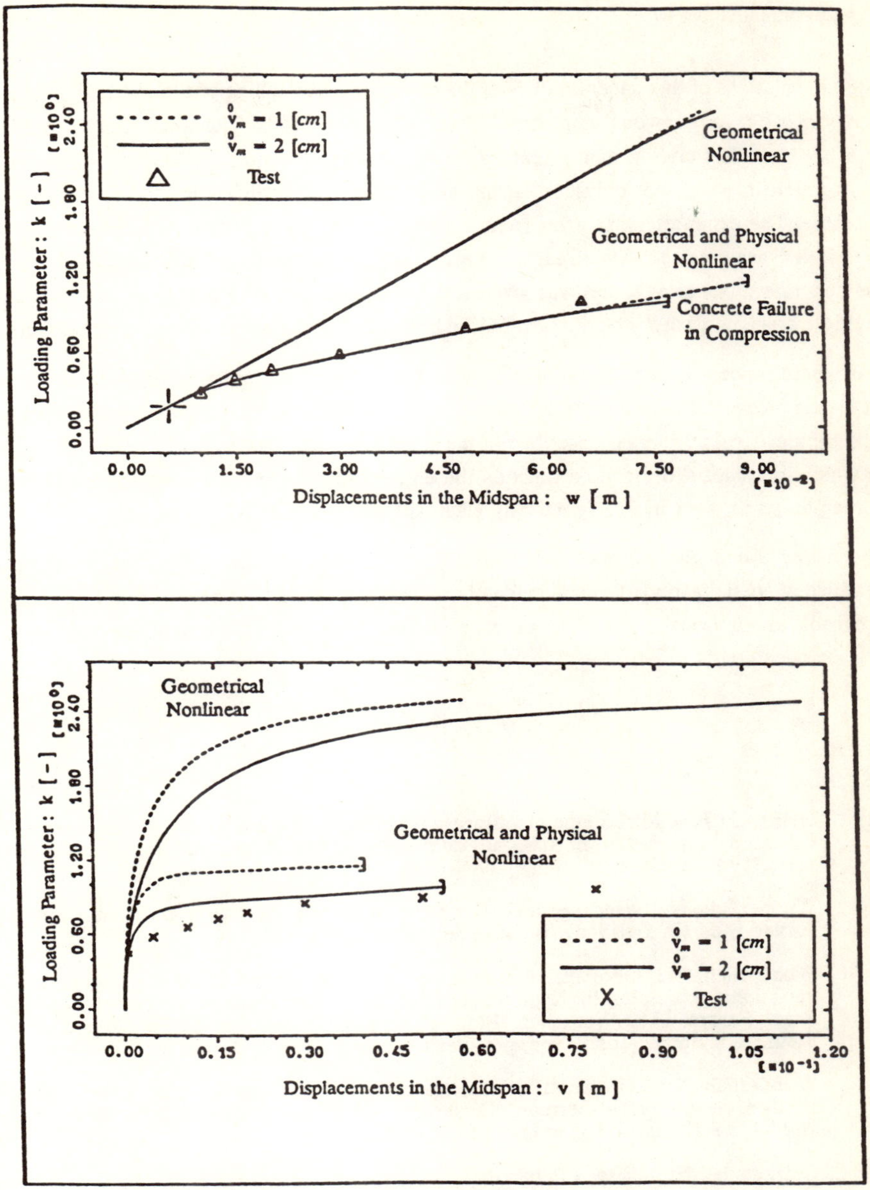

Fig.8: Load–Displacement Behaviour for Lateral Buckling
of a Reinforced Concrete Beam
a) Vertical Displacements w
b) Horizontal Displacements v

7. Conclusions

The formulations of this paper emphasize the constitutive modelling of the concrete in connection with a one-dimensional approach for spatially loaded rods. The work should contribute to a better understanding of the load carrying capacity of beams under combined bending and torsion in which not only normal but also shear stresses influence the development of cracking. The discrete cracks grow from initially arbitrary distributed microdefects and may appear in various directions already in lower load levels due to the low tensile strength of the concrete. Special attention was given also to unloading and reloading in the progressive multiple cracking to describe the post failure behaviour of the damaged concrete realistically.

The whole process is considered as a general damaging of the material and the decrease of its elastic properties is also taken into account. With suitable indicators the accumulated damaging under tension may be described in the framework of a generalized theory of plasticity in which in analogy to the yield function the entire damage function is composed of several damaging parts, also including existing elasto-plastic models.

The material and physically nonlinear theory and method for the analysis of structures assembled of spatially loaded beams made of concrete was realized in a corresponding computer program. In numerical simulations of test problems good performance and suitable accuracy was demonstrated.

8. References

[1] Heins, E.: Eine allgemeine nichtlineare Stabtheorie mit einem dreidimensionalen Rißmodell für Stahlbeton. Berichte aus dem Konstruktiven Ingenieurbau, Technische Universität München, 1991.

[2] Wunderlich, W., Beverungen, G.: Geometrically nonlinear theory and analysis of plane, curved rods (in German). Bauingenieur 52 (1977), 225-237.

[3] Wunderlich, W., Obrecht, H.: Large spatial deformations of rods using generalized variational principles. In: "Nonlinear Finite Element Analysis in Structural Mechanics". Proc. Europe-U.S. Workshop, Ruhr-Universität Bochum 1980, W. Wunderlich, E. Stein, K.J. Bathe (eds.), Springer-Verlag, Berlin (1981), 185-216.

[4] Wunderlich, W: Differential system and transfer-matrices for the bending theory of shells of revolution (in German). Dissertation TH Hannover 1966. Schriftenreihe: Lehrstuhl für Stahlbau TH Hannover, Heft 4 (1966).

[5] Cramer, H.: Numerische Behandlung nichtlinearer Probleme der Boden- und Felsmechanik mit elasto-plastischen Stoffgesetzen. Tech. Wiss. Mitteilungen 80-5, Inst. Konstr. Ing.-Bau, Ruhr-Universität Bochum, 1980.

[6] Chen, W.F., Han D.J.: Plasticity for Structural Engineers. Springer-Verlag, 1988.

[7] Eberhardsteiner, J., Meschke, G., Mang, H.: Triaxiales konstitutives Modellieren von Beton; zum Zwecke der Durchführung vergleichender Traglastanalysen dickwandiger Stahlbetonkonstruktionen mittels der Methode der Finiten Elemente. Institut für Festigkeitslehre der TU Wien, Juni 1987.

[8] De Borst, F.: Non-Linear Analysis of Frictional Materials. Dissertation, Technische Hochschule Delft, 1986.

[9] Han, D.J., Chen, W.F.,: Constitutive Modelling in Analysis of Concrete Structures. Journal of the Engineering Mechanics Division, ASCE, Vol. 113, EM4, 577-593, April 1987.

[10] Chen, W.F.: Plasticity in Feinforced Concrete. McGraw Hill, New York, 1982.

[11] Kupfer, H., Hilsdorf, H., Rüsch, H.: Behaviour of Concrete under Biaxial Stresses. Journal fo the Engineering Mechanics Division, ASCE, Vol.99, EM 4, 852-866, August 1973.

[12] Betten, J.: Elastizitäts- und Plastizitätslehre. Vieweg-Verlag, 1985.

[13] Willam, K.: Stoffgesetze und numerische Ansätze für Versagensvorgänge in Beton. In: "Nichtlineare Berechnungen im Konstruktiven Ingenieurbau", E. Stein, (Hrsg.), Springer-Verlag (1989), 324-342.

[14] Kotsowos, M.D., Newman, J.B.: A mathematical description of the deformational behaviour of concrete under complex loading. Magazine of Concrete Research, Vol. 31, No. 107, 77-90, June 1979.

[15] Walraven, J.C.: Experiments on shear transfer in cracks in concrete. Part II: Analysis of results. Report No. 5-79-10, Stevin Laboratory, Delft University of Technology, January 1979.

[16] Ramberg, W., Osgood, W.R.: Description of stress-strain curves by three parameters. NACA TN-902, 1943.

[17] Wunderlich, W.: Calculation of transfer-matrices, applied to the bending theory of shells of revolution. Proc. Intern. Symposium "The Use of Electronic Digital Computers in Structural Engineering." Newcastle upon Tyne (1966).

[18] Fletcher, R.: A new approach to variable metric algorithms. The Computer Journal, Vol. 13, Nr. 3, August 1970.

[19] Stein, E., Lambertz, K.H., Plank, L., Wunderlich, W., Cramer, H., Redanz, W.: Die DFGBIB des Schwerpunktes "Nichtlineare Berechnungen im KIB", eine Sammlung von Programmbausteinen für die Forschung. In: "Nichtlineare Berechnungen im Konstruktiven Ingenieurbau", E. Stein (Hrsg.), Springer-Verlag (1989), 649-686.

[20] Pauli, W.: Versuche zur Kippstabilität an praxisgerechten Fertigteilträgern aus Stahlbeton und Spannbeton. FB Konstruktiver Ingenieurbau der Technischen Hochschule Darmstadt, Institut für Massivbau, Dissertation 1990.

Crashworthiness Simulations of Composite Structures – a first Step with Explicit Time Integration

A.Matzenmiller †, K.Schweizerhof ‡

† Visiting scholar, University of California, Berkeley
Formerly CADFEM GmbH, Ebersberg, Germany
‡ Professor, Institut für Baustatik, University of Karlsruhe, Germany

Summary:
An overview over recent progress in the numerical simulations of composite damage is presented. The paper first concentrates on failure criteria for fiber-reinforced composites, indicating various failure mechanisms to model the post-failure behavior. Computational aspects for the crash worthiness analysis of laminated composite shell structures under low velocity impact are given next. Numerical results for the axial crushing of composite tubes and conical shells are presented and compared to experimental investigations. Finally the limits and merits of the implemented material and shell models are discussed.

1 Introduction

Low velocity impact loading on composite structures has received considerable attention especially for vehicle design. The failure behavior, energy dissipation and load carrying capacity of laminate tubes and cones, used as so-called "crash elements", is of primary interest for the safety analysis and occupant protection in automotives.

Our main purpose is to give an overview over recent work in crashworthiness analysis of composite structures. Little analysis for the crushing of laminated composites under low velocity impact has been carried out, although a number of researchers reported test results during the last decade [*Thornton 1979 and 1986; Hull 1982; Keal 1983; Price,Hull 1987; Maier 1990*]. Experimental investigations on fiber - reinforced tubes and cones indicate a wide range of material damage like matrix crushing, delamination and fiber breakage. The first two types of failure often show a rather slowly progressing failure with high energy dissipation, while the last failure type might initiate catastrophic collapse of the entire structure with little dissipation of kinetic energy. Since the microscopic failure behavior of laminates is still not completely understood, applicable mechanical models have not been proposed for the nonlinear analysis of composite structures. Nevertheless, first attempts towards the simulation of composite crash have been made [*Murray 1989;*

Matzenmiller, Schweizerhof, Maier 1990]. For a suitable mechanical model one has to resort to existing theories of failure prediction for laminated composites:

- a choice of an appropriate post-failure model like elasto-plasticity, damage mechanics or a simple degradation model;
- an efficient contact algorithm with the capability of eroding elements
- and a robust solution algorithm to monitor the frequently very jaggy time history response curve of forces.

Chapter 2 deals with failure phenomena of composite materials. The Hashin failure criteria [*Hashin 1980*] are chosen to account for the various modes of failure and are implemented in the Finite Element code LS-DYNA3D [*Hallquist 1990*] with various modifications. A short review of post-failure theories for laminated composites is outlined in chapter 3. Limitations of the different failure theories are briefly discussed. The main aspects of a numerical model for composite crash analysis are introduced in chapter 4. Numerical results are given in chapter 5 for two different shapes of crash elements.

2 Failure of Composite Materials

A homogenized continuum provides the basis for the constitutive model of each laminae. The usual rules of mixtures lead to the elasticity moduli and strength parameters of the unidirectional laminae, calculated from the properties of the fiber and resin data together with the volume ratios between fibers and matrix. Although the stress-strain response of unidirectional laminae is known to be very nonlinear [*Hahn, Tsai 1973*], the concept of linear elasticity is applied in order to determine the elasticity properties of the macro-mechanical model.

In contrast to integrated failure criteria for laminates, where the cross section of the composite structure is viewed as a homogeneous anisotropic body, greater universality is gained by checking failure in each individual layer of the laminate, since the failure criteria themselves do not depend on the stacking sequence or the layer thickness.

2.1 Failure Mechanisms

The direct calculation of failure loads and failure behavior on the basis of micromechanical models is far too complex to be carried out for the purpose of structural analysis of fiber reinforced composite structures. Even the rigorous derivation of the constitutive equations in form of a phenomenological parameter model from entirely micromechanical considerations is still an open field of research. A promising approach is to define failure criteria for critical stress or strain states in terms of averaged quantities. However, most failure criteria [*Hill 1948; Tsai 1971; Neale, Labossière 1989*] indicate only the initiation of failure due to critical combinations of stresses or strains. No predictions are generally made about the type of failure despite overwhelming experimental evidence, showing distinct failure mechanisms in fiber reinforced composites.

Practical composite structures are designed from stacks of unidirectional (UD-) laminae with different orientations. It is typical for many combinations of matrix and phase materials, that matrix cracking in individual layers can occur, well before total failure of the laminate cross section, caused by complete fiber rupture, is reached. While the occurrence of matrix cracking is important for the integrity of many composite structures, it might not immediately cause the collapse of it. For the integrity it is necessary to work with criteria, indicating the beginning and the various phenomena of failure within each phase, whereas for the final collapse in crash analysis it is of primary concern to rely on criteria for the ultimate resistance against total failure of the composite material. The latter question is most conveniently addressed by means of a homogenized continuum model. The different failure mechanisms due to tension or compression, caused by stresses in fiber direction or perpendicular to the reinforcement, have been cast into failure criteria with experimental evidence by a number of researchers e.g. [*Hashin 1980; Hahn, Erikson, Tsai 1982; Christensen 1987; Christensen, Swanson 1988*].

According to [*Puck 1969*] and [*Christensen, Swanson 1988*] at least fiber rupture and matrix/interphase failure need to be distinguished. In addition to these two forms of failure, some forms more related to the microstructure of the phase material might also occur. The latter are fiber buckling or kinking due to compression, loss of bond strength between fiber and matrix in combination with fiber pullout as well as delamination at the interface between individual layers of UD laminae.

Each failure mode indicating the loss of the tensile and/or compressive ultimate strength of the UD-laminae is quite different. They are most critical for the dissipation of kinetic energy in the mechanical model. The dominating failure mechanisms in composite structures take place under tensile and compressive stress states. Tensile failure is typical for cone shaped crash elements, resulting from hoop stresses as well as bending moments due to shape distortions. Thus, applicable failure criteria for the current issue must clearly distinguish between modes of failure for tension or compression as well as for failure in the direction of the principal axis of the material, since all four strength parameters differ up to more than an order of magnitude for the UD-laminae. The major failure mechanisms of laminates, often denoted as failure modes, are summarized and linked to their associated strength parameters:

I Matrix interface cracking due to tensile stresses, acting transverse to the fiber direction, in combination with shear. Failure depends mainly on the tensile strength and shear strength of the resin as well as on the fiber/matrix bond strength.

II Matrix crushing due to transverse compression and shear. Failure is mainly affected by the compressive and shear strength of the resin.

III Fiber breakage due to tensile loading is mainly influenced by the tensile strength of the fibers and the fiber/matrix volume ratio. However, uniaxial off-axis tests show, that even for small deviations of the loading direction from the fiber direction the failure load of the specimen is very sensitive with respect to the shear strength of the matrix material and to the fiber/matrix bond strength.

IV Fiber buckling and matrix crushing due to compression in fiber direction. The strength of the laminae is not only influenced by the compressive strength of the constituents but also by the elastic stiffness and by the shear strength of the matrix material. The latter becomes especially important for high volume ratios of fibers in comparison to the matrix material [*Rosen, Dow 1972*].

V Delamination of individual layers or separation of stacks of layers.

VI Debond of fiber and matrix or fiber pullout.

The first four failure modes report the failure of the unidirectional laminae, whereas mode **V** is a typical failure form of the laminate. Consequently, the laminae modes are accounted for in two-dimensional theories. Sufficiently accurate representation of the fifth mode requires a complete three dimensional setup of the constitutive equations and kinematics. Mode **VI** describes the problem of interfacial failure, common to all multiphase solids, consisting of a matrix material and some sort of reinforcements or aggregates. It only can be monitored by a micromechanical idealization, which is beyond the scope of the analysis of engineering structures.

The failure modes do not only indicate the type of failure. As already mentioned, they additionally determine the remaining transmittable strength of the damaged laminae in the post-failure regime. The remaining strength of the laminae, after failure is initiated, is most critical for the post-failure response and the crushing behavior of the structure.

Since the strength of the UD-laminae perpendicular to the fibers is governed by the resin and the bond strength, one refers to their associated failure modes also as the matrix modes. Consequently, mode **III** and **VI** are denoted as fiber modes.

2.2 Failure criteria of UD-laminae

The ultimate choice of an appropriate failure criterion for composite materials is still in discussion, despite the need for reliable analyses of fiber reinforced plastics. In view of the aforementioned requirements, the failure criteria in [*Hashin 1980*] is best suited for the purpose of postfailure analysis of composites. The x_1 -axis of the material coordinate system for the UD-laminae is assumed to coincide with the fiber direction. Hashin specifies

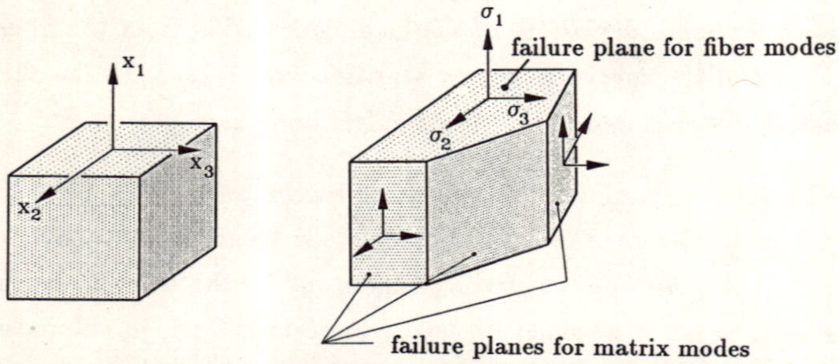

Figure 1: Coordinate system and failure planes of laminae

a three–dimensional failure criterion for a transversely isotropic material in terms of its integrity basis, given as:

$$I_1 = \sigma_{11} \tag{1}$$
$$I_2 = \sigma_{22} + \sigma_{33} \tag{2}$$
$$I_3 = \sigma_{23}^2 - \sigma_{22}\sigma_{33} \tag{3}$$
$$I_4 = \sigma_{12}^2 + \sigma_{13}^2 \tag{4}$$
$$I_5 = 2\,\sigma_{12}\sigma_{23}\sigma_{13} - \sigma_{22}\sigma_{13}^2 - \sigma_{33}\sigma_{12}^2 \tag{5}$$

The approximation of a failure criterion by means of a complete quadratic polynomial takes the form:

$$A_1I_1 + B_1I_1^2 + A_2I_2 + B_2I_2^2 + C_{12}I_1I_2 + A_3I_3 + A_4I_4 = 1 \tag{6}$$

Only the first four invariants up to quadratic order of stress components appear in eqn.(6). The constants A_3 and A_4 are determined in two simple shear tests.

$$A_3 = \frac{1}{\tau_T^2} \qquad A_4 = \frac{1}{\tau_A^2}$$

τ_T and τ_A represent the transverse and the axial shear strength. Failure is assumed to be caused by normal and shear stresses, acting on the failure planes (see fig.1). Fracture for the matrix modes occurs in a plane tangential to the fiber direction. The normal and shear stress components for the tangential plane are only functions of σ_{22} , σ_{33} , σ_{12} , σ_{13} and σ_{23} but not σ_{11}. Hence, the invariant I_1 has no influence on the stress components associated with the failure plane of the matrix modes.

Matrix mode:

$$A_m(\sigma_{22}+\sigma_{33})+B_m(\sigma_{22}+\sigma_{33})^2+\frac{1}{\tau_T^2}(\sigma_{23}-\sigma_{22}\sigma_{33})+ \\ +\frac{1}{\tau_A^2}\left(\sigma_{12}^2+\sigma_{13}^2\right)=1 \tag{7}$$

The failure plane for the fiber modes is assumed to be oriented perpendicular to the fibers. Only the $\sigma_{11}-$, $\sigma_{12}-$ and $\sigma_{13}-$ stress components contribute to fiber failure. Hence, the invariants I_2 and I_3 do not appear in the failure criteria.

Fiber mode:

$$A_f\sigma_{11}+B_f\sigma_{11}^2+\frac{1}{\tau_A^2}\left(\sigma_{12}^2+\sigma_{13}^2\right)=1 \tag{8}$$

Each of the primary failure criteria for matrix and fibers is next subdivided into tensile and compressive modes. The two sets of unknown coefficients for both cases are denoted as A_m^+ , B_m^+ , A_f^+ , B_f^+ and A_m^- , B_m^- , A_f^- , B_f^- respectively. They are related to the strength parameters, determined by simple uniaxial tension and compression tests. The following notation for the strength parameters is adopted:

X_t = tensile strength in fiber direction
X_c = compressive strength in fiber direction
Y_t = tensile strength transverse to the fibers
Y_c = compressive strength transverse to the fibers

Out of four uniaxial tests only four equations can be derived to quantify the eight unknowns. By physical reasoning and due to limited material data the contribution of distinct invariants to the various failure criteria is considered as insignificant. The failure criteria then become:

Tensile fiber mode:

$$\left(\frac{\sigma_{11}}{X_t}\right)+\frac{1}{\tau_A^2}\left(\sigma_{12}^2+\sigma_{13}^2\right)=1 \tag{9}$$

Compressive fiber mode:

$$\sigma_{11} = X_c \tag{10}$$

Tensile matrix mode:

$$\frac{1}{Y_t^2}\left(\sigma_{22}+\sigma_{33}\right)^2 + \frac{1}{\tau_t^2}\left(\sigma_{23}^2 - \sigma_{22}\sigma_{33}\right) + \frac{1}{\tau_A^2}\left(\sigma_{12}^2 + \sigma_{13}^3\right) = 1 \tag{11}$$

Compressive matrix mode:

$$\frac{1}{Y_c}\left[\left(\frac{X_c}{2\tau_t}\right)^2 - 1\right]\left(\sigma_{22}+\sigma_{33}\right) + \frac{1}{4\tau_T^2}\left(\sigma_{22}+\sigma_{33}\right)^2 + \\ + \frac{1}{\tau_T^2}\left(\sigma_{23}^2 - \sigma_{22}\sigma_{33}\right) + \frac{1}{\tau_A^2}\left(\sigma_{12}^2 + \sigma_{13}^2\right) = 1 \tag{12}$$

After introducing the plane stress assumptions into the failure criteria, they are applicable for the shell model used in the numerical analysis of chapter 5. In many practical applications the transverse and the axial shear strength are assumed to be equal and will be denoted as S_c. For the purpose of the computational implementation the failure modes e_f , e_c , e_m and e_d are introduced, which play the role of indicator functions. Their sign indicates either failure or a complete elastic response of the material.

Tensile fiber mode $\sigma_{11} > 0$:

$$e_m^2 = \left(\frac{\sigma_{11}}{Y_t}\right)^2 + \left(\frac{\sigma_{12}}{S_c}\right)^2 - 1 \begin{cases} \geq 0 & \text{failed} \\ < 0 & \text{elastic} \end{cases} \tag{13}$$

Compressive fiber mode $\sigma_{11} < 0$:

$$e_c^2 = \left(\frac{\sigma_{11}}{X_c}\right)^2 - 1 \begin{cases} \geq 0 & \text{failed} \\ < 0 & \text{elastic} \end{cases} \tag{14}$$

Tensile matrix mode $\sigma_{22} > 0$:

$$e_m^2 = \left(\frac{\sigma_{22}}{Y_t}\right)^2 + \left(\frac{\sigma_{12}}{S_c}\right)^2 - 1 \begin{cases} \geq 0 & \text{failed} \\ < 0 & \text{elastic} \end{cases} \tag{15}$$

Compressive matrix mode $\sigma_{22} < 0$:

$$e_d^2 = \left(\frac{\sigma_{22}}{2S_c}\right)^2 + \left(\left(\frac{Y_c}{2S_c}\right)^2 - 1\right)\frac{\sigma_{22}}{Y_c} + \left(\frac{\sigma_{12}}{S_c}\right)^2 - 1 \begin{cases} \geq 0 & \text{failed} \\ < 0 & \text{elastic} \end{cases} \tag{16}$$

2.3 Remarks on the Hashin-Failure Criteria :

(i) The case of simple shear in laminae coordinates is not treated as a separate failure form. Rather it is included in a combination with axial stress states in the above listed failure criteria. Failure of unconfined UD-laminae specimens, due to pure shear in the laminae coordinate system, takes place in the tangential failure plane. Both matrix modes are candidates for the related failure criterion.

(ii) In none of the four failure criteria an interaction term of the form $\sigma_{11}\sigma_{22}$ between the normal stress components is present. Although, the product of the invariants I_1 and I_2 in eq.(6) accounts for the interaction of biaxial stress states, it does not appear in eq.(7) and (8). This is a consequence of the assumption that only stress components, acting on the failure planes, contribute to the failure criteria for fiber and matrix modes.

(iii) The failure criterion of Tsai und Wu [*Tsai, Wu 1971*] for general anisotropic solids is a quadratic polynomial of the form:

$$F_{ij}\sigma_i\sigma_j + F_i\sigma_i = 1$$

The term F_{12} is the only nonvanishing interaction parameter of the plane stress case. If its dimensionless counterpart F_{12}^* is set equal to $-\frac{1}{2}$

$$F_{12}^* = \frac{F_{12}}{\sqrt{X_T X_c Y_T Y_c}} = -\frac{1}{2}$$

the analogy to the von Mises yield criterion may be recovered. For isotropic materials having equal strength in tension and compression, the significance of the coupling term F_{12} in the failure criterion is well known. However, if the anisotropy among the strength parameters is very pronounced, the nondimensional interaction coefficient F_{12}^* turns out to be small. In [*Narayanaswami, Adelmann 1977*] good agreement with experimental results is reported by specifying $F_{12} = 0$.

(iv) The graphical representation of the failure criteria is given in fig. 2 for the matrix modes and in fig. 3 for the fiber modes.

(vi) The tensile fiber mode criterion eq.(15) is extremely sensitive to shear stresses for practical strength parameters. This is in accordance with off-axis tests, where the fiber direction does not coincide with the principle tensile stress component σ_x of a specimen, subjected to uniaxial loading. If the angle between the fiber direction and

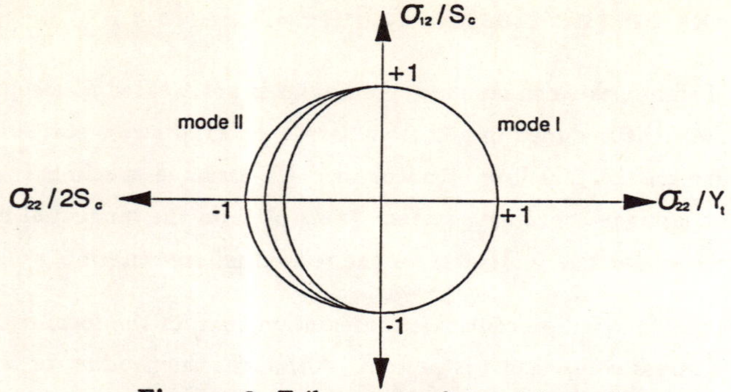

Figure 2: Failure curve for matrix modes

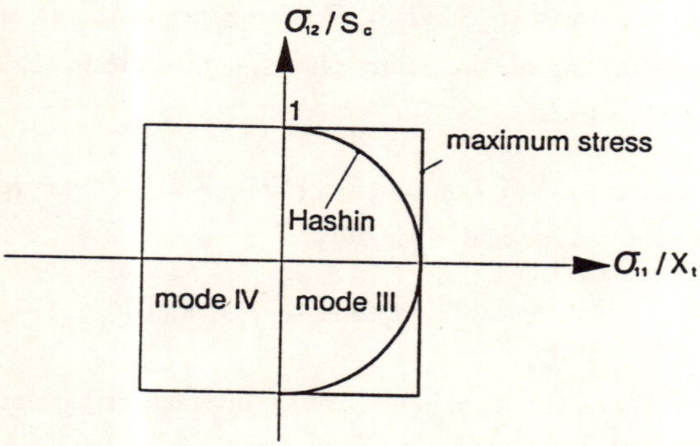

Figure 3: Failure criteria for fiber modes according to Hashin [*Hashin 1980*] and the maximum stress criterion.

the tensile stresses σ_x is denoted by θ, the fiber stresses σ_{11} and the shear stresses σ_{12} in principal material coordinates are:

$$\sigma_{11} = \sigma_x \cdot \cos^2\theta \quad ; \quad \sigma_{12} = -\sigma_x \cos\theta \sin\theta$$

Hence, the tensile fiber mode criterion becomes:

$$e_f^2 = \left(\frac{\sigma_x}{X_t}\right)^2 \cos^4\theta + \left(\frac{\sigma_x}{S_c}\right)^2 \cos^2\theta \sin^2\theta - 1 \tag{17}$$

For glass fiber/vinylester-resin the strength parameters X_t, X_c and S_c typically take the values:

$$X_t = 1120\, N/mm^2 \quad \text{and} \quad S_c = 77\, N/mm^2$$

(v) The large difference in magnitude between the tensile strength X_t and the shear strength S_c of practical laminae results in an extreme sensitivity of the tensile fiber

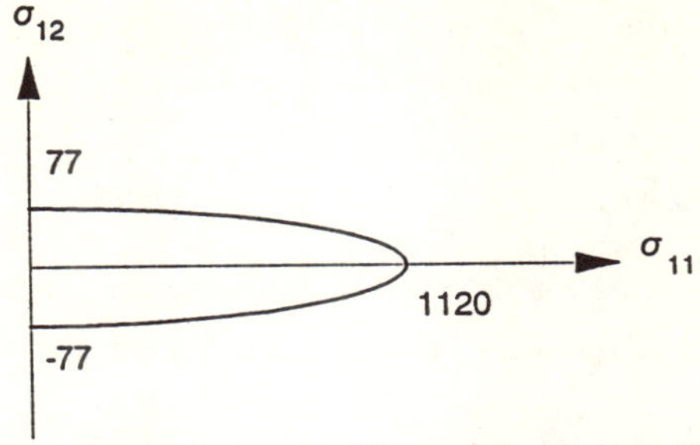

Figure 4: Failure criteria for tensile fiber mode of glass fiber/vinylester -resin.

mode criterion with respect to shear stresses. The shear contribution already amounts to 50% of e_f^2 according to eq.(17), if the angle θ between the loading direction and the fiber orientation is equal to 2.8°. For $\theta = 45°$ the fiber stresses contribute less than 0.5% to failure. Although, the sharp reduction of the load carrying capacity for composites in off-axis tests is well known, the predominant failure mode is rather shearing than fiber breakage. It is recommended to scale the shear contribution in eq.(13) by a factor β (with $0 \leq \beta \leq 1$) to account for the in situ-shear strength of laminates.

$$e_f^2 = \left(\frac{\sigma_{11}}{X_t}\right)^2 + \beta\left(\frac{\sigma_{12}}{S_c}\right)^2 - 1 \tag{18}$$

Since the in-situ shear strength S_L of the laminae in a laminate is considerably higher in comparison to the UD-laminae strength S_c , the scale factor

$$\beta = \left(\frac{S_C}{S_L}\right)^2$$

is significantly less than one.

3 Post-Failure Model For Composite Damage

General Features and Review of Earlier Work

Fiber reinforced composites generally show brittle failure characteristics on the microscale level, unless special care is taken to improve their ductility. The rigorous experimental investigation of the microstructural damage process, enabling the formulation of physically based constitutive laws, remains the ultimate objective in constitutive theory. However, as long as such a deducible approach has not become state-of-the-art in composite research, engineering practice still relies on phenomenological structure-parameter-models, fitted to experimental measures. Since any practical mesh size prevents the sufficiently accurate modeling of the extreme fine microstructural damage mechanisms, phenomenological models will always form the basis for the computational analysis of the damage evolution process.

The simplest computational artifice for the post-failure analysis of composite laminates consists of a combination of the so-called ply-discount method and a degradation rule for the material properties. The ply-discount method is based on the assumption of vanishing elastic moduli, if failure of the laminae is detected. Thus, only the remaining integer laminae contribute to the stiffness and remaining strength of the laminate cross-section. One of the earlier degradation assumptions for the stress-strain relationship during the post-failure response dates back to Petit et al. [*Petit, Waddoups, 1969*]. An analytical technique is proposed there, in order to determine the load-displacement response. The stress-strain relationship after failure is assumed to be linear in the softening part with stresses gradually diminishing along path (1) in the failed laminae - see Fig. 5.

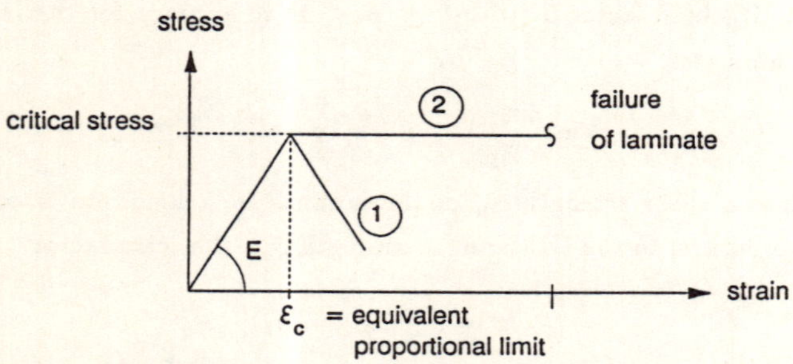

Figure 5: Simplified stress-strain diagram for the post-failure model.

Hahn et al. suggest in [*Hahn, Tsai 1974*] that any failed laminae carries its failure load in the post-critical range until the entire laminate cross-section fails. The resulting stress-strain relationship is shown as path (2) in fig. 5. This assumption has been reported to give improved results for the post-failure analysis of glass-fiber/epoxy-resin composites in comparison to the total failure hypothesis for laminae, where both, the strength and the stiffness, are set to zero after failure is encountered.

The gradient of the nominal stress-strain-diagram for the softening branch depends on numerous factors such as ductility of the matrix material, fracture toughness of the interface, pullout resistance of fractured fibers, fiber briding of cracks, processing flaws like misaligned fibers etc. If "snap-back" on the constitutive level is excluded, discontinuously vanishing stresses at failure form the lower bound for path (1).

A multiaxial generalization of degradation models - though physically correct - is not obvious; for an opposite example see [*Chang, Chang 1987*]. Continuous stiffness degradation is accounted for in [*Foerster, Kappe 1970*] after failure is indicated by the Puck-Schneider failure criteria - see [*Puck, Schneider 1969*] and [*Greenwood 1977*]. Transverse and shear failure are associated with matrix failure due to resin cracking. The transverse elasticity modulus, the shear modulus and Poisson's ratio gradually decrease to zero under increasing strains. In none of the above simple degradation models the inelastic strains and the unloading/reloading events are addressed. It is therefore not surprising, that plasticity theory is used for the numerical analysis of the nonlinear behavior of composites. In addition to a sound theoretical basis powerful computational algorithms are available for an elasto-plastic analysis.

An elastic, perfectly plastic material behavior for compression and linear elasticity for tension has been assumed in [*Zweben, 1978*] for the analysis of aramid fiber composite beams under three-point flexural loading. Elasto-viscoplasticity of the "Perzyna"-type in combination with an anisotropic yield criterion was applied to the analysis of laminated shells in [*Owen, Liu 1985*]. Elastoplasticity also formed the basis of composite analysis in [*Leewood, Doyle, Sun 1987*]. A coupled elasto-plastic damage concept has been proposed in [*Allix et.al. 1986*] for the nonlinear analysis of laminates. Stiffness degradation mainly effects the elastic modulus of the transverse fiber direction and shear modulus. The model in [*Allix et.al. 1986*] consists of an associated flow rule and anisotropic damage. Since the damage variables are governed by the transverse fiber stress and the shear stresses, the mathematical structure of the constitutive equations is compatible with Hashin's failure criteria.

Further insight into the post-failure phenomena of the laminae is gained by looking into the failure of structures, which is denoted as collapse in order to distinguish it from failure types of the material.

4 Collapse of Laminate Composite Tubes Under Impact Loading

A series of low velocity impact tests on various geometrical shapes of test specimens complement the post-failure prediction in [*Hahn, Tsai, 1974*], if the strength is interpreted as a threshold, corresponding to the resistance force in the specimen during impact. Although quantitative data in impact damage are rather limited, there is some indication for a sufficiently constant force-time history response of prismatic specimens under axial impact, provided ideal crash conditions [*Keal 1983; Price, Hull 1987; Czaplicki 1990; Maier 1990*] are given to justify the stress-strain-path (2) for the strength in compression. In contrast to the failure of laminate composites under statical loading there is neither a well defined failure state nor a sudden ("discontinuous") change from the integer to the failed material state of composite structure under low velocity impact.

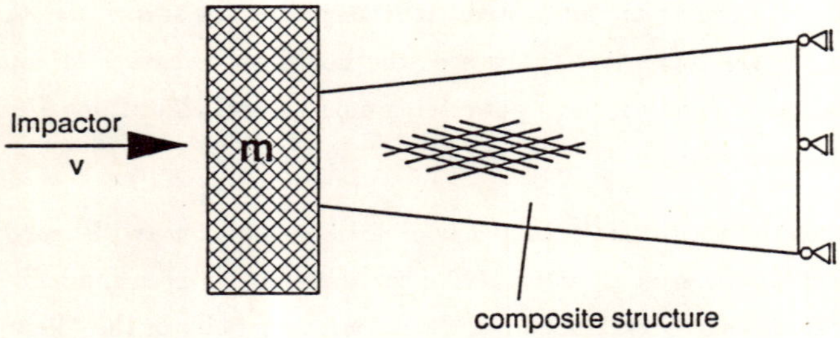

Figure 6: Impactor (mass m, velocity v_0) and target (laminated composite material).

The basic collapse phenomena have been investigated on the simple model of comparing tubes [*Keal 1983; Maier 1990*] under axial impact loading. Three different forms of collapse can be distinguished [*Price, Hull 1987*], depending on the wall-thickness, geometrical shape, stacking sequence of laminae, brittleness of material, "trigger type" [*Thornton, Jeryan 1988*] etc..

a) Tubes with small wall thickness collapse by progressive folding, accompanied by the formation of sharp hinges, similar to the collapse of metal or thermoplastic tubes.

b) Thick-walled tubes with reinforced ends fail by catastrophic fracture after the load reaches the compressive strength of the material.

c) Thick-walled tubes fail by progressive, crushing when the end of the tube is chamfered to create a local region of high stress, causing progressive deterioration of the materials. Failure is initiated at the chamfer, acting as a local imperfection. A highly fragmented zone then propagates down the tube. Away from the crash zone the material retains its original integrity.

As already pointed out, for given material properties, the wall-thickness becomes the decisive factor, whether the collapse form is of geometrical nature (buckling and folding as in case a)) or is due to material failure as in cases b) and c). This classification is well known for elastic-plastic buckling of metal structures and is standard design practice for shells, constructed of sufficiently ductile materials. Within the class of material induced collapse, form b) is fundamentally different from c), since its failure is sudden and uncontrollable, whereas the latter is progressive with a "predictable development". The damage zone in case b) is localized, as it is typical for brittle type of failure - see fig. 7.

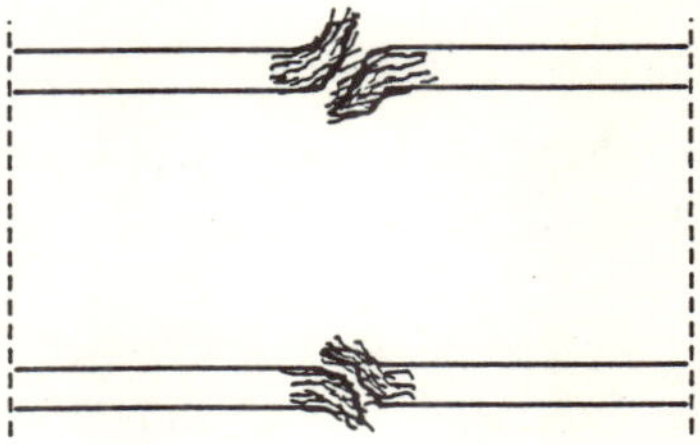

Figure 7: "Catastrophic collapse" of tube due to localized failure.

The damage zone in the case of progressive crushing , as outlined in case c), is depicted in fig. 8. The entire crash front constitutes of different states of damage.

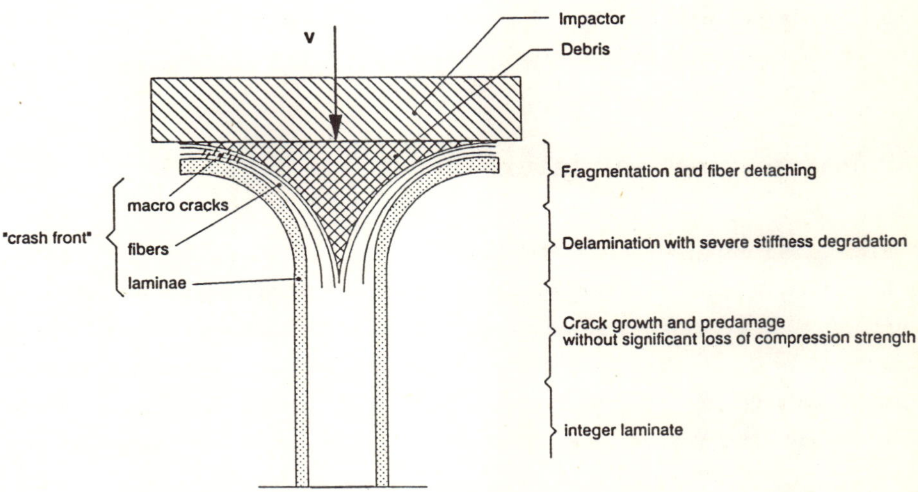

Figure 8: Crash front with progressive damage propagation in glass/fiber-epoxy laminae.

Severe stiffness degradation occurs in the crash front towards the contact zone between impactor and target, where a state of complete fragmentation up to the degree of pulverization - see for example [*Thornton 1988*] - is developed. As a consequence the stiffness properties of the homogenized continuum tend to vanish, a phenomenon known as erosion in high velocity impact problems. Erosion is defined as the total failure of the material. Material modeling for impact problems falls into three different categories:

1. **Volumetric response, or resistance to compressibility**: The equation of state expresses the relationship between pressure, volume and internal energy. However, this is insignificant for the current issue, but is of primary concern in hypervelocity impact dynamics.

2. **Resistance to distortion**: The strength parameters are the dominating variables and need to be carefully introduced into the constitutive equations - see chapter 2.

3. **Increasing inability to transmit stress as damage accumulates**: Tensile stress components tend to zero after a prescribed tensile fracture stress in the process of damage accumulation is reached. Total failure may be defined, if a critical value of damage K, calculated from the time integral, is exceeded.

$$K = \int_0^t (\sigma - \sigma_0)\, dt$$

The interested reader may be referred to an excellent review–article by [*Asay, Kerley 1987*] for further information on material behavior due to dynamic loading.

5 Numerical simulation

The numerical examples within the research project CARMAT 2000 [*Matzenmiller, Schweizerhof, Maier 1990*] are chosen for two purposes. First to compare the behavior of composite and steel cylinders in the specific application as crash elements in automotive structures, for it is referred to the above cited report. Second to investigate the applicability of a standard shell model for the simulation of the complex failure behavior of cylinders and cones made out of composite material under longitudinal impact loading.

Remark: The authors are aware of the fact that the failure of the crash elements is accompanied by delamination to a wide extent, which is observed also in the experiments [*Drumm, Klipfel, Maier 1989*]. Delamination due to normal stresses cannot be properly covered by the kinematical model used in standard shell theory, as normal stresses perpendicular to the shell reference surface are not present in the mechanical model. Shell models are intend to predict the coupled membrane and bending response of thin structures, which is often not the main effect in structures under such particular loading as investigated in this project. As far as a three-dimensional analysis demonstrated [*Maier 1990*] the stresses perpendicular to the shell surface are present and can be modeled using three-dimensional continuum theory and solid elements. However, for realistic applications such an analysis, which requires at least 4 ÷ 6 solid elements through the thickness of the shell, is not very efficient and involves considerable computational effort. Thus, the shell model is used throughout the numerical investigations despite its limitations. The aforementioned effects due to delamination are at least partially incorporated into the material model. The merits and limits of this approach will be discussed in the conclusions to the numerical examples.

5.1 Crashworthiness behavior of a composite cylinder

Within the research project a cylinder consisting out of 13 layers of glass-fiber vinylester laminae with a stacking sequence of $(+/-45°)_6 / 90°$ is analyzed. The geometrical quantities and the specific weight of the cylinder are taken directly from the corresponding experiment [*Müller 1989*]; the material data (table 1.) are determined via statical tests. The perfect geometry of the cylinder (radius $r = 23.4\,mm$, length $l = 100.16\,mm$, wall thickness $t = 2.6\,mm$) is perturbed by the addition of a geometrical imperfection with a maximum amplitude w_0 (ratio $w_0 / t = 0.2$).

Young's modulus in fiber direction	$E_x = 45847\,N/mm^2$
Poisson's ratio in plane ∥ to fiber direction	$\nu_{xy} = 0.26$
Young's modulus ⊥ to fiber direction	$E_y = 17506\,N/mm^2$
Poisson' ratio in plane ⊥ to fiber direction	$\nu_{yz} = 0.331$
Shear modulus in plane	$G_{xy} = 8631\,N/mm^2$
Transverse shear modulus ∥ to fiber direction	$G_{yz} = 6574\,N/mm$
Transverse shear modulus ⊥ fiber direction	$G_{xz} = 8631\,N/mm^2$
Shear strength	$S_c = 77\,N/mm^2$
Tensile failure strength in fiber direction	$X_t = 1120\,N/mm^2$
Compressive failure strength in fiber direction	$Y_c = 900\,N/mm^2$
Tensile failure strength ⊥ to fiber direction	$Y_t = 39\,N/mm^2$
Compressive failure strength ⊥ to fiber direction	$Y_t = 39\,N/mm^2$

Table 1: Material data for each UD-lamina of the composite cylinders

The application of the material data, taken from tests on a unidirectional lamina, for a composite laminate is depending on the assumption of a perfect homogeneous material for the complete stack of plys. Fabrication flaws and misalignments result in the failure of the ply at its weakest location. From the viewpoint of statistics it is rather unlikely to encounter the severest material defects of each lamina at the same cross section of a laminate. Thus it can be expected, that the in-situ strength for tension and compression of a single layer can be assumed to be larger within the laminate than tested as a single UD-layer. This assumption is checked by a numerical increase in strength by a factor of 1.5.

The degradation in nominal strength due to the diffusive crack formation in the crash front (and contact zone) and the delamination of stacks of laminae is taken into account by a softening factor. This factor is defined as the ratio of the average internal force in the test cylinder under impact loading - obtained in the experiment - to the analytical ultimate force under static loading [*Maier 1990*]. It is chosen to account for the apparent reduction of the laminae strength due to delamination. The average internal force in the experiment is 33 kN. The static analysis, based on a Tsai-Wu criterion as well as on a maximum stress criterion gives a value of 55 kN. Hence the estimate for the softening factor is calculated to $33/55 = 0.6$. For efficiency reasons within the explicit time integration scheme used, elements are removed. Thus "elimination" of an element takes place when the critical time step [*Courant etal. 1928*] has reached a quarter of the initial value, which is almost equivalent to 75% compression of the original element. A more realistic failure could be modeled, if the degradation in stiffness, representing a strongly crushed material, is also used for a decrease in the sound wave speed. This will be considered in future studies.

In order to get a controlled failure and to reduce the peak forces found for cylinders with perfect geometry, a trigger is introduced at the impacted end of the cylinder by using trigger angles $\alpha = 15°, 35°$ and $75°$. In the experimental investigations the trigger is made as shown in figure 9, whereas in the analysis the trigger is modeled using elements with a reduced thickness.

Discretization: The full cylinder is modeled by bilinear shell elements based on Reissner/Mindlin theory [*Hughes, Liu 1981; Belytschko, Lin, Tsay 1984; Hallquist, Benson 1986*]. To represent the appropriate deformation pattern the length of the elements close to the expected failure zone is reduced. Thus a mesh with 15 elements of $4mm$ length and 20 elements of $2mm$ length in longitudinal and 48 equally spaced elements in circumferential direction is used. The trigger is modeled as shown in fig. 9. For an angle of $\alpha = 35°$ the trigger length is calculated as $l_T = 4\,mm$; the thickness of the first two rows of elements is reduced to $t_1 = 1.00\,mm$ and to $t_2 = 1.8\,mm$ respectively. For an angle of $\alpha = 75°$ only the thickness of the first row of elements is reduced to $t_1 = 1.00\,mm$. A fur-

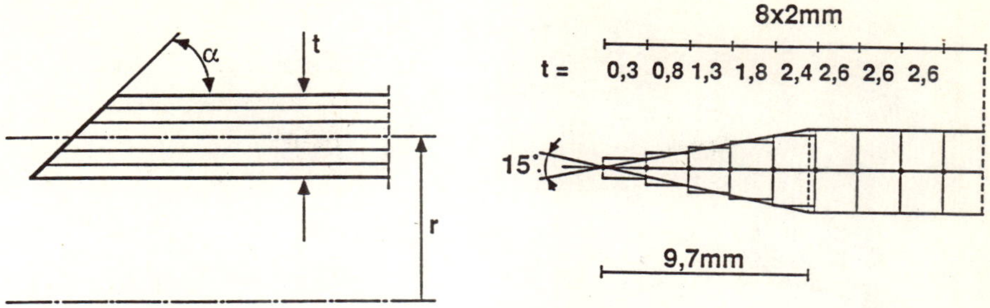

Figure 9: Trigger in the experiment and trigger discretization with finite elements

ther mesh refinement was tested; no mesh size dependency was found; mesh convergence seems to be reached to sufficient accuracy.

The laminate is modeled by 13 layers of uniform thickness (0.2 mm). Thus the integration points are chosen to be equidistant in the center of each layer and equal weighting factors for each point were used. The hourglass control within the uniformly reduced integrated elements follows the proposal of Belytschko and coworkers [*Belytschko, Ong, Liu, Kennedy 1984*] recommending "stiffness" control.

Loading and boundary conditions: The frontal end of the cylinder with the trigger can slide frictionless on the area of contact, which is achieved in the experiments by a smooth steel surface of the impactor. The impactor itself is assumed to be a rigid contact surface (mass $m = 19.83\,kg$) moving with an initial velocity of $v = 7.89\,mm/sec$ as in the experiments. In order to account for further imperfections of the experiment the angle of impact deviates from the cylinder axis by 0.573°.

Objective of the study and choice of parameters: The rather small radius/thickness ratio of $r\,/\,t = 23.4\,/\,2.6 = 9$ leaves imperfection sensitivity in the case of buckling out of consideration. As the elastic buckling stresses are far beyond the material strength, buckling is not issue too. The presented results mainly serve the purpose of a general validation of the introduced failure models for crashworthiness analysis. The following effects and quantities are of interest in this example:

- Maximum internal force and its complete time-history diagram. In particular a constant level of the internal force during the deformation process is to be achieved.
- Size of deformation compared to the dissipated energy to find the effectiveness of the crash element.

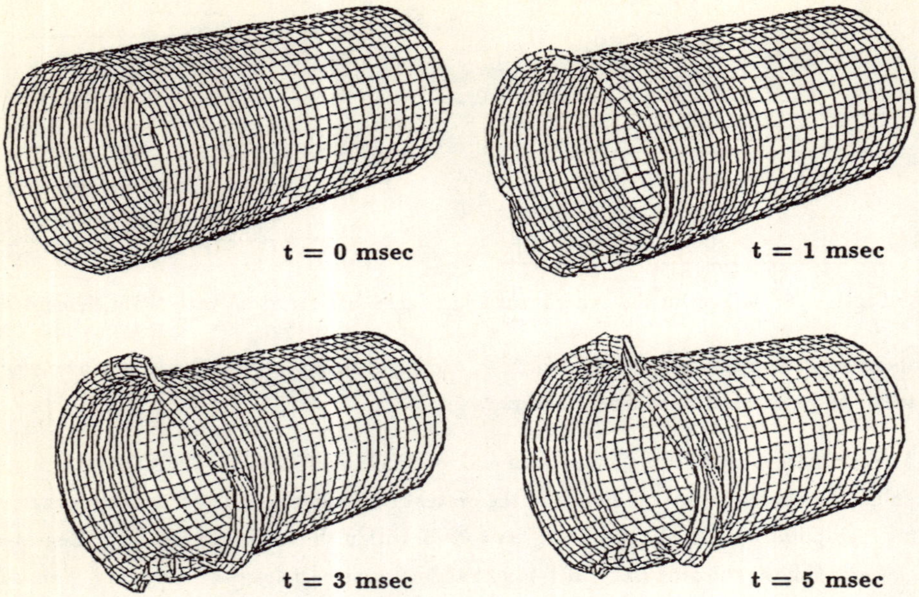

Figure 10: Deformation of composite cylinder under impact loading

- Influence of material strength on the maximum of the internal force (= resisting force) and on the load-deformation curve.
- Influence of the trigger on the gradient and on the maximum of the load-deformation curve. - Variations of trigger angles $\alpha = 15°, 35°, 75°$.
- Partial reduction of tensile strength in fiber direction after failure compared to full reduction after failure (total failure hypothesis).

Numerical results: The static ultimate load is found to be $P_{crit} \approx 49.2\,kN$. The corresponding failure mode is failure of the matrix under compression with an average value of $130\,N/mm^2$ for the axial membrane stress. For the dynamic analysis the finite element program LS-DYNA3D [*Hallquist 1990; Hallquist et.al. 1985*] is used, to which the failure models for the composite material were added. The explicit time integration scheme in LS-DYNA3D is well suited for highly nonlinear problems, involving wave propagation and contact, which are dominating the crashworthiness behavior. The run-time for the compression wave in the longitudinal direction is estimated to $T = 50\,\mu sec$ with an average sound speed of about $C = 4000\,m/sec$ for the laminate. The initial amplitude of the compression wave has a value of roughly $20 \div 25\,kN$ ($\sigma = 55 \div 70\,N/mm^2$), see fig. 11. Some test analyses showed that the failure occurred in an experimentally not observed

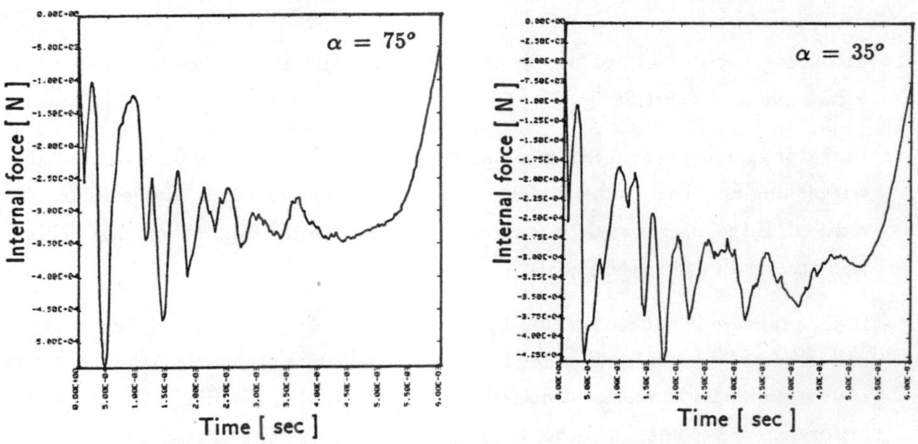

Figure 11: Force vs. time curve for trigger angles of 75° and 35°

fashion in the fiber tension mode, if the shear term as suggested in [*Hashin 1980*] is involved. Thus this term is replaced by the simple maximum stress criterion, which gives results rather close to the experimentally measured values.

Trigger angle	Max. internal force	At time
	[kN]	[m sec]
75°	54.1	0.46
35°	43.1	0.40
	42.8	1.75
15°	38.5	1.35
	41.3	2.30
Experiment	35.8	1.40
15°	37.0	2.20

Table 2: Maximum internal forces, time of occurrence dependent on trigger form

Maximum internal forces: The crushing behavior, observed in the experiments, is geometrically represented by a partial folding of the shell at the loaded end. Elements are compressed and folding (see fig. 10), all types of failure modes in the laminae are found. A strong relation between the developments of the folds and the oscillations in the time-history curve is obvious (see figs. 12 and 13). In table 2 the maxima of the internal forces are listed versus their time of occurrence compared to the results of the experiments. The following conclusions can be drawn:

- Increasing the strength of the ply compared to a simple layer leads to unrealistic values and is not justified.
- Larger trigger angles lead to larger maximum values of the internal force than smaller trigger angles. The gradient of the load time history curve is also considerably reduced, if the trigger angle α is changed from 75° to 15° (fig. 11 and 12). This is also observed in the experiments.
- If fiber rupture is not represented by a total failure of the layer (complete strength and stiffness loss) but is accounted for as a 50% reduction of the strength parameters – a coarse model to represent delamination in the case of bending – the force-time history curve becomes smoother than with a total reduction to zero (see fig. 12).

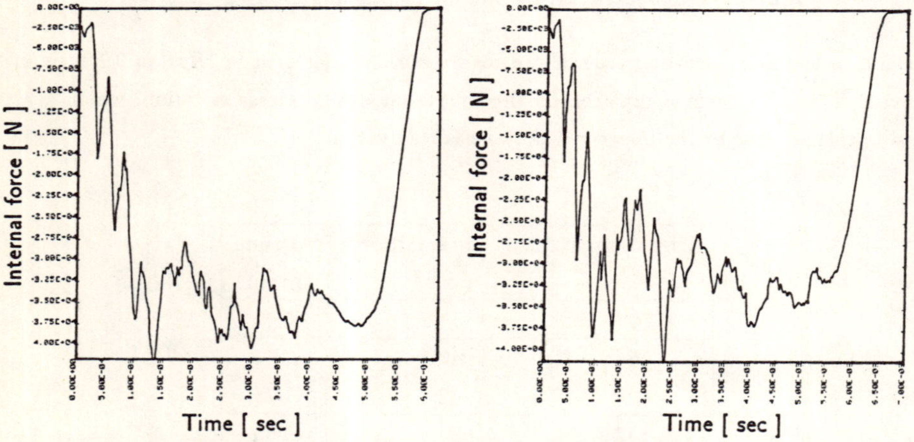

Total failure hypothesis — Partial failure ≡ 50% strength reduction

Figure 12: Force vs. time curve for a trigger angle of 15 °, total failure hypothesis vs. reduction of strength

- Mesh dependency is expected to be of concern for the current subject. The mesh dependence of the results is superimposed on the response due to folding and is known to cause local "spikes" in the response diagrams. However, for the current analysis it seems to be less pronounced, since all response curves did not change significantly after mesh refinement. The element size had just to be adapted to allow for the folding as described above. This was confirmed also by other numerical tests.

Force deformation curve and dissipated energy: The force deformation curve and the force-time history curve of the calculation with a trigger angle of 15° and the above mentioned partial reduction of tensile strength of the fiber in the case of failure shows good agreement with the experimental data. The final axial deformation is almost exactly predicted. A similarly good agreement can be found comparing the curves of the dissipated energy versus time and axial deformation, see figs. 13 and 14. The expected energy absorption [*Thornton, Jeryan 1988; Thornton, Edwards 1982*] is fully achieved.

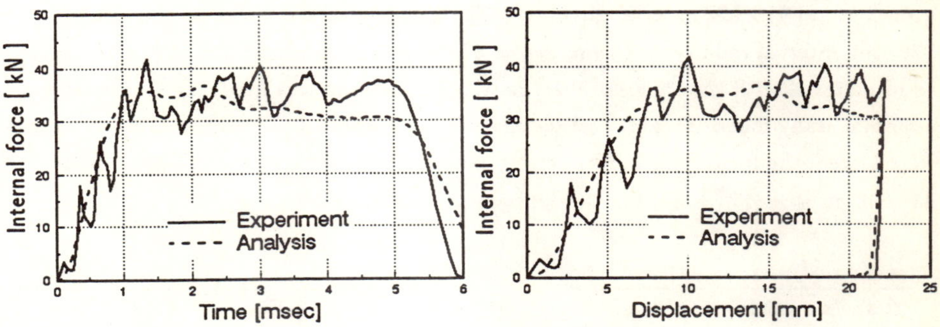

Figure 13: Force vs. time and displacement - comparison of analysis and experiment

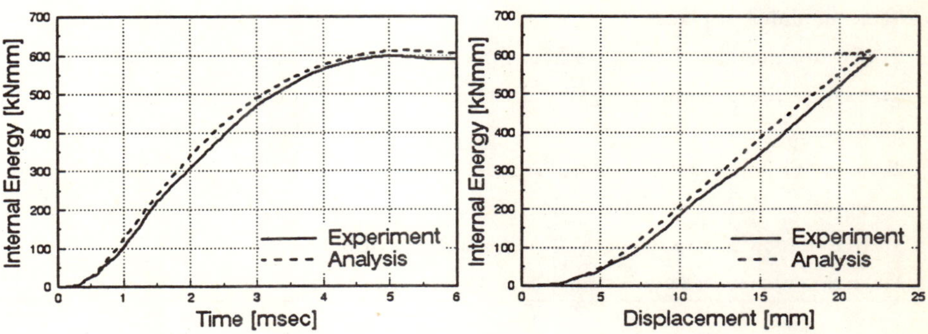

Figure 14: Energy vs. time and displacement - comparison of analysis and experiment

In figure 10 the development of the deformation and crushing of the cylinder by folding of the partially failed lamina is shown. Facing the limits of the kinematical assumptions in the shell theory used and the rather coarse material model, the results for the global quantities are surprisingly close to the experimental values.

5.2 Crashworthiness Behaviour of a Composite Conical Shell

The rather sharp gradient of the load deformation curve found for cylindrical specimens is known to be reduced by changing the shape into a conical form. The corresponding experiments [*Price, Hull 1987*] show a less steep gradient and a continuous increase in the average load, resulting in a smoother deceleration of the impacting vehicle than with cylindrical crash elements. For a better evaluation of the capabilities of the numerical analysis the finite element calculations are performed prior to the experiments.

The conical shell is designed not to fail by buckling, because of the low energy dissipation in this case. In addition the internal force, transmitted during the crash, should not grow beyond 100 kN in axial direction. The geometrical data chosen are: length l = $220\,mm$, internal radii $r_1 = 15\,mm$, $r_2 = 57.2\,mm$, wall thickness $t = 4.41\,mm$. Epoxy-resin material with a 55 % glass fiber/matrix volume ratio was used. The material data, experimentally found from statical tests, are listed in table 3. Several stacking variations were tested; the focus here is on a symmetrical laminate with 26 layers (90° / ± 10° / ± 45° / ± 80° / ± 45° ± 80° / ± 45°)$_S$; layer thickness $0.169\,mm$.

Young's modulus in fiber direction	E_x	=	$42300\,N/mm^2$
Poisson's ratio in plane ‖ to fiber direction	ν_{xy}	=	0.2729
Young's modulus ⊥ to fiber direction	E_y	=	$15600\,N/mm^2$
Poisson's ratio in plane ⊥ to fiber direction	ν_{yz}	=	0.345
Shear modulus in plane	G_{xy}	=	$6150\,N/mm^2$
Transverse shear modulus ‖ to fiber direction	G_{yz}	=	$5800\,N/mm$
Transverse shear modulus ⊥ fiber direction	G_{xz}	=	$6150\,N/mm^2$
Shear strength	S_c	=	$63\,N/mm^2$
Tension failure strength in fiber direction	X_t	=	$1120\,N/mm^2$
Compression failure strength in fiber direction	Y_c	=	$1020\,N/mm^2$
Matrix tension failure strength ⊥ to fiber direction	Y_t	=	$45\,N/mm^2$
Matrix compression failure strength ⊥ to fiber direction	Y_t	=	$142\,N/mm^2$

Table 3: Material data for each UD-lamina of the composite conical shell

Boundary conditions and loading: The impactor (mass 77.18 kg, velocity $v_0 = 15.56\,m/sec$) hits the cone at the end with the small radius. The chosen velocity is the standard value, used in crash tests for automotives. The boundary conditions of the numerical analysis are identical to the cylinder example, except that the radial outward deflection at the larger end of the cone is restrained.

Discretization, trigger: 48 equally spaced shell elements in circumferrential and 35 elements in longitudinal direction are used. The length of the elements in axial direction starts with $l_e = 2.00\,mm$ at the loaded end and is increased in each element row by 0.25 mm towards the other end. In some start-up analyses rather high internal forces were found for the perfect conical shell. Thus a trigger zone with the size equal to 4 element rows is modelled by linearly increasing the thickness from $t_1 = 1\,mm$ up to $t_4 = 4\,mm$. This reduces the peak internal forces considerably, despite observations in the experiments [*Price, Hull 1987*] indicate that for a controlled failure process no additional imperfections seem to be needed.

Numerical results: The resulting force-time history curve is given in figure 15, which is to a wide extend in good correlation to known experimental results. The strong oscillations, found in the numerical analysis, are also visible in experiments with test specimen of different plys, stacking sequences and geometrical shapes [*Maier 1990*]. The decrease in the average internal force after 2 msec, however, was not observed in these and other experiments [*Price, Hull 1987*]. This effect is a consequence of the overall bending, developing after the maximum load is reached, which is accompanied by an ovalization of

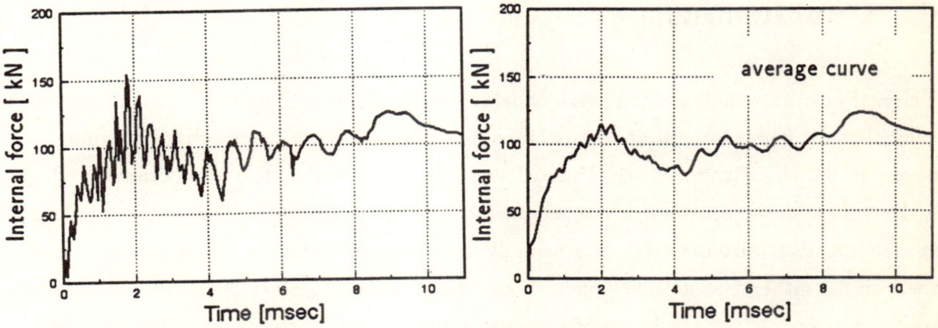

Figure 15: Force vs. time diagram, composite conical shell

the cross section, see fig. 16. The latter seems to be at least partially the (expected) result of the kinematical restrictions of the standard shell models, which do not allow a geometrical description of delamination. However we hope to get further insight into the real behavior, after the experiments are performed. Further numerical investigations

- also currently performed - are hampered by the fact, that even on vector computers a considerable amount of CPU time is needed for a single analysis.

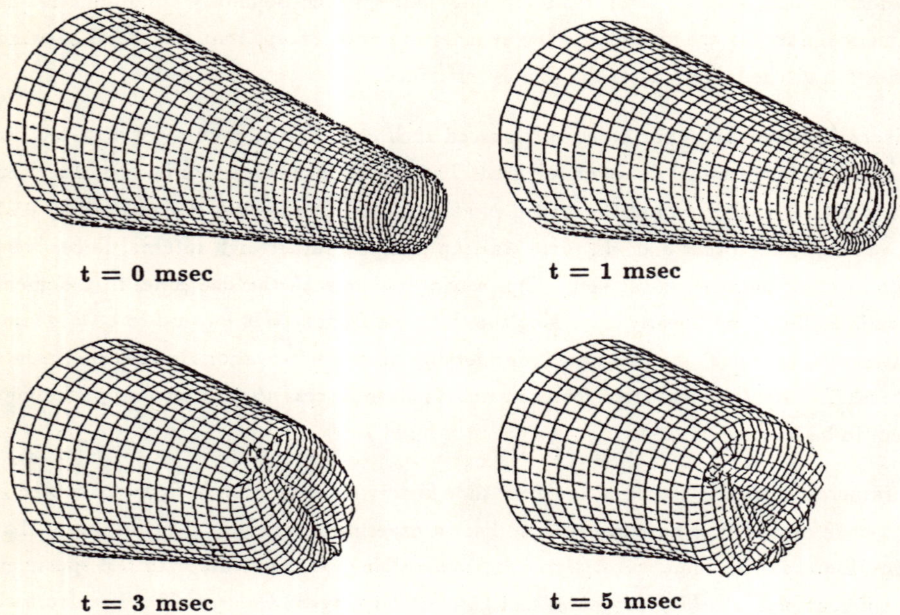

Figure 16: Deformation and crushing of composite conical shell

6 Conclusions

The rather simple shell and material model for the simulation of failure for cylindrical and conical shells under longitudinal impact loading gave results fairly close to experimentally observed values, if the numerical model was able to represent the local failure by local folding. Improvements towards damage theory, replacing the simple, "ad hoc" degradation models, possibly also material data from dynamical tests and advanced kinematics in the shell model either via improved shell elements including more variables – for a first step see [*Simo et.al. 1989*] – or layers of elements are necessary. For such an analysis, however, a rather massive computer power is required.

Acknowledgements: The research was conducted within the frame of the EUREKA project "CARMAT 2000", initiated and directed by Dr.-Ing. M.Maier, BASF AG, Ludwigshafen, Germany. The continuing support and cooperation of Dr.Maier and the BASF AG are gratefully acknowledged.

References

[1] Allix, O.; Gilletta, D.; Ladeveze, P.: Non Linear Mechanical Behavior of Laminates, in: Brittle Matrix Composites, ed. Brandt, A. M. and Marshall, I. H., pp. 227-240, Elsevier Applied Sciences, 1986.

[2] Asay, J.R.; Kerley, G.I.: The Response of Materials to Dynamic Loading, Int.J.Impact Eng., Vol.5, 1987, pp. 69-99

[3] Belytschko, T.; Lin, J.I.; Tsay, C.-S.: Explicit Algorithms for the Nonlinear Dynamics of Shells, CMAME, Vol. 42, 1984, pp. 225 - 251.

[4] Belytschko,T.; Ong J.S.-J.; Liu,W.K.; Kennedy,I.M.: Hourglass Control in Linear and Nonlinear Problems, Comp. Meth. Appl. Mech. Eng. 43(1984) 251-276.

[5] Chang, F. K.; Chang, K. Y.: Post Failure Analysis of Bolted Composite Joints in Tension or Shear-Out Mode Failure, J. Composite Materials, Vol. 21, pp. 809-823, 1987.

[6] Courant, R., Friedrichs, K. und Lewy, H.: Über die partiellen Differentialgleichungen der mathematischen Physik, Mathematische Annalen 100, 1928, pp. 32 - 74.

[7] Christensen, R. M.; Swanson, S. R.: Evaluation of a New Failure Criterion for Fibrous Composite Materials, in: Mechanics of Composite Materials, eds. G. J. Dvorak, N. Laws, AMD-Vol. 92, pp. 63-68, 1988.

[8] Christensen, R. M.: Tensor Transformations and Failure Criteria for the Analysis of Fiber Composite Materials , J. Composite Materials, Vol. 22, No. 6 pp. 874-897, 1988.

[9] Czaplicki, M. J.; Robertson, R.E.; Thornton, P. H.: Non-Axial Crushing of E-Glass/Polyester Pultruded Tubes, J. of Comp. Materials, Vol. 24, Oct. 1990, pp. 1077-1100.

[10] Drumm,R.; Klipfel,W.; Maier,M.: Prüfen des Impact- und Crashverhaltens von Polymer-Strukturen, Journal Materialprüfung, Vol. 31, 1989, Carl-Hanser-Verlag, Munich.

[11] Foerster, R.; Kappe, W.: Spannungs- und Bruchanalyse an Glasfaser/Kunststoff-Wickelkörpern, Kunststoffe, Vol.60, 1970 pp 1053-1059.

[12] Greenwood, J.H.: German Work on GRP Design, Composites, Vol.8, pp 175-184, 1977.

[13] Hahn, H. T.; Tsai, S. W.: Nonlinear Elastic Behavior of Unidirectional Composite Laminate, Journal of Composite Materials, Vol. 7, 1973, pp. 102 - 110.

[14] Hahn, H. T.; Tsai, S. W.: On the Behavior of Composite Laminates After Initial Failures, Journal of Composite Materials, Vol. 3, 1974, pp. 280 - 305.

[15] Hahn, H. T.; Erikson, J. B.; Tsai, S. W.: Characterization of Matrix/Interface Controlled Strength of Unidirectional Composites, in: Fracture of Composite Materials, eds. G. C. Shi and V. P. Tamuz, Martinus Nijhoff, pp. 197-214, 1982.

[16] Hallquist, J. O.: LS-DYNA3D User's Manual (Nonlinear Dynamic Analysis of Structures in Three Dimensions), Lawrence Livermore Software Technology Corporation, Livermore CA, 1990.

[17] Hallquist, J. O.; Benson, D. J.: A Comparison of an Implicit and Explicit Implementation of the Hughes-Liu Shell, in: Finite Element Methods for Plate and Shell Structures, eds.: T. J. R. Hughes und E. Hinton, Pineridge Press, Swansea, U.K., 1986.

[18] Hashin, Z.: Failure Criteria for Unidirectional Fiber Composites, Journal of Applied Mechanics, 1980, Vol. 47, pp. 329 - 334.

[19] Hill, R.: A Theory of the Yielding and Plastic Flow of Anisotropic Metals, Proc. Royal Soc. London, series A, Vol.193, 1948, pp.281-297

[20] Hughes, T. J. R.; Liu, W. K.: Nonlinear Finite Element Analysis of Shells, Comp. Meth. Appl. Mech. Eng., Vol 26, 1981, pp. 331 - 362.

[21] Hull, D.: Energy Absorption of Composite Materials under Crash Conditions, in "Progress in Science and Engineering of Composites", eds. T. Hayushi, K. Kawata and S. Umekawa, pp. 861 - 870, Tokio, 1982.

[22] Keal, R.: Post Failure Energy Absorbing Mechanism of Filament Wound Composite Tubes, Ph. D. thesis, Department of Metallurgy and Materials Science, University of Liverpool, Sept. 1983.

[23] Leewood, A.R.; Doyle, J.F.; Sun, C.T.: Finite Element Program for Analysis of Laminated Anisotropic Elastoplastic Materials, Comp.& Structures, Vol.25, 1987, pp. 749-758.

[24] Maier, M.: Experimentelle Untersuchung und numerische Simulation des Crashverhaltens von Faserverbundwerkstoffen. PhD Thesis, University Kaiserslautern, December 1990.

[25] Matzenmiller, A.; Schweizerhof, K.; Maier,M.: Numerische Simulation des Schädigungsverhaltens von Faserverbundwerkstoffen, Zwischenbericht zum CARMAT-2000 Projekt, CAD-FEM GmbH, Ebersberg, BASF AG, Ludwigshafen, 1990.

[26] Müller, M.: Experimentelle Untersuchung und Optimierung des Crashverhaltens von einfachen Strukturelementen aus Faserverbundwerkstoffen, Diploma Thesis, University Kaiserslautern, 1989.

[27] Murray, Y. D.: Theory and Verification of the Fiber Composite Damage Model Implemented in DYNA3D, Technical Report, APTEC Inc., San Jose, CA, December 1989.

[28] Narayanaswami, R.; Adelman, H. M.: Evaluation of the Tensor Polynomial and Hoffman Strength Theories for Composite Materials, Vol. 11, 1977, pp. 366 - 377.

[29] Neale, K. W.; Labossière, P.: A General Failure Theory for Fiber–Reinforced Composite Laminae, AMD–Vol 104 of the ASME, "Mechanics of Plastics and Plastic Composites", Editor: V. K. Stokes, 1989, pp. 113 - 122.

[30] Owen, D.R.J.; Guo Quiang Liu: Elastic-viscoplastic Analysis of Anisotropic Laminated Plates and Shells, Proceedings of the NUMETA 1985 Conference, Swansea, 1985.

[31] Petit, P. H.; Waddoups, M. E.: A Method of Predicting the Non–Linear Behavior of Laminated Composites, Journal of Composite Materials, Vol. 3, 1969, pp. 2 - 19.

[32] Price, J. N.; Hull, D.: Axial Crushing of Glass Fiber–Polyester Composite Cones, Composites Science and Technology, Vol. 28, 1987, pp. 211 - 230.

[33] Puck, A.: Festigkeitsberechnung an Glasfaser/Kunststoff–Laminaten bei zusammengesetzter Beanspruchung, Kunststoffe, Bd. 59, Heft 11, 1969.

[34] Puck, A.; Schneider, W.: On Failure Mechanisms and Failure Criteria of Filament--Wound Glass-Fiber/Resin Composites, Plastics and Polymers 37, 1969, pp. 33 - 44.

[35] Rosen, B.W.; Dow, N.F.: Mechanics of Failure of Fibrous Composites, in: Fracture - An Advanced Treatise, eds: Liebowitz, H., Academic Press, 1972, pp. 661-674.

[36] Simo, J. C.; Fox, D. D.; Rifai, M. S.: Geometrically Exact Stress Resultant Shell Models: Formulation and Computational Aspects of the Nonlinear Theory, in: Analytical and Computational Models of Shells, eds. Noor, A. K. und Belytschko, T. und Simo, J. C., Winter Annual Meeting of the ASME, ASME N. Y., CED–Vol. 3, 1989.

[37] Thornton, P. H.: Energy Absorption in Composite Structures, J. Composite Materials, Vol. 13, p. 247-262, 1979.

[38] Thornton, P. H.: The Crush Behavior of Glass Fiber Reinforced Plastic Sections, Composites Science and Technology, Vol. 27, 1986, pp. 199 - 223.

[39] Thornton P. H.; Jeryan, R. A.: Crash Energy Management in Composite Automotive Structures, Int. J. Impact Engng. Vol. 7, Nr. 2, pp. 167-180, 1988.

[40] Thornton, P. H.; Edwards, P. J.: Energy Absorption in Composite Tubes, J. Composite Materials, Vol. 6, 1982, pp. 521 - 545.

[41] Tsai, S.W.; Wu, E.M.: A General Theory of Strength for Anisotropic Materials, J.Composite Materials, Vol.5, 1971, pp. 73-96

[42] Zweben, C.: The Flexural Strength of Aramid Fiber Composites, J. Composite Materials, Vol. 12, pp. 422-430, 1978.

Local A Posteriori Error Estimates and Numerical Results for Contact Problems and Problems of Flow Through Porous Media

C. Y. Lee
J. T. Oden
M. Ainsworth

Texas Institute for Computational Mechanics
The University of Texas at Austin
Austin, Texas, U.S.A.

Summary

A new *a posteriori* error estimator is introduced and applied to variational inequalities occurring in problems of contact and flow through porous media. One issue in constructing element-wise *a posteriori* error estimates is the localization of the error. In the present work, this is accomplished by a special mixed formulation in which continuity conditions at interfaces are treated as constraints. An upper bound of the local energy norm of the error is obtained using a duality approach. This approach leads to error indicators which provide rigorous upper bounds of the element errors. A discussion of a compatibility condition for the well-posedness of the local error analysis problem is given. Three numerical examples are solved to check the compatibility of the local problems and convergence of the effectivity index both in a local and a global sense, with respect to local refinements.

1 Introduction

Consider the problem of finding a function u such that

$$-\Delta u \geq f \;:\; u \geq \varphi \qquad \text{in } \Omega$$
$$(\Delta u - f)(u - \varphi) = 0 \quad \text{a.e. in } \Omega$$

subject to $u = g$ on $\partial\Omega$, where $\Omega \in \mathbb{R}^2$ is assumed to be a bounded domain with Lipschitz boundary $\partial\Omega$ and φ is a given function. This problem is the strong form of the obstacle problem [3]. Similar formulations arise in the modeling of fluid flow through porous media [5] or the unilateral contact of elastic bodies [4]. The major difficulty in these problems is that there is a contact region or free boundary which is unknown *a priori.* This difficulty renders these types of problems nonlinear. However, these free-boundary problems can be formulated as variational inequalities which can be solved for u; then unknown contact region or boundary can be calculated *a posteriori.*

The uniqueness and existence of the solutions are proved in Ainsworth [1] for the case of a Signorini-Obstacle type variational inequalities, and in Kikuchi and Oden [4] for the case of Signorini type inequalities. The finite element discretization of variational inequalities is discussed in [4].

In the present work, adaptive finite element methods are assumed, the key component of which is an *a posteriori* error estimator which can be used to identify the elements where the approximation is poor. We attempt to develop an acceptable error estimator which satisfies the following criteria:

1. the estimated error provides an upper bound to the true error and is very close to the exact error

2. property 1) holds even element–wise

3. the error estimation technique is easily incorporated into existing finite element codes

4. the basis from which the estimator is derived is well founded mathematically

Property 2) is necessary for good adaptive FEMs while 1) and 2) should be checked by numerical experiments. In this paper, a new *a posteriori* error estimator for h-adaptive method is introduced. The mathematical proofs of our estimator are provided in [2]. Here we emphasize properties 1) and 2), and verify the quality of the *a posteriorii* error estimator when applied to the obstacle problem and to the flow through porous media problems, both being examples of problems characterized by variational inequalities. These model problems are taken from Oden and Kikuchi [5] and Elliot and Ockenden [3]. Details of theory and proofs of properties of variational inequalities for contact problems can be also found in Kikuchi and Oden [4].

2 Model Problem

Consider the following obstacle problem

Find u such that

$$\begin{aligned} -\Delta u \geq f\ ;\ u \geq \varphi \qquad & \text{in } \Omega \\ (\Delta u + f)(u - \varphi) = 0 \quad & \text{a.e. in } \Omega \end{aligned}$$

subject to $u = g$ on $\partial\Omega$. First, we define a non–empty closed convex set of admissible functions

$$K = \left\{ v \in H^1(\Omega) | \, v \geq \varphi \text{ in } \Omega,\ v = g \text{ on } \partial\Omega \right\}$$

where $\varphi \in C(\Omega)$ and $\varphi \leq 0$ on $\partial\Omega$. Assuming $\Delta u \in L^2(\Omega)$, we construct a weak form of the above problem by multiplying by $(v - \varphi)$ and $(u - \varphi)$, and integrating over Ω, to obtain

$$\int_\Omega (\Delta u + f)(v - \varphi) dx \geq 0 \qquad \forall\, v \in K$$

$$-\int_\Omega (\Delta u + f)(u - \varphi) dx = 0$$

where u denotes the true solution and v denotes an arbitrary function from K. By subtracting one equation from the other, we obtain

$$\int_{\Omega} -\Delta u(v-u)dx \geq \int_{\Omega} f(v-u)dx$$

Applying Green's formula

$$\int_{\Omega} \nabla u \cdot \nabla(v-u)dx \geq \int_{\Omega} f(v-u)dx \tag{1}$$

To simplify notation, we next introduce a bilinear form and a linear functional defined by

$$a:\ K \times K \to \mathbb{R},\quad a(u,v) \ \equiv \int_{\Omega} \nabla u \cdot \nabla v\ dx \tag{2}$$

$$f:\ K \to \mathbb{R},\quad f(v) \ \equiv \int_{\Omega} fv\ dx \tag{3}$$

Then the model problem can be recast in the general weak form as follows: Find

$$u \in K:\ J(u) \leq J(v) \qquad \forall\, v \in K \tag{4}$$

where

$$J \equiv \frac{1}{2}a(v,v) - f(v) \tag{5}$$

Equivalently,

$$\text{Find } u \in K:\ a(u, v-u) \geq f(v-u) \qquad \forall\, v \in K \tag{6}$$

We focus our attention on a general class of variational inequalities of which (6) is a typical member. To discretize (5), we introduce a finite dimensional subset $V_h \in H^1(\Omega)$ such that

$V_h = \{v_h \in C(\Omega) | v_h \in P_1(\Omega_k), \Omega_k =$ a finite element in a partition $\mathcal{Q}_h$ of $\Omega,\ v_h = g$ on $\partial\Omega\}$

For simplicity, linear triangular elements are used to construct the functions in V_h, since

1. linear triangular meshes can be locally refined without introducing constrained nodes
2. linear triangular elements provide for a simple implementation of the unilateral condition $v_h \geq \varphi$ in Ω (for example, when φ vanishes identically or is convex, we need only to set $v_h \geq \varphi$ at nodal points)

We define a finite dimensional non-empty closed convex subset of V_h according to

$$K_h \equiv \{v_h \in V_h | v_h(\boldsymbol{x}_\Delta) \geq \varphi(\boldsymbol{x}_\Delta) \quad \forall\, \Omega_k\}$$

with $\boldsymbol{x}_\Delta$ a nodal point. In this case $K \supset K_h$ when φ vanishes or is convex. But, in general, $K \not\supset K_h$, in which we can set $v_h \geq \varphi\, \forall\, \boldsymbol{x} \in \Omega_k$ with more constraints in the definition of K_h.

Finally, the discretized formulation is:

$$\text{Find } \ u_h \in K_h:\ J(u_h) \leq J(v_h) \qquad \forall\, v_h \in K_h \tag{7}$$

The convergence of u_h to u as $h \to 0$ is proven in [4]. In the numerical results which follow, the discrete inequalities corresponding to finite element models of (6) are solved by the projected SOR method [5].

3 Error Estimation

Define $K - \{u_h\} \equiv \{w | w = v - u_h \quad \forall\, v \in K\}$ and the error $e \equiv u - u_h$. By replacing u with $e + u_h$, eq. (6) becomes

$$a\,(e + u_h, v - e - u_h) \geq f\,(v - e - u_h) \text{ or}$$
$$a\,(e, v - e - u_h) \geq f\,(v - e - u_h) - a\,(u_h, v - e - u_h)$$

Let $w = v - u_h \in K - \{u_h\}$. Then, since v is arbitrary, w is also arbitrary, and we arrive at the following variational inequality for the error:

$$\text{Find } \ e \in K - \{u_h\} \ : \ a(e, w - e) \geq f(w - e) - a\,(u_h, w - e) \quad \forall\, w \in K - \{u_h\} \tag{8}$$

and it is easy to show that (8) is equivalent to the minimization problem

$$\text{Find } \ e \in K - \{u_h\} \ : \ J(e) \leq J(w) \qquad \forall\, w \in K - \{u_h\} \tag{9}$$

where

$$J(w) \equiv \frac{1}{2} a(w, w) - f(w) + a\,(u_h, w)$$

From (6), with $v = u_h \in K_h$, we have

$$a\,(u, u - u_h) \leq f\,(u - u_h)$$

$$a(e, e) = a(u, e) - a\,(u_h, e) \leq f(e) - a\,(u_h, e)$$

Therefore

$$\frac{1}{2}\, \|e\|_E^2 \equiv \frac{1}{2} \int_\Omega \nabla e \cdot \nabla e \; dx = \frac{1}{2} a(e, e) \leq -J(e) \tag{10}$$

To split the original global problem into local problems, we allow the global shape function to have discontinuities between elements, and use these discontinuities between elements as constraints by introducing a Lagrangian:

$$\begin{aligned} &L(w, \mu) \equiv J(w) - \langle \mu, [w] \rangle \\ &\text{where } \langle u, v \rangle \equiv \sum_k \int_{\partial \Omega_k} u(s) v(s) dx \end{aligned} \tag{11}$$

$[w] \equiv$ jump between elements which share a common edge.

For each element Ω_k, we define the elementwise admissible function set

$$V_k \equiv \left\{ v \in H^1(\Omega) | v \geq -u_h \text{ a.e. in } \Omega_k, \ v = 0 \text{ on } \partial\Omega_k \cap \partial\Omega \right\}$$

Clearly

$$L(w, \mu) \equiv J(w) \text{ when } w \in K - \{u_h\} \text{ and } [w] = 0$$

From (10) and (11), we construct a saddle point problem

$$
\begin{aligned}
-\frac{1}{2}\|e\|_E^2 \geq J(e) &= \inf_{w\in K-\{u_h\}} J(w) \\
&= \inf_{w\in K-\{u_h\}} \sup_{\mu\in R} L(w,\mu) \\
&\geq \sup_{\mu\in R} \inf_{w\in K-\{u_h\}} L(w,\mu) \\
&\geq \inf_{w\in K-\{u_h\}} L(w,\mu) \quad \forall\, \mu \in \mathbb{R}
\end{aligned}
$$

Integrating by parts with $\Delta u_h = 0$ in the linear element case

$$
\begin{aligned}
L(w,\mu) &= \frac{1}{2}a(w,w) - f(w) + a\left(u_h, w\right) - \langle \mu, [w]\rangle \\
&= \frac{1}{2}a(w,w) - f(w) + \sum_k k\left\langle \frac{\partial u_h}{\partial n_k}, w\right\rangle_{\partial\Omega_k} - \langle \mu, [w]\rangle \\
&= \sum_k J_k(w) - \left\langle \mu - \left\{\frac{\partial u_h}{\partial n_k}\right\}, [w]\right\rangle
\end{aligned}
\tag{12}
$$

where

$$
\begin{aligned}
J_k(w) &\equiv \frac{1}{2}a(w,w)_k - (r,w)_k + \frac{1}{2}\int_{\partial\Omega_k} R(s)w(s)ds \\
a(w,w)_k &\equiv \int_{\Omega_k} \nabla w\cdot\nabla w\; dx \\
(r,w)_k &\equiv \int_{\Omega_k} \left(f(x) + \Delta u_h\right) ds \\
R(s) &\equiv -\left[\frac{\partial u_h}{\partial n}\right] \text{ when } s\in \text{ interior boundary} \\
& \qquad g - \frac{\partial u_h}{\partial n} \text{ when } s\in \text{ exterior boundary}
\end{aligned}
$$

Here, we choose $\mu^* = \left\{\frac{\partial u_h}{\partial n}\right\}$ to make the coupled term vanish where $\{\ \}$ denotes the average between elements which share a common edge.

Finally, we have

$$
\|e\|_E^2 \leq -2\sum_k \inf_{w\in K-(u_h)} J_k(w) \tag{13}
$$

Now we have an upper bound which can be calculated from the following local minimization problem:

$$
\text{Find } w_k \in V_k :\ J_k(w_k) \leq J_k(v_k) \qquad \forall\, v_k\in V_k \tag{14}
$$

The importance of (14) is that the global problem (8) has been split into a set of local problems. But we now have to consider the well–posedness of these local problems.

Formally, the local problem is

$$-\Delta w_k \geq f,\ w_k \geq -u_h + \varphi$$
$$(-\Delta w_k - f)(w_k + u_h - \varphi) = 0 \qquad \text{in } \Omega_k$$

with the boundary condition

$$\frac{\partial w_k}{\partial n} = \begin{cases} -\frac{1}{2}\left[\frac{\partial u_h}{\partial n}\right] & \text{on interior boundary} \\ 0 & \text{on exterior boundary} \end{cases}$$

This is a Neumann type problem. As proved in [1], for a unique solution to exist to this boundary value problem, we must have

$$\int_{\Omega_k} f(x)dx - \frac{1}{2}\int_{\partial\Omega_k}\left[\frac{\partial u_h}{\partial n}\right] ds < 0 \tag{15}$$

But this condition is not satisfied in general and, without any modification of the local problem, either no solution or infinitely many solutions will exist. To overcome this difficulty, we add a constant δ_k to f to make the compatibility valid:

$$\delta_k = \frac{1}{\text{meas}(\Omega_k)}\left(\int_{\Omega_k} f(x)dx - \frac{1}{2}\int_{\partial\Omega_k}\left[\frac{\partial u_h}{\partial n}\right] ds\right) \tag{16}$$

Then the local problems are well posed whenever $\delta_k < 0$.

Next, we record a duality result of some importance.

Theorem 1

Let

$$M = \left\{\boldsymbol{p} \in \left[L^2(\Omega)\right]^2 | \nabla\cdot\boldsymbol{p} + f \leq 0 \text{ a.e. in } \Omega,\ \boldsymbol{n}\cdot\boldsymbol{p} \geq 0 \text{ on } \partial\Omega\right\}$$

and let

$$G(\boldsymbol{p}) \equiv -\frac{1}{2}\int_\Omega \boldsymbol{p}\cdot\boldsymbol{p}\, dx - \int_\Omega \varphi(f + \nabla\cdot\boldsymbol{p})dx + \int_{\partial\Omega}\varphi(g - \boldsymbol{n}\cdot\boldsymbol{p})ds$$

Then for any $v \in K$ and any $\boldsymbol{p} \in M$,

$$J(v) \geq J(u) = G(\nabla u) \geq G(\boldsymbol{p}) \text{ and } u \in M$$

Proof: See in [1] ∎

We next apply this theorem to each elementwise local problem (14); the result is:

$$-2J_k(w_k) \leq G_k(\boldsymbol{p}) \qquad \forall\, \boldsymbol{p} \in M_k \tag{17}$$

where,

$$M_k \equiv \left\{ \boldsymbol{p} \in \left[L^2(\Omega_k)\right]^2 |\nabla\cdot\boldsymbol{p}+f \leq 0 \text{ a.e. in } \Omega_k\ \boldsymbol{n}\cdot\boldsymbol{p} \geq \frac{1}{2}\left[\frac{\partial u_h}{\partial n}\right] \text{ on } \partial\Omega_k \right\} \tag{18}$$

$$G_k(\boldsymbol{p}) \equiv \int_{\Omega_k} \boldsymbol{p}\cdot\boldsymbol{p}\, dx + 2\int_{\Omega_k} \varphi(f+\nabla\cdot\boldsymbol{p})dx + 2\int_{\partial\Omega_k} \varphi\left(\boldsymbol{n}\cdot\boldsymbol{p} - \frac{1}{2}\left[\frac{\partial u_h}{\partial n}\right]\right) ds \ (\varphi \equiv \varphi - u_h) \tag{19}$$

With (13) and (16), we finally obtain

$$\|e\|_E^2 \leq \sum_k G_k(\boldsymbol{p}) \qquad \forall\, \boldsymbol{p} \in M_k \tag{20}$$

In view of this result, the local errors can be estimated by identifying the $\boldsymbol{p}$ which makes this upper bound closest to the true error.

Toward this end, consider the extremal condition of this dual problem

$$\boldsymbol{p} = \nabla w$$

Clearly M_k to which $\boldsymbol{p}$ belongs and $K - \{u_h\}$ to which w_k, the local solution, belongs, are different and it seems the two sets have nothing in common. But from the extremal condition, the true solution in both primal and dual problems is in both sets, and that is the common part of the two sets. Hence, even if w_k is not the common part, we expect w_k to be close to the true solution w. So we take the $\boldsymbol{p}$ as ∇w_k and check how good the assumption is. Moreover, from the formal form of the local problem, we also expect

$$f + \Delta w_k \simeq 0 \quad \text{in } \Omega_k \tag{21}$$

$$\frac{\partial w_k}{\partial n} - \frac{1}{2}\left[\frac{\partial u_h}{\partial n}\right] \simeq 0 \quad \text{on } \partial\Omega_k \tag{22}$$

so we assume only the first term is dominant and neglect the others. Thus, our final estimator is

$$\|e\|_E^2 \leq \sum_k \int_{\Omega_k} \nabla w_k \cdot \nabla w_k \, dx \tag{23}$$

where w_k is the solution of each local problem.

4 Numerical Examples

Three model problems are solved which provide typical examples of variational inequalities to verify the assumptions and to test the accuracy of the estimator. As noted earlier, the projected SOR method is used to solve both global and local problems, and triangular elements are employed. For the local problems, three mesh refinements are used to check convergence of the effectivity indices with respect to local refinement (see Fig. 1).

Note that in the calculation of local effectivity indices, zeros indicate that the exact error is zero.

4.1 Obstacle Problem

This problem has an exact solution which makes it possible to calculate the exact error and effectivity indices. The configuration and the numerical results are given in Figs. 2–7.

4.2 Dam Problem

This problem is to determine the free boundary between saturated and dry regions inside a dam with given inlet and outlet water levels. With the given water levels, the discharge can be calculated and is in effect given in the boundary conditions. The primitive form of the problem is of mixed type but, by the use of the Baiocchi transformation, a Dirichlet type is obtained. Since no exact solution is available, a fine mesh solution ($h = 5/40$) is taken to be the exact solution. This assumption was checked by computing the relative error, which is the ratio of the L_2 norm of the estimated error to the L_2 norm of the numerical solution, and we found that the relative errors were 11.37, 6.01, and 1.57 percent of the cases of $h = 5/5$, $5/10$, and $5/40$, respectively. This indicates that the $h = 5/40$ case is relatively close to the exact solution. The configuration of the problem and the resulting global effectivity indices for each case are given in Figs. 8–11.

4.3 Dam With a Sheet Pile Problem

The governing equation and the boundary conditions are the same as for the previous problem. In this case we do not know the discharge q *a priori.* This makes the problem of nonlinear even in the boundary conditions and requires the use of an iterative procedure to determine the discharge. Using the fact that the discharge at the tip of the sheet pile is equal to zero, we can solve for q by Newton's method. Hence, the solution scheme is: 1) first, solve for q by Newton's method with an initial guess, 2) determine the boundary conditions, then 3) use the same method as in the previous problem. However, the discharge will depend on the mesh size, i.e., we expect that the finer the mesh, the closer the discharge is to the exact value. Since no exact solution is available, we again solve the problem with a comparatively fine mesh, $h = 5/40$, and take this as the exact solution. This assumption was checked by computing the relative errors as was done in the previous problem.

This problem is solved with two different discharges. One is obtained with the coarse mesh of each problem, and the other is obtained with the finest mesh which should be the best approximation to the exact value. Computed relative errors were 18.42, 9.49, and 2.41 percent for the cases of $h = 5/5$, $5/10$, and $5/40$, respectively, with variable discharges. These were 16.83, 9.09, and 2.41 percent with the fixed discharge $q = 3.8890$ which was obtained by solving the finest mesh case ($h = 5/40$). This indicates we could assume that $h = 5/40$ case is relatively close to the exact solution.

The configuration of the problem and the resulting global effectivity indices for each case are given in Figs. 12–16.

5 Conclusions

Our principal goal is to check the accuracy of a new *a posteriori* error estimator using numerical examples. We made the assumption that in (18), only the first term is dominant. From our numerical results, the effectivity index is increasing as the local refinement gets finer. It is therefore likely that as we make the local mesh finer, the neglected terms become increasingly negligible relative to the first term. This tendency could also be observed in local effectivity indices.

As to the question of how good the error estimator is in the global sense, the effectivity indices are all less than 1.5. This level of global effectivity indices were unexpectedly low. Locally, the effectivity indices are between 1.0 and 2.0 and the distribution of the local effectivity indices become more regular as the local mesh becomes finer. As expected, indices are worst near the free boundary.

Acknowledgement: The support of this work by the National Science Foundation under Grant Number MSM-8805552 is gratefully acknowledge.

References

1. M. Ainsworth, "Analysis of a Signorini–Obstacle Type Variational Inequality," TICOM Report 91-1, Austin, Texas, 1991.

2. M. Ainsworth, J. T. Oden, and C. Y. Lee, "Local *A Posteriori* Error Estimators and Adaptive Algorithm for Finite Element Approximation of the Obstacle Problem," (preprint).

3. C. M. Elliot, and J. R. Ockenden, "Weak and Variational Methods for Moving Boundary Problems,"Pitman Pub., Boston, 1982.

4. N. Kikuchi, and J. T. Oden, "Contact Problems in Elasticity: A Study of Variational Inequalities and Finite Element Methods," *SIAM*, Philadelphia, 1988.

5. J. T. Oden, and N. Kikuchi, "Theory of Variational Inequalities, Flow Through Porous Media," *Int. J. Eng. Sci.*, **18**, pp. 1173–1284.

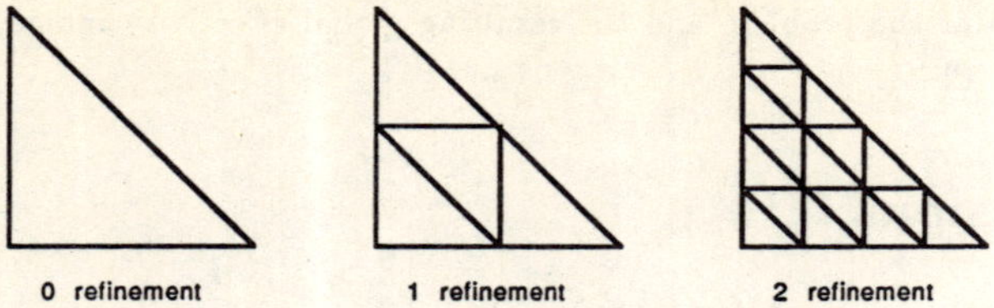

Fig.1 Local Refinements

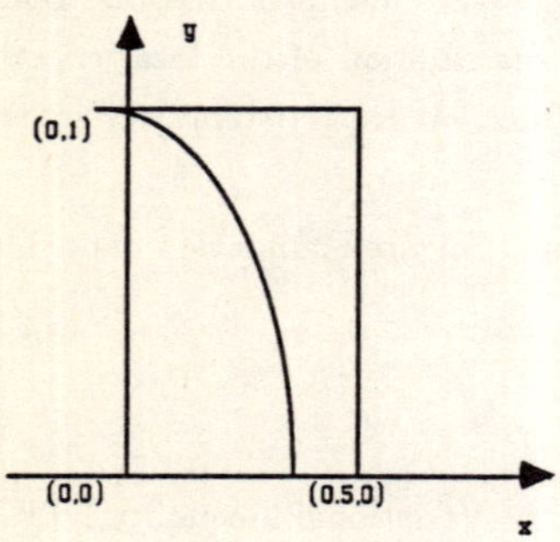

Fig.2 Configuration of Obstacle Problem

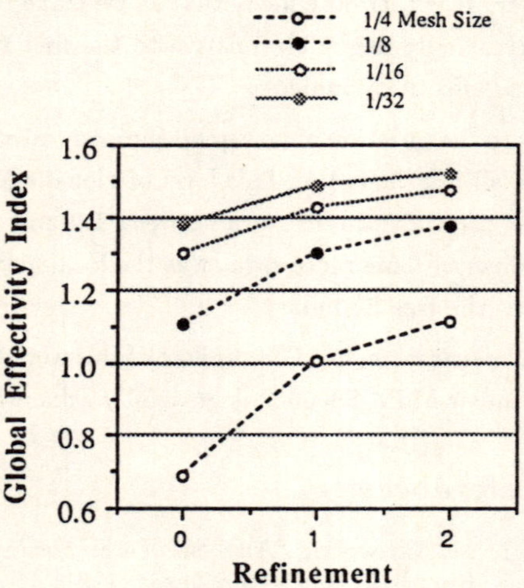

Fig.3 Obstacle Problem

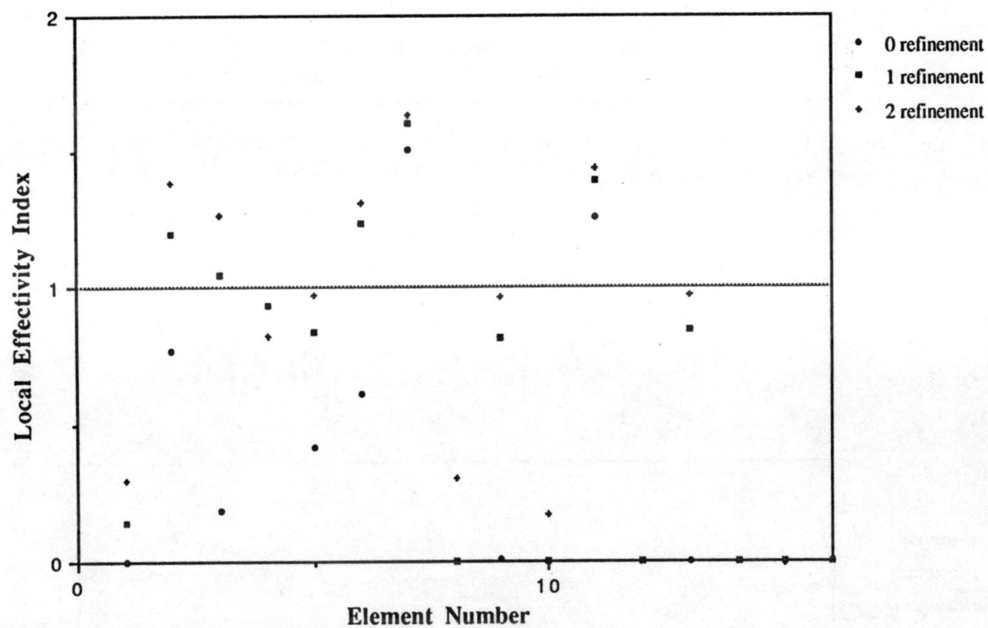

Fig.4 Obstacle Problem (h=1/4)

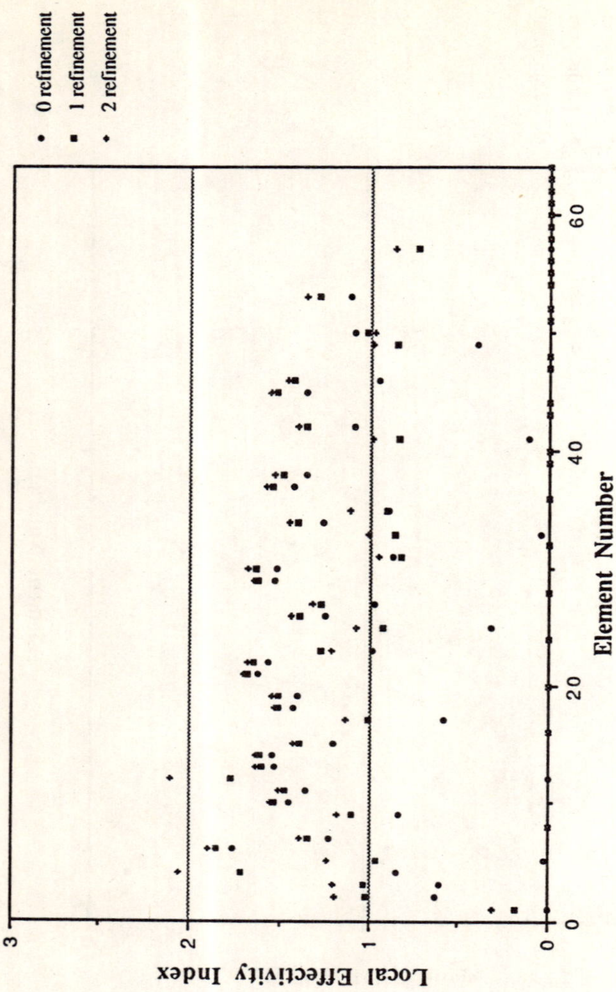

Fig.5 Obstacle Problem (h=1/8)

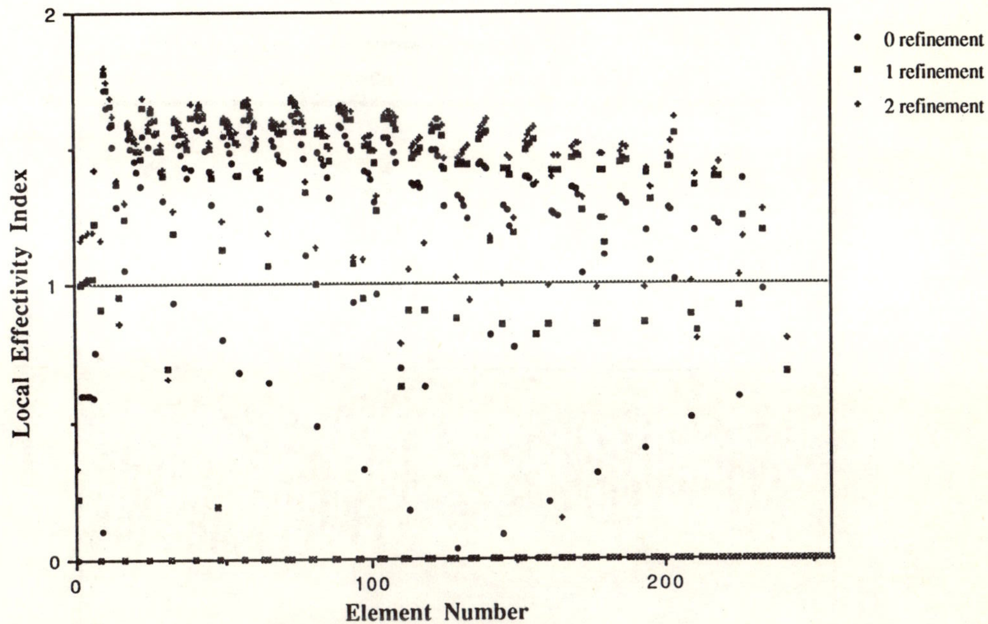

Fig.6 Obstacle Problem (h=1/16)

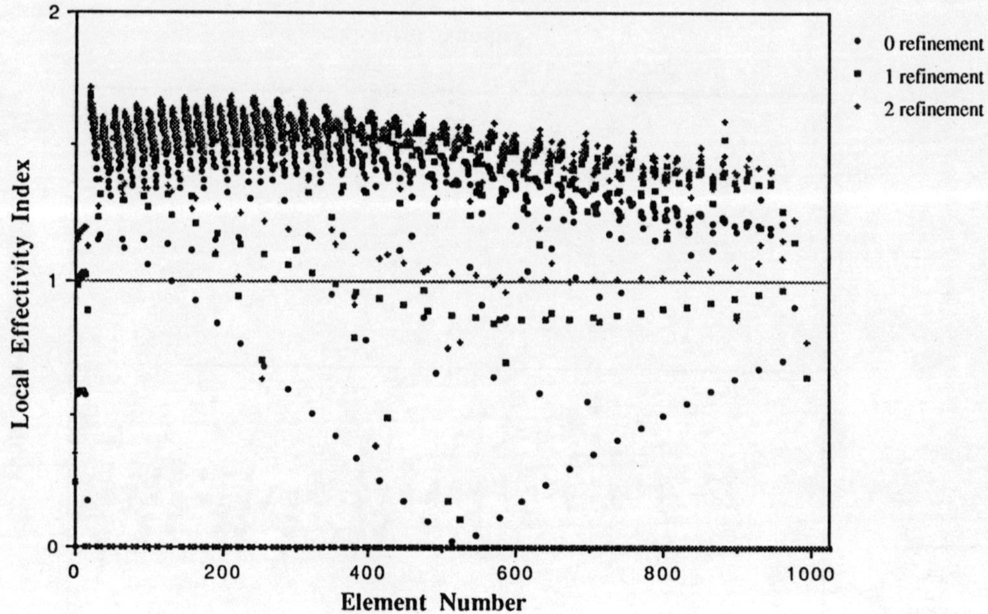

Fig.7 Obstacle Problem (h=1/32)

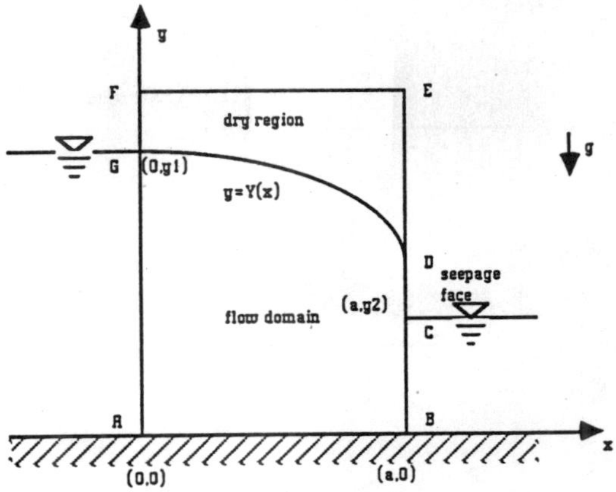

Fig.8 Configuration of Dam Problem

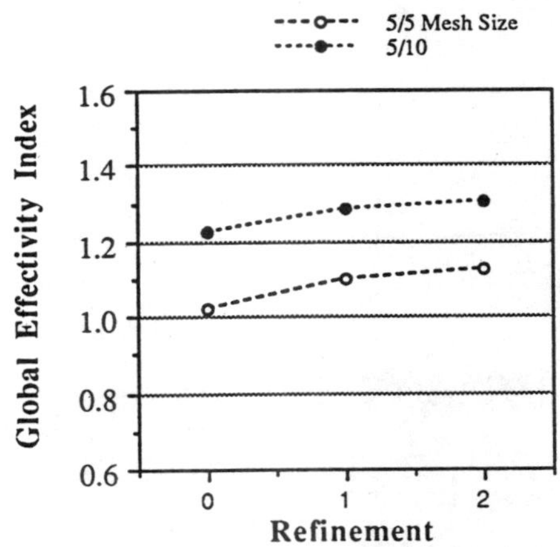

Fig.9 Dam Problem

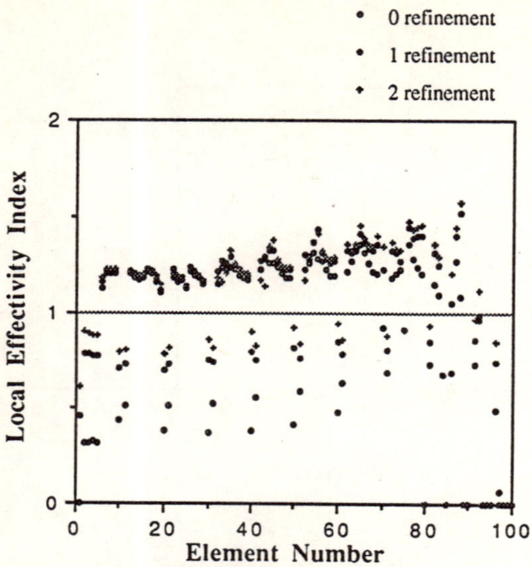

Fig.10 Dam Problem (h=5/5)

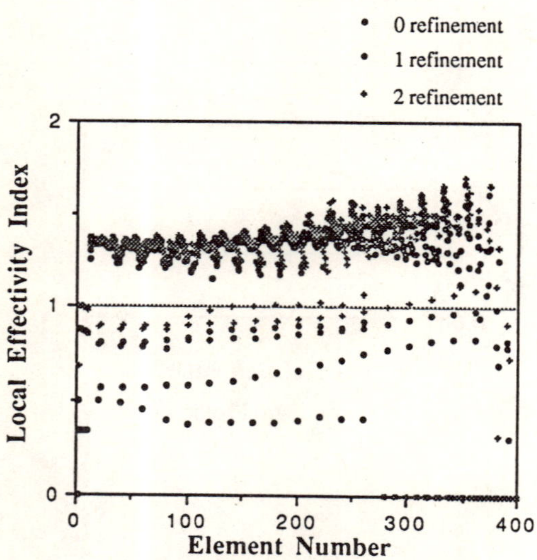

Fig.11 Dam Problem (h=5/10)

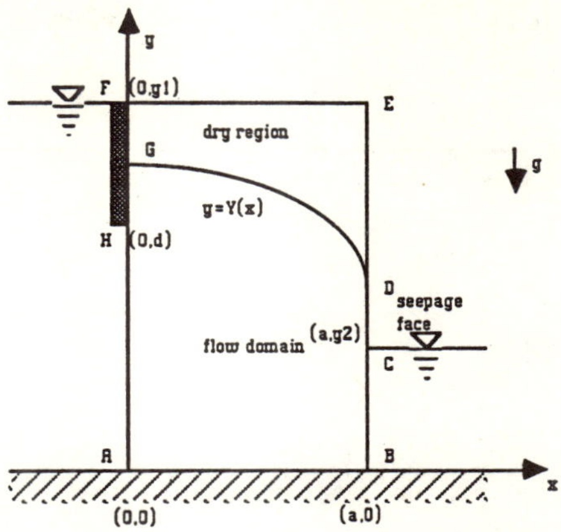

Fig.12 Configuration of Dam with a Sheet Pile

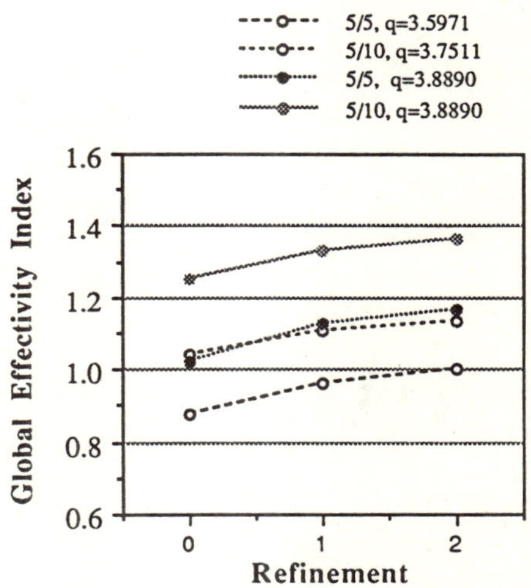

Fig.13 Dam with a Sheet Pile

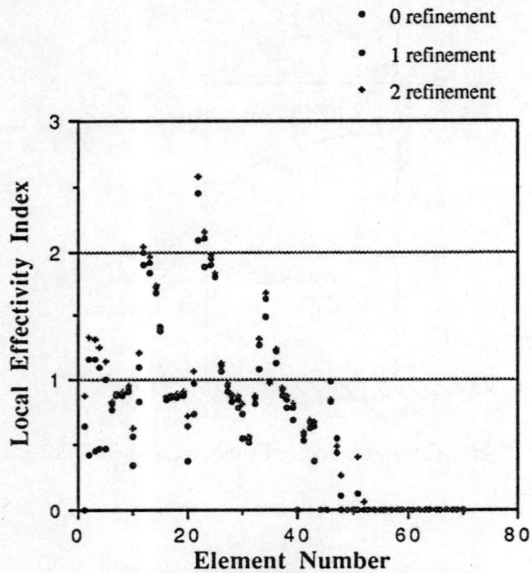

Fig.14 Dam with a Sheet Pile Problem
(q=3.5971, h=5/5)

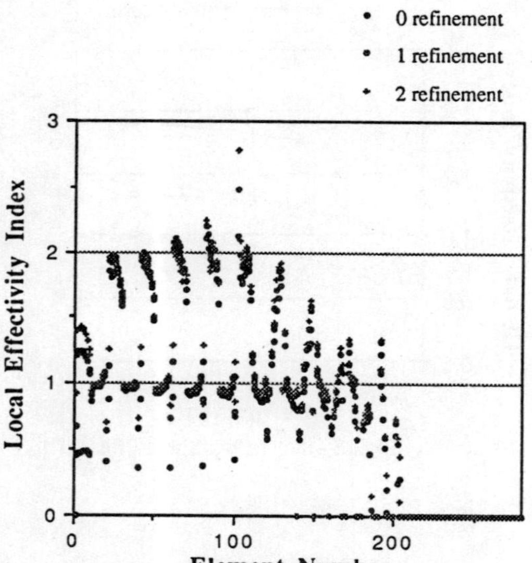

Fig.15 Dam with a Sheet Pile Problem
(q=3.7511, h=5/5)

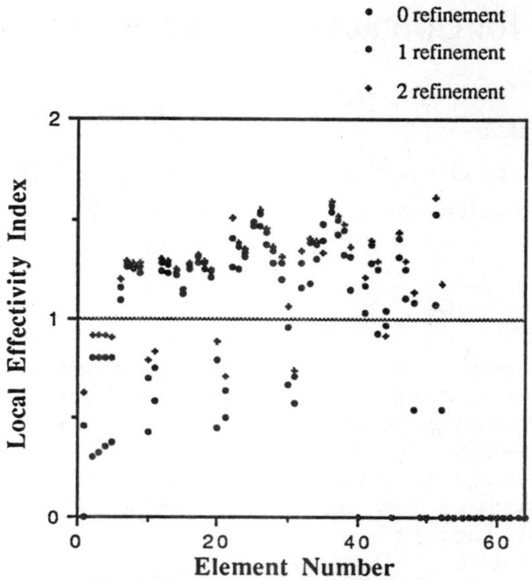

Fig.16 Dam with a Sheet Pile Problem
(q=3.8890, h=5/5)

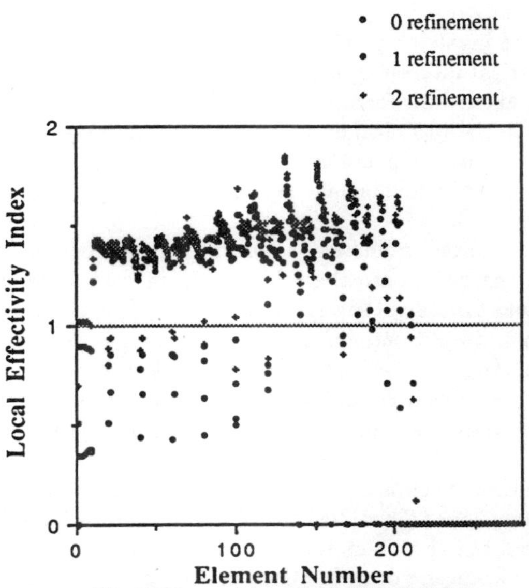

Fig.17 Dam with a Sheet Pile Problem
(q=3.8890, h=5/10)

On a Patch Test for Contact Problems in Two Dimensions

Robert L. TAYLOR and Panayiotis PAPADOPOULOS

Department of Civil Engineering, University of California at Berkeley, Berkeley, CA 94720, U.S.A.

Abstract

In this paper we propose a simple finite element completeness check in the spirit of the patch test originating in the work of Irons, with particular reference to frictionless contact problems in two dimensions. Attention is primarily focused on classical slideline treatments based on node-to-node or node-to-surface geometric assumptions. Representative element meshes are subjected to the proposed test via a penalty formulation and conclusions are drawn for both linear and quadratic order elements. In view of encountered shortcomings, a more general slideline methodology is advocated.

1 Introduction

Boundary-value problems involving mechanical contact between deformable bodies are of considerable practical interest in engineering applications. In the context of finite element methods, algorithms for the numerical simulation of contact phenomena originally appeared in the technical literature of the early seventies, [1,2]. In the following years until today the methodologies have substantially improved following closely the growth that finite elements have generally experienced within the realm of structural mechanics.

A pivotal objective in contact numerical algorithms is the accurate transmission of pressure forces from one body onto the other, coupled with observance of non-penetration. This two-fold task is not always an easy one. The geometric constraint of impenetrability is by nature *unilateral*, in the sense that it is enforced only on the parts of the body surfaces where contact actually takes place, while it remains inactive elsewhere. An additional complication stems from the fact that, in most cases, the exact location and extent of the contacting surfaces is not known to the analyst in advance.

Algorithms for finite element simulations of contact problems depend crucially on two tacitly related assumptions. These are the discretization of boundary surfaces potentially in contact and the approximation of pressure distribution within the regions of actual contact. Geometric considerations prevail in identifying the fictitious contact elements, whereas stability and accuracy guide, to a point, the choice of pressure fields.

In this paper we revisit classical slideline treatments based on the node-to-node and node-to-surface assumptions, [3,4], and attempt an assessment of their robustness through the introduction of a simple benchmark problem, referred to as a *patch test.* Standard arguments show that satisfaction of the patch test constitutes a necessary condition for algorithmic convergence. Failure to conform with its requirements typically results in loss of accuracy.

The constrained extremization problem of unilateral contact can be numerically approximated by means of formulations directly related to the Lagrange multiplier concept. Here, we employ a penalty formulation to free the space of admissible displacements from restrictions pertaining to contact constraints. The behavior of the tested algorithms is evaluated and explained, while shortcomings are identified for both linear and quadratic elements. The conclusions obtained are used to motivate alternative methodologies.

2 Classical slidelines: a brief review

Node-to-node and node-to-surface slideline treatments are used extensively in numerous research and commercial finite element programs that model contact phenomena. Here, the underlying assumption is that contact between interacting bodies will be accounted for with reference to the boundary nodal points only. The NIKE/DYNA codes at the Lawrence Livermore National Laboratory were among the first to adopt the above methodology and test the resulting numerical algorithms in large scale computations, [5].

The node-to-node method presupposes a one-to-one correspondence between boundary nodes on the contacting bodies. This is a geometrically limited assumption in that it can be easily violated, often to unacceptable level, when large body motions take place during the loading process. To its advantages one may recognize the obvious algorithmic simplicity and the unbiased treatment of the contacting surfaces. The assumed pressure field emanating from the node-to-node logic is essentially an equivalent force field concentrated at the nodes that are in contact, see Fig. 2.1.

The node-to-surface method permits a *slave* node to hit at a generic point across the opposite *master* surface in the sense of Fig. 2.2. This logic allows for arbitrarily large motions to be treated accurately, since changes in contact geometry can be directly accounted for by the algorithm at every given load step. It should be noted, however, that this procedure invokes bias between the two bodies as the nodes of the one (the slave) hit the surface of the other (the master). This is the 'one-pass' version of the method. The corresponding pressure field is shown in Fig. 2.3. Clearly, one can eliminate the geometric bias by reversing the role of master and slave surfaces and repeating the same process, thus obtaining the 'two-pass' version illustrated in Fig. 2.4.

In the above slideline treatments the non-penetration condition is enforced at discrete (nodal) points along the contacting surfaces. At these same points, we obtain contact forces, that is nodal forces desirably equivalent to an accurate approximation of the actual pressure field. In practice, tributary area methods are sometimes used to *a posteriori* recover a pressure field from the concentrated forces.

In the following section, we address the accuracy of pressure transmission along the contact surfaces, in the context of the above classical slideline treatments.

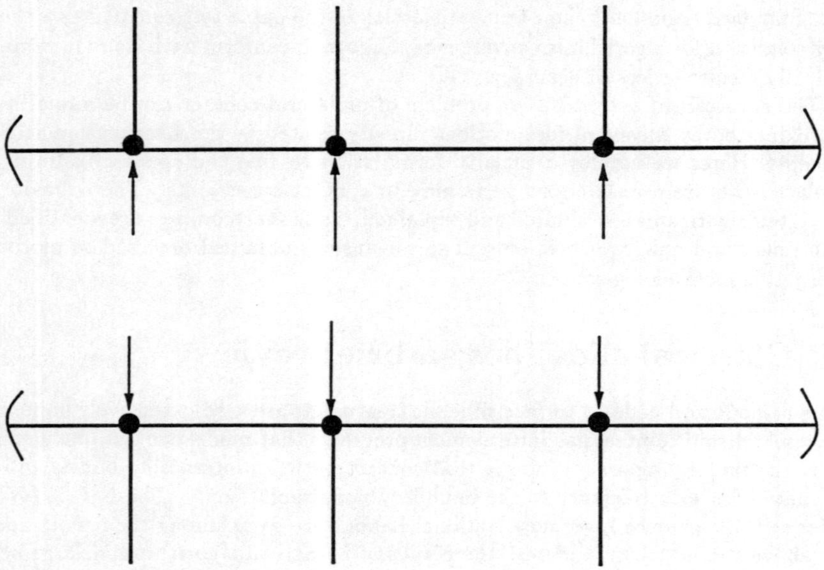

Fig. 2.1 Node-to-node contact

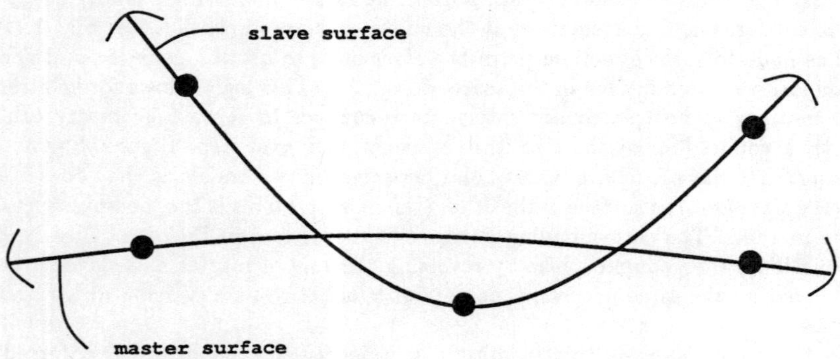

Fig. 2.2 Node-to-surface contact

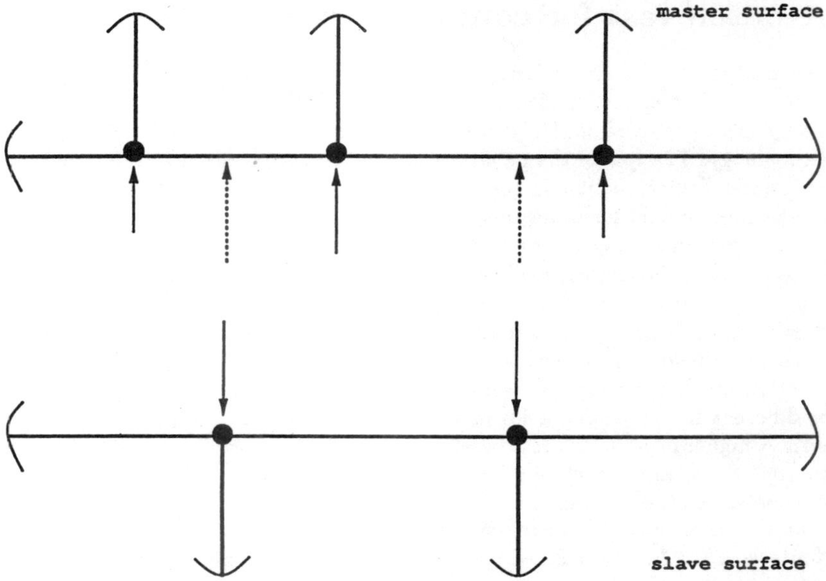

Fig. 2.3 Node-to-surface contact: one-pass algorithm

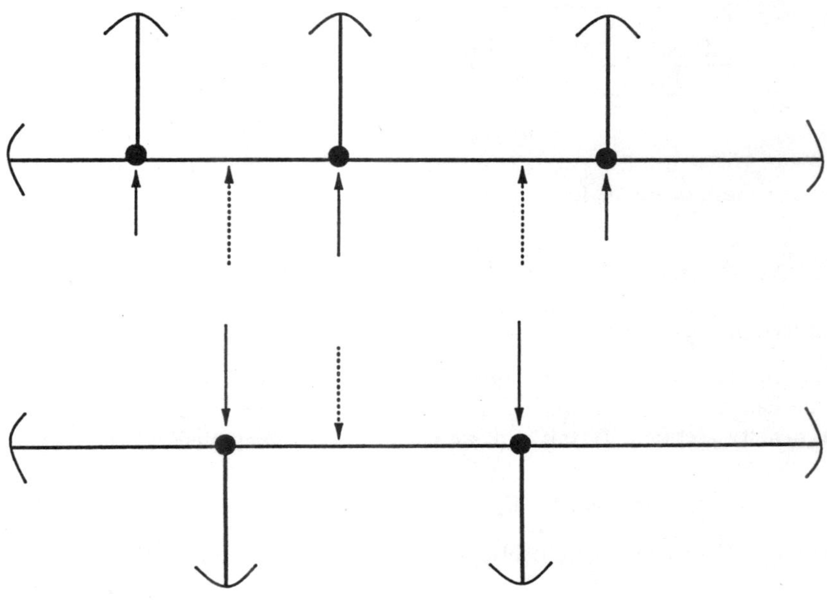

Fig. 2.4 Node-to-surface contact: two-pass algorithm

3 A patch test for contact problems

We propose a simple two-dimensional completeness test for finite element formulations of contact problems. The original idea of Irons, [6], extended to mixed finite elements by Zienkiewicz *et al.*, [7], is used here to assess the robustness of the contact methodology. The role of the patch test in convergence is discussed in detail by Zienkiewicz and Taylor, see [8]. In the spirit of the Irons patch test and its generalizations, the requirement for exact reproduction of a constant strain field gives way to a similar requirement pertaining to the (equivalent) pressure field, recovered by the algorithm along the contact surfaces.

Based on the above rationale, we introduce the problem of Fig. 3.1. A rectangular elastic punch is forced into a finite elastic foundation by means of a uniformly distributed force field. Since both bodies have the same material properties, the exact solution invokes constant pressure throughout the area of contact (this is also true for different material properties in the two bodies if a frictionless interface is assumed). A higher order patch test involving the application of linearly distributed force on top of the same body system may also be conducted. In this paper we restrict ourselves to the constant pressure test.

All computations have been performed within the environment of the fully nonlinear finite element program FEAP, which is partially documented in [8]. Small deformation linear and quadratic isoparametric elements in plane strain have been used throughout the analysis. The constraint conditions are treated by a classical penalty formulation.

The element meshes considered in this work are shown in Fig. 3.2. The following table conveniently summarizes the test results obtained.

Patch test results							
	Element meshes						
Pressure	A	B-1p	B-2p	C-1p	C-2p	D	E
Constant	p	p/f	p	f	p	p	f

p: pass f: fail

The above results are analyzed in the following section.

4 Evaluation of slideline treatments

4.1 Summary of the governing equations

We briefly summarize the equations governing the boundary-value problem of unilateral mechanical contact in the absence of friction, as formulated by means of a classical penalty method. To this end, we define the augmented energy functional Π_ϵ according to

$$\Pi_\epsilon(\mathbf{u}) = \Pi_{TPE} + \int_C \frac{1}{2}\epsilon h^2 \, d\Gamma , \tag{4.1}$$

where Π_{TPE} is the usual total potential energy for the bodies and $h(\mathbf{X}, \mathbf{u})$ is the

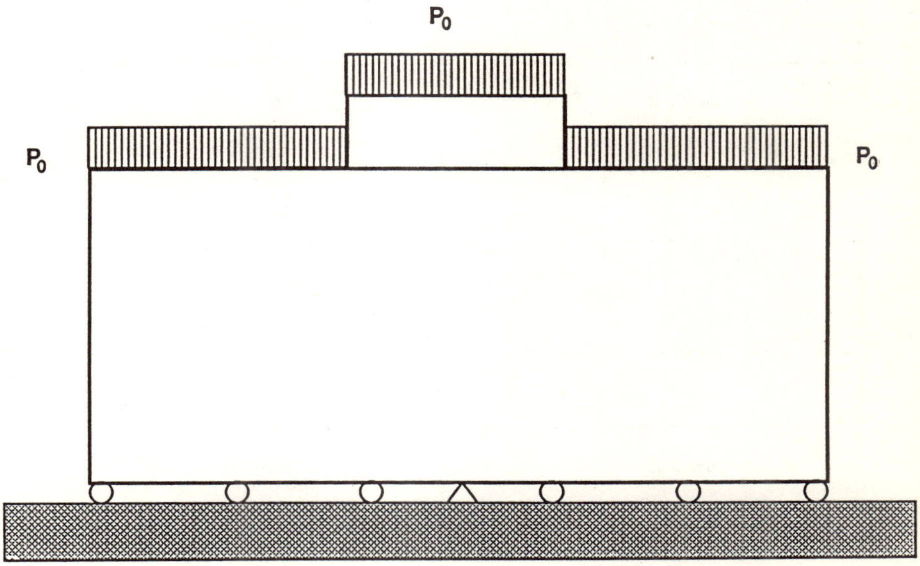

Fig. 3.1 Constant pressure patch test

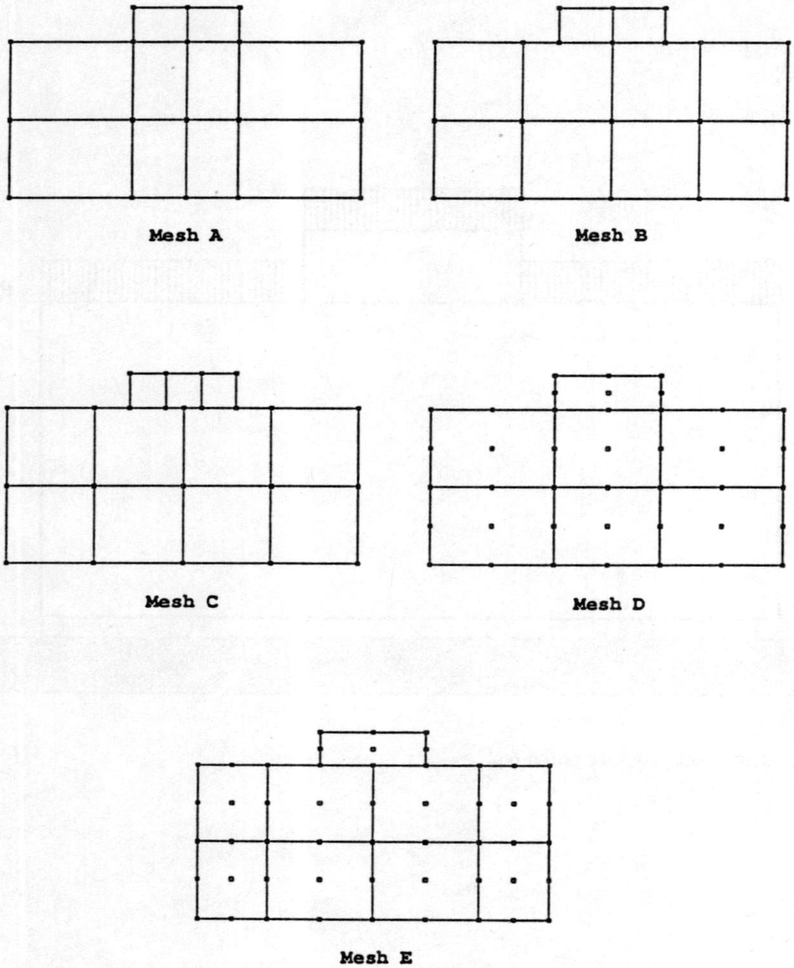

Fig. 3.2 Finite element meshes for contact patch test

local gap (penetration) function along the actual contact surface C, depending upon the boundary displacements of both bodies, $\mathbf{u}$, and the surface coordinates in the reference configuration, $\mathbf{X}$. The geometric constraint of non-penetration requires that

$$h = 0 \tag{4.2}$$

along the contact surface. A penalty parameter $\epsilon > 0$ controls the degree to which the above constraint is imposed. It is important to note that (4.2) becomes inactive if contact does not occur or release (i.e. positive pressure) is detected. Extremization of (4.1) with respect to the unconstrained displacement field V gives rise to the equation

$$\delta\mathbf{u}\cdot\mathbf{R} + \int_C \delta h\, \epsilon h\, d\Gamma = 0\ , \tag{4.3}$$

where $\mathbf{R}$ represents the residual force emanating from (unconstrained) extremization of the total potential energy functional. Clearly, the higher the value of ϵ, the more accurately (4.2) is enforced. The quantity ϵh can be regarded as the penalty force associated with the approximate satisfaction of (4.2).

A Lagrange multiplier formulation can be realized by introducing the functional

$$\Pi_L(\mathbf{u}\,, p) = \Pi_{TPE} + \int_C p\, h\, d\Gamma\ , \tag{4.4}$$

where the pressure field P is independently assumed at the outset. Finally, an augmented Lagrangian functional Π_{AL} combining a penalty with a classical Lagrange multiplier method can be defined according to

$$\Pi_{AL}(\mathbf{u}\,, p) = \Pi_{TPE} + \int_C \frac{1}{2}\epsilon h^2\, d\Gamma + \int_C p f(h)\, d\Gamma\ ,$$

with $f(0) = 0$ (e.g. $f(h) = h$). Standard arguments show that both penalty and augmented Lagrangian formulations converge to the Lagrange multiplier solution in the continuum setting as the penalty parameter approaches infinity. Thus, in what follows it is justified to consider all three formulations under the same line of thought.

4.2 Pressure distribution

The purpose of this subsection is to propose a simple analysis of the behavior exhibited by the slideline methods under consideration, when subjected to the patch test in conjunction with the family of Lagrange multiplier formulations outlined above. It will be shown that accurate representation of a constant pressure field is attainable by some methods, depending on how integration of the momentum balance equation (4.3) is performed along the contact surface. It should be noted that, given a particular slideline treatment, *a single mesh for which the patch test fails is sufficient to ascertain overall lack of robustness.*

In a discrete setting, the finite element approximation for the displacement field gives rise to an approximation of the local gap (penetration) function in the form

$$h = \mathbf{H}(\mathbf{X}\,, \mathbf{u})\,\mathbf{u} + \mathbf{H}_0(\mathbf{X})\ ,$$

where $\mathbf{H}$, $\mathbf{H}_0$ involve degrees of freedom associated, in principle, with surface nodes. With the additional assumption of constant pressure p_0 distributed throughout the region of contact, C, the penalty term in (4.3) becomes

$$\int_C \delta \mathrm{u}\, p_0\, \mathbf{H}\, d\Gamma\ . \tag{4.5}$$

It is the integration of this particular term that governs the behavior of the various slideline treatments. In view of the above conjecture, let us attempt to explain the patch test results obtained by the meshes of the previous section.

In meshes A though C, where linear elements are used, the integrand in (4.5) is a piecewise linear function in space. Clearly, the node-to-node contact of mesh A allows for nodal quadrature to be equivalent to the trapezoidal rule within each of the two symmetrically formed segments. Therefore, integration of (4.5) is exact and the patch test is passed.

In mesh B we distinguish between one- and two-pass algorithms. Moreover, within the one-pass procedure, there exist two alternatives; indeed, each of the two bodies may serve as the slave allowing the honor of being a master to the other. It is evident that the one-pass algorithm obtained by rendering the foundation master is practically equivalent to the two-pass algorithm. This is because, in this case, reversal of the master-slave characterization will not enhance the force distribution, but it will merely add stiffness to the center node. It follows that both formulations will pass the patch test, since, again, integration of (4.5) is performed by nodal quadrature within each of the segments on which the integrand in (4.5) is a linear function. The remaining one-pass algorithm unambiguously fails the patch test, because it attempts to integrate a piecewise linear function with only one quadrature point (corresponding to the center node).

Mesh C can be regarded as representative of a wide class of meshes encountered in contact finite element discretizations. Here, the distribution of nodes is arbitrary with respect to the corresponding nodes on the opposite surface. None of the two potential one-pass algorithms is successful. The rationale is the same as before; there exist four segments, in which the integrand of (4.5) is piecewise linear, and, if we consider the punch as slave (resp. master), there are four (resp. one) quadrature points to integrate it with. The two-pass algorithm has exactly as many nodal integration points as needed (i.e. five) for accurate integration of (4.5). Furthermore, elementary geometry dictates that the contact surface obtained in the context of the two-pass algorithm with arbitrary node distributions will be invariably flat, provided that contact conditions are detected for all boundary nodes. This is an undesirable feature whenever the algorithm is used to tie two meshes with arbitrary boundary node distribution, because it disallows accurate modeling of warping or bending along the tied surface.

Meshes D and E are constructed by 9-node quadratic elements, which, when undistorted, render the integrand of (4.5) piecewise cubic in a suitable local coordinate system. The three-point nodal quadrature of mesh D (Simpson's rule) corresponding to the node-to-node contact of three nodes is adequate for the purpose of satisfying the patch test. The same applies even if the contacting elements are distorted, although still flat (note that this case is entirely equivalent to having two *a priori connected* flat elements, which, allow, as is well known, exact transmission of constant surface forces).

Mesh E fails the patch test. This can be simply explained by the insufficient number of integration points (i.e. three points) involved in integrating the two piecewise cubic terms pertaining to the given mesh. The same test is similarly failed by one- or two-pass node-to-surface formulations.

Extensive numerical experimentation has confirmed the above findings. In summarizing, we claim that one-pass algorithms for linear elements and all nodal-based algorithms for quadratics fail, in general, the proposed patch test. Exceptions are limited and pertain to special meshes, which are, indeed, of little use whenever large body motions occur. On the other hand, the two-pass algorithm passes the test for all linear element meshes.

4.3 Solvability conditions for the discrete problem

A discrete contact algorithm is subject to solvability conditions, owing to the geometric constraint of impenetrability. The choice of pressure approximation is pivotal in rendering the problem solvable, as well as in providing a stable solution. As it becomes apparent, there are two diametrically opposed, but equally significant conditions to be met. First, the pressure approximation should be adequately low in order, or discretized in such a way that it be 'computable' given the number of available active displacement degrees of freedom, in the spirit of the *constraint counts* in [7]. Second, the same pressure should possess enough approximating power to accurately and stably represent the exact field.

By their construction, contact algorithms based on nodal logic (including both node-to-node and node-to-surface methods) satisfy the constraint counts. This is because, in this case, unknown pressures are always associated with active boundary nodes. Stability is not as easy to assess. Numerical experience with linear elements shows that the two-pass algorithm is generally stable (recall that it involves the maximum allowable number of independent pressure quantities!). One-pass algorithms are potentially unstable depending on the particular topology of the meshes, see for example mesh B of Fig. 3.2.

Keeping in mind the patch test along with the solvability requirements, we proceed in proposing an alternative approach to nodal treatments, which is conceptually more general and applicable to all order elements.

5 Contact segments: a general approach

The fundamental assumption of nodal contact is the discretization of the contact surface into contact elements, each of them consisting of a node on the slave surface and a corresponding node or element edge on the master surface. The resulting pressure field is buried into the nodal forces computed by the algorithm and is only artificially recovered. The computational significance of this approach must be appreciated, especially in light of its straightforward three-dimensional generalization. Here, we advocate an alternative, more general approach based on a continuous discretization of the contact surfaces into *contact segments*, along which a distributed pressure field is assumed. For an earlier attempt in the same direction, see [9].

In order to motivate this approach let us reconsider mesh B of Fig. 3.2. Two contact segments may be easily defined by the end points of the element edges along the contact surface. For each segment we make an approximation for the pressure field. It can be easily inferred that continuous piecewise linear pressure field and integration by trapezoidal rule, recovers the two-pass algorithm. Discontinuous piecewise constant pressure field (essentially equivalent to integrating (4.5) with one point quadrature within each segment) yields an alternative method. In both cases the patch test and the constraint counts are satisfied. However, the reduced order of the pressure field gives rise to stability problems, realized in the form of equi-oscillations in the computed pressure, as illustrated in Fig. 5.1. Note that this behavior is not necessarily suppressed by mesh refinement. Oden, [10], seems to reject all algorithms that employ discontinuous pressure fields for the Signiorini problem (i.e. when the foundation is rigid) and this is in accordance with our experience. It may be concluded that for linear elements the segments provide, on the one hand, a more general algorithmic framework, but, on the other hand, do not offer a more attractive alternative to the two-pass treatment.

For quadratic elements, the segmental logic provides new insight to the problem. Geometrically unbiased segments may now be defined as in Fig. 5.2, so that each one be associated with only a single element edge from each body. By way of continuous pressure approximation and Simpson's integration within a segment, the constant patch test is passed, contrary to the two-pass algorithm, while constraint counts and stability are not sacrificed. Moreover, contact calculations retain their local nature and may be performed within a finite element program employing a nodal logic with minimal additional effort. Details of this methodology may be found in [11].

Conclusion

In this work, robustness of contact algorithms has been addressed. Attention has been focused on pressure approximation and its algorithmic implications, rather than merely on satisfaction of the non-penetration condition. It has been shown that some nodally based algorithms fail to simulate exactly constant pressure conditions, owing to the inadequate approximation of the pressure distribution. Stability considerations are shown to further restrict the choice of pressure fields. Finally, a more general treatment is shown to emanate from continuous discretization of the contact surfaces and the associated pressures.

Acknowledgements

This work has been supported by the Lawrence Livermore National Laboratory under contract Nr. W-7405-ENG-48. The continuous interest of Dr. G.L. Goudreau of LLNL is especially acknowledged.

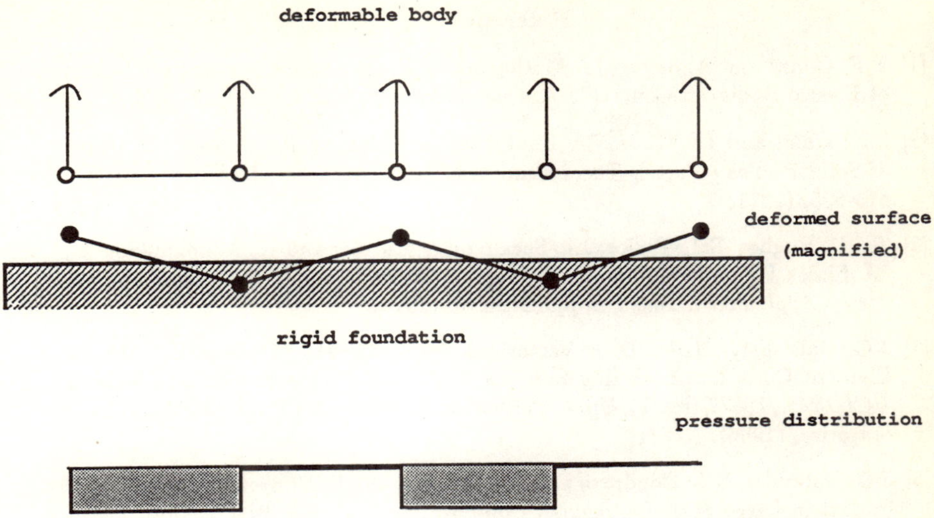

Fig. 5.1 Contact segments for linear elements: pressure oscillation due to discontinuous constant pressure approximation

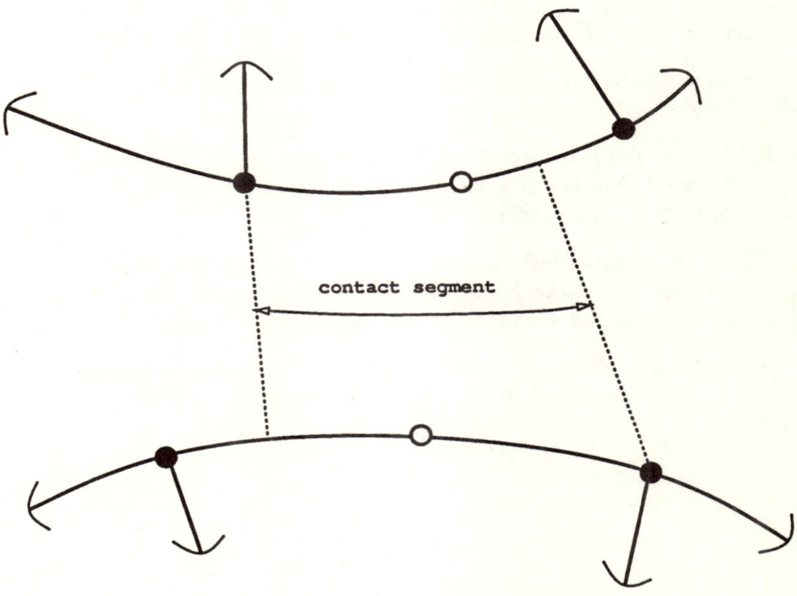

Fig. 5.2 Definition of contact segments for quadratic surfaces

References

[1] T.F. Conry and A. Seireg, "A Mathematical Programming Method for Design of Elastic Bodies in Contact", *J. Appl. Mech.*, 38, pp. 387-392, (1971).

[2] S.K. Chan and I.S. Tuba, "A Finite Element Method for Contact Problems of Solid Bodies - Part I. Theory and Validation", *Int. J. Mech. Sci.*, **13**, pp. 615-625, (1971).

[3] T.J.R. Hughes, R.L. Taylor, J.L. Sackman, A. Curnier and W. Kanoknukulchai, "A Finite Element Method for a Class of Contact-impact Problems", *Comp. Meth. Appl. Mech. Engr.*, 8, pp. 249-276, (1976).

[4] J.O. Hallquist, "NIKE2D: A Vectorized, Implicit, Finite Deformation, Finite Element Code for Analyzing Static and Dynamic Response of 2-D Solids", *UCID rep.*, 19677 (rev.1), Univ. California, Lawrence Livermore National Laboratories, (1986).

[5] J.O. Hallquist, G.L. Goudreau and D.J. Benson, "Sliding Interfaces with Contact-impact in Large Scale Lagrangian Computations", *Comp. Meth. Appl. Mech. Engr.*, **51**, pp. 107-137, (1985).

[6] B.M. Irons, "Numerical Integration Applied to Finite Element Methods", *Conference on Use of Digital Computers in Structural Engineering*, Univ. Newcastle, (1966).

[7] O.C. Zienkiewicz, S. Qu, R.L. Taylor and S. Nakazawa, "The Patch Test for Mixed Formulations", *Int. J. Num. Meth. Engr.*, 23, pp. 1873-1883, (1986).

[8] O.C. Zienkiewicz and R.L. Taylor, *The Finite Element Method*, 4th ed., Vol.1, McGraw-Hill, London, (1989).

[9] J.C. Simo, P. Wriggers and R.L. Taylor, "A Perturbed Lagrangian Formulation for the Finite Element Solution of Contact Problems", *Comp. Meth. Appl. Mech. Engr.*, 50, pp. 163-180, (1985).

[10] J.T. Oden, "Exterior Penalty Methods for Contact Problems in Elasticity", in *Europe-US Workshop: Nonlinear Finite Element Analysis in Structural Mechanics*, (ed. Bathe, Stein and Wunderlich), Springer, Berlin, (1980).

[11] P. Papadopoulos and R.L. Taylor, "A Mixed Formulation for the Finite Element Solution of Contact Problems", *UCB/SEMM Report*, 90/18, Univ. California, Berkeley, (1990).

An Interface Interpolation Scheme for Quadratic Convergence in the Finite Element Analysis of Contact Problems

Adrian Luis Eterovic and Klaus-Jürgen Bathe
Department of Mechanical Engineering, Massachusetts Institute of Technology, Cambridge MA 02139

Abstract

This paper presents an interpolation scheme for the kinematics of two-dimensional contact problems. The scheme satisfies sufficient conditions to achieve quadratic convergence with the Newton-Raphson iteration method. Within a Lagrange multiplier method framework, the first and second order contributions to the system matrix are obtained explicitly. The proposed formulation is intended for the solution of large strain contact problems.

1 Introduction

The development of solution methods for the finite element analysis of contact problems has received much attention during the past years. Interest in this research topic arises from both the theoretical difficulties in the numerical treatment of the governing equations and the broad range of potential applications of efficient contact analysis procedures.

There are two main ingredients in the specification of a solution method for contact problems. One is the choice of the global iteration method, which has to be selected so as to handle efficiently the resulting system of inequality contrained equations. The second is the local formulation of the contact interface conditions and the derivation of the corresponding contributions to the out-of-balance load vector and tangent stiffness matrix.

A number of global iteration strategies has been proposed and tested. Among these are Lagrange multiplier methods [1, 2, 3, 4, 5], penalty methods [6, 7, 8, 9], the perturbed Lagrangian method [10] and the augmented Lagrangian method [11, 12]. These procedures have been applied successfully to some finite deformation problems. However, such solutions can be difficult to obtain and a new formulation—the constraint function method—has been proposed in [13] to provide increased solution effectiveness.

Formulations of the kinematics of contact between two flexible bodies have been largely based on contact segments corresponding to the finite element discretization. Isoparametrically interpolated linear segments for the two- and three-dimensional cases are at present frequently used. An important feature in the analysis of finite deformation problems is the use of consistent tangent contact matrices. The derivation of these matrices has been addressed, for example, in references [14, 15, 16].

Isoparametric interpolation of the surface geometry and displacements, however, can lead to discontinuities of the corresponding contact stiffness across adjacent contact segments. These discontinuities can account for slow rates of convergence of the iteration method.

Our purpose in this paper is to present a kinematic formulation that satisfies sufficient conditions to achieve quadratic convergence with the Newton-Raphson iteration method. We briefly review the general conditions in Section 2. Within the framework of the standard Lagrange multiplier method, we derive in Section 3 the contact stiffness matrices. This is done with the purpose of identifying the relevant terms, which are the same for any Lagrange method and the constraint function method. We then proceed in Section 4 to derive these terms in the case of a general smooth interface interpolating curve, and based on the resulting expressions a particular interpolation scheme is developed in Section 5. We show that the necessary conditions require that in the interpolation the second derivatives be continuous across adjacent contact segments.

2 Conditions for quadratic convergence

We briefly consider in this section the conditions for quadratic convergence of the Newton-Raphson method in the solution of a system of nonlinear equations. First, we review a well-known result regarding convergence of the method (see for example reference [17]).

Theorem Consider the system $\mathbf{f}(\mathbf{x}) = \mathbf{0}$, where $\mathbf{f}$ is a C^1 map from $\Omega \subset R^n$ into R^n. Suppose that $\mathbf{f}(\mathbf{x}^*) = \mathbf{0}$ and $D\mathbf{f}(\mathbf{x}^*)$ is nonsingular. Then there exists a $\delta > 0$ such that for all $\mathbf{x}^{(0)} \in B(\mathbf{x}^*, \delta)$, the sequence generated by the iteration

$$\mathbf{x}^{(i+1)} = \mathbf{x}^{(i)} - [D\mathbf{f}(\mathbf{x}^{(i)})]^{-1}\mathbf{f}(\mathbf{x}^{(i)}) \tag{1}$$

is well defined, satisfies $\mathbf{x}^{(i)} \in B(\mathbf{x}^*, \delta)$ for all i, and converges to $\mathbf{x}^*$.

Proof Fix $M > \|[D\mathbf{f}(\mathbf{x}^*)]^{-1}\|$. Since $D\mathbf{f}$ is continuous, $[D\mathbf{f}]^{-1}$ is well defined and continuous on a neighborhood of $\mathbf{x}^*$. Let $\epsilon > 0$ be such that

$$\|[D\mathbf{f}(\mathbf{x})]^{-1} - [D\mathbf{f}(\mathbf{x}^*)]^{-1}\| \leq M - \|[D\mathbf{f}(\mathbf{x}^*)]^{-1}\| \quad \text{for all } \mathbf{x} \in B(\mathbf{x}^*, \epsilon) \tag{2}$$

It follows that

$$\|[D\mathbf{f}(\mathbf{x})]^{-1}\| \leq \|[D\mathbf{f}(\mathbf{x})]^{-1} - [D\mathbf{f}(\mathbf{x}^*)]^{-1}\| + \|[D\mathbf{f}(\mathbf{x}^*)]^{-1}\| \leq M \tag{3}$$

Then for $\mathbf{x}^{(i)} \in B(\mathbf{x}^*, \epsilon)$, $[D\mathbf{f}(\mathbf{x})]^{-1}$ is well defined and we can use (1) to write

$$\begin{aligned} \mathbf{x}^{(i+1)} - \mathbf{x}^* &= \mathbf{x}^{(i)} - \mathbf{x}^* - [D\mathbf{f}(\mathbf{x}^{(i)})]^{-1}\mathbf{f}(\mathbf{x}^{(i)}) \\ &= [D\mathbf{f}(\mathbf{x}^{(i)})]^{-1}[D\mathbf{f}(\mathbf{x}^{(i)})(\mathbf{x}^{(i)} - \mathbf{x}^*) - \mathbf{f}(\mathbf{x}^{(i)})] \end{aligned} \tag{4}$$

Since $D\mathbf{f}$ is continuous on $B(\mathbf{x}^*, \epsilon)$ and $\mathbf{f}(\mathbf{x}^*) = \mathbf{0}$ we have

$$\mathbf{f}(\mathbf{x}^{(i)}) = \left\{ \int_0^1 D\mathbf{f}(\mathbf{x}^* + s(\mathbf{x}^{(i)} - \mathbf{x}^*))\, ds \right\} (\mathbf{x}^{(i)} - \mathbf{x}^*) \tag{5}$$

Substituting this into (4)

$$\mathbf{x}^{(i+1)} - \mathbf{x}^*$$
$$= [D\mathbf{f}(\mathbf{x}^{(i)})]^{-1} \left\{ \int_0^1 \left[D\mathbf{f}(\mathbf{x}^{(i)}) - D\mathbf{f}(\mathbf{x}^* + s(\mathbf{x}^{(i)} - \mathbf{x}^*)) \right] ds \right\} (\mathbf{x}^{(i)} - \mathbf{x}^*) \tag{6}$$

and taking norms on both sides we obtain

$$\|\mathbf{x}^{(i+1)} - \mathbf{x}^*\|$$
$$\leq \|[D\mathbf{f}(\mathbf{x}^{(i)})]^{-1}\| \left\{ \int_0^1 \|D\mathbf{f}(\mathbf{x}^{(i)}) - D\mathbf{f}(\mathbf{x}^* + s(\mathbf{x}^{(i)} - \mathbf{x}^*))\| ds \right\} \|\mathbf{x}^{(i)} - \mathbf{x}^*\| \tag{7}$$

By continuity of $D\mathbf{f}$ there exists δ with $0 < \delta \leq \epsilon$ such that if $\mathbf{x}^{(i)} \in B(\mathbf{x}^*, \delta)$ then $\|D\mathbf{f}(\mathbf{x}^{(i)}) - D\mathbf{f}(\mathbf{x}^* + s(\mathbf{x}^{(i)} - \mathbf{x}^*))\| \leq \frac{1}{2}M^{-1}$. We have from (7) and (3) that $\|\mathbf{x}^{(i+1)} - \mathbf{x}^*\| \leq \frac{1}{2}\|\mathbf{x}^{(i)} - \mathbf{x}^*\|$ so $\mathbf{x}^{(i+1)} \in B(\mathbf{x}^*, \delta)$ and the conclusions follow. □

Corollary If, in addition to the assumptions of the Theorem there exists $L > 0$ such that

$$\|D\mathbf{f}(\mathbf{x}) - D\mathbf{f}(\mathbf{y})\| \leq L\|\mathbf{x} - \mathbf{y}\| \quad \text{for all } \mathbf{x}, \mathbf{y} \in B(\mathbf{x}^*, \delta) \tag{8}$$

then

$$\|\mathbf{x}^{(i+1)} - \mathbf{x}^*\| \leq \tfrac{1}{2}LM\|\mathbf{x}^{(i)} - \mathbf{x}^*\|^2 \tag{9}$$

i.e. the sequence $\{\mathbf{x}^{(i)}\}$ converges quadratically to $\mathbf{x}^*$.

Proof If (8) holds, then it follows from (7) and (3) that

$$\begin{aligned} \|\mathbf{x}^{(i+1)} - \mathbf{x}^*\| &\leq M\left\{ \int_0^1 L\|(1-s)(\mathbf{x}^{(i)} - \mathbf{x}^*)\| ds \right\} \|(\mathbf{x}^{(i)} - \mathbf{x}^*)\| \\ &= \tfrac{1}{2}ML\|(\mathbf{x}^{(i)} - \mathbf{x}^*)\|^2 \end{aligned} \tag{10}$$

as required. □

The Lipschitz condition (8), referred to as Lipschitz continuity, is stronger than continuity of $D\mathbf{f}$ on $B(\mathbf{x}^*, \delta)$, but weaker than differentiability. To illustrate its necessity, consider the following example.

Example Let $f(x) = x + x^{4/3}$. Then $f(0) = 0$, $f'(0) = 1$ and $f'(x) = 1 + 4/3x^{1/3}$, so f is C^1 and the Newton-Raphson sequence $\{x^{(i)}\}$ converges to $x^* = 0$. However

$$\lim_{i\to\infty} \frac{|x^{(i+1)}|}{|x^{(i)}|^2} = \lim_{i\to\infty} \frac{|(x^{(i)})^{-2/3}|}{|3 + 4(x^{(i)})^{1/3}|} = \infty \tag{11}$$

thus convergence is not quadratic. It can be easily shown that f'' is unbounded on any neighborhood of 0, does not exist at 0, and f' is not Lipschitz continuous on a neighborhood of 0.

3 Equilibrium equations

We consider the large-strain quasistatic response of a continuous system composed of several bodies undergoing contact conditions. For simplicity, we restrict ourselves to elastic materials and frictionless contact.

Since our objective is to identify the contribution to the equilibrium equations due to the contact constraints, we do not present details of the continuum discretization. Let $\mathbf{u}$ be the vector of displacement degrees of freedom, and let the discretized form of the total potential energy be given by [18]

$$\Pi(\mathbf{u}) = U(\mathbf{u}) - \mathbf{R}^T\mathbf{u} \tag{12}$$

where $U(\mathbf{u})$ is the strain energy and $\mathbf{R}$ is the load vector. We have assumed here displacement independent loading. Furthermore, let $\mathbf{g}(\mathbf{u})$ be the vector of contactor node gaps. We require that

$$\mathbf{g}(\mathbf{u}) \geq \mathbf{0} \tag{13}$$

i.e., no-interpenetration is allowed pointwise at the contactor nodes. The equilibrium problem then corresponds to

$$\min_{\mathbf{u}} \Pi(\mathbf{u}) \quad \text{such that} \quad \mathbf{g}(\mathbf{u}) \geq \mathbf{0} \tag{14}$$

Let $\boldsymbol{\lambda}$ be the vector of Lagrange multipliers corresponding to the constraints in (14). These multipliers correspond to the normal contact tractions on the target surfaces. Let

$$L(\mathbf{u}, \boldsymbol{\lambda}) = \Pi(\mathbf{u}) - \boldsymbol{\lambda}^T\mathbf{g}(\mathbf{u}) \tag{15}$$

be the corresponding Lagrangian function. The first order conditions for the solution of (14) are given by

$$[D_{\mathbf{u}}L(\mathbf{u}, \boldsymbol{\lambda})]^T = \mathbf{0} \tag{16}$$

$$\mathbf{g}(\mathbf{u}) \geq 0 \qquad \boldsymbol{\lambda} \geq \mathbf{0} \qquad \boldsymbol{\lambda}^T\mathbf{g}(\mathbf{u}) = 0 \tag{17}$$

Let $\mathbf{u}^*, \boldsymbol{\lambda}^*$ be a solution of (16–17). Denote by $\bar{\mathbf{g}}(\mathbf{u})$ those components of $\mathbf{g}(\mathbf{u})$ that correspond to nonzero components in $\boldsymbol{\lambda}^*$. Using (15) and (12) we have that $\mathbf{u}^*, \boldsymbol{\lambda}^*$ solve the system of nonlinear equations

$$[DU(\mathbf{u})]^T - [D\mathbf{g}(\mathbf{u})]^T\boldsymbol{\lambda} - \mathbf{R} = \mathbf{0} \tag{18}$$

$$\bar{\mathbf{g}}(\mathbf{u}) = \mathbf{0} \tag{19}$$

Here $[DU(\mathbf{u})]^T$ is the vector of internal nodal point forces. In the following we do not pursue the problem of identifying the components in $\bar{\mathbf{g}}(\mathbf{u})$, i.e. finding the set of active constraints, and concentrate instead upon the local properties of the system (18–19) around the solutions $\mathbf{u}^*$ and $\boldsymbol{\lambda}^*$.

According to the discussion in Section 2, in order to apply the Newton-Raphson method to solve the system (18-19) we need that the Jacobian matrix

$$\begin{bmatrix} D^2\Pi(\mathbf{u}) - \boldsymbol{\lambda}^T D^2\mathbf{g}(\mathbf{u}) & -[D\bar{\mathbf{g}}(\mathbf{u})]^T \\ -D\bar{\mathbf{g}}(\mathbf{u}) & \mathbf{0} \end{bmatrix} \tag{20}$$

evaluated at $\mathbf{u}^*$ and $\boldsymbol{\lambda}^*$ be nonsingular. This will be the case if $D^2\Pi(\mathbf{u}^*) - (\boldsymbol{\lambda}^*)^T D^2\mathbf{g}(\mathbf{u}^*)$ is positive definite and $D\bar{\mathbf{g}}(\mathbf{u}^*)$ has full rank. Assuming that the total potential energy is a strictly convex function of displacements, the first condition holds when $D^2\Pi(\mathbf{u})$ dominates the contact contribution $\boldsymbol{\lambda}^T D^2\mathbf{g}(\mathbf{u})$. The second requirement is automatically satisfied when the contact conditions are enforced at each contactor node.

Quadratic convergence to the solution, however, requires that the Jacobian matrix (20) satisfy the Lipschitz condition (8). Assuming that the discretized total potential energy function is sufficiently smooth, this condition is met provided the Hessian $D^2\mathbf{g}$ of the vector of gap functions is itself Lipschitz continuous.

Henceforth we seek an interpolation scheme for the kinematics of contact that satisfies the abovementioned requirements on $D^2\mathbf{g}$.

4 Contact kinematics

Consider a contactor node and the corresponding nodes on the target surface as indicated in Figure 1. Let $\mathbf{x}$ be the current coordinates of the contactor node and $\mathbf{a}_i$ with $i = 1, \ldots, n$ be the current coordinates of the nodes on the target surface. Suppose that $\mathbf{r}(s; \mathbf{a})$ is a curve that interpolates the target nodes, where s measures the arc length and $\mathbf{a} = (\mathbf{a}_1, \ldots, \mathbf{a}_n)$. We assume that the target body lies to the left of the direction of increase in s. In this section we derive the corresponding gap function g and its derivatives for this general interpolating curve. The resulting expression for D^2g is used in Section 5 to choose a specific form of $\mathbf{r}(s; \mathbf{a})$.

The unit tangent and normal vectors to the assumed target surface $\mathbf{r}(s; \mathbf{a})$ are given by

$$\mathbf{t}(s; \mathbf{a}) = \mathbf{r}'(s; \mathbf{a}) \tag{21}$$

$$\mathbf{n}(s; \mathbf{a}) = \mathbf{t}(s; \mathbf{a}) \times \mathbf{e} \tag{22}$$

where $()'$ indicates differentiation with respect to the parameter s, $\mathbf{e}$ is the unit vector pointing outwards from the plane in which all the nodes lie, and

× indicates cross product. The corresponding derivatives satisfy

$$\mathbf{t}'(s;\mathbf{a}) = k(s;\mathbf{a})\mathbf{n}(s;\mathbf{a}) \tag{23}$$

$$\mathbf{n}'(s;\mathbf{a}) = -k(s;\mathbf{a})\mathbf{t}(s;\mathbf{a}) \tag{24}$$

where $k(s;\mathbf{a})$ is the (signed) curvature of the target surface.

The "target point" is defined to be the closest point on the target surface to the contactor node. It is obtained by solving the problem

$$\min_{s} \tfrac{1}{2}\|\mathbf{x} - \mathbf{r}(s;\mathbf{a})\|^2 \tag{25}$$

Let $s^*(\mathbf{x},\mathbf{a})$ be the solution to (25) Assuming an interior solution, $s^*(\mathbf{x},\mathbf{a})$ satisfies identically the first order condition

$$[\mathbf{x} - \mathbf{r}(s^*;\mathbf{a})]^T\mathbf{t}(s^*;\mathbf{a}) = 0 \tag{26}$$

Let the function $f(s;\mathbf{x},\mathbf{a})$ be given by

$$f(s;\mathbf{x},\mathbf{a}) = [\mathbf{x} - \mathbf{r}(s;\mathbf{a})]^T\mathbf{n}(s;\mathbf{a}) \tag{27}$$

We define the gap function by

$$g(\mathbf{x},\mathbf{a}) = f(s^*(\mathbf{x},\mathbf{a});\mathbf{x},\mathbf{a}) \tag{28}$$

In view of equation (28) we have for the first derivatives of g

$$D_{\mathbf{x}}g = D_{\mathbf{x}}f + f'D_{\mathbf{x}}s^* \tag{29}$$

$$D_{\mathbf{a}}g = D_{\mathbf{a}}f + f'D_{\mathbf{a}}s^* \tag{30}$$

Differentiating equation (27) with respect to s we have

$$f'(s;\mathbf{x},\mathbf{a}) = -k(s;\mathbf{a})[\mathbf{x} - \mathbf{r}(s;\mathbf{a})]^T\mathbf{t}(s;\mathbf{a}) \tag{31}$$

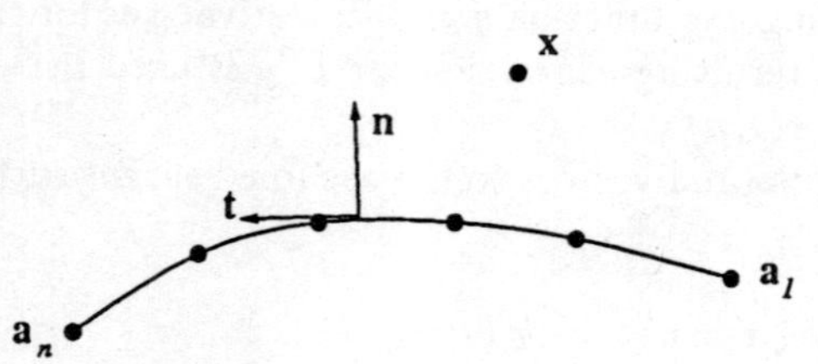

Figure 1: Contactor node and corresponding target nodes.

where we have used (21), (24), and the fact that $\mathbf{t}(s;\mathbf{a})^T\mathbf{n}(s;\mathbf{a}) = 0$ for all s. Evaluating (31) at $s = s^*(\mathbf{x},\mathbf{a})$, and using the first order condition (26) we obtain

$$f'(s^*;\mathbf{x},\mathbf{a}) = 0 \tag{32}$$

Differentiating (27) with respect to the nodal point coordinates yields

$$D_{\mathbf{x}}f = \mathbf{n}^T \tag{33}$$
$$D_{\mathbf{a}}f = (\mathbf{x}-\mathbf{r})^T D_{\mathbf{a}}\mathbf{n} - \mathbf{n}^T D_{\mathbf{a}}\mathbf{r} \tag{34}$$

Since $\mathbf{n}$ is a unit vector, each column of the matrix $D_{\mathbf{a}}\mathbf{n}$ is orthogonal to $\mathbf{n}$. It follows from (26) that when evaluated at s^*, equation (34) reduces to

$$D_{\mathbf{a}}f = -\mathbf{n}^T(D_{\mathbf{a}}\mathbf{r}) \tag{35}$$

Substituting equations (32), (33) and (35) into (29–30), we obtain for the first derivatives of the gap function

$$D_{\mathbf{x}}g = \mathbf{n}^T \tag{36}$$
$$D_{\mathbf{a}}g = -\mathbf{n}^T(D_{\mathbf{a}}\mathbf{r}) \tag{37}$$

To obtain the second derivatives, we have from equations (29), (30) and (32) that

$$D^2_{\mathbf{xx}}g = D^2_{\mathbf{xx}}f + (D_{\mathbf{x}}s^*)^T D_{\mathbf{x}}f' \tag{38}$$
$$D^2_{\mathbf{xa}}g = D^2_{\mathbf{xa}}f + (D_{\mathbf{x}}s^*)^T D_{\mathbf{a}}f' \tag{39}$$
$$D^2_{\mathbf{aa}}g = D^2_{\mathbf{aa}}f + (D_{\mathbf{a}}s^*)^T D_{\mathbf{a}}f' \tag{40}$$

Differentiating the identity (32) with respect to $\mathbf{x}$ and $\mathbf{a}$ and solving for $D_{\mathbf{x}}s^*$ and $D_{\mathbf{a}}s^*$ we have

$$D_{\mathbf{x}}s^* = -D_{\mathbf{x}}f'/f'' \tag{41}$$
$$D_{\mathbf{a}}s^* = -D_{\mathbf{a}}f'/f'' \tag{42}$$

Substitution in (38–40) yields

$$D^2_{\mathbf{xx}}g = D^2_{\mathbf{xx}}f - \frac{(D_{\mathbf{x}}f')^T D_{\mathbf{x}}f'}{f''} \tag{43}$$
$$D^2_{\mathbf{xa}}g = D^2_{\mathbf{xa}}f - \frac{(D_{\mathbf{x}}f')^T D_{\mathbf{a}}f'}{f''} \tag{44}$$
$$D^2_{\mathbf{aa}}g = D^2_{\mathbf{aa}}f - \frac{(D_{\mathbf{a}}f')^T D_{\mathbf{a}}f'}{f''} \tag{45}$$

Differentiating equations (33) and (34) and evaluating at s^* we have

$$D^2_{\mathbf{xx}}f = 0 \tag{46}$$
$$D^2_{\mathbf{xa}}f = D_{\mathbf{a}}\mathbf{n} \tag{47}$$
$$D^2_{\mathbf{aa}}f = -g(D_{\mathbf{a}}\mathbf{n})^T D_{\mathbf{a}}\mathbf{n} - (D_{\mathbf{a}}\mathbf{r})^T D_{\mathbf{a}}\mathbf{n} - (D_{\mathbf{a}}\mathbf{n})^T D_{\mathbf{a}}\mathbf{r} - \mathbf{n}^T D^2_{\mathbf{aa}}\mathbf{r} \tag{48}$$

while differentiating equation (31) gives

$$D_{\mathbf{x}}f' = -k\mathbf{t}^T \tag{49}$$

$$D_{\mathbf{a}}f' = -(\mathbf{x}-\mathbf{r})^T\mathbf{t}D_{\mathbf{a}}k - k(\mathbf{x}-\mathbf{r})^T D_{\mathbf{a}}\mathbf{t} + k\mathbf{t}^T D_{\mathbf{a}}\mathbf{r} \tag{50}$$

and

$$f'' = -k'(\mathbf{x}-\mathbf{r})^T\mathbf{t} - k(\mathbf{x}-\mathbf{r})^T\mathbf{t}' + k(\mathbf{r}')^T\mathbf{t} \tag{51}$$

Using expressions (23–24) and evaluating at $s = s^*$ we have

$$D_{\mathbf{a}}f' = k\mathbf{t}^T[gD_{\mathbf{a}}\mathbf{n} + D_{\mathbf{a}}\mathbf{r}] \tag{52}$$

and

$$f'' = k(1 - kg) \tag{53}$$

Substitution in (43–45) leads finally to

$$D^2_{\mathbf{xx}}g = -\frac{k}{1-kg}\mathbf{t}\mathbf{t}^T \tag{54}$$

$$D^2_{\mathbf{xa}}g = D_a\mathbf{n} + \frac{k}{1-kg}\mathbf{t}\mathbf{t}^T[gD_{\mathbf{a}}\mathbf{n} + D_{\mathbf{a}}\mathbf{r}] \tag{55}$$

$$\begin{aligned} D^2_{\mathbf{aa}}g = & -g(D_{\mathbf{a}}\mathbf{n})^T D_{\mathbf{a}}\mathbf{n} - (D_{\mathbf{a}}\mathbf{r})^T D_{\mathbf{a}}\mathbf{n} - (D_{\mathbf{a}}\mathbf{n})^T D_{\mathbf{a}}\mathbf{r} - \mathbf{n}^T D^2_{\mathbf{aa}}\mathbf{r} \\ & -\frac{k}{1-kg}[gD_{\mathbf{a}}\mathbf{n} + D_{\mathbf{a}}\mathbf{r}]^T\mathbf{t}\mathbf{t}^T[gD_{\mathbf{a}}\mathbf{n} + D_{\mathbf{a}}\mathbf{r}] \end{aligned} \tag{56}$$

Equations (36–37) and (54–56) give the required derivatives Dg and D^2g in terms of the gap g, the tangent and normal vectors $\mathbf{t}$ and $\mathbf{n}$, the curvature k, and the matrices $D_{\mathbf{a}}\mathbf{r}$, $D_{\mathbf{a}}\mathbf{n}$ and $D^2_{\mathbf{aa}}\mathbf{r}$.

5 Interpolation scheme

Given the target nodes $\mathbf{a}_1, \ldots, \mathbf{a}_n$, we choose to interpolate separately each target segment defined by the two consecutive nodes $\mathbf{a}_j$ and $\mathbf{a}_{j+1}$. We specify at each target node a slope and a curvature. This will yield continuity across adjacent segments of the second derivatives of the gap function with respect to the contactor and target node positions and finite jumps of the third derivatives. It can be shown that this is sufficient for Lipschitz continuity of the second derivatives.

The interpolation function over a target segment will therefore have to satisfy six independent conditions. The slope and curvature at each target node are tied to the positions of that node and of the two neighboring target nodes. Given $\mathbf{a}_{j-1}$, $\mathbf{a}_j$ and $\mathbf{a}_{j+1}$, we set the slope and the curvature at node $\mathbf{a}_j$ to correspond to the slope and curvature at $\mathbf{a}_j$ of a circle passing through $\mathbf{a}_{j-1}$, $\mathbf{a}_j$ and $\mathbf{a}_{j+1}$, as indicated in Figure 2. (Note that the end segments need to be treated differently.)

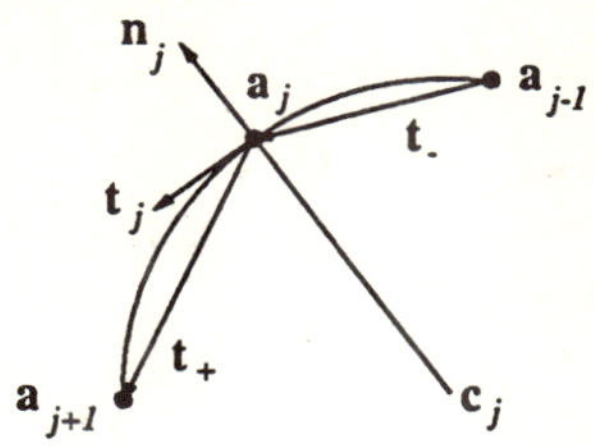

Figure 2: Choice of slope and curvature at target node $\mathbf{a}_j$.

Defining $\mathbf{t}_-$ and $\mathbf{t}_+$ by

$$\mathbf{t}_- = \mathbf{a}_j - \mathbf{a}_{j-1} \tag{57}$$

$$\mathbf{t}_+ = \mathbf{a}_{j+1} - \mathbf{a}_j \tag{58}$$

it can be shown that when $\mathbf{t}_- \times \mathbf{t}_+ \neq \mathbf{0}$ the position vector $\mathbf{c}_j$ of the center of the circle is given by

$$\mathbf{c}_j = \mathbf{a}_j + \tfrac{1}{2}\left[(\mathbf{t}_+ \cdot \mathbf{t}_+)\mathbf{t}_- + (\mathbf{t}_- \cdot \mathbf{t}_-)\mathbf{t}_+\right] \times \frac{(\mathbf{t}_+ \times \mathbf{t}_-)}{\|\mathbf{t}_+ \times \mathbf{t}_-\|^2} \tag{59}$$

We then have for the normal and tangent vectors

$$\mathbf{n}_j = \pm \frac{(\mathbf{c}_j - \mathbf{a}_j)}{\|\mathbf{c}_j - \mathbf{a}_j\|} \qquad \mathbf{t}_j = \mathbf{e} \times \mathbf{n}_j \tag{60}$$

where the sign is chosen so that $\mathbf{t}_j \cdot (\mathbf{a}_{j+1} - \mathbf{a}_{j-1}) > 0$. The signed curvature is given by

$$k_j = \frac{1}{\mathbf{n}_j \cdot (\mathbf{c}_j - \mathbf{a}_j)} \tag{61}$$

When $\mathbf{t}_- \times \mathbf{t}_+ = \mathbf{0}$ we take $\mathbf{t}_j = \mathbf{t}_-/\|\mathbf{t}_-\| = \mathbf{t}_+/\|\mathbf{t}_+\|$, $\mathbf{n}_j = \mathbf{t}_j \times \mathbf{e}$ and $k_j = 0$.

Consider the interpolation problem for the segment bounded by the nodes $\mathbf{a}_j$ and $\mathbf{a}_{j+1}$. In Figure 3 we show the local coordinate system used to define the interpolating curve $\phi(r)$, where r measures distance along the vector $\mathbf{t} = \mathbf{a}_{j+1} - \mathbf{a}_j$ from its midpoint.

This curve is constructed according to

$$\begin{aligned} \phi(r) = (\tan\alpha_j)\psi_j(r) + (\tan\alpha_{j+1})\psi_{j+1}(r) \\ + k_j\left(1+\tan^2\alpha_j\right)^{3/2}\theta_j(r) + k_{j+1}\left(1+\tan^2\alpha_{j+1}\right)^{3/2}\theta_{j+1}(r) \end{aligned} \tag{62}$$

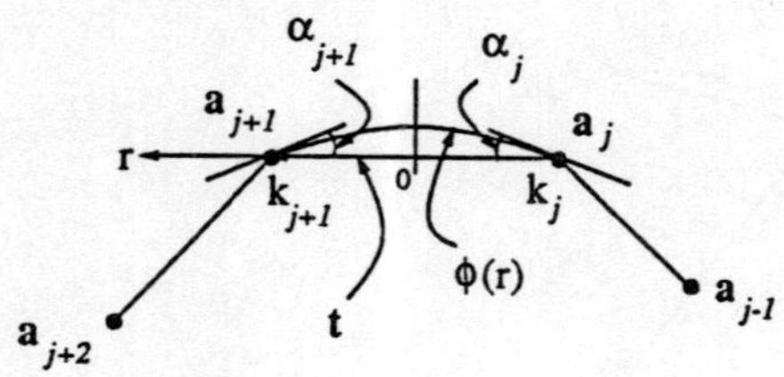

Figure 3: Local coordinate system over a target segment.

where

$$\tan\alpha_j = \frac{(\mathbf{t}_j \times \mathbf{t}) \cdot \mathbf{e}}{\mathbf{t}_j \cdot \mathbf{t}} \tag{63}$$

$$\tan\alpha_{j+1} = -\frac{(\mathbf{t}_{j+1} \times \mathbf{t}) \cdot \mathbf{e}}{\mathbf{t}_{j+1} \cdot \mathbf{t}} \tag{64}$$

and

$$\psi_j(r) = \frac{1}{2a^4}(a/2 - r)^3(a/2 + r)(5a + 6r) \tag{65}$$

$$\psi_{j+1}(r) = -\frac{1}{2a^4}(a/2 - r)(a/2 + r)^3(5a - 6r) \tag{66}$$

$$\theta_j(r) = \frac{1}{2a^3}(a/2 + r)^2(a/2 - r)^3 \tag{67}$$

$$\theta_{j+1}(r) = \frac{1}{2a^3}(a/2 + r)^3(a/2 - r)^2 \tag{68}$$

In these definitions $a = \|\mathbf{t}\|$. The function $\phi(r)$ so defined satisfies

$$\phi(-a/2) = 0 \tag{69}$$

$$\phi(a/2) = 0 \tag{70}$$

$$\phi'(-a/2) = \tan\alpha_j \tag{71}$$

$$\phi'(a/2) = -\tan\alpha_{j+1} \tag{72}$$

$$\phi''(-a/2) = k_j\left(1 + \tan^2\alpha_j\right)^{3/2} \tag{73}$$

$$\phi''(a/2) = k_{j+1}\left(1 + \tan^2\alpha_{j+1}\right)^{3/2} \tag{74}$$

Since the signed curvature for ϕ is given by

$$k(r) = \frac{\phi''(r)}{(1 + [\phi'(r)]^2)^{3/2}} \tag{75}$$

equations (73) and (74) ensure that

$$k(-a/2) = k_j \tag{76}$$
$$k(a/2) = k_{j+1} \tag{77}$$

Therefore the curve $\phi(r)$ satisfies all six required conditions.

6 Concluding remarks

Our objective in this paper was to develop the out-of-balance load vector and stiffness matrix contributions of a general contact interface interpolation and, based on the continuity requirements for quadratic convergence of the Newton-Raphson iteration, present a "conforming" interpolation scheme. We have shown that having a continuous normal to the contact surface is not sufficient for the contact stiffness matrices to be Lipschitz continuous. Actually, the interpolation has to provide Lipschitz continuity of the target surface curvature.

Since the discontinuities referred to appear only at the boundary points of the target segments, the rate of convergence would be affected whenever one or more of the contactor nodes lie in a neighborhood of these points in the converged solution. When this is the case an interpolating curve with continuous second derivatives can in general only ensure superlinear convergence, while piecewise linear surfaces clearly do not provide in general the required continuity conditions.

Equations (54–56) show the contribution of the contact formulation to the structural stiffness. Three special cases are of interest. When the target surface is rigid and flat (i.e. has zero curvature), all terms vanish. When we allow for a rigid target surface that is curved, only the term in (54) has to be included. In the more general case of a flexible target surface, finally, all terms in (54–56) have to be added.

We would like to emphasize that the kinematic description of Sections 4 and 5 remains valid when applied to other Lagrangian solution methods, and in particular is also suitable for the constraint function method given in reference [13].

References

[1] A. Francavilla and O. C. Zienkiewicz. A note on numerical computation of elastic contact problems. *International Journal for Numerical Methods in Engineering*, 9:913–924, 1975.

[2] J. R. Hughes, R. L. Taylor, J. L. Sackman, A. Curnier, and W. Kanoknukulchai. A finite element method for a class of contact-

impact problems. *Computer Methods in Applied Mechanics and Engineering*, 8:249–276, 1976.

[3] K. J. Bathe and A. B. Chaudhary. A solution method for planar and axisymmetric contact problems. *International Journal for Numerical Methods in Engineering*, 21:65–88, 1985.

[4] A. B. Chaudhary and K. J. Bathe. A solution method for static and dynamic analysis of three-dimensional contact problems with friction. *Computers & Structures*, 24:855–873, 1986.

[5] B. Nour-Omid and P. Wriggers. A two-level iteration method for solution of contact problems. *Computer Methods in Applied Mechanics and Engineering*, 54:131–144, 1986.

[6] N. Kikuchi and Y. J. Song. Remarks on relations between penalty and mixed finite element methods for a class of variational inequalities. *International Journal for Numerical Methods in Engineering*, 15:1557–1561, 1980.

[7] J. T. Oden. Exterior penalty methods for contact problems in elasticity. In W. Wunderlich, E. Stein, and K. J. Bathe, editors, *Nonlinear Finite Element Analysis in Structural Mechanics: Proceedings of the Europe-U.S. Workshop*. Springer-Verlag, July 1981.

[8] J. T. Oden and N. Kikuchi. Finite element methods for constrained problems in elasticity. *International Journal for Numerical Methods in Engineering*, 18:701–725, 1982.

[9] J. O. Hallquist, G. L. Goudreau, and D. J. Benson. Sliding interfaces with contact-impact in large-scale Lagrangian computations. *Computer Methods in Applied Mechanics and Engineering*, 51:107–137, 1985.

[10] J. C. Simo, P. Wriggers, and R. L. Taylor. A perturbed Lagrangian formulation for the finite element solution of contact problems. *Computer Methods in Applied Mechanics and Engineering*, 50:163–180, 1985.

[11] R. Glowinski, P. Le Tallec, and M. Vidrascu. Augmented Lagrangian techniques for solving frictionless contact problems in finite elasticity. In P. G. Bergan, K. J. Bathe, and W. Wunderlich, editors, *Finite Element Methods for Nonlinear Problems: Proceedings of the Europe-U.S. Symposium*. Springer-Verlag, August 1985.

[12] P. Wriggers, J. C. Simo, and R. L. Taylor. Penalty and augmented Lagrangian formulations for contact problems. In Middleton J. and G. N.

Pande, editors, *Proceedings of the International Conference on Numerical Methods in Engineering—Theory and Applications, NUMETA '85*, pages 97–106, January 1985.

[13] A. L. Eterovic and K. J. Bathe. On the treatment of inequality constraints arising from contact conditions in finite element analysis. *Computers & Structures*, 1991. In press.

[14] P. Wriggers and J. C. Simo. A note on tangent stiffness for fully nonlinear contact problems. *Communications in Applied Numerical Methods*, 1:199–203, 1985.

[15] P. Wriggers. On consistent tangent matrices for frictional contact problems. In G. N. Pande and J. Middleton, editors, *Proceedings of the International Conference on Numerical Methods in Engineering—Theory and Applications, NUMETA '87*, July 1987.

[16] H. Parisch. A consistent tangent stiffness matrix for three-dimensional non-linear contact analysis. *International Journal for Numerical Methods in Engineering*, 28:1803–1812, 1989.

[17] D. P. Bertsekas. *Constrained Optimization and Lagrange Multiplier Methods*. Academic Press, New York, 1982.

[18] K. J. Bathe. *Finite Element Procedures in Engineering Analysis*. Prentice-Hall, Englewood Cliffs, N.J., 1982.

On the Formulation and Numerical Treatment of Finite Deformation Frictional Contact Problems

T. A. LAURSEN & J. C. SIMO

Abstract

A formulation is presented in which the finite deformation frictional contact of a deformable body with a rigid obstacle is considered. In this discussion the notation and formulation of the finite deformation problem *without* friction, introduced in Simo & Laursen [1991], is first reviewed, leading to a new discussion of how the Coulomb friction constraint can be added to this description within the bounds of the finite deformation theory. Once the mathematical description of the constrained problem is set forth, it is discussed how penalty and augmented Lagrangian techniques are used to enforce the constraints in practice (as discussed in Simo & Laursen [1991]). A development of the virtual work of contact, with its linearization, is subsequently given for the special case of two-dimensional problems. Finally, the utility of the resulting treatment is demonstrated through two finite deformation examples.

1. Introduction.

The finite element treatment of frictional contact problems has been of increasing interest in the past decade, as evidenced by a growing body of work on the subject. Within the realm of the infinitesimal theory, substantial contributions to the understanding of frictional problems can be found, for example, in Campos, Oden & Kikuchi [1982], Oden & Pires [1984], and Gallego & Anza [1989], where formulations and implementations of classical and non-classical variants of the Coulomb friction law are discussed and numerically demonstrated. More recently, attention has also been directed toward the more difficult problem of frictional contact in *large deformations*, with implementations discussed in Ju & Taylor [1988], Kikuchi & Oden [1988], Chapter 12, and Wriggers, Vu Van, & Stein [1991].

In most of these references the penalty method is used for the enforcement of contact constraints, with the Lagrange multiplier method being another popular technique. The attributes of these two methods for solving constrained problems are well-documented, with the penalty method having the advantages of not introducing additional variables and maintaining definiteness of the equation system, albeit at the cost of possible penalty sensitivity. In the case of the frictional contact problem, the penalty method also provides a convenient framework within which integration algorithms developed for elastoplasticity (see, for example, Simo & Hughes [1991]) can be applied to the evolution of the frictional stress, subject to a slip (yield) condition given by the Coulomb law. This idea, discussed by Giannakopoulos [1989] and contained in many implementations (for example, Wriggers, Vu Van, & Stein [1991]), makes the penalty regularization particularly useful for frictional problems. More recently, even the problem of ill-conditioning has been addressed for frictional contact through the development of an augmented Lagrangian extension of the standard penalization, as given by the authors in Simo & Laursen [1991].

What remains, however, is for a sufficiently general formulation for frictional contact to be developed such that arbitrary discretizations, large deformation effects, multiple bodies in contact, and problems in both two and three dimensions are readily handled. Toward this end, this paper will present a framework for the treatment of the frictional contact of one deformable body with a rigid obstacle in large deformations. In particular, the formulation presented here is suitable for either two or three spatial dimensions and results in contact expressions for which augmented Lagrangian techniques discussed in Simo & Laursen [1991] are easily applied. Through a presentation of the finite element equations in two dimensions, it is shown how the formulation allows for a sensible and straightforward treatment of the contact contribution to the virtual work statement, leading to a closed form expression for both the residual and its linearization (tangent stiffness), needed for Newton-Raphson solution of the global equation system. Finally, numerical examples will be given which demonstrate the usefulness of the proposed formulation. It is felt that through the presentation of this one body problem, a framework is developed within which extensions to multiple body contact are naturally made and implemented in both two and three dimensions.

2. The Rigid Obstacle Problem in Finite Deformations.

In this section we develop the mathematical description of the problem we wish to consider, the motion of a deformable body in the presence of a rigid obstacle, with the contact between body and obstacle being governed by a Coulomb friction law. In so doing we set the stage for section 3, in which the finite element equations will be developed, enabling numerical treatment of this system. It is noteworthy that the development of sections 2 and 3 is somewhat novel in that all definitions and formulations are made in the continuum setting, without particular reference to the finite element discretization ultimately introduced. As such, our development will apply equally well to arbitrary discretizations and in this sense is quite general.

2.1. Notation.

We consider a deformable body, denoted by Ω in its reference configuration (see Figure 2.1). For the purposes of this discussion, Ω is taken to be an open subset of either $\mathbb{R}^2$ or $\mathbb{R}^3$. Points in Ω are denoted by $\boldsymbol{X}$ and are mapped onto the current configuration $\varphi(\Omega)$ via the deformation $\varphi : \Omega \rightarrow \mathbb{R}^2$ (or $\mathbb{R}^3$). We thus denote points in the current configuration by $\boldsymbol{x} = \varphi(\boldsymbol{X})$. The body undergoes the deformation φ in the presence of a rigid obstacle, which we shall assume for simplicity to be stationary in the reference frame of interest (although this need not in general be the case). The presence of this obstacle restricts the deformation of Ω to a subset of the ambient space called the *admissible region*, the interior of which we shall denote by K. In this paper we assume that the boundary of K, $\partial\mathsf{K}$, is smooth (at least C^3). We prescribe a part of $\partial\Omega$ as Γ, and consider this to be a part of the outer boundary of the body which shall contain all material points of contact with the obstacle surface ($\partial\mathsf{K}$) at all times. We take γ to be the image of Γ over φ.

Finally, we define a function g over the spatial domain, which has the property that $g < 0$ in K, $g = 0$ on $\partial\mathsf{K}$, and $g > 0$ elsewhere. Clearly, for the position of a point $\boldsymbol{X} \in \Omega$ to be admissible, $g(\varphi(\boldsymbol{X})) \leq 0$. We refer to g as the *gap function*, but defer its precise definition until section 3.

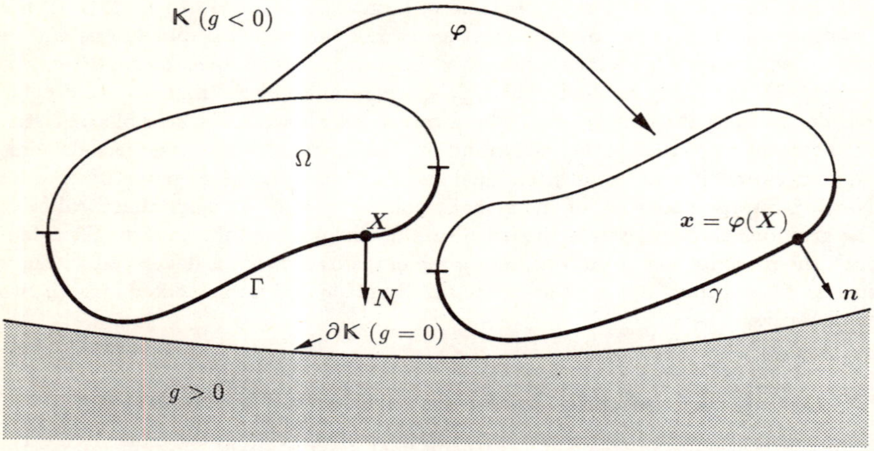

Figure 2.1. Notation for the rigid obstacle problem.

2.2. Review of the frictionless obstacle problem.

With the above notation in hand, we proceed to review a statement of the static frictionless contact problem previously given in Simo & Laursen [1991]. We begin with statements of the equilibrium equations and the boundary conditions:

$$\left.\begin{aligned} \text{Div}\boldsymbol{P} + \boldsymbol{f} &= \boldsymbol{0} \text{ in } \Omega, \\ \boldsymbol{P}\boldsymbol{N} &= \bar{\boldsymbol{t}} \text{ on } \Gamma_\sigma, \\ \varphi &= \bar{\varphi} \text{ on } \Gamma_\varphi, \end{aligned}\right\} \tag{2.1}$$

where

$\boldsymbol{P}$:= first Piola – Kirchhoff stress tensor

$\boldsymbol{f}$:= prescribed body force

$\boldsymbol{N}$:= outward normal in the reference configuration and Γ_σ and Γ_φ are the parts of $\partial\Omega$ upon which the traction ($\bar{t}$) and displacement ($\bar{\varphi}$) are prescribed. It is to be noted that we say nothing about the constitutive law governing $\boldsymbol{P}$; the development which follows makes no assumptions as to what the material behavior may be, other than to say that the description must be consistent with large deformation constitutive theory.

We next proceed to state the contact conditions, which amount to a restriction on the admissible deformations $\varphi(\cdot)$ of the reference configuration Ω:

For all $\boldsymbol{X} \in \Gamma$, and all admissible deformations $\boldsymbol{x} = \varphi(\boldsymbol{X})$, we require that:

$$\left.\begin{aligned} g(\boldsymbol{x}) &\leq 0, \\ t_N = \boldsymbol{\nu} \cdot \boldsymbol{PN} &\geq 0, \\ t_N g(\boldsymbol{x}) &= 0, \\ \boldsymbol{t}_T = -\boldsymbol{PN} + t_N \boldsymbol{\nu} &= \boldsymbol{0}, \end{aligned}\right\} \tag{2.2}$$

where

$\boldsymbol{\nu} := -\boldsymbol{n}(\boldsymbol{x})$, where $\boldsymbol{n}$ is the outward normal to the body in the current configuration

$t_N :=$ the normal contact pressure (note the sign change)

$\boldsymbol{t}_T :=$ the frictional traction (again note a sign change)

Inspection of equations (2.2) reveals that $(2.2)_1$ furnishes the kinematic restriction of impenetrability, while $(2.2)_2$ requires the contact pressure to be compressive. Equation $(2.2)_3$ mandates that this contact pressure may only be nonzero where contact occurs ($g = 0$), while $(2.2)_4$ assures that the response is frictionless. The reason for the introduction of $\boldsymbol{\nu}$, a unit vector with opposite sense of $\boldsymbol{n}$, will be made clear in section 3. For the present, we remark that condition $(2.2)_4$ will be replaced by an additional set of restrictions corresponding to the Coulomb friction law in the next section. Satisfaction of the governing equations (2.1) subject to the contact restrictions (2.2) constitutes solution of the frictionless rigid obstacle problem.

2.3. Introduction of the frictional constraint.

To introduce the frictional constraint into this problem we next consider a motion of the body $\varphi(\cdot, t) : \Omega \to \mathsf{R}^3$, where $t \in [0, T]$ denotes time. Thus, we consider a point $\boldsymbol{X} \in \Gamma$, with current position $\boldsymbol{x} = \varphi(\boldsymbol{X}, t)$. Letting $\boldsymbol{y}$ denote any spatial point of $\partial\mathsf{K}$, we define for $\boldsymbol{x} \in \gamma$ a point $\bar{\boldsymbol{y}} \in \partial\mathsf{K}$, where $\bar{\boldsymbol{y}}$ is the point of $\partial\mathsf{K}$ for which $\|\boldsymbol{x} - \boldsymbol{y}\|$ is minimum. Clearly, if contact occurs at $\boldsymbol{x} = \varphi(\boldsymbol{X}, t)$, $\boldsymbol{x} = \bar{\boldsymbol{y}}$, while otherwise the two points are distinct. Figure 2.2 may be consulted for a graphical illustration of these ideas.

Remark: Concerning the role played by the time variable in this section, we emphasize that the Coulomb friction constraint we shall introduce is *rate independent*. Hence, if attention is restricted to the quasistatic problem, as we do here, $t \in [0, T]$ need not coincide with the actual time. In fact, t can be taken as any parameter indexing subsequent configurations. ■

Having made the definition of $\bar{\boldsymbol{y}}(t)$ for $\boldsymbol{x} = \boldsymbol{\varphi}(\boldsymbol{X}, t)$ with $\boldsymbol{X} \in \Gamma$, we return to the assumption introduced in section 2.1; namely, that $\partial\mathsf{K}$ is smooth (at least C^3). We exploit this assumption in the following way: for a 2-D (3-D) problem, we assume $\boldsymbol{y}$ takes a parametric form $\boldsymbol{y} = \boldsymbol{y}(\xi)$ ($\boldsymbol{y} = \boldsymbol{y}(\xi, \eta)$ in 3-D) in some open subset of $\partial\mathsf{K}$ containing $\bar{\boldsymbol{y}}(t)$, where the mapping $\boldsymbol{y}(\xi)$ ($\boldsymbol{y}(\xi, \eta)$) is such that the third partial derivative (derivatives) exist. In a finite element setting, the domain $\mathsf{I} \subset \mathsf{R}$ (or $\mathsf{I} \subset \mathsf{R}^2$) such that $\xi \in \mathsf{I}$ (or $(\xi, \eta) \in \mathsf{I}$) is naturally defined via the isoparametric map.

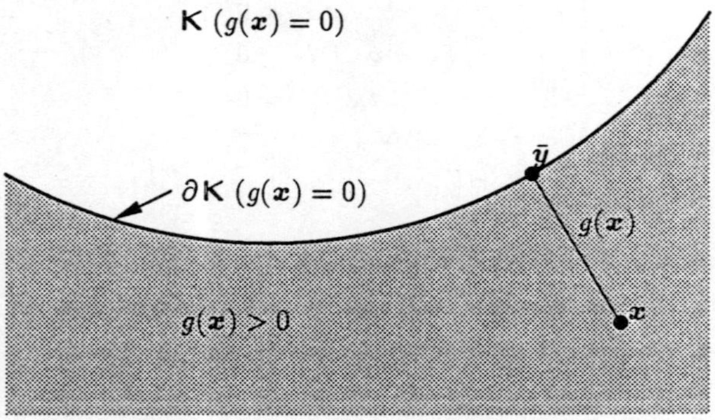

Figure 2.2. Definition of the closest point $\bar{\boldsymbol{y}}$ to a point $\boldsymbol{x} \in \gamma$, shown in two dimensions.

Now let $\bar{\xi}(t) \in \mathsf{I}$ (or $(\bar{\xi}(t), \bar{\eta}(t)) \in \mathsf{I}$) be the point in the domain I of the parametrization for which $\bar{\boldsymbol{y}}(t) = \boldsymbol{y}(\bar{\xi}(t))$ (or $\bar{\boldsymbol{y}}(t) = \boldsymbol{y}(\bar{\xi}(t), \bar{\eta}(t))$). Note that this point is a function of time. In addition, we may define a basis at each point in an open set about $\bar{\boldsymbol{y}}(t)$ as follows:

$$\textbf{2-D} \quad \left.\begin{aligned} \boldsymbol{e}_1(\xi) &= \boldsymbol{y}_{,\xi}(\xi) \\ \boldsymbol{e}_2 &= \text{unit vector into the plane} \\ \boldsymbol{e}_3(\xi) &= \frac{\boldsymbol{e}_1(\xi) \times \boldsymbol{e}_2}{\|\boldsymbol{e}_1(\xi)\|} \end{aligned}\right\} \tag{2.3}$$

$$\textbf{3-D} \quad \left.\begin{aligned} \boldsymbol{e}_1(\xi, \eta) &= \boldsymbol{y}_{,\xi}(\xi, \eta) \\ \boldsymbol{e}_2(\xi, \eta) &= \boldsymbol{y}_{,\eta}(\xi, \eta) \\ \boldsymbol{e}_3(\xi, \eta) &= \frac{\boldsymbol{e}_1(\xi, \eta) \times \boldsymbol{e}_2(\xi, \eta)}{\|\boldsymbol{e}_1(\xi, \eta) \times \boldsymbol{e}_2(\xi, \eta)\|} \end{aligned}\right\} \tag{2.4}$$

In each case, a set of dual basis vectors may be computed such that $\boldsymbol{e}^i \cdot \boldsymbol{e}_j = \delta^i_j$ (i, j=1-3). We now particularize these bases to their values at $\bar{\boldsymbol{y}}$, and make the following definitions:

$$\textbf{2-D} \quad \left.\begin{aligned} \boldsymbol{\tau}_1 &= \boldsymbol{e}_1(\bar{\xi}(t)) \\ \boldsymbol{\nu} &= \boldsymbol{e}_3(\bar{\xi}(t)) \end{aligned}\right\} \tag{2.5}$$

$$\textbf{3-D} \quad \left.\begin{aligned} \boldsymbol{\tau}_1 &= \boldsymbol{e}_1(\bar{\xi}(t), \bar{\eta}(t)) \\ \boldsymbol{\tau}_2 &= \boldsymbol{e}_2(\bar{\xi}(t), \bar{\eta}(t)) \\ \boldsymbol{\nu} &= \boldsymbol{e}_3(\bar{\xi}(t), \bar{\eta}(t)) \end{aligned}\right\} \tag{2.6}$$

The dual bases at $\bar{\boldsymbol{y}}$ are similarly defined and denoted; for example, in 2-D we define $\boldsymbol{\tau}^1$ such that $\boldsymbol{\tau}^1 \cdot \boldsymbol{\tau}_1 = 1$ and $\boldsymbol{\tau}^1 \cdot \boldsymbol{\nu} = 0$. We note that in both the 2-D and 3-D cases $\boldsymbol{\nu}$ is its own dual. Although not explicitly specified, the bases in (2.5) and (2.6) are functions of time because of the dependence of $\bar{\boldsymbol{y}}$ on the motion $\boldsymbol{\varphi}(\boldsymbol{X}, t)$ of the material point $\boldsymbol{X} \in \Gamma$.

With all this notation at hand we return to consideration of any point $\boldsymbol{X} \in \Gamma$ and write the expression of its tangential velocity with respect to the obstacle. Holding X fixed, the determination of $\bar{\boldsymbol{y}}$ becomes dependent on time; thus, with a slight abuse of notation we write $\bar{\boldsymbol{y}}(\boldsymbol{X}, t)$, $\bar{\xi}(\boldsymbol{X}, t)$, and $\bar{\eta}(\boldsymbol{X}, t)$. The time derivative of the function $\bar{\boldsymbol{y}}(\boldsymbol{X}, t)$ is the rate of change of the *projection* of $\boldsymbol{\varphi}(\boldsymbol{X}, t)$ on $\partial\mathsf{K}$, which we define as the *tangential relative velocity* $\boldsymbol{V}_T$:

$$\textbf{2-D} \quad \left.\begin{aligned} \boldsymbol{V}_T(\boldsymbol{X}, t) &= \frac{d}{dt}[\bar{\boldsymbol{y}}(\boldsymbol{X}, t)] \\ &= \boldsymbol{y}_{,\xi}(\bar{\xi})\dot{\bar{\xi}}(\boldsymbol{X}, t) \\ &= \dot{\bar{\xi}}(\boldsymbol{X}, t)\boldsymbol{\tau}_1 \end{aligned}\right\} \tag{2.7}$$

$$\textbf{3-D} \quad \left.\begin{aligned} \boldsymbol{V}_T(\boldsymbol{X}, t) &= \frac{d}{dt}[\bar{\boldsymbol{y}}(\boldsymbol{X}, t)] \\ &= \boldsymbol{y}_{,\xi}(\bar{\xi}, \bar{\eta})\dot{\bar{\xi}}(\boldsymbol{X}, t) + \boldsymbol{y}_{,\eta}(\bar{\xi}, \bar{\eta})\dot{\bar{\eta}}(\boldsymbol{X}, t) \\ &= \dot{\bar{\xi}}(\boldsymbol{X}, t)\boldsymbol{\tau}_1 + \dot{\bar{\eta}}(\boldsymbol{X}, t)\boldsymbol{\tau}_2 \end{aligned}\right\} \tag{2.8}$$

Expressed in words, $\boldsymbol{V}_T(\boldsymbol{X}, t)$ is the tangential component of the relative material velocity between the point $\boldsymbol{X} \in \Gamma$ and the adjacent obstacle.

Using this definition, we may set forth the Coulomb friction law:

$$\left.\begin{aligned} \Phi &:= \|\boldsymbol{t}_T\| - \mu t_N \leq 0, \\ \boldsymbol{V}_T &= \zeta \frac{\partial}{\partial \boldsymbol{t}_T}\Phi = \zeta \frac{\boldsymbol{t}_T}{\|\boldsymbol{t}_T\|}, \\ \zeta &\geq 0, \\ \zeta\Phi &= 0. \end{aligned}\right\} \tag{2.9}$$

where $\boldsymbol{t}_T$ is as given in $(2.2)_4$, resolved in the basis we have introduced such that $\boldsymbol{t}_T = t_T^1 \boldsymbol{\tau}_1$ in 2-D and $\boldsymbol{t}_T = t_T^1 \boldsymbol{\tau}_1 + t_T^2 \boldsymbol{\tau}_2$ in 3-D.

Here, equation $(2.9)_1$ represents the Coulomb, or slip condition, with Φ being the *slip function*, analogous to the yield function in plasticity. Equation $(2.9)_2$ gives the slip law, which together with $(2.3)_3$ ensures that all slip occurs in the opposite direction of the applied tangential traction. Finally, equation $(2.9)_4$ requires that slip only occurs when $\Phi = 0$; otherwise *perfect stick* prevails. We also note that as a consequence of these equations $\boldsymbol{t}_T = \mathbf{0}$ if $t_N = 0$.

If equations (2.9) are used to replace $(2.2)_4$, we obtain the contact conditions for the frictional problem, which must hold, for all time, for all $\boldsymbol{X} \in \Gamma$. Satisfaction of these contact constraints while also satisfying equations (2.1) constitutes solution of the frictional contact problem.

3. Finite Element Treatment of the Frictional Contact Problem.

With the problem definition of section 2 in hand, we next proceed to develop a finite element procedure by which the problem may be solved numerically. In order to eliminate unwanted restrictions on the solution space which result from the presence of the contact conditions (see Kikuchi & Oden [1988] for a detailed discussion), we first introduce a penalty regularization for the contact conditions. At the same time, we shall also briefly discuss the manner in which this regularization can be extended to an *augmented Lagrangian* framework, as given in Simo & Laursen [1991]. Following this development, it will be discussed how the regularized Coulomb friction law may be integrated in time to produce an update scheme for the contact tractions. Finally, all these ingredients will be put together, using the two dimensional case as a model problem, to develop the expression of *virtual work* (and its linearization) corresponding to this formulation.

3.1. Penalty and augmented Lagrangian enforcement of constraints.

We begin by defining the gap function $g(\boldsymbol{x})$ as promised in section 2.1. Crucial to this definition is the concept of the closest point projection, as implied heavily by the discussion of section 2.3:

$$\left.\begin{aligned} g &= |g| \,\operatorname{sign}(g) \\ |g| &= \|\boldsymbol{x} - \bar{\boldsymbol{y}}\| = min_{\boldsymbol{y} \in \partial \mathsf{K}} \|\boldsymbol{x} - \boldsymbol{y}\| \\ \operatorname{sign}(g) &= \begin{cases} 1 \text{ if } (\bar{\boldsymbol{y}} - \boldsymbol{x}) \cdot \boldsymbol{\nu}(\bar{\boldsymbol{y}}) \geq 0 \\ -1 \text{ otherwise} \end{cases} \end{aligned}\right\} \tag{3.1}$$

As defined previously, $\bar{\boldsymbol{y}}$ is the *closest point projection* of $\boldsymbol{x} = \varphi(\boldsymbol{X}, t)$ onto $\partial\mathsf{K}$, and $\boldsymbol{\nu}(\bar{\boldsymbol{y}})$ is the outward normal to the obstacle at $\bar{\boldsymbol{y}}$. Expressed in words, $|g|$ is the distance between $\boldsymbol{x}$ and $\bar{\boldsymbol{y}}$, and $\operatorname{sign}(g)$ is given by the admissibility of $\boldsymbol{x}$. Additionally, we now see the reason for the introduction of $\boldsymbol{\nu}$ in (2.2): for points $\boldsymbol{X} \in \Gamma$ such that $\boldsymbol{x} = \varphi(\boldsymbol{X}, t)$ is in contact with $\partial\mathsf{K}$ ($g(\boldsymbol{x}) = 0$) at time t, $\boldsymbol{\nu}(\bar{\boldsymbol{y}})$ is indeed the negative of $\boldsymbol{n}(\boldsymbol{x})$. For points not in contact ($g(\boldsymbol{x}) < 0$) this is no longer true, but in this case $\boldsymbol{t}(\boldsymbol{x}) = \boldsymbol{PN}(\boldsymbol{X}) = \mathbf{0}$, making this detail insignificant. It will become apparent in what follows that $\boldsymbol{\nu}$ is more convenient for our implementation than is $\boldsymbol{n}$.

We now regularize the normal contact conditions $(2.2)_{1-3}$ by the introduction of the following penalization:

$$\left.\begin{aligned} t_N &= \epsilon_N \langle g \rangle \\ \langle g \rangle &= \begin{cases} g \text{ if } g \geq 0 \\ 0 \text{ otherwise} \end{cases} \end{aligned}\right\} \tag{3.2}$$

Equations (3.2) shall hereafter replace $(2.2)_{1-3}$. The quantity ϵ_N, the *normal penalty parameter*, causes conditions $(2.2)_{1-3}$ to be exactly satisfied as it tends to infinity. *Penalty sensitivity* refers to the possible difficulty in choosing ϵ_N large enough to adequately enforce constraints without causing ill-conditioning; more shall be said about this at the end of the section.

Turning to the frictional constraint, we note immediately that our definition of $\boldsymbol{V}_T$ possesses a very desirable quality; it *transforms objectively* under superposed spatial rigid body motions. By inspection of (2.7) and (2.8) we see that the components $V_T^1 = \dot{\bar{\xi}}$ and $V_T^2 = \dot{\bar{\eta}}$ are unaltered by a change of frame of the form $\boldsymbol{x}^+ = \boldsymbol{Q}(t)\boldsymbol{x} + \boldsymbol{c}(t)$, where $\boldsymbol{Q}(t)$ represents a proper orthogonal transformation and $\boldsymbol{c}(t)$ a rigid translation. Additionally, the bases $\boldsymbol{\tau}^i$ transform according to $\boldsymbol{\tau}^{i^+} = \boldsymbol{Q}\boldsymbol{\tau}^i$. Because of these facts we have the result that $\boldsymbol{V}_T^+(\boldsymbol{X},t) = \boldsymbol{Q}\boldsymbol{V}_T(\boldsymbol{X},t)$. It should be carefully noted that this objectivity of $\boldsymbol{V}_T$ holds despite the fact that the material velocity field $\boldsymbol{V}(\boldsymbol{X},t) = \dot{\boldsymbol{\varphi}}(\boldsymbol{X},t)$ is *not* objective. We exploit the objectivity of $\boldsymbol{V}_T$ to construct an objective frictional regularization, which we now give:

$$\left.\begin{aligned} \Phi &:= \|\boldsymbol{t}_T\| - \mu t_N \leq 0, \\ \boldsymbol{V}_T - \zeta \frac{\boldsymbol{t}_T}{\|\boldsymbol{t}_T\|} &= \frac{1}{\epsilon_T} \overset{\triangledown}{\boldsymbol{t}}_T, \\ \zeta &\geq 0, \\ \zeta \Phi &= 0. \end{aligned}\right\} \tag{3.3}$$

where

$$\overset{\triangledown}{\boldsymbol{t}}_T = \begin{cases} \dot{t}_T^1 \boldsymbol{\tau}_1 \text{ in 2-D} \\ \dot{t}_T^1 \boldsymbol{\tau}_1 + \dot{t}_T^2 \boldsymbol{\tau}_2 \text{ in 3-D} \end{cases} \tag{3.4}$$

Equations (3.3) are seen to be identical to equations (2.9) with the exception of $(3.3)_2$, which is a penalization of the slip rule $(2.9)_2$ that is exact as $\epsilon_T \to \infty$. The objective rate $\overset{\triangledown}{\boldsymbol{t}}_T$ introduced in $(3.3)_2$ and defined in (3.4) corresponds to the standard notion of the Lie derivative, and ensures that the proposed evolution law is frame indifferent (as can be verified). This definition also proves to be convenient for later implementation.

The regularization for the frictional obstacle problem is now complete, as summarized by equations (3.2) and (3.3). It is these expressions which we shall hereafter develop and work into the proposed finite element scheme. Before leaving this section, however, it is worthwhile to discuss how *augmented Lagrangians* can be incorporated into these definitions, with the goal of reducing the aforementioned *penalty sensitivity*. The normal and frictional contact laws can be rewritten in this context as follows:

$$t_N = \langle \lambda_N + \epsilon_N g \rangle \tag{3.5}$$

and

$$\left.\begin{aligned} \Phi &:= \|\boldsymbol{t}_T\| - \mu t_N \leq 0, \\ \boldsymbol{V}_T - \zeta \frac{\boldsymbol{t}_T}{\|\boldsymbol{t}_T\|} &= \frac{1}{\epsilon_T}(\overset{\triangledown}{\boldsymbol{t}}_T - \overset{\triangledown}{\boldsymbol{\lambda}}_T), \\ \zeta &\geq 0, \\ \zeta\Phi &= 0. \end{aligned}\right\} \tag{3.6}$$

where the objective rate (i.e., the Lie derivative) $(\overset{\triangledown}{\cdot})$ is as defined in (3.4). When integrated in the manner discussed in the next section, these expressions lead to a time-stepping algorithm for frictional stress which includes the effects of *estimated Lagrange multipliers* λ_N and $\boldsymbol{\lambda}_T$. Together with update formulas for these multipliers, this helps define powerful iterative schemes which enable very accurate satisfaction of constraints for *finite* penalties. Due to the length of this development the augmented Lagrangian approach will not be discussed further; the interested reader is referred to Simo & Laursen [1991] for full details.

3.2. Time integration of the frictional constraint.

Returning to (3.3), we next wish to integrate the frictional evolution laws with respect to time to obtain a time-stepping algorithm for the frictional tractions. We do this with the idea that the motion of our body in the presence of the obstacle will be solved for *incrementally* in time; thus the problem can be broken down into a series of subintervals $[t_n, t_{n+1}]$ such that in each subinterval we know all information (displacements, stresses, etc.) at t_n and wish to find these quantities at t_{n+1}.

Accordingly, we shall apply a *backward Euler* scheme to equations (3.3), mimicking a common technique for integrating elastoplastic constitutive relations (see Simo & Hughes [1991]). In doing so, we also appeal to definitions (2.7) and (2.8) for $\boldsymbol{V}_T$ and (3.4) for $\overset{\triangledown}{\boldsymbol{t}}_T$. Performing this operation leads to the following:

$$\left.\begin{aligned} \Phi_{n+1} &:= \|\boldsymbol{t}_{T_{n+1}}\| - \mu t_{N_{n+1}} \leq 0, \\ t^1_{T_{n+1}} &= t^1_{T_n} + \epsilon_T \left[\bar{\xi}_{n+1}(\boldsymbol{X}) - \bar{\xi}_n(\boldsymbol{X}) - \Delta\zeta \frac{t^1_{T_{n+1}}}{\|\boldsymbol{t}_{T_{n+1}}\|}\right] \\ t^2_{T_{n+1}} &= t^2_{T_n} + \epsilon_T \left[\bar{\eta}_{n+1}(\boldsymbol{X}) - \bar{\eta}_n(\boldsymbol{X}) - \Delta\zeta \frac{t^2_{T_{n+1}}}{\|\boldsymbol{t}_{T_{n+1}}\|}\right] \quad \text{(if 3D)} \\ \Delta\zeta &\geq 0, \\ \Delta\zeta\Phi_{n+1} &= 0. \end{aligned}\right\} \tag{3.7}$$

In these expressions, subscripts of the form $(\cdot)_n$ refer to quantities at time t_n (known data), while $(\cdot)_{n+1}$ refers to a quantity at time t_{n+1} and represents an unknown value to be found. The problem posed in (3.7) is readily solved by the introduction of a *trial state* and subsequent *return map*; substantial discussion and analysis of this type of algorithm is given in Simo & Hughes [1991]. In order to expedite our development we refer directly to Box 3.1, which outlines the integration algorithm for the contact tractions.

Box 3.1. Integration algorithm for contact tractions

1. Known : $\boldsymbol{t}_n(\boldsymbol{X}),\ \bar{\xi}_n(\boldsymbol{X})$
2. Compute trial state :

$$t_{N_{n+1}} = \epsilon_N \langle g_{n+1} \rangle$$

$$t^{1^{\text{trial}}}_{T_{n+1}} = t^1_{T_n} + \epsilon_T[\bar{\xi}_{n+1}(\boldsymbol{X}) - \bar{\xi}_n(\boldsymbol{X})]$$

$$t^{2^{\text{trial}}}_{T_{n+1}} = t^2_{T_n} + \epsilon_T[\bar{\eta}_{n+1}(\boldsymbol{X}) - \bar{\eta}_n(\boldsymbol{X})] \quad \text{(if 3D)}$$

$$\Phi^{\text{trial}}_{n+1} = \|\boldsymbol{t}^{\text{trial}}_{T_{n+1}}\| - \mu t_{N_{n+1}}$$

3. Return if necessary :

IF ($\Phi^{\text{trial}}_{n+1} \leq 0$) THEN

$$t^i_{T_{n+1}} = t^{i^{\text{trial}}}_{T_{n+1}} \qquad \text{(stick)}$$

ELSE (return map)

$$t^i_{T_{n+1}} = \mu t_{N_{n+1}} \frac{t^{i^{\text{trial}}}_{T_{n+1}}}{\|\boldsymbol{t}^{\text{trial}}_{T_{n+1}}\|} \qquad \text{(slip)}$$

ENDIF

4. Complete contact traction description :

$$\boldsymbol{t}_{n+1}(\boldsymbol{X}) = \begin{cases} t_{N_{n+1}} \boldsymbol{\nu}_{n+1}(\bar{\boldsymbol{y}}) - t^1_{T_{n+1}} \boldsymbol{\tau}_{1_{n+1}}(\bar{\boldsymbol{y}}) \text{ if 2-D} \\ t_{N_{n+1}} \boldsymbol{\nu}_{n+1}(\bar{\boldsymbol{y}}) - t^1_{T_{n+1}} \boldsymbol{\tau}_{1_{n+1}}(\bar{\boldsymbol{y}}) - t^2_{T_{n+1}} \boldsymbol{\tau}_{2_{n+1}}(\bar{\boldsymbol{y}}) \text{ if 3-D} \end{cases}$$

It is important to emphasize what is involved in developing equation (3.7) and subsequently Box 3.1. Given a point $\boldsymbol{X} \in \Gamma$, we first evaluate (on the basis of $\varphi_{n+1}(\boldsymbol{X})$) what the closest point $\bar{\boldsymbol{y}}$ is. In the integration of the rate equation $(3.3)_2$ discussed above, the identification of $\boldsymbol{X}$ is *held constant.* We may think of the t^i_T and t_N as being values associated with the point $\boldsymbol{X}$, and of the bases $\boldsymbol{\tau}^i$ and $\boldsymbol{\tau}_i$ as being associated with $\bar{\boldsymbol{y}}$. With this conceptualization, and because of the special definitions of $\boldsymbol{V}_T$ and $\overset{\triangledown}{\boldsymbol{t}}_T$ made earlier, our discrete integration algorithm given in Box 3.1 is frame indifferent. It also has the advantages of being quite simple to perform and amenable to exact linearization as will be seen in the next section.

3.3. Formulation and linearization of the virtual work of contact.

With the penalization and integration scheme in hand for the contact conditions, we now return to the governing equations and construct a weak form of the boundary value problem we have posed. Multiplying $(2.1)_1$ by admissible variations, integrating over Ω, and applying integration by parts in the usual manner gives:

$$\left.\begin{aligned} G(\varphi,\delta\varphi) &:= \int_{\Omega} \boldsymbol{P}\cdot \mathrm{Grad}[\delta\varphi]\, d\Omega - \int_{\Omega} \boldsymbol{f}\cdot\delta\varphi\, d\Omega - \int_{\Gamma_\sigma} \bar{\boldsymbol{t}}\cdot\delta\varphi\, d\Gamma \\ &= \int_{\Gamma} \boldsymbol{t}(\boldsymbol{X},t)\cdot\delta\varphi(\boldsymbol{X})\, d\Gamma \end{aligned}\right\} \tag{3.8}$$

which must hold for all $\delta\varphi$ with $\delta\varphi = 0$ on Γ_φ and for all time t. The quantity we have defined in $(3.8)_1$ as $G(\varphi,\delta\varphi)$ is recognized as the usual sum of the internal virtual work and the virtual work of the applied body forces and tractions for the quasistatic problem; this contribution to the finite element equations is attained in the standard way and will not be discussed further.

The right hand side of $(3.8)_2$, on the other hand, is the virtual work of the contact tractions, to which we now devote our full attention. For simplicity, in what follows we shall consider the *two dimensional* case. Particularized to time t_{n+1}, the algorithmic approximation $\boldsymbol{t}_{n+1}(\boldsymbol{X})$ to $\boldsymbol{t}(\boldsymbol{X},t_{n+1})$ is as given in Box 3.1, while $\delta\varphi(\boldsymbol{X})$ is the (material) variation at $\boldsymbol{X}\in\Gamma$, which we shall now investigate. We may write $\delta\varphi(\boldsymbol{X})$ as follows:

$$\delta\varphi(\boldsymbol{X}) = [\boldsymbol{\nu}\cdot\delta\varphi(\boldsymbol{X})]\boldsymbol{\nu} + [\boldsymbol{\tau}_1\cdot\delta\varphi(\boldsymbol{X})]\boldsymbol{\tau}^1 \tag{3.9}$$

We now introduce an approximation for $\boldsymbol{\tau}_1\cdot\delta\varphi(\boldsymbol{X})$ which we shall employ in the subsequent formulation and implementation. This approximation takes the following form:

$$\delta\varphi(\boldsymbol{X})\cdot\boldsymbol{\tau}_1(\bar{\boldsymbol{y}}) \approx \|\boldsymbol{\tau}_1(\bar{\boldsymbol{y}})\|^2\delta\bar{\xi} \tag{3.10}$$

where $\delta\bar{\xi}$ is the variation in $\bar{\xi}$ caused by the variation in $\varphi(\boldsymbol{X},t)$. This approximation has two important properties: first, it becomes *exact* as $g(\varphi(\boldsymbol{X},t))\to 0$, and second, it gives the virtual work an extremely compact form which, when linearized, leads to a *completely symmetric* system of equations in the *frictional stick* case.

To see these points, we first consider the equation by which $\bar{\xi}$ is determined, given $\boldsymbol{X}\in\Gamma$. By virtue of the fact that $\bar{\boldsymbol{y}}(\boldsymbol{X},t) = \boldsymbol{y}(\bar{\xi}(\boldsymbol{X},t))$ is the *closest point projection* of $\boldsymbol{x} = \varphi(\boldsymbol{X},t)$ onto $\partial\mathsf{K}$, we have:

$$[\varphi(\boldsymbol{X},t) - \boldsymbol{y}(\bar{\xi}(\boldsymbol{X},t))]\cdot\boldsymbol{\tau}_1 = 0. \tag{3.11}$$

This equation defines $\bar{\xi}$. By taking variations of (3.11), one may verify:

$$\left.\begin{aligned} \delta\bar{\xi} &= \frac{\delta\varphi(\boldsymbol{X})\cdot\boldsymbol{\tau}_1}{\|\boldsymbol{\tau}_1\|^2(1+g\kappa)} \\ \kappa &= \frac{1}{\|\boldsymbol{\tau}_1\|^2}\boldsymbol{\nu}\cdot\boldsymbol{y}_{,\xi\xi}(\bar{\xi}) \end{aligned}\right\} \tag{3.12}$$

An examination of $(3.12)_1$ will show that as $g\to 0$ the approximation in (3.10) does indeed become exact. Interestingly, the quantity κ in $(3.12)_2$ has the physical interpretation of being the *curvature* of the obstacle surface at $\bar{\boldsymbol{y}}$.

Having made the approximation in (3.10), we substitute it into (3.9), and then evaluate the integrand of $(3.8)_2$ using (3.9) and the results in Box 3.1. Doing so gives:

$$-\int_\Gamma \boldsymbol{t}(\boldsymbol{X})\cdot\delta\boldsymbol{\varphi}(\boldsymbol{X})\,d\Gamma \approx -\int_\Gamma (t_N\boldsymbol{\nu}\cdot\delta\boldsymbol{\varphi} - t_T^1\|\boldsymbol{\tau}_1\|^2\delta\bar{\xi})\,d\Gamma := G_c(\boldsymbol{\varphi},\delta\boldsymbol{\varphi}) \tag{3.13}$$

where the fact that all quantities are evaluated at time t_{n+1} is understood. Equation (3.13) gives the contact contribution to the virtual work. Also needed for the finite element solution of this nonlinear problem is a tangent stiffness (Hessian), obtained by taking another variation of (3.13). This calculation, while somewhat lengthy, is nonetheless straightforward and renders the following result (with the symbol $\Delta(\cdot)$ recognized as being an abbreviation for the directional derivative):

$$\left.\begin{aligned}\Delta G_c(\boldsymbol{\varphi},\delta\boldsymbol{\varphi}) &= \left.\frac{d}{d\epsilon}\right|_{\epsilon=0} G_c(\boldsymbol{\varphi}+\epsilon\Delta\boldsymbol{\varphi},\delta\boldsymbol{\varphi})\\ &= \Delta G_{c_N}(\boldsymbol{\varphi},\delta\boldsymbol{\varphi}) + \Delta G_{c_T}(\boldsymbol{\varphi},\delta\boldsymbol{\varphi})\end{aligned}\right\} \tag{3.14}$$

where the expressions for $\Delta G_{c_N}(\boldsymbol{\varphi},\delta\boldsymbol{\varphi})$ and $\Delta G_{c_T}(\boldsymbol{\varphi},\delta\boldsymbol{\varphi})$ may be attained by direct calculation and are given by:

$$\left.\begin{aligned}\Delta G_{c_N}(\boldsymbol{\varphi},\delta\boldsymbol{\varphi}) = \int_\Gamma \Big[&\epsilon_N(\boldsymbol{\nu}\cdot\delta\boldsymbol{\varphi})(\boldsymbol{\nu}\cdot\Delta\boldsymbol{\varphi}) + \\ &\frac{t_N\kappa}{(1+g\kappa)\|\boldsymbol{\tau}_1\|^2}(\boldsymbol{\tau}_1\cdot\delta\boldsymbol{\varphi})(\boldsymbol{\tau}_1\cdot\Delta\boldsymbol{\varphi})\Big]\,d\Gamma\end{aligned}\right\} \tag{3.15}$$

and

$$\left.\begin{aligned}&\Delta G_{c_T}(\boldsymbol{\varphi},\delta\boldsymbol{\varphi}) = \\ &\int_\Gamma \Bigg\{\frac{\Delta t_T^1}{1+g\kappa}\delta\boldsymbol{\varphi}\cdot\boldsymbol{\tau}_1 \\ &+ \frac{t_T^1}{\|\boldsymbol{\tau}_1\|^2(1+g\kappa)^2}\Big[(\delta\boldsymbol{\varphi}\cdot\boldsymbol{y}_{,\xi\xi}(\bar{\xi}))(\Delta\boldsymbol{\varphi}\cdot\boldsymbol{\tau}_1) + (\Delta\boldsymbol{\varphi}\cdot\boldsymbol{y}_{,\xi\xi}(\bar{\xi}))(\delta\boldsymbol{\varphi}\cdot\boldsymbol{\tau}_1)\Big] \\ &- \frac{t_T^1}{\|\boldsymbol{\tau}_1\|^4(1+g\kappa)^3}\Big[(1-2g\kappa)\boldsymbol{\tau}_1\cdot\boldsymbol{y}_{,\xi\xi}(\bar{\xi}) + \\ &\qquad g\boldsymbol{\nu}\cdot\boldsymbol{y}_{,\xi\xi\xi}(\bar{\xi}))\Big](\delta\boldsymbol{\varphi}\cdot\boldsymbol{\tau}_1)(\Delta\boldsymbol{\varphi}\cdot\boldsymbol{\tau}_1)\Bigg\}\,d\Gamma\end{aligned}\right\} \tag{3.16}$$

where

$$\Delta t_T^1 = \begin{cases}\dfrac{\epsilon_T\Delta\boldsymbol{\varphi}\cdot\boldsymbol{\tau}_1}{\|\boldsymbol{\tau}_1\|^2(1+g\kappa)} & \text{(stick)}\\[2ex] -\dfrac{\mu\,\mathrm{sign}(t_T^1)\epsilon_N\Delta\boldsymbol{\varphi}\cdot\boldsymbol{\nu}}{\|\boldsymbol{\tau}_1\|} - \dfrac{\mu t_N\,\mathrm{sign}(t_T^1)\boldsymbol{\tau}_1\cdot\boldsymbol{y}_{,\xi\xi}(\bar{\xi})}{\|\boldsymbol{\tau}_1\|^5(1+g\kappa)}\Delta\boldsymbol{\varphi}\cdot\boldsymbol{\tau}_1 & \text{(slip)}\end{cases} \tag{3.17}$$

In writing the above expressions, we remember that all quantities vary with $\boldsymbol{X}$ through identification of closest points $\bar{\boldsymbol{y}}(\boldsymbol{X})$. It is important to realize that (3.14)-(3.17) represent an *exact linearization* of the contact virtual work given in (3.13), *independent* of the finite element discretization. This linearization yields asymptotically quadratic rates of convergence if a Newton-Raphson scheme is used to solve (3.8) (provided $G(\varphi, \delta\varphi)$ is also exactly linearized). The linearization is fairly simple in the sense that because the contact integral is evaluated in the *reference configuration*, *only the integrand* varies with the motion. Finally, we note that the linearization is completely symmetric if *frictional stick* occurs, but otherwise includes a nonsymmetric contribution in regions of Γ that undergo sliding. This final result is in accordance with previously reported treatments of friction in a discrete setting (see, for example, Wriggers, Vu Van, & Stein [1991]).

The description of the two dimensional frictional obstacle problem is now complete, with the exception of the finite element discretization and numerical quadrature of (3.13), (3.15), and (3.16) . For the sake of brevity we we omit these details, with the comment that discrete (matrix) expressions are attained by mere substitution of the finite element approximations for φ and $\delta\varphi$ into (3.13) and (3.14)-(3.17). The integration is performed by dividing the integral over Γ into a series of subintegrals over element edges Γ_e, and then applying numerical quadrature over each of these. For instance, in the second numerical example where biquadratic elements are used for the 2-D discretization, Simpson's rule was used to perform the quadrature, as suggested, for example, in Kikuchi & Song [1981].

A Final Remark. It is emphasized that the machinery introduced in sections 2 and 3 to introduce and treat the frictional constraint for this one body problem is completely amenable to finite deformation multiple body problems as well. In particular, the main difference in treating this broader class of problems is the realization that the identification of $\bar{\boldsymbol{y}}$ for a point $\boldsymbol{X} \in \Gamma$ now depends on the motion of a second body, whose surface contains a material point currently at $\bar{\boldsymbol{y}}$. Accordingly, when we formulate the virtual work for such a problem, we must account for the virtual work associated with *both bodies* induced by the presence of the contact traction $\boldsymbol{t}(\boldsymbol{X}, t)$. The linearization of the equations is carried out as discussed herein, accounting for the motion of the second body, and interestingly *also* produces a symmetric Hessian for the the frictional stick case in 2-D. Finally, it is emphasized that the 3-D statement of virtual work and its linearization were omitted from this treatment only for the sake of brevity, as they are performed exactly analogously to the development presented here. All these points will be addressed in future work by the authors. ■

4. Numerical Examples.

In this section we shall present two large deformation calculations, in order to demonstrate the applicability of the formulation we have presented. These calculations were performed using the finite element program FEAP, as described most recently in Zienkiewicz & Taylor [1989]. The finite deformation plasticity formulation used for the material description has been set forth in Simo [1988]. The first of these two problems, a problem of thin film indentation, was performed using 4-node (bilinear) continuum elements, while the second, the drawing of a wire, employed 9-node (biquadratic) elements. The indentation problem is frictionless, providing an illustration of the type of problem discussed in section

2.2, while the wire drawing problem is done both with and without friction, providing an example of the frictional problem introduced in section 2.3. The fact that exactly the same continuum formulation of the contact problem is used for both problems, and does not depend on element order, is felt to be a particular strength of the formulation we have introduced.

4.1. Thin film indentation.

In the first problem we explore the microindentation of a very thin film on a substrate. The mechanics of this process, although heavily studied experimentally (see, for example Stone, LaFontaine, Alexopoulos, Wu & Li [1988]), are not completely explained as yet by a solid analytical description. Because of the extreme complexities introduced by nonlinear material response, large deformations, and contact nonlinearities, the finite element method provides an attractive opportunity to investigate a problem which is very difficult to handle analytically. Understanding of this problem is quite important to industrial applications where very thin films are a key component, such as in the design of magnetic storage disks.

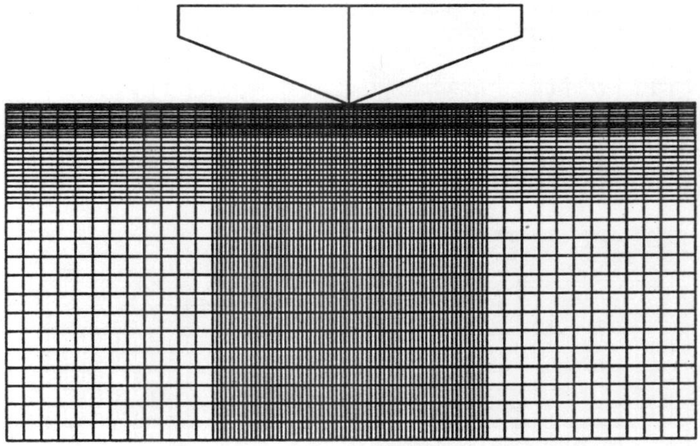

Figure 4.1. Mesh for the axisymmetric thin film indentation problem, reflected though the axis of symmetry.

In this simulation, we investigate the indentation of a 1 μm- thick film on a substrate, with the maximum depth of indentation equal to one film thickness. Mimicking experiments of this type, the calculation models the process of a rigid indenter being pushed into the film-substrate system and then withdrawn. Because the particular goal of this problem is to investigate the response of a film which is harder than its substrate, we use nominal material properties for silicon (bulk modulus=95.3 GPa, shear modulus=49.7 GPa, and yield stress=4410 MPa) for the film, and properties for aluminum (bulk modulus=74.4

GPa, shear modulus=28.5 GPa, yield stress=485 MPa) for the substrate. No hardening is presumed for the silicon, while the (isotropic) hardening rate for the aluminum is taken to be 146 MPa. This axisymmetric problem has been run with and without friction, with the result being that for the geometry considered, the presence of friction makes very little difference in the calculated response. Thus, we present only frictionless results here, which may be thought of as an application of the formulation presented in section 2.1.

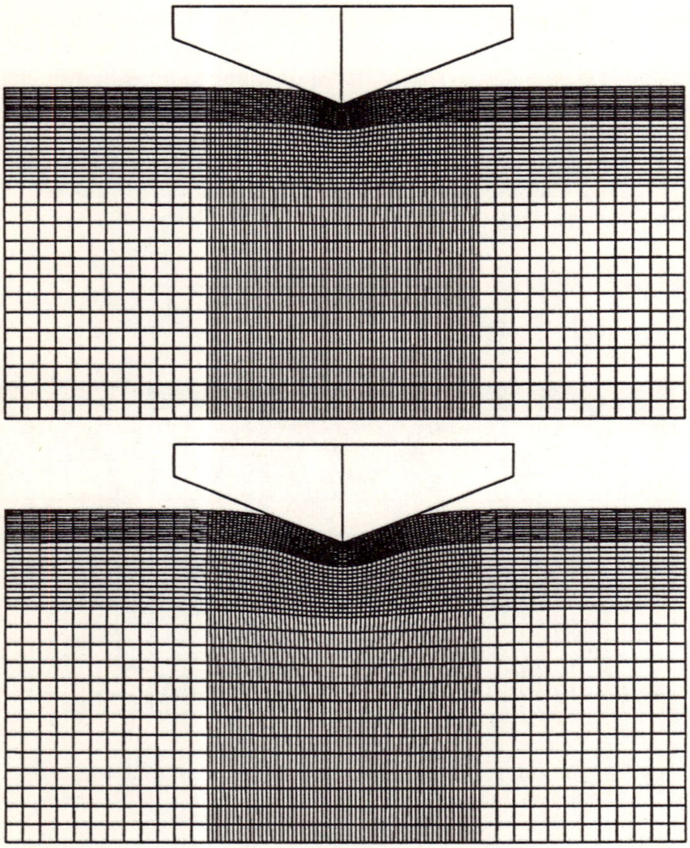

Figure 4.2. Deformed configurations for the axisymmetric thin film indentation problem, shown for indentation depths of one-half and one film thickness.

In Figure 4.1 we see the finite element mesh for this problem, which contains 1628 4-node continuum elements. The boundary conditions are prescribed as fixed on the outer surfaces of the film-substrate mesh, with the island of material chosen large enough such

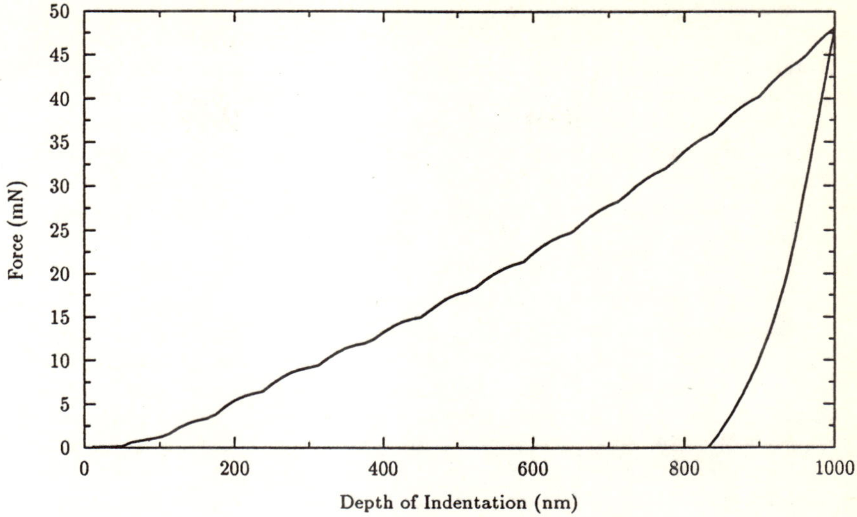

Figure 4.3. Computed force vs. depth of indentation for the thin film indentation problem.

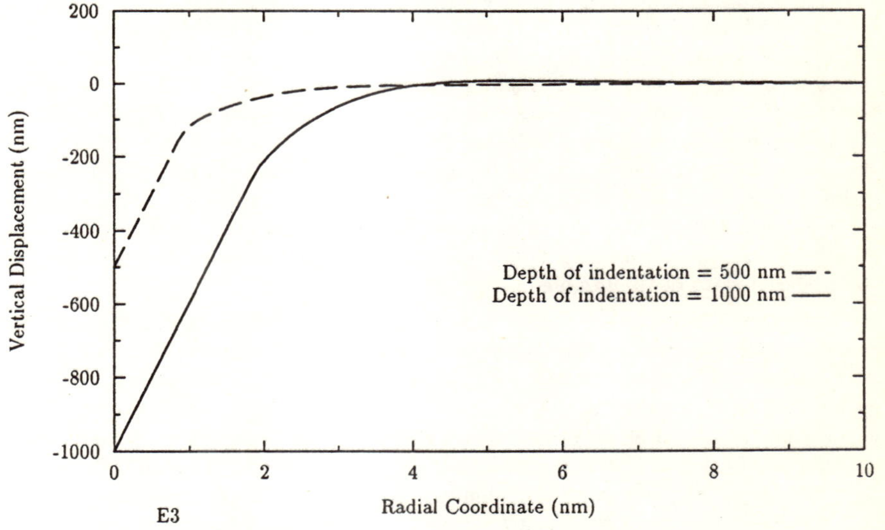

E3

Figure 4.4. Computed surface profiles for the thin film indentation problem.

that the results obtained are not sensitive to these boundary conditions. The motion of the rigid indenter (obstacle) is prescribed such that full insertion of the indenter is accomplished in 80 load steps, while unloading is achieved in 100 more load steps. Figure 4.2 shows deformed meshes for indentation depths of one-half and one film thickness, while Figure 4.3 shows the computed load-depth of indentation curve for this configuration during both insertion and withdrawal of the punch. A further example of the useful information attainable from a calculation of this type is given in Figure 4.4, which shows the computed profiles of the film surface at two different indentation depths. In examining these profiles, one can see that because the film is harder than the substrate the material adjacent to the actual contact area is pulled down, rather than displaying the pileup effect seen in indentation of softer films (such as aluminum).

We can see this distinction between harder films and softer ones even more clearly by viewing Figures 4.5 and 4.6, which show the computed plastic region for the configuration discussed in this paper (Figure 4.5), as compared to that of another in which the indentation of a soft film (aluminum) over a *relatively harder* substrate (silicon) is simulated (Figure 4.6). The difference between these two cases is readily apparent, and together with the other features of this problem, provides a brief demonstration of how the contact methodologies presented in this paper have application to a tangible industrial problem.

4.2. Drawing of a wire, with and without friction.

In the axisymmetric problem we now present, a cylindrical sample with properties K(bulk modulus) $= 63.84 \times 10^9$, G(shear modulus) $= 26.12 \times 10^9$, and σ_Y(yield stress) $=$ 31×10^6 is drawn through a doubly-curved die. The hardening law for the material is taken to be linear, with the hardening modulus equal to $\frac{G}{100}$, and the discretization for the cylinder is accomplished using 80 9-node continuum elements. The die wall restricting the material is parameterized in a piecewise fashion as a quadratic function of a scalar parameter ξ. As can be seen from Figure 4.7, the billet, with initial length .254 and initial radius .0508, is drawn (pulled) through a distance equal to twice its initial length, with the result that at the end of this process the upper end of the billet has been reduced to one-half its original diameter. This loading is accomplished using displacement control on the upper surface of the sample.

Two calculations were done, one without friction and one with a frictional coefficient of 0.1. In the frictionless case, the normal penalty was taken to be $\epsilon_N = 10^{14}$, while in the frictional calculation the penalties $\epsilon_N = 10^{13}$ and $\epsilon_T = 10^9$ were used. In addition, the augmented Lagrangian scheme alluded to briefly in this paper and described fully in Simo & Laursen [1991] was employed to ensure accurate enforcement of constraints. In this way, an analyst can avoid time-consuming iterations to find optimal combinations of penalty and time step; the augmented Lagrangian supplement to the penalty regularization allows the problem to be completed using a much wider range of penalty parameters than was possible previously. In these calculations, 400 steps were utilized to produce the solutions, with 4 augmentations performed per time step.

As can be seen from Figure 4.7, the frictional ($\mu = 0.1$) and frictionless cases produce a slight but discernible difference in the deformation pattern. The distinction is more

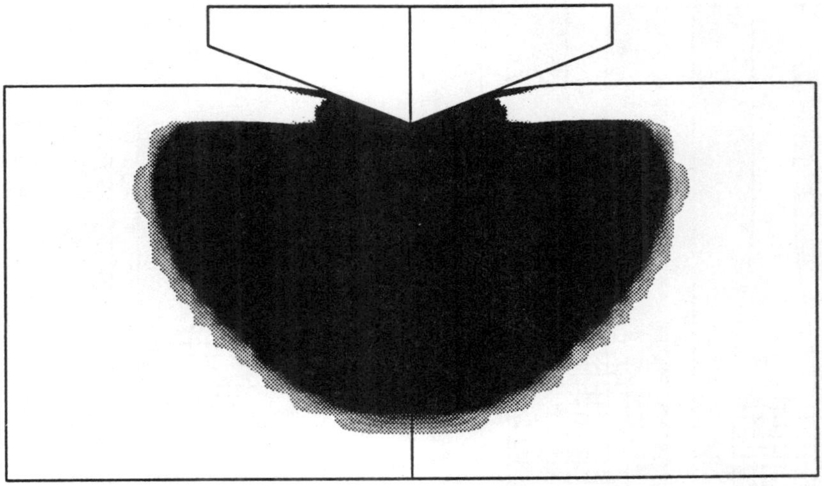

Figure 4.5. Computed plastic region for a silicon film on an aluminum substrate (plastic region shown in black and surrounding gray).

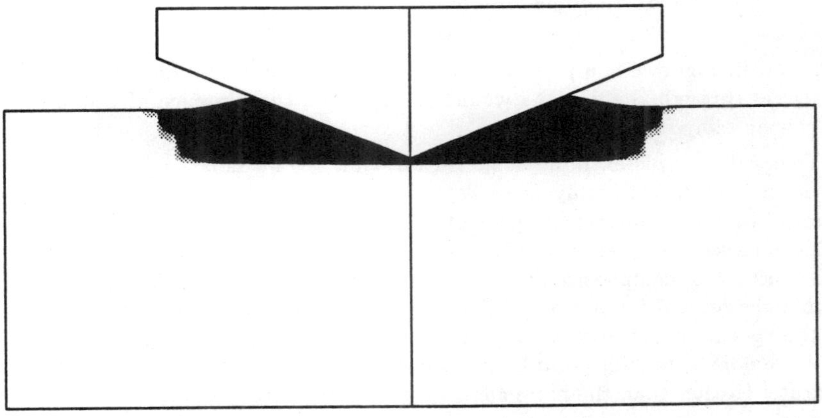

Figure 4.6. Computed plastic region for an aluminum film on a silicon substrate (plastic region shown in black and surrounding gray).

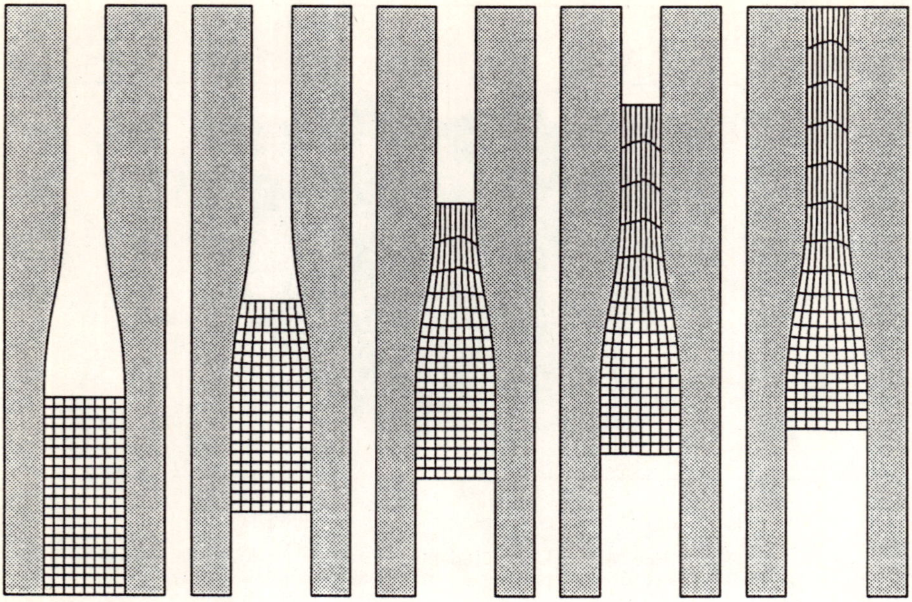

Figure 4.7. Mesh and computed deformations for the frictionless and frictional ($\mu = 0.1$) wire drawing problems. In each state, the frictionless solution is shown left of the centerline, while the frictional one is shown on the right.

readily seen in Figure 4.8, a plot of the total force required to perform the drawing as the billet moves through the die. As would be expected, the presence of friction makes the wire drawing more difficult.

To conclude, this example demonstrates that the formulation we have developed is useful in the solution of truly *large deformation* frictional contact problems, even with substantial material nonlinearity present. The performance of a contact treatment in the presence of these effects relies crucially on careful definition and linearization of the contact virtual work. It is reemphasized that the development of these features given in this paper is completely general for two dimensions, and does not rely on the order of element used or on the specific parametrization of the obstacle surface (as long as it is smooth). This example demonstrates this point by providing an obstacle with curvature and a body with biquadratic (rather than the more common bilinear) discretization.

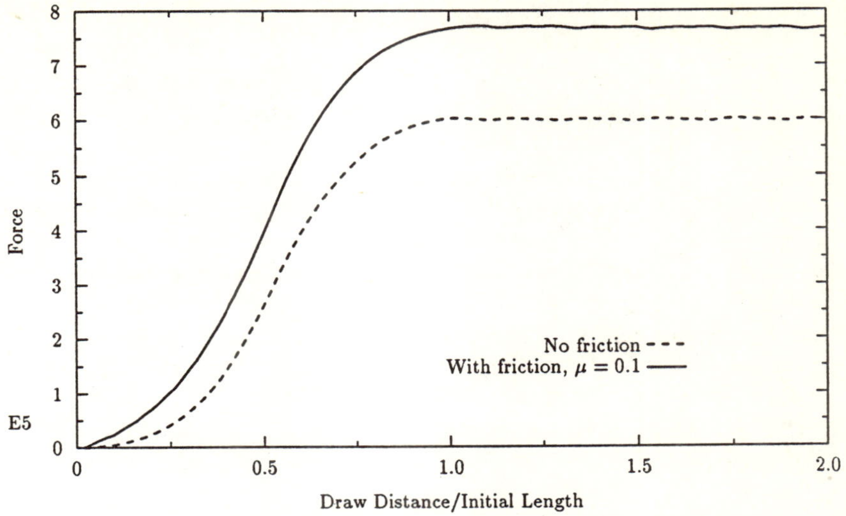

Figure 4.8. Computed force vs. displacement for the frictionless and frictional ($\mu = 0.1$) wire drawing problems.

5. Summary and Conclusions.

In this paper we have sought to give as general a statement of the *large deformation* Coulomb friction problem as possible, for the special case where a deformable body contacts a rigid obstacle. In the course of this discussion, we have developed and penalized the normal and frictional constraints in a manner amenable to either two or three dimensional problems. As a special case, we have considered the 2-D problem and formulated the variational equation *and its linearization.* An important consequence of our approach is that it is completely general; matrix expressions for the contact residual and stiffness may be attained by mere substitution into the expressions of section 3.3 and subsequent integration. An example of the type of problem which can be solved using this approach has been given as well. It is felt that the conceptualization of the frictional contact problem introduced in this paper is an important step toward the development of a similar framework for large deformation *multi-body* problems.

Acknowledgement: Support for this work was provided by IBM and Lawrence Livermore National Laboratory under contract nos. 2DJA-0348 and 2DJA-792 with Stanford University; this support is gratefully acknowledged.

References

Campos, L.T., J.T. Oden & N. Kikuchi, [1982], "A Numerical Analysis of a Class of Contact Problems with Friction in Elastostatics," *Computer Methods in Applied Mechanics and Engineering*, **34**, 821-845.

Gallego, F.J. & J.J. Anza, [1989], "A Mixed Finite Element Model for the Elastic Contact Problem," *International Journal for Numerical Methods in Engineering*, **28**, 1249-1264.

Giannakopoulos, A.E., [1989], "The Return Mapping Method for the Integration of Friction Constitutive Relations," *Computers and Structures*, **32**, 157-167.

Ju, J.-W. & R.L. Taylor, [1988], "A Perturbed Lagrangian Formulation for the Finite Element Solution of Nonlinear Frictional Contact Problems," *Journal of Theoretical and Applied Mechanics*, supplement to vol. 7, 1-14.

Kikuchi, N. & Y.J. Song, [1981], "Penalty/Finite-Element Approximations of a Class of Unilateral Problems in Linear Elasticity," *Quarterly of Applied Mathematics*, **39**, 1-22.

Kikuchi, N. & J.T. Oden, [1988], *Contact Problems in Elasticity: A Study of Variational Inequalities and Finite Element Methods,* SIAM, Philadelphia.

Oden, J.T. & E.B. Pires, [1984], "Algorithms and Numerical Results for Finite Element Approximations of Contact Problems with Non-Classical Friction Laws," *Computers and Structures*, **19**, 137-147.

Simo, J.C., [1988], "A Framework for Finite Strain Elastoplasticity Based on the Multiplicative Decomposition and Hyperelastic Relations. Part II: Computational Aspects," *Computer Methods in Applied Mechanics and Engineering*, **68**, 1-31.

Simo, J.C. & T.J.R. Hughes, [1991], *Elastoplasticity and Viscoplasticity: Computational Aspects*, Springer-Verlag, Berlin.

Simo, J.C. & T.A. Laursen, [1991], "An Augmented Lagrangian Treatment of Contact Problems Involving Friction," submitted to *Computers and Structures.*

Stone, D., W.R. LaFontaine, P. Alexopoulos, T.-W. Wu, & C.-Y. Li, [1986], "An Investigation of Hardness and Adhesion of Sputter-Deposited Aluminum on Silicon by Utilizing a Continuous Indentation Test," *Journal of Materials Research*, **3**, 141-147.

Wriggers, P., T. Vu Van & E. Stein, [1990], "Finite Element Formulation of Large Deformation Impact-Contact Problems with Friction," *Computers and Structures*, **37**, 319-331.

Zienkiewicz, O.C. & R.L. Taylor, [1989], *The Finite Element Method. Volume 1: Basic Formulation and Linear Problems*, McGraw-Hill, London.

A Computational Study of the Residual Stress Distribution in Thermally Loaded Beams of Arbitrary Cross-Section on Frictional Support

F. D. Fischer*, E. Hinteregger* and F. G. Rammerstorfer**

*Institute of Mechanics, University for Mining and Metallurgy, Franz-Josef- Str. 18, A-8700 Leoben, Austria

**Institute of Lightweight Structures and Aerospace Engineering, Vienna University of Technology, Gußhausstr. 27-29/317, A-1040 Vienna, Austria

Summary

A beam of length 100 m with the profile of a railway rail lies on a cooling bed during the cooling process after hot-rolling. The analytical solution for the distribution of the frictional forces along the beam leads to problems which are up to now unsolved. Therefore, the Finite Element Method is used to compute both the residual stresses in longitudinal direction and the deformations of such a horizontally lying long beam. The frictional forces between the beam and the cooling bed have great influence on the remaining residual stress state. A cold beam shows substantial differences in the residual stress distribution between the middle and the end part.

Introduction

The computation of the residual stresses in hot-rolled beams with abitrary cross-sections after cooling down is a highly nonlinear problem. Several thermal heat transfer conditions varying with time must be taken into account to investigate different cooling strategies. Special emphasis is necessary to analyse the radiation at the beginning of the cooling process as well as the convective heat transfer between the beam and the supporting construction [1]. In this paper the cross-section of the beam has the profile of a railway rail of the type UIC 54 E. Using the Finite Element Method the transient temperature distribution over the cross-section can be computed as a two dimensional heat conduction problem with the appropriate heat transfer coefficients corresponding to the different cooling strategies. With these temperature fields it is possible to compute the residual stresses in a thermo-elstic-plastic analysis. Different cooling conditions lead to very different residual stress states. The results of several calculations are discussed in detail elsewhere [1]. Fig. 1 shows typical residual axial stresses remaining after cooling down under consideration of thermal interactions of neighbouring rails.

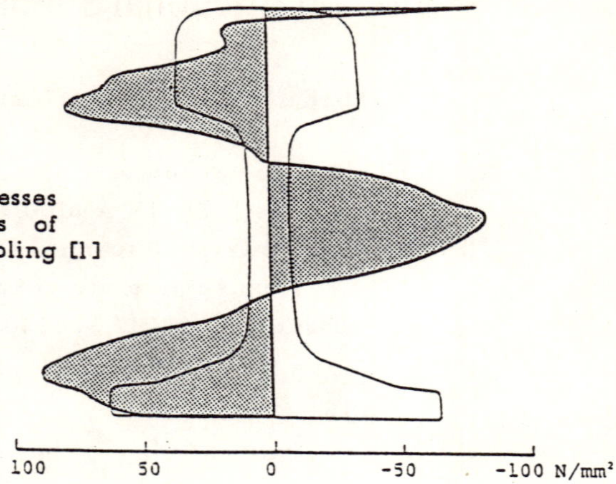

Fig. 1: Longitudinal residual stresses [N/mm^2] along the axis of symmetry at the end of cooling [1]

<u>Dependence of the residual stresses on the frictional support</u>

The production length of a rail depends on the profile and the transport facilities. In this paper a 100 m long rail is investigated. For computing the residual stresses of a railway rail on the cooling bed it is supposed that the rail is initially straight. Since the head and the base of the rail cool down in distinct manner a temperature moment arises. Soon after the beginning of the cooling down process the first distortions at the rail ends appear. The rail deflects with transversal movements of the end portions. This leads to frictional forces between the horizontally lying rail and the supporting construction of the cooling bed.

It is assumed that the deflection of the rail at the cooling bed is not restricted by neighbouring rails. After cooling down the ends diverge in a considerable manner from the initially straight axis, but the main part in the middle of the rail remains straight and keeps its original position. This is due to the fact that after a certain distance away from the rail end the bending moment caused by the frictional forces compensates the moment due to the thermal stresses.

The principal mechanism which also appears in an elastic beam on a rough surface with dead weight and thermal load is described in detail in [2].

The temperature moment m_Θ for a temperature distribution independent on the x-direction is calculated due to [3] as

$$m_\Theta(t) = \int_A E\, \alpha T(y,z,t)\, z\, dy\, dz \,. \tag{1}$$

E is the Young's modulus assumed to constant over A. α is the linear coefficient of thermal expansion and T(y,z,t) is the temperature depending on the time t and the position (y,z). By considering possible friction force distributions along the rail axis one finds slip regions, i. e. intervals in which the transverse deflection w(x) due to the transient temperature moment $m_\Theta(t)$ varies with time. In these regions the transverse load per unit length, p(x), is governed by Coulomb's friction law

$$|p(x)| = \mu q \quad . \tag{2}$$

μ is the friction coefficient and q is the dead load of the rail per unit length acting in y-direction. p(x) changes its sign in those points $x_i(t)$ where the velocity just vanishes and the movement turns its direction:

$$\frac{\partial w}{\partial t}(x = x_i(t)) \quad := \quad \dot{w}(x_i) = 0 \,, \quad i = 1, 2, \ldots . N(t) \quad . \tag{3}$$

The rail lying on the cooling bed cannot be regarded as purely elastic. The decomposition of the total strain tensor ε gives

$$\varepsilon = \varepsilon_{el} + \varepsilon_{pl} + \varepsilon_{th} \quad . \tag{4}$$

ε_{el}, ε_{pl}, and ε_{th} are the elastic, plastic and the thermal strains. Approximating the problem by an uniaxial stress state according to Bernoulli's hypothesis

$$\varepsilon = -z\frac{\partial^2 w(x,t)}{\partial x^2} + C(x,t) \,. \tag{5}$$

the resultant bending moment with respect to the y-axis is given by

$$M = \int_A z\, \sigma\, dA = \int_A z\, E\, (-z\, \frac{\partial^2 w}{\partial x^2} + C - \varepsilon_{pl} - \alpha T)\, dA \quad . \tag{6}$$

M must be equilibrated by moments due to mechanical loads, which in our case are simply the distributed forces p(x). Assuming E to be constant over the cross section we obtain from equ. (6)

$$\frac{\partial^2 w}{\partial x^2} = -\frac{1}{EJ_y}(M + M_{pl} + m_\Theta) \tag{7}$$

with

$$M_{pl} := \int_A z\, E\, \varepsilon_{pl}\, dA \quad . \tag{8}$$

The question arises if regions $[a,b]_j$ along the beam axis exist in which adhesive friction appears, i. e.

$$\frac{\partial w}{\partial t} = 0 \qquad \forall\, x \in [a,b]_j \tag{9}$$

within intervals of finite length. In such adhesive friction intervals equ. (9) in combination with equ. (7) leads to

$$\frac{\partial M}{\partial t} + \frac{\partial M_{pl}}{\partial t} = -\frac{\partial m_{\Theta}}{\partial t} \quad . \qquad \forall\, x \in [a,b]_j \tag{10}$$

The temperature field and, therefore, m_{Θ} are assumed to be independent of x. Hence, the condition (10) requires that within adhesive friction regions the time variation of the sum of the bending moment M due to the mechanical loads p(x) and M_{pl} must be independent of x, too.

$$\frac{\partial^2 (M + M_{pl})}{\partial x\, \partial t} = 0 \qquad \forall\, x \in [a,b]_j \quad . \tag{11}$$

Changing the sequence of differentiation we can conclude from (11):

$$\frac{\partial^2 (M + M_{pl})}{\partial t\, \partial x} = \frac{\partial}{\partial t} Q + \frac{\partial^2 M_{pl}}{\partial t\, \partial x} = 0 \qquad \forall\, x \in [a,b]_j \quad . \tag{12}$$

(Q being the transverse shear resultant.)
If we assume a monotonic development of the process, starting with $m_{\Theta}(t=0) = 0$, $w(x, t = 0) = 0$, $M_{pl}\ (x, t = 0) = 0$, and, hence, $Q\ (t = 0) = 0$, condition (12) in combination with equ. (11) leads to

$$Q = 0 \qquad \forall\, x \in [a,b]_j \tag{13}$$

and, hence,

$$\frac{\partial Q}{\partial x} = -p(x) = 0 \qquad \forall\, x \in [a,b]_j \tag{14}$$

as long as no plastification appears in $[a,b]_j$, i. e. $M_{pl} = 0 \quad \forall\, x \in [a,b]_j$.
Since in the regions which remain straight the same temperature stress and strain situation appears in each cross-section, i. e. $\forall\, x \in [a,b]_j$ we may assume that even in the elatic-plastic range M_{pl} is independent of x and, hence, $\frac{\partial M_{pl}}{\partial x} = 0$ and equ. (14) remains valid.
From (14) we can conclude that regions which are at rest are not held by adhesive friction forces $p(x) \neq 0$ but are subjected to the following conditions:

$$M(x) + M_{pl}(x) = -m_{\Theta}, \tag{15}$$

$$Q(x) = 0, \tag{16}$$

$$p(x) = 0 \tag{17}$$

for all $x \in [a,b]_j$.

Let us now consider where such regions might appear for $\partial m_\theta / \partial t \neq 0$. For this purpose consider Fig. 2. Because of (15) the end portions must not be at rest for $t > 0$. One inner region - due to symmetry requirements in the middle of the beam - could be at rest since (15) - (17) can be fulfilled by any number $N > 1$ of tightly connected load intervals according to equ. (2) and (3) as long as $L(t) := x_N(t)$ is smaller than half the complete length of the beam.

Fig. 2 shows the friction force system of a cooling beam placed on a continuous support.

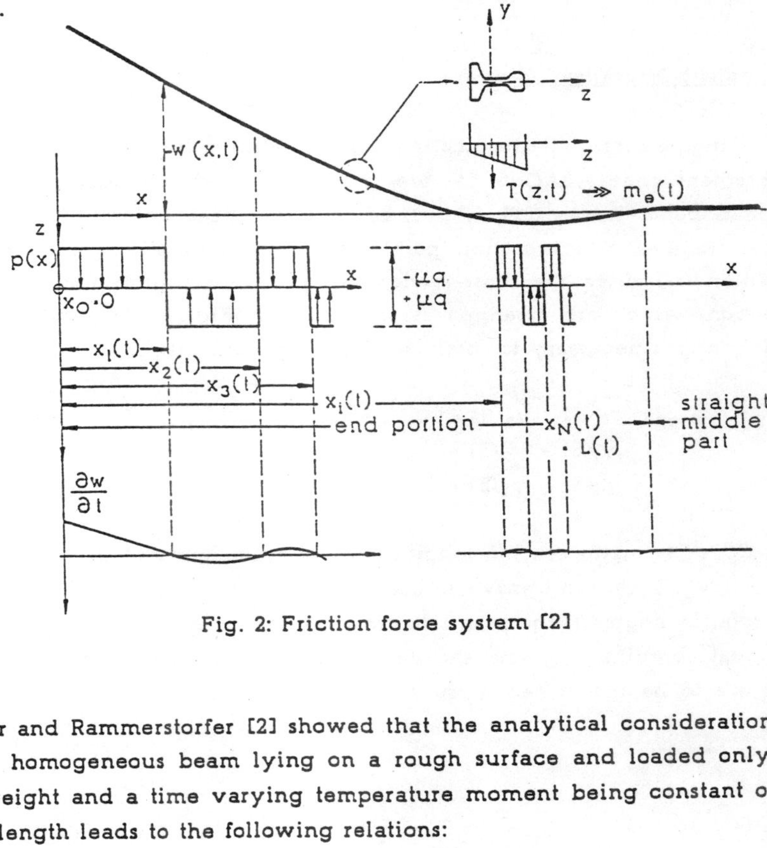

Fig. 2: Friction force system [2]

Fischer and Rammerstorfer [2] showed that the analytical consideration of the elastic homogeneous beam lying on a rough surface and loaded only by its deadweight and a time varying temperature moment being constant over the beam length leads to the following relations:

$$2\sum_{i=1}^{N-1} (-1)^{i-1}\, x_i = (-1)^N x_N \quad . \tag{18}$$

$$2\sum_{i=1}^{N-1} (-1)^{i-1}\, x_i^2 = (-1)^N x_N^2 - 2\, m_\theta / (\mu q) \quad , \text{ and} \tag{19}$$

$$2\sum_{i=1}^{N-1} (-1)^{i-1} [\Lambda(i)(x_k - x_i)^3 - (x_N - x_i)^2 (3x_k - 2x_N - x_i)]\, x_i = \frac{3}{\mu q}\, m_\theta\, (x_N - x_k)^2 \tag{20}$$

with $k = 1, 2, \ldots, N-1$ and the logical function $\Lambda(i) = 1$ for $i \leq k$, $\Lambda(i) = 0$ for $i > k$.

With (20) - (22) N-1 differential equations plus two algebraic equations (the equilibrium equations) exist. This differential/algebraic system of N+1 equations for N unknowns $x_1(t), \ldots, x_N(t)$ and as the $(N+1)^{th}$ the number of unknowns N itself has not been solved up to now. Therefore, the Finite Element Method is applied to this problem.

Finite-Element-Discretization

For modeling the rail on the cooling bed quadratic beam elements of the Finite Element code ABAQUS [4] are used. They behave according to the Timoshenko beam theory [5], i. e. they include shear deformations. As the allowable number of integration points for each thermally loaded element cross-section is limited the cross-section of the rail is represented by 6 sub-cross-sections which are arranged as indicated in Fig. 3. To assure a plane cross-section it is necessary to establish the nodal constraints:

$$u_1(S) = u_2(S) = \ldots = u_6(S) \quad . \qquad (21.1)$$

$$v_1(S) = v_2(S) = \ldots = v_6(S) \quad . \qquad (21.2)$$

$$\varphi_1(S) = \varphi_2(S) = \ldots = \varphi_6(S) \quad . \qquad (21.3)$$

$u_1(S), u_2(S), \ldots, u_6(S)$ are the longitudinal displacements of the nodes, $v_1(S), v_2(S), \ldots, v_6(S)$ are the transverse displacements and $\varphi_1(S), \varphi_2(S), \ldots, \varphi_6(S)$ are the rotation angles of the propriate cross-section at the node S of the beams. In the local coordinate system the deformations of the single element cross-sections are to be prescribed in such a way that they act like the total rail

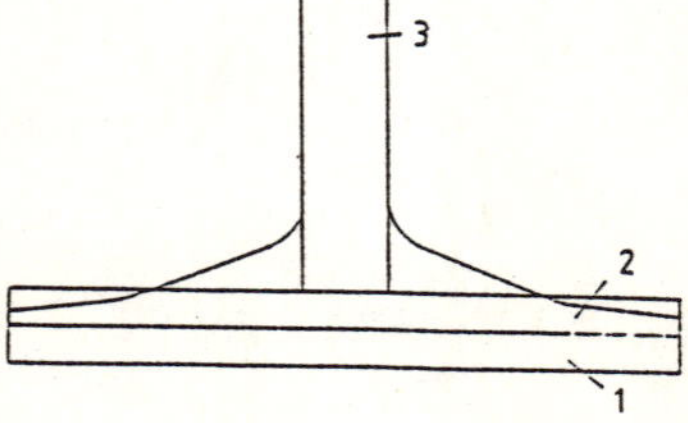

Fig. 3: Rail cross-section represented by 6 single beam cross-sections

cross-section. Obviously the sum of the single cross-sectional areas of all 6 beams as well as the total moment of inertia must coincide with the original rail.

One half of the 100 m long rail on the cooling bed is taken into account due to symmetry. The transfer of the frictional forces is discretized by transverse truss elements as shown in Fig. 4. They represent the frictional action of the supporting steel beams of the cooling bed.

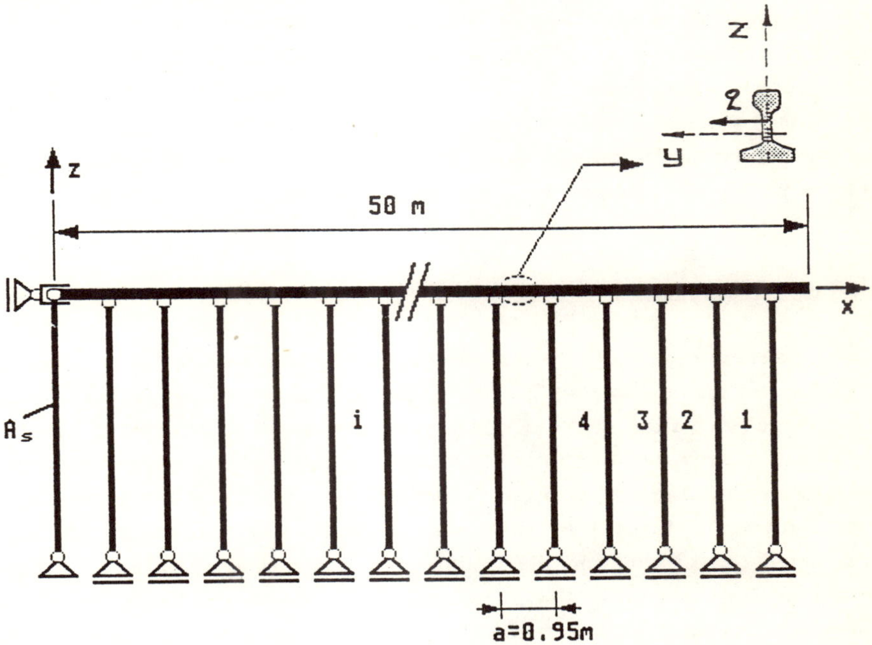

Fig. 4: Finite-Element-model of a rail on the cooling bed

The greatest value of the frictional force P(x) can be determined by $|P(x)| \leq \mu q a$ if the coefficient of friction and the dead load are known and a is the distance between neighbouring supporting beams (a= 0.95 m). This is the greatest possible normal force in the trusses, too. Using a fictitious elasto-plastic behaviour of the trusses the yield stress $\sigma_{T.i.y}$ is defined by

$$\sigma_{T.i.y} = \frac{\mu q a}{A_S} \quad . \tag{22}$$

A_s is the area of each truss. The Young's modulus of the truss material is chosen to be very high so that the trusses behave almost rigidly before the adhesive limit is reached. Sliding corresponds with plastifying of the fictitious material. By this model the sign of the frictional force - tensile or compressive force in the truss - is found automatically.

Adhesive friction coefficients μ (between rail and not lubricated supporting beam) could be found by measurements to be in the range of 0.38 and 0.60. The mean value is 0.45. According to the computed temperature fields distinct transient temperature distributions T(t,y,z) are prescribed in 32 integration points of the rail cross-section. The temperature field is constant along the axis and, therefore, independent on x. Using this model the development of longitudinal residual stresses due to cooling can be computed 'globally' with good approximation. In general the residual stress state is multi-axial. In this paper the problem is approximated by uniaxial stress state assumptions.

Transversal Displacements w(x,t)

Fig. 5 and 6 show the time-dependent variation of the transversal displacements w(x,t) of a 100 m long rail which is straight at the beginning of the cooling process. Negative values of the displacements at the rail ends point to bending over the rail base. After 3 minutes of cooling the rail end reaches its maximum amplitude in the range of -18.5 cm (curve 5 in Fig. 5). Then the tendency of the displacements turns into the other direction as far as a maximum displacement of 16.3 cm from the initial possition is reached after 11 minutes (curve 2 in Fig. 6). During the phase transformation (austenite - perlite) the temperature moment m_{Θ} decreases and the displacements become smaller once again and the motion turns back. After 20 minutes cooling the rail reaches the maximum deflection over base (curve 4 in Fig. 6). Curve 7 in Fig. 6 shows the final deformations. It can be seen from Fig. 5 and 6 that in a middle region of approximately 2 x 23 m length during the whole cooling procedure no displacements occur. Only the end portions deflect. It can be concluded that rails with a total length of less than approx. 54 m deflect transversally in each point during the cooling process. If this length is exceeded a region in the middle of the rail exists which is not deflected.

If the cold rail is lifted from the cooling bed, of course, an elastic spring-back appears and the computation shows a residual deflections along the total length of the rail. Nearly a parabolic curve can be observed as the rail deflects without the influence of the frictional forces. The difference in the final deflection between the end and the middle part of the 100 m long rail amounts to 4.708 m. This results are in good accordance with measurements.

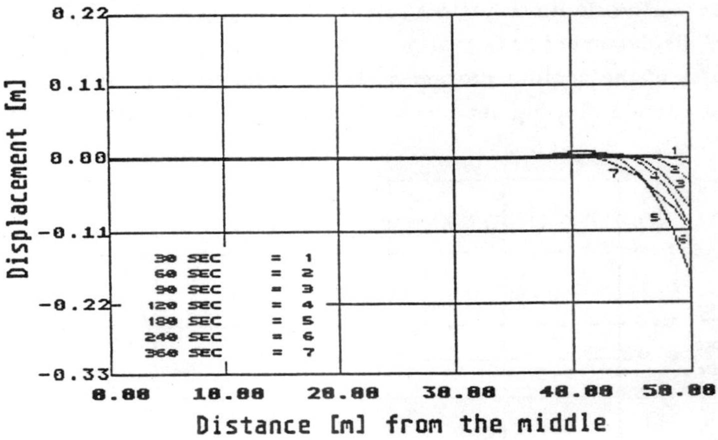

Fig. 5: Displacements of a rail on the cooling bed (0 - 6 min)

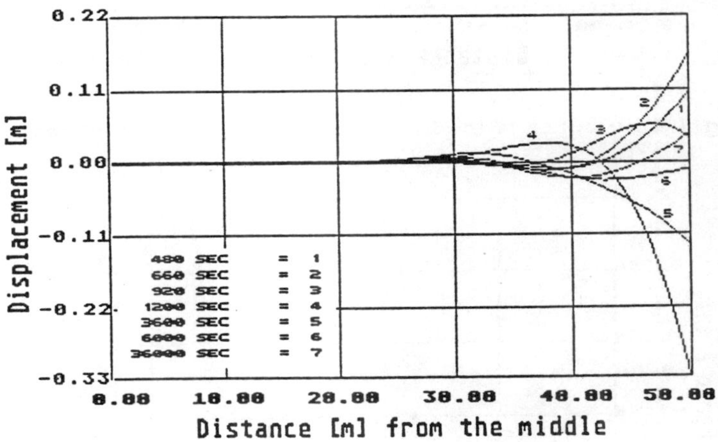

Fig. 6: Displacements of a rail on the cooling bed (8 min - 10 h)

Distribution of the frictional forces P(x,t)

Due to cooling of the rail on the cooling bed frictional forces between the rail and the supporting construction are avoken. The development of the frictional forces during the cooling down is shown in Fig. 7 - 9 where the normal forces $P_i(x,t)$ in the trusses (see Fig. 4) are plotted. If the normal stress $\sigma_{T,i}$ in the trusses reaches the fictitious yield stress (P_{max}, $|P_{min}|$ = 2.25 N) (see equ. (22), changes of displacements w(x,t) appear.

In the course of the cooling process a "broadening" of the distribution as well as several changes in the tendency of the direction of the frictional forces

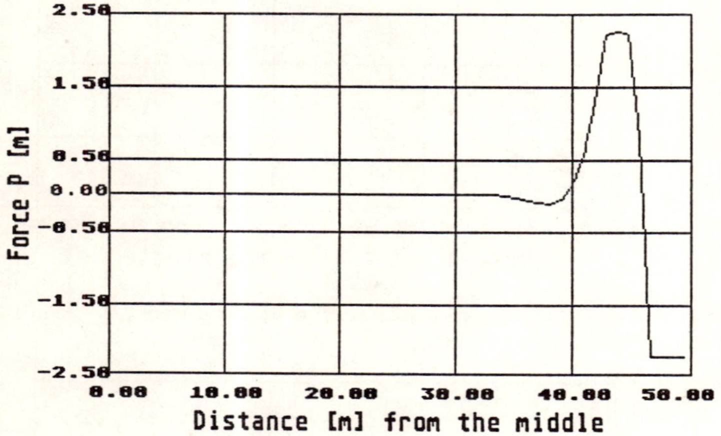

Fig. 7: Distribution of the frictional forces (after 1 min of cooling)

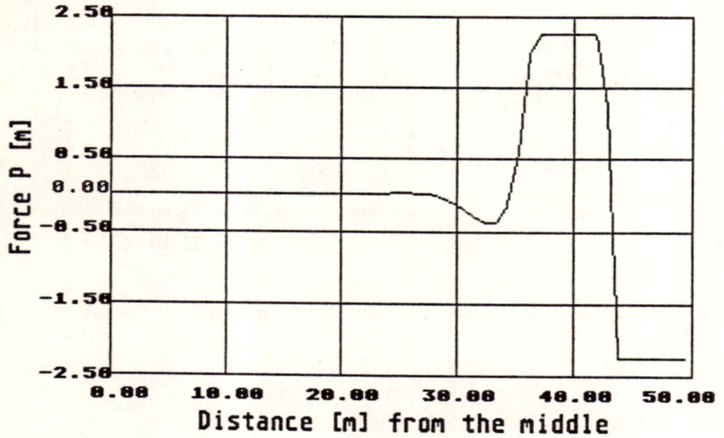

Fig. 8: Distribution of the frictional forces (after 3 min of cooling)

occur. Again it could be observed that along the middle portion of the 100-m-rail no frictional forces arise, in accordance with the analytical findings, since there are no displacements. Using this Finite Element model with a fictitious elastoplastic behaviour of the trusses very small deformations in the truss elements occur due to their elastic behaviour even if the adhesive limit is not reached. Therefore, the P(x)-curves do not show a really distinct "jump pattern" between P_{max} and P_{min} as in the analytical solution, [2].

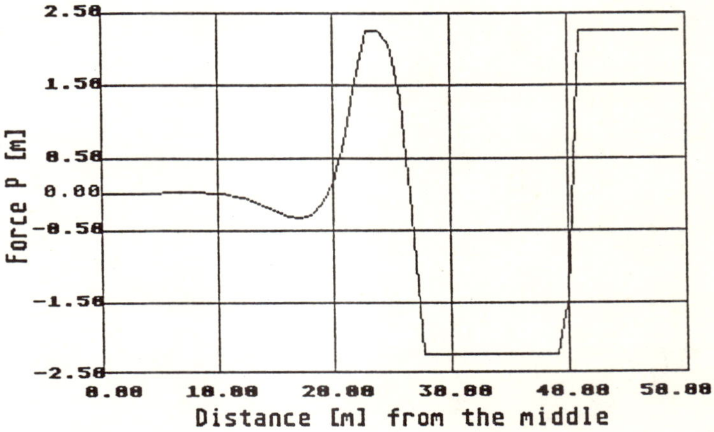

Fig. 9: Distribution of the frictional forces (after 10 h of cooling)

Longitudinal residual stresses $\sigma_x(x,y,z)$ in a cooled down rail

After cooling and lifting of the rail from the cooling bed a certain level of residual stresses still remains with dominant longitudinal stresses. They can be computed with good approximation by this beam-truss-system. The results are plotted for some distinct points of the rail cross-section in Figs. 10 - 12.
It can be seen that the residual stress distributions in the cross-sections along the longitudinal axis of the rail differ significantly.For example, the longitudinal stress of the peripheral base turned towards the cooling bed (curve 1 in Fig. 11) varies between -90.8 N/mm^2 in the middle of the rail and -141.7 N/mm^2 near the end of the rail. This is an increase of the stress magnitude of 56 %. Curve 2 in the same figure shows a completely different trend. In the mid portion of the rail compressive stresses in the range of -30.5 N/mm^2 prevail and near the rail end tensile stresses in the range of 25.4 N/mm^2 can be found. Curve 1 in Fig. 12 shows a qualitatively different trend. The maximum tensile stresses occur approximately 9 m away from the raii end and the stresses decrease towards the mid and the end.

The comparison of these results obtained for a rail cooled down without frictional influence with that obtained for the end portion of the 100-m-rail cooled under frictional conditions shows good agreement. The residual stresses in the middle of the frictional supported rail correspond with the stresses in a rail which is forced to remain straight during cooling down due to the contact to the neighbouring rails [1].

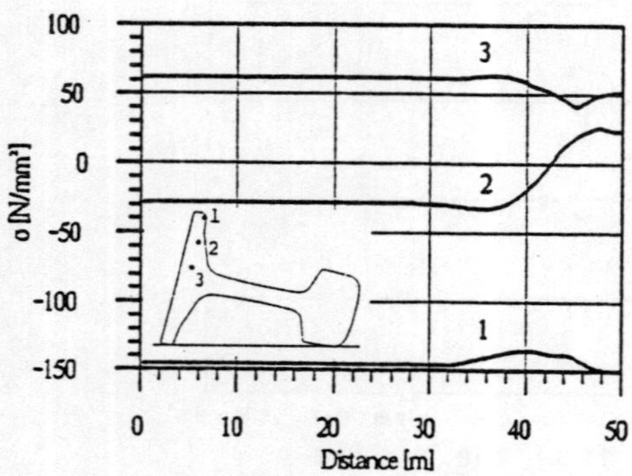

Fig. 10: Distribution of longitudinal residual stresses (middle of the rail at 0 m, rail end at 50 m)

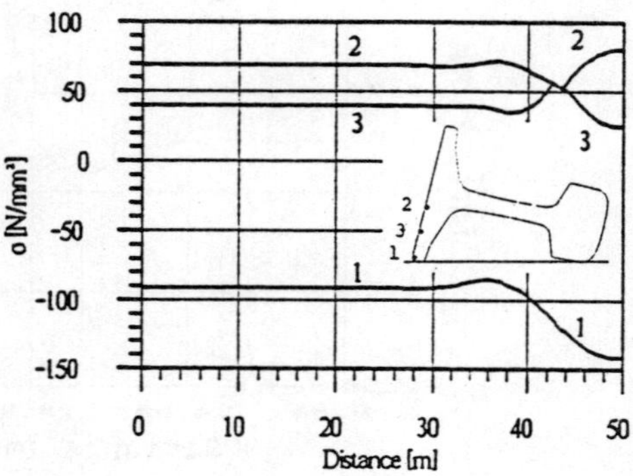

Fig. 11: Distribution of longitudinal residual stresses (middle of the rail at 0 m, rail end at 50 m)

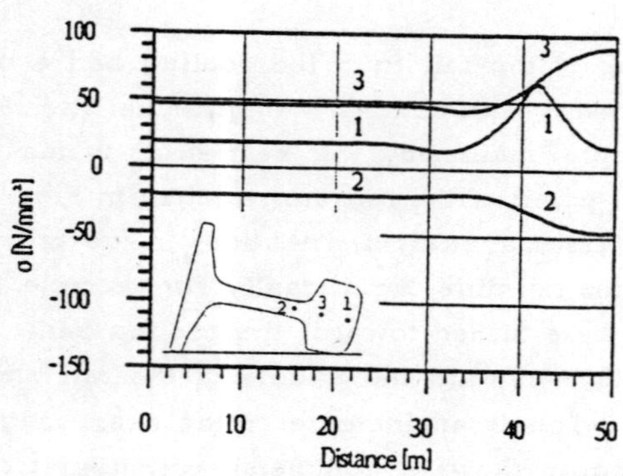

Fig. 12: Distribution of longitudinal residual stresses (middle of the rail at 0 m, rail end at 50 m)

The presented results illustrate that measured values of residual stresses must be interpreted in a distinct manner as it has been done in the past. Formerly measured residual stresses $\sigma(y,z)$ were only assigned to a specific point of the rail cross-section assuming that the measuring point is sufficiently far away from the rail end. According to the presented results the dependence of the residual stresses on the longitudinal coordinate of the measuring point has to be taken into account, too, $\sigma_x = \sigma_x(x,y,z)$. For the usual measuring a reference specimen is cut out of the beam. The assumption that the measured residual stresses are constant over the whole length except a small end part in the range of approximately two beam heights is not longer valid.
This may be one reason for the great scatter of reported residual stress values on the same y-z-coordinates of the cross-section. Even values with different signs can be found in literature. It seems that it has not been considered on which part of the rail length the specimen has been taken from.
An early work [6] deals with the measurement of residual stresses of unstraightened rails. Further measurement results can be found in [7,8,9]. Plotting all the known values of measured residual stress in one diagram Fig. 13 shows the field of measured and computed longitudinal residual stresses, respectively, along the axis of symmetry.

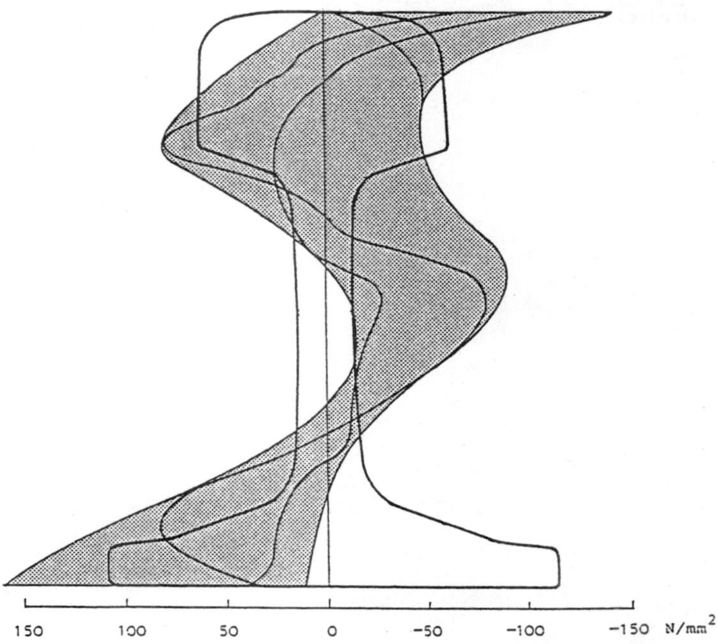

Fig. 13: Scatter of the longitudinal residual stress along the axis of symmetry taken from literature

References

1 Hinteregger, E.: Eigenspannungen und Verformungen in Schienen nach dem Walzen vor dem Richten. Dissertation, Montanuniversität Leoben 1990

2 Fischer, F. D. and F. G. Rammerstorfer: The Thermally Loaded Heavy Beam on a Rough Surface. Proc. 8th Symposium on "Trends in Applications of Mathematics to Mechanics", F. Ziegler edt., Longman Higher Education & Reference, Burnt Mill, UK, 1991

3 Parkus, H.: Thermoelasticity, Wien, New York: Springer-Verlag 1976

4 ABAQUS, Version 4.7, Hibbit, Karlson und Sorensen, Providence, Rhode Island, USA, 1988

5 Timoshenko, S. P.: Strength of Material, Part III, Advanced Theory and Problems, 3. Edition, Princeton: D. Van Nostrand Co 1956

6 Meier, H.: Eigenspannungen in Eisenbahnschienen, Organ f. d. Fortschritt des Eisenbahnwesens 91 (1936) 320-329

7 Konjuchov, A. D., D. M. Rabinovic, I. J. Vinokurov and V. S. Serebrjakov: Einfluß der Technologie auf die Restspannungen in durchgehärteten Schienen, STAL in Deutsch 10 (1969) 1048-1051

8 Asbeck, H. O. and M. Heyder: Eigen- und Richtspannungen in walzneuen Schienen, Eisenbahntechnische Rundschau 26 (1977) 217-222

9 Kolmogorov, S. V., Y. D. Makarov and O. N. Mikhailov: Formation of residual stresses in heat treatment hardened rail, Steel in the USSR 17 (1987) 271-273